GENOME DUPLICATION

D0160957

GENOME DUPLICATION

Melvin L. DePamphilis

Stephen D. Bell

LONDON AND NEW YORK

Garland Science
Vice President: Denise Schanck
Editor: Elizabeth Owen
Editorial Assistant: David Borrowdale
Production Editor: Georgina Lucas
Illustrator: Oxford Designers & Illustrators
Cover Design: Andrew Magee
Copyeditor: Nik Prowse
Typesetting: Georgina Lucas
Proofreader: Joanne Clayton
Indexer: Liza Furnival

Melvin L. DePamphilis, Ph.D., is currently a member of the Senior Biomedical Research Service at the National Institutes of Health (NIH) in Bethesda, Maryland, where he heads a research group investigating DNA replication and gene expression during animal development. After receiving his Ph.D. in Biochemistry from the University of Wisconsin in 1970 for his work on bacterial motility, Dr. DePamphilis carried out post-doctoral research at the University of Wisconsin on enzyme mechanisms and then at Stanford University Medical School on viral DNA replication. He continued his studies on DNA replication in animal viruses as a professor at Harvard Medical School, extending them to frog eggs, mouse embryos, and cultured mammalian cells. He later joined the Roche Institute of Molecular Biology and then moved to the NIH.

Stephen D. Bell, Ph.D., is the Professor of Microbiology at the Sir William Dunn School of Pathology at Oxford University where he heads a research group investigating DNA replication and cell division in archaeal cells. Prior to this he was a group leader at the Medical Research Council Cancer Cell Unit in Cambridge, UK. After receiving his Ph.D. in Genetics from the University of Glasgow in 1992 for his work on mechanisms of transposition in bacteria, Dr. Bell performed post-doctoral studies at the University of Glasgow and later at Cambridge University on gene expression mechanisms in primitive eukaryotic cells and in archaea.

©2011 by Garland Science, Taylor & Francis Group, LLC

ISBN 978-0-4154-4206-0

This book contains information obtained from authentic and highly regarded sources. Reprinted material is quoted with permission, and sources are indicated. A wide variety of references are listed. Reasonable efforts have been made to publish reliable data and information, but the authors and the publisher cannot assume responsibility for the validity of all materials or for the consequences of their use.

All rights reserved. No part of this publication may be reproduced, stored in a retrieval system, or transmitted in any form or by any means—graphic, electronic, or mechanical, including photocopying, recording, taping, or information storage and retrieval systems—without permission of the copyright holder.

Published by Garland Science, Taylor & Francis Group, LLC, an informa business, 270 Madison Avenue, New York NY 10016, USA, and 2 Park Square, Milton Park, Abingdon, OX14 4RN, UK.

Printed in the United States of America
15 14 13 12 11 10 9 8 7 6 5 4 3 2 1

Visit our web site at http://www.garlandscience.com

Library of Congress Cataloging-in-Publication Data

DePamphilis, Melvin L.
Genome duplication / Melvin L. DePamphilis, Stephen D. Bell.
p. ; cm.
Includes bibliographical references and index.
ISBN 978-0-415-44206-0
1. DNA replication. I. Bell, Stephen D. II. Title.
[DNLM: 1. Gene Duplication. 2. Genome. QU 475]
QP624.5.R48D47 2010
572.8'645--dc22

2010032858

QP
624.5
.R48
D47
2011

Mixed Sources
Product group from well-managed forests, controlled sources and recycled wood or fiber
www.fsc.org Cert no. SW-COC-002985
©1996 Forest Stewardship Council

PREFACE

This book is dedicated to the premise that nothing is more fundamental to Life than the ability to reproduce. That ability begins with the simple fact that all living organisms store their blueprint for life (the genome) in deoxyribonucleic acid (DNA). DNA is the longest, thinnest biological molecule and the machinery that replicates it must complete the job quickly and accurately if the organism is to flourish. Over the past 60 years, science has elucidated the structure and organization of the genome (Chapter 1) and the molecular mechanisms that reproduce it in a variety of biological systems. Accordingly, the goal of *Genome Duplication* is to explain how genomes are duplicated in living organisms, from the simplest cell to the most complex multicellular organism, and how genome duplication impacts upon and is regulated by cellular activity. *Genome Duplication* is not intended to be an encyclopedia, but rather to identify the underlying concepts and principles that govern genome duplication in a world rich in biodiversity.

The development of concepts and principles that apply to all living organisms through the study of a few well-chosen paradigms is possible, because all living organisms can be classified into three domains: bacteria, archaea, and eukarya (Chapter 2). This book focuses on the intestinal bacterium *Escherichia coli* and the soil bacterium *Bacillus subtilis*, the archaeon *Sulfolobus solfataricus*, the single-cell eukaryotes *Saccharomyces cerevisiae* (budding yeast) and *Schizosaccharomyces pombe* (fission yeast), and multicellular eukaryotes *Drosophila melanogaster* (fruit fly), *Xenopus laevis* (African clawed frog), and the mammals *Mus musculus* (mouse) and, of course, *Homo sapiens*. Many examples from other organisms, as well as from bacterial and animal viruses and episomes, are included to illustrate specific points and provide historical context.

Not only do all living organisms use the same molecule to encode their genome, they all use the same mechanism to replicate it—semi-conservative DNA replication by the replication-fork mechanism (Chapters 3 through 6). Remarkably, the sequence of events in the assembly and activation of replication forks is essentially the same throughout the three domains of life. To paraphrase Jacques Monod's famous dictum, what is true for replication forks in bacteria is also true for replication forks in elephants. Thus, genome duplication is arguably the most highly conserved of biological processes and, as such, it is a window into biological evolution. Not surprisingly, the subject of evolution appears throughout the book, but particularly in Chapters 2 and 15.

Genomes do not exist as simple DNA molecules in bacteria, archaea, or eukarya, because DNA is frequently modified chemically and invariably organized into a DNA–protein complex (Chapter 7). These modifications protect the DNA from damage, compact it into a tiny space, and regulate expression of its genetic information. Thus, each time a cell divides, it must dismantle these modifications in order to duplicate its genome, and then restore them as DNA replication proceeds. Moreover, as genome duplication occurs in eukarya, the two sibling molecules (sister chromatids) are embraced by a unique ring structure that remains until they are separated during cell division.

During the course of evolution, the nature of replication origins changed from unique DNA sequences that are essential for assembly and activation of their cognate replication machinery to ill-defined sequences that are determined more by epigenetic than genetic parameters (Chapters 8 through 10). This transition from rigidly defined replication origins to conditionally defined replication origins allowed the evolution of complex, multicellular organisms. The machinery that converts DNA replication origins into two replication forks that travel in opposite directions is impressive (Chapter 11), involving 20 different proteins in bacterial cells and 66 in human cells.

To be productive, all organisms must link cell division tightly to genome duplication. In eukarya, duplication of the nuclear portion of their genome is generally (but not always) restricted to once and only once per cell division (Chapter 12). When a cell has committed itself to dividing, it either completes the process or dies. However, plants and invertebrates frequently employ multiple rounds of genome duplication in the absence of cell division to facilitate post-mitotic growth. Only bacteria and probably some archaea can successfully launch a second set of replication forks from the same replication origin before the first set of forks have terminated.

Eukarya contain a series of feedback loops driving the cell-division cycle in one direction. Checkpoints (Chapter 13) ensure that one phase is completed before the next phase begins. Bacteria contain at least two checkpoints, one to ensure that a cell has produced sufficient materials to support cell division (the restriction checkpoint) and one that senses DNA damage during DNA replication. Eukarya sense DNA damage before, during, and after the DNA replication has taken place. Eukaryotic cells also detect stalled replication forks, prevent entry into mitosis before DNA replication is completed, and prevent cell division before mitosis is completed. If a cell fails to correct the problem, then it is programmed to die.

All of this leads to the understanding and treatment of human disease (Chapter 14). Remarkably, with the exception of diseases caused by prions and RNA viruses, all of the remaining infectious diseases are caused by either microorganisms, DNA viruses, or RNA viruses with a DNA intermediate in their life cycle. This means that differences in the mechanisms by which DNA is duplicated in microbes and viruses compared with mammals offer therapeutic targets. In fact, current antiviral therapies target the viral genome duplication almost exclusively. Mistakes in genome duplication as well as infection by DNA viruses can also result in noninfectious diseases, such as cancer, a leading cause of death worldwide. DNA replication proteins not only provide useful biomarkers for the early detection of cancer, but most of the current cancer therapies target critical proteins that function at replication forks or that regulate genome duplication. Mistakes in genome duplication also are associated directly with a number of genetic disorders, including those associated with neurodegenerative diseases, such as Huntington's disease, spinocerebellar ataxia, and fragile X syndrome. Thus, it is not hyperbole to state that understanding how genomes are duplicated, how the process of genome duplication is regulated, and how genome duplication differs among the various forms of life is critical to the understanding and treatment of human disease.

Perhaps now one can appreciate how the subject of this book—genome duplication—links together all living organisms, their origins, their evolution, and their future. Unfortunately, the literature on this subject can be intimidating, and students often become lost in the labyrinth of mechanisms created by a diverse group of biosystems and described with a Byzantine nomenclature that challenges professionals as well as novices. *Genome Duplication* attempts to correct these problems, and with a basic, college-level knowledge of biology and chemistry, the reader will find this book both comprehensive and comprehensible. That said, the authors welcome any corrections or suggestions from the reader that would improve this book in a future edition.

ACKNOWLEDGMENTS

This book would not have been possible without the steadfast support, guidance, and encouragement of Elizabeth Owen, David Borrowdale, and Georgina Lucas at Garland Science, and the unending patience and love of Margarete DePamphilis and Rachel Bell, the women with whom we share our lives. We also thank Tom and Roger Kornberg for the dedication of this book to their father, Arthur, and for relating the importance of their mother, Sylvy, to his career.

The literature citations within each chapter are limited by the format of this book, but we acknowledge the incredible history of a field that spans at least a century of time and the dedicated effort of thousands of men and women whose curiosity, imagination, and hard work has revealed both the complexity of the subject and the unifying pillars upon which it rests.

Finally, we give a special thanks to all those who helped us with advice and materials.

Steve Anderson (Rutgers, The State University of New Jersey, USA); Robert Bambara (University of Rochester, USA); Deepak Bastia (Medical University of South Carolina, USA); Stephen P. Bell (Massachusetts Institute of Technology, USA); James Berger (University of California, Berkeley, USA); Anja Bielinsky (University of Minnesota, USA); Julian Blow (University of Dundee, UK); Eric Boye (Oslo University Hospital, Norway); Peter Burgers (Washington University School of Medicine, USA); Brian Calvi (Indiana University, USA); Tony Carr (University of Sussex, UK); Mike Cashel (National Institute of Child Health and Human Development, USA); David Clark (National Institute of Child Health and Human Development, USA); Nick Coleman (University of Cambridge, UK); Susan Cotterill (St George's, University of London, UK); Rebecca Curley (Wellcome Trust Centre for Cell Biology, University of Edinburgh, UK); Rob de Bruin (University College London, UK); John Diffley (Cancer Research UK London Research Institute, UK); Nick Dyson (Massachusetts General Hospital Cancer Center, USA); Arturo Falaschi (International Centre for Genetic Engineering and Biotechnology, Italy); Ellen Fanning (Vanderbilt University, USA); David Frick (University of Wisconsin, USA); Errol Friedberg (University of Texas Southwestern Medical Center, USA); Michael Fry (Technion – Israel Institute of Technology, Israel); Ian Grainge (University of Newcastle, Australia); Alan Grossman (Massachusetts Institute of Technology, USA); Crisanto Gutierrez (Centro de Biología Molecular Severo Ochoa, Spain); Samir Hamdan (Harvard Medical School, USA); Barbara Hamkalo (University of California, Irvine, USA); Joyce Hamlin (University of Virginia, USA); Lea Harrington (Wellcome Trust Centre for Cell Biology, University of Edinburgh, UK); Rob Hoeben (Leiden University Medical Center, Netherlands); Joel Huberman (Roswell Park Cancer Institute, USA); Erik Johansson (Umeå University, Sweden); Reid Johnson (University of California, Santa Barbara, USA); Jon Kaguni (Michigan State University, USA); Geoff Kapler (Texas A&M Health Science Center, USA); Paul Kaufman (University of Massachusetts, USA); Stephen Kearsey (University of Oxford, UK); Rolf Knippers (University of Konstanz, Germany); Robert Kuchta (University of Colorado, USA); Thomas Kunkel (National Institute of Environmental Health Sciences, USA); Chrissie Lee (National Institute of Child Health and Human Development, USA); Robert Lehman (Stanford University School of Medicine, USA); Tomas Lindahl (Cancer Research UK London Research Institute, UK); Stuart Linn (University of California, Berkeley, USA); David MacAlpine (Duke University Medical Center, USA); Ken Marians (Memorial Sloan-Kettering Cancer Center, USA); Francesca Mariotti (Wellcome Trust Centre for Cell Biology, University of Edinburgh, UK); Charles McHenry (University of Colorado, USA); Roger McMacken (Johns Hopkins University, USA); Paul Modrich (Howard Hughes Medical Institute, Duke University, USA); Andrew Murray (Harvard University, USA); Christopher Pearson (SickKids Research Institute, Toronto, Canada); Yves Pommier (National Cancer Institute, USA); Nick Rhind (University of Massachusetts, USA); Charles Richardson (Harvard Medical School, USA); Etienne Schwob (Centre National de la Recherche Scientifique, France); Polina Shcherbakova (University of Nebraska Medical Center, USA); José Sogo (ETH Hönggerberg, Switzerland); Paul Speight (University of Sheffield, UK); Bruce Stillman (Cold Spring Harbor Laboratory, USA); Sandra Weller (University of Connecticut Health Center, USA); Sam Wilson (National Institute of Environmental Health Sciences, USA); Marc Wold (University of Iowa, USA); Roger Woodgate (National Institute of Child Health and Human Development, USA).

NOMENCLATURE

Conventions in the naming of genes and proteins varies considerably among the different organisms as to when to use CAPITALS, Title Case, and *italics*, but for the sake of simplicity Title Case is used for the names of all proteins and *italicized Title Case* for the names of all genes.

Names of individual proteins are in Title Case, but the names of protein complexes are in CAPITALS. Thus, the eukaryal MCM helicase consists of six proteins termed Mcm2, Mcm3, Mcm4, Mcm5, Mcm6, and Mcm7. Unfortunately, the names of genes, proteins, and protein complexes can be very confusing, not only because different names are used to describe the same protein (gene) in different organisms, but because nomenclature developed over time as information became available.

For a detailed comparison of nomenclature for proteins involved in genome duplication in mammals, budding yeast, fission yeast, flies, amphibians, plants, and archaea, see the Appendix in DePamphilis, M.L. (ed.) (2006) *DNA Replication and Human Disease* (Cold Spring Harbor Laboratory Press, Cold Spring Harbor, NY). Nomenclature for all genes of individual species are available via the following websites.

Escherichia coli: http://ecolihub.org/

Sulfolobus solfataricus: http://www-archbac.u-psud.fr/projects/sulfolobus/

Saccharomyces cerevisiae, *Saccharomyces* Genome Database: http://www.yeastgenome.org

Schizosaccharomyces pombe, "PombePD" BioKnowledge Library: http://www.proteome.com

Drosophila melanogaster, "FlyBase": http://flybase.bio.indiana.edu

Xenopus laevis, "Xenbase": http://www.xenbase.org

Homo sapiens, NCBI database: http://www.ncbi.nlm.nih.gov/entrez/query.fcgi?CMD=search&DB=gene

the HUGO Gene Nomenclature Committee: http://www.genenames.org/

"HumanPSD" & "GPCR-PD", BioKnowledge Library: http://www.proteome.com

Arabidopsis thaliana Database: http://www.arabidopsis.org

Arabidopsis thaliana project: http://mips.helmholtz-muenchen.de/plant/athal/

STYLE

A hyphen (-) is used to join parts of protein names, such as human DNA polymerase-α or DNA Pol-α•primase, and *E. coli* DNA polymerase-I or DNA Pol-I.

A dot (•) is used to denote protein•protein interactions, such as Cdk1•CcnB protein kinase, and H2A•H2B dimer, rather than the conventional hyphen in an effort to avoid ambiguities between names of proteins (which are often hyphenated) and protein interactions.

A slash (/) is used to separate alternative names. For example, Cdk1/Cdc25/Cdc2/CdkA are names for the same protein or gene in mammals, budding yeast, fission yeast, and the plant arabidopsis, respectively.

The figures in this book are freely available in jpeg format at http://www.garlandscience.com

DEDICATION

This book is dedicated to the memory of Arthur Kornberg, biochemist, teacher, mentor, acknowledged father to the field of DNA replication, paternal father to three sons, and *pater familias* to a multitude of scientists who had the good fortune of knowing him. Arthur and his wife, Sylvy Kornberg, shared their love for research, science, and family, and worked together in the Microbiology Department that Arthur founded in 1953 at Washington University, St. Louis. In 1959, Arthur Kornberg, along with Sylvy and colleagues from St. Louis, established a new Department of Biochemistry at Stanford University. Also in 1959, and just six years after James Watson and Francis Crick reported the structure of DNA, Arthur Kornberg received the Nobel Prize in Physiology or Medicine. The prize was awarded for his discovery of the enzymatic mechanism of DNA synthesis and for the identification and characterization of the first enzyme that synthesizes duplex DNA in a template-directed manner. Arthur Kornberg's contributions to DNA enzymology are monumental, and Paul Berg, in a tribute to his mentor, friend, and colleague of more than fifty years, wrote that Arthur Kornberg's "achievements influenced a generation of biochemists to undertake problems as seemingly intractable as gene expression, signal transduction, intracellular protein transport, and many others. The ability to clone, amplify, and sequence genes that constitutes the "biological revolution" followed in large measure from the identification and purification of polymerases, ligases, nucleases, and related enzymes that emerged from these studies." Arthur Kornberg was truly a legend in his own time, and we all miss him.

Paul Berg, "Arthur Kornberg", Proceedings Of The American Philosophical Society Vol. 153, No. 4, December 2009.

Arthur Kornberg (1918–2007) and his wife Sylvy Ruth Kornberg (1917–1986) in their laboratory in the Department of Biochemistry, Stanford University, CA, 1959.

CONTENTS IN BRIEF

CONTENTS IN DETAIL

Chapter 1
GENOMES

The **genome** is the sum total of all the **genetic** information required to reproduce a particular organism. A genome contains all of the **genes** that encode the proteins and RNA molecules necessary for building and maintaining an organism, as well as all of the genetic information required to regulate the expression of these genes so that their products appear at the correct time during development, and in the correct cell type. All of this genetic information – information inherited from one generation to the next – constitutes the genome. Genomes are found both in cellular organisms and in the **viruses** and **bacteriophages** that replicate in them.

What all cellular organisms share in common is the molecule that encodes their genome: the double-stranded form of deoxyribonucleic acid commonly known as **duplex DNA**, double-stranded DNA, or simply DNA. Genomes of viruses, on the other hand, consist of either double-stranded DNA (dsDNA) or single-stranded DNA (ssDNA) as well as double-stranded or single-stranded ribonucleic acid (RNA). Cellular organisms can also harbor **episomes**, small duplex DNA molecules that replicate independently of the cell's chromosomes. In **bacteria**, episomes are also referred to as plasmids. Bacterial **plasmids** are circular dsDNA molecules that replicate autonomously in their host cell. Their size varies from 1 to over 400 kilobase pairs (kbp), and cells contain from one to hundreds of copies, with a typical copy number of between 1 and 20. The reason that viruses and episomes cannot reproduce themselves outside of their host cell – and are therefore not alive – is that part of their genome is encoded by the genome of their host cell. One bizarre parasitic organism, the prion, depends completely on the DNA genome of the host. Prions are infectious particles composed only of protein that propagate by refolding abnormally into a structure that is able to convert a normal cellular protein into an abnormal cellular protein (the prion).

Genomes need not be contained within a single, huge molecule. In fact, genomes greater than approximately 10 kbp are generally segmented into molecules of smaller size, and one or more segments of a genome may be stored in different places. Among the eukarya, most of the genome is stored within the nucleus, but a relatively tiny portion is stored within the mitochondrion in animals and additionally in the chloroplast in plants. Each of these DNA molecules encodes genes specific for the function of the organelle that harbors it. Since these organelles are responsible for generating ATP, their small DNA molecules (16–2600 kbp) are nevertheless important for the survival of the organism.

DNA does not exist alone in nature; it exists only in association with proteins. DNA in all cellular organisms is organized into a DNA–protein complex called chromatin in which the acidic (i.e. negatively charged at neutral pH) DNA is wrapped around a stable complex of basic (i.e. positively charged at neutral pH) proteins. These basic proteins are called histones in eukarya, histones or histone-like proteins in archaea, and histone-like proteins in bacteria. They bind DNA tightly to form a structure called the **nucleosome**. Although the composition and organization of chromatin differs between bacterial, archaeal, and eukaryal chromosomes, the principle is the same: compacting DNA into a small space while still providing a dynamic structure that permits variation in the expression of its genes. Many other proteins also bind to chromatin. Some of them bind to the DNA, some to other proteins.

The small genomes of bacteria, archaea, viruses, and episomes are stored in a single piece of chromatin called a **chromosome**. These chromosomes are usually, but not always, circular. The much larger genomes of eukarya are stored in many individual chromosomes. Eukaryal chromosomes are normally linear and terminated at each end by a unique structure called the telomere. When chromosomes are duplicated during cell division, the two sibling chromosomes are called **sister chromatids**. The number of copies of each chromosome per cell, referred to as **ploidy**, can vary with bacterial and archaeal species ranging from examples of **haploid** organisms (containing a single copy of their genome) to **polyploid** organisms (many copies of their chromosome). Single-cell eukaryotes, such as yeast, can exist in both a haploid and a **diploid** (two copies of the genome) state. **Gametes** (eggs and sperm) of plants and animals are haploid, but most of their other cells are diploid. Some specialized cells of plants and animals are polyploid, but they are terminally differentiated and do not undergo cell division.

GENOME COMPOSITION

DNA

The modern science of genetics, the study of heredity and variation in living organisms, began with the observations by Gregor Mendel (published in 1866, but largely ignored for the next 35 years) that organisms inherit traits in discrete units. Two concepts emerged from Mendel's work. The "Law of Segregation" states that when individuals within a species reproduce, each gamete receives only one copy of each hereditary unit. The "Law of Independent Assortment" states that each hereditary unit assorts independently of the others during gamete formation. Mendel's hereditary units are what we now call genes, and the second law is true only for genes on different chromosomes.

The discovery of what genes are and how they are duplicated each time a cell reproduces itself began in 1869 when Friedrich Miescher, a student in the laboratory of Felix Hoppe-Seyler at the University of Tübingen, Germany, discovered DNA. At that time, it was thought that cells consisted largely of protein, but Miescher noted the presence of something in pus cells that "cannot belong among any of the protein substances known hitherto." Since it was derived entirely from the cell's nucleus, he named it nuclein. Subsequently, Miescher discovered that nuclein could be obtained from many other cells, and that it was unusual in that it contained phosphorus as well as carbon, oxygen, nitrogen, and hydrogen. Miescher's work was not published until 1871, because Hoppe-Seyler insisted on first confirming its reproducibility. Miescher died in 1895, long before the significance of his discovery was appreciated. By 1893, another student of Hoppe-Seyler, Albrecht Kossel, had discovered that nuclein contained four nucleic acid bases, for which he received the Nobel Prize in Physiology or Medicine in 1910. Nevertheless, the credit for laying the foundation for determining the structure of nucleic acids belongs to Phoebus Levene. At the beginning of the twentieth century, Levene and his coworkers at the Rockefeller Institute in New York City elucidated the structures of both sugar components of the nucleic acids, established the nature of the nucleosides and nucleotides, and proposed a structure for nucleic acids that consisted of a linear combination of nucleotides with the correct internucleotide linkage. However, the biological significance of DNA remained unrecognized until 1944. The complexity of proteins and polysaccharides had long been established by that time, but the chemistry of nucleic acids suggested that they were simple polymers of a tetranucleotide repeat. It is doubtful that anyone seriously thought that such a simple chemical substance as DNA could encode all the genetic information required to reproduce a living organism.

The first step toward identification of DNA as the genetic material was taken in 1928 with the discovery of genetic transformation of bacteria by the British microbiologist Frederick Griffith. In trying to understand the nature of pneumonia, a leading killer of the day, Griffith observed that if he inoculated mice with a nonvirulent form of *Streptococcus pneumoniae* together with a heat-killed virulent form of this bacterium, the nonvirulent form readily became virulent. Griffith's discovery provided a way to identify the chemical nature of genetic information. But the true nature of the genetic material awaited the breakthrough work of Oswald Avery, Colin MacLeod, and Maclyn McCarty at the Rockefeller Institute. They succeeded in purifying pneumococcus transforming factor in 1944 and identified it as DNA! Understandably, this discovery was met with a great deal of skepticism. DNA was considered by many to be too simple a molecule to house our genes. More likely, it was thought, DNA was simply the molecule upon which proteins were organized in such a way as to encode genetic information. The debate continued until 1952 when Alfred Hershey and Martha Chase demonstrated that when bacteriophage T2 infects *Escherichia coli* most of the phage DNA enters the cell while most of the phage protein does not. This discovery, when taken together with the Nobel Prize-winning work of James Watson and Francis Crick, based on the X-ray crystallographic data of Rosalind Franklin and the base-pairing rules of Erwin Chargaff, finally settled the question. The structure of DNA was a double helix, consisting of two intertwined polymers of deoxyribonucleotides in which every adenine in one strand is paired with a thymine in the other strand, and every guanine in one strand is paired with a cytosine in the other strand. It is now clear that DNA exists in nature either as ssDNA, or as dsDNA, and the sequence composition of DNA varies extensively from one species to another. However, without exception, the genomes of all cellular organisms, as well as their episomes, are composed of dsDNA. Watson and Crick concluded in their landmark publication in 1953 that "It has not escaped our notice that the specific [base] pairing we have postulated immediately suggests a possible copying mechanism for the genetic material [DNA]": a prophetic statement, if ever there was one.

The role of proteins in this story is not over. During the past 50 years, the field of epigenetics has come into its own. Epigenetics refers to heritable changes either in the visible characteristics of an organism (phenotype), or in the pattern of gene expression during an organism's development that are caused by mechanisms other than changes in DNA sequence. For example, totipotent stem cells differentiate into various pluripotent cell lineages that in turn differentiate into the myriad of specialized cells that comprise an adult organism. In mammals, examples would include the development of neurons, muscles, skin, and blood. Epigenetic changes come about through proteins that bind to, and in some cases chemically modify, DNA. DNA does house the genes, but the proteins encoded by these genes regulate the activity of genes. The answer to the question of whether DNA or protein is the heritable material has come to include both.

Composition Of DNA

The structure of DNA in cellular organisms is unique and universal, but its composition is not. DNA is an unbranched polymer consisting of four major subunits called deoxyribonucleoside monophosphates (**dNMPs**).

Each dNMP consists of one of four bases: adenine, guanine, cytosine, and thymine. Adenine and guanine are purines, cytosine and thymine are pyrimidines. **Purines** are aromatic, heterocyclic bases that consist of a six-member pyrimidine ring fused to a five-member imidazole ring (Figure 1-1B). **Pyrimidines** are aromatic, heterocyclic bases that contain two nitrogen atoms in a six-member ring. Each dNMP consists of a five-carbon sugar whose 1′-C is linked by an **N-glycosidic bond** to one of the nitrogens in either a purine or pyrimidine base, and whose 5′-C is linked by a **phosphodiester bond** to a phosphate (Figure 1-1A). Deoxyribonucleoside diphosphates (**dNDPs**) are identical to dNMPs except that they contain an additional phosphate linked to the dNMP phosphate by a **phosphoanhydride bond**. Deoxyribonucleoside triphosphates (**dNTPs**) contain a pyrophosphate linked to the dNMP phosphate via a phosphoanhydride bond (Figure 1-1A). The names and abbreviations for each of the bases, nucleosides, and nucleotides found in DNA are in Table 1-1. Thus, DNA is composed of four different deoxyribonucleosides [deoxyadenosine (dA), deoxyguanosine (dG), deoxycytidine (dC), and deoxythymidine (dT)] and four different deoxyribonucleotides (dAMP, dGMP, dCMP, and dTMP). The four dNMPs can be arranged in any sequence, allowing DNA to encode an almost endless number of genes.

The composition of DNA differs from its close relative, RNA, in only two critical ways. First, the sugar moiety in RNA is ribose instead of 2′-deoxyribose (Figure 1-1A), thus giving rise to the names deoxynucleotide for DNA and nucleotide for RNA. Second, RNA contains uracil in place of thymine (5-methyluracil) (Figure 1-1B; Table 1-1). RNA and DNA both contain adenine, guanine, and cytosine.

DNA STRUCTURE

Primary Structure

The primary structure of a biopolymeric molecule refers to the exact specification of its atomic composition and the chemical bonds connecting those atoms. For a typical unbranched, uncross-linked biopolymer such as

Table 1-1. Names And Abbreviations For Nucleosides And Nucleotides

	Bases	Nucleosides	Nucleotides		
			Monophosphate	Diphosphate	Triphosphate
Purine	Adenine (A)	Adenosine (A)	AMP	ADP	ATP
		Deoxyadenosine (dA)	dAMP	dADP	dATP
	Guanine (G)	Guanosine (G)	GMP	GDP	GTP
		Deoxyguanosine (dG)	dGMP	dGDP	dGTP
Pyrimidine	Cytosine (C)	Cytidine (C)	CMP	CDP	CTP
		Deoxycytidine (dC)	dCMP	dCDP	dCTP
	Thymine (T)	Thymidine or deoxythymidine (dT)	dTMP	dTDP	dTTP
	Uracil (U)	Uridine (U)	UMP	UDP	UTP

Nucleosides and nucleotides containing ribose have no prefix, but those containing deoxyribose carry the prefix *deoxy-*, which is abbreviated to 'd'. Since thymine never appears with ribose, deoxythymidine is often abbreviated simply as T. Nucleotides are written using the name of the nucleoside followed by the number of phosphates (e.g. adenosine monophosphate).

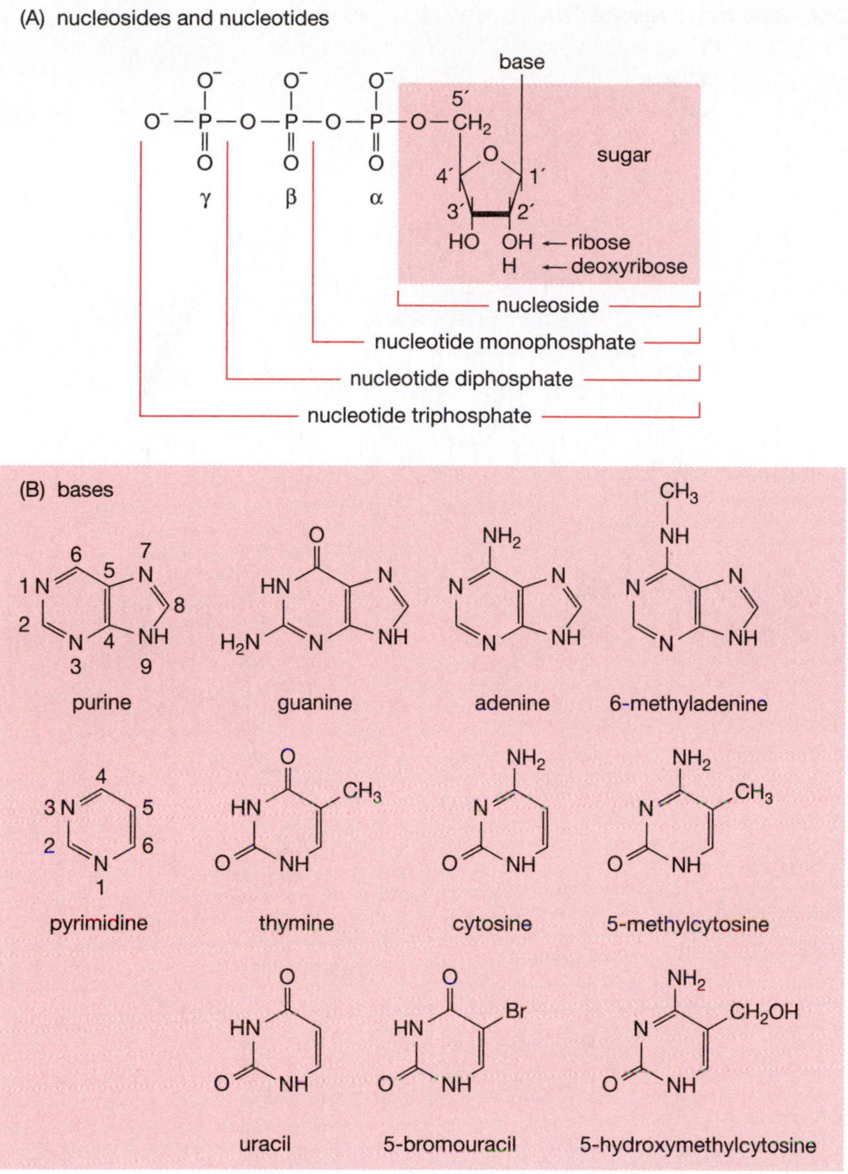

(A) nucleosides and nucleotides

base

sugar

ribose

deoxyribose

nucleoside

nucleotide monophosphate

nucleotide diphosphate

nucleotide triphosphate

(B) bases

purine

guanine

adenine

6-methyladenine

pyrimidine

thymine

cytosine

5-methylcytosine

uracil

5-bromouracil

5-hydroxymethylcytosine

Figure 1-1. Primary structure of DNA and RNA. (A) Structure of nucleosides and nucleotides. The carbon atoms in ribose are numbered 1′, 2′, 3′, 4′, and 5′ to distinguish them from the numbered atoms in purines and pyrimidines. RNA contains ribose. DNA contains deoxyribose. (B) Purine and pyrimidine bases commonly found in DNA genomes. RNA always contains uracil in place of thymine. 5-Bromouracil is a chemically modified form of uracil that can substitute for thymine.

DNA, RNA, and protein, the primary structure is equivalent to specifying the sequence of its monomeric subunits. For example, the dNMPs in DNA and the NMPs in RNA are linked ‘head-to-tail’ with the phosphate linked to the 5′-C of one nucleoside and the 3′-C of the next (Figure 1-2A). Moreover, RNA can be linked directly to DNA, as in the example of RNA-primed nascent DNA chains. Such sequences can be represented as stick diagrams (Figure 1-2B), although the primary structure of DNA and RNA (i.e. their nucleotide sequence) is generally represented simply as a sequence of letters indicating the appropriate base (for example, ACTG = 5′-ACTG-3′ = 5′-adenosine-cytidine-thymidine-guanosine-3′).

Secondary Structure

Secondary structure refers to the three-dimensional form of local segments within the polymer. In nucleic acids, secondary structure is defined both by hydrogen bonding between bases, and by nonelectrostatic interactions that promote bases to stack one on top of the other. For example, DNA in cellular genomes consists of two polydeoxyribonucleotide strands in

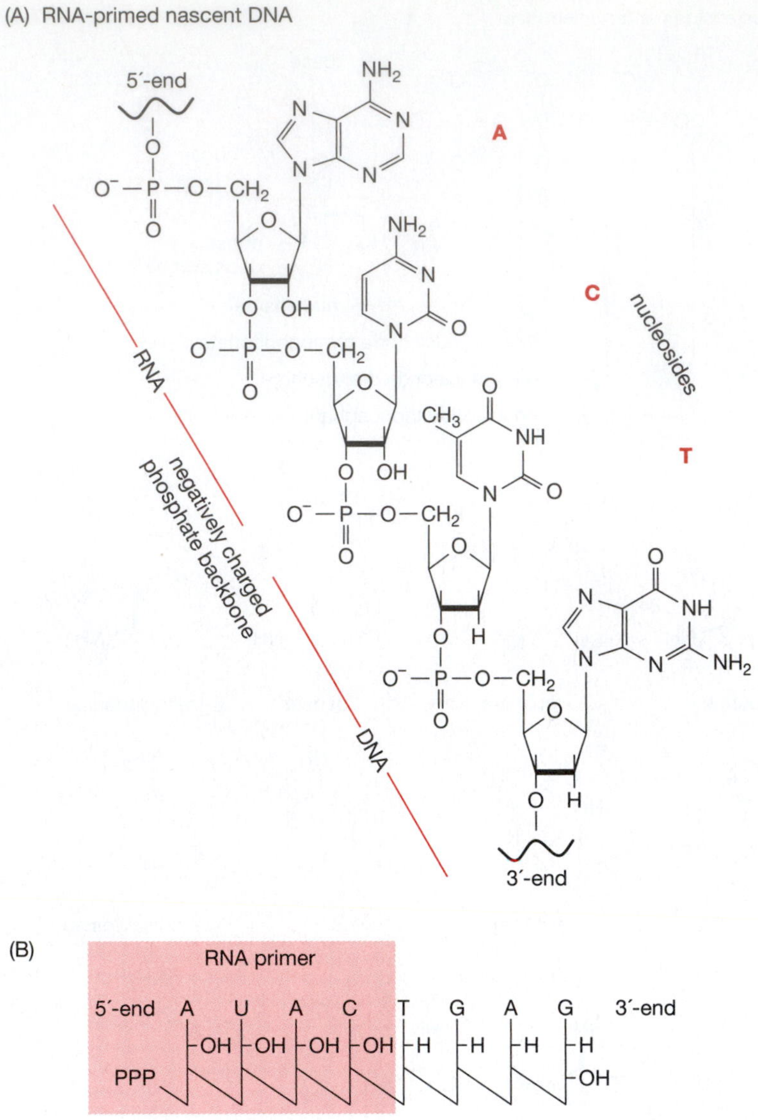

Figure 1-2. Phosphodiester covalent links between 3′- and 5′-hydroxyl groups on the ribose and deoxyribose sugar moieties in RNA, DNA, and RNA-p-DNA chains. (A) RNA-p-DNA chain of the type found during DNA replication. One oxygen of each phosphoryl group is negatively charged at neutral pH, thereby creating a polyanionic molecule. (B) Abbreviated format for structure in (A). Note the 5′→3′ polarity of the molecule.

which the sugar-phosphate linkages form the backbone of a structure that is held together by **hydrogen bonds** between base pairs (Figure 1-3). Two features of **base-pairing** interactions have profound consequences for the mechanism by which the DNA sequence is copied. First, adenine in one strand always pairs with thymine in the other, and guanine in one strand always pairs with cytosine in the other (Watson–Crick base pairs). Therefore, one strand of DNA is not an exact copy of the other strand; it is a complementary copy of the other strand. Moreover, since only specific pairs of bases are allowed, it follows that if the sequence of bases on one chain is given, then the sequence on the other chain is determined automatically. The fact that GC base pairs form three hydrogen bonds whereas AT base pairs form only two means that GC-rich DNA melts at higher temperatures than AT-rich DNA. Therefore, AT-rich regions are easier to unwind during genome duplication.

The second feature of DNA secondary structure is that the two polydeoxynucleotide strands are aligned in opposite directions (**antiparallel**) so that the 5′-end of one strand is opposite the 3′-end of the other strand. Since all **DNA polymerases** synthesize DNA only in one direction (5′→3′), it follows that DNA synthesis on one template strand must occur in the direction opposite to DNA synthesis on the complementary

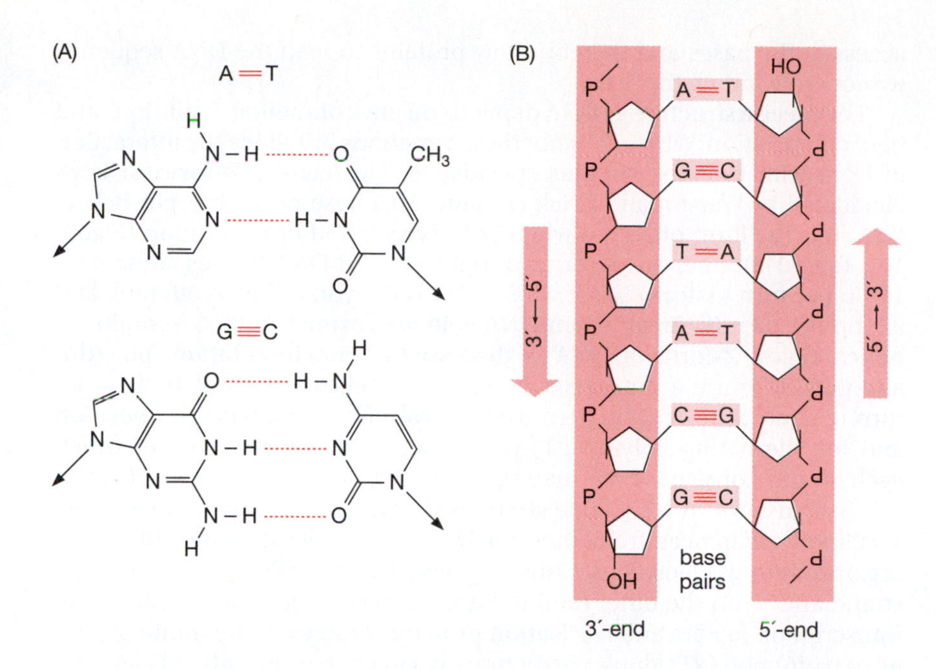

Figure 1-3. DNA secondary structure. (A) Watson–Crick base pairs in dsDNA contain two hydrogen bonds (dashed lines) between adenine and thymine, and three hydrogen bonds between guanine and cytosine. Arrows indicate where deoxyribose is attached. (B) Two polydeoxyribonucleotide strands with their sugar-phosphate backbones (pink panels) aligned antiparallel to one another (5′→3′:3′←5′).

template strand. One consequence of base-pairing and **base-stacking** interactions is that each of the two strands must coil around the same axis, with both strands forming right-handed helices, thereby giving rise to the term double helix (Figure 1-4). The sugar-phosphate backbones of each strand lie on the outside of the helix, giving it a polyanionic surface that acts like an ion-exchange resin for positively charged proteins. The purine-pyrimidine base-pairing rules produce a DNA structure with a fairly uniform diameter of about 20 Å. When the relative sizes of individual atoms are considered, space-filling models (Figure 1-4B) reveal that the base pairs are neatly stacked one on top of the other with the presence of a major and a minor groove in the helix. In B-form DNA (see below) the minor groove (12–13 Å) is about 68% the width of the major groove (18–19 Å). Since the bases are inside the helix and the phosphates are outside, these grooves provide

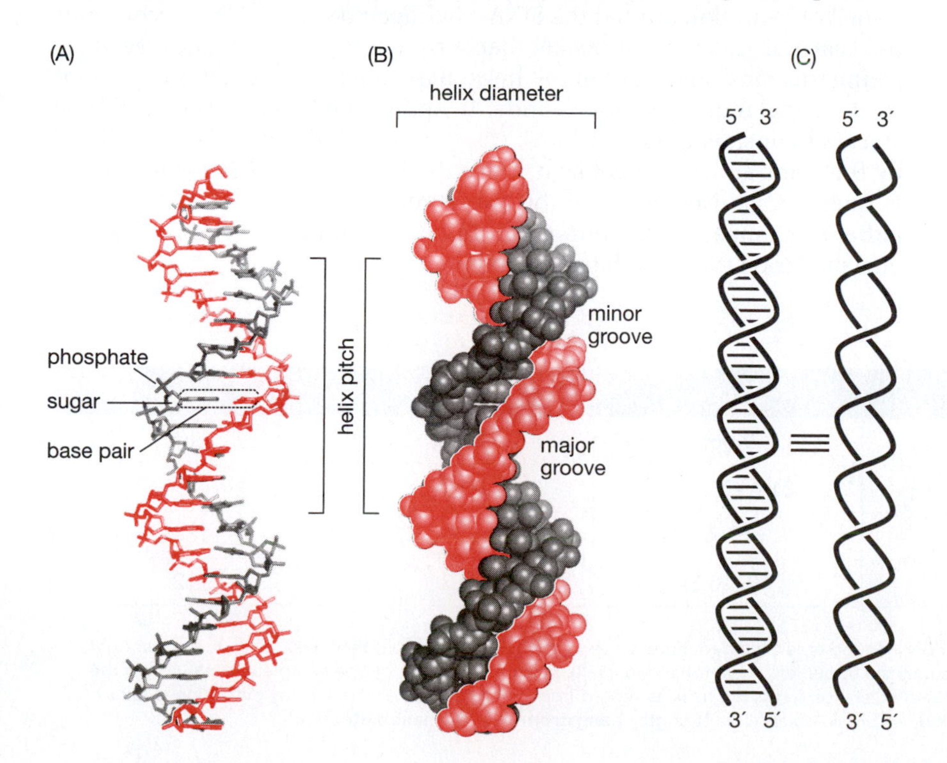

Figure 1-4. The three-dimensional structure of DNA can be represented either by sticks showing the direction of each covalent bond (A) or by spheres representing the space occupied by each atom (B). In these images of B-form DNA, only P, O, N, and C are indicated, hydrogens have been omitted for simplicity. Also for the sake of simplicity, a simple cartoon of DNA structure (C) is often used, and often without indicating base pairs. A right-handed helix wraps about its axis in the direction of the fingers on the right hand. A left-handed helix wraps in the opposite direction. Handedness (or chirality) is a property of the helix, not of the perspective: a right-handed helix cannot be turned or flipped to look like a left-handed one unless it is viewed in a mirror.

access to the bases and thereby allow proteins to read the DNA sequence without unwinding the helix.

The precise structure of DNA depends on environmental conditions and base composition. Whether or not these variations influence the interaction of DNA with proteins remains speculative. The 'classic' **B-form** of DNA elucidated by Watson and Crick contains 10.5 base pairs (bp) per helical turn. It is the form observed when DNA is hydrated or in solution (Figure 1-4; Table 1-2). Alternative forms of right-handed DNA helices containing 11 bp per turn (**A-form**) exist when DNA is dehydrated with ethanol, and 9.3 bp per turn (**C-form**) when DNA is in approximately 60% humidity. A noncanonical Z-form of DNA is discussed below. In solution, poly(dA) and poly(dT) form a duplex structure with a helical repeat of 10.0 bp per turn in contrast to 10.5 bp per turn for DNA of random base composition and for alternating poly(dA-dT) (a synthetic dsDNA molecule in which each strand consists of alternating A and T bases). Poly(dA):poly(dT) (a synthetic dsDNA in which one strand is dA and one strand is dT) exhibits structural differences from random DNA that make it more difficult to organize into nucleosomes. Thus, regions of native DNA with As on one strand and Ts on the other tend to be more accessible to interaction with transcription factors and replication proteins. Moreover, the minor groove in poly(dA):poly(dT) duplex structures is larger than in native DNA, and therefore more accommodating to interaction with proteins. For example, one subunit of the origin-recognition complex in fission yeast (Orc4) binds tightly and specifically to the minor groove in the asymmetric A:T-rich DNA regions that are common to fission yeast DNA-replication origins. Finally, RNA:DNA hybrids in which one strand of the duplex is RNA and one strand is DNA adopt the A-form. Since RNA:DNA duplexes form at DNA replication forks each time RNA-primed DNA synthesis is initiated during DNA replication, the transition from an A-form (RNA:DNA hybrid duplex) to a B-form (duplex DNA) structure may trigger the release of one enzyme and the loading of another.

The stability of duplex DNA comes both from hydrogen bonds between the bases in the two strands (base-pairing), and from the hydrophobic and van der Waals interactions between bases in the same strand (base stacking). In addition, the helix is covered with water molecules that form a shell of hydration around the DNA. Hydrogen bonds in DNA account for 2–3 kcal mol^{-1}, somewhat weaker than most hydrogen bonds because of the geometric constraints within the helix. Base-stacking interactions account for 4–15 kcal mol^{-1} per dinucleotide. To appreciate the importance of base stacking, consider a stack of coins. The position of any one coin is stabilized by the coins above and below it. Once the DNA double helix is formed, it is quite stable. Nevertheless, the two strands can be unwound (melted) either by changes in environmental conditions, or by the action of specific enzymes and DNA-destabilizing proteins.

Table 1-2. DNA Secondary Structure

Parameter	A-form	B-form	C-form	Z-form
Helix sense	Right	Right	Right	Left
Helix pitch	28 Å	34 Å (35.7 Å)	31 Å	45 Å
Helix diameter	23 Å	20 Å		18 Å
Base pairs per turn	11	10 (10.5)	9.3	12

Helix sense refers to the fact that helices are either right-handed or left-handed. If the curve of the helix moves from the lower left to the upper right, then it is right-handed. If it moves from the lower right to the upper left, it is left-handed. **Helix pitch** is the length of one complete turn (360°) of the helix measured along the helix axis. **Helix diameter** is the distance across the helix as viewed from one end. Data are from X-ray diffraction studies of DNA fibers. Data in parentheses are for DNA in solution of physiological ionic strength. 1 **angstrom** (Å) = 0.1 **nanometer** (nm).

The two strands of DNA separate easily when the pH is more than 12 or less than 2. This results from ionization of the bases with concomitant disruption of the hydrogen bonds. In addition, the shell of hydration is disrupted which destabilizes base stacking. However, prolonged exposure to acid results in loss of purine and pyrimidine bases by cleavage of glycosidic bonds. Since phosphodiester bonds become more sensitive to hydrolysis at sites where a base is lost, strong acid will degrade DNA. In contrast, the DNA backbone is relatively stable in alkali. RNA, on the other hand, is slowly degraded in strong acid and rapidly degraded in strong alkali, the result of nucleophilic attack by the 2′-OH on the adjacent 3′-PO_4 group to produce cyclic 2′-(3′-)NMPs that resolve into a mixture of 2′-NMPs and 3′-NMPs. This bit of chemistry allowed identification of RNA-p-DNA **covalent bonds** in newly synthesized DNA.

The two strands of DNA separate as the temperature is raised (**denaturation**) and reanneal as the temperature is lowered (**renaturation**). The melting temperature (T_m) is defined as the temperature at which 50% of a DNA molecule forms a stable double helix and 50% exists as single strands. The T_m depends on the length of the molecule, its nucleotide composition, the type and concentration of positively charged cations and molecules present, and the pH. The classic formula, which works well for long DNA molecules, is T_m = 69.3+0.41(G+C%) in 0.2 M Na^+, pH 7.0 (Marmur and Doty, 1962; for short oligodeoxynucleotides, see Owczarzy *et al.*, 1997). The greater the GC content, the more stable the DNA.

Tertiary Structure

Tertiary structure describes atomic positions in three-dimensional space. One of the most critical properties of DNA is the fact that it is not topologically rigid like a steel rod, but flexible like a strand of spaghetti *al dente*; it can bend, coil, and fold-up upon itself. DNA topology, as it relates to genome duplication, takes the form of DNA bending and DNA superhelical turns.

Bent DNA

DNA bending can be induced in solution either by association with DNA-binding proteins, or by its sequence composition. 'Bent DNA' itself is a sequence-directed curvature of the DNA helix that is associated frequently with replication origins in bacteria, viruses, yeast, and mammals. In random-sequence DNA, a fragment of 150–200 bp resists deformation. For example, it will not form circles *in vitro*. However, if this length of DNA consists of a series of about 10 runs of three to four As each separated by Cs and Gs so that they are phased 10 bp apart, then circles will form that can be sealed using appropriate enzymes and that can be visualized by electron microscopy. Thus, bent DNA regions have the inherent ability to either facilitate or to discourage binding of proteins from specific sites in the genome. For example, bent DNA can facilitate interaction between proteins bound to specific sites that are far apart in terms of intervening base pairs by forcing the intervening sequence to bend and thus greatly shorten the actual distance between the two sites.

Supercoiled DNA

Coiling is what happens when a flexible fiber is wrapped around a central axis (Figure 1-5A). If the ends of the fiber are then immobilized, either by attaching them to an immobile structure (Figure 1-5B) or to each other to form a circle (Figure 1-5D), and the central axis is then removed, the coils will remain, because the fiber cannot unwind about its own central axis. Now consider the flexible fiber as duplex DNA. When the pipe is withdrawn, the DNA, which already consists of two flexible fibers coiled about one another, will now be coiled around itself. These additional coils are called supercoils.

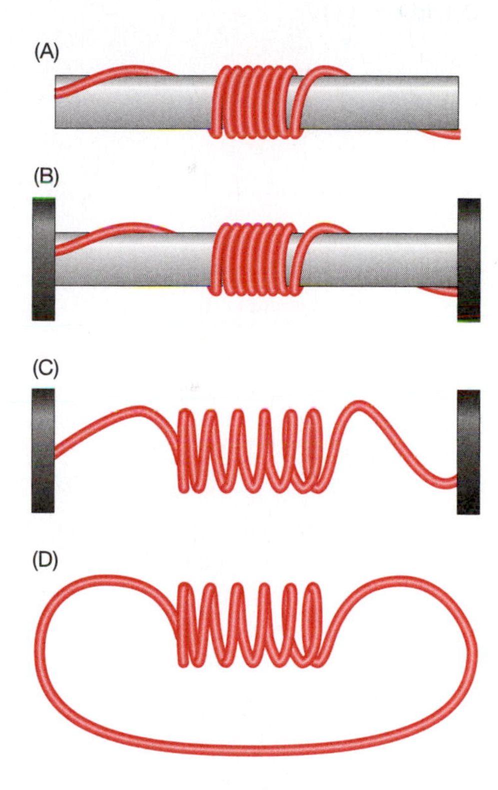

Figure 1-5. Coiling results from wrapping a flexible fiber around a central axis. Take, for example, wrapping a wire around a pipe (A). If the ends of the wire are then immobilized by attaching them either to an immobile structure (B, the black bars), or to each other (D), the coils are trapped. They will remain even after the pipe is removed (C, D). Panel C is analogous to the situation with long linear DNA molecules coiled around nucleosomes and attached to components of the nucleus. Panel D is analogous to circular DNA molecules. All cellular organisms contain enzymes called topoisomerases that allow DNA to undergo such topological changes.

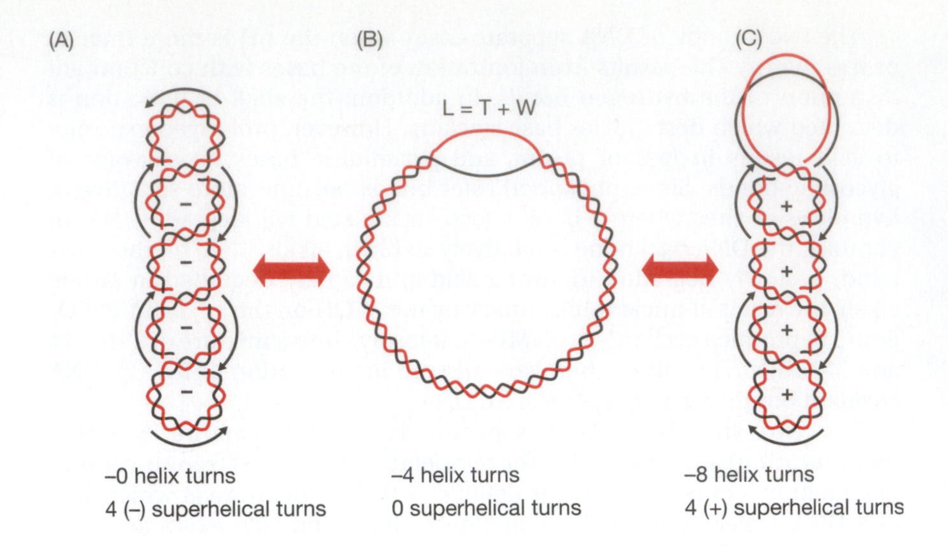

Figure 1-6. Three topologically equivalent DNA molecules. (A) Circular, double-stranded, covalently closed B-form DNA molecule containing four negative superhelical turns. Negative superhelical turns are in the same direction as right-handed helical turns in B-form DNA (indicated by the red arrow crossing over the black arrow). (B) The same molecule shown in (A), but without four of its right-handed turns in its DNA helix, as indicated by the ssDNA bubble. This molecule can now lie flat on a planar surface. (C) The same molecule shown in (B), but absent four more helical turns. This change is compensated by four positive superhelical turns. Positive superhelical turns are in the 'left-handed direction' (indicated by the red arrow crossing over the black arrow). (Adapted from Schvartzman, J.B. and Stasiak, A. (2004) *EMBO Rep.* 5: 256–261. With permission from Macmillan Publishers Ltd.)

Supercoiling is a fundamental property of the helical structure of DNA. To understand this property, consider a circular B-form DNA molecule that is covalently closed (no nicks or gaps in either strand). In such molecules, one strand cannot be rotated about the other without causing the helix to twist upon itself (Figure 1-6). This twisting is called '**supercoiling**'. When DNA is forced to lie flat on a surface the number of times one strand crosses the other is defined as the **linking number** (L). There are two kinds of crossover: those within the dsDNA helix itself (1 per 10.5 bp in the case of B-DNA), and those that occur when the dsDNA becomes coiled (when the entire helix is twisted about itself). L is equal to the number of twists about the helical axis (T) plus the writhe (W) of the helical axis through three-dimensional space. The topological state of a DNA molecule can be defined by the simple equation, L = T+W. **Negative superhelical turns** are right-handed coils (Figure 1-6A). **Positive superhelical turns** are left-handed coils (Figure 1-6C). As long as the DNA molecule is part of a topologically closed system (one that does not permit it to rotate about its own helical axis), then any changes in the number of helix turns must be compensated by the same number of superhelical turns in the *opposite direction.*

Imagine that one strand of the DNA helix is broken to allow free rotation about the phosphodiester bond in the opposite strand. Now imagine that the two strands of the circular molecule are separated (unlinked) by four turns of the helix, and then the helix is resealed. The resulting DNA would now be under-wound by four helical turns (Figure 1-6B). If the unpaired bases are then rejoined via hydrogen bonding – without breaking and rejoining one of the two sugar-phosphate backbones – the linking number would be maintained by adding four negative superhelical turns (Figure 1-6A). Superhelical turns in DNA are also called supercoils. Note that the direction of the superhelical turns in negative supercoils is the same as the direction of the helical turn. Conversely, if further DNA unwinding increases the number of unlinked bases to eight helical turns – again, without breaking and rejoining one of the two strands – the linking number would be maintained by the appearance of four positive superhelical turns (Figure 1-6C), because the direction of positive superhelical turns is opposite to the direction of helical turns in B-DNA. Thus, the three DNA structures in Figure 1-6 are topologically equivalent.

SUPERHELICITY

The *Gedanken* experiment described above reveals that topological changes in DNA can make DNA unwinding either more or less difficult, and therefore the presence of DNA supercoils will either inhibit or promote any process, such as DNA replication, that requires DNA unwinding. There are four mechanisms by which nature either introduces or removes superhelical turns into DNA. Three of these mechanisms (chromatin assembly and disassembly, DNA topoisomerases, and DNA unwinding) are discussed in detail in later chapters.

Superhelicity Generated By Nucleosome Assembly

Eukaryotic chromatin consists of repeating units called nucleosomes in which DNA is wrapped around an octamer of histones. When nucleosomes in a chromosome constructed from a covalently closed circular DNA molecule are disassembled in solution, the DNA contains one negative superhelical turn for each nucleosome that was present, revealing the presence of one left-handed rotation around each histone octamer. These results demonstrate that chromatin assembly is facilitated by negative supercoils in DNA, because they are topologically equivalent to negative **toroidal** turns, and negative toroidal turns are easily wrapped around the histone octamer (Figure 1-7). Similarly, negative supercoils facilitate binding of origin-recognition proteins, the first step in the assembly of pre-replication complexes. Conversely, negative superhelical turns are generated by chromatin disassembly, a process that occurs during genome duplication.

Interestingly, X-ray crystallographic analysis of nucleosomes concluded that DNA makes about 1.75 left-handed rotations around each histone octamer. Based on the model in Figure 1-5, one would therefore expect that release of the proteins from chromatin should produce about 1.75 negative superhelical turns for each nucleosome. This difference between the expected superhelicity of 1.75 turns per nucleosome (as predicted from the considerations illustrated in Figure 1-6) and the observed value of 1 is referred to as the linking number **paradox**. It reflects the fact that DNA in crystals contains 10 bp per helical turn whereas DNA in physiological solution contains 10.5 bp per helical turn (see Table 1-2), a difference that is likely due to differences in the extent of DNA hydration.

Superhelicity Generated By DNA Gyrases

All cellular organisms possess enzymes called topoisomerases that allow DNA to change its topology. Most topoisomerases act by providing a swivel about which DNA can rotate around its own helical axis. These enzymes can relax DNA by allowing it to rid itself of superhelical energy. However,

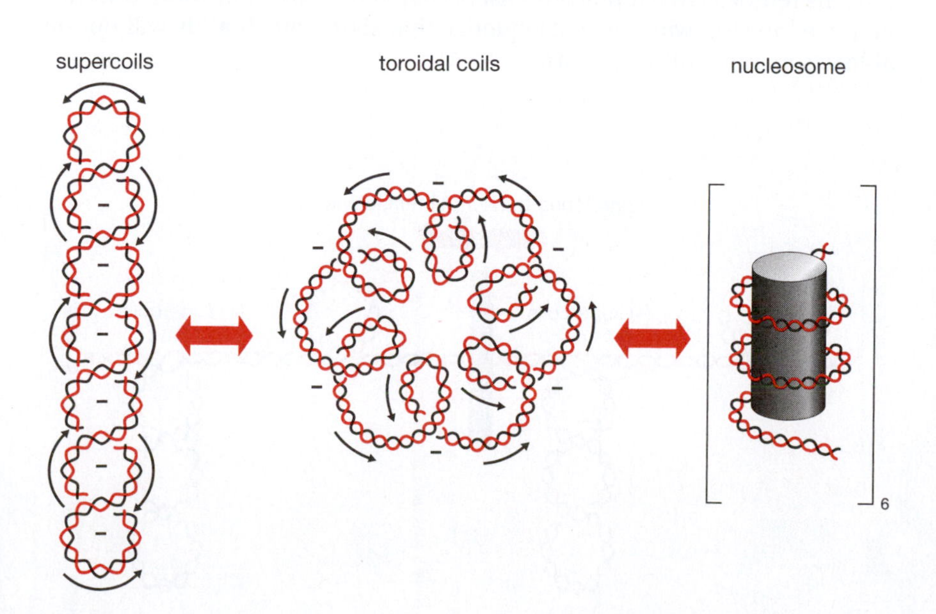

Figure 1-7. Nucleosome assembly creates supercoils. A single negative supercoil is equivalent to a single negative toroidal coil. When DNA wraps around a histone octamer [(H2A, H2B, H3, H4)$_2$] to form a nucleosome, it creates one negative toroidal coil for each nucleosome. Thus, the circular DNA pictured here with six negative superhelical turns is the topological equivalent of six nucleosomes. Note that negatively supercoiled DNA is a right-handed superhelix in a circular, covalently closed duplex, but a left-handed superhelix when seen in toroidal form (as in nucleosomes). The toroidal coil is equivalent to a wire wrapped around a cylindrical core (Figure 1-5). (Adapted from Schvartzman, J.B. and Stasiak, A. (2004) *EMBO Rep.* 5: 256–261. With permission from Macmillan Publishers Ltd.)

bacteria encode a unique topoisomerase called **DNA gyrase** that can generate negative supercoils. As with nucleosome assembly, DNA wraps around the DNA gyrase, but in a positive direction rather than a negative one. A positive supercoil is then converted directly into a negative one by passing a DNA segment through a transient double-strand break. Reversal of this procedure relaxes DNA. Each round of supercoiling is driven by a conformational change induced by ATP binding, and ATP hydrolysis then permits a new cycle to begin. Archaea contain a unique **reverse DNA gyrase** that introduces positive rather than negative superhelical turns into the genome. Positive superhelical turns inhibit DNA melting, whereas negative superhelical turns promote DNA melting (discussed below). Thus, the action of reverse DNA gyrase helps to stabilize duplex DNA structure under environmental conditions that would otherwise denature it, such as high temperatures. Eukarya lack DNA gyrases; they add and subtract supercoils through nucleosome assembly, DNA replication, and topoisomerase activity.

Superhelicity Generated By Transcription

Based on the *Gedanken* experiment in Figure 1-6, one can readily understand that any process that unwinds the DNA helix invariably generates DNA supercoils. For example, DNA transcription by RNA polymerase and its cofactors melts the DNA at the site of transcription so that it can synthesize RNA that is complementary to one of the two DNA templates. As the transcription machine moves along the DNA, the nascent RNA is released from its template, and the DNA duplex rapidly re-forms. Assuming that the DNA cannot rotate freely about its helical axis, positive supercoils will be generated in front of the site of transcription, and an equal number of negative supercoils will be generated behind (Figure 1-8). In this way, the DNA linking number remains constant.

Superhelicity Generated By Replication

DNA replication, like DNA transcription, must unwind the DNA and therefore generate superhelical turns in front of the replication machine as replication proceeds. However, the presence of single-stranded regions in each of the two newly replicated template strands allows free rotation about the phosphodiester linkage to remove the negative supercoils that would have otherwise accumulated behind the replication fork. The positive supercoils that are generated in front of the replication machine, however, must be removed by a topoisomerase. Otherwise, DNA replication forks will be forced to stop with the consequence that **sister chromatids** will not be able to separate during mitosis.

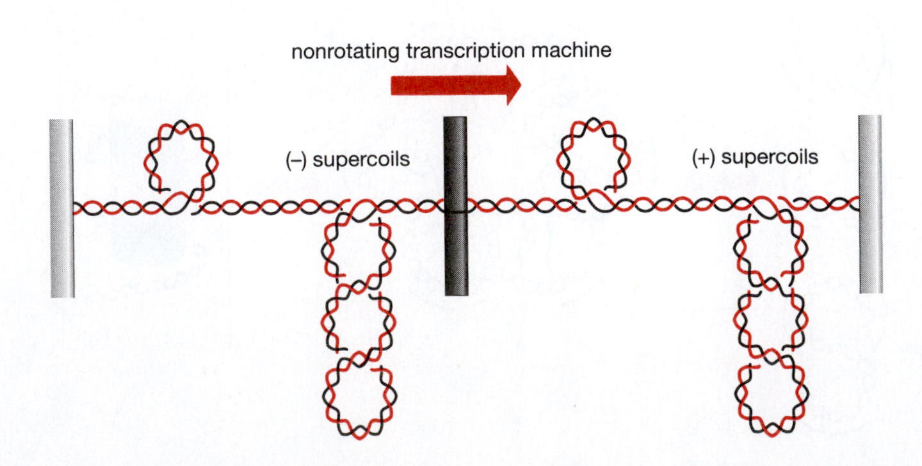

Figure 1-8. Transcription generates supercoils. The transcription machine is represented by a rod. When this rod cannot rotate around the helical axis of the DNA template, then as transcription proceeds, positive supercoils will be generated in front of the site of transcription and an equal number of negative supercoils will be generated behind. The DNA is attached to a hypothetical immobile structure (e.g. nuclear structure) that prevents it from rotating freely about its helical axis. The same would be true for covalently closed, circular DNA molecules. (Adapted from Wang, J.C. (2002) *Nat. Rev. Mol. Cell Biol.* 3: 430–440. With permission from Macmillan Publishers Ltd.)

NONCANONICAL DNA STRUCTURES

Negative supercoiling facilitates both DNA folding and compaction as well as the untwisting of DNA. Moreover, the torsional energy stored in supercoiled DNA is proportional to the square of the number of supercoils. This means that supercoiling DNA provides an enormous reservoir of energy that can be used to promote assembly of chromatin and interaction between DNA replication proteins and DNA. In addition, negative supercoiling can promote formation of noncanonical DNA structures, some of which have been implicated as regulatory elements either in DNA replication or in gene expression. One sobering thought in considering these noncanonical DNA structures is how they would affect the process of genome duplication.

Z-DNA

Z-DNA is a novel form of duplex DNA in that it is a left-handed helix (Figure 1-9) and therefore decidedly different from any of its right-handed counterparts (Table 1-2). Z-DNA occurs with GC-rich sequences (especially alternating C and G), and it is stabilized by the presence of 5-methylcytosine, polycations such as spermine and spermidine, and negatively supercoiled DNA. For example, in a circular covalently closed DNA with three negative supercoils, changing one turn of the DNA helix from right-handed to left-handed would eliminate two negative supercoils and the energy of negative supercoiling would then stabilize a small segment of Z-DNA. Antibodies prepared against Z-DNA stain specific regions of chromosomes, and these regions correlate with transcriptionally active genes. DNA transcription generates negative superhelical turns behind the transcription machine (Figure 1-8), and these negative supercoils could stabilize GC-rich sequences in a Z-DNA conformation. Since Z-DNA cannot form nucleosomes, this transition may serve to keep promoters accessible to transcription factors. Furthermore, several Z-DNA-specific binding proteins have been identified, strongly supporting the hypothesis that Z-DNA plays a biological role in regulating gene expression. When a region of Z-DNA exists within a larger B-form DNA molecule, the junction between the B- and Z-forms of DNA probably consists of three to four partially unwound base pairs. Such regions may facilitate DNA unwinding by initiator proteins, and therefore might serve as components of DNA replication origins.

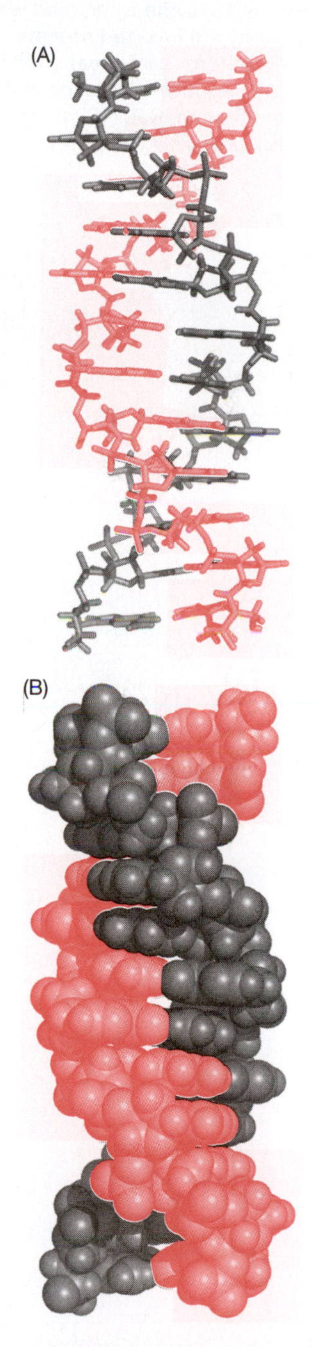

Figure 1-9. Z-DNA is a left-handed helix in which only a major groove is evident. One helical turn (12 bp) is shown as a stick model (A) and as a space-filling model (B). Hydrogen atoms are omitted, and all atoms in a single strand of the helix are colored the same.

Cruciform DNA

DNA commonly contains a large number of sites in which the base sequence is repeated either directly, or as a mirror image, or as an inversion. In **direct repeats**, the same sequence is repeated with the same polarity, whereas **mirror repeats** occur when the same sequence is repeated with mirror-image polarity, and **inverted repeats** occur when the same sequence is repeated with opposite polarity (Figures 1-10A and 1-12A). Therefore, inverted repeats are palindromes; they read the same forward or backward (remember that the two strands of DNA are antiparallel). Palindromic sequences can be represented either as a linear duplex DNA structure or as a cruciform structure (Figure 1-10B). The cruciform conformation induces a sharp kink in the DNA molecule that is localized at the apex of the extruded helix. This kink requires a loop of three to four unpaired bases. In contrast, the bend at the base of the structure does not require the presence of unpaired bases.

Inverted repeats can exist either in an extended linear conformation or a cruciform conformation, depending on conditions. Conversion of a linear inverted repeat into a cruciform requires DNA unwinding, and the rate of DNA unwinding depends on base composition, superhelical density, temperature, and ionic strength. The extended linear conformation is favored in the presence of low salt and low superhelical density. The cruciform conformation is favored in the presence of high salt and high

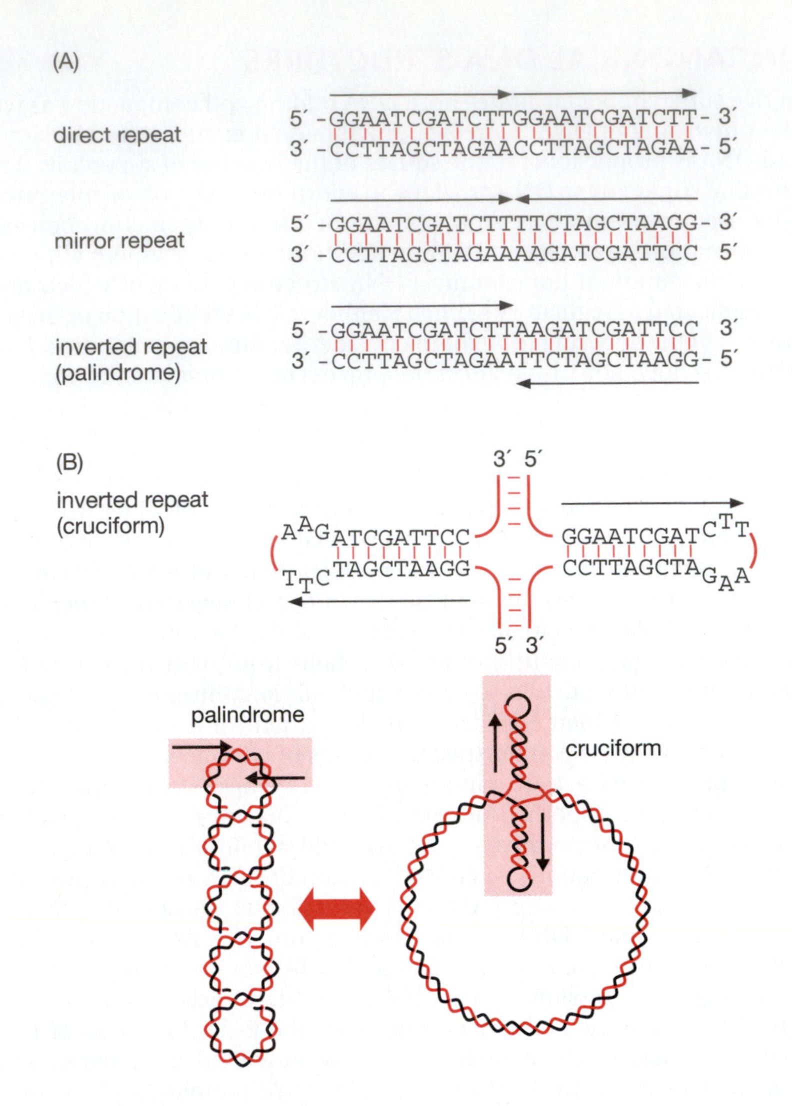

Figure 1-10. Repeated sequences in DNA. (A) Three types of repeated sequences exist in DNA. Arrows indicate the nature of the repeat: direct (a 5′→3′ sequence is repeated in the same direction), mirror (a 5′→3′ sequence is repeated but with the opposite polarity), and inverted (a 5′→3′ sequence is repeated but on the complementary strand and with opposite polarity). Inverted repeats are palindromes; the sequence is the same when read with the same polarity. (B) Inverted repeats can adopt a cruciform conformation that is stabilized by negative DNA supercoils. (Adapted from Sinden, R.R. (1994) *DNA Structure and Function*. With permission from Elsevier.)

negative superhelical density. Negative supercoils promote melting at the center of symmetry and stabilize the extruded structure. Approximately one negative superhelical turn is relaxed for every 10.5 bp of cruciform that is extruded. Therefore, the greater the negative superhelical density, the more easily cruciforms can form.

Cruciforms are more easily generated from AT-rich inverted repeats than from GC-rich repeats, because CG base pairs contain three hydrogen bonds whereas AT base pairs contain only two, and therefore CG base pairs require more energy to melt. Thus, it is not surprising that cruciforms are extruded more easily at higher temperatures than at lower ones. However, as temperatures rise above about 41°C, the rate of cruciform extrusion slows down, because other sequences in the DNA begin to melt (unwind). This reduces the negative superhelical density of the molecule and the major driving force behind the formation of cruciforms. Once formed, however, cruciforms are quite stable even when the DNA is relaxed by a topoisomerase, because the reverse reaction, returning to the extended linear conformation, requires introduction of one negative supercoil for every 10.5 bp of cruciform arm. Cruciforms, like duplex DNA, are stabilized by positively charged molecules such as divalent cations and spermidine.

Inverted repeats are widely distributed in the chromosomal DNA of many eukaryotes, and cruciform structures have been implicated as a component of DNA replication origins. Inverted repeats are present at the replication origins of phage, plasmids, mitochondria, eukaryotic virus, and mammalian cells, and have been associated with amplified genes. Moreover, they are functionally important for the initiation of DNA replication in bacteria and their plasmids as well as in mammalian viruses. Antibodies that appear specific for cruciform DNA not only stain nuclei, they can prevent initiation of DNA synthesis, suggesting that DNA replication may be regulated by cruciform-specific binding proteins. It is worth noting that if cruciform structures are involved in initiation of DNA replication, the act of replication would convert them back into linear DNA, thus providing a mechanism that would restrict origin activation to once per cell-division cycle.

Triplex DNA

Formation of triplex DNA (molecules consisting of three DNA strands) can occur at DNA regions containing only purines on one strand and only pyrimidines on the other. A third strand of DNA can then lie in the major groove of the DNA double helix (Figure 1-11A) where it is stabilized by formation of **Hoogsteen base pairs**, a minor variation of base-pairing that applies the N7 position of the purine base (as a hydrogen-bond acceptor) and C6 amino group (as a donor) (Figure 1-11A, B). Since poly-purine:polypyrimidine tracts are widespread in mammalian genomes, the potential for triple-helix formation is great.

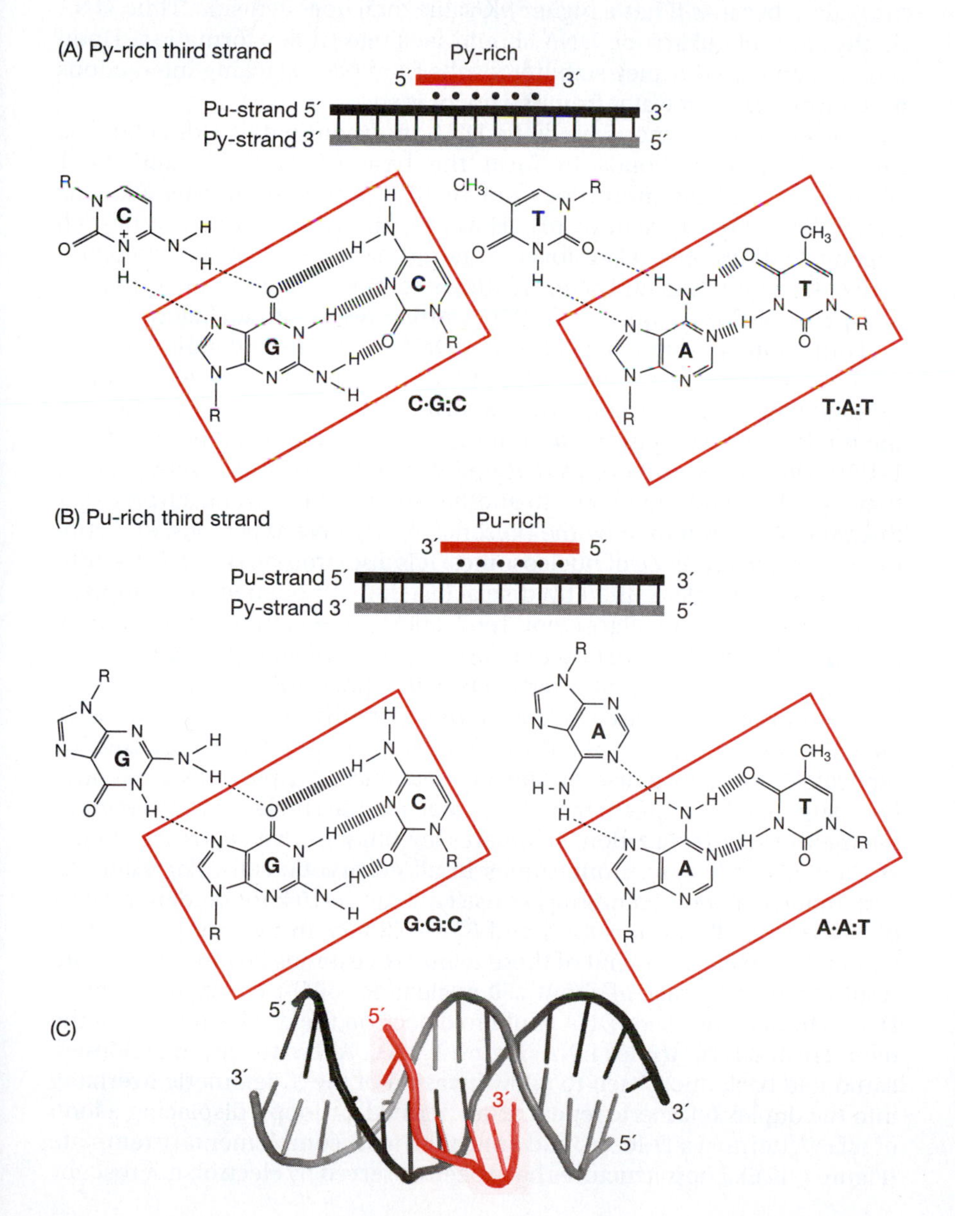

Figure 1-11. Triplex DNA. Four Hoogsteen base pairs can form in triple helix DNA, depending on whether the third strand is purine (Py)-rich (A) or pyrimidine (Pu)-rich (B). The four traditional Watson–Crick base pairs (G:C, A:T) are encompassed by a red box. The H-bonds forming Hoogsteen base pairs are indicated by dotted lines. Note that Hoogsteen base pairs form between two guanines or two adenines, as well as between cytosine and guanine, and between thymine and adenine. Protonation at N3 on the Hoogsteen cytosine is required for C to pair with G:C. (C) A DNA strand (black) containing a stretch of purines (GACT**GAGAGA**) and a complementary DNA strand (gray) containing a stretch of pyrimidines (CTGA**CTCTCT**) form a double helix in solution. A pyrimidine-rich strand (red; **CTCTCT**) lies in the major groove parallel, in terms of its 5′→3′ orientation, to the purine-rich strand, forming Hoogsteen base pairs with the purine-rich strand of the duplex DNA. (Panel C adapted from Asensio, J.L., Brown, T. and Lane, A.N. (1999) *Structure* 7: 1–11. With permission from Elsevier.)

Triplex-forming oligonucleotides bind in the major groove of duplex DNA via two Hoogsteen hydrogen bonds formed between the base of the third strand and the purine base of the duplex DNA. Third-strand binding occurs more slowly than traditional Watson–Crick base pairing, but once formed, triplexes are stable. Duplex DNA undergoes helical distortions as a consequence of triplex DNA formation and these distortions can trigger endogenous recombination and repair mechanisms in the cell. The presence of divalent cations augments triplex stability, as do naturally occurring polyamines such as spermine, spermidine, and putrescine, by reducing electrostatic repulsion among the three phosphate backbones.

Triplex DNA exists in two motifs that are distinguished by the orientation of the third strand with respect to the purine-rich strand in the duplex DNA. Pyrimidine-rich oligonucleotides bind parallel to the polypurine target sequence (Figure 1-11A), whereas purine-rich oligonucleotides bind antiparallel to the polypurine target (Figure 1-11B). Thymines or protonated cytosines of the pyrimidine-rich oligonucleotide bind to A:T or G:C Watson–Crick base pairs, respectively, forming the canonical base triads T·A:T and C·G:C. This motif is favored by acidic conditions, since N3 protonation of the third-strand cytosine is required for proper Hoogsteen bonding to the N7 of duplex guanine. However, 5′-methylcytosine alleviates this limitation, because it has a higher *p*K value than does cytosine. Thus, DNA methylation of eukaryotic DNA should facilitate triplex formation. Upon binding, increased triplex stability results from base-stacking interactions and the exclusion of water from the major groove.

The purine-rich motif binds to DNA in an antiparallel direction via reverse Hoogsteen bonds to form the base triads G·G:C and A·A:T (Figure 1-11B). This motif does not require protonation, thus allowing triplex formation at neutral pH. However, the guanines in purine-rich oligonucleotides can also form G-tetrad structures at physiological concentrations of salt and acid, creating a competition between formation of triplex DNA and G-quadruplex DNA structures described below.

Formation of intramolecular triplex DNA can occur at regions containing both a purine oligonucleotide and a mirror repeat symmetry. Mirror repeats allow intramolecular triplex structures to form if one strand of the duplex DNA contains a continuous sequence of purine bases (Figure 1-12B). However, this requires that one stretch of the DNA is unwound so that a single strand can anneal to another stretch of DNA. Not surprisingly, this process is promoted by the presence of negative supercoils, and suppressed by the presence of nucleosomes. It is also promoted by DNA synthesis on one template, thereby displacing the complementary template. Such 'single-strand displacement' replication occurs at the mitochondrial DNA (mtDNA) replication origin to form a D-loop (Figure 10-4) and during replication of adenovirus and certain bacterial phage (Figure 10-2).

Intramolecular triplex DNA structures have been shown to occur *in vivo*, and have been suggested to affect genome duplication by causing replication forks to pause at specific sequences. Triplex DNA has also been suggested as a mechanism for protecting the single-stranded ends of telomeres from degradation. Telomeres are unique DNA sequences found at the ends of linear chromosomes in all eukaryotic cells. For example, vertebrate telomeres commonly consist of long (1–10 kbp) tandem repeats of 5′-TTAGGG-3′ in one strand and 3′-AATCCC-5′ in the complementary strand. However, the 3′-end of these telomeres extends beyond the 5′-end, resulting in an average of about 150 nucleotides of G-rich single-stranded DNA. This purine-rich ssDNA could fold back on the duplex portion of the telomere and form triplex DNA (Figure 1-12D). Alternatively, the telomere could fold back upon itself to allow invasion of the 3′-telomeric overhang into the duplex telomeric repeat array (termed a t-loop), displacing a loop of ssDNA (termed a D-loop), and annealing to its complementary template (Figure 1-12E). Loop structures have been observed by electron microscopy

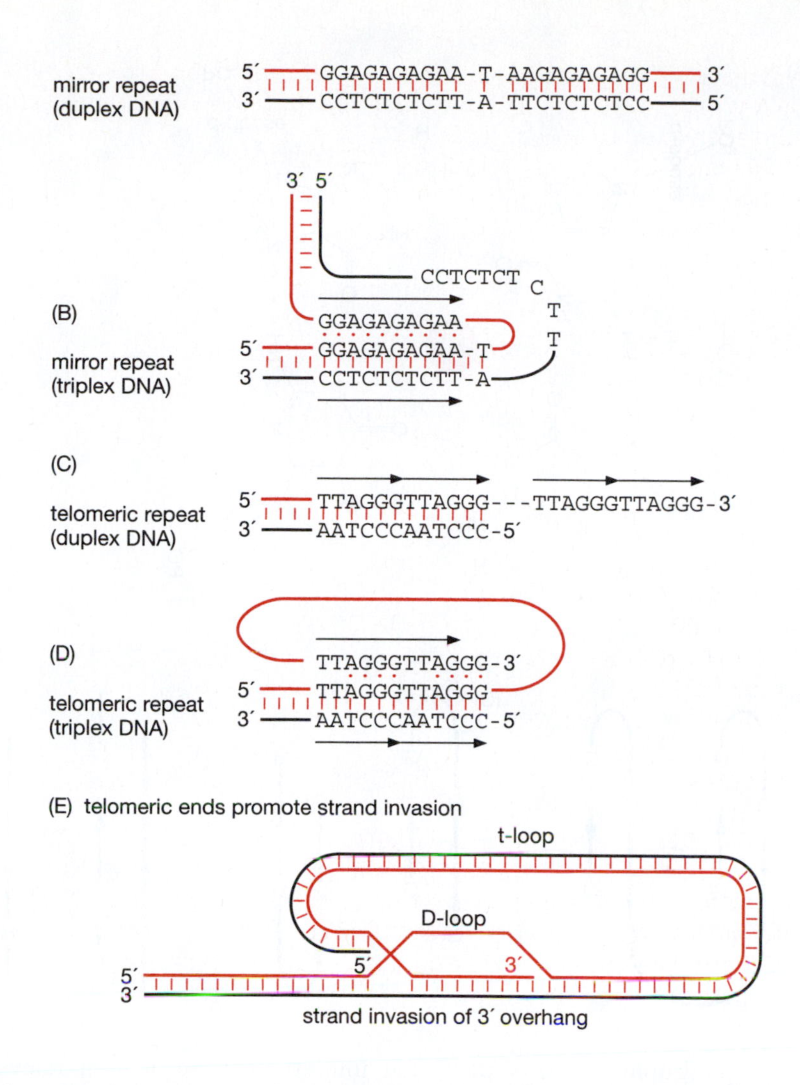

Figure 1-12. Formation of intramolecular triplex structures. Mirror repeat sequences (A) in duplex DNA facilitate formation of triplex DNA structures (B). The 3′-ends of vertebrate telomeres are G-rich single-stranded DNA regions consisting of 5′-TTAGGG-3′ repeats (C). These purine-rich ssDNA regions could be stabilized either by formation of triplex DNA (D) (data from Veselkov, A.G., Malkov, V.A., Frank-Kamenetskll, M.D. and Dobrynin, V.N. (1993) *Nature* 364: 496), or by invading part of the telomere and reannealing with its complementary strand, referred to as a t-loop (E) (data from Palm, W. and de Lange, T. (2008) *Annu. Rev. Genet.* 42: 301–334.) Red lines indicate Watson-Crick base pairs. Red dots indicate Hoogsteen base pairs.

of telomeres from mammals, trypanosomes, ciliates, and plants. In both cases, the advantage to the cell is to protect the 3′-ends of its telomeres from degradation, although the strand-invasion mechanism simply exchanges one region of ssDNA for another. Although triplex DNA can form from telomere oligonucleotides in solution, it has not been identified *in vivo*. In contrast, electron microscopy analysis of psoralen-crosslinked telomeric DNA purified from human and mouse cells reveals large terminal loops at their telomeres, part of which is ssDNA, suggesting that these terminal loops formed by a strand-invasion mechanism.

Quadruplex DNA

Arguably the most Byzantine noncanonical DNA structures detected so far occur at G-rich sequences in either DNA or RNA in the form of four-stranded structures called **quadruplex DNA** or G-quadruplexes. At their core is the G-tetrad formed by Hoogsteen base pairings and stabilized by a monovalent metal ion such as K^+ or Na^+ in its central axis (Figure 1-13A). G-quadruplexes consist of a minimum of two G-tetrads stacked one on top of the other and held together by loops of DNA arising from the intervening mixed-sequence nucleotides that are not involved in the G-tetrads themselves (Figure 1-13B–D). The net result is a four-stranded structure with the bases on the inside and the sugar-phosphate backbones on the outside. X-ray crystallographic analysis of G-rich DNA sequences and molecular modeling has revealed a wide variety of orientations, stoichiometries, and geometries in which G-quadruplexes can be assembled, but the question remains as to whether or not such structures actually form *in vivo*.

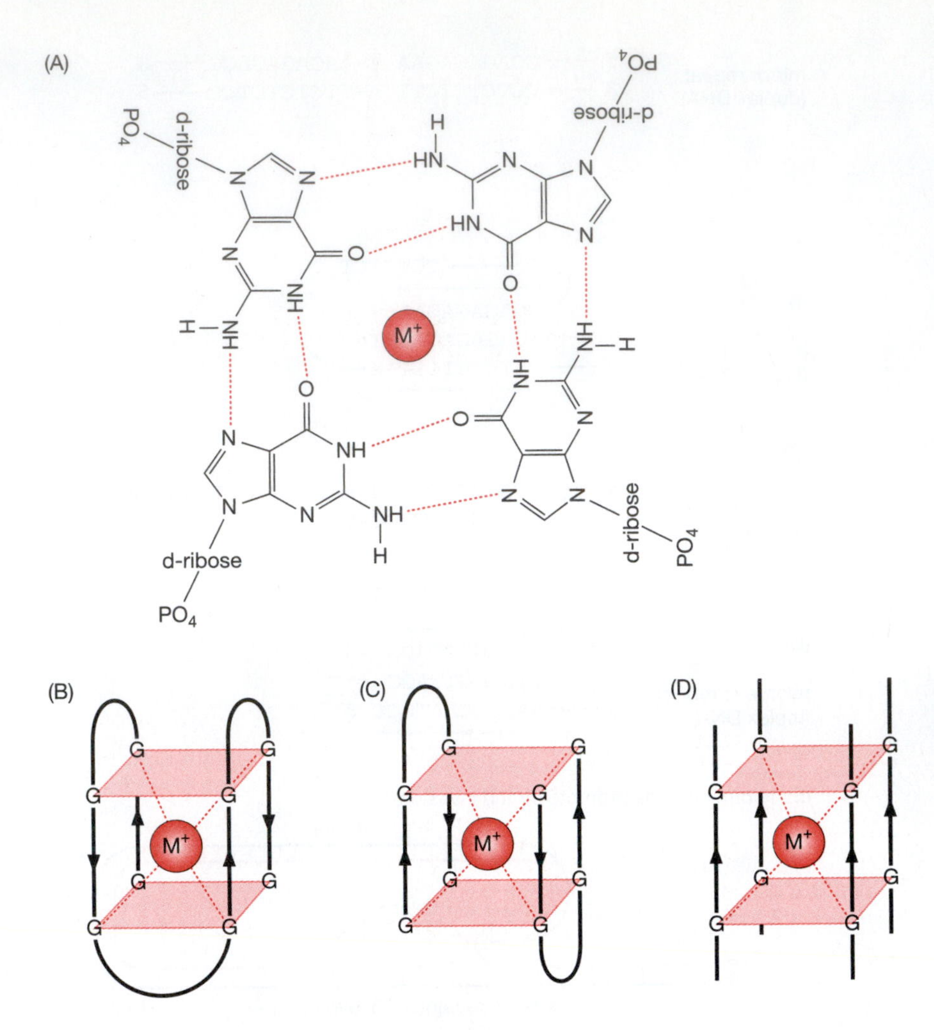

Figure 1-13. G-quadruplex DNA. (A) Four deoxyguanosines can form a tetrad using Hoogsteen base-pairing rules (Figure 1-11). Such G-tetrads are stabilized by the presence of a central monovalent metal ion. Dashed red lines are H bonds. (B) When two or more G-tetrads are stacked one on top of the other, they form a four-stranded DNA structure called a G-quadruplex in which the G-tetrads are held together by the intervening nucleotides, and the phosphates are all on the outside forming a highly anionic surface. At least three different orientations are possible consisting either of parallel (D, four-molecular quadruplex) or antiparallel (B, monomolecular quadruplex; C, bimolecular quadruplex) alignments of the two DNA strands with DNA loops either at the same (B) or opposite (C) poles of the structure. Each shaded rectangle represents a G-tetrad linked together by intervening T residues (vertical lines). Arrows indicate polynucleotide strand polarity (5′→3′).

G-quadruplexes are of particular interest because of their relevance to telomeres, the long G-rich sequences found at the ends of eukaryotic chromosomes. For example, vertebrate telomeres commonly are composed of d(TTAGGG) repeats, and the 3′-end of these telomeres extends beyond the 5′-end, resulting in an average of 150 nucleotides of ssDNA. This single-stranded telomeric repeat can fold into five or six four-repeat bimolecular quadruplexes (Figure 1-14) that may protect the telomeric ends from degradation. This structure does not depend on DNA supercoiling. Drugs that stabilize G-quadruplex formation *in vitro* inhibit telomere maintenance in mammalian cells, suggesting that telomeres form G-quadruplexes *in vivo*, as well as *in vitro*. Potential quadruplex sequences have also been identified in nontelomeric genomic DNA, particularly in nuclease-hypersensitive

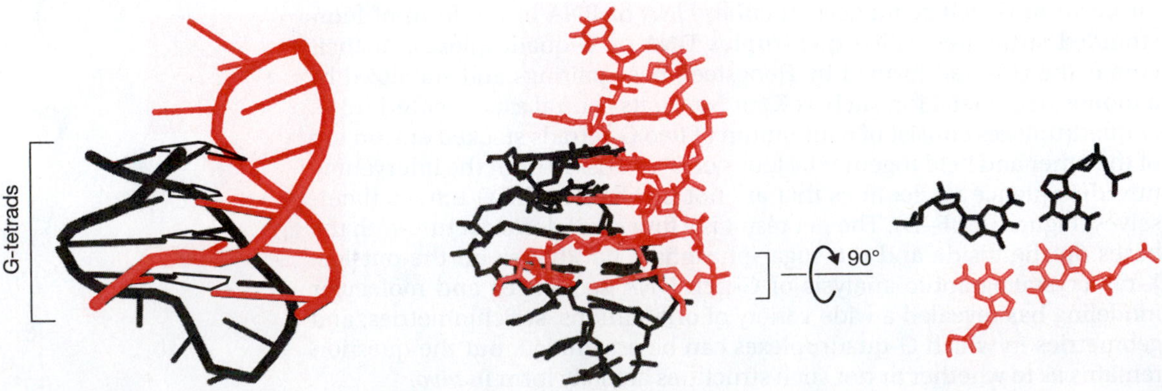

Figure 1-14. The *Oxytricha* telomeric sequence (GGGGTTTTGGGG) crystallized in solution as a G-quadruplex structure with a stack of four G-tetrads separated by loops of four Ts at the top and bottom of image. Left: the path of the two sugar-phosphate backbones, and the direction of their bases. Dotted lines indicate the four G-tetrads. Each loop each contains four Ts. Middle: stick model of the positions of all atoms with the exception of hydrogens. *Oxytricha* telomeric DNA [$(G_4T_4)_n$] is in the antiparallel form with loops at opposite poles (Figure 1-13C). Right: a single G-tertrad is extracted from the image in the middle panel and rotated 90° to allow a view from above the tetrad. (Images generated from PDB file 1D59 using Pymol. Data from Kang, C., Zhang, X., Ratliff, R., Moyzis, R. and Rich, A. (1992) *Nature* 356: 126–131.)

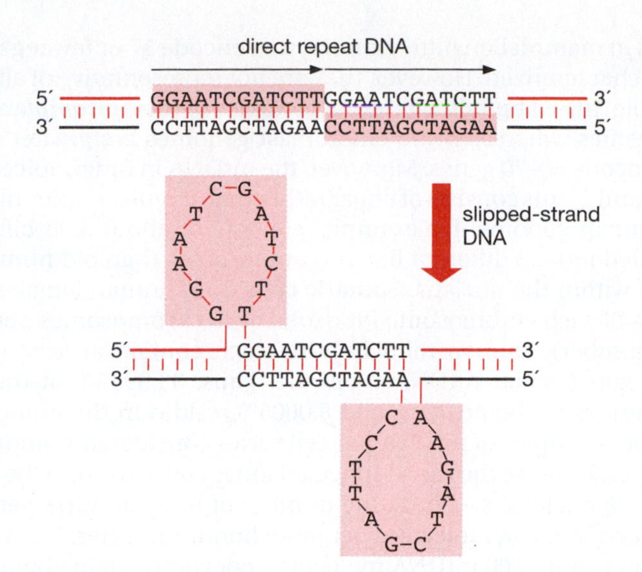

Figure 1-15. Slipped-strand DNA can form from a region comprising two direct repeats. One loop is located at the 5′-end of the direct repeats while the other is located at the 3′-end. (Adapted from Sinden, R.R., Pytlos-Sinden, M.J. and Potaman, V.N. (2007) *Front. Biosci.* 12: 4788–4799. With permission from Frontiers in Bioscience.)

promoter regions. Here supercoils would be expected to facilitate formation of quadruplex structures, as they have been demonstrated to do so with triplex structures.

Slipped-Strand DNA

Complementary strands of direct repeats can anneal in a misaligned ('slipped') arrangement along the DNA helix axis. In principle, this slipped-strand DNA may form in almost any direct repeat, but in practice such structures have been detected only in short DNA repeats that are unstable. To form slipped-strand DNA, the direct repeat region must unwind and one strand of one copy of a direct repeat must pair with the complementary strand of another copy of a second direct repeat (Figure 1-15). Slipped-strand DNA structures do not appear to be stable, because there is considerable loss of helix stability from the loss of hydrogen-bonding and base-stacking interactions of DNA within the two loops. However, certain factors may compensate. Loop–loop interaction can occur as the looped-out strands in opposite DNA strands are complementary, and structures with loop–loop interactions have been reported for various other DNA structures. Moreover, in the known examples of slipped-strand DNA structures, the sequences involved contain some degree of inverted repeat symmetry, in addition to direct repeat symmetry. This inverted repeat symmetry can produce DNA hairpins in the looped-out arms of slipped-strand structures, which would contribute to the stability of these DNA structures. Finally, as with other noncanonical DNA structures, unwinding the DNA double helix is facilitated by negatively superhelical DNA.

GENOME CHARACTERISTICS

Chromosome Shape, Number, And Distribution

The size, shape, and distribution of genomes in cellular organisms vary enormously (Table 1-3). With few exceptions, the entire genome in bacteria and archaea, two of the three domains of life, is contained in a single DNA molecule that is generally, but not always, circular rather than linear. Only the larger genomes found in eukarya are segmented, with the number of different chromosomes varying from three in fission yeast to a fern species (*Ophioglossum reticulatum*) reported to have 630 different chromosomes. In addition, whereas the bulk of eukaryotic genomes resides within the nucleus, a tiny fraction is contained in the mitochondria of animals or the chloroplasts of plants. Of these, only the small (<20 kbp) circular DNA molecules in mammalian mitochondria that encode 37 or fewer genes have been well characterized. However, they are not representative of all mtDNA. For example, the 6 kbp mitochondrial genomes of the *Apicomplexa* encode only five genes, whereas some chloroplast genomes are greater than 200 kbp and encode 50–70 genes. Moreover, the mtDNA in fungi, apicomplexan parasites, and plants consists of linear rather than circular duplex molecules.

The human genome, for example, consists of about 3.08 billion base pairs divided into 23 different linear chromosomes (haploid number) that are stored within the nucleus. **Somatic cells** from human females contain two copies of each chromosome for a total of 46 chromosomes per nucleus (diploid number). Human mitochondria each contain at least one copy of the 16 kbp circular mtDNA molecule. Thus, 99.9995% of the human genome resides in the nucleus, and 0.0005% resides in the mitochondria. The number of copies of mtDNA per cell varies considerably among different cell types, because they are duplicated independently from the chromosomes in the nucleus, and both the number of mitochondria per cell and the number of mtDNA molecules per mitochondrion varies. Human sperm each contain about 100 mtDNA molecules, oocytes contain about 100 000, and somatic cells contain from 1000 to 10 000 copies of mtDNA.

DNA Length

DNA is an extraordinarily long, remarkably stable, and easily replicated molecule. Yet it is sufficiently flexible that it can be compressed into a space only a tiny fraction of its length. This is because all DNA molecules are only approximately 20 Å in diameter (the equivalent of ~ 6 bp), but their length varies enormously. The human genome is 3.08 billion (10^9) bp in size. Therefore, a single diploid human cell would contain 6.16 billion bp of DNA, or 6.3 pg of DNA, which is the equivalent of 2.09 m of DNA, where DNA mass (pg) = genome size (bp)/(0.978×10^9 bp pg^{-1}); DNA length (m) = [genome size (bp)](0.34×10^{-9} m bp^{-1}); 1 micrometer (μm) = 2941 bp; 10^6 base pairs = 1 Mbp = 0.34 mm (Sinden, 1994; Dolezel *et al.*, 2003). This is the amount of DNA that must be duplicated every time a human cell divides. Since each diploid cell contains 46 chromosomes, a single human chromosome will contain, on average, a DNA molecule almost 5 cm in length (~147 Mbp). In fact, the largest human chromosome contains 245 Mbp and the smallest contains 47 Mbp (Table 1-3). That means that the genome-duplication machinery must be able to replicate single DNA molecules that are centimeters long and contain millions of base pairs.

Now consider the problem of genome duplication as a single-cell human embryo develops into an adult. A typical adult weighs about 75 kg, which is the equivalent of about 29 trillion cells or about 60 trillion m of DNA (number of cells per organism = [body mass (g)/2]/[1.3×10^{-9} g $cell^{-1}$]). To appreciate the magnitude of this number, compare it with the minimum distance between the Earth and the Sun (146 billion m; 91 million miles). Thus, the total amount of DNA in a single human adult will stretch to the Sun and back about 200 times! Remarkably, humans do not have the largest genome on the planet. That honor goes to some amphibians, fish, insects, plants, and even protozoa. The amount of DNA in cellular genomes varies considerably, from less than 1 Mbp in some bacteria and archaea to greater than 100 000 Mbp in some animals and plants (Table 1-3).

C-Value Paradox

The amount of DNA in a single cell is often much greater than required to encode all of the genes. For example, only about 2% of the DNA in a human cell contains the 22 000 genes that encode proteins. Thus, there is ample storage space for future deposits of genetic information, making DNA an excellent 'gene bank'. Haploid DNA content (historically referred to as the C-value) cannot be translated directly into genome size, because some

Table 1-3. Examples Of Genome Size, Shape, And Segmentation

Organism, Organelle, Chromosome		Common name	Genome size (Mbp)	Chromosomes (1N)
Mitochondria	*Homo sapiens*	Human	0.016	1 circular
	Schizosaccharomyces pombe	Fission yeast	0.020	1 circular
	Candida parapsilosis	Yeast-like fungi	0.031	1 linear
	Saccharomyces cerevisiae	Budding yeast	0.086	1 circular
Chloroplasts	*Spinacia oleracea*	Spinach	0.151	
	Zea mays	Maize (corn)	0.570	
	Cucumis melo	Muskmelon	2.6	
Bacteria	*Mycoplasma genitalium* (smallest bacterial genome)	Parasitic bacterium	0.580	1 circular
	Borrelia burgdorferi	Spirochete bacterium	0.935	1 linear
	Escherichia coli	Intestinal bacterium	4.6	1 circular
	Bacillus subtilis	Sporulating soil bacterium	4.2	1 circular
	Streptomyces coelicolor	Soil bacterium	8.7	1 linear
Archaea	*Aeropyrum pernix*	Hyperthermophilic archaeon	1.67	1 circular
	Methanothermobacter thermoautotrophicus	Methanogen archaeon	1.75	1 circular
	Pyrococcus abyssi	Hyperthermophilic archaeon	1.77	1 circular
	Sulfolobus solfataricus	Thermophilic archaeon	2.99	1 circular
Yeast	*Saccharomyces cerevisiae*	Budding yeast	12.2	16 linear
	Sc chromosome 4 (largest)		1.53	
	Sc chromosome 1 (smallest)		0.23	
	Schizosaccharomyces pombe	Fission yeast	13.8	3 linear
	Sp chromosome I (largest)		5.7	
	Sp chromosome III (smallest)		3.5	
Nematodes	*Meloidogyne graminicola* (smallest animal genome)	Root-knot nematode	29	
	Caenorhabditis elegans	Roundworm	100	10 linear
Insects	*Drosophila melanogaster*	Fruit fly	180	4 linear
	Dm chromosome 3 (largest)		68.8	
	Dm chromosome 4 (smallest)		4.3	
Birds	*Gallus gallus*	Chicken	1050	20 linear
Animals	*Mus musculus*	Mouse	2600	20 linear
	Xenopus laevis	African clawed frog	3100	18 linear
	Homo sapiens	Human	3080	23 linear
	Hs chromosome 1 (largest)		245	
	Hs chromosome 21 (smallest)		47	
	Triturus vulgaris	Common newt	24 300	24
	Protopterus aethiopicus (largest animal genome)	Marbled lungfish	130 000	38
Algae	*Chlamydomonas reinhardtii*	Unicellular green algae	100	17 linear
	Ostreococcus tauri (smallest plant genome)	Unicellular green algae	11.5	18 linear
Plants	*Arabidopsis thaliana*	Arabidopsis	129	5 linear
	Oryza sativa	Rice	389	12
	Cucumis melo	Muskmelon	860	12
	Spinacia oleracea	Spinach	989	6
	Zea mays	Maize	2500	5
	Populus trichocarpa	Poplar tree	480	19
	Fritillaria assyriaca (largest plant genome)	Fritillaria (flowering plant)	124 000	24

Data are taken from various on-line databases (www.genomesize.com/, www.genomesonline.org/gold.cgi, www.kew.org/cval/homepage.html) and from Gregory (2005). Genome sizes are based either on DNA sequence data (where available), or on DNA mass. 1 million (10^6) base pairs (bp)=1 megabase pair (Mbp)=1000 (10^3) (kilobase pairs, kbp).

organisms contain large amounts of mitochondrial or chloroplast DNA (particularly among the protozoa), and some organisms are highly polyploid (particularly among plants). When these facts are taken into account, the range in genome size among all organisms varies by 600 000-fold (Figure 1-16). The smallest genome appears to be in the parasitic microsporidium *Encephalitozoon intestinalis* (0.0023 pg DNA), and the largest is in the free-living amoeba *Chaos chaos* (1400 pg DNA). Animal genomes range from 0.03 to 0.1 pg among the nematodes to 112–132 pg among the lungfish. Plant genomes range from 0.01 pg in the algae to more than 100 pg in the angiosperms.

One is tempted to assume that the amount of DNA per cell is related somehow to the complexity of the organism. Although it is true that complex organisms such as eukarya do have larger genomes than simple organisms such as bacteria and archaea, there exists a great deal of variation in the range of genome sizes within a class of organisms (Figure 1-16). For example, there is an approximate 100-fold difference between the smallest amphibian genome and the largest. Yet, the overall body plan and metabolism of these animals is quite similar. Even more remarkable is the fact that some insects, mollusks, fish, and amphibians have larger genomes than humans. Thus, we have a paradox. If all (or most) of the DNA in an organism's genome encodes the proteins required to build the organism, then why does a newt need eight times as many genes as a frog? In fact, it doesn't. The number of genes is fairly similar within each group; what varies is the amount of DNA, not the number of genes.

Genetic Complexity Paradox

Only about 10% of the genome in vertebrates encodes proteins. It is the amount of DNA that does not encode protein that increases dramatically with the complexity (sophistication) of the organism (Table 1-4). The differences in genome sizes within a class of organisms are due almost entirely to differences in the amount of this noncoding DNA, the function of which is unknown.

Humans are certainly anatomically different and appear to be more sophisticated organisms than chimpanzees and mice, but the number of genes in these three mammals is quite similar. Even more striking is the fact

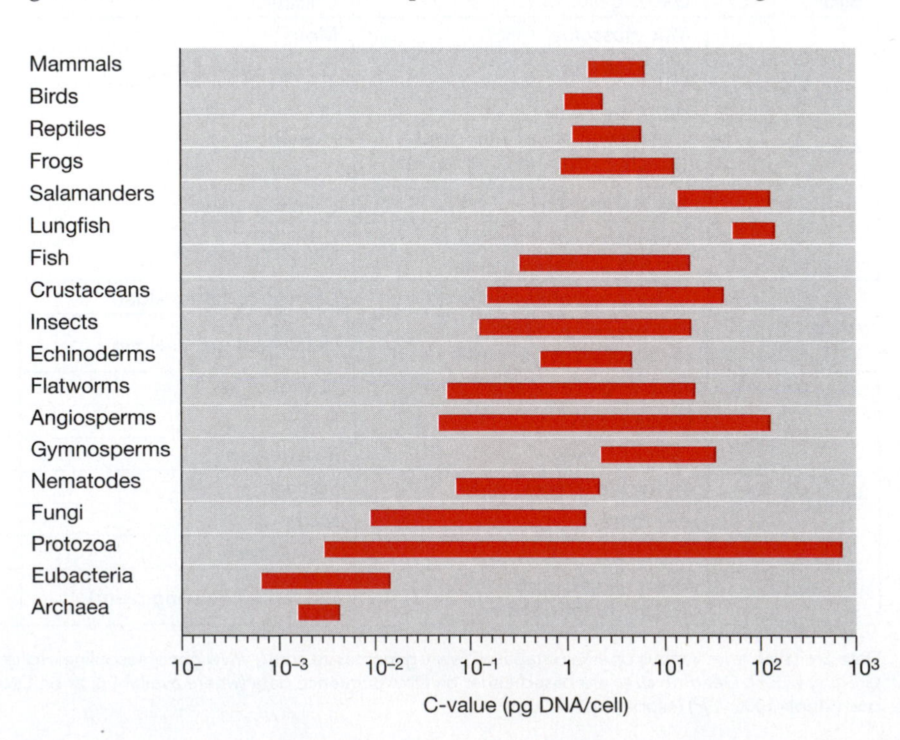

Figure 1-16. The haploid amount of DNA per cell (C-value) in various organisms. 10^{12} picogram (pg) = 1 g; 1 pg of DNA = 978 Mbp. (Adapted from Gregory, T.R. (ed.) (2005) *The Evolution of the Genome*. With permission from Elsevier.)

Table 1- 4. Genome Size And Number Of Genes

Species	Common name	Genome size (Mbp)	Genes	Genes per Mbp
Homo sapiens	Human	3080	22 000	7
Pan troglodytes	Chimpanzee	3350	23 000	7
Mus musculus	Mouse	2600	22 000	8
Populus trichocarpa	Poplar tree	480	45 000	94
Oryza sativa	Rice	389	41 000	105
Arabidopsis thaliana	Arabidopsis	129	26 500	205
Caenorhabditis elegans	Roundworm	100	19 500	195
Escherichia coli	Bacterium	4.6	4700	1021

that we share more than 97% of our DNA sequence and almost all of our genes with the chimpanzee, and about 88% of our genes with rodents. Moreover, humans are definitely more sophisticated than trees and vegetables, but the genomes in these plants contain 60% more genes than does the human genome! Even the simplest plants, such as *Arabidopsis*, have as many genes as mammals, despite the fact that its genome is 20-times smaller. In brief, the size of genomes and the number of genes can vary enormously with little relationship to biological complexity.

Two discoveries over the past decade suggest how these paradoxes will likely be resolved. First, a single gene can produce more than one protein or RNA product. Humans have about 22 000 structural genes but an estimated 100 000 proteins. This feat is accomplished through the mechanism of RNA splicing. After transcription, the multiple exons that comprise a single gene can be combined in different ways to produce different proteins. Therefore, the biological complexity of an organism is not a simple function of the number of different genes it contains. Second, protein-encoding genes comprise less than 5% of the human genome, but the remaining DNA is clearly not the 'junk' it was once thought to be. Instead, the remaining DNA encodes an assortment of sequences that regulate the expression of protein-encoding genes. Differences in the extent to which gene activity is regulated do appear to correlate with biological complexity. For example, 9% of all genes in *Homo sapiens* encode transcription factors, whereas transcription factors account for only 5% of the genes in *Drosophila* and 3% of the genes in *Saccharomyces cerevisiae.*

If there is a lesson in these paradoxes, it is this: either the presence or absence of a small number of genes, or a difference in the pattern of gene expression, is what distinguishes one species from another. Therefore, all of the genome must be duplicated accurately and completely, not simply that portion of the genome that houses the genes for proteins, transfer RNAs, and ribosomal RNAs, but the 'junk' DNA as well.

SUMMARY

- The genomes of all cellular organisms are composed of duplex DNA, two strands of a polymer (Figure 1-2) containing four different deoxyribonucleotide subunits (Table 1-1; Figure 1-1) twisted into a double helix (Figure 1-4).
- Under physiological conditions, duplex DNA adopts the canonical B-form (Table 1-2). The helix is stabilized by stacking bases within each

strand, by hydrogen bonds between bases of complementary strands, and by exclusion of water molecules. The molecule is hydrophobic on the inside, but hydrophilic on the outside as a consequence of the external position of its phosphates. Thus, it is water soluble and binds tightly to basic proteins.

- The structure of DNA is not uniform; it depends on nucleotide composition and sequence as well as environmental conditions such as solvent, salt, and pH. Thus, noncanonical DNA structures may exist at various sites throughout genomes, and they, as well as variations in DNA sequences, will either facilitate or inhibit genome duplication.
- Noncanonical DNA structures include Z-DNA (Figure 1-9; Table 1-2), cruciform DNA (Figure 1-10), triplex DNA (Figure 1-11; 1-12), quadruplex DNA (Figure 1-13; 1-14), and slipped-strand DNA (Figure 1-15). These unusual forms of DNA are promoted by the presence of negative superhelical turns whose energy can be used to melt and re-form DNA structures. In addition, telomere DNA can form terminal loops (Figure 1-12E). If these noncanonical structures exist *in vivo* as they do *in vitro*, they present unique problems for the machinery of genome duplication.
- DNA has a polarity, because the sugar residues are linked together by 5′→3′ phosphodiester bonds (Figure 1-2). In duplex DNA, the two strands are antiparallel; the 5′-end of one is opposite to the 3′-end of the other (Figure 1-3). This imposes a severe constraint on the mechanism for genome duplication, because all DNA polymerases synthesize DNA only in one direction (5′→3′).
- The number of purine bases equals the number of pyrimidine bases. The number of adenosines equals the number of thymidines, and the number of guanosines equals the number of cytidines. Base modifications are possible, particularly methylation of A in bacteria and C in multicellular eukarya (Figure 1-1). However, the four dNMPs are not uniformly distributed in DNA. Thus, nucleotide composition and sequence varies considerably throughout the genome.
- DNA does not exist in cells as a naked molecule, but as a nucleoprotein complex termed chromatin. In both archaea and eukarya, chromatin consists of repeating units (nucleosomes) consisting of a histone octamer around which is coiled about 147 bp of DNA. An analogous repeating subunit exists in bacteria by substituting HU protein for histones.
- Dissociating nucleosomes and unwinding the DNA helix both generate superhelical turns in DNA, although in opposite directions. Dissociating nucleosomes produces one negative superhelical turn for each nucleosome that is disassembled. Unwinding DNA produces one positive superhelical turn for each 10 bp unwound. The exact numbers depend on the secondary structure of DNA under the conditions in which these events occur. The intrinsic energy in positive superhelical turns makes duplex DNA more difficult to unwind and therefore more resistant to replication. Negative superhelical turns have the opposite effect. Negative superhelical turns promote transcription, replication, and protein binding, as well as the formation of noncanonical structures.
- A genome is all of the genetic information required to duplicate an organism. This includes the tiny fraction encoded by mtDNA. Genomes in cellular organisms consist of from less than 1 Mbp to more than 100 000 Mbp of duplex DNA. Those in mitochondria, bacteria, and archaea are generally, but not exclusively, single unbranched circular molecules. The larger genomes found in eukarya are segmented into DNA–protein complexes called chromosomes. Based on sequence analysis, the DNA in a single chromosome consists of a single unbranched linear DNA molecule. The longest human DNA molecule is about

245 Mbp or about 8.3 cm long! Thus, the mechanism for duplicating genomes must deal with single DNA molecules of remarkable length. It is only because DNA is a highly flexible molecule that these extremely long lengths of DNA can be coiled around histones, and then folded and condensed into tiny volumes such as eukaryotic nuclei.

- DNA as the genetic material presents us with two important paradoxes. One is the apparent lack of correspondence between the amount of DNA per genome and the complexity of eukaryotic organisms. The other is that at least 90% of vertebrate DNA does not encode for proteins, ribosomal RNA, or transfer RNA. What it contributes is still a mystery. Nevertheless, all of the genome must be duplicated before a cell can divide. Whether or not all of the genome must be duplicated accurately for an organism to survive and reproduce remains to be seen.

REFERENCES

Additional Reading

Bacolla, A. and Wells, R.D. (2004) Non-B DNA conformations, genomic rearrangements, and human disease. *J. Biol. Chem.* 279: 47411–47414.

Bates, A.D. and Maxwell, A. (2005) *DNA Topology*. Oxford University Press, Oxford.

Neidle, S. (2007) *Principles of Nucleic Acid Structure*. Academic Press, London.

Neidle, S. and Balasubramanian, S. (2006) *Quadruplex Nucleic Acids.* RSC Biomolecular Sciences, Cambridge.

Ohyama, T. (2005) DNA *Conformation and Transcription*. Landes Bioscience, Austin, TX.

Portugal, F.H. and Cohen, J.S. (1977) *A Century of DNA*. MIT Press, Cambridge, MA.

Rich, A. and Zhang, S. (2003) Timeline: Z-DNA the long road to biological function. *Nat. Rev. Genet.* 4: 566–572.

Sinden, R.R. (1994) *DNA Structure and Function*. Academic Press, London.

Sinden, R.R., Pytlos-Sinden, M.J. and Potaman, V.N. (2007) Slipped strand DNA structures. *Front. Biosci.* 12: 4788–4799.

Literature Cited

Dolezel, J., Bartos, J., Voglmayr, H. and Greilhuber, J. (2003) Nuclear DNA content and genome size of trout and human. *Cytometry A* 51: 127–128.

Gregory, T.R. (ed.) (2005) *The Evolution of the Genome.* Elsevier/Acadamic Press, Burlington, MA.

Marmur, J. and Doty, P. (1962) Determination of the base composition of deoxyribonucleic acid from its thermal denaturation temperature. *J. Mol. Biol.* 5: 109–118.

Owczarzy, R., Vallone, P.M., Gallo, F.J., Paner, T.M., Lane, M.J. and Benight, A.S. (1997) Predicting sequence-dependent melting stability of short duplex DNA oligomers. *Biopolymers* 44: 217–239.

Sinden, R.R. (1994) *DNA Structure and Function*: Academic Press, London.

Chapter 2
THREE DOMAINS OF LIFE

In this book we will describe how the ability to duplicate DNA, the genetic material of all cells, is essential to the propagation of life itself. As will become apparent in the following chapters, an immense amount of molecular detail has been established regarding the nature, function, and interplay of the enzymes that have the task of copying the famous double helix. This treasure trove of information has been obtained by painstaking studies performed in hundreds of laboratories around the world over the last 60 years. As is so often the way in science, studies in simple model systems have been of pivotal importance in establishing the mechanistic paradigms of the DNA replication process; these studies have in turn informed the design and interpretation of studies in more complex systems. This approach has the advantage of ensuring that diverse systems are studied and thus it is possible to tease apart general, shared, or conserved principles from species-specific esoterica. In addition, the study of diverse systems, from viruses, through microbes, to humans, allows us to trace the lines of descent of the machinery of DNA replication during evolution. Comparing the replication machinery of human cells to that of much more organizationally simple microbial systems allows us to observe the process by which evolution has embellished the core machineries of replication with increasingly sophisticated means of ensuring appropriate regulation of that core process. Therefore, to understand the importance of the differences between the DNA replication apparatus of different organisms, we need to understand the way in which life has evolved on Earth.

IN THE BEGINNING...

During the formation of the solar system it is believed that Earth was formed by the accretion of meteorites and related material. Radiometric dating of ancient rocks places the age of Earth at approximately 4.5 billion years. Ancient Earth would have been a remarkably inhospitable environment with little or no atmosphere and thus unprotected from constant bombardment by debris left over from the formation of the solar system. This bombardment, combined with the energies released during accretion, would mean that the entire planet was an immense molten sphere. However, eventually the early Earth reached a size where it generated sufficient gravity to retain a primitive atmosphere that probably included water vapor. This in turn led to cooling of the planet and solidification of the crust. It is thought that this occurred between 100 and 150 million years after the formation of Earth. Ongoing collisions with asteroids would have led to the release of more steam from the crust and resulted in the generation of clouds of water vapor in the atmosphere. As the planet cooled further, rain fell and thus gave rise to oceans. By this stage (about 4.2 billion years ago), the atmosphere contained ammonia, methane, carbon dioxide, nitrogen, and water vapor. The first evidence for cellular life dates to between 400 and 700 million years later, approximately 3.8–3.5 billion years ago. This evidence is in the form of rock formations that resemble modern-day stromatolites, natural features that are formed by colonies of cyanobacteria. In support of a similar origin for the ancient stromatolite-like rock formations, electron microscopy has provided evidence for cyanobacteria-like structures within these rocks. A second piece of evidence for ancient life is found in the presence of kerogen deposits dating to approximately 3.8 billion years ago. Kerogen is a complex mixture of high-molecular-weight organic compounds that is thought to be produced by the decay of organic matter. It is not only useful for dating the origins of life on Earth: when subjected to appropriate heating in the Earth's crust, kerogens transform and release crude oil and natural gas.

Clearly we've taken quite a jump here; from the formation of presumably sterile oceans 4.2 billion years ago to evidence for recognizable cellular life around 3.5 billion years ago. What events occurred in the intervening 700 million years?

Prebiotic Era

As mentioned above, the atmosphere of primitive Earth was rich in simple carbon- and nitrogen-based compounds such as ammonia, methane, carbon dioxide, and nitrogen gas. How did these simple compounds lead to the generation of complex macromolecules such as proteins, lipids, carbohydrates, and nucleic acids that are the essence of living cells? A series of classical experiments performed by Stanley Miller in Harold Urey's laboratory yielded significant insights into the solution of this problem. Miller set up mixtures of water, methane, ammonia, and hydrogen gas in a reducing atmosphere and subjected this mixture to electrical discharges and elevated temperatures. Remarkably, after only a few days (a time span notably shorter than several hundred million years) examination of the reaction mixture revealed the synthesis of a diverse range of complex organic molecules, including the nitrogenous bases that form part of present-day nucleotides, amino acids (the building blocks of proteins), and a variety of sugars. Miller's experiments have recently been challenged by a re-evaluation of the reducing potential of the early Earth atmosphere: more recent studies suggest that it was a significantly less reducing environment than in the conditions used by Miller, and repeating Miller's experiments in mildly reducing conditions results in far fewer organic compounds being generated. However, it has been proposed that reducing conditions analogous to those used by Miller might be found in the vicinity of deep-sea hydrothermal vents. These features are formed when fissures in the Earth's surface reach the sea bottom, producing superheated water and localized deposition of a range of minerals. Such conditions may seen inimical to life, but in fact even today organisms, such as the evocatively named archaeon *Pyrodictium occultum*, thrive in these 'black smoker' environments. Such environments may, therefore, have provided the necessary conditions for the generation of organic molecules that are the building blocks of life. A number of potential routes have been proposed to facilitate the transition from pools of these precursor molecules to more complex polymers. For example, Günter Wächtershäuser has championed a role for the metal sulfides that are found abundantly at hydrothermal vents acting as surfaces to both trap and catalyze reactions between these precursor molecules.

Such chemical evolution could have facilitated generation of more complex molecules and polymers related to those found in present-day biological systems. Such trapping of organic molecules on surfaces would prevent their diffusion and loss into the surrounding ocean; it also conceptually serves as a step toward the compartmentalization that is a hallmark of present-day membrane-enclosed cells. Although scenarios such as those described above provide a potential source of complex molecules, they do not describe living systems. However, they set the stage for the generation

Table 2-1. Characteristics Of Living Systems

• Able to reproduce independently of other biological systems
• Consisting of one or more cells
• Encoding genes in DNA
• Able to obtain and use energy
• Responsive to the environment
• Capable of evolving

of complex nucleic acids, such as RNA. A growing body of data supports the hypothesis that RNA can act as both information carrier and engine for its own duplication, one of six defining characteristics of living systems (Table 2-1).

Like DNA (Figure 1-3), RNA is a nucleic acid containing a backbone of alternating sugar (ribose in the case of RNA) and phosphate moieties and with a nitrogenous base attached to the 1′ position of each ribose (Figure 2-1). In modern RNA there are four bases – A, G, C, and U – that can interact with each other in the following combinations of Watson–Crick base pairs: A = U and G ≡ C and additionally G = U. This means that the sequence of one RNA strand can direct the formation of a second complementary strand. This, in turn, can lead to a mechanism for copying and thus replicating the original sequence. In addition, base pairing also allows an

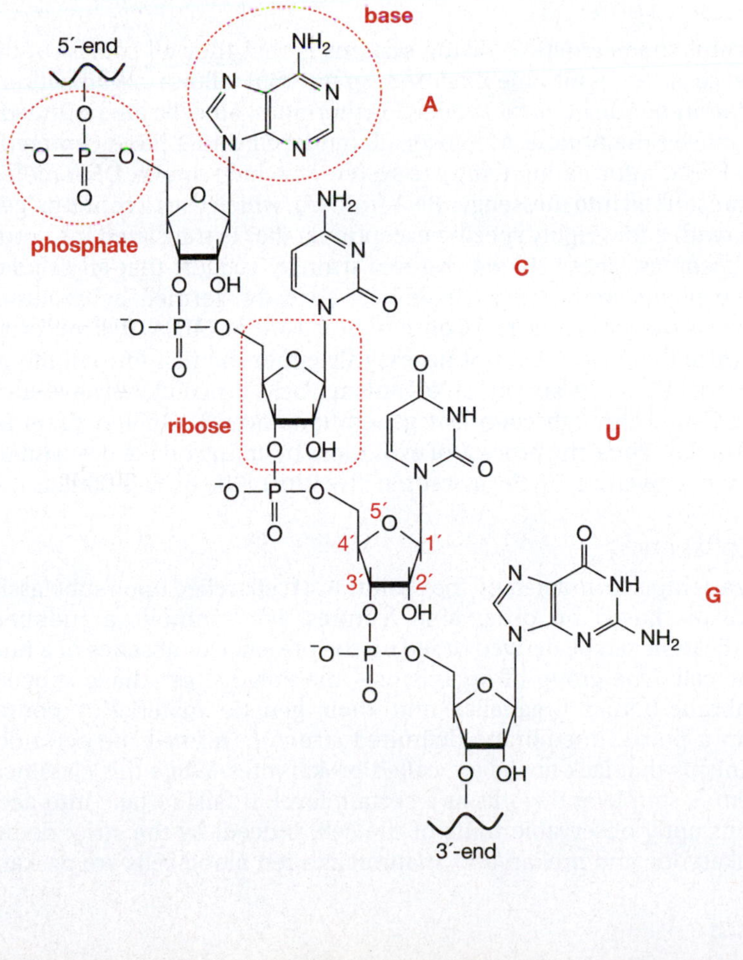

Figure 2-1. Diagram of an RNA chain with the phosphate, ribose sugar, and nitrogenous base groups indicated. The carbon positions of the central ribose sugar are numbered in red.

individual single-strand RNA molecule to fold back on itself, generating complex three-dimensional shapes. Therefore, the sequence of bases of an RNA molecule defines its pattern of folding and consequently its shape and thus its function.

These properties of RNA molecules were recognized in the late 1960s and led Francis Crick and Carl Woese to propose that RNA had the potential to act as an enzyme. This bold prediction was subsequently vindicated by work from Thomas Cech and Sidney Altman that revealed that RNA molecules could indeed act as enzymes (dubbed **ribozymes**), a discovery for which Altman and Cech received the Nobel Prize in Chemistry in 1989. Significantly, among the reactions that RNA can catalyze is the synthesis of new RNA in a template-dependent reaction. Thus, it is now widely believed that there was an RNA world around 3.8–3.6 billion years ago, a world in which RNA acted both as template and replication machine as well as potentially synthesizing other biomolecules. However, RNA is a labile molecule and thus has limited stability. Over evolutionary time, DNA was selected as a more stable carrier of genetic information. Similarly, while some ribozymes still exist, the majority of catalysis in present-day cells is mediated by more stable and potentially more versatile proteins. Strikingly, however, the ability to translate RNA sequence into protein sequence in **ribosomes** is, in fact, dependent on a ribozyme: formation of a **peptide bond** in the active site of the ribosome is catalyzed by an adenine base in the RNA component of the ribosome (see Figure 2-2 in color plate section).

Present-day ribosomes are combinations of RNA and protein. However, the proteins play structural, not catalytic, roles. Thus, the ribosome may be a snapshot of a key transitional step in evolution when proteins were becoming an important component of the accessory architecture of sophisticated RNA-based machines. The preservation of the RNA-based catalysis in ribosomes is testament to the efficiency of the reaction.

Cellular Life

A second characteristic of living systems is that they all consist of one or more cells, a membrane-enclosed space that allows organization and regulation of biochemical events. Furthermore, all cells are bounded by a lipid bilayer membrane. Moreover, all modern cellular life is centered on a DNA-based information template system in which duplex DNA molecules are transcribed into messenger RNA (mRNA), which uses a common genetic code (with a few highly specific exceptions) that is translated into proteins by ribosomes. These shared features strongly suggest that all cellular life on the planet arose from a common ancestor, termed by evolutionary biologists the Last Universal Common Ancestor or **LUCA**. It is important to recognize that LUCA was not necessarily either the first form of life or the simplest. LUCA was 'simply' an organism whose descendants have survived and adapted through countless generations over 3.5 billion years to the present day. Thus, the process of evolution, by acting on the descendants of LUCA, has given rise to the awesome diversity of life on this planet.

Phylogenetics

Early attempts to understand the evolution of cells relied upon subclassifying organisms based on observable features. For example, a rudimentary classification can be derived based on the presence or absence of a nucleus in the cell. One group of organisms – the eukaryotes – have subcellular, membrane-bound organelles and their genetic material is contained within a porous membrane-delimited structure termed the cell nucleus. Organisms that lack nuclei are called prokaryotes. While this classification system is simple and works at a certain level, it fails to take into account less instantly observable traits of the cell. Indeed, by the strict definition of eukaryotic and prokaryotic, mammalian red blood cells are prokaryotic

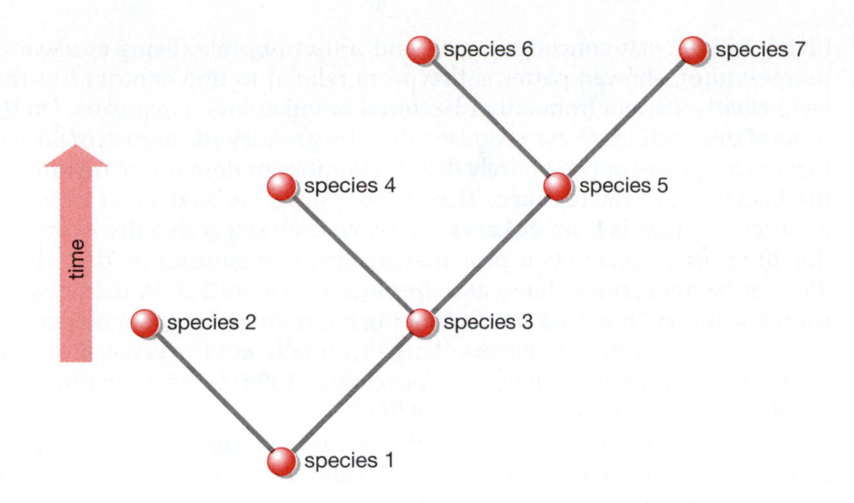

Figure 2-3. An example of a phylogenetic tree. In this illustration species 1 is the last common ancestor. As time passes, some species become extinct, e.g. species 2 and 4. Yet other species split to give rise to new lineages: for example, species 5 yields species 6 and 7.

whereas all other cells in the human body are eukaryotic. Fortunately, with the development of the science of molecular systematics, it is possible to glean a more sophisticated and accurate view of the classification and degree of relatedness of organisms, a process termed phylogeny.

Central to Darwin's theory of evolution by natural selection is the generation of diversity within a population. Some individuals will be better fit to their environment and will therefore have a reproductive advantage. Thus, the process of evolution can be envisaged as a series of bifurcations in which a parental species gives rise to two daughter species. Over time a species can either be maintained, become extinct, or undergo another bifurcation and thus yield further, novel, progeny. Such events can be visualized as a tree-like structure termed a phylogenetic tree (Figure 2-3). Phylogenetic trees plot the appearance and disappearance of species as a function of time.

This concept of vertical inheritance with modification is central to evolutionary theory. It also makes a key prediction. If extant species can be compared meaningfully, it should be possible to extract the rules of relationship and lines of descent of present-day lineages. Such analyses can be performed by comparison of the sequence of universally conserved genes, the classical example of which is the gene for the RNA component of the small subunit of the ribosome (ribosomal RNA, or rRNA). These analyses are based on the assumption that random alterations to the sequence of the rRNA gene occur with a fixed probability over time. Thus, two organisms with only one change in the sequence between their rRNAs are likely to be more closely related to one another – that is, share a more recent common ancestor – than two organisms with 10 sequence differences between their rRNAs. This approach was pioneered by Carl Woese and colleagues in the late 1970s and yielded a staggering discovery. Woese was performing fingerprint analysis of rRNAs. This was the state of the art prior to the development of nucleic acid sequencing methodologies (Figure 2-4). To examine the sequence composition of rRNAs, the isolated RNA was subjected to digestion with ribonuclease T1 (an enzyme that cuts RNA only after G residues). Separation of the digestion products revealed a pattern that was diagnostic of the sequence composition of the rRNA.

Using this approach with rRNA isolated from a range of bacteria and eukaryotes, Woese was able to discern signature patterns corresponding to either bacterial or eukaryotic species. The real excitement started when he examined the rRNA from a range of microorganisms that had been classified as prokaryotes based on their gross morphology. Remarkably, these organisms, a methanogen (methane-generating organism), a halophile

Figure 2-4. One of Woese's original films showing the separation of oligonucleotides (dark gray spots). Each spot was circled with a marking pen and its oligoribonucleotide sequence indicated in red ink. (Courtesy of Carl Woese, University of Illinois.)

(living in high salt concentrations), and a thermophile (living at elevated temperature), showed patterns that were related to one another but that were clearly distinct from either bacterial or eukaryotic organisms. On the basis of this finding, Woese proposed that the **prokaryotic** branch of life was in fact comprised of two entirely distinct families or domains of organisms, the bacteria and the **archaea**. These two groups are as distinct from one another as either is from **eukarya**. Thus, Woese's staggering discovery was that there are three, not two, principal lineages of organisms or "domains of life": the bacteria, the archaea, and the eukarya (Figure 2-5). With increasing sophistication of nucleic acid sequencing methodologies and the discovery of many more archaeal species, Woese's initially controversial proposals were fully vindicated and it became possible in 1990 for Woese to propose a topology for the phylogenetic "tree of life."

As can be seen from Figure 2-5, Woese's tree places LUCA at the divergence point of the bacterial lineage and a super-lineage that subsequently split to give rise to the eukarya and the archaea. As will become apparent in this book, this sister-grouping of eukarya and archaea reflects a closer relationship between the gene-transcription, DNA-repair, and DNA-replication machineries of these organisms than to those of bacteria.

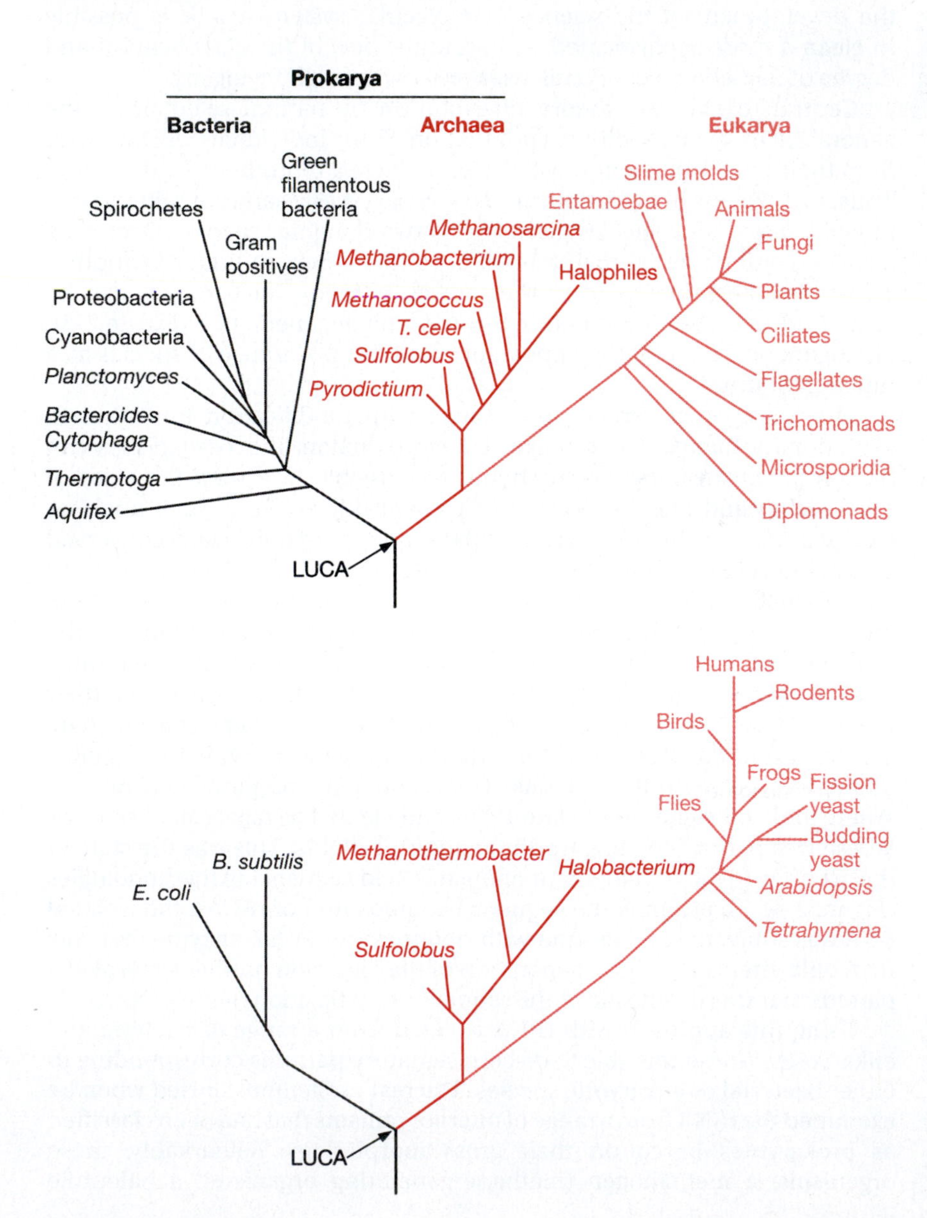

Figure 2-5. The "Universal Tree of Life" proposed by Carl Woese in 1990. Examples of organisms in the three domains of life; bacteria, archaea, and eukarya are shown in gray, red, and pink, respectively. The lower tree indicates the relationships between the most highly studied representative or model organisms that have provided a picture of genome duplication.

CHARACTERISTICS OF THE THREE DOMAINS

Many of the metabolic pathways that comprise the vast majority of any organism's gene repertoire are common between archaea and bacteria, and the size and structure of archaeal cells are indistinguishable from those of Gram-positive bacteria, except that archaea differ from most bacteria in the composition of their cell membrane and cell wall. The majority of genes that distinguish archaea from bacteria are involved in DNA replication, DNA transcription, and RNA translation. Of these, the DNA-replication machinery appears most different between the two domains. Most archaeal proteins participating in DNA replication are more similar in sequence to those found in eukarya than to analogous replication proteins in bacteria. However, archaea have only a subset of the eukaryal replication machinery, apparently needing fewer polypeptides and structurally simpler complexes. For example, of the 14 proteins that form pre-replication complexes on eukaryal genomes, only two homologs of these proteins are required to form pre-replication complexes on archaeal genomes (Table 2-2). Similarly, of the 25 different proteins required to drive eukaryotic replication forks, only 14 homologs of these proteins appear to be required at archaeal replication forks. Striking is the fact that none of the bacterial DNA replication genes appear to have found their way into archaea. The archaeal replication apparatus also contains features not found in other organisms owing, in part, to the broad range of environmental conditions, some extreme, in which members of this domain thrive. In terms of transcription, archaea possess a 12–14 subunit RNA polymerase (RNAP) closely related to the eukaryotic nuclear RNAPs and clearly more complex than the bacterial RNA polymerase. Similarly, the transcription factors that recruit the archaeal RNA polymerase to transcription start sites have close relatives in eukaryotes and not in bacteria. Archaea also differ from bacteria in some ribosomal proteins and translation initiation factors, but most rRNAs,

Table 2-2. Summary Of Characteristic Features Of Bacteria, Archaea, and Eukarya.

Property	Prokarya		Eukarya
	Bacteria	Archaea	
Nuclei	No	No	Yes
Organelles	No	No	Yes
Internal membranes	Rare[a]	Rare[a]	Common
Cell size	<5 μm	<5 μm	10–100 μm
Genome size	0.6–13 Mbp	0.5–6 Mbp	0.6–4 000 000 Mbp
Chromosome topology	Circular (usually)	Circular	Linear
Chromatin proteins	No histones	Histones (some species)	Histones
Lipids	Ester-linked	Ether-linked	Ester-linked
DNA transcription	One RNA polymerase: α, β, β', ω subunits	One RNA polymerase: >12 subunits	Three RNA polymerases: each >12 subunits
DNA replication Origin recognition Replicative helicase Helicase loader	 DnaA DnaB DnaC	 Orc1/Cdc6 Mcm Orc1/Cdc6	 ORC(1-6)+Cdc6 Mcm(2-7) ORC(1-6), Cdc6, Cdt1

[a]Cyanobacteria (bacterial domain of life) and halophilic archaea have some internal membrane systems.

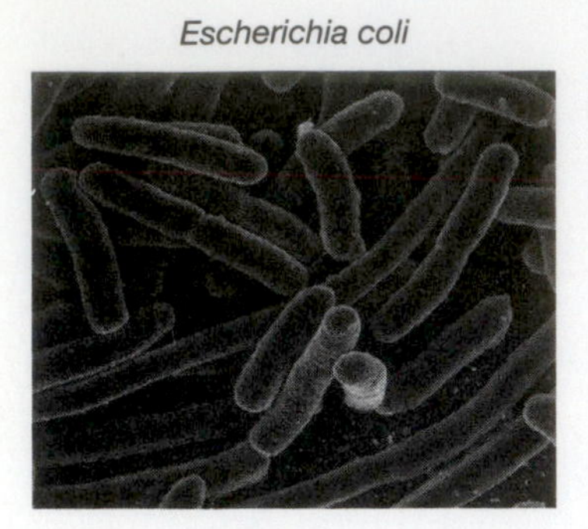

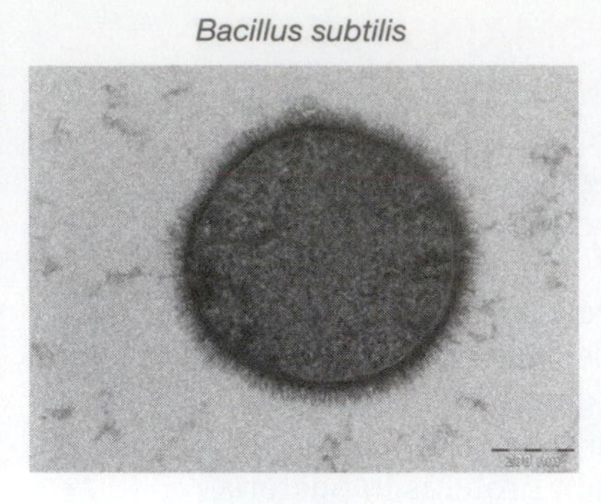

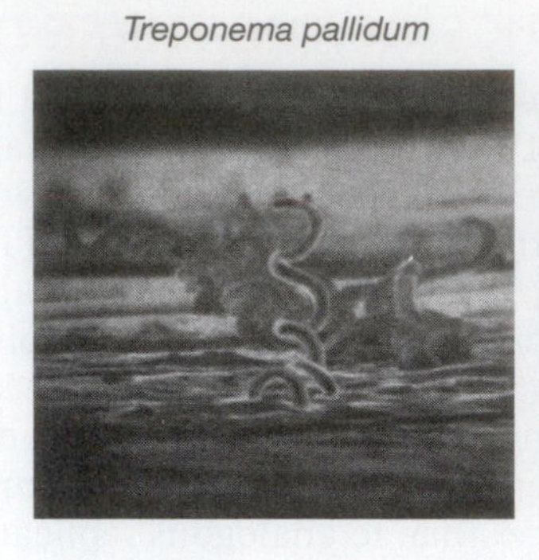

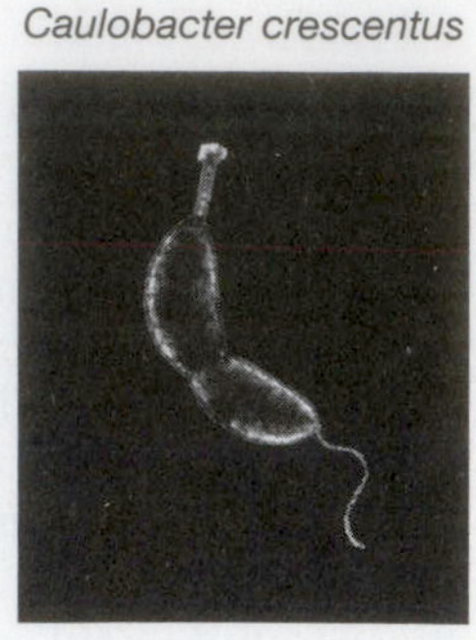

Figure 2-6. Representative bacterial species. Most of what is known about genome duplication is derived from *Escherichia coli* and *Bacillus subtilis* with supporting information from other Gram-negative enterobacteria. *B. subtilis* is a rod-shaped bacterium like *E. coli*. Shown is a cross section through *B. subtilis* to reveal the absence of internal cellular structure. (*Caulobacter crescentus* image courtesy of Yves Brun, Indiana University.)

ribosomal proteins, elongation factors, amino acid-charging enzymes, and transfer RNAs are common to both bacteria and archaea.

Bacteria

Bacteria are generally small, single-celled organisms (Figure 2-6). The membrane is surrounded by a cell wall composed of sugar and protein groups. Within the cell, the DNA is organized into a structure called a nucleoid that is not physically separated from the rest of the cell by a membrane, as is the case in eukaryotic nuclei (Figure 2-7). Within the DNA itself, genes are often arranged in operons, where several genes are cotranscribed as a single **poly-cistronic mRNA**. Because there is no physical division between DNA and cytoplasm, the mRNA can be immediately translated by ribosomes, thus transcription and translation are coupled. The majority of characterized bacteria have circular chromosomes, although some species such as Streptomycetes have linear genomes. As will be seen in Chapter 8, this linearity has interesting implications for the way in which these chromosomes are replicated.

Archaea

Although the first archaea to be identified lived in rather bizarre ecological niches, it has become apparent that archaeal species are found in almost every environment on the planet, from Antarctic surface waters at near-freezing temperatures to the vicinity of hydrothermal vents on the ocean floor where species such as *Pyrodictium* thrive in temperatures in excess

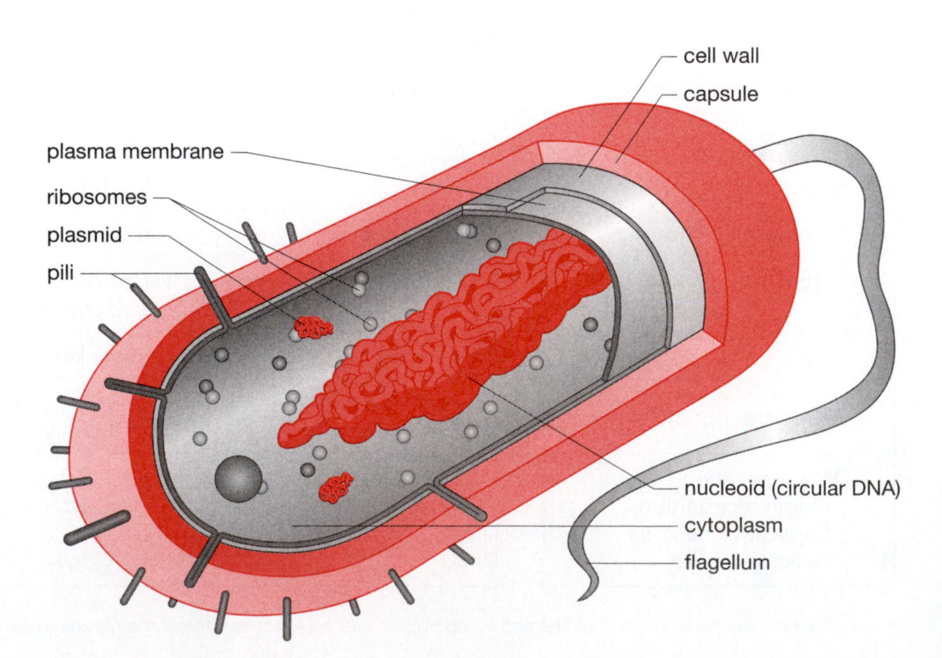

Figure 2-7. Organization of a typical bacterium. The chromosome forms a densely packaged structure in the cytoplasm, termed a nucleoid.

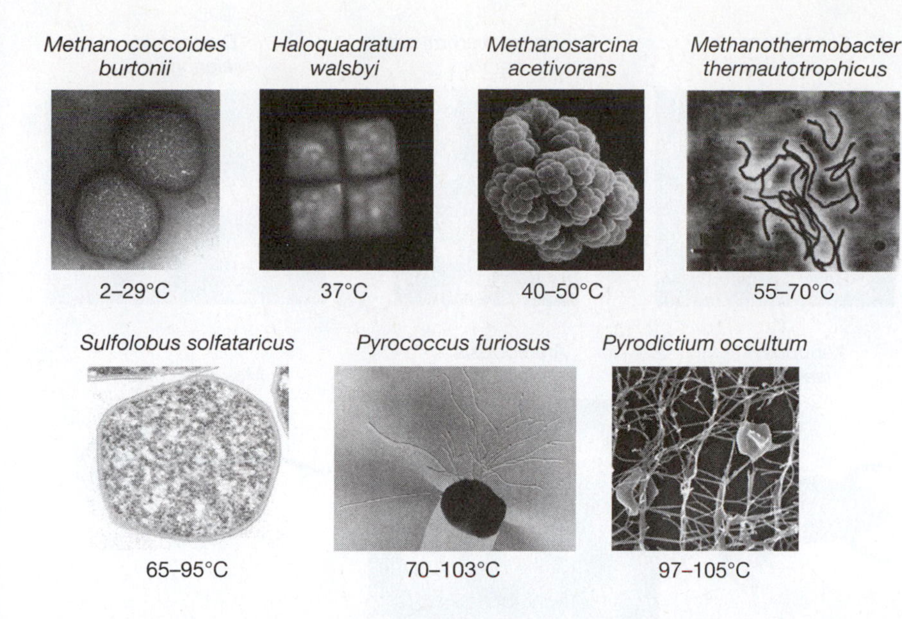

Figure 2-8. Representative archaeal species with their optimal growth temperatures indicated below the panels. (*Methanococcoides burtonii* image courtesy of Ricardo Cavicchioli, University of NSW. *Methanosarcina acetivorans* image courtesy of Herbert Fang, University of Hong Kong. *Sulfolobus solfataricus* image courtesy of the late Wolfram Zillig, Max Planck Institute of Biochemistry. *Pyrococcus furiosus* image courtesy of Karl Stetter, Universität Regensburg. *Pyrodictium occultum* image from Rieger, G., Rachel, R., Hermann, R. and Stetter, K.O. (1995) *J. Struct. Biol.* 115: 78–87. With permission from Elsevier.)

of 100°C (Figure 2-8). Sampling of the biodiversity of microbial life in the ocean suggests that archaea are, after bacteria, the second most abundant form of life on Earth.

Archaea are morphologically classified as prokaryotes. They do not possess nuclei or (with a few specific exceptions) other membrane-bound subcellular structures, and so the gross morphology of an archaeon is well represented by that of a bacterium (Figures 2-6 and 2-7). At the crude organizational level, archaeal genomes resemble those of bacteria. All archaeal chromosomes characterized to date are circular and contain many poly-cistronic transcription units. However, the archaeal RNA polymerase has at least 12 subunits, all of which are related to components found in eukaryotic RNA polymerase and many of which have no counterparts in the simpler bacterial enzyme. As will be seen in later chapters, the archaeal DNA replication system is highly related to, but much simpler than, that found in eukaryotes.

Eukaryotes

Eukaryotic cells (Figure 2-9), like the organisms from which they are derived (Figure 2-10), are organizationally complex. The majority of DNA within

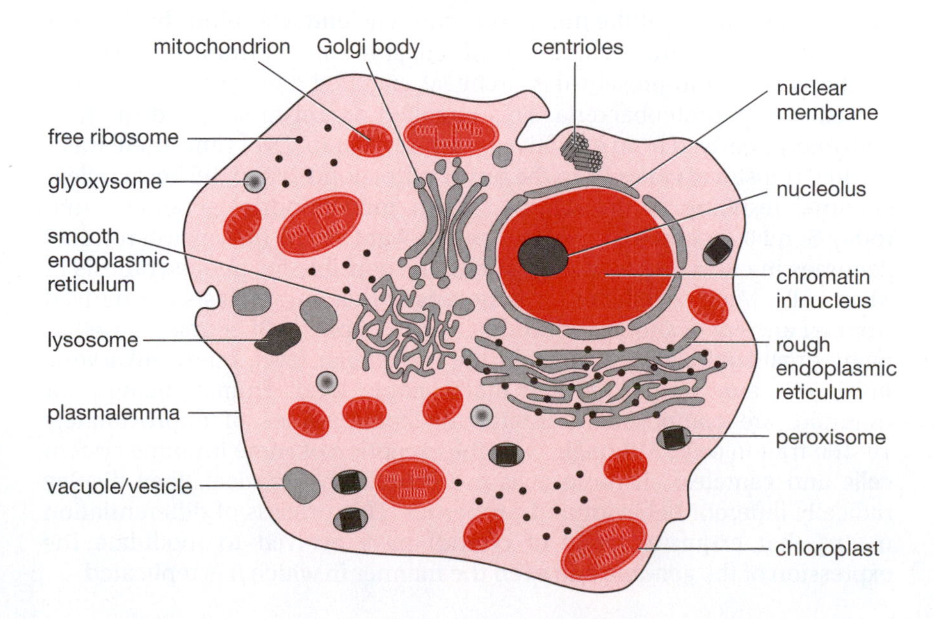

Figure 2-9. Cartoon of the organization of an idealized eukaryotic cell. Note that chloroplasts are restricted to photosynthetic organisms.

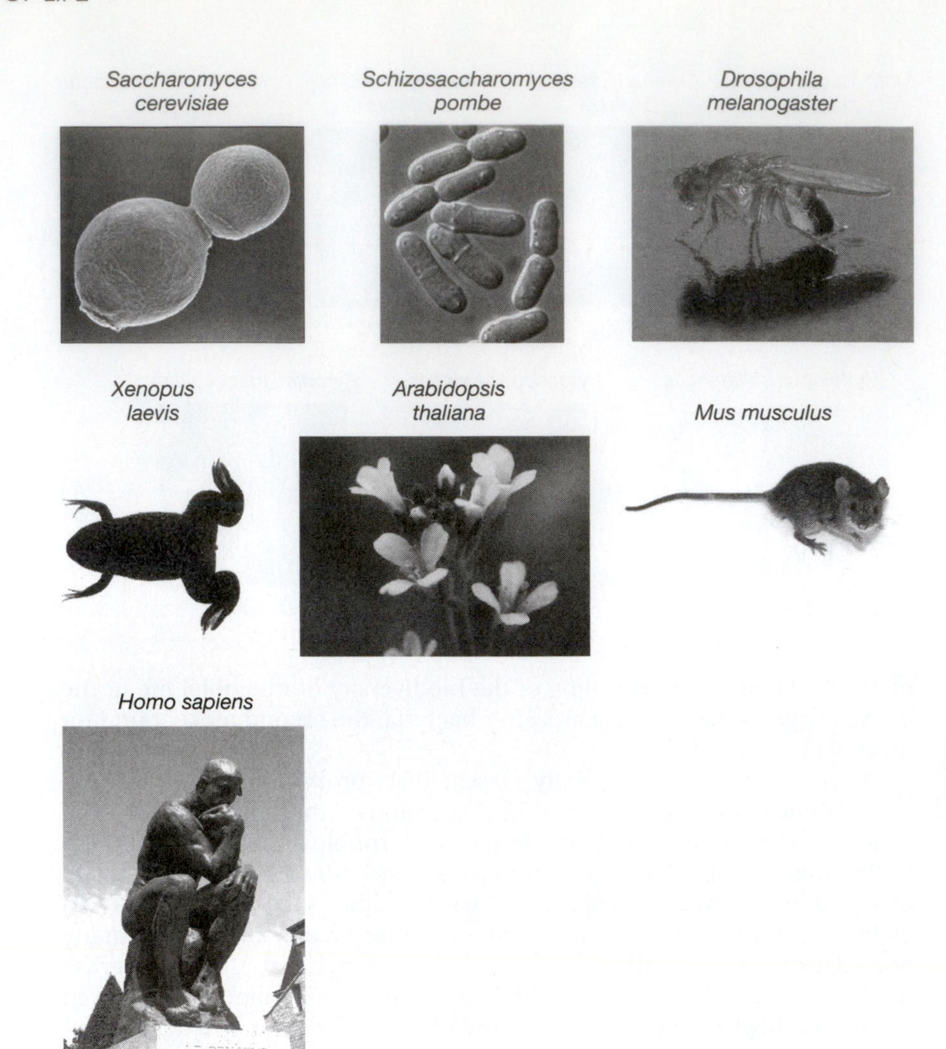

Figure 2-10. Representative eukaryotic organisms used to study genome duplication. (*Saccharomyces cerevisiae* image courtesy of Yoshifumi Jigami, National Institute of Advanced Industrial Science and Technology. *Schizosaccharomyces pombe* image courtesy of Nicholas Rhind, University of Massachusetts Medical School. *Drosophila melanogaster* image courtesy of Andre Karwath under Creative Commons Attribution-Share Alike 2.5 License. *Arabidopsis thaliana* image courtesy of Alberto Salguero under Creative Commons Attribution-Share Alike 3.0 License. *Mus musculus* courtesy of George Shuklin under Creative Commons Attribution-Share Alike 1.0 Generic License.)

eukaryotic cells is found within the nucleus, a membrane-bound organelle where it is highly compacted by association with DNA-binding proteins called histones into the form of chromatin (Chapter 7). Eukaryotic cells, however, contain additional DNA in some of their organelles. Mitochondria, for example, possess double-stranded DNA molecules (a circle of 16 569 bp in humans). These mitochondrial genomes are thought to reflect the evolutionary source of the mitochondrion. The **endosymbiont hypothesis** proposes that at the generation of eukaryotes, a primitive organism (perhaps related to present-day archaea) engulfed a bacterium related to present-day α-proteobacteria. This engulfed organism supplied the host with energy derived from oxidative phosphorylation. Over time, there was a gradual transfer of genes from the proto-mitochondrion to the host nuclear genome, resulting in the stripped-down mitochondrial genome found today. Similar arguments have been put forward for the presence of residual genomes in chloroplasts, the organelles responsible for photosynthesis in plant cells. More specifically, it is proposed that chloroplasts are derived from an ancient cyanobacterium-like endosymbiont. Of course, as well as being organizationally complex at the subcellular level, many eukaryotic organisms are complex at the macroscopic level. Human beings, for example, are composed of interacting communities of approximately 10–100 trillion cells. Although, with the exception of some immune system cells and gametes, all these cells are genetically identical, they display radically different behavior and properties. This process of differentiation means that exquisite levels of control have evolved to modulate the expression of the genome and even the manner in which it is replicated.

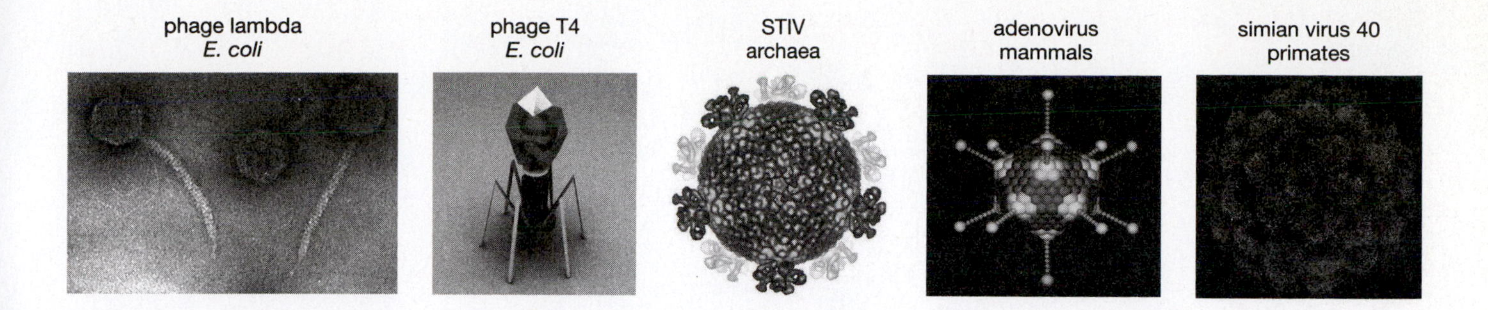

Figure 2-11. Examples of some of the viruses that have been used to study DNA replication in bacteria, archaea, and mammals. Note that for historical reasons viruses that infect bacteria are normally called bacteriophage, or phage for short. (Phage lambda image courtesy of Ross Inman, University of Wisconsin-Madison. STIV image courtesy of Mark Young, TBI, Montana State University. Simian virus 40 image courtesy of Phoebus87 under Creative Commons Attribution-Share Alike 3.0 License.)

VIRUSES: UBIQUITOUS PARASITES

Bacteria, archaea, and eukaryotes are all living beings: they are entities capable of independent reproduction, metabolism, and evolution (Table 2-1). But preying upon these cellular organisms are forms of the ultimate parasite: the viruses (Figure 2-11). Viruses are small self-contained packages of protein and nucleic acid that are utterly dependent on host cells for their replication and propagation, often to the host's detriment. As will be seen in later chapters, viruses have been invaluable tools for understanding the concepts that underpin cellular processes. In many cases, the way that a virus hijacks a cell's machinery gives insight into the way that that machinery is normally controlled. Additionally, viruses sometimes possess simplified machineries that are nonetheless closely related to the host's apparatus and thus can serve as model systems to understand the related, but more complex, host system.

As well as being useful tools for molecular biologists, an increasing body of data is suggesting that viruses have played a pivotal role in shaping the content of host cell genomes and, as will be detailed in Chapter 15, may even have contributed to some key steps in evolution.

SUMMARY

- The Earth was formed 4.5 billion years ago.
- The first evidence for cellular life dates to 3.8–3.5 billion years ago.
- The first living systems were probably based on RNA genomes.
- RNA can be an information source and an enzyme.
- All extant living organisms have a common ancestor: LUCA.
- All cellular organisms can be classified into one of three domains of life: bacteria, archaea, and eukarya.
- Viruses parasitize all three domains of life and may have played key roles in transitions in evolution.

REFERENCES

Additional Reading

Barton, N.H et al (2007) *Evolution.* Cold Spring Harbor Laboratory Press, New York.

Woese, C., Kandler, O., and Wheelis, M. (1990) Towards a natural system of organisms: proposal for the domains Archaea, Bacteria, and Eukarya. *Proc. Natl. Acad. Sci.* 817(12): 4576–4579.

Chapter 3
REPLICATION FORKS

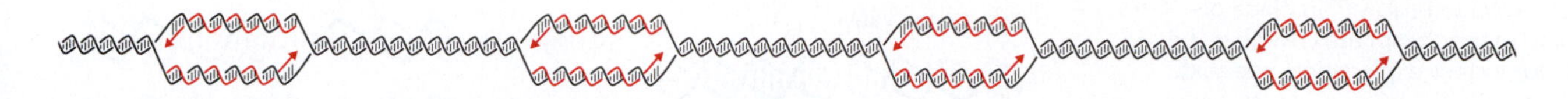

The previous chapters presented two fundamental characteristics of all living organisms: they can be grouped into three domains (bacteria, archaea, and eukarya) and they use the duplex form of DNA (dsDNA) to encode their genomes. This chapter presents two more unifying characteristics: they replicate their genomes semi-conservatively, and they do so using the replication-fork mechanism, a mechanism used by many bacteriophages, bacterial plasmids, and animal viruses, as well as by cellular organisms. To be sure, other strategies for genome duplication also exist in nature, but they are confined to the genomes of some bacteriophage, bacterial plasmids, animal and plant viruses, and to the tiny fraction of the eukaryotic genome contained in mitochondria. These non-replication-fork mechanisms, as well as the effects imposed on DNA replication by DNA modification, chromatin structure, and nuclear organization, are described in subsequent chapters.

DNA REPLICATION

Three years after the structure of DNA was solved, the first DNA polymerase (an enzyme that synthesizes DNA) was purified from the *E. coli* bacterium, a discovery for which Arthur Kornberg received the Nobel Prize in Physiology or Medicine in 1959. *E. coli* DNA polymerase I reads the sequence of one strand of dsDNA while synthesizing the complementary strand. This discovery, which marked the first step in understanding how the copying mechanism works, was followed rapidly by evidence that DNA does, in fact, replicate semi-conservatively, with one strand serving as a template upon which a new complementary DNA strand is synthesized. Therefore, when a cell divides, the DNA in each of the two progeny cells contains one old and one new DNA strand. Thus, the first essential concept of genome duplication was established: the two antiparallel strands are separated (unwound) and a specific enzyme (DNA polymerase) synthesizes a complementary copy of each strand.

Given the structure of DNA, there are only three possible modes by which it can be replicated: conservative, semi-conservative, and dispersive (Figure 3-1). In **conservative replication**, the parental DNA molecule is duplicated without changing its structure or composition. The circular ssDNA genomes found in some viruses can be considered to replicate conservatively, although they pass through a transient dsDNA intermediate that is replicated semi-conservatively (Figure 8-8). In **semi-conservative replication**, the two strands of the double helix are unwound, and a complementary copy of each strand is synthesized. This mode is used to duplicate the genomes of all cellular organisms, as well as the dsDNA genomes of all viruses and plasmids. In **dispersive replication**, newly replicated sections of DNA are dispersed among old unreplicated DNA such that the newly replicated portions are covalently linked to the unreplicated

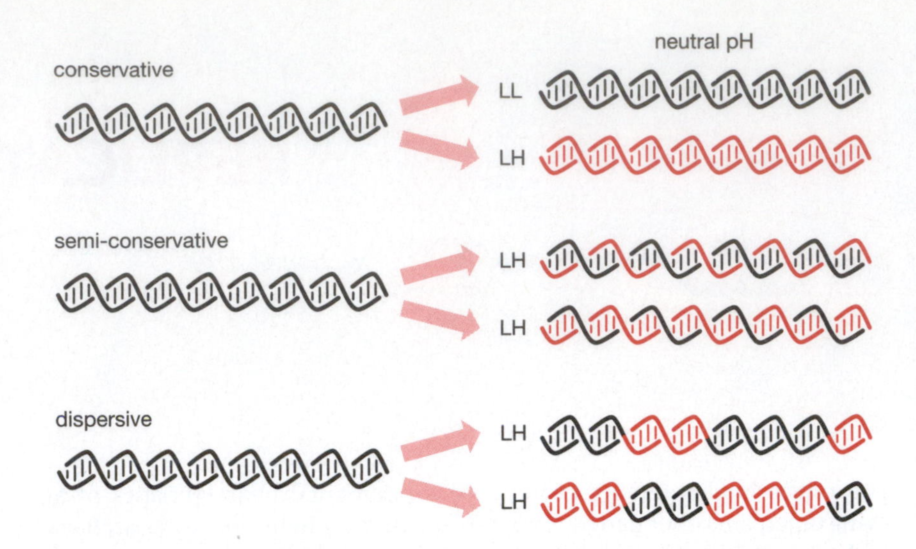

Figure 3-1. Distinguishing the three possible modes for duplicating DNA under conditions where dsDNA is stable (neutral pH, 0.2 M salt). Parental DNA is black and contains elements of normal density. Nascent DNA is red and contains elements of heavy density [e.g. bromodeoxyuridine monophosphate (BrdUMP) in place of thymidine monophosphate (TMP)]. H, heavy; L, light; for further explanation, see text.

portions. This mode of replication has been observed only with parvovirus genomes.

The three modes of DNA replication can be distinguished experimentally by incorporating a density-labeled nucleotide analog such as bromodeoxyuridine triphosphate (BrdUTP) (Figure 3-1) into nascent DNA synthesized during the first round of cell division. If replication is either semi-conservative or dispersive, then newly replicated DNA will contain 50% of the normal isotope (light DNA) and 50% of the dense isotope (heavy DNA). Thus, all of the newly replicated native DNA will be of hybrid density (LH). However, if DNA replication is conservative, then half of the resulting DNA will be completely light (LL) and half will be completely heavy (HH) (Figure 3-1).

Semi-Conservative DNA Replication

In 1958, Matthew Meselson and Franklin Stahl demonstrated that bacterial genomes replicated semi-conservatively. They propagated *E. coli* for several hours in medium containing $^{15}NH_4Cl$ (ammonium chloride containing the

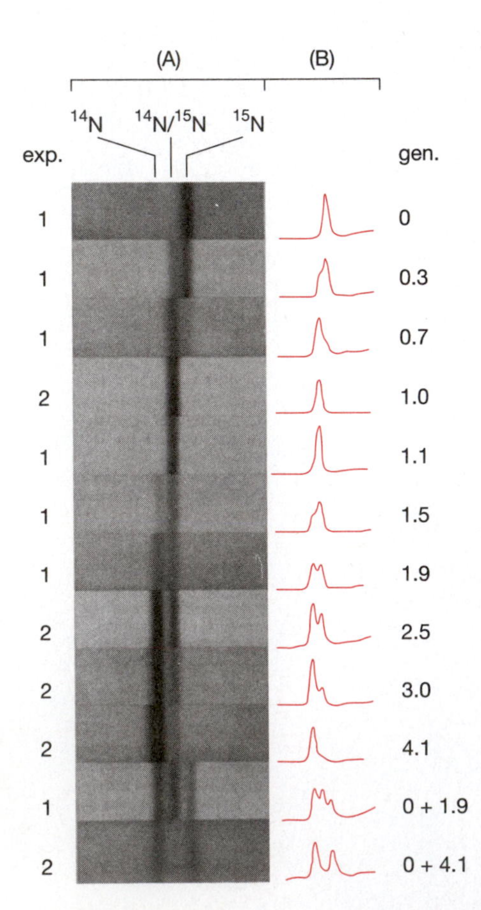

Figure 3-2. Density-gradient equilibrium centrifugation of *E. coli* DNA at neutral pH reveals the semi-conservative nature of genome duplication. In a high-speed centrifuge, CsCl sediments away from the axis of revolution to form a density gradient that remains constant when the centrifugal force driving the salt in one direction and the diffusion forces dispersing the salt in all directions reach equilibrium. Under these conditions, DNA forms a band at its buoyant density (typically 1.701 g cm^{-3}). Since ^{14}N-DNA is less dense than ^{15}N-DNA by 0.014 g cm^{-3}, they separate into two bands (^{14}N and ^{15}N, respectively). Ultraviolet (UV) absorption photographs (A) reveal the positions of DNA in the centrifuge tube, and densitometer tracings of the UV images (B) provide quantifiable results. When bacterial lysates were prepared at various times after addition of an excess of ^{14}N-labeled substrates to cells proliferating in ^{15}N-labeled substrates, newly replicated DNA became lighter in density. The number of cell divisions [generation times (gen.)] that occurred are indicated. The degree of labeling of a species of DNA corresponded to the relative position of its band between the bands of fully labeled and unlabeled DNA shown in the lower-most frame, which serves as a density reference. Mixtures of the 0+1.9 and the 0+4.1 generation samples confirmed the presence of DNA with an intermediate (hybrid) density (^{14}N/^{15}N). When allowance is made for the relative amounts of DNA in the three bands, the intermediate band is 50±2% of the distance between the ^{14}N and ^{15}N bands. Therefore, the hybrid DNA band contains half light DNA and half heavy DNA. (exp. = experiment; gen. = generation) (From Meselson, M. and Stahl, F.W. (1958) *Proc. Natl. Acad. Sci. USA* 44: 671–682. With permission from the National Academy of Sciences.)

heavy isotope of nitrogen) in order to make the density of all DNA strands heavier than normal (Figure 3-2, 0 generations). The bacteria were then transferred to medium containing the normal nitrogen isotope ($^{14}NH_4Cl$). As the cells continued to grow and divide for one generation (45 min), all of the DNA shifted in density from heavy (fully substituted with ^{15}N) to hybrid density in which half of the nitrogen in the DNA band was ^{15}N and half was ^{14}N (Figure 3-2, 1 generation). However, after 1.9 generations two bands were seen, one corresponding to hybrid $^{14}N/^{15}N$-DNA, and the other to DNA molecules made entirely of ^{14}N. By the time four generations of cells were produced, all of the DNA contained predominantly ^{14}N (Figure 3-2, 4.1 generations).

These data showed that DNA replication did not occur by a conservative mode, because only one DNA band was observed after one cell generation. A conservative mode would produce two different dsDNA molecules after one generation, one containing only ^{15}N and one containing only ^{14}N. The single $^{14}N/^{15}N$-DNA band detected after one generation was consistent either with semi-conservative or with dispersive DNA replication. However, the two bands seen after two generations were consistent only with semi-conservative replication. Dispersive DNA replication would produce only hybrid $^{14}N/^{15}N$-DNA after two generations, whereas semi-conservative replication would produce one hybrid dsDNA band and one light dsDNA band composed entirely of ^{14}N-DNA. Since the conditions of these experiments maintained the DNA in its duplex form, these results demonstrated that *E. coli* duplicated its chromosome semi-conservatively (Figure 3-1). See also Box 3-1.

Semi-conservative replication can be distinguished from dispersive replication during the first round of cell division by denaturing the DNA before subjecting it to density analysis (Figure 3-3). Under these conditions, semi-conservative DNA replication produces a completely light strand (L) and a completely heavy strand (H), but dispersive replication produces only ssDNA of hybrid density (LH). This test is critical to distinguish semi-conservative DNA replication from damaged DNA that was repaired in the presence of the heavy isotope, or that underwent various forms of recombination during a single cell division. In the latter cases, DNA will contain patches of newly synthesized density-labeled DNA dispersed among unreplicated DNA and linked covalently to unreplicated strands. The greater the extent of DNA repair or recombination, the greater the chance that these phenomena will be mistaken for semi-conservative DNA replication when assayed under neutral pH conditions.

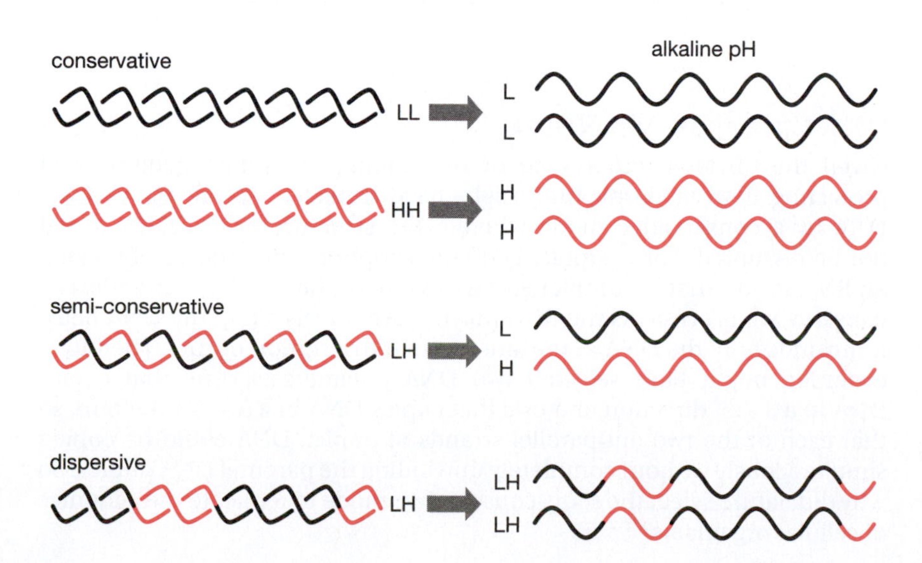

Figure 3-3. Distinguishing the three modes for duplicating DNA from DNA repair at alkaline pH. Density labeling pattern for each of the three possible modes of DNA replication before (left) and after (right) DNA is denatured. Parental DNA (black) contains elements of normal density. Nascent DNA (red) contains elements of heavy density (e.g. BrdUMP in place of TMP).

Box 3-1. Evidence for Semi-Conservative DNA Replication

Although the bacterial experiment is often cited, semi-conservative genome duplication was first demonstrated in 1957 by Herb Taylor using autoradiography to visualize the distribution of radiolabeled nucleotides in the two sibling chromatids of newly duplicated chromosomes (Taylor *et al.*, 1957). Root cells of the broad bean plant (*Vicia faba*) were allowed to incorporate ^{3}H-thymidine (^{3}H-TdR) into their DNA during the 8 h period of replicative DNA synthesis (S phase) in their cell-division cycle. The cells converted ^{3}H-TdR into ^{3}H-dTTP, a substrate for DNA polymerase. The root cells were then transferred to a solution containing colchicine but not ^{3}H-TdR. Colchicine prevents cells from exiting metaphase, a time during mitosis when chromosomes are greatly condensed and the two sibling chromatids of each newly duplicated chromosome are linked together at their centromere (cen). Thus, individual chromosomes could be visualized by squashing the cells on a glass slide and then staining their DNA with an appropriate dye. The sites of ^{3}H-TdR incorporation were visualized by covering the slide with a photographic emulsion. Tritium (^{3}H) emits low-energy β particles that activate the silver halide in the emulsion. After about 2 weeks, a sufficient number of silver halide ions form such that subsequent development and fixation turn the radiated silver halide into visible black grains.

The results revealed that both chromatids in each newly duplicated chromosome were labeled along their entire length (panel A). When cells that had been treated in this manner were allowed to proceed into the next cell-division cycle in the absence of ^{3}H-TdR, each chromosome contained one labeled and one unlabeled chromatid, except where a reciprocal sister chromatid exchange (SCE) had occurred. However, cells that transited the first cell division before exposure to colchicine underwent a second round of genome duplication in the absence of ^{3}H-TdR. These cells were arrested by colchicine when they attempted to undergo a second round of mitosis, and their chromosomes contained one radiolabeled chromatid lying beside one unlabeled chromatid (panel B). Thus, each chromosome appeared to contain a single DNA molecule that underwent semi-conservative replication each time a cell divided.

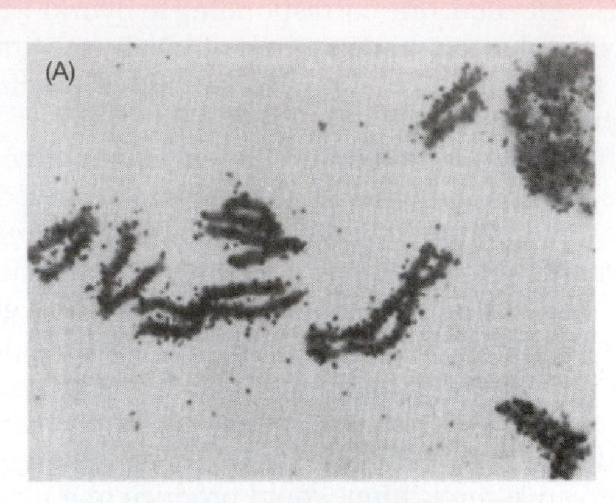

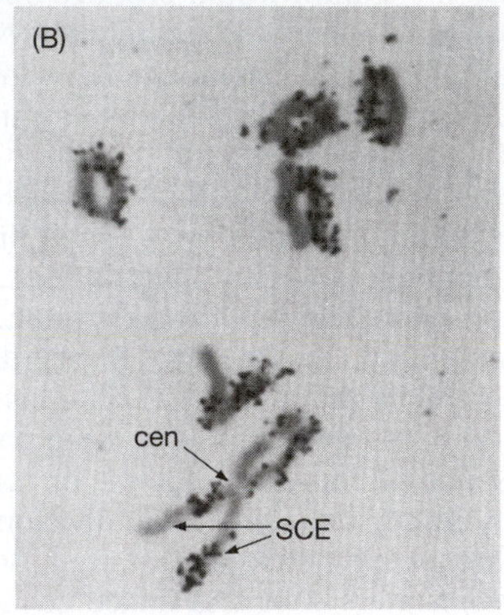

(From Taylor, J.H., Woods, P.S. and Hughes, W.L. (1957) *Proc. Natl. Acad. Sci. USA* 43: 122–128. With permission from the National Academy of Sciences.)

DNA Replication And Natural Selection

Given the obvious importance of maintaining an intact genome and protecting it from harm, one might have expected nature to replicate DNA by a conservative mode wherein the structure of the DNA would not be disrupted. For example, gene transcription, the process of making an RNA strand that is complementary to one of the two DNA strands, is a conservative process. As the RNA polymerase travels along the DNA helix, it unwinds only the DNA at the site where transcription occurs. Therefore, evolution might have selected two DNA polymerases, one that copies DNA in a 3′→5′ direction and one that copies DNA in a 5′→3′ direction, so that each of the two antiparallel strands of duplex DNA could be copied simultaneously without completely unwinding the parental DNA duplex. So why did nature select the semi-conservative mode to replicate the genomes of cellular organisms?

The simplest explanation is that nature does not design solutions to problems, nature selects solutions from the tools that are available. Given the fact that all DNA polymerases synthesize DNA only in the 5′→3′ direction, the primordial DNA polymerase must have appeared during evolution prior to the three domains of life. As mutations within this enzyme occurred, more efficient and more specialized versions were selected for various roles in the duplication and repair of genomes in bacteria, archaea, and eukarya. Thus, the problem of replicating duplex DNA was more easily solved by a semi-conservative mode that allowed the existing families of enzymes to copy the template.

DNA Replication Forks

With the discovery of the structure of DNA, the mechanism for duplicating the genetic material became apparent: unzip the double helix to expose the bases so that DNA polymerase can make a complementary copy of each strand by forming the correct base pair (Figure 1-3) and then link the newly inserted deoxynucleotides together. If DNA synthesis occurs concurrently on both template strands, this process produces a Y-shaped molecule referred to as a replication fork (Figure 3-4). Thus, the selective advantage of this mechanism is genomic stability, because it produces a minimum amount of ssDNA as a transient intermediate. ssDNA is hypersensitive to chemical and enzymatic degradation as well as to mechanical shear. In addition, since chromatin assembly requires dsDNA, replication forks allow rapid assembly of newly replicated DNA into chromatin. This further protects the genome while simultaneously maintaining it in a highly condensed form that can be stored within the tiny space of a bacterial or archaeal cell or the nucleus of a eukaryotic cell.

The Replication-Fork Paradox

DNA is comprised of two strands arranged in an antiparallel configuration (5′→3′:3′←5′; Figures 1-3 and 1-4), but DNA polymerases synthesize DNA only in one direction (5′→3′). Moreover, DNA polymerases, in contrast to RNA polymerases, cannot initiate polynucleotide synthesis *de novo*; they can only add nucleotides to the 3′-OH terminus of a polynucleotide that is annealed to a DNA template. Thus, the antiparallel 3′→5′ template strand can readily be copied because the 3′-OH end of the growing nascent DNA strand acts as a 'primer' for DNA synthesis. Accordingly, this template strand is termed the 'leading-strand template', because the primer can be extended continuously in one direction. But how is the complementary 5′→3′ template strand (the 'lagging-strand template') copied at the same time as the leading-strand template?

The solution to this paradox emerged in 1966 with the discovery by Reiji Okazaki that DNA replication in *E. coli* involves the transient synthesis of short nascent DNA fragments that are then joined together to make long

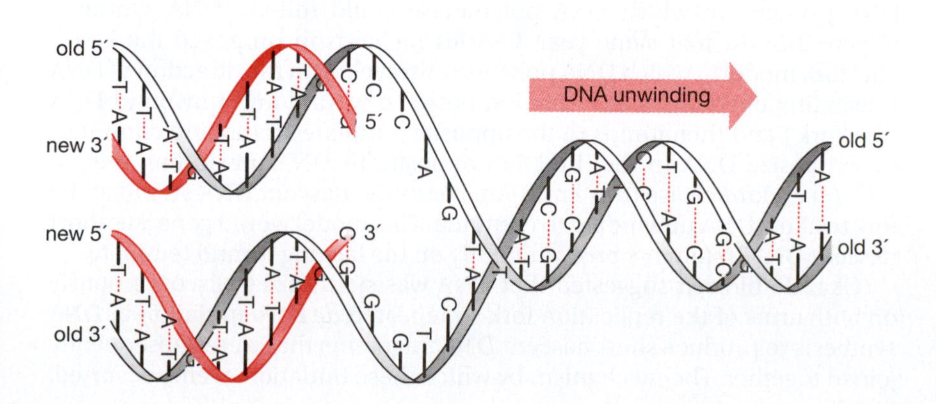

Figure 3-4. The replication fork is the site where dsDNA is unwound and a complementary copy of each of the two strands is synthesized concurrently. New (nascent) DNA strands are in red. Old (template) DNA strands are in black. Complementary DNA strands are antiparallel to one another.

Box 3-2. Discovery of Okazaki Fragments

Okazaki fragments (O.f.) were discovered by pulse-labeling *E. coli* with ^{3}H-thymidine under conditions that greatly slowed the rate of cell growth and division (Sakabe and Okazaki, 1966). *E. coli* were cultured at their optimum temperature of 37°C for several generations in the presence of ^{14}C-thymidine to uniformly label their DNA with ^{14}C (black line). The cells were then cooled to 20°C and pulse-labeled with ^{3}H-thymidine for 10 s to label nascent DNA (red line) under conditions where a slower rate of DNA replication might reveal the presence of transient intermediates (A). The doubling time for *E. coli* is approximately 40 min at 37°C and approximately 250 min at 20°C. In pulse-chase experiments (B, C), a vast excess of unlabeled thymidine was added to cells that had been pulse-labeled for 10 s at 20°C, and the incubation continued for the times indicated. Total cellular DNA was then isolated and fractionated by sedimentation in alkaline sucrose gradients to completely denature it. The amount of acid-insoluble radioactivity was quantified in each fraction of the gradient that could be made soluble by treatment with deoxyribonuclease (an enzyme that specifically digests DNA). Under these conditions, most of the ^{3}H-DNA appeared first as fragments approximately 50 to approximately 5000 nucleotides in length that then rapidly became long ^{3}H-DNA fragments, consistent with a role as transient intermediates in DNA replication. m, minutes. The amount of radiation detected from the ^{14}C and ^{3}H isotopes was reported as 'counts per minute' (CPM).

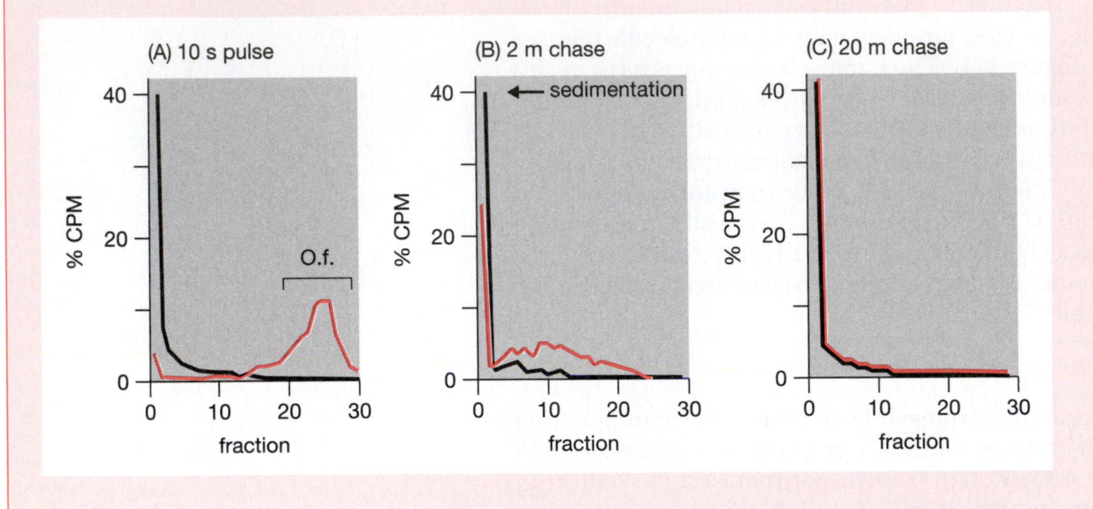

nascent DNA chains (Box 3-2). Thus began a lengthy debate over whether Okazaki's fragments were genuine intermediates in DNA replication or experimental artifacts. This debate faded away as Okazaki fragments were identified during genome duplication in a wide variety of biological systems, and under a wide variety of experimental conditions both *in vitro* and *in vivo*. The mechanism of DNA replication, however, remained a mystery.

To explain the existence of short transient nascent DNA intermediates at replication forks, three basic models were proposed. In 1969, Ed Haskell and Cedric Davern proposed the pre-fork synthesis model in which staggered single-strand breaks occurred in front of the fork to provide DNA primers on which DNA polymerase could initiate DNA synthesis (Figure 3-5). In that same year, Charles Richardson proposed the knife-and-fork model in which DNA polymerase traveling in the direction of DNA unwinding encounters the junction between wound and unwound DNA (the 'fork') and then jumps to the opposite template strand and continues to synthesize DNA in the direction opposite to DNA unwinding (Figure 3-6). An endonuclease (the 'knife') then cuts the nascent DNA strand at the fork to allow DNA unwinding to continue. This model would generate short nascent DNA fragments predominantly on the lagging-strand template.

Okazaki himself suggested that DNA was synthesized discontinuously on both arms of the replication fork by repeated *de novo* initiation of DNA synthesis to produce short nascent DNA fragments that were subsequently joined together. The mechanism by which these initiation events occurred,

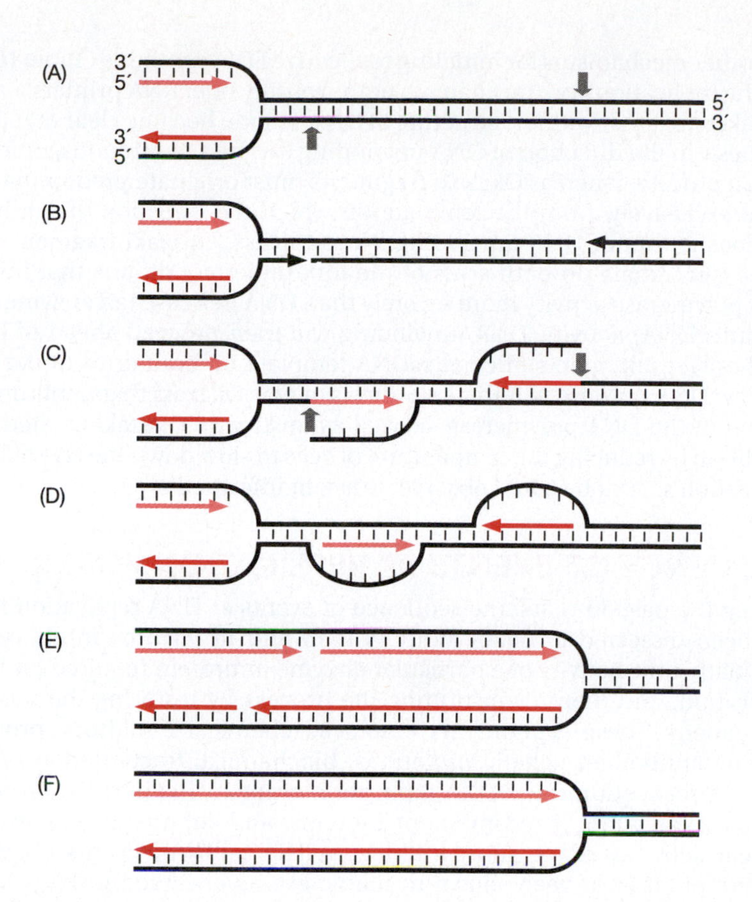

Figure 3-5. The pre-fork synthesis model proposed that staggered single-strand breaks occur in front of a replication fork (A, B, gray arrows) to provide DNA polymerase with 3′-OH-terminated primers (black arrowheads) on which to initiate DNA synthesis (C, red arrows). DNA polymerase copies the template strand opposite the primer. As the DNA in front of the polymerase is unwound, the DNA behind the nascent strand is cut again (C). This allows rejoining of the parental DNA strand (D). Rejoining would be facilitated by a 5′→3′ exonuclease that excises part of the nascent strand to allow reannealing of the parental templates. The break in the templates would be sealed by DNA ligase. The short nascent DNA fragments are then extended by DNA polymerase (E) and joined to the ends of the long growing nascent strands (F). This model accounts for short nascent DNA strands originating from both arms of the replication fork.

however, was unknown at the time. In the end, two key discoveries proved that Okazaki's model was correct in principle, if not in detail. The first was that Okazaki fragments contain a short oligoribonucleotide covalently linked to their 5′-end (Figure 3-7A). This distinguished replicative DNA synthesis from DNA synthesis involved in repair or recombination of DNA, because replicative DNA synthesis begins on the 3′-end of an RNA primer (also called initiator RNA), rather than on the 3′-end of a DNA primer. The second was the discovery of a novel class of enzymes termed DNA primases whose sole purpose was to allow DNA polymerase to initiate replicative DNA synthesis at a great diversity of sequences in ssDNA templates. Thus, replicative DNA synthesis could begin almost anywhere on a ssDNA template. The frequency of initiation in various cells, viruses, and episomes that use the replication-fork mechanism ranges from once every 30 or so nucleotides to once every 5000 nucleotides (Figure 3-7B). Later studies on genomes from bacteriophages, animal viruses, and episomes would reveal

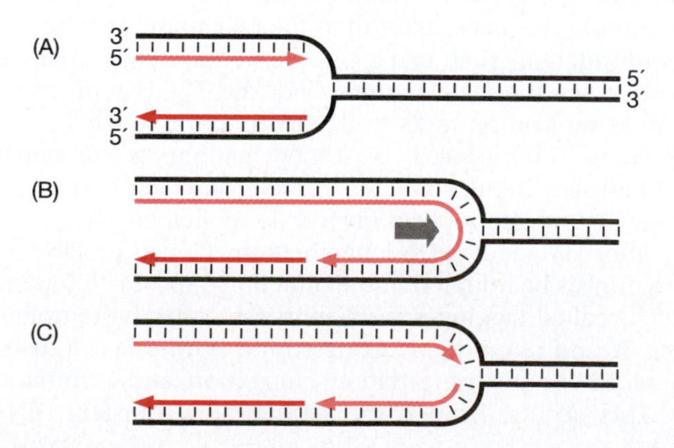

Figure 3-6. The knife-and-fork model proposed that DNA polymerase (A, arrowheads) jumps from one template to the other when it encounters a replication fork (B). An endonuclease (gray arrow) then cuts the nascent DNA strand at the site of the strand switch (B, C). The 3′-OH end of the cut strand (C, arrowhead) then acts as a primer that can be extended further by DNA polymerase on both template strands. DNA helicase could then unwind DNA further and the process repeated. This model accounts for short nascent DNA strands originating from one arm of the fork, but not the other.

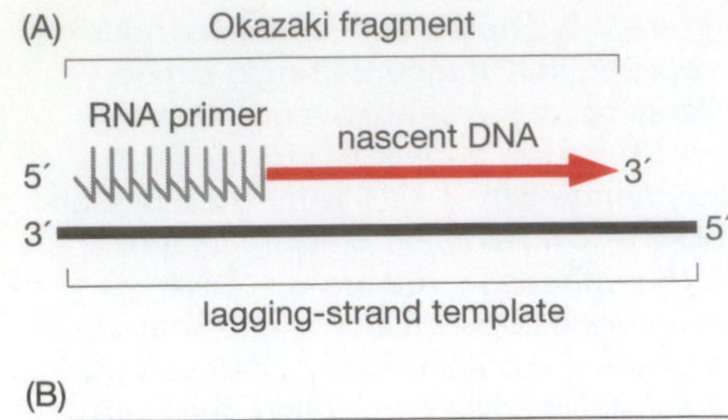

(B)

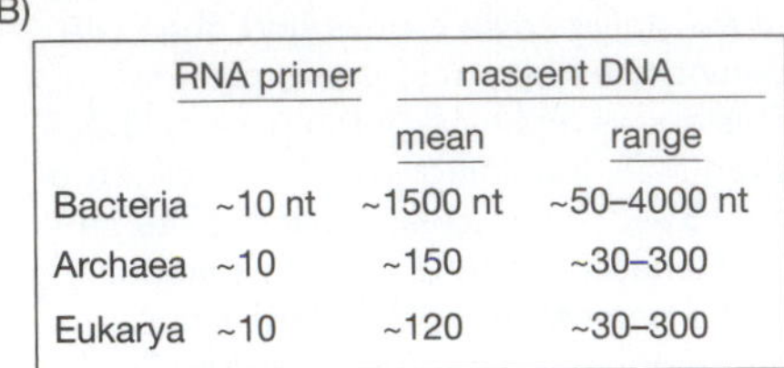

	RNA primer	nascent DNA	
		mean	range
Bacteria	~10 nt	~1500 nt	~50–4000 nt
Archaea	~10	~150	~30–300
Eukarya	~10	~120	~30–300

Figure 3-7. (A) Okazaki fragments consist of short nascent DNA chains (red) with an oligoribonucleotide (RNA primer, gray) covalently linked to their 5′-ends. (B) Except for Okazaki fragments from eukarya, the lengths of RNA primers are based on the properties of DNA primases *in vitro*. Data for eukarya reflect *in vivo* analysis of Okazaki fragments from mammalian cells and papovaviruses. nt, nucleotides.

four other mechanisms for initiating replicative DNA synthesis (Table 10-1), but the replication-fork mechanism employs only one: RNA primers.

Taken together with the structure of DNA, it soon became clear that DNA synthesis in the direction of DNA unwinding can be a continuous, uninterrupted process, whereas Okazaki fragments must originate predominantly, if not exclusively, from the template on which the direction of synthesis is opposite that of DNA unwinding. Nevertheless, Okazaki-fragment synthesis can initiate on both arms of the fork under conditions that inhibit DNA polymerase activity more severely than **DNA helicase** (an enzyme that unwinds DNA) activity. DNA unwinding will then proceed ahead of DNA synthesis, resulting in sufficient ssDNA template on both arms of the fork to allow DNA primase to initiate synthesis of new Okazaki fragments downstream of the DNA polymerase, as well as upstream. Okazaki created this condition by reducing the temperature of cells to slow down the rate of DNA replication so that he could observe transient intermediates.

SEQUENCE OF EVENTS AT REPLICATION FORKS

During the past 50 years, the sequence of events at DNA replication forks has been dissected in detail, both *in vitro* and *in vivo*, by inhibiting or eliminating the activity of a particular enzyme or protein required for DNA replication, and then reconstituting the process by restoring the missing component. These experiments employed chemical inhibitors, protein-specific antibodies, genetic mutations, biochemical fractionation of cell extracts, reconstitution of DNA replication events using purified proteins, ectopic expression of recombinant proteins, and suppression of specific protein activities using short interfering RNA (siRNA). Events observed *in vitro* often have been shown to mimic events observed *in vivo*. These experiments were carried out, for the most part, using either bacterial or eukaryal DNA and proteins, or those of the viruses and plasmids that replicate in these cells. Similar experiments with fully reconstituted replication forks have not yet been carried out with archaeal proteins, but the replication proteins that have been identified and characterized so far in archaea are either homologs or orthologs of their eukaryal counterparts. Therefore, it is reasonable to assume that many of the mechanistic details of replication of archaeal genomes will closely resemble the replication of eukaryal genomes.

At a first approximation the sequence of events at replication forks is virtually identical throughout the three domains of life, as well as in the dsDNA genomes of bacteriophages, animal viruses, and plasmids. DNA unwinding and DNA synthesis on the leading-strand template normally occur concomitantly, an important linkage that avoids the potential catastrophe of generating large amounts of ssDNA that would be more sensitive to breakage or modification than dsDNA. Nevertheless, for simplicity, the sequence can be considered as five sequential events (Figure 3-8).

The rate-limiting step at DNA replication forks is **DNA unwinding**. DNA synthesis cannot take place faster than the two strands of the double helix can be unwound. Only then can a complementary copy of DNA template be synthesized. As the DNA is unwound, the 3′-OH end of the nascent DNA strand at replication forks is extended continuously by a replicative DNA polymerase. This process is termed **leading-strand synthesis**, and the ssDNA template being copied is termed the **leading-strand template**, because this arm of the replication fork is replicated ahead of the other arm. Only after sufficient ssDNA has been produced by DNA unwinding can DNA synthesis be initiated in the direction *opposite* to fork movement. Hence, this is called **lagging-strand synthesis**, and the template is called the **lagging-strand template**. Lagging-strand synthesis is a discontinuous process; it requires repeated initiation, elongation, and termination of DNA synthesis. This occurs through the synthesis of transient, RNA-primed,

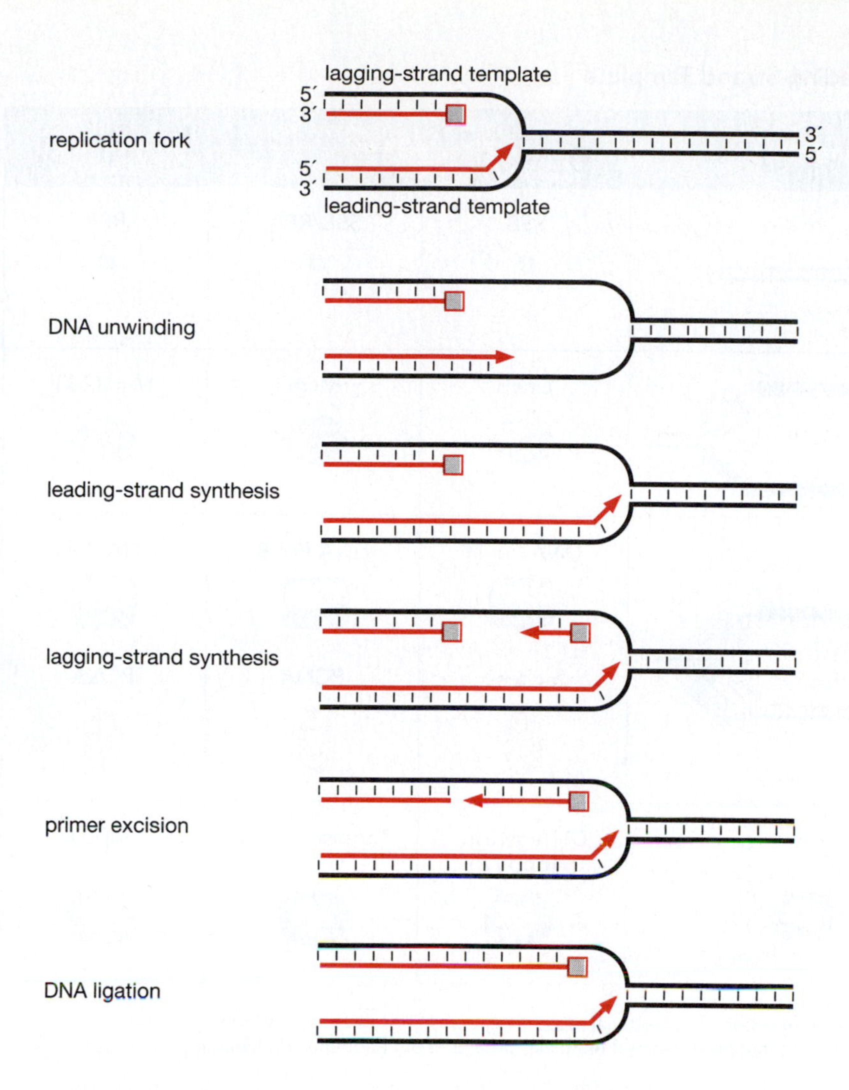

Figure 3-8. Sequence of events at a DNA replication fork. Nascent DNA (red) is synthesized from the RNA primer (box) at its 5′-end to its 3′-end (arrowhead). Since the two strands of DNA are antiparallel, the polarity of the nascent DNA strand is opposite to that of its template strand (black). Thus, DNA synthesis on the two arms of the fork occurs in opposite directions.

nascent DNA chains commonly referred to as Okazaki fragments in honor of their discoverer. Synthesis of the RNA primer is the rate-limiting step in Okazaki-fragment metabolism. Once the primer is synthesized, DNA synthesis begins immediately and rapidly extends the primer until it reaches the 5′-end of the nascent DNA strand further downstream. One byproduct of RNA-primed DNA synthesis is that part or all of the RNA primer used to initiate synthesis of the previous Okazaki fragment still exists at the 5′-end of the nascent DNA chain downstream. This primer must be removed before the new Okazaki fragment can be joined to the long growing nascent DNA strand. Otherwise, the newly replicated genome would contain fragments of RNA embedded throughout. **Primer excision** does not occur independently; it is linked to DNA synthesis. Finally, DNA ligation covalently links the 3′-end of the Okazaki fragment to the 5′-end of the growing nascent DNA strand by a phosphodiester bond. The half-life of Okazaki fragments is approximately 1 s in bacteria and approximately 1 min in eukarya. The length of Okazaki fragments is 50–4000 nucleotides in bacteria and 30–300 nucleotides in archaea and eukarya. The reasons for these differences are discussed in subsequent chapters.

DNA Unwinding And Leading-Strand Synthesis

As the lagging-strand template is exposed by the action of DNA helicase, it rapidly associates with a ssDNA-binding protein to produce replication forks with ssDNA on one arm (the lagging-strand template) and dsDNA on the other (the leading-strand template). For historical reasons, these

Table 3-1. Sequence Of Events On The Leading-Strand Template

Event	DNA structure	Proteins		
		Bacteria	Archaea	Eukarya
ssDNA binding	3′ 5′ 5′ 3′	SSB	SSB/RPA	RPA
DNA unwinding	3′ 5′ 5′ 3′	DnaB	Mcm	Mcm(2-7)
Leading-strand synthesis	3′ 5′ 5′ 3′	DNA Pol-III β-clamp	DNA Pol-B PCNA	DNA Pol-ε PCNA
DNA relaxation	−1 helical turn; +1 supercoil	DNA gyrase, Topo IV	Topoisomerase	Topo I

Nascent DNA is represented as a red line and RNA primers as gray bars. The bacterial replicative helicase travels along the lagging-strand template, as shown. Archaeal and eukaryal helicases travel along the leading-strand template. β-clamp is a homodimer. Proliferating cell nuclear antigen (PCNA) is a homotrimer in most organisms.

proteins are termed SSB (ssDNA-binding protein) in bacteria, SSB or RPA (replication protein A) in archaea, and RPA in eukarya (Table 3-1). Their function is to prevent reannealing of complementary DNA sequences, either between the two template strands, or within the template itself. For example, DNA contains various types of repeated sequences, some of which can form secondary structures (cruciforms) that impede DNA synthesis (Figure 1-10). ssDNA-binding proteins also interact directly and specifically with other proteins at replication forks.

DNA unwinding is carried out by a unique enzyme called the **replicative DNA helicase**. In cellular organisms this is a hexameric protein complex that passes one of the two template strands through its center as it travels in the direction of the replication fork, melting the two DNA strands as it does. Most, but not all (e.g. herpes simplex virus), replicative DNA helicases are hexameric structures that generally, but not always (e.g. T7 bacteriophage), derive the energy required for disrupting the double helix from hydrolysis of ATP into ADP and inorganic phosphate. Bacterial replicative DNA helicases (termed DnaBs for historical reasons) travel along the lagging-strand template. Archaeal and eukaryal replicative DNA helicases [termed minichromosome maintenance (Mcm) proteins] travel along the leading-strand template. Under normal conditions, DNA unwinding and leading-strand DNA synthesis occur concomitantly, thereby minimizing the amount of ssDNA in front of the leading-strand DNA polymerase. However, selective inhibition of DNA synthesis does not prevent DNA unwinding from continuing in the absence of DNA synthesis and generating large amounts of ssDNA. Under such conditions, replicative DNA polymerases can initiate DNA synthesis downstream of the 3′-end of the nascent DNA strand, resulting in Okazaki fragments appearing on both arms of the replication fork.

Leading-strand DNA synthesis is carried out by DNA polymerase III in bacteria, by a B-family DNA polymerase in archaea, and by DNA polymerase-ε in eukarya. Each of these enzymes contains a 3′→5′ exonuclease activity that ensures accurate base pairing between the nascent strand and the template strand by excising any misincorporated nucleotide, a process termed **proofreading**. However, in each case, the ability of the leading-strand DNA polymerase to extend the primer without dissociating from the template (its **processivity**) is greatly facilitated by association with another protein that assembles into either a doughnut-shaped dimer (the β-clamp in bacteria) or trimer (proliferating cell nuclear antigen, or PCNA, in archaea and eukarya) with the newly replicated dsDNA template running through its center.

DNA Relaxation

As base pairs are melted and the double helix is unwound, superhelical turns will be introduced in order to compensate for the change in topology. In formal terms, the linking number (the number of times one strand crosses the other) remains constant (Figures 1-6 and 1-8). For each 10 bp of DNA melted by DNA helicase, one right-handed (i.e. negative) turn of the helix is unwound, and one left-handed (i.e. positive) superhelical turn will appear at the site of DNA unwinding (Table 3-1; Figure 3-9A). This occurs only when

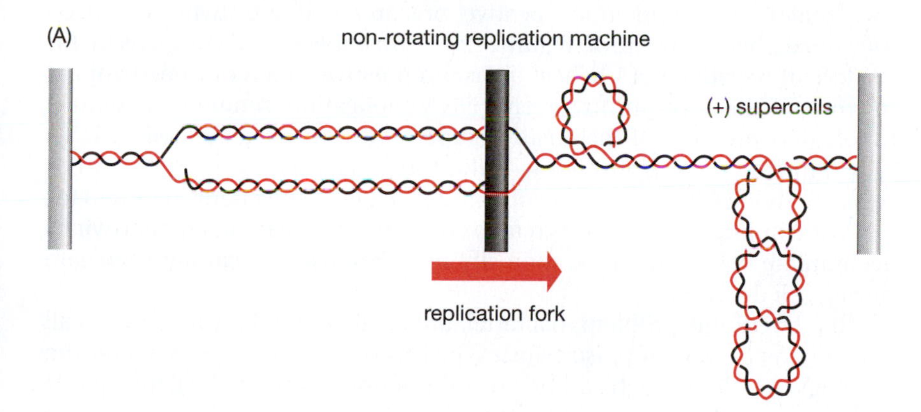

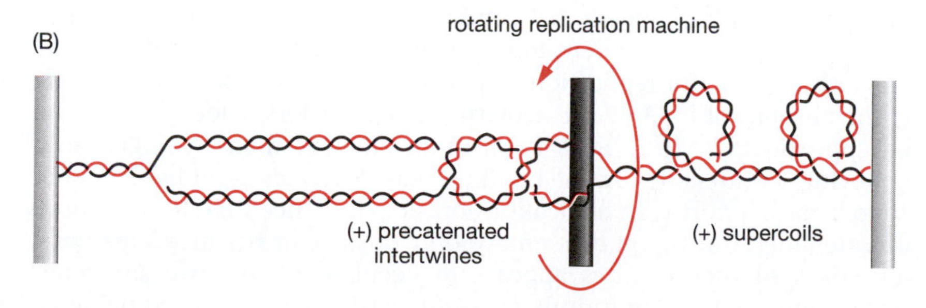

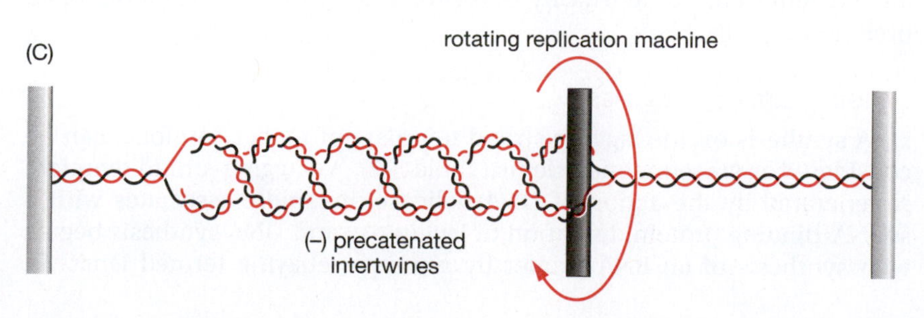

Figure 3-9. Supercoiling at DNA replication forks is a consequence of unwinding negative helical turns in a DNA molecule that cannot rotate freely about its axis. Here linear DNA is attached to a hypothetical immobile structure (e.g. nuclear structure). (A) When the replisome (rod) does not rotate, DNA unwinding produces positive supercoils in front of the fork. (B) If the replisome rotates in the direction opposite to unwinding, then some of the positive superhelical turns can be redistributed as positive precatenated intertwines behind the replication fork. (C) If the replisome rotates in the same direction as DNA unwinding, then for each negative helical turn removed, one negative precatenated intertwine could appear behind the fork. (Adapted from Wang, J.C. (2002) *Nat. Rev. Mol. Cell Biol.* 3: 430–440. With permission from Macmillan Publishers Ltd.)

one DNA strand cannot freely rotate about the other. For example, a circular, covalently closed, dsDNA molecule cannot freely rotate about its own helical axis, whereas a circular dsDNA that contains a single phosphodiester bond interruption in one strand can use the phosphodiester bond in the complementary strand as a swivel to relieve topological tension. In theory, linear DNA molecules can rotate about their axis during DNA unwinding, thereby potentially alleviating topological tension, but in actuality rotation is severely restricted. This is, in part, a simple consequence of the length of the DNA molecules themselves. Try untwisting the strands of a long rope from one end; the longer the rope, the more difficult it is to unwind the strands. In addition, DNA is organized into nucleoprotein complexes (i.e. chromosomes) that are attached to multiple sites in the nucleoid of bacteria and archaea, and in the nuclei of eukarya. Since DNA that is attached to something cannot rotate freely, these attachment sites define topologically constrained domains or 'DNA loops' within long linear molecules.

In principle, positive superhelical turns generated by DNA unwinding will be distributed either in front of the replication fork as supercoils, or behind the fork as positive precatenated intertwines, depending on whether or not the DNA replication machine at the fork (the **replisome**) rotates. If the replisome does not rotate, then positive supercoils are generated in front of the fork as negative DNA helical turns are removed (Figure 3-9A). However, if the replisome rotates in a direction opposite to DNA unwinding, then some of the positive superhelical turns can be redistributed as positive precatenated intertwines behind the replication fork (Figure 3-9B). This allows the fork to continue advancing. Alternatively, if the replisome rotates in the same direction as DNA unwinding, then for each negative helical turn that is unwound, one negative precatenated intertwine could be introduced behind the fork (Figure 3-9C). Since this would compensate for the loss of negative DNA helical turns, no positive supercoils need appear in front of the fork. In any event, DNA replication generates a serious topological problem. If the positive superhelical turns generated by DNA unwinding are not removed (i.e. if the torsional stress in the DNA is not relaxed), then DNA unwinding will grind to a halt as the energy required for DNA unwinding becomes too great to continue. If precatenated intertwines accumulate behind the fork, then sibling chromosomes cannot segregate during cell division.

To address this problem, bacterial, archaeal, and eukaryal genomes all encode one or more topoisomerases, a class of enzymes that can alter the topology of DNA through addition or subtraction of superhelical turns and/or precatenated intertwines (Tables 4-7–4-9). They do this by cleaving the phosphodiester backbone of DNA and then resealing it after the supercoils have unraveled. Two different topoisomerases (DNA gyrase and Topo IV) are both required for replication-fork progression in the bacterium *E. coli*; in the absence of DNA gyrase activity, replication-fork velocity is reduced by approximately 70%. Topo IV is more efficient than DNA gyrase at removing precatenanes. In the budding yeast *S. cerevisiae*, either Topo I or Topo II can support DNA replication-fork activity. Since Topo II can remove precatenanes, but Topo I can remove only positive or negative supercoils, formation of precatenanes appears to occur when positive supercoils are not removed. In the fruit fly *D. melanogaster*, Topo I is essential in all proliferating cells.

Lagging-Strand Synthesis

DNA synthesis on the lagging-strand template of replication forks can be considered as six sequential events (Table 3-2). As lagging-strand template is generated by the action of DNA helicase, it rapidly associates with a ssDNA-binding protein. Initiation of lagging-strand DNA synthesis begins with synthesis of an RNA primer by a unique enzyme termed DnaG in

Table 3-2. Sequence Of Events On The Lagging-Strand Template

Event	DNA structure	Proteins		
		Bacteria	Archaea	Eukarya
Template	 3′ 5′	SSB	SSB/RPA	RPA
Initiation	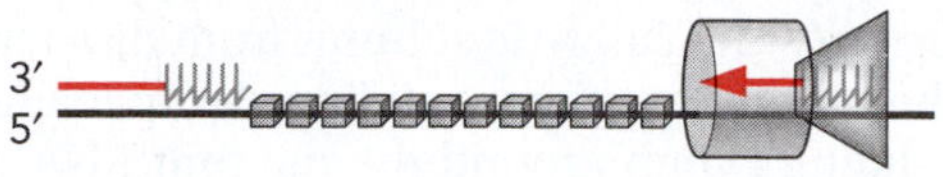 3′ 5′	DNA Pol-III DnaG	DNA Pol-B Primase	DNA Pol-α Primase
Polymerase handoff[a]	 3′ 5′	None	DNA Pol-B PCNA RFC	DNA Pol-δ PCNA RFC
Elongation[a] and proofreading	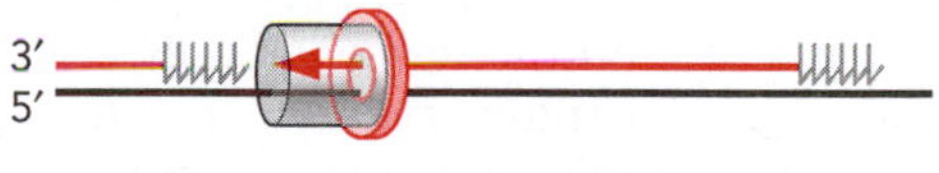 3′ 5′	DNA Pol-III β-clamp Sliding clamp loader	DNA Pol-B PCNA	DNA Pol-δ PCNA
Primer excision[b]	 3′ 5′	RNase HI DNA Pol-I (5′→3′ exo nuclease)	RNase HII FEN1	RNase HII FEN1 DNA2
Gap-filling[b]	 3′ 5′	DNA Pol-I DNA Pol-III	DNA Pol-B	DNA Pol-δ
Ligation	 3′ 5′	DNA lig	DNA lig I	DNA lig I

Nascent DNA is represented as a red line and RNA primers as gray bars.
[a]The bacterial sliding clamp loader and the archaeal and eukaryal RFC analogs are five-subunit complexes that load PCNA onto DNA. The *E. coli* sliding clamp loader is part of DNA Pol-III holoenzyme. RFC is not part of the replication-fork machine.
[b]Primer excision occurs with concomitant DNA synthesis (gap-filling).

bacteria and DNA primase in archaea and eukarya. This primer is then rapidly extended by a replicative DNA polymerase (Pol-III, a B-family DNA polymerase, or Pol-α in bacteria, archaea, and eukarya, respectively). The initiation site for RNA-primed DNA synthesis is selected stochastically from many possible sites distributed throughout the genome. DNA primase exhibits little sequence specificity, but it is highly specific for length. Regardless of the source of DNA primase, RNA primers are invariably 3 to 13 nucleotides in length, beginning with either ATP or GTP and ending at a randomly chosen nucleotide.

In bacteria, DNA Pol-III exists as a complex of 10 different proteins called the Pol-III holoenzyme (Table 4-6). This complex includes a sliding-clamp activity that prevents the catalytic subunit from dissociating from the DNA as it extends the primer:template, a sliding clamp loader that is required to assemble the sliding clamp around the nascent RNA primer, and a 3′→5′ exonuclease activity that facilitates proofreading of the newly synthesized DNA. If the last base pair formed by the catalytic subunit is incorrect, the 3′→5′ exonuclease excises the incorrect base, allowing the catalytic subunit to try again. Consequently, the Pol-III holoenzyme can rapidly and accurately synthesize long stretches of DNA on either the leading-strand or lagging-strand template. The resulting Okazaki fragments are an average of 1–2 kbp in length.

Where bacteria use the same type of DNA polymerase to replicate both leading- and lagging-strand templates, eukaryotes use three DNA polymerases. DNA primase is tightly bound to DNA Pol-α, the replicative eukaryotic DNA polymerase that initiates DNA synthesis on the RNA primer. However, this DNA polymerase•DNA primase complex lacks both the processivity to extend the primer more than 10–20 nucleotides, and the 3′→5′ exonuclease activity necessary to proofread the accuracy of its synthesis. Therefore, Pol-α•primase, as this complex is called, quickly hands-off to another DNA polymerase that is more appropriate for synthesis of long nascent DNA strands. DNA Pol-δ, like its leading-strand counterpart DNA Pol-ε, utilizes the PCNA sliding clamp to ensure a high degree of processivity, and it contains a 3′→5′ exonuclease proofreading activity. As with *E. coli*, loading the sliding clamp requires the services of a sliding clamp loader to assemble it around the nascent dsDNA. Like the *E. coli* sliding clamp loader, the eukaryotic sliding clamp loader, termed replication factor C (RFC), is a five-protein complex that is indispensable for assembling the replisome, but, in contrast to the *E. coli* sliding clamp loader, RFC is not an integral part of the replisome. This switch in DNA polymerases permits Pol-α•primase to disengage quickly from the primer:template and initiate a new Okazaki fragment before the old one has been completed. It also allows selection of leading- and lagging-strand replicative DNA polymerases that are tailored for their respective jobs.

Archaea also possess multiple DNA polymerases, although precise roles for individual enzymes have not yet been established. All archaea contain DNA polymerases belonging to the B-family of DNA polymerases, a family that includes the eukaryotic replicative DNA polymerases α, δ, and ε (Table 4-3). The euryarchaea also encode DNA polymerases belonging to family D. The archaeal Pol-D polymerase preferentially extends DNA from an RNA primer, and one of its subunits possesses 3′→5′ exonuclease activity that can excise mispaired nucleotides and therefore may provide a proofreading function. It has been suggested that Pol-B replicates the leading strand and Pol-D replicates the lagging strand. The crenarchaea do not encode a D-family polymerase, but they do encode multiple B-family polymerases, suggesting that different B-family polymerases play different roles on the leading- and lagging-strand templates.

DNA elongation continues until the DNA polymerase encounters the 5′-end of the nascent DNA strand downstream. Before the new Okazaki fragment can be joined to the one that preceded it, the RNA primer on the downstream nascent DNA strand must be excised completely. Primer excision in bacteria and eukarya is linked with DNA synthesis. This not only ensures that excision of the RNA primer is complete, it ensures that the initial dNMPs used to extend the primer are removed as well as the RNA. The error frequencies (the frequency at which the wrong nucleotide is incorporated) for DnaG, DNA primase, DNA Pol-III, and DNA Pol-α are much greater than for the highly processive replicative DNA polymerases Pol-III holoenzyme, Pol-δ, and Pol-ε. Thus, in eukarya, primer excision removes the first 10–20 nucleotides of the nascent RNA-primed DNA strand. This not only increases the fidelity of DNA replication on the lagging-strand template, but it maintains the integrity of newly replicated DNA by ensuring that ssDNA gaps are not left behind by the excision process.

In bacteria, this job is carried out by RNase HI and DNA Pol-I. However, only Pol-I is essential; inactivation of Pol-I results in the accumulation of Okazaki fragments, whereas inactivation of RNase H in *E. coli* has little effect on genome duplication. RNase H is an endonuclease that can degrade the RNA component of an RNA:DNA duplex (RNA annealed to DNA), but it cannot remove the ribonucleotide at the RNA-p-DNA junction that joins the 5′-end of the nascent DNA strand to the 3′-end of its RNA primer. Pol-I is unique in that it contains both 5′→3′ and 3′→5′ exonuclease activities in addition to DNA synthesis activity. Therefore, it can, in principle, excise the RNA primer with its 5′→3′ exonuclease activity while concomitantly extending the 3′-OH end of the nascent Okazaki fragment (i.e. fill in the gap left by the RNA primer) by displacing Pol-III. Moreover, Pol-I could fill in the gap accurately by using its 3′→5′ proofreading exonuclease activity. However, given the fact that *E. coli* RNA primers are rapidly excised *in vivo*, leaving only a small fraction of the Okazaki fragments with one to three ribonucleotides remaining at their 5′-ends, it appears that Pol-I, the most abundant DNA polymerase activity in *E. coli*, excises RNA primers, including the RNA-p-DNA junction, before Pol-III has completed the elongation phase. Such is not the case in eukaryotes. In eukaryotes, RNA primer excision is tightly coupled with Okazaki-fragment synthesis. The PCNA•Pol-δ complex displaces the RNA primer as well as part of the DNA to which it is attached. PCNA then orchestrates the FEN1 nuclease to release the displaced RNA-p-DNA oligonucleotide and degrade it. This process is probably facilitated by RNase HII and by DNA2, a protein with both helicase and nuclease activities. Given that archaea have homologs of the proteins used by eukarya for RNA primer excision, it is reasonable to assume that the mechanism for RNA primer excision used in eukarya also occurs in archaea.

Once the RNA primer, including the RNA-p-DNA junction, is excised and the 3′-end of the Okazaki fragment extended to meet the 5′-end of the long nascent DNA strand, the final step in lagging-strand synthesis occurs. Formation of a covalent bond between the two pieces of nascent DNA is carried out by the replicative **DNA ligase**, an enzyme that is highly specific for re-forming a single phosphodiester bond between a 5′-P-dNMP and adjacent 3′-OH-dNMP on a DNA template.

Conservation Of Replication-Fork Events

Remarkably, the events at replication forks are virtually identical among cellular organisms and the viruses and episomes that inhabit them. However, a number of significant differences lie in the details of the proteins that carry out analogous functions in bacteria versus archaea and eukaryotes. Viral genomes less than 10 kbp, such as papovaviruses [e.g. simian virus 40 (SV40)] and papillomaviruses, rely almost exclusively on their host cell to provide most of the proteins necessary to drive their replication forks (Table 3-3). At most, small viral genomes produce only one or two replication-fork proteins, and these are required to assemble replication forks at specific sites in the viral genome (replication origins). In some cases, this protein is also

Table 3-3. Viral Paradigms For Replication Forks And Their Host's Counterpart

E. coli 4.6 Mbp	Phage λ 48 kbp	Phage T7 40 kbp	Phage T4 169 kbp	*H. sapiens* 3080 Mbp	SV40 5.2 kbp	HSV 152 kbp
DnaB	DnaB	T7 gp4	T4 gp41	Mcm(2-7)	SV40 T-antigen	UL5,8,52
SSB	SSB	T7 gp2.5; SSB	T4 gp32	RPA	RPA	UL29
DnaG	DnaG	T7 gp4	T4 gp61	Primase	Primase	UL5,8,52
DNA Pol-III	DNA Pol-III	T7 gp5	T4 gp43	DNA Pol-α DNA Pol-δ DNA Pol-ε	DNA Pol-α DNA Pol-δ	UL30
β-clamp	β-clamp	Thioredoxin	T4 gp45	PCNA	PCNA	UL42
γ-complex	γ-complex	None	T4 gp44,62	RFC	RFC	None
RNase HI	RNase HI	T7 gp6	T4 gp33.2	RNase HII	RNase HII	UL30
DNA Pol-I	DNA Pol-I	T7 gp6	T4 gp33.2	FEN1	FEN1	UL12?
DNA ligase	DNA ligase	T7 gp1.3 or host ligase	T4 ligase	DNA lig I	DNA lig I	DNA lig I

Bacteriophage λ, T4, and T7 replicate in *E. coli*. Animal viruses SV40 and HSV replicate in human cells. Proteins are frequently named after their gene (e.g. T7 gp4 is bacteriophage T7 gene 4 protein). HSV genomes contain two unique regions: the long unique region (U_L) contains 56 genes, and the short region contains 12 genes.

the replicative DNA helicase. Larger viral genomes, such as bacteriophage T4 and herpes simplex virus (HSV), encode at least four proteins that are required to drive their replication forks, although lysogenic viruses such as bacteriophage λ that integrate into the host genome still rely extensively on their host for DNA replication proteins. Viruses that encode their own replication proteins are invariably lytic viruses that kill their host cell during the process of reproducing themselves.

ORGANIZATION OF REPLICATION FORKS

Replication-Fork Configuration

The advantage of the replication-fork mechanism is that both templates are replicated concurrently. Yet because the two strands of DNA have opposite polarities, the leading- and lagging-strand DNA polymerases travel in opposite directions; only the leading-strand polymerase travels in the same direction as DNA unwinding. In the classic replication fork-configuration, the events on the leading and lagging templates are pictured as if they operate independently of one another (Figure 3-10A). If this were the case in bacteria, then the lagging-strand Pol-III could travel 3 kbp or more in the opposite direction from the leading-strand Pol-III. DnaG (the DNA primase) could be lost from the replication fork each time it completes an RNA primer and must then locate another initiation site. DNA helicase could generate excess amounts of lagging-strand template, because DnaG is not localized at the replication fork. Moreover, if the velocity of DNA helicase exceeded that of leading-strand synthesis by Pol-III, then excess leading-strand template would be generated as well.

In 1975, Bruce Alberts suggested a simple solution to this dilemma that was based on his analysis of bacteriophage T4 DNA replication. He suggested that DNA replication occurred within a large protein complex termed a replisome, which in the case of T4 phage consisted of an eight-protein complex that could replicate phage T4 DNA at rates of 24 kbp/min^{-1}. The DNA polymerases catalyzing leading- and lagging-strand synthesis, as well as the T4 DNA primase, are anchored to the replicative DNA helicase.

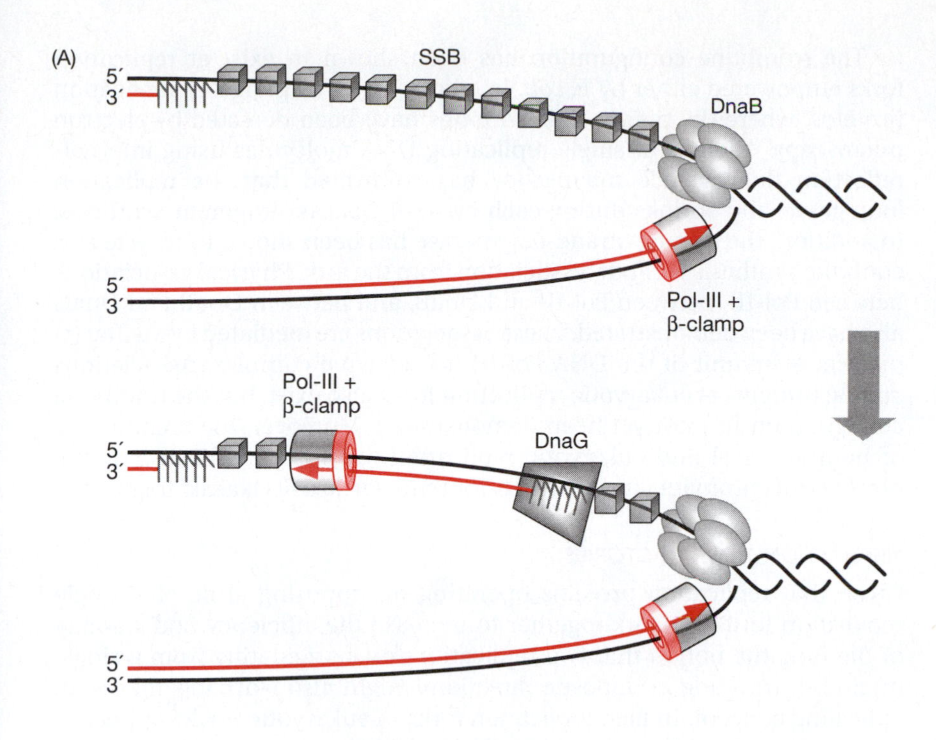

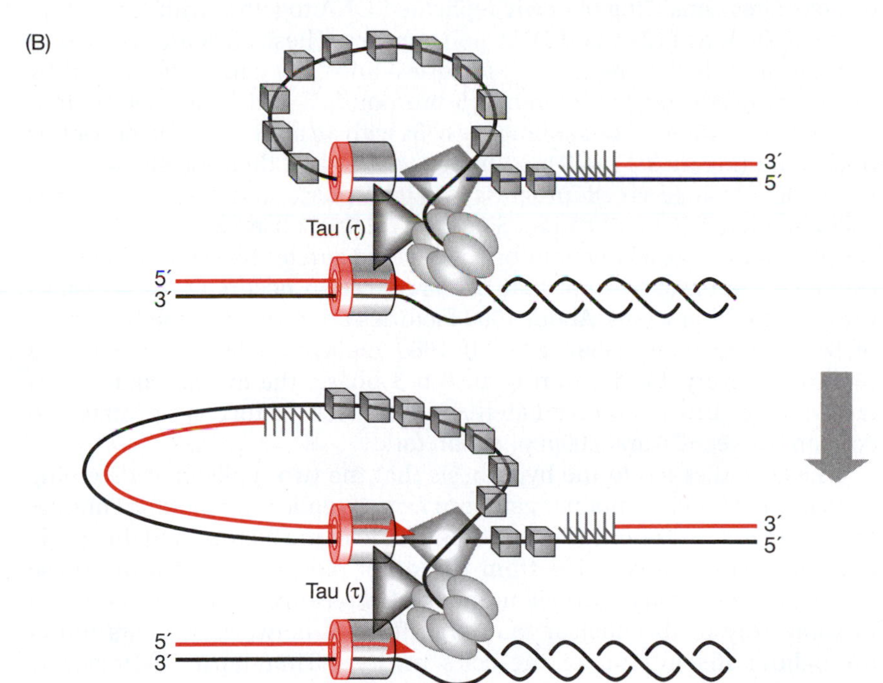

Figure 3-10. Classic (A) and trombone (B) configuration for DNA replication forks in bacterial genomes. DNA unwinding and leading-strand synthesis occur concomitantly followed by lagging-strand synthesis. Proteins and symbols are described in Tables 3-1 and 3-2. In the trombone model, Tau (τ) links two Pol-III enzymes together as well as to DnaB. DnaG is bound to DnaB. In principle, the two Pol-III enzymes can face either in the same direction or in opposite directions. For an animated model of a replication fork in action, visit www.cofc.edu/~zimmermana/replication1_2.swf. (Data from Hamdan, S.M, Richardson, C.C. (2009) *Annu. Rev. Biochem.* 78: 205–243.)

In this configuration (shown for *E. coli* proteins in Figure 3-10B), DNA is drawn into the helicase, producing a ssDNA loop at replication forks. Because the lagging-strand polymerase is physically linked to its leading-strand counterpart, the lagging-strand template is being moved in the direction opposite to the direction of nascent strand synthesis. To overcome this apparent clash, the polymerase loads onto the newly synthesized RNA primer and then extrudes behind it the newly formed duplex DNA. Once the Okazaki fragment is completed, the lagging-strand polymerase disengages and is recycled back to the next primer that is deposited by primase. Since the structure of these replication forks is reminiscent of a slide trombone, this configuration of replication forks was dubbed the trombone model.

The trombone configuration has been shown to exist at replication forks empowered either by *E. coli*, by phage T4, or by phage T7 replication proteins, where the predicted DNA loops have been detected by electron microscopy. Analysis of single replicating DNA molecules using internal-reflection fluorescence microscopy has confirmed that the replication loop grows and shrinks during each cycle of Okazaki-fragment synthesis. In addition, the lagging-strand polymerase has been shown to recycle and continue synthesis without dissociation from the fork. Physical associations between Pol-III, between Pol-III and DnaB, and between DnaB and DnaG also have been demonstrated. These associations are mediated by the Tau (τ) protein, a subunit of the DNA Pol-III holoenzyme complex. Associations among proteins at eukaryotic replication forks also exist, but the trombone configuration has not yet been demonstrated. Moreover, the architecture of both archaeal and eukaryotic replication forks is complicated by the presence of chromatin and by the 10-fold smaller size of Okazaki fragments.

Replication-Fork Factories

Given that replication proteins operating on opposing arms of a single replication fork can work together to increase the efficiency and stability of the fork, the notion that two replication forks originating from a single origin but traveling in opposite directions might also work together is an appealing concept. In fact, replication forks in eukaryotic nuclei appear as discrete sites consisting of newly replicated DNA together with replication proteins such as PCNA and DNA polymerases. These sites are referred to as replication foci or replication factories, and they can be visualized by incorporating the nucleotide analog 5-bromouracil (BrdU; Figure 1-1B) into nascent DNA followed by staining the cells with an antibody specific for this nucleotide (Figure 3-11A). The number of sites and their locations within the nucleus change as cells progress through S phase, and then disappear as cells complete DNA replication. Since the original observations by Hiromu Nakamura in 1986, a large number of detailed studies have confirmed and extended the basic concept that DNA replication occurs at discrete sites throughout the nucleus. About 1000 factories are formed at the beginning of S phase in mammalian cells. If 1000 replication factories complete replication every 45 min during an 8 h S phase, the average number of factories per cell is approximately 10 000. Each factory is estimated to contain between 20 and 200 replication forks.

These studies led to the hypothesis that the two replisomes diverging from a single origin remain together to form a replication factory. Parental DNA is drawn into one side of each replisome in the factory, and the newly replicated DNA is extruded from the other side (Figure 3-11B). These factories (which vary in their number of replisome pairs) are attached in some way to the nuclear matrix/scaffold, a network of fibers found throughout the interior of the cell's nucleus. Time-lapse photography of specific DNA sequences in live yeast cells that have been labeled to allow detection of specific DNA sequences shows that the formation of replication factories is a consequence of DNA replication itself. Moreover, sister replication forks diverging from the same origin remain associated with each other within a replication factory until the entire replication unit (the replicon) is duplicated.

Considerable biochemical and cytological data also support the hypothesis that replication forks generated from the same origin remain associated with each other and remain within a replication factory until they terminate. For example, the SV40 genome is contained in a small circular DNA molecule with a single replication origin. The replisome at each replication fork contains one hexameric DNA helicase that unwinds the DNA. Electron microscopy reveals that the two SV40 DNA helicases at sister replication forks associate with one another and thereby pull the DNA

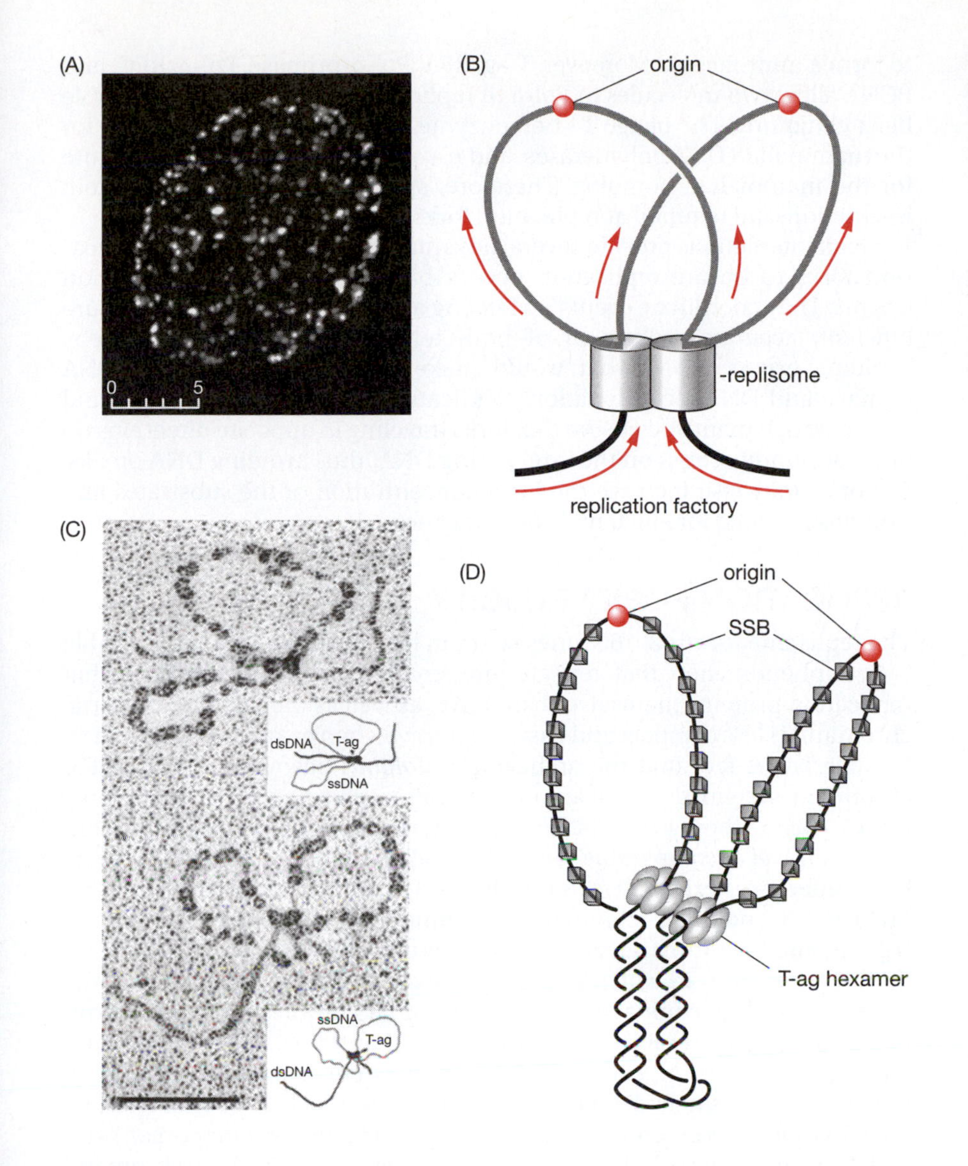

Figure 3-11. Replication factories. (A) Human diploid fibroblast cells were labeled with 5-bromodeoxyuridine for 5 min, stained with an anti-BrdU monoclonal antibody, and then sliced into thin sections and viewed in a confocal microscope. The resulting images display a section through the interior of the nucleus. The bar represents 5 μm in 1 μm divisions. (B) Two cellular replisomes bound together to produce a replication factory. Parental DNA (black line) is drawn through the replication factory in the direction of the arrows (red) to produce two sister chromatids (gray lines). (C) SV40 T-antigen (T-ag) binds to the SV40 replication origin where it assembles autonomously into two hexameric DNA helicases (Figure 11-3) and then unwinds the SV40 chromosome bidirectionally. Electron micrographs show the two DNA helicases bound together with the supercoiled dsDNA portion of the SV40 molecule wound into a tight rope downstream of the helicases and the unwound ssDNA portion spread out over the supporting grid. The ssDNA has been coated with *E. coli* SSB to prevent renaturation. (D) Cartoon interpretation of the images in C. ((A) from Ma, H., Siegel, A.J. and Berezney, R. (1999) *J. Cell Biol.* 146: 531–542. With permission from The Rockefeller University Press; (B) adapted from Kitamura, E., Blow, J.J. and Tanaka, T.U. (2006) *Cell* 125: 1297–1308. With permission from Elsevier; (C) from Wessel, R., Schweizer, J. and Stahl, H. (1992) *J. Virol.* 66: 804–815. With permission from the American Society for Microbiology.)

through the helicase rather than each helicase traveling independently along the DNA (Figure 3-11C, D).

The situation in bacteria is less clear. By tracking the progression of sister replication forks with respect to genetic loci in live cells, David Sherratt and colleagues revealed that sister replisomes in *E. coli* separate and move to opposite cell halves shortly after initiation, migrating outward as replication proceeds and both returning to midcell as replication termination approaches. These data are not consistent with models in which replisomes act within a fixed site. They conclude that independent replication forks follow the path of the compacted chromosomal DNA, with no structure other than DNA anchoring the replisome to any particular cellular region.

The biological significance of replication-fork factories remains speculative. The major difficulty is the lack of a functional dependency between two replication forks (one fork may stall while the other one continues on) and the apparent absence of these factories during *in vitro* DNA replication. Bacteriophages, bacterial plasmids (including plasmids using an *E. coli* replication origin), and animal virus DNA can be replicated *de novo* using purified proteins in a soluble experimental setting. No requirement for attachment to nuclear matrix/scaffold proteins has been reported. Nevertheless, as illustrated with SV40 T-antigen (T-ag), the replisomes at two replication forks within a single replication unit (i.e. replicon) may associate to form a mini-factory. Moreover, T-ag, RPA, Pol-α•primase, Pol-δ, RFC, and PCNA allow two molecules of Pol-δ to replicate both strands of the double helix conjointly. The phage T4 holoenzyme complex cannot substitute for the mammalian DNA polymerases, and *E. coli* DNA ligase cannot substitute for the mammalian homolog. Therefore, species-specific protein•protein interactions are required at replication forks.

Such interactions provide several advantages. They synchronize replication forks to ensure replication occurs bidirectionally from replication origins. They coordinate events on leading and lagging templates to ensure efficient, accurate replication of both templates concurrently, thereby avoiding excess ssDNA that would make the fork vulnerable to DNA damage and DNA recombination. Replication factories would help avoid DNA entanglement and ensure that forks traveling in opposite directions do not place undue stress on the intervening DNA, thus avoiding DNA breaks. Factories may also increase the local concentration of the substrates and proteins required for efficient DNA replication.

REPLICATION-FORK VELOCITY

The replisome travels 20–60 times faster in bacteria than in eukarya (Table 3-4), a phenomenon that reflects inherent differences in the rates that replication proteins themselves can move, as well as the fact that bacterial chromatin is less compact and easier to unravel than eukaryotic chromatin. Notable is the fact that the archaeon *Sulfolobus* has a less complicated chromatin system than eukaryotes, but its replication forks move only twice as fast as those in yeasts. Remarkably, the replisomes among different eukarya travel at comparable rates of 0.5–3 kbp min^{-1}, despite the fact that large differences exist between the size and organization of their genomes (Tables 1-3 and 1-4). Mammalian genomes are methylated at GC-rich regions, and they are organized into regions of highly compacted chromatin (heterochromatin) in which genes are transcriptionally repressed, as well as regions of less compacted, more accessible chromatin (euchromatin) in which genes are actively transcribed (Chapter 7). Heterochromatin replicates later during cell division than does euchromatin. In contrast, the genomes of single-cell eukarya such as yeast lack DNA methylation, contain much less heterochromatin than found in multicellular organisms, and lack the critical linker histone (H1) that is involved in chromatin compaction. Flies encode histone H1, but have very limited amounts of DNA methylation. Frog eggs, wherein many DNA replication studies have been done, are transcriptionally inactive and contain an open form of chromatin that is devoid of histone H1. The fact that all of these organisms replicate their genome at about the same rate suggests that their replisomes are fundamentally similar, and that fork velocity is limited by

Table 3-4. Velocities (kbp min^{-1}) Of DNA Replication Forks *In Vivo*

Bacteria	Archaea	Budding yeast	Fission yeast	Flies	Frogs	Mammals
Escherichia coli	*Sulfolobus acidocaldarius*	*Saccharomyces cerevisiae*	*Schizosaccharomyces pombe*	*Drosophila melanogaster*	*Xenopus laevis*	*Homo sapiens*
60	5.3	2.9 (0.5–6)	2.8 (2–4.5)	2.6	0.5 (0.3–1.1)	1.5 (0.6–3.6)

Mean values given with ranges in parentheses for nuclear genomes. Fork rates vary during S phase and at different sites in the chromatin. Human mitochondrial DNA forks travel, on average, at 0.27 kbp min^{-1}. Data are shown for *E. coli* (Kornberg and Baker, 1992), Archaea (Duggin *et al.*, 2008), yeast (Antequera, 2004; Feng *et al.*, 2006; Heichinger *et al.*, 2006), flies (Blumenthal *et al.*, 1974; MacAlpine *et al.*, 2004), frogs (Walter and Newport, 1997), and mammals (Berezney *et al.*, 2000; Conti *et al.*, 2007). Bacteriophages λ, T7, and T4 have fork rates comparable to those of *E. coli*.

the inherent speed at which DNA unwinding and DNA synthesis can occur concomitantly.

NUCLEOTIDE POOLS

Both the velocity and fidelity of DNA and RNA synthesis at replication forks depend on the concentration and composition of nucleoside triphosphate pools. In 1962, Peter Reichard demonstrated that dNTPs are synthesized from their corresponding NDPs through the action of ribonucleoside diphosphate reductase (Figure 3-12), more commonly known as ribonucleotide reductase (RNR). The resulting dNDPs are converted into their corresponding dNTPs by a ubiquitous nonspecific nucleoside diphosphate kinase, and then incorporated into DNA by the action of a DNA polymerase. RNR is a heterotetramer composed of two large and two small subunits. Its activity and substrate specificity are regulated by binding of different nucleoside triphosphates to two allosteric sites on the large subunit, thus preventing dNTP over-accumulation and achieving an optimal ratio of the four deoxyribonucleotides. In eukaryotic cells, several additional regulatory mechanisms exist. Transcription of the genes encoding the large subunits increases several fold when cells enter S phase, and the small subunit is translocated from the nucleus to the cytoplasm. Both of these events would generate higher levels of RNR activity, presumably to support a higher rate of dNTP synthesis. Similarly, Sml1, a protein that inhibits RNR activity, is phosphorylated and inactivated by the DNA damage-response pathway, thus preparing the way for DNA synthesis during DNA repair. In summary, four separate pathways converge to regulate RNR activity: allosteric regulation by dNTPs, transcriptional regulation of large subunit levels, regulation of subcellular localization, and regulation of specific inhibitors. Together they ensure that cells maintain dNTPs at levels that are just sufficient to support DNA replication. For example, yeast contain fewer than 1.8×10^6 dNTP molecules per cell, which is less than the amount needed to copy the 1.2×10^7 bp yeast genome. Thus, replication-fork velocities are limited by the cellular concentrations of dNTPs.

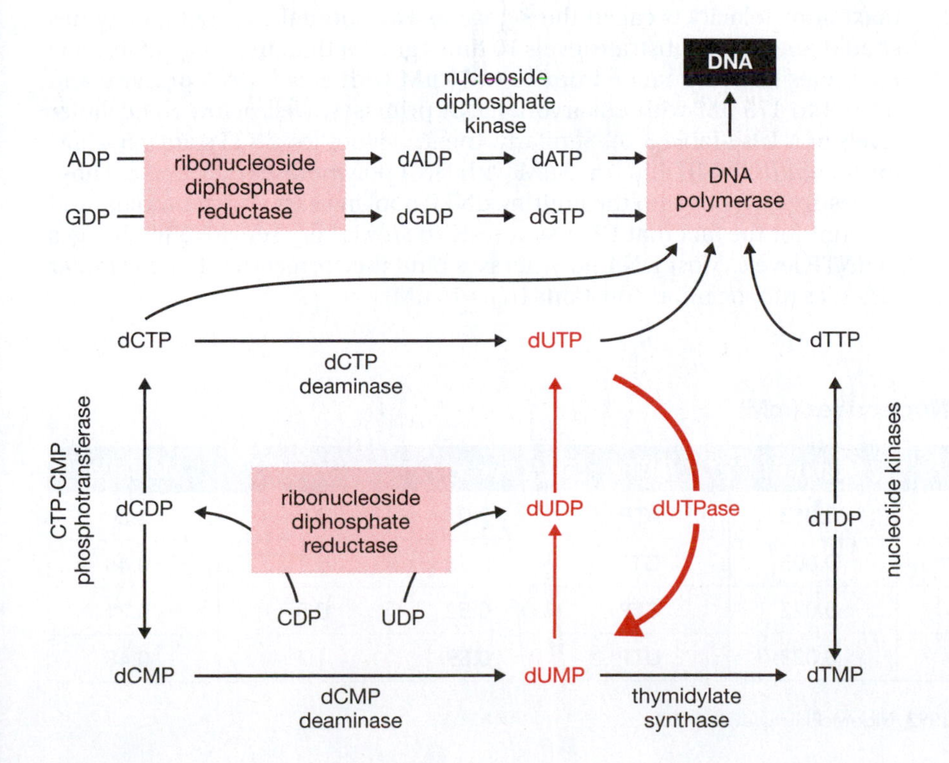

Figure 3-12. Synthesis of dNTP precursors for DNA synthesis. dUTP synthesis and degradation are indicated in red. (Data from Kornberg, A. and Baker, T.A. (1992) *Deoxyribonucleic Acid Replication*, 2nd edn.)

Cells have evolved mechanisms to limit dNTP accumulation to support replication fidelity and yet maintain dNTP levels above the minima required for efficient DNA replication and repair. Limiting the size of the dNTP pool increases the fidelity of DNA replication. Artificially expanding the dNTP pool increases mutagenesis by stimulating extension of DNA strands that have misincorporated the wrong nucleotide. For example, BrdU (Figure 1-1B) is converted by the cell into BrdUTP and then incorporated in place of dTTP. The extent of incorporation depends on the intracellular ratio of BrdUTP to dTTP, and this ratio can be increased by addition of 5-fluorouracil (a specific inhibitor of thymidylate synthase). Incorporation of BrdU is commonly used to identify newly synthesized DNA either by the change in its density due to the presence of BrdUMP, or by reaction with antibodies directed specifically against BrdU. Note, however, that the use of 5-fluorouracil to stimulate BrdU incorporation into nascent DNA also increases the level of dUTP (Figure 3-12).

Another example is misincorporation of dUTP into DNA in place of dTTP. dUMP is a normal precursor in the synthesis of dTTP (Figure 3-12). However, dUMP is also converted into dUTP by nucleoside mono- and diphosphate kinases (Figure 3-12). In addition, dUTP is produced by deamination of dCTP. Normally, dUTP is rapidly dephosphorylated to dUMP by dUTPase, and therefore does not enter DNA. However, conditions that increase the dUTP/dTTP ratio result in the appearance of dUMP in place of dTMP in nascent DNA. This can occur *in vitro* when DNA is replicated in the presence of a high dUTP/dTTP ratio, or *in vivo* either when thymidylate synthase is inhibited in order to stimulate incorporation of BrdU in place of dTMP, or when dUTPase is inhibited (e.g. genetic mutation). One consequence of dUTP misincorporation into nascent DNA is that uracil-N-glycosylase excises the uracil base, thereby generating abasic sites that result in subsequent mutagenesis or DNA breakage during DNA replication. These abasic sites are also sensitive to alkaline cleavage, resulting in fragmentation of nascent DNA, an artifact in which DNA damage and repair can be mistaken for Okazaki fragments. Therefore, caution must be exercised in experiments in which dTTP pools are reduced so as to increase the concentration of a labeled DNA precursor.

The concentration of substrate needed to drive an enzyme at half of its maximum velocity is called the K_m value. For optimal efficiency, enzymes should operate at substrate levels 10 times greater than their K_m values. The K_m values for NTPs range from 10 to 50 μM with *E. coli* DNA primase and from 8 to 175 μM with eukaryotic DNA primases, well below the cellular levels of NTPs (Table 3-5). Similarly, the K_m values for dNTPs are 0.5–4 μM for *B. subtilis* Pol-III and 0.5–5 μM with DNA polymerases α, δ, and ε. These values are on a par with the limiting dNTP pool in yeast and mammals, and account for the fact that DNA synthesis *in vivo* is very sensitive to changes in dNTP levels. Most DNA polymerases bind the correct dNTP with similar affinities [dissociation constants (K_d) ≈10 μM].

Table 3-5. Cellular Concentrations Of Nucleotides (mM)

dNTP	Bacteria	Yeast	Mammals	NTP	Bacteria	Yeast	Mammals
dATP	0.18	0.016	0.013	**ATP**	3.0	3.0	2.8
dGTP	0.12	0.012	0.005	**GTP**	0.92	0.7	0.48
dCTP	0.07	0.014	0.022	**CTP**	0.52	0.5	0.21
dTTP	0.08	0.030	0.023	**UTP**	0.89	1.7	0.48

Data are for asynchronous cells (Kornberg & Baker, 1992; Nick-McElhinny *et al*, 2010).

Deoxyribonucleotide pools are not equimolar; they have a characteristic asymmetry, with dGTP typically the smallest (Table 3-5). In budding yeast, for example, the dGTP pool is threefold smaller than the dTTP pool. dNTP pools also fluctuate during cell division, increasing several fold when eukaryotic cells begin DNA replication by markedly increasing their rate of dNTP synthesis to compensate for the rapid consumption of dNTPs at replication forks. Nevertheless, cells undergoing DNA replication are very sensitive to dNTP depletion. When dNTP synthesis is blocked by addition of hydroxyurea (a specific inhibitor of ribonucleotide reductase), replication stops well before any dNTP pool is exhausted.

SUMMARY

- All three domains of life replicate their DNA semi-conservatively by the replication-fork mechanism. Moreover, the sequence of events at replication forks in the chromosomes of all cells, as well as the viruses and episomes that proliferate in them, is essentially the same, and the proteins that drive them are highly conserved in function, if not in amino acid sequence.
- DNA at replication forks is unwound by DNA helicases: engines that use the energy derived from binding, hydrolysis, and release of ATP.
- ssDNA-binding proteins do not melt DNA; they prevent the two complementary DNA strands from reannealing.
- Given the structure of duplex DNA (two complementary antiparallel strands) and the unidirectionality of DNA polymerases (nascent DNA strands are elongated only in the 5′→3′ direction), DNA synthesis is continuous on the antiparallel 3′←5′ leading-strand template and discontinuous via the repeated initiation, elongation, and joining of Okazaki fragments on the 5′→3′ lagging-strand template. Okazaki fragments are short, transient RNA-primed nascent DNA chains. They are about 10 times shorter in archaea and eukarya than in bacteria.
- DNA synthesis begins with the synthesis of an RNA primer by the enzyme DNA primase. A replicative DNA polymerase extends the RNA primer by addition of dNTPs to its 3′-OH end. Bacteria and archaea have one or more replicative DNA polymerase (depending on the speies in question); eukarya have three.
- The processivity of DNA polymerases is greatly enhanced by their association with sliding clamps. Sliding clamps are circular molecules that are loaded onto duplex DNA by special protein complexes called sliding clamp loaders. Sliding clamps act as assembly platforms for various nucleic acid-processing enzymes, including enzymes that remove RNA primers from DNA.
- Leading- and lagging-strand progression are coupled in bacteria and some of their phage through protein•protein interactions that form a stable replication machine or replisome. The trombone model provides a molecular basis for this coupling. Something similar is presumed to exist in eukarya. In addition, replisomes operating at sister replication forks can aggregate into stationary 'factories' that pull the DNA through one end and extrude the two sibling chromatids out the other end.
- Replication-fork velocity differs significantly among the three domains of life. Bacterial forks are fastest by far, followed by archaeal forks and then eukaryal forks. The differences result both from the inherent properties of the replisome and from the structure of the nucleoprotein complex (chromatin) that must be replicated.
- Both the velocity and fidelity of DNA and RNA synthesis at replication forks depend on the concentration and composition of nucleoside triphosphate pools.

REFERENCES

Additional Reading

Burgers, P.M.J. and Seo, Y.-S. (2006) Eukaryotic DNA replication forks. In *DNA Replication and Human Disease*, DePamphilis, M.L. (ed.), pp. 105–120. Cold Spring Harbor Laboratory Press, Cold Spring Harbor, NY.

DePamphilis, M.L. and Wassarman, P.M. (1980) Replication of eukaryotic chromosomes: a close-up of the replication fork. *Annu. Rev. Biochem.* 49: 627–666.

Johnson, A. and O'Donnell, M. (2005) Cellular DNA replicases: components and dynamics at the replication fork. *Annu. Rev. Biochem.* 74: 283–315.

Marians, K.J. (1992) Prokaryotic DNA replication. *Annu. Rev. Biochem.* 61: 673–719.

Waga, S. and Stillman, B. (1998) The DNA replication fork in eukaryotic cells. *Annu. Rev. Biochem.* 67: 721–751.

Literature Cited

Antequera, F. (2004) Genomic specification and epigenetic regulation of eukaryotic DNA replication origins. *EMBO J.* 23: 4365–4370.

Berezney, R., Dubey, D.D. and Huberman, J.A. (2000) Heterogeneity of eukaryotic replicons, replicon clusters, and replication foci. *Chromosoma* 108: 471–484.

Blumenthal, A.B., Kriegstein, H.J. and Hogness, D.S. (1974) The units of DNA replication in *Drosophila melanogaster* chromosomes. *Cold Spring Harb. Symp. Quant. Biol.* 38: 205–223.

Conti, C., Sacca, B., Herrick, J., Lalou, C., Pommier, Y. and Bensimon, A. (2007) Replication fork velocities at adjacent replication origins are coordinately modified during DNA replication in human cells. *Mol. Biol. Cell* 18: 3059–3067.

Duggin, I.G., McCallum, S.A. and Bell, S.D. (2008) Chromosome replication dynamics in the archaeon *Sulfolobus acidocaldarius*. *Proc. Natl. Acad. Sci. USA* 105: 16737–16742.

Feng, W., Collingwood, D., Boeck, M.E., Fox, L.A., Alvino, G.M., Fangman, W.L., Raghuraman, M.K. and Brewer, B.J. (2006) Genomic mapping of single-stranded DNA in hydroxyurea-challenged yeasts identifies origins of replication. *Nat. Cell Biol.* 8: 148–155.

Heichinger, C., Penkett, C.J., Bahler, J. and Nurse, P. (2006) Genome-wide characterization of fission yeast DNA replication origins. *EMBO J.* 25: 5171–5179.

Kornberg, A. and Baker, T.A. (1992) *DNA Replication*, 2nd edn. W. H. Freeman and Company, New York.

MacAlpine, D.M., Rodriguez, H.K. and Bell, S.P. (2004) Coordination of replication and transcription along a *Drosophila* chromosome. *Genes Dev.* 18: 3094–3105.

Meselson, M. and Stahl, F.W. (1958) The replication of DNA in *Escherichia coli*. *Proc. Natl. Acad. Sci. USA* 44: 671–682.

Nick-McElhinny S.A., Watts B.E., Kumar D., Watt D.L., Lundstrom E.B., Burgers P.M., Johansson E., Chabes A., and Kunkel T.A. (2010) Abundant ribonucleotide incorporation into DNA by yeast replicative polymerases. *Proc. Natl. Acad. Sci. USA.* 107: 4949–4954.

Sakabe, K. and Okazaki, R. (1966) A unique property of the replicating region of chromosomal DNA. *Biochim. Biophys. Acta* 129: 651–654.

Taylor, J.H., Woods, P.S. and Hughes, W.L. (1957) The organization and duplication of chromosomes as revealed by autoradiographic studies using tritium-labeled thymidine. *Proc. Natl. Acad. Sci. USA* 43: 122–128.

Walter, J. and Newport, J.W. (1997) Regulation of replicon size in *Xenopus* egg extracts. *Science* 275: 993–995..

Chapter 4

REPLICATION PROTEINS: LEADING-STRAND SYNTHESIS

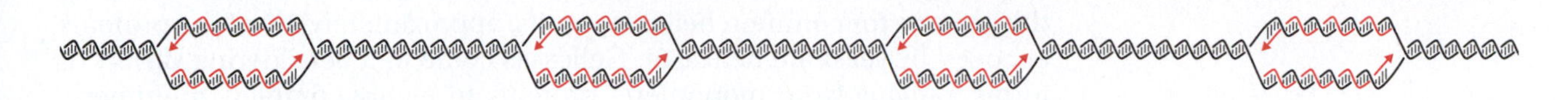

The preceding chapter summarized the mode of DNA replication (semiconservative), the structure of replication forks, the sequence of events by which they replicate DNA, and the proteins involved. In this chapter and its sequel, these topics are examined in detail by first looking at the events on the leading-strand template (Table 3-1) and then on the lagging-strand template (Table 3-2) of a fork already in progress. The replicative DNA helicase is already unwinding DNA so that the replicative DNA polymerases can synthesize complementary copies of both strands. As ssDNA appears at the fork, ssDNA-binding proteins rapidly associate with it, simultaneously preventing complementary sequences from reannealing and facilitating DNA unwinding and DNA synthesis. The leading-strand DNA polymerase is already engaged in DNA synthesis. It is already associated with a sliding clamp that facilitates its ability to travel long distances without dissociating from the template. Topoisomerases are busy removing the superhelical stress that results from unwinding DNA. How these proteins got there in the first place will be discussed in Chapters 5 and 11, but first we will focus on the primary goal of the replication fork: duplicating the sequence of millions of base pairs of DNA.

To this end, bacteriophages and animal viruses have proved invaluable. When these parasites infect cells, they produce large amounts of replication proteins encoded by their own genomes, and they induce their host to produce more of the replication proteins needed for viral genome duplication that are not encoded by the virus. Moreover, genetic mutations were easier to produce in viruses than in their host cell. This simplified the identification, purification, and characterization of replication proteins, and thereby opened the door to understanding the biochemistry of replication proteins. More recently the tools of recombinant DNA technology and DNA sequencing, and improvements in electron microscopy and X-ray crystallography, have allowed entrance into the age of structural biology. The outcome of these advances has been a renewed appreciation of the remarkable similarities among proteins that carry out the same function but that are encoded by very different organisms. Moreover, these proteins do not operate independently of one another, but rather function together as a single replication machine or replisome in which protein•protein interactions can be quite species-specific.

DNA HELICASES

The central organizing feature of the double helix lies in the hydrogen bonds that hold the base pairs together. Yet these very bonds must be broken to allow the two DNA strands to act as templates for the synthesis of two new sibling molecules. Melting base pairs is energetically unfavorable under normal cellular conditions and requires a specialized enzyme called a helicase to catalyze the process. Consequently, replication forks cannot advance any faster than the replicative DNA helicase can unwind the DNA duplex.

To melt base pairs, all helicases require the energy released during the reaction cycle of binding, hydrolysis, and release of a nucleotide triphosphate that is usually, but not always, ATP. Since every cellular process that involves either DNA or RNA also involves modulation of nucleic acid structure, it is not surprising that helicases are extremely abundant in cells. Based on searching the genome for common helicase motifs, approximately 2% of the proteins encoded by yeast are helicases. Helicases come in a bewildering variety of forms, ranging from monomeric proteins to highly complex, multimeric machines. With the exception of the trimeric herpesvirus DNA helicase, helicases that unwind DNA during cellular genome duplication (Table 4-1) fall into the latter category: they are ring-shaped multimeric assemblies with molecular weights of over 300 000 Da. These ATP-fueled machines are members of the AAA+ family of proteins, a large and functionally diverse superfamily of NTPases characterized by a conserved nucleotide-binding and catalytic site called the AAA+ domain.

To melt DNA effectively, the helicase must translocate directionally along the DNA strand. Since the two strands of DNA have opposite polarities, but DNA unwinding at replication forks proceeds in one direction, helicases travel either in a 3′→5′ or in a 5′→3′ direction along one of the two arms of the replication fork. Interestingly, the bacterial replicative helicase, typified by *E. coli* DnaB, moves 5′→3′ along a DNA strand, whereas the archaeal and eukaryotic replicative helicase, the MCM protein complex, moves 3′→5′ (Figure 4-1). This has the intriguing consequence that replication forks in bacteria have the helicase on the opposite strand from the helicase in archaea and eukaryotes. As discussed below, this may have important consequences for the coordination of events during DNA synthesis.

Table 4-1. Replicative DNA Helicases

Genome	Protein	Preferred NTP or dNTP	Direction	Function
E. coli	DnaB	ATP	5′→3′	Replication
S. solfataricus	Mcm	ATP, dATP	3′→5′	Replication
S. cerevisiae	Mcm(2-7)	ATP	3′→5′	Replication
S. pombe	Mcm(2-7)	ATP	3′→5′	Replication
H. sapiens	Mcm(2-7)	ATP	3′→5′	Nuclear DNA replication
H. sapiens	Twinkle	UTP	5′→3′	mtDNA replication
Phage T7	gp4	TTP	5′→3′	Helicase-primase, T7 DNA replication
Phage T4	gp41	GTP	5′→3′	T4 DNA replication
Phage SPP1	gp40	ATP	5′→3′	SPP1 DNA replication
SV40	Large T-ag	ATP	3′→5′	SV40 DNA replication
BPV	E1	ATP	3′→5′	BPV DNA replication
HSV	UL5, UL8, UL52	ATP	5′→3′	Helicase-primase complex, HSV DNA replication

Bacteriophages T7 and T4 replicate in *E. coli*. SPP1 replicates in *B. subtilis*. Simian virus 40 (SV40), bovine papillomavirus (BPV), and herpes simplex virus (HSV) replicate in mammalian cells.

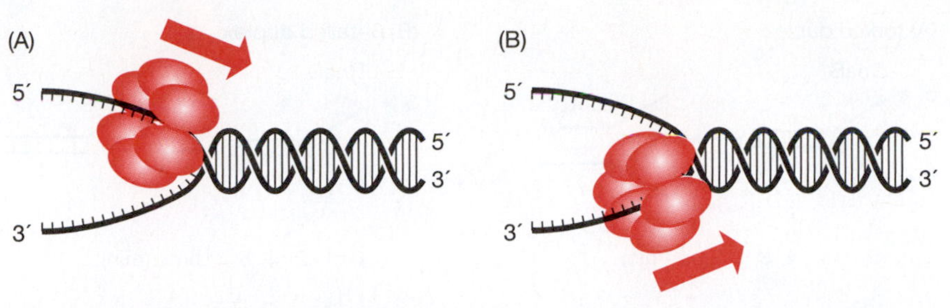

Figure 4-1. DNA helicase polarity. (A) Hexameric DnaB helicase (red spheres) translocates 5′→3′ along the lagging-strand template at *E. coli* DNA replication forks, unwinding dsDNA in its path. (B) Hexameric MCM helicases accomplish the same task by translocating 3′→5′ along the leading-strand template in archaea and eukarya.

The DnaB Family Of Helicases

The best-understood cellular replicative helicases are members of the *E. coli* DnaB family. DnaB oligomerizes into a homohexameric ring that has been observed by electron microscopy to form either sixfold or threefold symmetry states. The DnaB ring unwinds DNA at the replication fork by encircling the 5′-lagging-strand template, and then translocating along the template in the 5′→3′ direction while occluding the 3′-leading-strand template. The crystal structure of a DnaB monomer reveals two domains separated by a flexible linker. In the presence of ATP, these monomers assemble into hexamers with an outer diameter of approximately 115 Å and a height of approximately 75 Å (Figure 4-2). The diameter of the central channel varies between approximately 25 Å and approximately 50 Å, depending on the orientations of the C-terminal domains. Thus, the central channel of DnaB family helicases is wide enough to accommodate a 20 Å duplex DNA molecule.

The fact that DnaB hexamers contain a central channel large enough to accommodate a dsDNA molecule suggests that DnaB may be able to mount dsDNA and translocate along its surface without melting it. This indeed is the case. DnaB can thread itself onto the 5′-end of ssDNA and then move 5′→3′ until it encounters dsDNA (Figure 4-3). If the substrate mimics a replication fork, DnaB melts the DNA, but if the substrate is missing one arm of the fork, then DnaB does not melt it. Instead, the helicase continues to travel along the dsDNA portion of the molecule, as evidenced by the fact that it can displace a protein bound to the dsDNA at a specific site downstream. Hence, DnaB helicase can travel along a dsDNA molecule as well as a ssDNA molecule, presumably by passing the dsDNA through its central channel.

While the DnaB paradigm is useful, the properties of replicative DNA helicases do vary. For example, HSV DNA helicase consists of three different subunits (polypeptides) and is presumed to function as a trimer, although a hexameric structure containing two copies of each subunit has not been ruled out. The bacteriophage T7 DNA helicase is unusual in that it uses TTP instead of ATP as its preferred fuel. Nevertheless, the helicase domain comprising the C-terminal half of the molecule includes the six conserved sequence motifs characteristic of the DnaB family of replicative helicases. Both electron microscopic and crystallographic data reveal that the T7 helicase forms both hexamers and heptamers in the presence of TTP. The central channel of the heptamer varies from 24 to 28 Å in diameter along its length, large enough to accommodate dsDNA, whereas the hexamer

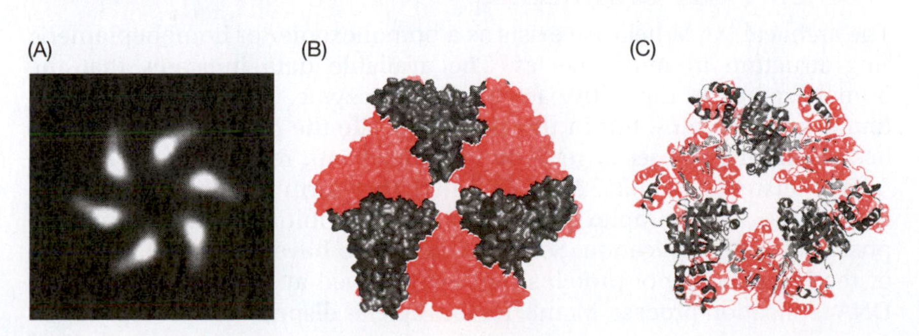

Figure 4-2. DnaB family of DNA helicases are homohexameric structures with a central channel. (A) Negatively stained electron micrograph of *E. coli* DnaB. X-ray crystallographic structure of bacteriophage SPP1 helicase gp40 (another DnaB family helicase) shown as a surface representation (B) and a ribbon representation (C). Alternating subunits in the homohexameric assembly are colored gray and red for the sake of clarity. Similar results have been obtained with DnaB (Bailey *et al.*, 2007). Images generated using Pymol from PDB file 3BGW. ((A) from Yang, S., Yu, X., VanLoock, M.S. *et al.* (2002) *J. Mol. Biol.* 321: 839–849. With permission from Elsevier.)

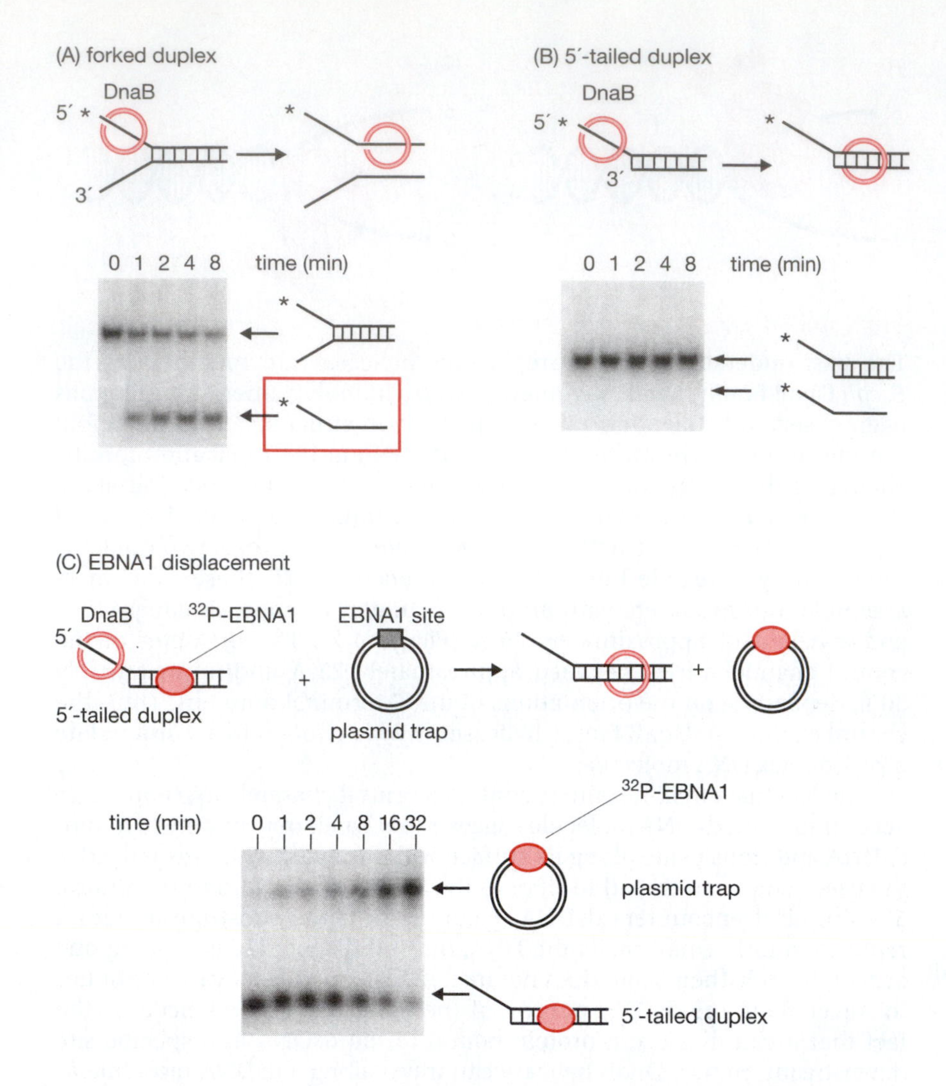

Figure 4-3. DnaB helicase can travel on dsDNA as well as on ssDNA. (A) DnaB melts a forked duplex DNA substrate in which the 5′-end of one strand is radiolabeled (*), releasing it into solution (red box). DNA molecules were separated according to size and shape by electrophoresis through a polyacrylamide gel. (B) The radiolabeled strand is not displaced if the 3′-tail is absent. (C) A ^{32}P-labeled EBNA1 DNA-binding protein bound to the dsDNA segment of the 5′-tailed duplex is released by DnaB and trapped (so it cannot rebind) by binding to an excess of plasmid DNA that also contains the EBNA1-binding site. Therefore, under these conditions, DnaB can mount dsDNA and translocate down its length. (From Kaplan, D.L. and O'Donnell, M. (2002) *Mol. Cell* 10: 647–657. With permission from Elsevier.)

appears to be able to accommodate only a single strand of DNA through its central orifice. T7 helicase exists as a hexamer when bound to ssDNA and as a mixture of heptamer and hexamer when not bound to ssDNA. Thus, the switch from heptamer to hexamer may provide a ring-opening mechanism for assembly of the helicase around ssDNA. A similar situation appears to exist with the bacteriophage SPP1 DNA helicase. Here the hexamer has two tiers: a threefold symmetric N-terminal tier and a sixfold symmetric C-terminal tier. The central channel changes from 17 to 42 Å, suggesting that dsDNA can enter one end, but only ssDNA can come out the other. Given the differences between DnaB and MCM helicases, and the variation among viral helicases, it is possible that the mechanism of DNA unwinding at replication forks varies among the three forms of life.

The MCM Family Of Helicases

The archaeal MCM helicase exists as a homohexamer or homoheptameric ring structure in most species. The available data indicates that the homohexamer is the active form of the enzyme. *Mcm* genes are also found in eukaryotes, but in this domain of life the progenitor *Mcm* gene has undergone a series of gene-duplication events, resulting in six distinct (albeit related) subunits (Mcm2–Mcm7) that form a stable complex in all eukarya. This complex is required for both initiation and elongation phases of DNA replication. Some metazoa also have additional members of the MCM family of proteins that are involved at different steps in the DNA replication process. Mcm8, for example, is dispensable for replication initiation, but it is important for the elongation step of DNA synthesis. The gross morphologies of archaeal and eukaryal MCMs are superficially similar to the bacterial DnaB-type helicases, forming doughnut shapes with central channels through which nucleic acids can potentially pass, but the mechanisms by which MCM helicases effect DNA unwinding are currently a topic of intense investigation.

Some MCM helicases can use either ATP or dATP to catalyze unwinding of artificial oligonucleotide substrates including DNA–RNA hybrids, as long as the substrate has a 3′-extended ssDNA tail. Presumably, this tail threads through the center hole in the helicase, thereby allowing the helicase to load onto the DNA substrate. Individual MCM subunits have at least two distinct DNA-binding sites. The first, found in an N-terminal domain, appears to help hold the MCM helicase on the DNA. A second DNA-binding site resides in the middle of the ATPase motor domain of the protein. This site is in the form of a hairpin-like structure that changes position during ATP binding, hydrolysis, and release, thereby providing the power stroke that drives the motor along the DNA. This mechanism of action of the MCM helicase appears similar to that employed by helicases found in a number of animal virus genomes.

Each of the six different subunits that comprise the Mcm(2-7) complex in eukaryotes is conserved among all eukarya, and each is required for DNA replication in budding yeast, fission yeast, and frog egg extract. Furthermore, recent work has revealed that the purified Mcm(2-7) complex possesses helicase activity *in vitro*. This point had been highly controversial for many years, but biochemical studies by Tony Schwacha revealed that robust helicase activity could be detected under highly defined buffer conditions. The six Mcm subunits are closely related to one another, with the greatest homology residing within a central 200-amino acid region related to the AAA+ family of ATPases. However, no individual subunit exhibits ATPase activity, suggesting that two or more Mcm proteins are required to form an active site for ATP hydrolysis. In fact, binary combinations of subunits Mcm(2,6), Mcm(3,7), and Mcm(4,7) exhibit ATPase activity by virtue of the fact that one subunit provides the ATP-binding site while the other subunit provides a critical arginine residue required for ATP hydrolysis. Analysis of ATP-binding-site mutants suggests that Mcm4, Mcm6, and Mcm7 are the catalytic subunits, and Mcm2, Mcm3, and Mcm5 are the regulatory subunits in the MCM helicase. In fact, a complex containing only Mcm4, Mcm6, and Mcm7 (Figure 4-4) exhibits DNA helicase activity, whereas Mcm(2,3,5) does not, suggesting that Mcm2, Mcm3, and Mcm5 either regulate MCM helicase activity, or facilitate loading the helicase onto DNA. A common architecture is predicted for Mcm(2-7) in which Mcm2 interacts with Mcm5 and Mcm6, Mcm3 interacts with Mcm5 and Mcm7, and Mcm4 interacts with Mcm6 and Mcm7 (Figure 4-4). Perhaps other cellular proteins are involved as well. Thus, eukarya may contain multiple forms of MCM helicases with unique properties and functions.

Rotary Engine Activity Of DNA Helicases

The simplest model for helicase action is that the ring-shaped helicase is loaded onto DNA that is already partially melted by the action of initiator proteins, such that a single strand of DNA passes through the central channel of the hexamer. This channel contains potential DNA-binding sites that are contributed by the individual subunits of the helicase. As the wave of NTP binding, hydrolysis, and release travels around the ring, DNA is passed from one subunit to another. This transfer, coupled with the NTP-induced rotation between subunits, enables the helicase to walk forward one nucleotide per NTP hydrolyzed. In the simplest model, this forward motion of the helicase acts as a molecular bulldozer, unwinding the dsDNA ahead of it.

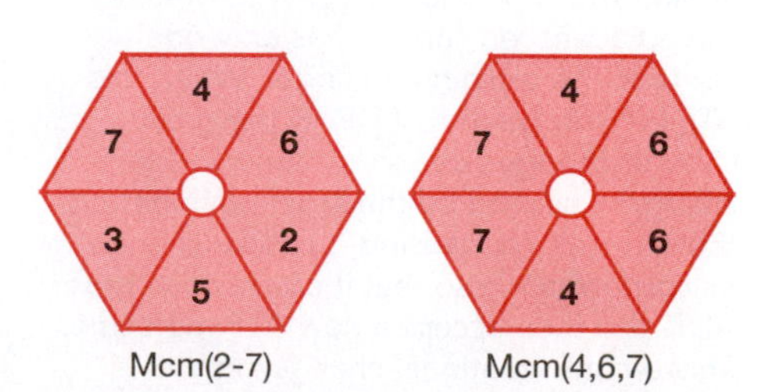

Figure 4-4. Subunit arrangement in the native MCM helicase, Mcm(2-7), and in the catalytically active subcomplex of Mcm(4,6,7).

One characteristic of AAA+ proteins is their ability to communicate with one another through shared NTP molecules. Each subunit possesses a cleft on one face of the protein that binds ATP via specific side chains in conserved amino acid sequences called the Walker A, Walker B, Sensor 1, and Sensor 2 motifs. The crystal structure of the T7 helicase provides an example (see Figure 4-5 in color plate section).

This provides a mechanism to propagate ATP-dependent conformational changes around the circle of subunits through endless waves of ATP binding, hydrolysis to ADP and phosphate, release of the ADP, and then rebinding of ATP (Figure 4-6). In T7 helicase, for example, binding of NTP by a subunit induces a 15° rotation of that subunit relative to the plane of the ring. How this mechanism operates when in motion can be deduced from the high-resolution crystal structure of bovine papillomavirus (BPV) E1 helicase in a complex with ssDNA. As with other replicative helicases, BPV E1 forms a hexameric structure. The structure of BPV E1 in complex with DNA reveals that the β-hairpins in the motor domain of this protein that bind DNA form a rising staircase around the central channel of the helicase (see Figure 4-7 in color plate section). As subunits within the hexameric helicase bind ATP, hydrolyze the ATP, and finally release the ADP, the positions of the β-hairpins slowly descend the staircase, allowing the helicase to walk along the ssDNA. Since the relative positions of the staircase elements would be dependent on the status of the nucleotide bound by the subunit, as the ripple of ATP binding, hydrolysis, and release passes around the ring-shaped helicase, sequential repositioning of the hairpins drives the enzyme along DNA.

How Helicases Unwind DNA

Three different models for DNA helicase action have been proposed. The simplest mechanism for unwinding DNA during DNA replication begins with loading two ring-shaped hexameric helicases onto a bubble of ssDNA such that a single strand of DNA passes through the central cavity of each hexamer and the other strand is placed outside the ring of protein (Figure 4-8). The mechanisms involved in helicase loading are discussed in Chapter 10. Since helicases travel in only one direction along a strand of ssDNA, and duplex DNA consists of two antiparallel strands, this means that the helicase on one strand will travel in the direction opposite to that of the helicase on the complementary strand. Therefore, a minimum of one helicase must be loaded onto each strand in order to effect bidirectional DNA replication.

The process of DNA unwinding can be viewed in one of two ways. Either the DNA remains stationary and each helicase acts independently by traveling away from one another (Figure 4-8A), stripping off the complementary strand as one might strip the casing from a wire, or the helicases remain stationary and the DNA is drawn through the helicase complex (Figure 4-8B). In the second case, the two helicases physically interact with one another. As the hexamers are held together, they cannot move apart, and so the end result is that ever-increasing loops of ssDNA

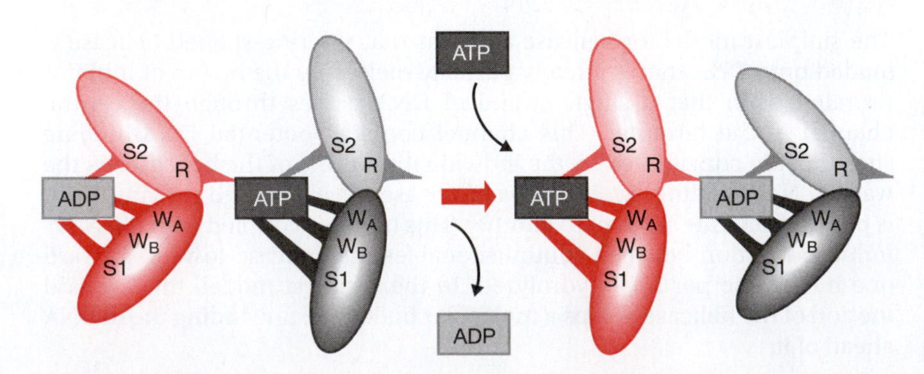

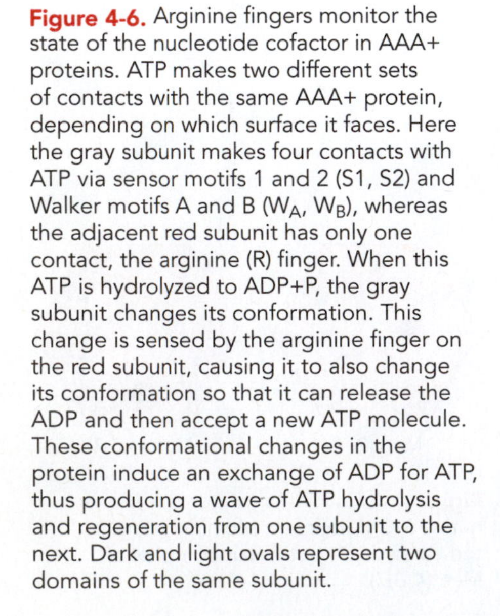

Figure 4-6. Arginine fingers monitor the state of the nucleotide cofactor in AAA+ proteins. ATP makes two different sets of contacts with the same AAA+ protein, depending on which surface it faces. Here the gray subunit makes four contacts with ATP via sensor motifs 1 and 2 (S1, S2) and Walker motifs A and B (W_A, W_B), whereas the adjacent red subunit has only one contact, the arginine (R) finger. When this ATP is hydrolyzed to ADP+P, the gray subunit changes its conformation. This change is sensed by the arginine finger on the red subunit, causing it to also change its conformation so that it can release the ADP and then accept a new ATP molecule. These conformational changes in the protein induce an exchange of ADP for ATP, thus producing a wave of ATP hydrolysis and regeneration from one subunit to the next. Dark and light ovals represent two domains of the same subunit.

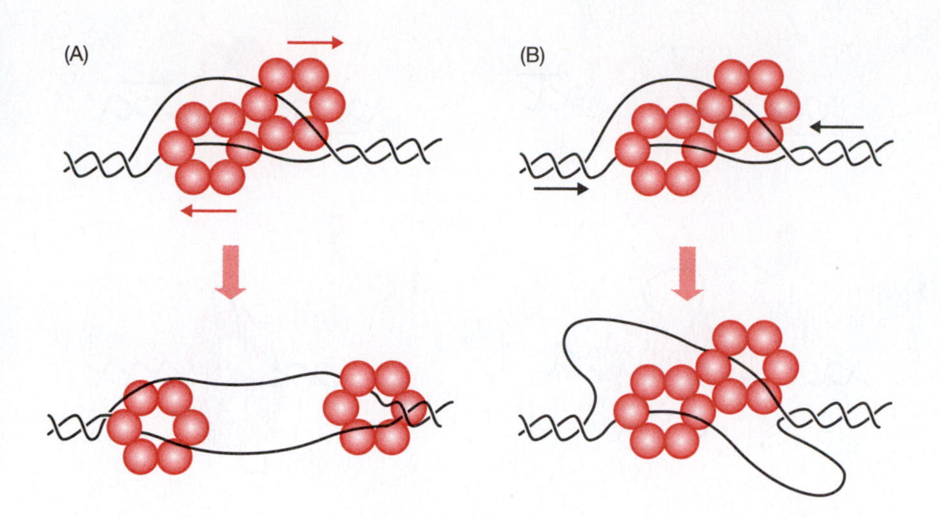

Figure 4-8. Needle-and-thread model for helicase action. Each helicase hexamer is loaded onto a single strand of DNA. (A) Two helicases travel independently in opposite directions. (B) Two helicases remain physically associated, and DNA is pulled into them.

are generated like two rabbit ears emerging from the helicase complex. Papillomavirus E1 DNA helicase appears to act in this manner.

An alternative model postulates that duplex DNA enters one side of the helicase and two separated DNA strands emerge (Figure 4-9). This model recognizes that helicase monomers consist of two domains, a large domain that is required for the helicase to travel along the DNA (the helicase motor domain) and a smaller domain that is required to melt base pairs (the helicase ploughshare domain). As base pairs are melted, one strand of DNA passes straight through the hexameric assembly, while the other strand emerges from the side of the helicase through a pore located between the two domains. Thus, a feature within the enzyme assembly acts as a ploughshare to split the two strands apart and direct them to their distinct exit pores. Note that in a variant of the model, the ploughshare could be a separate polypeptide that interacts with the helicase and acts to split the two strands apart. As the motor domain drags the ploughshare along, the melted region is extended. As with simple steric exclusion models (Figure 4-8), two variations are presented, one in which the two hexamers move independently, and one in which the two hexamers remain associated.

In the third model, the two helicases may each encircle a DNA duplex and, by acting together, directionally pump DNA toward one another. Rotation of the DNA as it passes through this static protein complex will cause the DNA to unwind and thereby extrude two loops of ssDNA between them (Figure 4-10A). Alternatively, each helicase latches onto a previously melted bubble of ssDNA produced during the process of initiation of DNA replication, encircles one of the two ssDNA loops, and then pulls it through the helicase, thereby extending the melted region (Figure 4-10B).

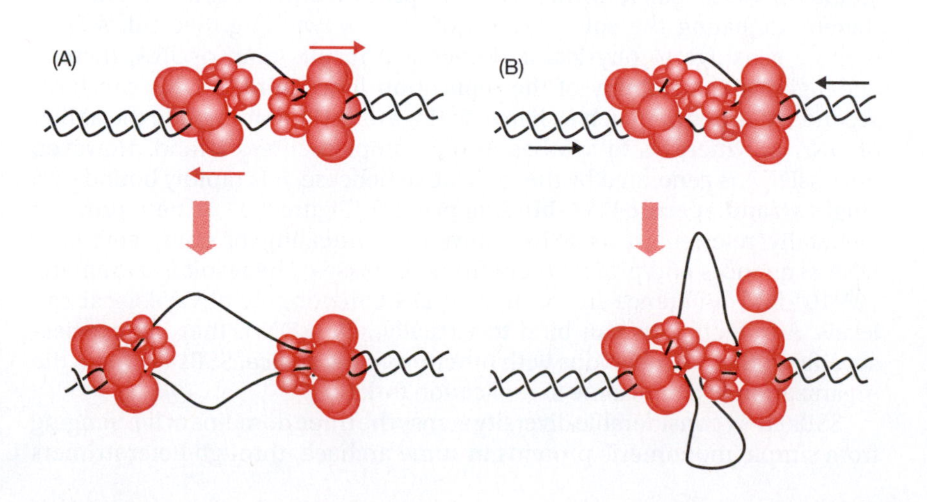

Figure 4-9. Ploughshare model for helicase action. The six large spheres represent the helicase motor, and the six small spheres represent the helicase ploughshare. dsDNA enters through the helicase motor, and one strand emerges from the center hole in the helicase and the other strand emerges between the motor and ploughshare. (A) The two helicases function independently, as in Figure 4-8A. (B) The two helicases are physically associated, as in Figure 4-8B.

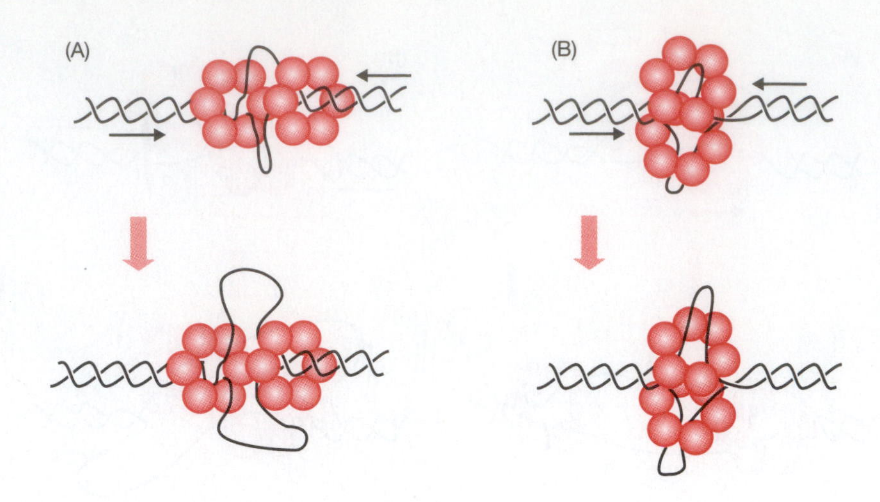

Figure 4-10. DNA pump model for helicase action. (A) Pair of helicases acting as dsDNA pumps, each drawing dsDNA into the space between the two hexamers from which loops of ssDNA extrude. (B) Pair of helicases acting as ssDNA pumps, each drawing a loop of ssDNA through its central pore.

The available data indicate that the DnaB and BPV E1 helicases use the simple steric exclusion needle-and-thread mode of helicase action (Figure 4-8). In contrast, structural studies of the SV40 large T-antigen helicase suggest that it utilizes the ploughshare model (Figure 4-9). As for the MCM helicases, the available biochemical data strongly support that both the archaeal Mcm complex and the eukaryal Mcm(4,6,7) subcomplex of the Mcm(2-7) helicase can efficiently melt DNA using a simple steric exclusion needle-and-thread mechanism. However, *in vivo*, the eukaryal MCM helicase operates within the context of a larger protein complex that includes Cdc45 and the four GINS proteins (discussed in Chapter 5). The MCM helicase may work as a rotary pump (Figure 4-10), with the Cdc45•GINS complex acting as a ploughshare.

Consistent with this model, single Mcm(2-7) heterohexamers can be loaded onto dsDNA *in vitro* in a reaction that is dependent on ORC, Cdc6, Cdt1, and ATP hydrolysis. Once loaded, the Mcm(2-7) single hexamers are converted into stable, head-to-head double hexamers connected via their N-terminal rings. The DNA runs through a central channel in the double hexamer, allowing the Mcm(2-7) complexes to slide passively along the DNA in either direction. These data are consistent with the two-step model discussed in chapter 11 in which Mcm(2-7) double hexamers are first assembled onto dsDNA at replication origins and then later converted into an active DNA helicase.

SINGLE-STRAND-SPECIFIC DNA-BINDING PROTEINS

Regardless of their mechanism of action, DNA helicases produce ssDNA, and ssDNA generates problems of its own. First, the two complementary strands of ssDNA can re-form dsDNA simply by reannealing to one another, thereby defeating the sole purpose of DNA unwinding. Second, ssDNA is more sensitive to physical and chemical insults than dsDNA, thereby endangering the stability of the replication fork. Third, ssDNA can form secondary structures such as hairpins (Figure 1-10) that impede the ability of DNA polymerases to synthesize the complementary strand. However, once ssDNA is generated by the replicative helicase, it is rapidly bound by a **single-strand-specific DNA-binding protein** (Figure 4-11). These proteins (generally referred to as SSBs) prevent reannealing of complementary DNA sequences and protect them from nucleases. The result is a dramatic 10^3–10^4-fold difference in the rate of DNA melting at physiological salt levels. Although SSBs can bind to virtually any ssDNA, they are species-specific in their associations with other proteins. Hence, SSBs facilitate the organization of a replisome at replication forks.

SSBs show considerable diversity across the three domains of life, ranging from simple monomeric proteins in some archaea, through heterotrimers

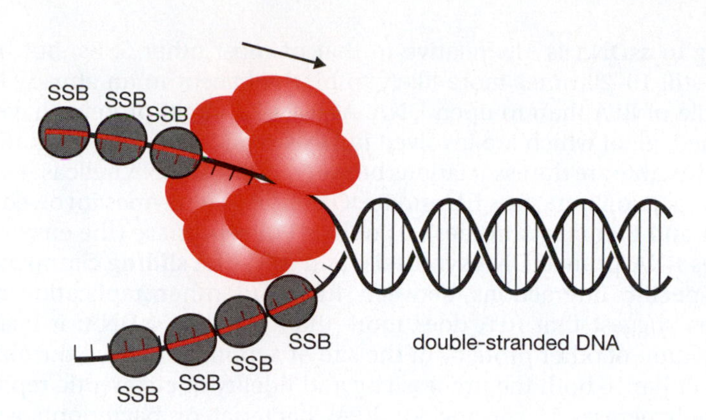

Figure 4-11. Replication fork with helicase on one template strand and single-strand-specific DNA-binding proteins (SSBs) on both template strands.

in eukaryotes and other archaea, to homotetramers in bacteria (Table 4-2). Bacterial SSBs bind tenaciously and specifically to ssDNA. Part of the reason is that they bind in a highly cooperative manner; at low ratios of protein to DNA, some ssDNA molecules are nearly saturated with protein while others in the same population contain none. Each 19-kDa *E. coli* SSB monomer comprises an N-terminal DNA-binding domain and a flexible C-terminal tail that binds such heterologous proteins as the Chi (χ) subunit of *E. coli* DNA polymerase III, *E. coli* DNA polymerases II and V, the DNA repair enzyme exonuclease I, DNA helicases PriA and RecQ, and uracil DNA glycosylase. Mutations within this C-terminal tail reveal that these interactions are critical to genome duplication and cell division. *E. coli* SSB binds ssDNA *in vitro* in two major modes. At low salt concentrations, an SSB dimer contacts approximately 35 nucleotides, whereas at high salt concentrations all four subunits of an SSB tetramer contact approximately 65 nucleotides of ssDNA (see Figure 4-12 in color plate section). One of these is most likely used in the cell.

Replication protein A (RPA) is the major single-stranded DNA-binding protein in all eukaryotic organisms examined. RPA is a heterotrimeric protein composed of subunits of approximately 70, 32, and 14 kDa. As with SSB, RPA binds nonspecifically to ssDNA and interacts with and/or modifies the activities of many different proteins. However, the cooperativity of RPA

Table 4-2. Composition Of ssDNA-Binding Proteins In Representative Species

Species	Protein	Organization	No. of OB folds
E. coli	SSB	Homotetramer	1 per monomer
Phage T4	gp32	Monomer	1
Phage T7	gp2.5	Homodimer	1 per monomer
Eukarya	RPA	Heterotrimer: RPA 70, 32, and 14	6 (of which 4 bind DNA)
HSV	UL29/ICP8	Homodimer	1 OB-like fold
Sulfolobus solfataricus	SSB	Monomer	1
Methanococcus	RPA	Monomer	4
Methanosarcina	RPA1 RPA2 RPA3	Homodimer/tetramer Homodimer Homodimer	4 per monomer 2 per monomer 2 per monomer
Pyrococcus	RPA	Heterotrimer: RPA 31, 32, and 14	3

OB folds, oligonucleotide-/oligosaccharide-binding folds; RPA, replication protein A (see text for explanations).

binding to ssDNA is low relative to that of most other SSBs, but human RPA is still 10–20 times more likely to bind adjacent to an already bound molecule of RPA than to open DNA. At least 23 target proteins have been identified, all of which are involved in DNA replication, recombination, or repair. Notable are the associations between RPA and DNA helicases such as SV40 T-ag, papillomavirus E1, and RecQ helicases (enzymes involved in the Werner and Bloom syndromes), as well as Pol-α•primase (the enzyme that initiates RNA-primed DNA synthesis) and RFC (the **sliding clamp loader**). Such specific interactions between RPA and other replication-related proteins suggest that RPA does more than stabilize ssDNA: it facilitates organization of other proteins at the site of action. Moreover, the ability of RPA to enhance both the processivity and fidelity of eukaryotic replicative DNA polymerases is specific for RPA: bacterial or bacteriophage SSBs cannot substitute for RPA.

SSB and RPA have similar affinities for DNA. The challenge with analyzing binding of these proteins is that their affinity is so high that measurements tend to be stoichiometric at physiological salt conditions. At very low concentrations of DNA and physiological salt conditions binding constants are in the range of 10^{-9}–10^{-10} M. The affinity of human RPA for polydeoxythymidine is high and the occluded binding site is up to 30 nucleotides, depending on which of three different modes of DNA binding are engaged.

Despite their diversity in subunit composition (Table 4-2), SSBs share a common architectural feature in the form of oligonucleotide-/oligosaccharide-binding folds (OB folds), although the number of such folds per SSB varies. OB folds are commonly employed in the recognition of ssDNA. They comprise a five-stranded β-sheet coiled to form a closed barrel-like structure that is capped by an α-helix located between the third and fourth β-strands. These folds employ aromatic residues to stack with, and thus stabilize, exposed nucleotide bases in the single-stranded DNA (Figure 4-13).

In RPA, OB folds are designated **DNA-binding domains** (DBDs) A–F (Figure 4-14). The RPA70 subunit has four OB folds (DBDs F, A, B, and C) connected by flexible, unstructured polypeptide linkers, suggesting that RPA can adopt a wide variety of conformations when binding to DNA. The crystal structure of the full RPA has not been determined, possibly because the molecule adopts multiple conformations on ssDNA. DBD F contains a basic cleft that interacts with both ssDNA and proteins involved in repair of DNA damage such as p53, Rad51, and BRCA2. DBD F is connected to the high-affinity DNA-binding core of RPA (DBD A and DBD B) by a long flexible linker. DBD A and DBD B interact directly with eight to ten nucleotides of DNA

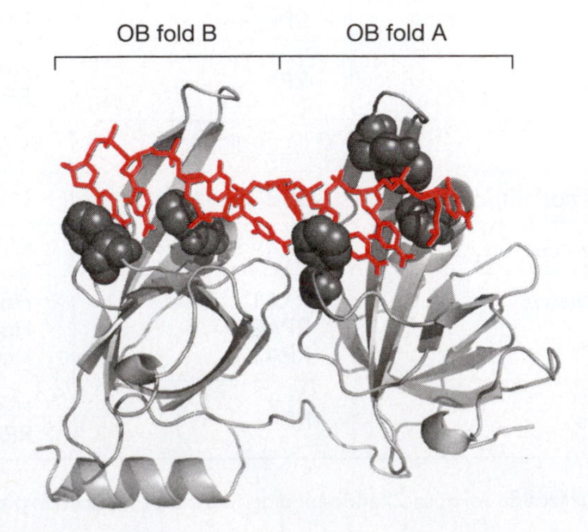

Figure 4-13. A complex between the ssDNA-binding domain (OB folds A and B) of human RPA70 subunit (gray) and eight nucleotides of ssDNA (red). Two OB folds (gray ribbons and threads) are visible. The atoms of aromatic residues in RPA that stack with bases (think of base-pair stacking in duplex DNA) are represented as dark gray spheres. Images generated using Pymol with PDB file 1JMC. (Data from Bochkarev, A., Pfuetzner, R.A., Edwards, A.M. and Frappier, L. (1997) *Nature* 385: 176–181.)

(Figure 4-13) and are both necessary and sufficient for high-affinity binding. DBD C is required for heterotrimeric complex formation. It interacts weakly with DNA. RPA32 and RPA14 each have one OB fold (D and E, respectively), and are required for RPA complex formation. DBD D interacts weakly with DNA. DBD E interacts with single-stranded telomeric DNA sequences. RPA32 is phosphorylated during the cell cycle and in response to cellular DNA damage. The N-terminal domain of RPA70 functions by interacting with the phosphorylation domain of RPA32, blocking undesirable interactions with the core DNA-binding domain of RPA.

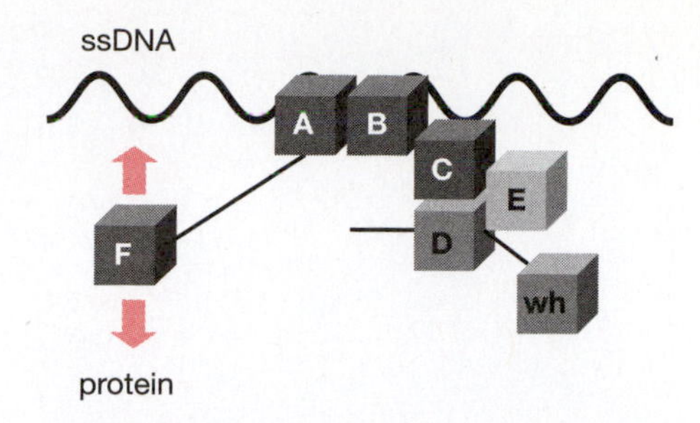

Figure 4-14. Model for RPA bound to ssDNA. RPA is a very flexible heterotrimeric ssDNA-binding protein consisting of RPA70 (DBDs F, A, B, and C), RPA32 [DBD D and the winged helix (wh) domain], and RPA14 (DBD E). Domain F appears to bind either ssDNA or one of several target proteins. (Data from Binz, S.K. and Wold, M.S. (2008) *J. Biol. Chem.* 283: 21559–21570.)

DNA POLYMERASES

DNA polymerases are essential for maintaining the integrity of genomes, as well as for duplicating them. Thus, it is not surprising that *E. coli* encodes five different DNA polymerases, only two of which are required for DNA replication (replicative DNA polymerases), and eukarya encode up to 15 DNA polymerases, only four of which are required for DNA replication (Table 4-3). The remainder are required either for specialized DNA repair processes, or to allow synthesis to proceed past various types of DNA damage that block replication-fork progression. What they all share in common is the fact that none of them can initiate DNA synthesis *de novo*: DNA polymerases can only extend DNA from a primer containing a free 3′-OH group that is annealed to a DNA template. During DNA replication, the primer is synthesized on the template by DNA primase. During DNA repair, the primer is the 3′-OH end of a broken DNA chain annealed to a DNA template. The only exception is **terminal deoxyribonucleotidyl transferase** (TdT). This enzyme can catalyze addition of ribo- as well as deoxyribonucleotides to the 3′-terminus of DNA in the absence of a template. TdT is expressed in lymphoid cells where it is involved in joining exons together during antibody gene recombination. The archaeal DNA primase also has TdT activity. Its function is unknown, but given the fact that most archaea lack polymerase X family members (discussed below), the archaeal primase's TdT activity may allow it to repair DNA damage as well as to initiate RNA-primed DNA synthesis. Notable is the fact that although eukaryal DNA primases do require a template, they are very error-prone, to the extent that they can even incorporate dNTPs as well as rNTPs, highlighting the evolutionary relationship between archaeal and eukaryal DNA primases. Telomerase, a DNA polymerase that maintains the ends of linear chromosomes, was originally thought to be a terminal nucleotidyl transferase, because it extends the 3′-OH termini of DNA, but this enzyme provides its own template in the form of an RNA component (Figure 6-10).

Replicative DNA polymerases must first recognize their appropriate template site, and then release from that site and synthesize millions of nucleotides worth of DNA. DNA synthesis must be accurate to maintain genomic stability, but not so accurate that genetic mutations are eliminated. To this end, those replicative DNA polymerases responsible for the bulk of DNA synthesis invariably have the ability to detect and correct errors in base pairing. However, some basal level of misincorporation may be necessary, or else the ability of a species to evolve by natural selection would be compromised. With the striking exception of *E. coli* Pol-II, the accuracy of nucleotide incorporation (**fidelity**) of most DNA polymerases is low compared with replicative DNA polymerases. All replicative DNA polymerases (apart from eukaryotic Pol-α) contain an associated 3′→5′ exonuclease activity that allows the enzyme to proofread the last base pair incorporated. If it is a mismatch, then the exonuclease activity can excise it and allow the polymerase to try again. Only DNA bacteriophages and animal viruses with genomes of approximately 40 kbp or more encode their own replicative DNA polymerase, and the properties of these viral enzymes are quite similar to those of either bacterial Pol-III or eukaryotic Pol-δ and Pol-γ.

Table 4-3. DNA Polymerases In Bacteria And Mammals

	DNA polymerase	Associated exonuclease	Error rate ($\times 10^{-5}$)	Processivity (nucleotides)
Bacterial replicative enzymes	Pol-I	3′→5′ 5′→3′	≤0.6	
	Pol-I (3′ exo⁻)	5′→3′	2.5	
	Pol-III	3′→5′	0.1–1	
Human replicative enzymes	Pol-α (alpha), nuclear	None	10	
	Pol-δ (delta), nuclear	3′→5′, inactivated	≤2	
	Pol-ε (epsilon), nuclear	3′→5′, inactivated	≤2	
	Pol-γ (gamma), mitochondrial	3′→5′, inactivated	≤1	~500
Bacterial repair enzymes	Pol-II	3′→5′	≤0.2	
	Pol-II (exo⁻)	None	4.8	
	Pol-V	None	10–100	
	Pol-IV	None	10–100	
Human repair enzymes	Pol-β (beta)	None	23	~10
	Pol-λ (lambda)	None	90	
	Pol-ζ (zeta)	None	130	
	Pol-θ (theta)	None	240	~75
	Pol-ν (nu)	None	350	~100
	Pol-κ (kappa)	None	580	
	Pol-η (eta)	None	3500	
	Pol-ι (iota)	None	72 000	

Error rate is the number of misincorporated nucleotides per 100 000 bp replicated. Data are for purified enzymes (Arana *et al.*, 2008; Rattray and Strathern, 2003; Zhong *et al.*, 2006). Processivity is the number of nucleotides synthesized following a given primer binding event and thus is a measure of the ability of replicative DNA polymerases to continuously incorporate nucleotides without dissociating from the primer:template.

The DNA Polymerase I Paradigm

The paradigm for DNA polymerases is *E. coli* DNA polymerase I, the first DNA polymerase to be characterized, for which Arthur Kornberg was awarded the Nobel Prize in Physiology or Medicine in 1959. Pol-I, as it is commonly known, was identified by its ability to catalyze incorporation of ^{14}C-TTP into DNA. It could incorporate at least 20 nucleotides into a growing chain without dissociating from the template molecule. As the enzyme was purified, it became apparent that it had additional activities. Remarkably, even highly purified Pol-I had the ability to degrade DNA exonucleolytically, as well as to synthesize it. Pol-I could degrade one strand of duplex DNA from either a 3′→5′ or a 5′→3′ direction. In contrast, DNA synthesis could proceed only in a 5′→3′ direction. It transpired that gentle treatment of purified DNA Pol-I with protease resulted in separation of a 323-amino acid domain containing the 5′→3′ exonuclease activity from a 604-residue protein (termed the Klenow fragment) containing the polymerization and 3′→5′ exonuclease activities. The tight association of DNA polymerase and 3′→5′ exonuclease activities is important for the fidelity of the enzyme. Obviously, a DNA polymerase must be able to copy DNA accurately; errors made during replication will, if uncorrected, result in base-pair mismatches that could potentially lead to genetic mutations. However, errors inevitably arise by accidental misincorporation of the wrong base by the polymerase. For *E. coli* DNA Pol-I, errors arise at a frequency of approximately one mistake for every 100 000 bases copied. If a derivative of DNA Pol-I that lacks the associated 3′→5′ exonuclease activity is substituted for the wild-type enzyme, then the error rate increases about 100-fold.

The structural basis of these activities has been revealed from X-ray crystal structures of *E. coli* and *Thermus aquaticus* DNA Pol-I in complex with DNA substrates. Their structure has been likened to a right hand gripping the primer:template substrate (see Figure 4-15 in color plate section). The palm of the hand contains the active site of the enzyme, the fingers help bind the incoming nucleotide and the template ahead of the enzyme, while the thumb domain grips the duplex DNA emerging from the active center. Although there are a number of phylogenetically distinct families of DNA polymerases, this right-hand gross topography is common to all DNA polymerases.

Another feature common to all DNA polymerases is that they use two metal ions to mediate catalysis (Figure 4-16). Metal ion A makes interactions with the 3′-OH of the growing DNA strand. This interaction is thought to lower the pK_a of the hydroxyl group and helps in its attack of the α-phosphate of the incoming dNTP. Both metal ions may additionally act to stabilize the intermediate that occurs during phosphodiester bond formation. Catalysis proceeds by an SN_2-type chemical reaction whereby the α-phosphate goes through a transition state involving five different bonds. In the final stage of the reaction, metal ion B facilitates the exit of the pyrophosphate cleaved off during phosphodiester bond formation.

The crystal structure of the Klenow fragment with DNA also illuminates the mechanism for proofreading. The catalytic site for polymerization and the proofreading site, both present in the Klenow fragment, are separated

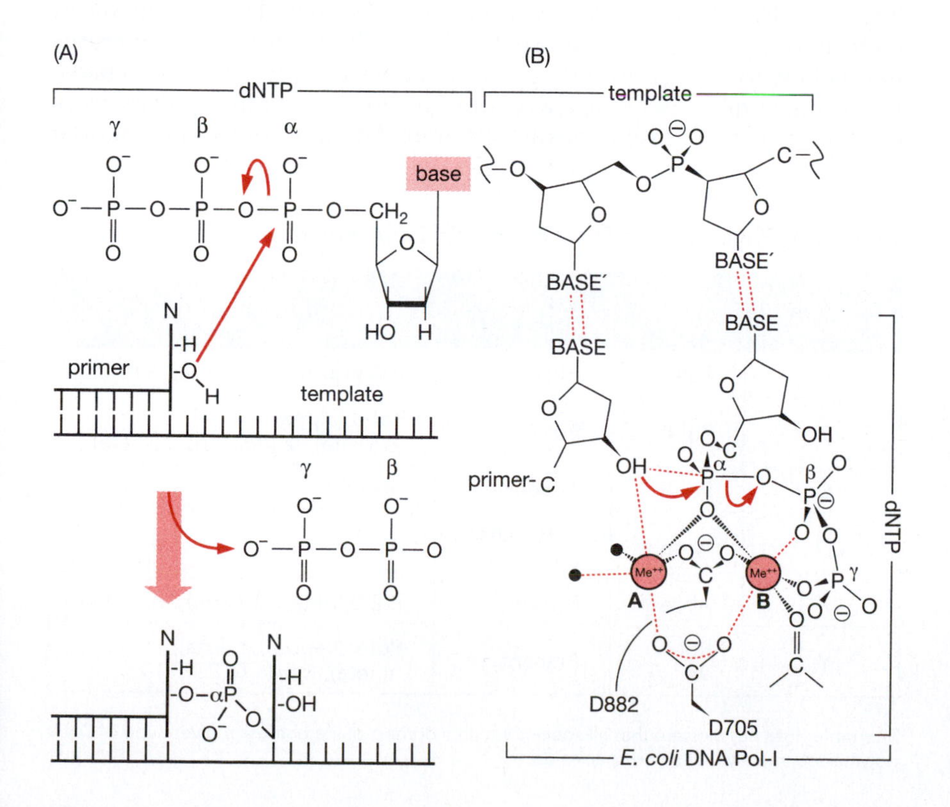

Figure 4-16. The two-metal-ion mechanism for DNA polymerase catalysis. (A) DNA polymerases extend oligonucleotide primers annealed to DNA templates by catalyzing a nucleophilic attack by the 3′-OH group of the primer on the α-phosphate of the dNTP with concomitant release of pyrophosphate. (B) The active site in *E. coli* Pol-I features two divalent metal ions (A and B) that together with two conserved aspartate residues (D882 and D705) form a stable penta-coordinated (also called trigonal bipyramidal) transition state. The positively charged divalent metal ions form ionic bonds (indicated as '–' charge distributed along dotted lines) and coordinate-covalent bonds (indicated as shaded triangles) with the free carboxyl group of aspartates 882 and 705. Metal ion A activates the primer's 3′-OH for attack on the α-phosphate of the dNTP. Metal ion B plays the dual role of stabilizing the negative charge that builds up on the leaving oxygen and chelating the β- and γ-phosphates. Contributions from a main chain carbonyl group complete the coordination of Metal ion B.

by about 30 Å, indicating that substantial movements would be required to position the DNA substrate at either site. Therefore, a competition exists between the two sites for the 3′-OH end of the primer. The exonuclease site preferentially binds a single-stranded 3′-end whereas the polymerization site prefers a Watson–Crick base-paired dsDNA substrate. Thus, if the wrong nucleotide is incorporated by accident, the resulting mismatched base pair will destabilize the duplex and favor recognition of the 3′-end by the exonuclease site (see Figure 4-17 in color plate section). This competition is augmented by the fact that polymerization is blocked by incorporation of a mispaired nucleotide: it is thought that this arises because the altered geometry of a mispair results in inappropriate positioning of the 3′-end for incorporation of the next nucleotide.

DNA Polymerase Families

Although *E. coli* Pol-I played a key role in defining many of the mechanistic principles in the polymerization of DNA, it soon became apparent that Pol-I was not the only DNA polymerase encoded by *E. coli*. Indeed, Pol-I does not perform the main task of genome duplication; its principal roles during DNA replication are in completing Okazaki-fragment synthesis and excision of RNA primers. The task of replicating the *E. coli* genome is taken by Pol-III. The situation in human cells is even more complex where at least 15 DNA polymerases have been identified. On the basis of amino acid sequence homologies, DNA polymerases can be grouped into six families: A, B, C, D, X, and Y (Table 4-4). DNA Pol-I falls into family A. Family B contains the main replicative polymerases of eukaryotes (Pol-α, -δ, and -ε) as well as the principal polymerase responsible for mutagenesis in the cell (Pol-ζ). Several archaeal and a number of bacteriophage enzymes are also members of family B. Bacterial Pol-III is in a distinct family (family C), while family D is unique to the euryarchaeal branch of the archaea. Families X and Y contain polymerases whose principal roles are in DNA repair and translesion DNA synthesis, respectively.

Family A

Bacterial DNA Pol-I is the archetypal member of family A DNA polymerases, and a number of bacteriophages such as T3 and T7 encode DNA polymerases that belong to this family, but, so far, none have been found in archaea. Interestingly, relatives of the bacteriophage group of family A polymerases are found in modern-day eukaryotic mitochondria. The mitochondrial

Table 4-4. Phylogenetic Distribution Of DNA Polymerase Families.

Polymerase family	Bacteria	Archaea	Eukarya
A	Pol-I	Absent	Pol-γ (gamma), -ν (nu), -θ (theta)
B	Pol-II	Pol-B	Pol-α (alpha), -δ (delta), -ε (epsilon), -ζ (zeta)
C	Pol-III	Absent	Absent
D	Absent	Euryarchaea only	Absent
X	Absent	Present	Pol-β (beta), -λ (lambda), -μ (mu)
Y	Pol-IV, -V	Present	Pol-κ (kappa), -ι (iota), -η (eta), Rev1, TdT

This table does not indicate that all species within a domain of life possess a given type of DNA polymerase, merely that some species do.

replicative DNA polymerase-γ (Pol-γ) is encoded within the nuclear genome and imported into mitochondria. Pol-γ is a heterodimeric enzyme in which the larger catalytic subunit (PolγA) is associated with another protein, the PolγB subunit, that enhances the enzyme's processivity. In contrast with other family A DNA polymerases (Pol-θ and Pol-ν), PolγA has a proofreading ability.

Animals and plants have three additional members of family A that are typified by Pol-θ, an enzyme consisting of a DNA polymerase domain fused to a DNA helicase domain. These proteins do not appear to be essential, because *Drosophila* mutants lacking the Pol-θ ortholog are viable. However, these strains are hypersensitive to a range of DNA-damaging agents, particularly those that cause interstrand cross-links. It is likely therefore that Pol-θ plays a role in DNA repair processes. Human Pol-ν has been implicated in DNA repair and in translesion DNA synthesis. Pol-ν discriminates poorly against misinsertion of dNTP opposite template thymine or guanine. Nevertheless, the processivity of both Pol-θ and Pol-ν is comparable to the family B replicative DNA polymerases Pol-δ and Pol-ε in the absence of PCNA.

Family B

Family B DNA polymerases are ubiquitous. Examples have been found in all three domains of life as well as in a broad range of viruses that prey on both eukaryotes and prokaryotes. Among the bacteria, Pol-B family members are typified by DNA Pol-II, an enzyme that is restricted primarily to the γ-proteobacteria. Pol-II is believed to play a role in restarting DNA replication following DNA damage, although *E. coli* strains that lack the gene encoding Pol-II are viable and show no overt phenotypes during normal growth.

In contrast to the rather limited distribution of family B DNA polymerases in bacteria, all archaeal genomes sequenced to date possess at least one family B member. These enzymes possess proofreading activity, and they are capable of interaction with the sliding clamp, PCNA. One intriguing feature of Pol-B in the archaea is its ability to sense uracil in the template ahead of the active site and cause the enzyme to stall. Uracil is not a normal component of DNA, but spontaneous deamination of cytosine followed by a tautomeric shift generates uracil bases in DNA. In addition, changes in nucleotide metabolism can result in dUTP incorporation in place of dTTP (Figure 3-12). Whereas cytosine normally base-pairs with guanine, the preferred partner for uracil is adenine. Thus, if left uncorrected, following passage of the replication fork, a C→U transition would result in the conversion of a G:C base pair to an A:T. This is a particular problem for hyperthermophiles as a 10°C increase in temperature results in a 5.7-fold increase in the rate of spontaneous deamination. The structural basis for this read-ahead function has been resolved with the observation that the archaeal family B polymerases have a small pocket that lies ahead of the active site (Figure 4-18). This pocket binds uracil and forces the polymerase to stall four bases before the uracil. This presumably leads, by

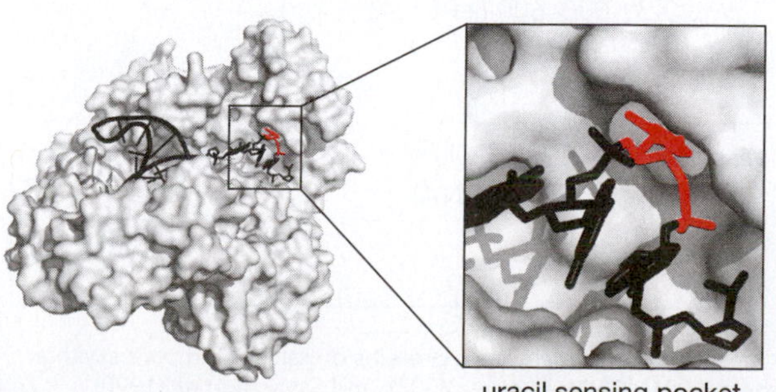
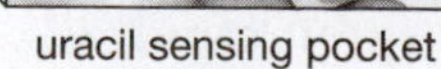

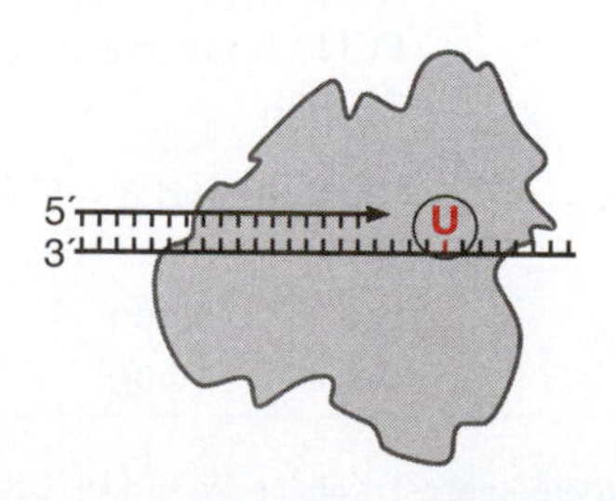

Figure 4-18. Uracil-sensing function of archaeal family B polymerases. Archaeal Pol-B contains a special pocket in its N-terminal domain that captures uracil bases in the template strand four bases ahead of the catalytic site. The structure of a trapped uracil-containing template has been determined and reveals the basis of this behavior. The uracil is shown in red and is clearly inserted into a pocket on the polymerase. (Data from Firbank, S.J., Wardle, J., Heslop, P.J. *et al.* (2008) *J. Mol. Biol.* 381: 529–539.)

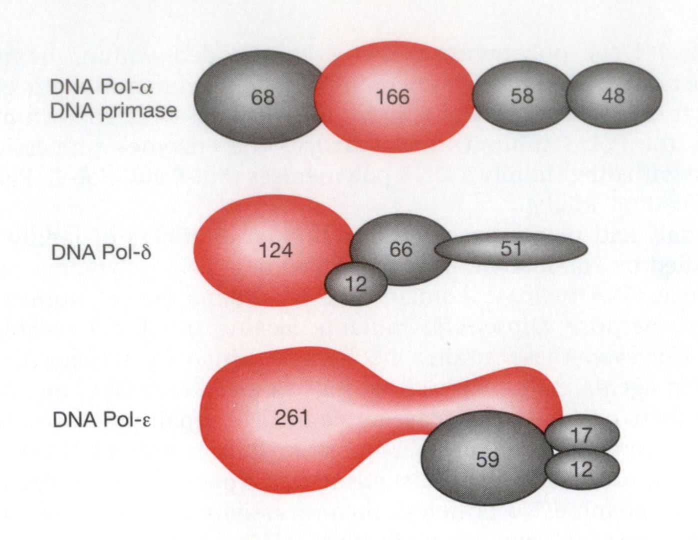

Figure 4-19. Eukaryotic family B DNA polymerases are multisubunit complexes. The catalytic subunit (red) is the largest. The 48 and 58 kDa subunits of Pol-α•primase constitute the DNA primase. The third subunit of Pol-δ is extremely elongated in shape and the catalytic subunit of Pol-ε is a two-domain polypeptide. Interactions with the other subunits are localized to the C-terminal domain. The numbers refer to the sizes of the subunits in the human enzymes. (Data from Burgers, P.M.J. and Seo, Y.-S. (2006) Eukaryotic DNA replication forks. In *DNA Replication and Human Disease* (DePamphilis, M.L., ed.), pp. 105–120.)

an as yet unresolved mechanism, to the eventual removal of uracil from the template strand.

All eukarya possess three B-type polymerases that are required for DNA replication (α, δ, and ε) and a fourth enzyme (ζ) that plays a role in mutagenesis. The three replicative DNA polymerases exist as multisubunit complexes (Figure 4-19), although the number and size of the subunits can vary among species (Table 4-5).

In eukarya, all replicative DNA synthesis is initiated by Pol-α•primase, an enzyme comprised of four different subunits. The largest subunit contains the DNA polymerase catalytic activity and is associated with the 68–79 kDa B-subunit. This B-subunit appears to have a regulatory role; it is differentially phosphorylated during the cell cycle. The two smallest subunits form the DNA primase, with the primase catalytic activity residing in the 48 kDa subunit. Pol-α is unique among replicative DNA polymerases, in that it lacks a proofreading 3′→5′ exonuclease activity, and it is about 10 times less processive than the other eukaryotic replicative DNA polymerases in the presence of a sliding clamp (Table 4-5). This makes it unsuited to the task of genome duplication, but well suited for the specific task of *de novo*

Table 4-5. Eukaryotic Replicative DNA Polymerases

Enzyme	Subunits (kDa)[a]	Function	Processivity (nucleotides)	Velocity (kb min^{-1}, 30°C)
Pol-α•primase	166 (167) 68 (79) 58 (62) 48 (48)	DNA synthesis Stability and regulation DNA primase stability DNA primase synthesis	10–30	~1
Pol-δ	124 (125) 66 (40) 51 (55) 12	Catalysis and 3′→5′ exonuclease Stability PCNA attachment Stability	5–10 (–PCNA) 500–600 (+PCNA)	~7
Pol-ε	261 (256) 59 (78) 17 (23) 12 (22)	Catalysis and 3′→5′ exonuclease ? dsDNA binding dsDNA binding	~60 (–PCNA) ~400 (+PCNA)	~2

[a]Sizes of mammalian and budding yeast (parentheses) proteins are shown. DNA polymerase rates for *S. cerevisiae* proteins on ssDNA templates with a single DNA primer were made in the presence of RPA, RFC, and PCNA. Data from Burgers (1991), Chilkova *et al.* (2007), and Syvaoja *et al.* (1990).

initiation of RNA-primed DNA synthesis. Thus, Pol-α•primase initiates RNA-primed DNA synthesis on both strands of replication origins. Once the RNA-primed DNA chain reaches about 30 nucleotides in length, the weakly processive Pol-α releases from the primer:template and hands-off to Pol-ε. Pol-ε then extends the 3′-end of the long growing nascent DNA strand on the leading-strand template. The same process is repeated for Okazaki-fragment synthesis on the lagging-strand template, except that Pol-α•primase hands-off to Pol-δ. Pol-δ extends Okazaki fragments on the lagging-strand template until DNA ligase I can join them to the 5′-end of the long growing nascent DNA strand. The advantage of this division of labor is that Pol-δ can be tailored to interact with the proteins that excise RNA primers and join Okazaki fragments together, and Pol-ε tailored to travel long distances without interruption.

Genetic evidence supporting this model comes from the use of Pol-δ and Pol-ε mutants that show an asymmetric mutator phenotype for certain mispaired bases. Together with the known direction of fork movement in yeast, this polymerase mutator asymmetry allows an unambiguous strand assignment of both polymerases. However, the catalytic domain of Pol-ε is not essential for yeast survival although mutant cells grow poorly. The non-essential nature of the Pol-ε catalytic domain suggests that leading-strand DNA synthesis can also be completed by Pol-δ. Historically, complete replication of SV40 genomes has been achieved *in vitro* using purified replication proteins that included both Pol-α and Pol-δ, but not Pol-ε. The ability of the cellular replication fork to utilize Pol-δ when Pol-ε is not available reflects the similarity of these two enzymes and the advantage of what is referred to in evolution as functional redundancy. The fact that Pol-ε is not required for replication forks generated by the SV40 replication origin may well reflect the association (or lack thereof) between cellular DNA polymerases and different DNA helicases [in this case, Mcm(2-7) and T-antigen].

Human DNA Pol-δ and Pol-ε also consist of four subunits, and as with Pol-α the catalytic activity for DNA synthesis resides in the largest subunit. In addition, the catalytic subunits of both Pol-δ and Pol-ε also exhibit a 3′→5′ exonuclease proofreading activity. In the case of Pol-δ, the second largest subunit bridges to the third subunit, p51, which contains an interaction motif for the sliding clamp, PCNA. The ability of Pol-δ to replicate long stretches of DNA is dependent on PCNA, to ensure processive synthesis (Table 4-5). The fourth subunit of Pol-δ appears to help stabilize the complex. Pol-ε also contains four subunits and is sensitive to PCNA, but not nearly to the extent displayed by Pol-δ. In the presence of accessory proteins, PCNA increases the processivity of Pol-δ manyfold, but PCNA has comparatively little effect on Pol-ε. The structure of the entire Pol-ε complex, visualized by cryo-electron microscopy in Figure 4-20, suggests that one of the roles of the three small Pol-ε subunits is to encircle the dsDNA, perhaps serving as an inbuilt sliding clamp and possibly explaining the relatively high processivity of Pol-ε in the absence of PCNA.

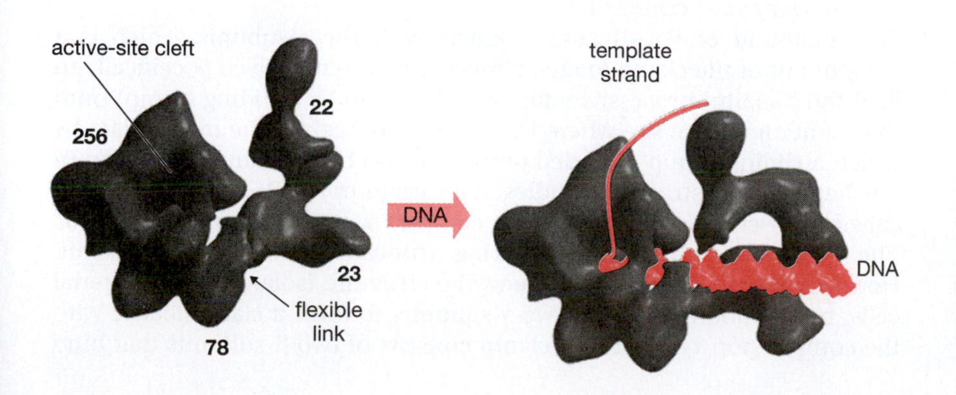

Figure 4-20. Structure of *S. cerevisiae* DNA Pol-ε by cryo-electron microscopy. A change in the relative orientations of the Pol-ε subunits is made possible by the flexible connection between them that would allow access of the DNA to the active site (left). A return of the tail to its normal conformation after entry of dsDNA into the active-site cleft would result in close interaction of the nucleic acid with the extended tail domain (right). This mode of interaction with DNA would explain the intrinsic processivity of Pol-ε. The minimum length of dsDNA required for processive DNA synthesis by Pol-ε (~40 bp) corresponds to the distance from the central portion of the catalytically active domain to the end of the extended tail domain formed by the three subunits. (From Asturias, F.J., Cheung, I.K., Sabouri, N. *et al.* (2006) *Nat. Struct. Mol. Biol.* 13: 35–43. With permission from Macmillan Publishers Ltd.)

The velocities of yeast Pol-α, Pol-δ, and Pol-ε *in vitro* (Table 4-5) are only 25% as fast as the bacterial replicative DNA polymerase, Pol-III, but they are comparable to the velocities of yeast replication forks *in vivo* (Table 3-4), suggesting that the overall rate of DNA synthesis in yeast is determined by the rate at which yeast replicative DNA polymerases can travel. Intriguingly, processivities of these three enzymes do not appear to correlate with their velocities (Table 4-5). PCNA•Pol-δ and PCNA•Pol-ε are each about 50-fold more processive than Pol-α•primase, but PCNA•Pol-δ is about three times faster than PCNA•Pol-ε. Given that PCNA•Pol-ε is responsible for DNA synthesis on the leading-strand template, these comparisons suggest that processivity is more important than speed when nature selects a DNA polymerase to carry out the bulk of DNA replication.

Family C

DNA polymerases belonging to family C are found exclusively in bacteria, where they constitute the principal DNA replication polymerase. The prototype is *E. coli* Pol-III, a heterotrimeric DNA polymerase consisting of an α-subunit that catalyzes DNA synthesis, an ε-subunit that is a 3′→5′ exonuclease, and a θ-subunit that stimulates the activity of the ε-subunit. Sequence analysis of the *E. coli* Pol-III α-subunit has failed to reveal significant homology with other DNA polymerases, leaving it unclear whether or not the molecular architecture of Pol-III is shared by other DNA polymerases. However, the recent elucidation of the crystal structure of the α-subunit from the bacterium *T. aquaticus* demonstrates that this enzyme possesses the classic right-handed configuration. Furthermore, the architecture of the catalytic domain is similar to that of the Pol-X family of DNA repair polymerases. Family X polymerases have been known for some time to be part of the β-nucleotidyltransferase structural superfamily and are clearly structurally and evolutionarily unrelated to polymerases of the A, B, and Y families. This complicated situation of distinct phylogenetic distributions of the various families of DNA polymerases becomes even more intriguing with the observation that some archaea encode novel DNA polymerases that belong specifically to family D.

As with eukaryotic replicative DNA polymerases, the bacterial replicative DNA polymerase III core enzyme is a multisubunit assembly consisting of three different proteins. The α-subunit contains the catalytic site for DNA polymerase activity. The ε-subunit is a 3′→5′ exonuclease that is facilitated by the θ-subunit in providing a proofreading function. The Pol-III core enzyme spontaneously assembles into a stable **holoenzyme** consisting of 10 different proteins (Table 4-6), but since several are present in two copies, it contains a total of 17 proteins that together with DnaB and DnaG are referred to as the replisome, the multiprotein complex that powers the replication fork (Figure 4-21). Together, these proteins provide extensive processivity and velocity to the polymerase, accounting for the remarkably high speed of bacterial replication forks *in vivo* (Table 3-4).

The holoenzyme concept

The α-subunit of Pol-III core interacts with the τ-subunit, which is a component of the clamp loader, a mechanism that evolved specifically to load the β-clamp processivity factor (referred to as a sliding clamp) onto the primer:template site where DNA synthesis begins. The mechanism by which a sliding clamp is loaded onto DNA has been gleaned largely from biochemical and structural studies of a variant of the Pol-III clamp loader called the γ-complex, a five-protein complex with a composition of $\gamma_3\delta\delta'$. The γ-subunit is a naturally occurring, truncated form of the τ-subunit. However, within the Pol-III holoenzyme structure isolated from bacterial cells, two τ-subunits replace two γ-subunits to form a clamp loader with the composition $\tau_2\gamma\delta\delta'$. The β-clamp consists of two β-subunits that bind

Table 4-6. *E. coli* DNA Polymerase III Holoenzyme

<table>
<tr><th>Subunit</th><th>Organization</th><th colspan="2">Function</th></tr>
<tr><td>α</td><td rowspan="3">Pol-III core</td><td colspan="2">DNA polymerase catalytic subunit, interacts with τ-subunit</td></tr>
<tr><td>ε</td><td colspan="2">3′→5′ exonuclease (proofreading)</td></tr>
<tr><td>θ</td><td colspan="2">Stimulates exonuclease</td></tr>
<tr><td>τ</td><td>Linker protein</td><td colspan="2">Dimerization of Pol-III core; links Pol-III core to DnaB; helps load β-clamp</td></tr>
<tr><td>γ</td><td rowspan="3">γ-complex</td><td>ATPase</td><td rowspan="3">ATP-dependent loading of the β-clamp onto the 3′-OH primer terminus</td></tr>
<tr><td>δ</td><td>Binds β-clamp</td></tr>
<tr><td>δ'</td><td>Stator, stimulates γ ATPase</td></tr>
<tr><td>χ</td><td></td><td colspan="2">Links Pol-III holoenzyme to SSB</td></tr>
<tr><td>ψ</td><td></td><td colspan="2">Links χ-subunit to γ-subunit; connects χ to clamp loader</td></tr>
<tr><td>β</td><td>β-clamp</td><td colspan="2">Processivity factor (sliding clamp), binds to α-subunit</td></tr>
</table>

Enzyme complex	Subunit composition	Processivity (nt)	Velocity (nt min^{-1}, 30°C)
Pol-III holoenzyme	$\alpha, \varepsilon, \theta, \tau, \gamma, \delta, \delta', \chi, \psi, \beta$	>5000	24 000
Pol-III*	$\alpha, \varepsilon, \theta, \tau, \gamma, \delta, \delta', \chi, \psi$	200	
Pol-III core	$\alpha, \varepsilon, \theta$	10	~5000

The holoenzyme is 792 kDa in size. nt, nucleotides. From Kim *et al.* (1996), McHenry (1991), and Wu *et al.* (1992).

to the α-subunit only after they are assembled onto DNA by the clamp loader. During the loading reaction, the β-clamp is bound to the δ-subunit of the clamp loader. The τ-subunit links two Pol-III cores together, and it further links the Pol-III core to the DnaB helicase. The χ-subunit identifies the lagging-strand template of the replication fork through association with SSB that is bound to ssDNA. The ψ-subunit links the χ-subunit to the clamp loader via its association with the γ-subunit.

The replisome concept

The existence of a biochemically and physically demonstrative holoenzyme leads directly to the concept of a replisome; two copies of the replicative DNA polymerase physically attached to both arms of the replication fork, to each other, and to the replicative DNA helicase (Figure 4-21). One copy of the Pol-III core is locked onto the growing Okazaki fragment by its association with a sliding clamp (two β-subunits) that encircles the duplex DNA behind it, and one copy is similarly locked onto the growing end of the leading strand. At the same time, both copies of Pol-III are held together by the sliding clamp loader ($\tau_2\gamma\delta\delta'$). In fact, the sliding clamp loader orchestrates the events that assemble and maintain the replisome. It is responsible for assembly of the sliding clamp onto the newly replicated DNA duplex so that the DNA polymerase can form a stable complex with the primer:template. It is responsible for keeping leading-strand synthesis and lagging-strand synthesis at the same place by holding onto both DNA polymerases. It is responsible for keeping both DNA polymerases attached to the DnaB helicase, thereby linking DNA synthesis with DNA unwinding. Finally, the sliding clamp loader is attached to SSB on the lagging-strand template by a bridge formed from the χ- and ψ- subunits, a feature that helps it guide the

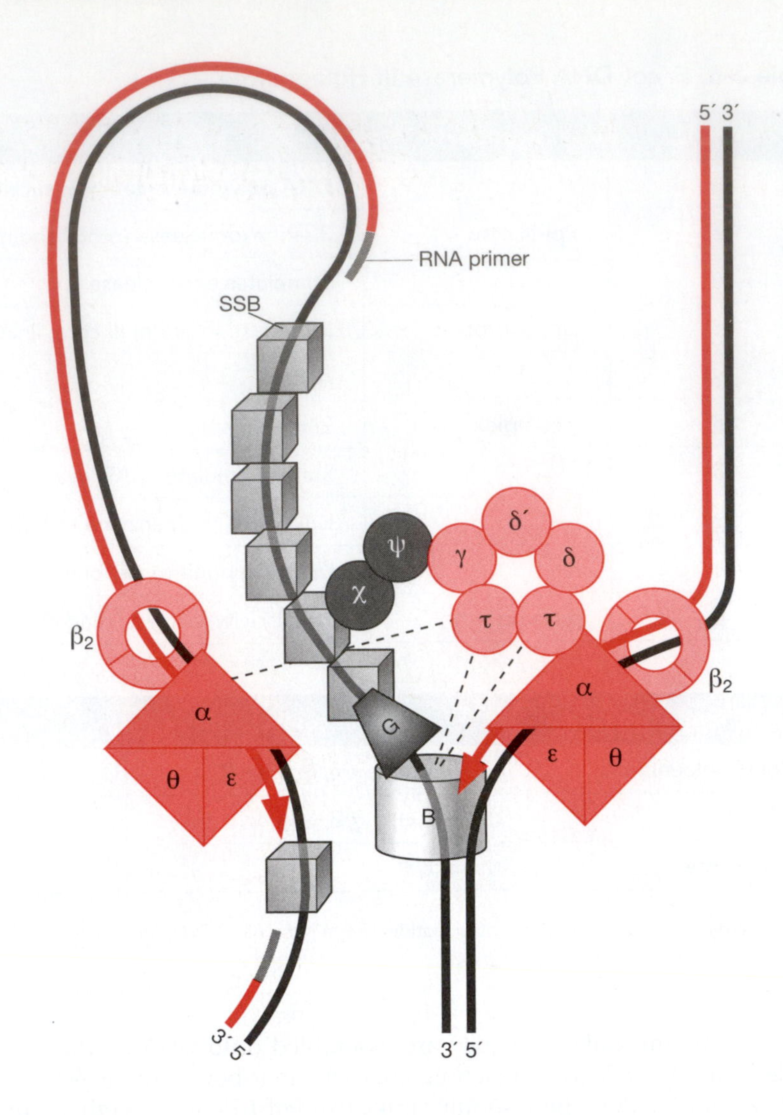

Figure 4-21. The bacterial replisome. The machine that powers replication forks in *E. coli* consists of the DNA polymerase III holoenzyme linked to the replicative DNA helicase (DnaB) and DNA primase (DnaG). Primase (G) interacts with both helicase (B) and SSB. The tau (τ)-subunit interacts with both helicase and the Pol-III α-subunit (indicated by contact lines). The stoichiometry and major protein•protein contacts are indicated. Parental DNA is black, newly synthesized DNA is red, and RNA primers are gray. (Adapted from McHenry, C.S. (2003) *Mol. Microbiol.* 49: 1157–1165. With permission from John Wiley & Sons, Inc.)

DnaG primase to its template as rapidly as DNA helicase can unwind the DNA (discussed in Chapter 5).

The Pol-III holoenzymes of *E. coli* and *T. aquaticus* contain two copies of the Pol-III core enzyme: one replicates the leading strand and the other the lagging strand. However, this homodimeric situation is not found in all bacteria. Rather, studies of *B. subtilis* revealed that this organism has two family C polymerases: $DnaE_{BS}$, a **homolog** of the *E. coli* Pol-III α-subunit, and Pol-C. An elegant series of genetic studies led to the proposal that Pol-C and $DnaE_{BS}$ play roles in leading- and lagging-strand synthesis, respectively. Whether or not the holoenzyme concept and the replisome concept apply to all bacteria, as well as to archaea and eukarya, remains to be demonstrated.

Family D

Members of the euryarchaeal kingdom of the archaea, in addition to having at least one family B DNA polymerase, possess a unique family of DNA polymerases called family D. These enzymes have two subunits, DP1 and DP2. DP1 is a 3′→5′ exonuclease activity, and DP2 is a DNA polymerase. The Pol-D polymerases are stimulated by PCNA through its association with DP1. Based on its biochemical properties, Pol-D may be the lagging-strand DNA polymerase. It may also initially extend the primer on the leading strand before transferring the role of leading-strand DNA polymerase to Pol-B.

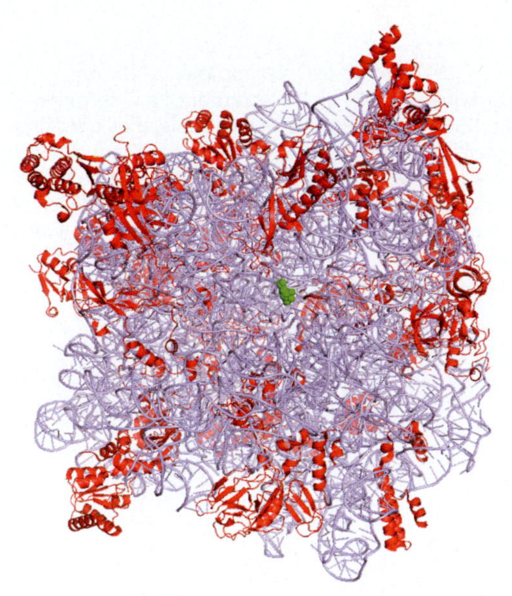

Figure 2-2. Structure of the ribosome, the RNA–protein complex that mediates translation of RNA transcripts of genes into proteins. The RNA chains (blue) reveal the incredible three-dimensional complexity of the ribosome. Clearly, the majority of the ribosome is composed of RNA with proteins (red) essentially decorating its surface. The catalytic site of the ribosome (green) lies buried in the center, far from the protein components of the ribosome.

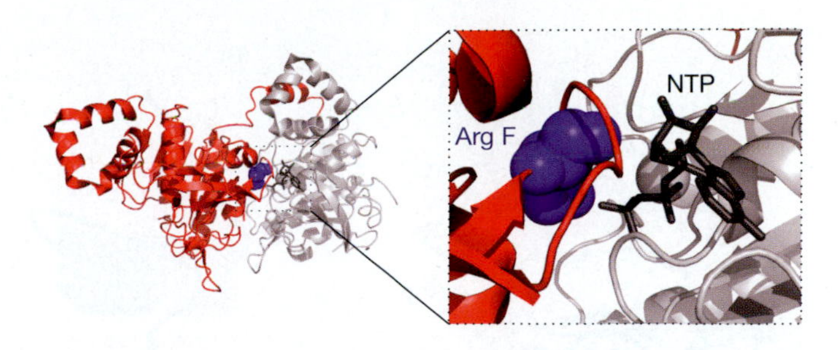

Figure 4-5. Two subunits (red and gray) of bacteriophage T7 helicase with a nonhydrolyzable analogy of NTP (black) sandwiched between them. The NTP is cradled in the active site (cleft) of the gray subunit, as the positively charged arginine finger (Arg F; blue spheres) from the red subunit caresses its 5′-phosphate. Spheres represent carbon, nitrogen, oxygen, and phosphorous atoms. Coiled ribbons are α-helices, and flat ribbons are β-strands; the remaining loops are shown as threads. (Data from Toth, E.A., Li, Y., Sawaya, M.R. *et al.* (2003) *Mol. Cell* 12: 1113–1123.)

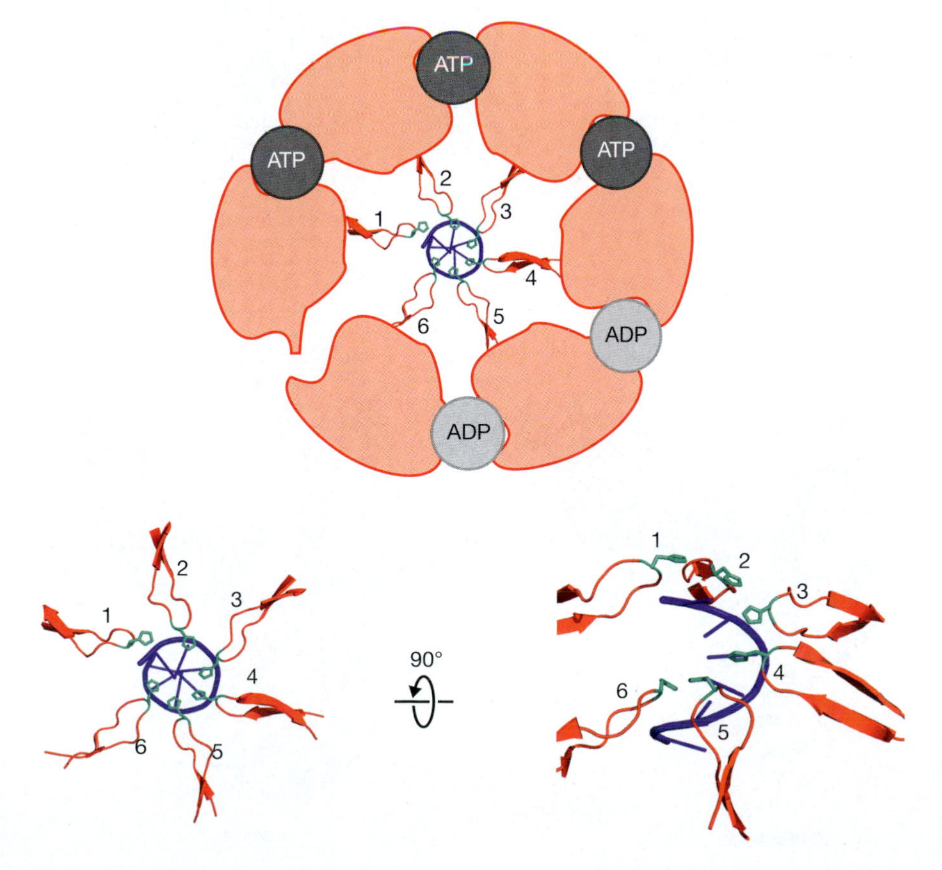

Figure 4-7. DNA binding within the bovine papillomavirus (BPV) E1 helicase. The upper panel is a view down the central channel of the helicase with ssDNA (dark blue) in the center. With one exception, the main bodies of each of the six helicase subunits (tan) contain either ATP or ADP bound between each of the monomers; one interface lacks any nucleotide. Each monomer contains a β-hairpin (red) structure that interacts with the DNA. A histidine residue (light blue) within the β-hairpin interacts with the phosphate of an adjacent nucleotide. The lower panels show two different views (rotated 90°) of the β-hairpin–DNA interactions. A β-hairpin consists of two β-strands that lie adjacent to one another in the amino acid sequence, but whose orientations in the protein structure are antiparallel, with the N-terminus of one strand adjacent to the C-terminus of its neighbor. A β-strand is a stretch of five to ten amino acids whose peptide backbones are almost fully extended. The two β-strands in a β-hairpin are linked by a short loop of two to five amino acids. (Data from Enemark, E.J. and Joshua-Tor, L. (2006) *Nature* 442: 270–275.)

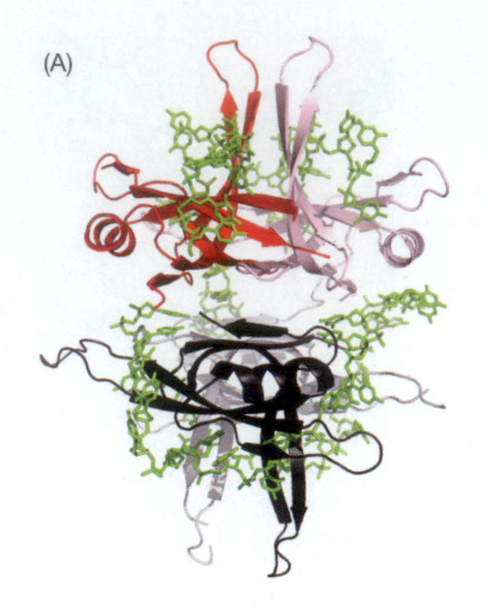

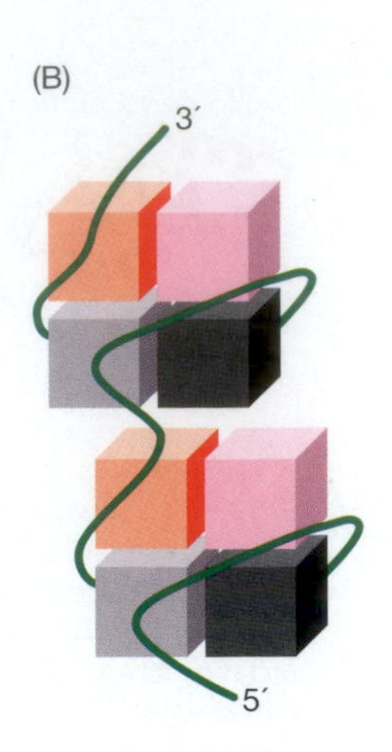

Figure 4-12. *E. coli* SSB•ssDNA complex. (A) X-ray crystallographic structure of a tetramer of a C-terminal-truncated SSB protein bound to two molecules of $(dC)_{35}$. Each protein monomer in the tetramer is given a different color. DNA is shown in green. (B) Model based on these and other data in which approximately 65 nucleotides of ssDNA interact with a cluster of four SSB subunits. (Data from Roy, R., Kozlov, A.G., Lohman, T.M. and Ha, T. (2007) *J. Mol. Biol.* 369: 1244–1257.)

Figure 4-15. *E. coli* DNA Pol-I has a 'right-handed' configuration. Polypeptide chains (gray) fold about the newly synthesized DNA strand (red) on its template strand (blue), as if the polymerase were someone's right hand.

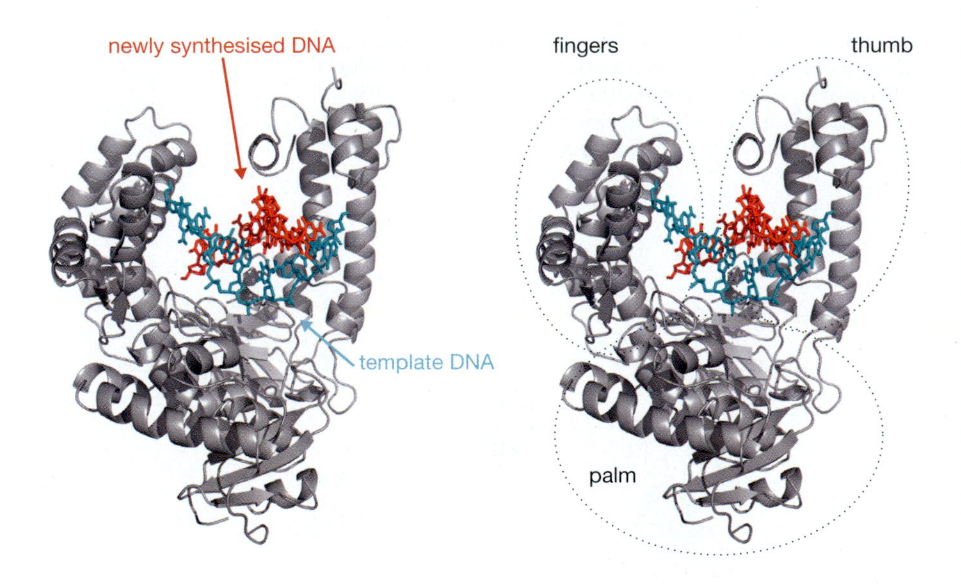

Figure 4-17. Comparison of elongation and editing complexes of the DNA Pol-I Klenow fragment. Polypeptide chains (gray) fold about the newly synthesized DNA strand (red) on its template strand (blue), placing the 3′-end of the nascent DNA strand either in the catalytic site for DNA synthesis (elongation) if the base pairing is correct, or in the editing site if the base pairing is incorrect. Images generated using PDB files 1TAU and 1KLN.

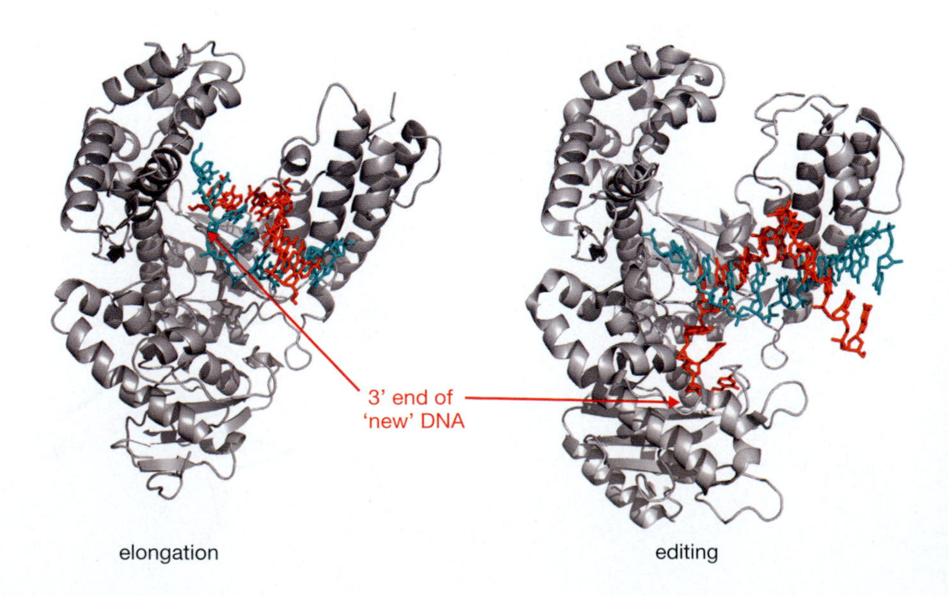

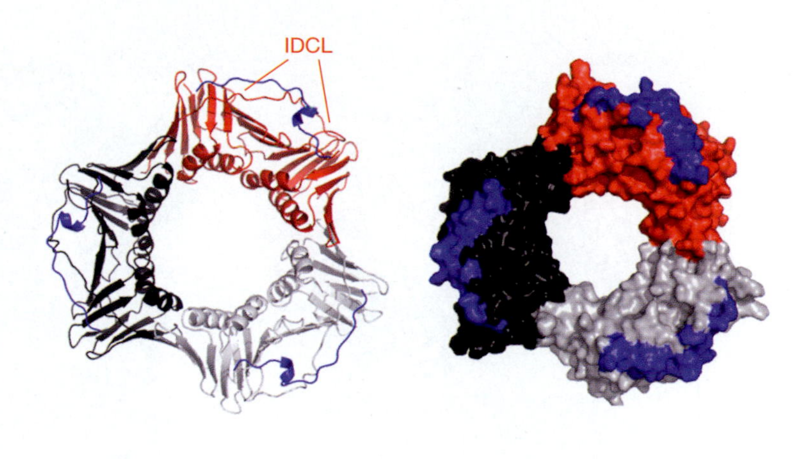

Figure 4-23. Human PCNA consists of three identical subunits. (A) Ribbon representation of PCNA shows the three subunits as black, gray, and red. A PCNA-interacting-protein (PIP) motif peptide (blue) is shown bound to each subunit. PCNA monomers have two similar globular domains, linked by a long, flexible interdomain connector loop (IDCL) that acts as the binding site for a PIP motif peptide. For clarity, the IDCL is indicated on the red subunit only but it is, of course, present on all three subunits. The three monomers are arranged head-to-tail to form a ring wherein the inner surface is formed by positively charged α-helices that associate with DNA, and an outer surface is composed of β-sheets. (B) Surface representation of PCNA. The overall dimensions of PCNA are similar to the bacterial β-clamp with an external diameter of 80 Å and an inner cavity of approximately 35 Å. The cross-sectional diameter of B-form dsDNA is 20 Å (Table 1-2). Images generated using Pymol with PDB file 1AXC.

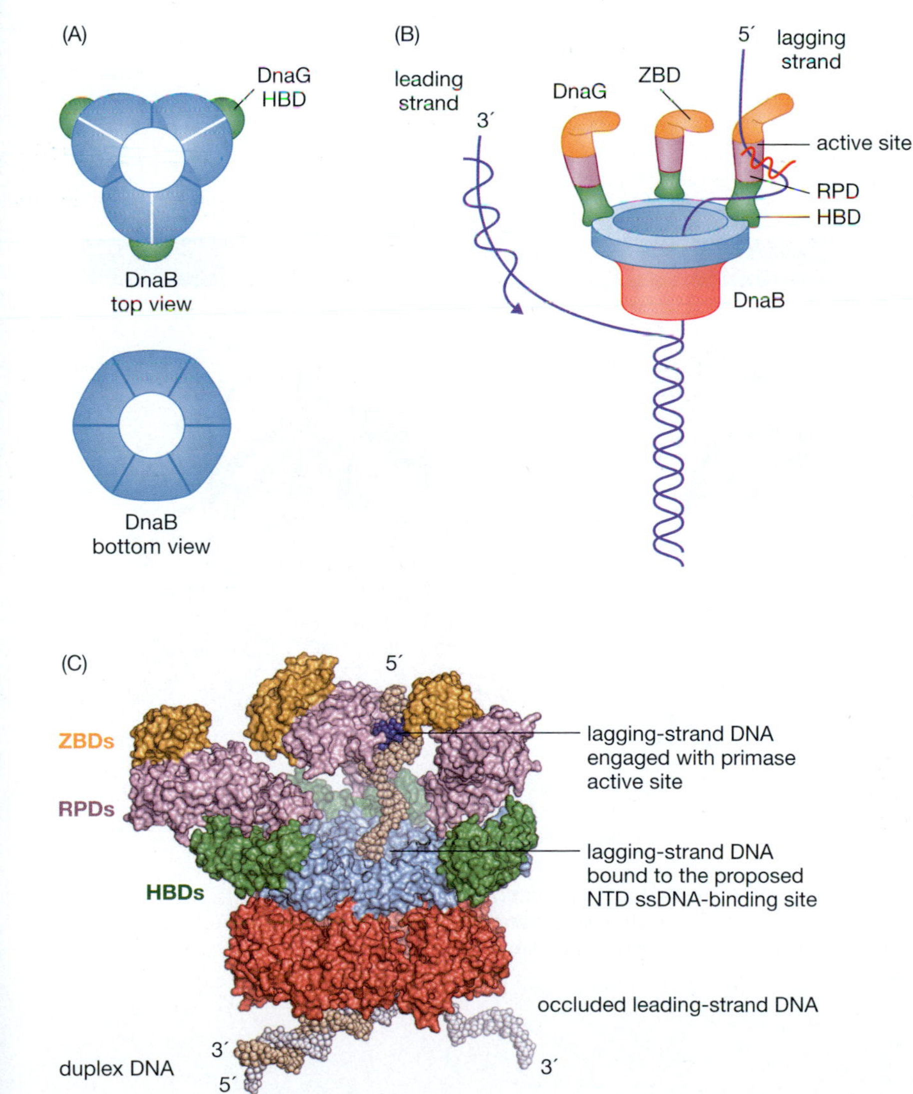

Figure 5-3. *E. coli* DNA helicase–DNA primase complex. (A) The DnaB helicase exists as a hexamer of six identical subunits. X-ray crystallographic analysis reveals that the side of DnaB facing the unwound DNA has sixfold symmetry, as expected. However, the opposite side of DnaB exhibits threefold symmetry: three DnaB dimers, each of which contacts the helicase-binding domain (HBD) of a single DnaG primase molecule. (B) Thus, each DnaB hexamer can bind three molecules of DnaG, each of which contains a single HBD, an RNA polymerase domain (RPD), and a zinc-binding domain (ZBD). The lagging-strand template threads through the center hole in the helicase hexamer and is then captured by the active site of the RPD in one of the DnaG molecules. This event may be facilitated by a ZBD from a different DnaG. The RNA primer is then extruded toward the outside of the complex. (C) Surface representation of the DnaB–DnaG complex described with the DnaG ZBDs in gold, the RPDs in lavender, and the HBDs in green. The HBD associates with the N-terminal domain (blue) of two DnaB monomers. dsDNA is fed into the C-terminal end (red) of the DnaB helicase, and one strand passes through the hexamer with a 5′→3′ polarity, while the other strand is displaced. The DnaB hexamer has an outer diameter of 115 Å and a height of 75 Å. The diameter of the central channel is approximately 50 Å, wide enough to accommodate dsDNA (~20 Å). [(A) and (B) adapted from Marians, K.J. (2008) *Nat. Struct. Mol. Biol.* 215: 125–127. With permission from Macmillan Publishers Ltd. (C) from Bailey, S., Eliason, W.K. and Steitz, T.A. (2007) *Science* 318: 459–463. With permission from the American Association for the Advancement of Science.)

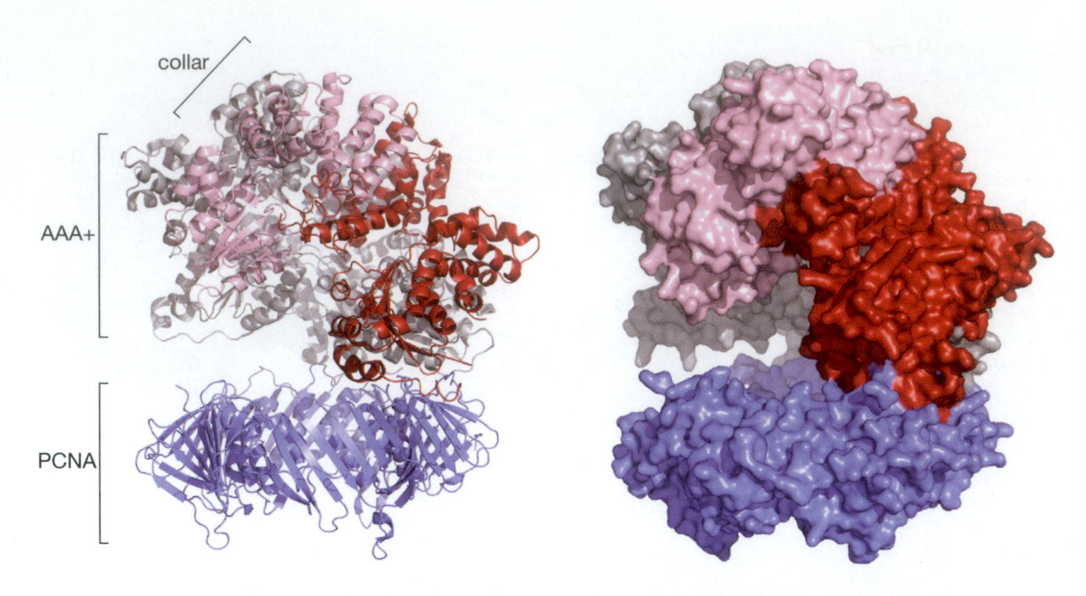

Figure 5-7. The crystal structure of a complex of yeast RFC [one large RFC subunit (red) and four small RFC subunits (pink and gray)] bound to a PCNA ring (blue). The positions of the AAA+ domains and collar domains (analogous to those in the γ-complex) are indicated. The figure was prepared using Pymol and PDB file 1SXJ. (Data from Bowman, G.D., O'Donnell, M. and Kuriyan, J. (2004) *Nature* 429: 724–730.)

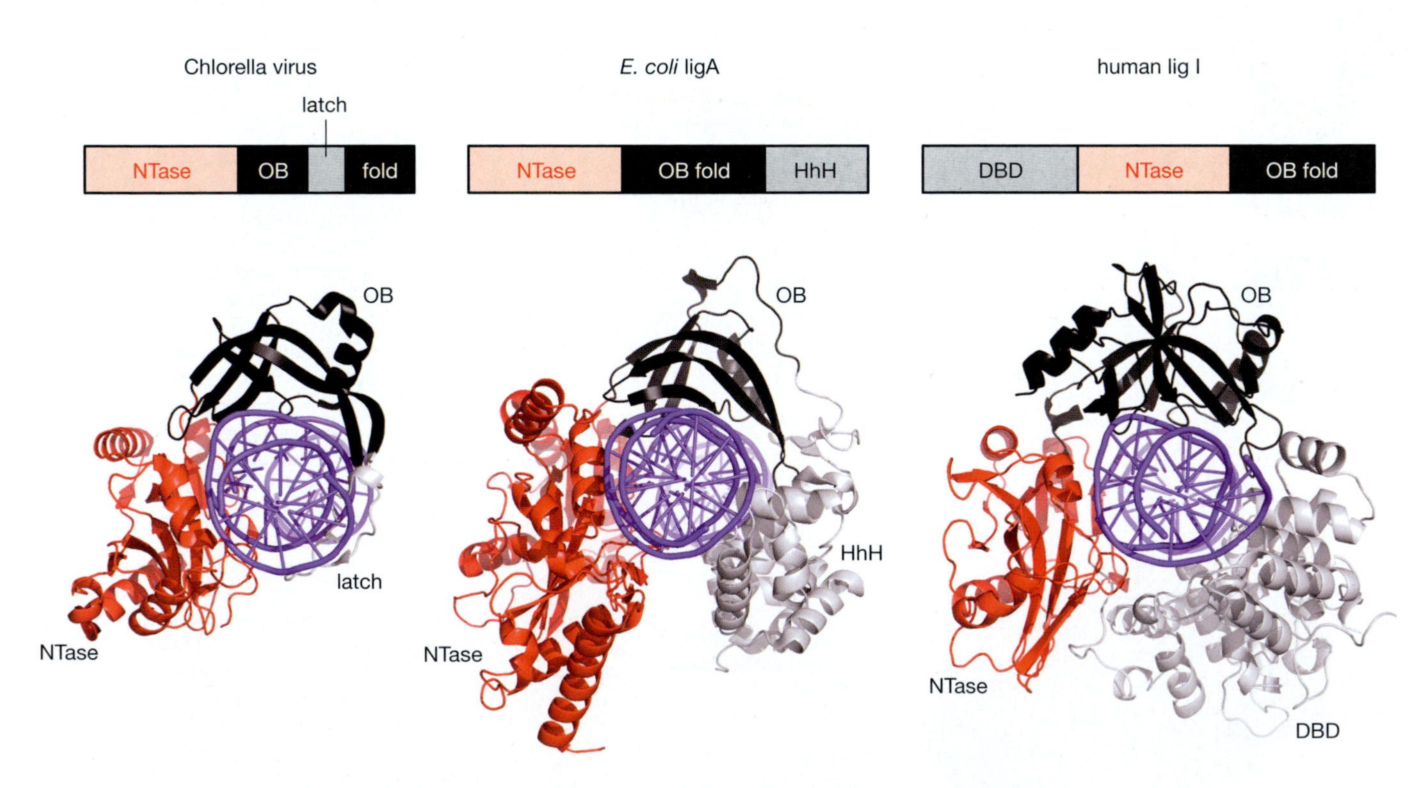

Figure 5-13. Crystal structures of DNA ligases bound to nicked DNA substrates. The locations of the domains are highlighted in the linear protein sequence and color-coded in both cartoon and structure. DBD, DNA-binding domain; HhH, helix-hinge-helix domain; NTase, nucleotidyltransferase; OB fold, oligonucleotide-/oligosaccharide-binding fold domain.

DNA Polymerase Fidelity And Molecular Evolution

The very act of duplicating the genome creates the basis for genomic instability, and the simple fact that natural selection has not completely suppressed the accumulation of errors during genome duplication attests to the theory that biological evolution relies upon these errors. The spontaneous mutation rate is one in 10^9 bp in *E. coli*, and one in 10^9–10^{11} bp in human cells. These mutations provide the mechanism by which natural selection operates. Remarkably, a significant portion of these errors are caused by misincorporation of the wrong nucleotide when DNA polymerases copy DNA, both during genome duplication and when damaged DNA is repaired.

The fidelity with which DNA synthesis occurs depends both on the ability of the polymerase to incorporate the correct nucleotide, and on the presence or absence of a 3′→5′ exonuclease that can excise misincorporated nucleotides. A comparison of the equilibrium dissociation constants for insertion of dCTP opposite dGMP reveals that the binding affinity for dCTP (13±5 μM, range 2–55 μM) varies little among DNA polymerases, whereas the intrinsic rate constant for nucleotide insertion varies over 10^5-fold. Thus, low-fidelity polymerases bind the correct nucleotide with high affinity, but insert it slowly. Mutation rates for bacterial DNA polymerases that lack associated exonuclease activity vary from 0.1 to 100 mistakes per 100 000 bp replicated (Table 4-3). Variation among human enzymes is even greater (between 10 mistakes and 72 000 mistakes per 100 000 bp replicated). However, these numbers improve markedly with addition of an exonuclease activity that can degrade one strand of duplex DNA beginning at its 3′-OH terminus. The same enzyme with its associated 3′→5′ exonuclease activity has a 4–24-fold lower error rate than in the absence of its proofreading activity (Table 4-3).

In principle, error rates in cellular organisms could be decreased by increasing polymerase fidelity through increased proofreading. However, this comes at the cost of increased nucleotide turnover. For example, a mutation in bacteriophage T4 DNA polymerase that suppresses mutations is temperature-sensitive for growth in *E. coli*, because, at the higher temperature, the increase in exonucleolytic proofreading leads to depletion of the nucleotide precursor pool, thereby reducing the rate of DNA synthesis. However, the limited accuracy of the replicative polymerases may not be the main source for the accumulation of errors as cells divide. In eukarya, translesion synthesis by Pol-ζ in response to spontaneous DNA damage appears to be a leading source of mutations.

Spontaneous mutation rates of *S. cerevisiae* strains lacking Pol-ζ are half those of wild-type strains, suggesting that the function of Pol-ζ during DNA replication is to bypass naturally occurring damage in the cell. Disruption of the Pol-ζ catalytic subunit is tolerated in *Drosophila*, *Arabidopsis*, and in vertebrate cell lines under some conditions, but not during mouse embryonic development. It is likely that the large mammalian genome inevitably accumulates damage from natural causes during the trillions of cell divisions required to produce an adult from a single fertilized egg. Bypass of this damage by Pol-ζ is preferable to the alternative: stalling of the replication machine that leads to cellular **apoptosis**. Given the size of the genome and the number of genome duplications required, this event would occur too frequently to be compatible with animal development.

When the wrong nucleotide is accidentally incorporated by a DNA polymerase, bacterial and eukaryotic cells employ a surveillance system called **mismatch repair** to recognize and repair these errors. The mismatch repair machinery distinguishes the newly synthesized strand from the template strand. In *E. coli*, the two strands are distinguished by the fact that methylated base pairs are separated at replication forks. The parental strand retains the methylation mark whereas the nascent strand must wait for the methylation machinery to recognize a hemimethylated site and restore it to a fully methylated site (Chapter 7). In eukaryotes, the sites in nascent DNA on the lagging-strand template where DNA ligase must act appear to direct mismatch repair to the correct strand, but how nascent DNA on the leading-strand template is recognized is not clear. The fact that the damage-detection and damage-repair systems are as complex as the replication machinery itself highlights the importance evolution has attached to DNA fidelity.

The goal of DNA replication and DNA repair is to achieve a balance between genomic stability and genetic mutation that allows species both to survive and to evolve.

SLIDING CLAMPS

DNA polymerases replicating the leading strand are faced with the daunting task of extending DNA by many thousands or even millions of nucleotides at a stretch. In other words, the enzyme needs to be highly processive; it should synthesize DNA continuously without disengaging from the template. Even lagging-strand DNA synthesis needs to incorporate from 30 to 300 or 400 nucleotides for each Okazaki fragment at eukaryal replication forks, and up to 4000 or 5000 nucleotides for each Okazaki fragment at bacterial replication forks. However, the processivity of the DNA polymerases responsible for initiation of RNA-primed DNA synthesis (Pol-III core in bacteria, and Pol-α in eukarya) is intrinsically low (between 1 and 30 nucleotides per DNA binding event). Thus, either a more processive polymerase is needed for extensive DNA synthesis, or an accessory factor is required to maintain template commitment. This accessory factor is called the sliding clamp. Its importance is highlighted by the fact that cells contain a special clamp loader that is required to attach the sliding clamp to the right spot at the right time. In bacteria, both the sliding clamp (β) and sliding clamp loader (γ-complex) are part of the macromolecular complex called the DNA polymerase III holoenzyme (Table 4-6; Figure 4-21). In eukarya, the sliding clamp loader (RFC) is not tightly bound to the replication machine (Figure 5-8).

Bacteria

In bacteria, the sliding clamp is a homodimer called the β-clamp. Its crystal structure reveals the strikingly elegant way in which the clamp functions. The clamp forms a circular molecule with DNA passing through the 35-Å diameter central cavity (Figure 4-22). Its name comes from the fact that it can slide along dsDNA or an RNA:DNA duplex, but it cannot slide along ssDNA. With this discovery, the question arose as to how the β-clamp is loaded onto duplex DNA far from the ends of a chromosome. This dilemma is particularly problematic in light of the observation that the β-clamp is very stable with a K_d of less than 50 nM, and a half-life on DNA of about

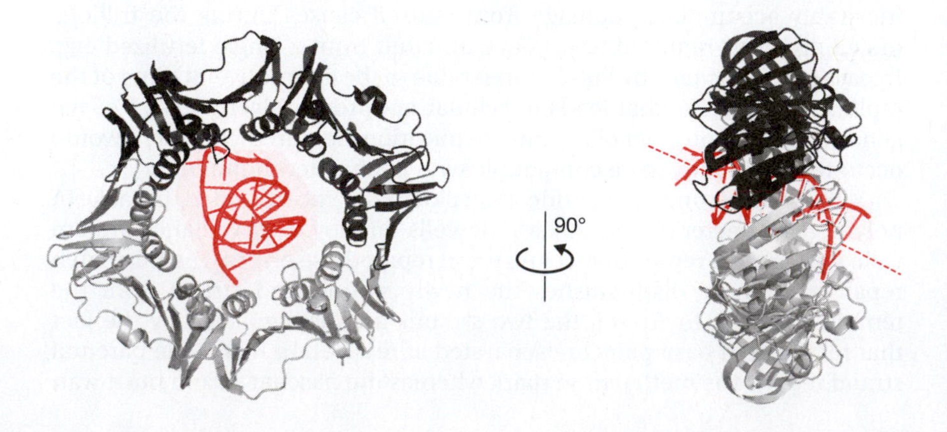

Figure 4-22. The *E. coli* β-clamp consists of two identical subunits (gray and black). An end-on view of DNA passing through the central hole is indicated. The external diameter is approximately 80 Å and the inner hole has a diameter of 35 Å. Note that the DNA is not perpendicular to the clamp, but rather passes through the highly positively charged central cavity at an angle of 22°. Images generated using Pymol with PDB file 3BEP.

100 min at 37°C. Clearly, an active clamp-loading mechanism must exist. In fact, as discussed in Chapter 5, sliding clamp loaders have been identified in all three domains of life.

Archaea And Eukarya

Both archaea and eukarya encode a protein called proliferating cell nuclear antigen (PCNA) that functions as a sliding clamp. However, while PCNA displays a gross morphology very similar to the β-clamp, its subunit composition is quite different. PCNA is a trimer (see Figure 4-23 in color plate section) that is composed of three identical subunits in eukarya, but in some archaea, such as *Sulfolobus solfataricus*, it is composed of three different, albeit related, subunits.

All sliding clamps have the capacity to enhance the processivity of their partner polymerases, but sliding clamps also interact with other proteins as well. For example, PCNA interacts with a diverse spectrum of DNA replication and repair factors, including endonucleases, ligases, helicases, and chromatin assembly factors. Thus, sliding clamps can be viewed as assembly platforms for a variety of DNA-based processes. Intriguingly, the various partner proteins have a common mechanism for interacting with the clamp. Typically the client protein will have a short motif, termed the PCNA-interacting-protein (PIP) motif, conforming to the consensus QXX(M/L/I)XX(F/Y)(F/Y). This interacts with a PCNA monomer via a strand that connects the two lobes of a monomer (Figure 4-23), termed the interdomain connector loop (IDCL). Thus, in principal, a PCNA trimer can interact simultaneously with three partner proteins, acting as a tool belt to couple a range of factors to a translocating DNA polymerase. Indeed, in studies of the heterotrimeric *Sulfolobus* PCNA, it was found that the distinct PCNA subunits had preferred interaction partners. Therefore, it is possible that PCNA can direct the assembly of a stereochemically defined array of accessory factors on the ring. An example of this organization is seen during completion of Okazaki-fragment synthesis (Ringmaster; Chapter 5).

DNA TOPOISOMERASES

Changes to DNA topology are an inevitable consequence of unwinding the double helix; removal of helical turns in duplex DNA must be compensated by addition of superhelical turns of the opposite orientation so that the total number of times that one strand of DNA wraps around the other remains constant. Hence, DNA replication results in the accumulation either of positive superhelical turns in the DNA in front (downstream) of a replication fork, or of precatenated intertwines behind (upstream) of a replication fork whose orientation can be either positive or negative, depending on whether or not the replication machine rotates and, if so, in which direction (Figure 3-9). If superhelical turns in front of the fork are not removed (relaxed), then the further DNA unwinding will cease and DNA synthesis at the fork will stop. If precatenated intertwines are not removed before replication terminates, then the two sibling molecules will remain intertwined, thereby preventing segregation of sister chromatids during cell division. Similar topological problems arise from the transcription, recombination, and repair of DNA, as well as from DNA replication and its assembly into chromatin. To resolve these problems, a family of enzymes evolved called topoisomerases.

DNA topoisomerases are nature's solution to 'knotty' problems. They alter DNA topology through addition or subtraction of superhelical turns and **catenated** intertwines. They accomplish this feat by unlinking the two strands of DNA so that one strand can rotate about the other. Topoisomerases are classified into two groups. Type I enzymes are monomeric proteins that transiently break a singel strand of DNA, thereby allowing the unbroken strand to freely rotate about the phosphodiester bond opposite the nick. Type II enzymes are dimeric or tetrameric enzymes that break

Table 4-7. Type IA DNA Topoisomerases

Bacteria	Archaea	Yeasts	Flies	Mammals	Activities		
					Relax supercoils	Add supercoils	Catenate/ decatenate
Topo I					(–) only		
Topo III	Topo III	Topo III	Topo IIIα	Topo IIIa	Hyper (–) only		Nicked or gapped DNA
Topo IIIβ			Topo IIIβ	Topo IIIβ			
Reverse DNA gyrase	Reverse DNA gyrase				(–)	ATP→(+)	

(–) and (+) indicate direction of superhelical turns. In the presence of ATP, reverse DNA gyrase forms positive supercoils, whereas DNA gyrase (Table 4-9) forms negative supercoils. Reverse gyrase is a unique hyperthermophile-specific DNA topoisomerase that induces positive supercoiling. It is a modular enzyme composed of a topoisomerase IA and a helicase domain, which cooperate in the ATP-dependent positive supercoiling reaction. Adapted from Champoux (2001) and Wang (2002).

both strands concurrently, thereby allowing one sibling chromosome to pass through the other. These two categories are further divided into subgroups A and B. Members of the same subfamily are structurally and mechanistically similar, whereas those of different subfamilies are distinct.

Type IA Enzymes

Type IA topoisomerases (Table 4-7) only relax negatively supercoiled DNA, and therefore they cannot relieve topological stress in front of replication forks. The enzyme operates in the following way. First, it binds to a site on duplex DNA and disrupts (melts) five to ten base pairs. Then a single phosphodiester link within one of the two strands within this region is broken and the 5′-end of the cleavage site is covalently linked to a unique site within the enzyme (Figure 4-24). This allows type IA enzymes to pass the uncut strand through the DNA break by an enzyme-bridging mechanism in which the DNA ends that are created at the break are bridged by the topoisomerase, and movements of the enzyme-bound DNA ends relative to each other mediate the opening and closing of a DNA gate (Figure 4-25). The

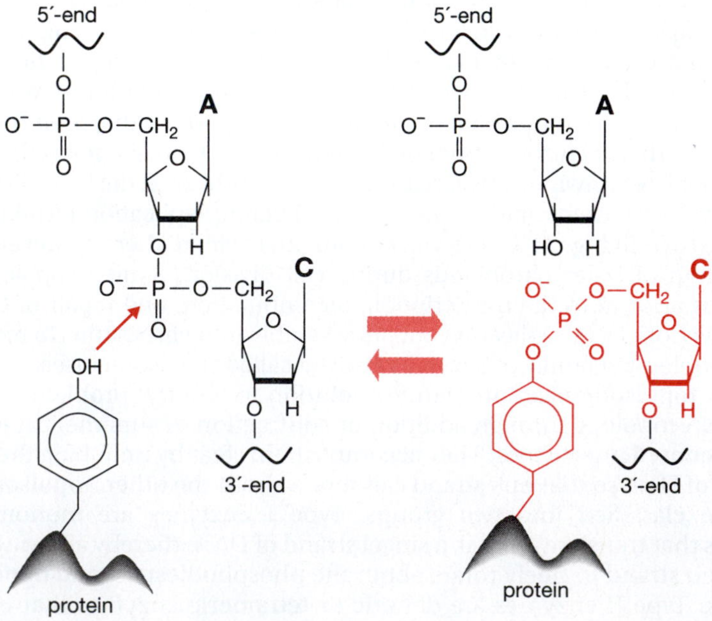

Figure 4-24. Topoisomerases break a phosphodiester bond in the DNA backbone with concomitant formation of a covalent bond between a tyrosine residue and the phosphate. Cleavage is a transesterification reaction in which the target NpN phosphodiester is attacked by a tyrosine of the enzyme. The reaction is reversible. With type IA and type II enzymes, a 3′-OH is the leaving group and the active-site tyrosyl becomes covalently linked to a 5′-PO_4 group, as shown. In the reaction catalyzed by type IB enzymes (not shown), a 5′-OH is the leaving group and the active-site tyrosyl becomes covalently linked to a 3′-PO_4 group.

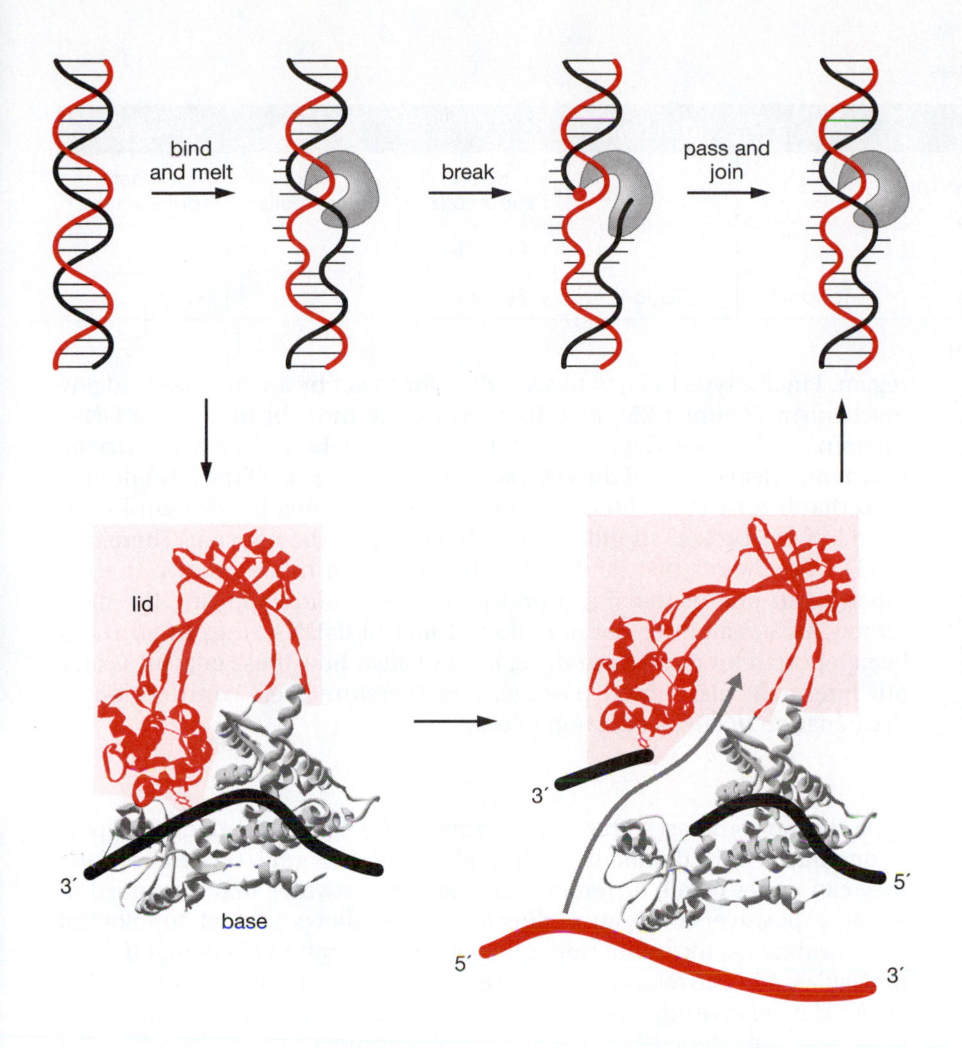

Figure 4-25. Type IA DNA topoisomerases have a characteristic toroidal shape with a central hole large enough to accommodate either ssDNA or dsDNA. The enzyme binds to DNA, inducing localized strand separation, and then introduces a transient break in one strand (black line). The 5′ end of the broken DNA strand is covalently bound to the active-site tyrosyl group (red circle/tyrosine molecule, see Figure 4-24) in the 'lid' of this monomeric enzyme, and the 3′ end is bound noncovalently to the 'base' of the enzyme. This prevents the DNA from rotating about its axis. Lifting the lid opens a gate in the DNA through which the other (non-broken) strand can pass (red line), and then the topoisomerase religates the broken strand. Thus, the DNA linking number changes by increments of one (ΔLk = 1) for each catalytic event. (Adapted from Champoux, J.J. (2002) *Proc. Natl. Acad. Sci. USA* 99: 11998–12000. With permission from the National Academy of Sciences.)

single-strand break is then resealed, and the enzyme then either releases the DNA or begins another catalytic cycle.

Type IA topoisomerases use the same mechanism to catenate or decatenate DNA in two independent molecules. The passed strand can be either ssDNA or dsDNA. The broken strand must be single stranded in the region of the active site, but may be double stranded at all other points. The enzyme appears to introduce a transient break across from the nick or gap. The reaction pathway is reversible, allowing the enzyme to either form or resolve catenated intertwines.

The energy required to pass one strand through the other comes from the negative superhelical turns becoming negative helical turns. The less negatively supercoiled the DNA, the more difficult it is for this reaction to occur. Therefore, the proficiency of the enzyme progressively decreases during the course of relaxing negative superhelical DNA. Since positive superhelical turns make DNA more difficult to unwind at the site where the enzyme binds, type IA enzymes do not relax positively supercoiled DNA.

Type IB Enzymes

Type IB topoisomerases (Table 4-8) relax both negatively and positively supercoiled DNA, allowing them to relieve topological stress in front of replication forks. Type IA and IB enzymes differ in three important ways. Type IA enzymes form a covalent bond with the 5′-end of the DNA cleavage site, whereas type IB enzymes form a covalent bond with the 3′-end of the DNA cleavage site. Type IA enzymes cut within a single-stranded DNA region, whereas type IB enzymes only cut within a double-stranded DNA

Table 4-8. Type IB DNA Topoisomerases

Bacteria	Archaea	Yeasts	Flies	Mammals	Activities		
					Relax supercoils	Add supercoils	Catenate/ decatenate
Topo V					(+) or (–)		
		Topo I	Topo I	Topo I	(+) or (–)		

region. Finally, type IA enzymes are thought to act by an enzyme-bridging mechanism (Figure 4-25), type IB enzymes are thought to act by a DNA-rotation mechanism (Figure 4-26). When a DNA-bound type IB enzyme transiently cleaves one of the DNA strands, only the side of the DNA double helix that is upstream of the nick – the side containing the protein-linked 3′-end of the broken strand – is tightly bound to the enzyme. Therefore, the DNA segments that flank a transient nick can rotate relative to each other about one of the single phosphodiester bonds opposite the nick. Although catenation or decatenation of nicked dsDNA rings *in vitro* has been reported for type IB enzymes, it is not clear how these enzymes carry out intermolecular strand passage, and therefore their contribution to decatenation *in vivo* is questionable.

Type II Enzymes

Type II topoisomerases (Table 4-9) catalyze the ATP-dependent transport of one intact DNA double helix through another (Figure 4-27). Therefore, they can either create or remove catenated intertwines that are coiled in either a positive or negative direction. This allows several topological transformations, including catenation and decatenation of circular dsDNA molecules, and the relaxation of either positively or negatively supercoiled DNA. The relative efficiencies of a given type II enzyme in catalyzing these reactions depend on the structural features of the DNA substrates and the enzyme•DNA complexes. DNA gyrase, for example, is unique in that a 140-bp DNA segment wraps around the enzyme, in a right-handed

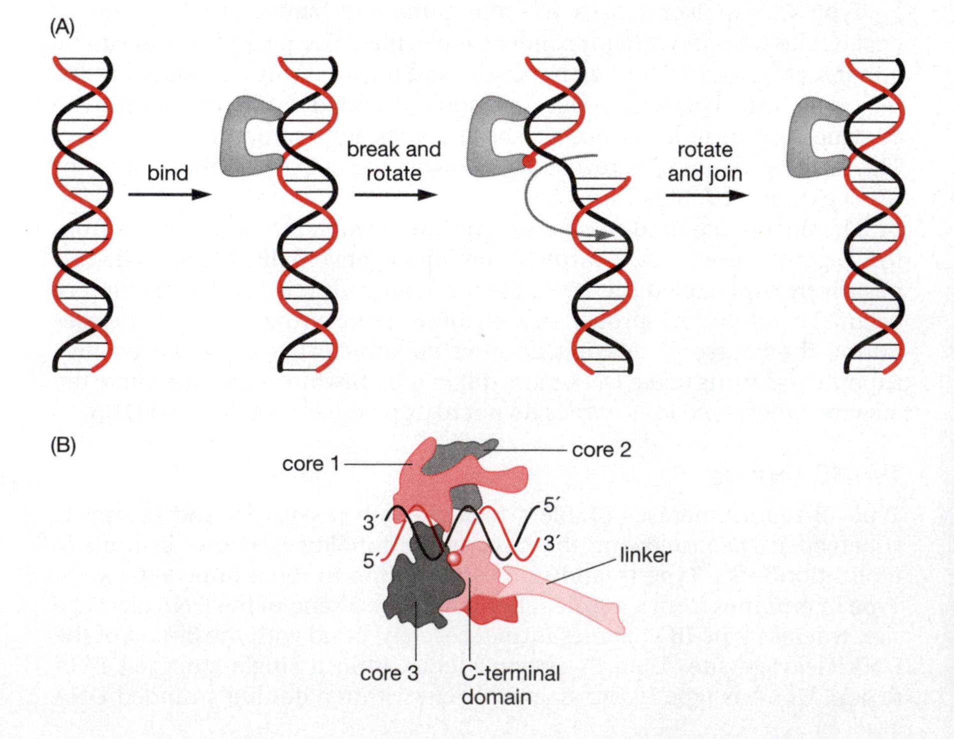

Figure 4-26. In contrast to type IA DNA topoisomerases, type IB enzymes change the linking number by allowing the rotation of one DNA strand around the other. (A) The protein first binds the DNA and cleaves one strand. The 3′-end of the broken DNA strand is covalently linked to the active-site tyrosyl of the enzyme (red circle). However, interaction between the enzyme and the 5′-end of the broken DNA strand is mostly ionic, so it permits rotation of that DNA segment relative to the protein until religation occurs. This allows multiple strand-passage events for each strand breakage and rejoining event. The average change in linking number varies with torque and the type of enzyme. (B) Section through the enzyme revealing how DNA passes through the center of the 'C' structure, thereby making contact with various core subdomains as well as the COOH-terminus of the protein. (Adapted from Wang, J.C. (2002) *Nat. Rev. Mol. Cell Biol.* 3: 430–440. With permission from Macmillan Publishers Ltd.)

Table 4-9. Type II DNA Topoisomerases

Bacteria	Archaea	Yeasts	Flies	Mammals	Activities		
					Relax supercoils	Add supercoils	Catenate/ decatenate
Gyrase	Gyrase				(+); (–)	ATP→(–)	Inefficient
Topo IV					Inefficient, (+)>(–)		Efficient
		Topo II	Topo II	Topo IIα			Efficient
				Topo IIβ			
	Topo VI				(+); (–)		Efficient

DNA gyrase requires ATP to introduce negative supercoils into DNA. Topo VI from archaea can be considered to be a type IIB enzyme (the only one known so far) because it has the same catalytic properties as type IIA enzymes, but differs in its DNA-binding site.

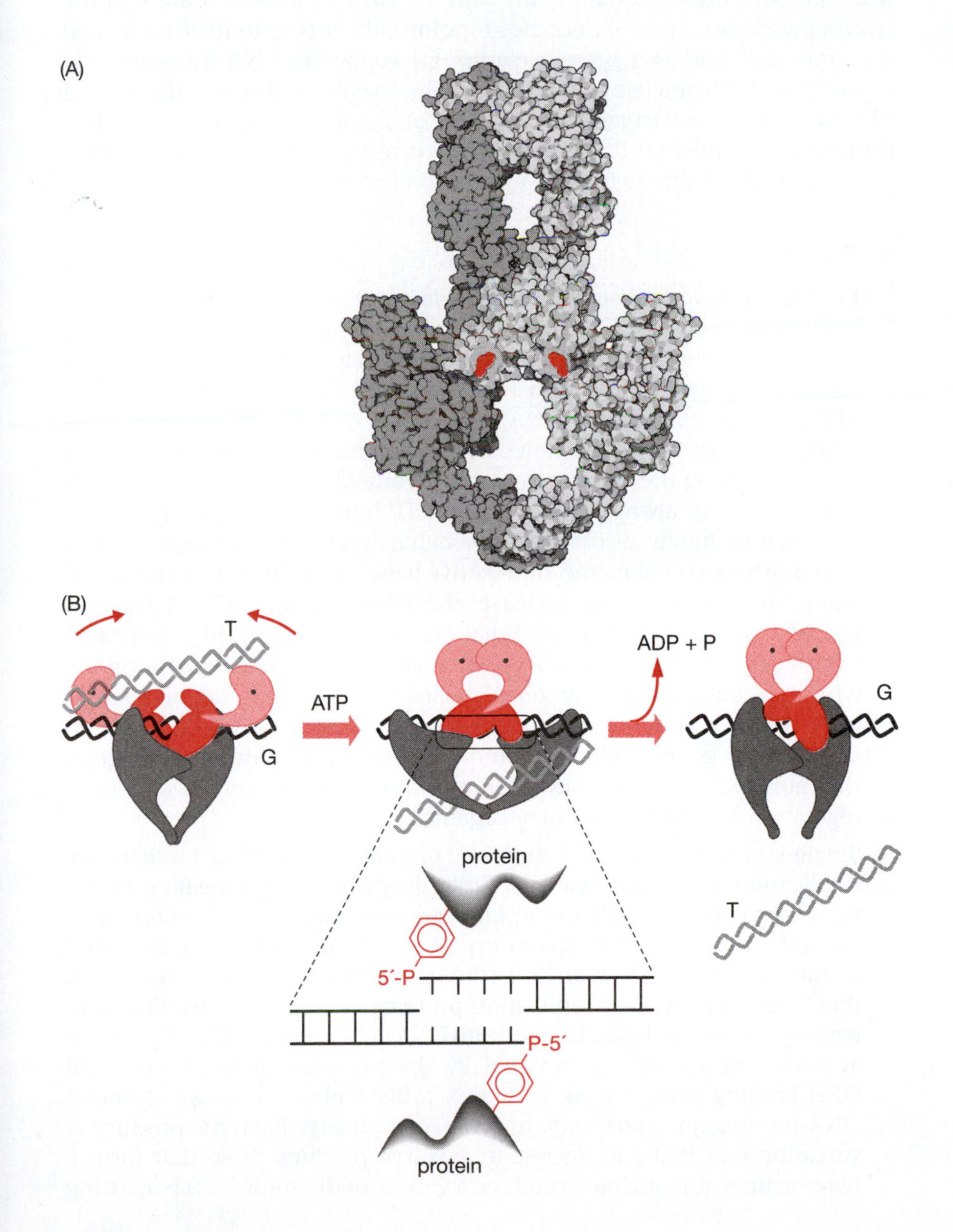

Figure 4-27. Type IIA topoisomerases are multidomain proteins that are required for dsDNA breakage and rejoining. (A) Eukaryotic type II topoisomerases are homodimers (A_2), while prokaryotic type II enzymes are heterodimers (A_2B_2). This space filling structure is modeled after two sets of coordinates, taken from human and yeast crystallographic data. (B) The duplex DNA segment that contains the enzyme-mediated DNA gate (G-segment) is depicted as a dark gray DNA helix. The DNA being passed through the G-segment (T-segment) is depicted as a light gray DNA helix. The opening and closing of the gate is driven by ATP binding and hydrolysis that occurs at the ATP binding site (•) in each molecule. ((A) adapted from Goodsell, D.S. (2002) *The oncologist* 7: 381–382. With permission from AlphaMed Press. (B) adapted from Wang, J.C. (2002) *Nat. Rev. Mol. Cell Biol.* 3: 430–440. With permission from Macmillan Publishers Ltd.)

orientation, to position the DNA segment to be transported (T-segment) and the DNA segment to be transiently cleaved (the gate- or G-segment) in a particular way. This right-handed wrapping of DNA around the enzyme is closely related to its preferential relaxation of positively supercoiled DNA, as well as its ability to introduce negative supercoils into a relaxed DNA ring or loop. During DNA replication, DNA Topo IV in bacteria and Topo II in eukarya are primarily responsible for resolving any precatenanes that form behind replication forks, because temperature-sensitive mutations in these enzymes produce cells with topologically intertwined chromosomes at the restrictive temperature.

The mechanism by which topoisomerase II passes one dsDNA molecule through another can be envisioned as six discrete steps, as follows. (1) Topo II initiates its catalytic cycle by binding to DNA. (2) In the presence of a divalent cation, Topo II generates a transient double-stranded DNA break and establishes a covalent enzyme–DNA cleavage complex (Figure 4-25). The break produces 4 bp protruding 5′-ends that are complementary and therefore can reanneal to facilitate rejoining of the broken DNA. (3) Upon binding two ATP molecules, the enzyme undergoes a conformational change that triggers passage of a second double helix through the double-stranded DNA break. Concomitant with the structural reorientation of the enzyme, topoisomerase II becomes topologically linked to its nucleic acid substrate and acts as a protein clamp. (4) Following DNA translocation, Topo II reseals the nucleic acid break. (5) The enzyme hydrolyzes the second ATP molecule, which triggers the opening of the protein clamp and confers the potential to release DNA. (6) Topo II returns to its original conformation and regains the ability to initiate a new round of catalysis.

SUMMARY

- Replicative DNA helicases consist of six protein subunits arranged as a circle with a hole in the center. In some cases, the hole appears too small for dsDNA to pass through, but large enough for ssDNA. In other cases, the helicase can clearly translocate along dsDNA as well as ssDNA. The basic mechanism of DNA unwinding is to thread one template strand through the center hole, and then translocate along this strand while extruding the other strand to the outside. The energy for this is usually, but not always, derived from ATP binding and hydrolysis, and the hexamer functions like a rotary engine drawing the template into its jaws one base pair at a time. Replicative helicases differ in two important ways. Bacterial replicative helicases translocate along the 5′→3′ lagging-strand template, whereas archaeal and eukaryal replicative helicases translocate along the 3′→5′ leading-strand template. Furthermore, while the bacterial and archaeal helicases are composed of a single gene product (homohexamers), the eukaryal helicase is composed of six different gene products (heterohexamer). This difference suggests that eukaryal helicases may have multiple roles or may be subject to regulation not found in simpler organisms.
- Single-strand-specific DNA-binding proteins are critical for genome duplication for several reasons. First, they prevent reannealing as the two complementary DNA templates are unwound at replication forks. Second, they eliminate secondary structure in both templates that would otherwise act as an impediment to DNA synthesis, and third, they interact with other replication proteins to coordinate leading- and lagging-strand synthesis. Hence, both DNA helicase and DNA polymerase activities are strongly stimulated by the presence of single-stranded DNA-binding proteins. As with replicative helicases, single-stranded DNA-binding proteins vary in their complexity. Bacteria produce a single protein that can aggregate, eukarya produce three that form a heterotrimer, and archaea produce a variety of different SSBs consisting

of from one to three different proteins. As with replicative DNA helicases, archaea exhibit features akin to both bacteria and eukarya. Some archaea produce a single SSB and some produce RPA-like complexes. The heterotrimeric RPA produced by eukarya appears to fulfil multiple functions by interacting with a variety of different proteins.

- One or more **replicative DNA polymerases** is encoded by virtually every genome over 40 kbp. Bacteria produce two DNA polymerases that are essential for DNA replication. Pol-III extends the RNA primer and is processive over long stretches of template, and Pol-I excises RNA primers. Archaea produce one or two, depending on the species. Bacterial Pol-III functions as part of a large stable complex of proteins referred to as the holoenzyme. With the addition of DNA helicase, DNA primase, and a second copy of the Pol-III core enzyme a replisome is assembled at replication forks that can coordinate leading- and lagging-strand DNA synthesis with DNA unwinding. Presumably, the same is true for other domains of life, although such complexes have been difficult to demonstrate. Archaeal DNA B polymerases have a special proofreading function that searches for uracil in the template. At high temperatures, deamination of cytosine produces uracil that can be misincorporated in place of thymidine. Eukarya produce three different replicative DNA polymerases for nuclear DNA replication, one for initiating DNA synthesis on RNA primers (Pol-α), one for extending leading-strand DNA (Pol-ε), and one for extending lagging-strand DNA (Pol-δ). They also produce a specific DNA polymerase for mitochondrial DNA replication (Pol-γ). Replicative DNA polymerases, in general, are highly processive and contain a $3'\rightarrow5'$ exonuclease that proofreads what the polymerase has just inserted, a function that greatly reduces the rate of base-pairing errors. Pol-α is the exception. Eukarya also produce at least eight different DNA polymerases that are specifically involved in DNA repair.
- The processivity of DNA polymerases is greatly enhanced by their association with **sliding clamps**. The bacterial sliding clamp is a homodimer, whereas both the archaeal and eukaryal sliding clamps are homotrimers. All three clamps form a circle with a central hole large enough to accommodate dsDNA. Once loaded onto nascent DNA, the appropriate DNA polymerase can attach and processively carry out DNA synthesis.
- **Topoisomerases** release the topological tension generated by DNA unwinding, as well as by a number of other cellular processes such as chromatin assembly, DNA transcription, and DNA recombination. Hence the need for a multitude of topoisomerases. However, DNA replication requires only two types of topoisomerases. One that can relax positive superhelical turns in front of replication forks and one that can relax catenated intertwines as two replication forks approach one another from opposite directions (Chapter 6). These requirements are provided by DNA gyrase and Topo IV in bacteria, and by Topo I and Topo II in eukarya. These enzymes all work by breaking either one or both DNA strands in a reaction that leaves the enzyme covalently attached to one end of the DNA. Reverse DNA gyrase provides a unique function in archaea by introducing positive supercoils that resist DNA melting at high temperatures.

REFERENCES

Additional Reading

Burgers, P.M. (2009) Polymerase dynamics at the eukaryotic DNA replication fork. *J. Biol. Chem.* 284: 4041–4045.

Corbett, K.D. and Berger, J.M. (2004) Structure, molecular mechanisms, and evolutionary relationships in DNA topoisomerases. *Annu. Rev. Biophys. Biomol. Struct.* 33: 95–118.

Fanning, E., Klimovich, V. and Nager, A.R. (2006) A dynamic model for replication protein A (RPA) function in DNA processing pathways. *Nucleic Acids Res.* 34: 4126–4137.

Hubscher, U., Maga, G. and Spadari, S. (2002) Eukaryotic DNA polymerases. *Annu. Rev. Biochem.* 71: 133–163.

Kunkel, T.A. and Burgers, P.M. (2008) Dividing the workload at a eukaryotic replication fork. *Trends Cell Biol.* 18: 521–527.

Modrich, P. (2006) Mechanisms in eukaryotic mismatch repair. *J. Biol. Chem.* 281: 30305–30309.

Moldovan, G.L., Pfander, B. and Jentsch, S. (2007) PCNA, the maestro of the replication fork. *Cell* 129: 665–679.

Wang, J.C. (2002) Cellular roles of DNA topoisomerases: a molecular perspective. *Nat. Rev. Mol. Cell Biol.* 3: 430–440.

Literature Cited

Arana, M.E., Seki, M., Wood, R.D., Rogozin, I.B. and Kunkel, T.A. (2008) Low-fidelity DNA synthesis by human DNA polymerase theta. *Nucleic Acids Res.* 36: 3847–3856.

Bailey, S., Eliason, W.K. and Steitz, T.A. (2007) Structure of hexameric DnaB helicase and its complex with a domain of DnaG primase. *Science* 318: 459–463.

Burgers, P.M. (1991) *Saccharomyces cerevisiae* replication factor C. II. Formation and activity of complexes with the proliferating cell nuclear antigen and with DNA polymerases delta and epsilon. *J. Biol. Chem.* 266: 22698–22706.

Champoux, J.J. (2001) DNA topoisomerases: structure, function, and mechanism. *Annu. Rev. Biochem.* 70: 369–413.

Chilkova, O., Stenlund, P., Isoz, I., Stith, C.M., Grabowski, P., Lundstrom, E.B., Burgers, P.M. and Johansson, E. (2007) The eukaryotic leading and lagging strand DNA polymerases are loaded onto primer-ends via separate mechanisms but have comparable processivity in the presence of PCNA. *Nucleic Acids Res.* 35: 6588–6597.

Kim, S., Dallmann, H.G., McHenry, C.S. and Marians, K.J. (1996) Coupling of a replicative polymerase and helicase: a tau-DnaB interaction mediates rapid replication fork movement. *Cell* 84: 643–650.

McHenry, C.S. (1991) DNA polymerase III holoenzyme. Components, structure, and mechanism of a true replicative complex. *J. Biol. Chem.* 266: 19127–19130.

Rattray, A.J. and Strathern, J.N. (2003) Error-prone DNA polymerases: when making a mistake is the only way to get ahead. *Annu. Rev. Genet.* 37: 31–66.

Syvaoja, J., Suomensaari, S., Nishida, C., Goldsmith, J.S., Chui, G.S., Jain, S. and Linn, S. (1990) DNA polymerases alpha, delta, and epsilon: three distinct enzymes from HeLa cells. *Proc. Natl. Acad. Sci. USA* 87: 6664–6668.

Wang, J.C. (2002) Cellular roles of DNA topoisomerases: a molecular perspective. *Nat. Rev. Mol. Cell Biol.* 3: 430–440.

Wu, C.A., Zechner, E.L. and Marians, K.J. (1992) Coordinated leading- and lagging-strand synthesis at the *Escherichia coli* DNA replication fork. I. Multiple effectors act to modulate Okazaki fragment size. *J. Biol. Chem.* 267: 4030–4044.

Zhong, X., Garg, P., Stith, C.M., Nick McElhinny, S.A., Kissling, G.E., Burgers, P.M. and Kunkel, T.A. (2006) The fidelity of DNA synthesis by yeast DNA polymerase zeta alone and with accessory proteins. *Nucleic Acids Res.* 34: 4731–4742.

Chapter 5

REPLICATION PROTEINS: LAGGING-STRAND SYNTHESIS

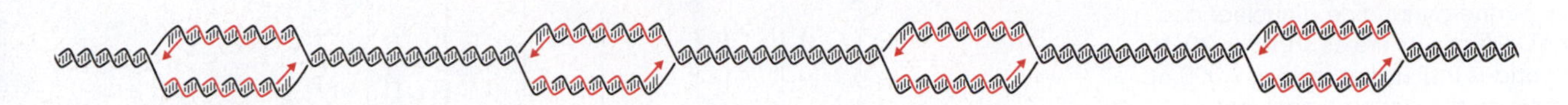

The preceding chapter focused on the proteins already engaged in DNA synthesis at active replication forks: helicases, ssDNA-binding proteins, DNA polymerases, sliding clamps, and topoisomerases. This chapter focuses on structure of RNA-primed nascent DNA and the proteins required to carry out the initiation, elongation, and termination of Okazaki-fragment synthesis on the lagging-strand template. The reason for splitting replication-fork proteins into leading- and lagging-strand events is to emphasize a simple but critical concept that applies to bacteria and eukarya, and most likely to archaea as well: initiation of Okazaki-fragment synthesis at replication forks is the same as initiation of DNA synthesis at replication origins. In other words, DNA synthesis at cellular replication origins begins by synthesizing the first Okazaki fragment on each strand. However, the first Okazaki fragment is extended continuously until it collides with a replication fork coming in the opposite direction. Although the replication-fork mechanism is also utilized by some bacteriophages, bacterial plasmids, and animal viruses, these genomes often employ a different mechanism to begin DNA synthesis at their replication origins than do cellular organisms (Figure 10-1).

RNA-PRIMED NASCENT DNA

DNA polymerases cannot initiate DNA synthesis *de novo*, they simply extend the 3′-ends of oligonucleotides. Therefore, the key to understanding *de novo* initiation of DNA synthesis at replication forks was to discover the nature of the primer. The answer was revealed in the chemistry of Okazaki fragments; they contain a short oligoribonucleotide covalently linked to their 5′-end (Figure 5-1). These RNA-primed nascent DNA chains have been identified in the replicating chromosomes of bacteria, archaea, yeast, and mammalian cells. Ironically, they are most difficult to characterize in simple organisms such as bacteria, because DNA synthesis is so fast that Okazaki fragments are difficult to isolate, and the mechanism for removing **RNA primers** is quite efficient, leaving only a small proportion of Okazaki fragments with only one to three ribonucleotides remaining at their 5′-ends. Success in bacterial systems came with the development of *in vitro* DNA replication systems that utilize purified proteins. Okazaki-fragment synthesis *in vivo* has been characterized most extensively in the replicating chromosomes of SV40 and polyomavirus. These viral chromosomes replicate in the nuclei of mammalian cells, relying completely on their host to provide all of the replication machinery except the DNA helicase, and they replicate about 60 times more slowly than bacterial chromosomes.

Since DNA replication was thought to begin at specific sites in most, if not all, genomes, the question naturally arose as to whether or not RNA primers are synthesized at specific sites on the DNA template. The answer came from several types of analyses. The 5′-terminal ribonucleotide on Okazaki

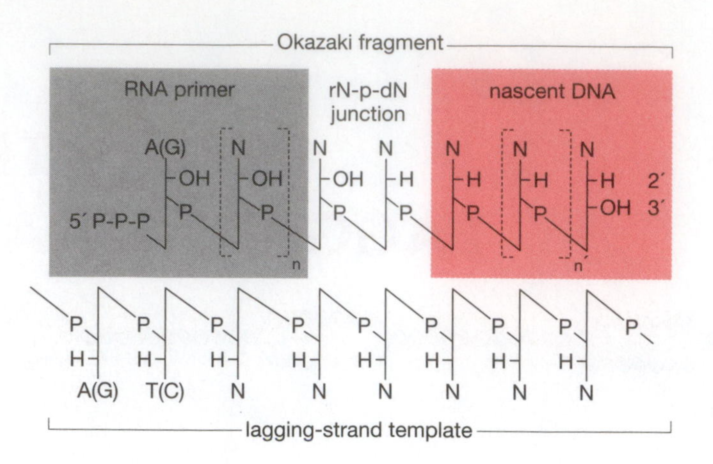

Figure 5-1. Okazaki fragments are short transient 5′-ppp-RNA-p-DNA-OH-3′ oligonucleotides that originate predominantly, if not exclusively, from the lagging-strand template. Their sequence is complementary to that of the template upon which they were made. RNA primers are typically about 10 residues in length and always begin with either ATP or GTP. Initiation sites in bacteria and eukarya begin at purine-pyrimidine dinucleotides. Nascent DNA chains are as short as 20 or 30 nucleotides (nt) and as long as 5000 nt, depending on the genome analyzed.

fragments synthesized *in vivo* was identified by radiolabeling short nascent DNA chains with [α-^{32}P]GTP using an enzyme (guanylyltransferase) that 'caps' the 5′-ends of mRNAs in eukaryotic cells with a single GMP molecule (Figure 5-2A). This reaction occurs only at the 5′-ends of RNA, and only if the 5′-terminal nucleotide begins with a di- or triphosphate (NDP or NTP). The identity of the labeled nucleotide can then be determined by degrading the RNA either with RNA-specific endonucleases or with alkali, agents to which the DNA moiety is resistant. Since only the first ribonucleotide incorporated by an RNA polymerase retains its 5′-triphosphate moiety, the capping assay identifies only the first ribonucleotide synthesized, not 5′-ribonucleotides that result from RNA degradation. Results from such experiments reveal that RNA primers always begin with either ATP or GTP at their 5′-end, a conclusion that is confirmed by the properties of purified DNA primases described below. In the case of mammalian cells, more than 80% of the RNA primers begin with ATP and the remainder with GTP.

To determine whether or not the transition from RNA to DNA occurs at specific sites in the template, advantage was taken of the properties of the rN-p-dN junction (Figure 5-1). The 3′-end of the RNA primer is linked covalently to the 5′-end of the nascent DNA chain via a phosphodiester bond. Given the mechanism by which DNA polymerases extend oligonucleotide primers (Figure 4-16), the phosphate linking the RNA and DNA moieties is the α-phosphate from the first dNTP added to the 3′-OH end of the RNA primer, and the nascent DNA chain ends with a 3′-OH terminal dNMP. Therefore, the rN-p-dN dinucleotide can be identified by a procedure called nearest-neighbor analyses. [α-^{32}P]dNTPs are incorporated into nascent DNA under conditions that allow faithful DNA replication *in vitro*. Okazaki fragments are then isolated and incubated in alkali which causes the 5′-phosphoryl group from the DNA to transfer to either the 2′- or 3′-hydroxyl group on the 3′-terminal ribonucleotide (Figure 5-2B). The resulting mixture of 2′(3′) rNMPs is then converted entirely into the 3′-OH isomer and fractionated by chromatography. If a single [α-^{32}P]dNTP is used to radiolabel nascent DNA, then the frequency at which this 5′-^{32}P appears with each of the four possible rNMPs at rN-p-dN junctions is revealed. The results of such experiments revealed that all 16 possible rN-p-dN junctions exist in Okazaki fragments at the frequencies expected from the nucleotide composition of the template. In other words, the transition from RNA to DNA synthesis does not occur at any preferred DNA sequence, a conclusion again confirmed by the properties of purified DNA primases described below.

Do Okazaki fragments mature to specific lengths? The relative number of Okazaki fragments of different lengths has been determined by labeling the 5′-terminal nucleotide at the rN-p-dN junction that is unique to Okazaki fragments, and then fractionating the labeled DNA fragments either by sucrose gradient sedimentation or by gel electrophoresis. First, an unlabeled phosphate is added enzymatically to the 5′-ends of any DNA chains that may have resulted from DNA fragmentation during the experiment. Then the 5′-ends of Okazaki fragments are 'unmasked' by treating the DNA with alkali to degrade the RNA primers as well as any RNA present. This produces a 5′-terminal hydroxyl group on the Okazaki-fragment DNA moiety (Figure

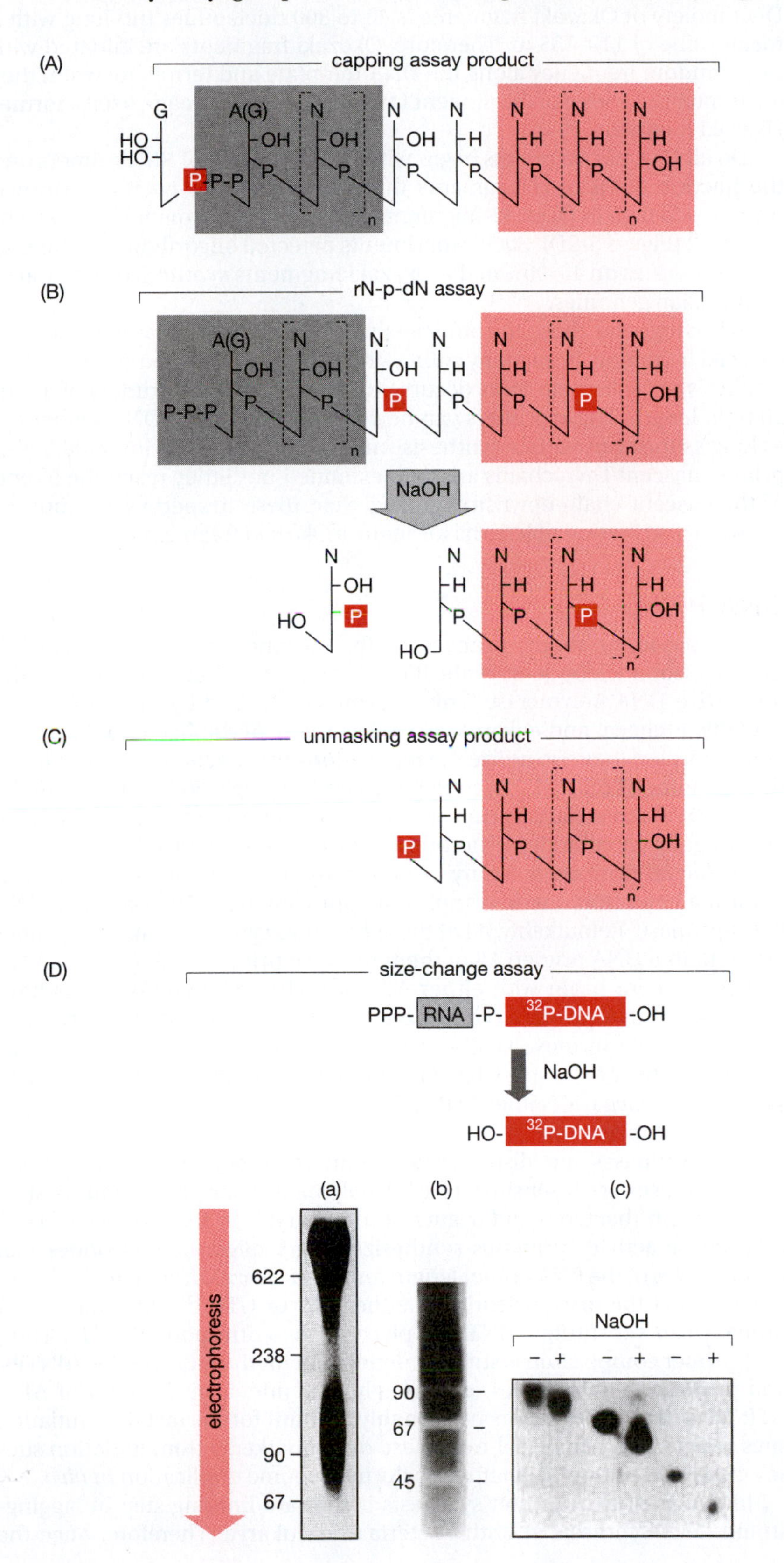

Figure 5-2. Properties of Okazaki fragments. (A) 5′-Terminal pppN nucleotides on Okazaki fragments can be radiolabeled (boxed P means ^{32}P) using an mRNA 'capping' enzyme. (B) The nucleotides present at the rN-p-dN junction can be determined by synthesizing Okazaki fragments in the presence of one [α-^{32}P]dNTP and then transferring the ^{32}P at the rN-p-dN junction from the dNMP to the NMP during alkaline hydrolysis of the RNA primer. (C) 5′-Terminal dNMPs in Okazaki fragments can be radiolabeled by removing the RNA with alkali and then enzymatically adding a phosphate to the 'unmasked' 5′-OH terminal dNMP. (D) The fraction of short nascent DNA chains (^{32}P-DNA) that contain an RNA primer can be determined by the fraction of ^{32}P-DNA fragments that migrate slightly faster during gel electrophoresis following RNA degradation by alkaline hydrolysis. Three segments were cut from a gel that corresponded to oligonucleotides of 45, 67, or 90 residues in length. One aliquot was incubated with NaOH (+) and one was not (–); both were then fractionated by alkaline gel electrophoresis, and ^{32}P-DNA chains detected by autoradiography. The difference in ^{32}P-DNA size corresponded to loss of RNA primers of approximately 10 nt. (Adapted from DePamphilis, M.L. (1995) *Methods Enzymol.* 262: 628–669. With permission from Elsevier. Additional data from Ogawa, T. and Okazaki, T. (1980) *Annu. Rev. Biochem.* 49: 421–457.)

5-2C). The 5′-terminal nucleotide of the unmasked nascent DNA chains can then be radiolabeled enzymatically with ^{32}P. Each nascent DNA fragment then contains the same amount of radioactivity, regardless of its length. Such experiments revealed that Okazaki fragments are not of unique length, but a continuum of lengths. In mammalian nuclei, for example, the DNA moiety of Okazaki fragments is 40 to 300 nucleotides (nt) long with a mean value of 110–135 nt. Therefore, Okazaki fragments are initiated with near-random frequency along the DNA template and terminate when they encounter the 5′-end of the nascent DNA strand downstream, itself a former Okazaki fragment.

Do all Okazaki fragments begin with RNA? The size of RNA primers and the fraction of Okazaki fragments that carry them has been determined from the change in Okazaki-fragment length upon treatment with alkali or RNase T2 (Figure 5-2D). Such experiments detected oligoribonucleotides of 6 to 12 residues on 40–90% of the Okazaki fragments synthesized on viral or mammalian genomes.

Taken together, these and other analyses of the chemistry and structure of Okazaki fragments from many sources reveal that *de novo* initiation of DNA synthesis at replication forks occurs on the ends of RNA primers of about 10 nt in length. These primers can be synthesized at most DNA sequences as long as they can initiate synthesis with ATP or GTP. Once initiated, RNA-primed nascent DNA chains are then extended until they reach the 5′-end of the nascent chain downstream. Together, these properties account for the wide distribution of lengths for mature Okazaki fragments.

DNA PRIMASE

DNA primase (or simply primase) is the enzyme required to synthesize the oligoribonucleotide at replication forks that is then extended by the replicative DNA polymerase. This enzyme is encoded by the genomes of bacteria, archaea, and eukarya, as well as many of the bacteriophage and animal viruses that also utilize the replication-fork mechanism to duplicate their genome (Table 5-1). Bacterial primase is a single 66 kDa polypeptide, but archaeal and eukaryal primase is a heterodimer comprised of a catalytic and a regulatory subunit with a combined size of approximately 70–110 kDa. The eukaryal primase is tightly bound to the two subunits of DNA Pol-α to form a heterotetrameric complex of approximately 350 kDa called DNA Pol-α•primase. Remarkably, all of these enzymes synthesize an RNA primer rather than a DNA primer. All of them produce primers of 3 to 15 residues, and all of them begin with either ATP or GTP. Where comparisons have been made, RNA primers synthesized *in vitro* by purified DNA primases are comparable to those synthesized *in vivo* by the replication machine. Primases appear to be specialized RNA polymerases (all which initiate RNA synthesis *de novo* with either ATP or GTP) that evolved to fill a unique niche in genome duplication.

DNA primases are distinguished from RNA polymerases in several ways. Primases are insensitive to RNA polymerase-specific inhibitors such as rifampicin (bacteria) and α-amanitin (eukarya). In the absence of DNA polymerase activity, primases synthesize longer oligoribonucleotides that are multiples of the RNA primer length found on Okazaki fragments. Primase requires that the first nucleotide is either ATP or GTP, but thereafter DNA primase can substitute a dNTP in place of the corresponding NTP. Thus, RNA primer composition is strongly dependent on the ratio of NTPs/dNTPs, and primase-initiation sites are strongly dependent on the ratio of ATP/GTP. These and other factors presumably account for the fact that initiation sites selected *in vitro* by Pol-α•primase differ markedly from initiation sites selected on the same DNA template during genome duplication *in vivo*.

Initiation of RNA primer synthesis is the rate-limiting step in lagging-strand DNA synthesis in both bacteria and eukarya. Therefore, once the

Table 5-1. RNA Primers Synthesized By DNA Primase

Genome	Genes	RNA primer		
		Composition	Mean length	Start site[a]
Purified primases from genomes that replicate in bacteria				
E. coli	DnaG	pppAGN$_{8-10}$	10–12	3′-GTC
Phage T7	gp4	pppACN$_{2-3}$	4–5	3′-CTG
Phage T4	gp61	pppACN$_{2-3}$	4–5	3′-TTG
Plasmid ColE2	rep	pppAGA	2–3	3′-GTCT
Purified primases from genomes that replicate in eukarya				
Budding yeast	Pri1, Pri2	ppp(A/G)N$_{7-9}$	8–10	
Fruit fly	Pri1, Pri2	ppp(A/G)N$_{7-14}$	8–15	
Murine	Prim1, Prim2	ppp(A/G)N$_{8-10}$	9–11	
Simian	Pol-α•primase	ppp(A/G)N$_{1-8}$	6–8	3′-CTTT
Human	Prim1, Prim2	ppp(A/G)N$_{6-13}$	11–14	
HSV	UL52	pppGG(N)$_{8-10}$	10–12	3′-AGCCCTCCCA
RNA primers synthesized *in vivo*				
SV40 in simian cells	Pol-α•primase	pppAN$_{5-8}$	6–9	3′-(A/G)T(T/C)

[a]The first one or two nucleotides in the primer-recognition site (underlined) are not copied into the primer.

primer is synthesized, the Okazaki fragment is rapidly completed and joined to the long growing nascent DNA strand. DNA primase binds ssDNA (K_d<100 nM) and then slides along the DNA in order to find a place to initiate primer synthesis. The rate-determining step for primer synthesis occurs at or prior to the first phosphodiester bond formed. Reported frequencies, however, vary from 0.001 to 0.2 events s^{-1}. The rate at which primase synthesizes RNA is about 0.0002 kb min^{-1}, far slower than Pol-α synthesizes DNA (Table 4-5). Clearly, the apparent rate of RNA primer synthesis depends strongly on experimental conditions. Nevertheless, throughout both *in vivo* and *in vitro* experiments in both bacteria and eukarya, as well as with bacteriophage and animal virus enzymes, the rate of Okazaki-fragment synthesis is limited by the rate of RNA primer synthesis.

To accommodate the needs of repeated initiation events on the lagging-strand template, primases must be promiscuous in their selection of initiation sites. Given the sizes of genomes (Table 1-3), *E. coli* requires approximately 5000 primase initiation events and human cells require approximately 30 000 000. Such demands require a low degree of sequence specificity, not only to provide flexibility in where initiation events can occur, but to avoid excess demands on the genetic requirements for evolution of species. Isolated DNA primases initiate synthesis of short oligoribonucleotides on ssDNA templates when provided with all four NTPs. Specific recognition sites have been identified for primases encoded by bacteria, bacteriophages, bacterial plasmids, and large animal viruses, but not for eukaryotic primases. When coupled with Pol-α, primase strongly prefers pyrimidine-rich sequences. However, start sites mapped in SV40 DNA replicating in simian cells were not recognized by the host Pol-α•primase, revealing that the state of DNA affects initiation-site selection by Pol-α•primase. For example, *in vitro*, bare ssDNA templates can adopt various secondary structures that are restricted *in vivo* by chromatin and the small target size (≤300 nt) of the Okazaki-fragment initiation zone on lagging-strand templates at replication forks.

How is primase promiscuity attained at the molecular level? The crystal structures of both the bacterial primase, DnaG, and the archaeal primase have been solved. Although clearly unrelated to one another, both enzymes share the feature of having active sites that are exposed on the surface of the enzyme and, thus, easily accessible by DNA; quite distinct from the deep channels and grooves found in other DNA and RNA polymerases. This agrees nicely with the facts that primases have broad substrate specificities and must disengage from the template-DNA strand after the RNA primer has been synthesized.

INITIATION OF OKAZAKI-FRAGMENT SYNTHESIS

The single most striking difference between Okazaki fragments synthesized by bacterium- and bacteriophage-encoded enzymes and those produced by archaeal and eukaryal enzymes is their mean length, which differs by about 10-fold; Okazaki fragments in bacteria, bacteriophages, and bacterial plasmids are generally 1–2 kb long whereas those produced in archaea and eukarya are generally 0.1–0.2 kb long. A likely explanation lies in the fact that bacterial replication forks are 20–60 times faster that those in archaea and eukaryotes (Table 3-4). Since initiation of Okazaki fragments is the rate-limiting step in lagging-strand synthesis, the faster the fork advances, the greater the amount of lagging-strand template that is produced per unit time, and therefore the distance between primase-initiation events will increase, with the result that the mean length of Okazaki fragments will increase.

A second characteristic of Okazaki fragments is their broad distribution of lengths about the mean, regardless of the genome from which they originate. In bacteria, the range is from 50 to 4000 nt, and in archaea and eukarya it is from 30 to 300 nt. This variation represents mature Okazaki fragments (those waiting for RNA primer excision and DNA ligation), because the same length distribution exists when either DNA Pol-I or DNA ligase is inactivated in *E. coli*, and when SV40 replicating DNA is treated with various DNA polymerases to complete the synthesis of Okazaki fragments. Therefore, Okazaki-fragment synthesis is not initiated at some predetermined distance from the 5′-end of the downstream nascent DNA strand, but at many different, stochastically chosen, sites upstream of the long nascent DNA strand on the lagging-strand template. This generates a population of short RNA-primed nascent DNA chains of widely varying lengths, consistent with the weak template specificity exhibited by DNA primase. Thus, the broad distribution of Okazaki-fragment lengths is determined by the frequency at which Okazaki fragments are initiated.

Some studies of Okazaki fragments synthesized during budding yeast, fly, and papovavirus DNA replication have detected the presence of discrete intermediate sizes of (p)ppRNA-DNA chains of approximately 30, 60, 125, and 240 nt in length, suggesting that Okazaki fragments are synthesized in a discontinuous manner by joining together smaller precursors of approximately 30 nt. However, the distribution of these intermediates in yeast is not altered by inactivating DNA ligase I, suggesting that these intermediates represent natural pauses during the extension of RNA primers that are observed under some conditions, but not under others.

The frequency of Okazaki-fragment initiation is determined by two parameters: the length of the lagging-strand template and the frequency of RNA primer synthesis. The fact that Okazaki fragments are not uniform in length means that one or both of these parameters varies from one replicating DNA molecule to another. The length of template is determined by the rate of DNA unwinding, which in turn is limited both by the intrinsic speed of the replicative DNA helicase and by the structure of the DNA•protein complex through which it travels. *In vivo*, DNA exists as a nucleoprotein complex called chromatin, the primary functions of which are to protect DNA from chemical and enzymatic attack and to condense the DNA into a size small enough to be stored in the cell. Although bacterial chromatin does not appear to impose a significant obstacle to the bacterial replicative helicase (DnaB), eukaryal chromatin is another matter. Chromatin in these organisms is organized into repeated units called nucleosomes consisting of histone octamers around which is wrapped the DNA. The center-to-center distance from one nucleosome to the next is 160–230 bp. Since this distance is strikingly similar to the mean size of Okazaki fragments in archaea and eukarya, it has been proposed that the length of the lagging-strand template may be restricted by the ability of these cells to disassemble nucleosomes. In fact, ssDNA gaps at replication forks in SV40 and in yeast genomes are approximately 220 nt in length. However, nucleosomes are only found in a subset of archaea, and even those, such as *Sulfolobus*, that have an entirely distinct chromatin packaging system have Okazaki fragments of 30–300 nt in length. It appears unlikely therefore that chromatin architecture directly influences Okazaki-fragment size.

The frequency of RNA primer synthesis on a DNA template is determined by the concentration of the replication factors involved. Studies with purified DNA and purified *E. coli* replication proteins reveal that the length of Okazaki fragments can vary from approximately 0.5 to more than 7 kb depending on the relative concentrations of DNA primase (DnaG), DNA helicase (DnaB), DNA polymerase (Pol-III holoenzyme), and their nucleotide substrates. As the ratio of DNA primase to DNA helicase is increased, the sizes of Okazaki fragments are decreased, because the frequency of initiation events per unit length of template is increased. In bacteria, DnaG is concentrated at replication forks by binding to DnaB. DnaG mutants that have an altered interaction with DnaB produce Okazaki fragments of altered length, even though their capacity to synthesize RNA primers and their ability to interact with the Pol-III holoenzyme remain unaffected. These and other experiments demonstrate that the ability of DNA primase to associate with DNA helicase at replication forks determines the frequency of initiation events on the lagging-strand template.

The numerical ratio of long nascent DNA chains to Okazaki fragments in SV40 replicating DNA is less than one. Therefore, these replication forks contain, on average, less than one Okazaki fragment per fork. In fact, the frequency of replication forks with ssDNA on one arm and dsDNA on the other is quite low, as judged by electron microscopic analysis of replicating DNA from various organisms. Moreover, the ssDNA gaps at replication forks in the chromosomes of both SV40 and yeast are the same size as the maximum length of Okazaki fragments (200–300 nt). This observation implies that there is never more than one Okazaki fragment per fork. Taken together, these observations reveal that initiation of Okazaki-fragment synthesis is the rate-limiting step in lagging-strand DNA synthesis. Once this occurs, the fragment is rapidly extended and ligated to the long nascent DNA strand, thereby removing it from the population faster than new Okazaki fragments can be initiated. Nevertheless, replication forks are capable of initiating replication more than once on the lagging-strand template, particularly in bacterial systems where more than two replicative DNA polymerases can exist (discussed below) or in the presence of elevated primase levels.

THE HELICASE-PRIMASE PARADOX

The intrinsic properties of helicases and primases are paradoxical. Regardless of which arm of the fork is encircled by the replicative DNA helicase (Figure 4-1), it always travels in the direction of fork movement. On the other hand, DNA primase (like DNA polymerase) can only polymerize nucleotides in the 5′→3′ direction. Therefore, on the lagging-strand template, primase is always traveling in the direction opposite to that of the helicase. If these two enzymes operate independently of one another, then DNA unwinding would quickly run far ahead of initiation of Okazaki-fragment synthesis, because DNA primase action is the rate-limiting step in Okazaki-fragment metabolism. Both bacteria and archaea/eukarya resolved this paradox by linking together the actions of these two proteins.

Bacteria

In bacteria, structural studies reveal that three molecules of the DnaG primase are bound by a single hexamer of the DnaB helicase (see Figure 5-3 in color plate section). Interestingly, binding of the DnaG molecules stabilizes the hexameric structure of DnaB, thereby contributing to the processivity of the helicase. Moreover, the catalytic sites of DnaG primase are positioned directly above the central channel of DnaB helicase, thus guiding the ssDNA produced by the DNA helicase to the active site on the DNA primase. This, therefore, provides a very efficient mechanism for unwinding DNA with concomitant initiation of RNA primer synthesis.

Some bacterial viruses, such as *E. coli* bacteriophage T7, have taken the coupling of primase and helicase one step further by fusing together the primase and helicase domains into a single polypeptide. Thus, the full-length T7 gp4 protein is actually an ortholog of both DnaG and DnaB; it has a DnaG-like primase domain at its N-terminus and a DnaB-like helicase domain at its C-terminus. However, here again a paradox arises. T7 gp4 helicase translocates along the lagging-strand template in the 5′→3′ direction. However, since all DNA primases synthesize RNA in a 5′→3′ direction, they must translocate along the lagging-strand template in the 3′→5′ direction. Thus, it would appear that T7 gp4 will be torn in two as its helicase and primase domains try to move in opposite directions along the same DNA strand! Of course, this doesn't happen. Rather, the act of primer synthesis forces a transient stalling of the helicase activity while the primer is synthesized. The same constraints apply to DnaB and DnaG at bacterial replication forks.

Based on the *E. coli* and bacteriophage T7 models, the following sequence of events occurs at replication forks unwound by a 5′→3′ helicase. A stretch of lagging-strand template in the order of 50–4000 nt is exposed by helicase action and presented to the primase. The helicase pauses while primase synthesizes a short oligoribonucleotide on this template in the direction opposite to that in which the helicase travels. This results in a loop of DNA generated between the helicase and the primase (Figure 3-10B). Primase is displaced from the primer:template by the replicative DNA polymerase [Pol-III in the case of *E. coli*, T7 Pol (T7 gp5) in the case of bacteriophage T7] that uses the short oligoribonucleotide as an RNA primer on which to initiate DNA synthesis, and then the replicative DNA helicase resumes DNA unwinding.

Archaea And Eukarya

By analogy with the bacterial DnaB•DnaG interaction, a direct interaction between DNA helicase and DNA primase might be anticipated in archaea and in eukarya. However, there is a key and fundamental difference between bacterial and both archaeal and eukaryal replicative DNA helicases: they travel in opposite directions. DnaB and T7 gp4 are 5′→3′ helicases, but Mcm and Mcm(2-7) are both 3′→5′ helicases (Figure 4-1 and Table 4-1). Therefore, MCM helicases translocate along the leading-strand template rather than the lagging-strand template. Indeed, no clear evidence for a direct interaction between helicase and primase exists in either archaea or eukarya. These organisms have resolved the helicase-primase paradox through the introduction of a unique group of proteins called GINS (which stands for Go, Ichi, Ni, and San, the numbers five, one, two, and three in Japanese).

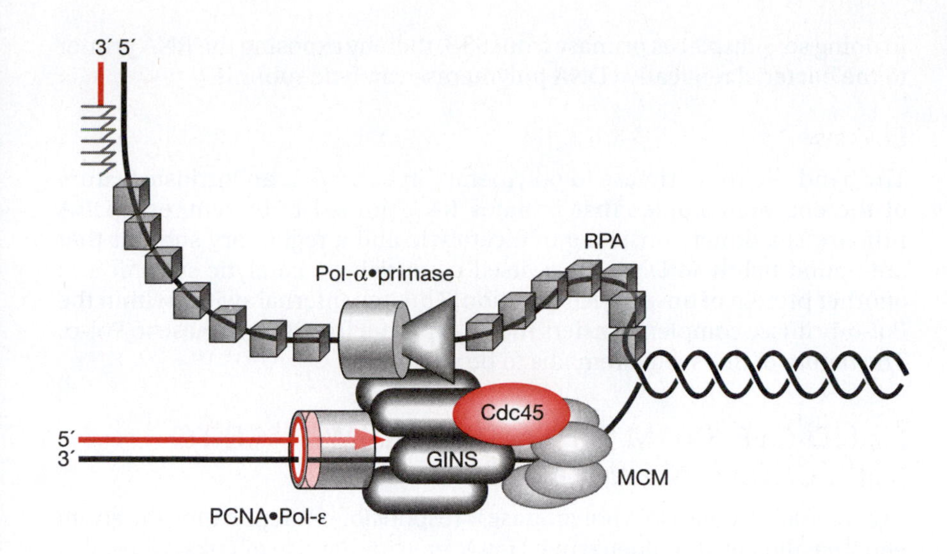

Figure 5-4. CMG helicase complex (Cdc45•MCM•GINS) appears to coordinate initiation events on the lagging-strand template by physically linking together the lagging-strand DNA polymerase (Pol-α, gray cylinder) and the leading-strand DNA polymerase (PCNA•Pol-ε) with DNA primase (trapezoid) and the MCM helicase. This allows the MCM helicase to translocate 3′→5′ on the leading-strand template, while simultaneously presenting the lagging-strand template to DNA primase (gray trapezoid).

GINS is a complex of four paralogous proteins [Sld5, Psf1 (Partner of Sld5), Psf2, and Psf3]. The four subunits are related to one another, indicative of their derivation from a common ancestor, and present-day archaea have simplified GINS complexes that are either homotetramers or dimers of heterodimers. GINS interacts directly with Mcm in archaea and is a component of a eukaryotic protein complex containing Cdc45, Mcm(2-7), and GINS (dubbed the CMG complex). The CMG complex exhibits a more robust 3′→5′ helicase activity than the isolated Mcm(2-7), suggesting that this 11-protein complex is the true replicative helicase in eukarya. Furthermore, although no direct MCM•primase interactions have been detected, both human and archaeal GINS have been shown to physically and functionally interact with their cognate primases. Thus, it is likely that GINS provides the linkage between the MCM helicase and DNA synthesis on the leading- and lagging-strand templates in archaea and eukarya (Figure 5-4).

The evolution of the 3′→5′ helicase in archaea and eukarya solved the dilemma faced by bacteria. The primase is presented to the opposite arm of the fork, the lagging-strand template, by the GINS complex. This allows the helicase to unwind DNA continuously, because the primase can now pull the lagging-strand template toward itself as it synthesizes an RNA primer. The result is a ssDNA loop between the RNA primer site and the replication fork. With the addition of the replicative DNA polymerase, we have the trombone model (Figure 3-10B) that had been previously suggested for replication forks in the chromosomes of bacteria and bacteriophages.

HAND-OFF FROM PRIMASE TO POLYMERASE

Bacteria

Once an RNA primer has been synthesized, primase must disengage so that a polymerase can extend the primer, and the primase can continue to travel with the helicase in order to initiate a new RNA primer further upstream on the lagging-strand template. However, studies of *E. coli* DnaG primase yielded the startling observation that it remains tightly bound to the primer. While this property may serve both to stabilize the short RNA:DNA duplex and to prevent its degradation by a nuclease, it inhibits displacement of the primase by the replicative DNA polymerase. Nature solved this problem with SSB, a protein that is required for primase to maintain its grip on the primer. One subunit of the *E. coli* Pol-III holoenzyme (Table 4-6 and Figure 4-21), the χ-subunit, competes with DnaG primase for binding to SSB, and, in doing so, χ displaces primase from SSB, thereby exposing the RNA primer to the bacterial replicative DNA polymerase catalytic subunit.

Eukarya

The hand-off from primase to polymerase in eukarya is an intrinsic feature of the enzyme complex that initiates RNA-primed DNA synthesis. DNA primase is a dimer consisting of a catalytic and a regulatory subunit that are bound tightly to DNA Pol-α, itself comprising a catalytic subunit and another protein of unspecified function. Thus, an internal switch within the Pol-α•primase complex transfers the RNA primer from the primase to Pol-α. The nature of this switch remains to be elucidated.

HAND-OFF FROM INITIATOR POLYMERASE TO REPLICATOR POLYMERASE

In bacteria, only one DNA polymerase is responsible for replicating the entire genome, the Pol-III holoenzyme. However, with the rise of eukaryotes, this job was distributed among three different DNA polymerases, Pol-α•primase, PCNA•Pol-δ, and PCNA•Pol-ε. Since only Pol-α•primase can initiate DNA synthesis *de novo*, a transition must occur on both leading- and lagging-strand templates where Pol-α•primase is replaced either by PCNA•Pol-δ or by PCNA•Pol-ε. This hand-off from one eukaryotic polymerase to another is facilitated by two events: one that promotes the release of Pol-α from the primer:template, and one that promotes binding of either Pol-δ or Pol-ε to the primer:template.

Mammalian Pol-α can utilize RNA primers about 30 times faster than primase can synthesize them. Once DNA synthesis begins, Pol-α processively extends the RNA primer by approximately 10 dNMPs, and then Pol-α begins to dissociate from the primer:template as it incorporates the next 10 dNMPs, thereby allowing RNA-p-DNA primers to become accessible to other DNA polymerases. This transition is marked by a change in Pol-α from a form that is insensitive to aphidicolin to one that is sensitive. Aphidicolin is a tetracyclic diterpene that selectively inhibits replicative DNA polymerases by competing primarily with dCTP incorporation. At some point in the synthesis of RNA-p-DNA primers, Pol-α•primase apparently changes conformation so that the dCTP-binding site in Pol-α becomes accessible to aphidicolin. This change may occur when the structure of the primer:template duplex changes from the A-form characteristic of RNA:DNA duplexes to the B-form characteristic of DNA duplexes (Table 1-2). This transition from an A-form to a B-form duplex may also trigger the hand-off from Pol-α either to Pol-δ on the lagging-strand template, or to Pol-ε on the leading-strand template by stimulating release of Pol-α from the primer:template.

Stimulating this transition further is loading the eukaryotic sliding clamp (PCNA) onto the RNA-p-DNA primer as soon as the primer is long enough. Since the dimensions of the doughnut-shaped sliding clamp are similar from bacteria to human, properties of the *E. coli* β-subunit can be extended to human PCNA and Pol-δ. The *E. coli* γ-complex (the RFC homolog) alone can assemble the β-clamp (the PCNA homolog) onto a primer with a length of 10–16 nt annealed to a DNA template, depending on steric hindrance from secondary structure. The Pol-III holoenzyme requires 22 nt in order to assemble the β-clamp onto DNA at replication forks. By analogy, the 20–40 nt-long RNA-p-DNA primers synthesized by Pol-α•primase at replication forks should be sufficient for loading the PCNA sliding clamp. Once the sliding clamp is in place, the affinity of either Pol-δ or Pol-ε for this PCNA•DNA primer:template complex is greatly increased. Since Pol-ε is part of the pre-initiation complex (Table 11-4 and Figure 11-15) at replication origins, Pol-ε will be loaded onto the first

RNA-primed nascent DNA strand synthesized at replication origins, and thus takes over leading-strand DNA synthesis. Apparently, Pol-δ is recruited to active replication forks where it takes over lagging-strand DNA synthesis (i.e. Okazaki-fragment synthesis). The high processivity of PCNA•Pol-δ and PCNA•Pol-ε complexes prevents further exchange with another polymerase until DNA synthesis is completed.

SLIDING CLAMP LOADERS

Sliding clamps play a truly central role in the genome-duplication process for all three domains of life, and therefore it is not surprising that all three domains produce a special five-protein complex to load the sliding clamp at internal sites along the chromosome (Table 3-2). The mechanism by which these sliding clamp loaders operate has been revealed from a combination of biochemical and structural studies.

Bacteria

The β-clamp can only be loaded onto dsDNA or RNA:DNA duplexes that contain a 3′-OH terminated primer, thereby targeting β-clamp assembly to the 3′-ends of Okazaki fragments, leading-strand synthesis, and sites where DNA repair is needed. In *E. coli*, the clamp loader is part of the DNA Pol-III holoenzyme where it resides within a seven-subunit complex with the composition of $\gamma\tau_2\delta\delta'\chi\psi$ (Table 4-6). The χ- and ψ-subunits are not essential for clamp loading. Rather they play roles in stabilizing the complex and in the release of primase from the RNA primer•ssDNA•SSB complex, as discussed above. The clamp loader's ATPase activity is associated with the 47 kDa γ- and the 71 kDa τ-subunits. Interestingly, these polypeptides are encoded by the same gene, *dnaX*. They are produced by a programmed frame-shift. Thus, the N-terminal region of τ is identical to that of γ; the difference between these proteins lies in the τ-specific C-terminal extension. The primary role of this extension lies in defining the architecture of the Pol-III holoenzyme and in coupling it to the DnaB helicase (Figure 4-21). Therefore, it is possible to reconstitute clamp loader activity in the form of $\gamma_3\delta\delta'$ (referred to as the γ-complex; Table 4-6). This feat allowed the clamp loader's crystal structure to be solved (Figure 5-5). The five subunits $\gamma_3\delta\delta'$ are arranged in a gapped circle with subunits being joined by a collar at their C-terminal domains. The N-terminal domain of each subunit contains an AAA+ domain and forms a stack with a right-handed helix. The gap lies between subunits δ and δ′.

When the clamp loader binds three ATP molecules, it can then interact stably with the β-clamp (in a manner analogous to PCNA•RFC in Figure 5-8). This in turn levers open one of the two interfaces between the β-clamp homodimer by at least 20 Å, permitting access of DNA. On binding to DNA, the ATP•γ-complex•β-clamp protein complex then hydrolyzes the three ATP molecules, releasing the β-clamp from the clamp loader, and allowing the β-clamp to reseal around the DNA substrate.

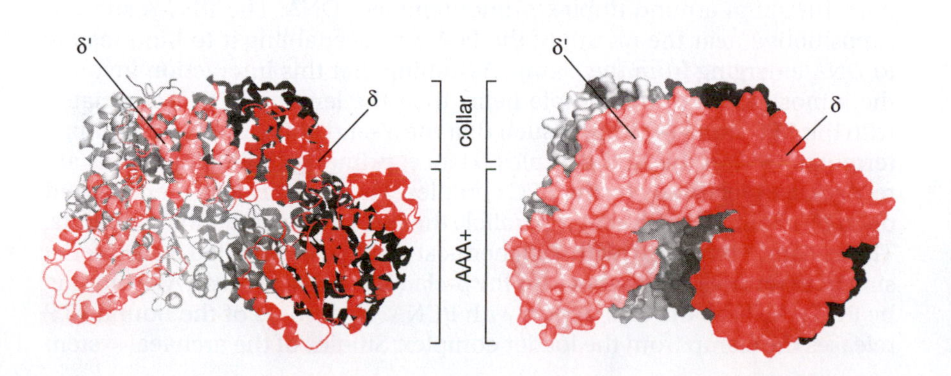

Figure 5-5. Crystal structure of the *E. coli* γ-complex: δ′-subunit (pink), δ-subunit (red), and three γ-subunits (three shades of gray). The C-terminal domains of the subunits form a circular collar that supports an asymmetric arrangement of the N-terminal ATP-binding domains of the γ-motor and the structurally related domains of the 'δ′-stator' and the 'δ-wrench'. A clear gap is observed between the AAA+ domains of δ and δ′. Figure prepared using Pymol from PDB file 1JR3. (Data from Jeruzalmi, D., O'Donnell, M. and Kuriyan, J. (2001) *Cell* 106: 429–441.)

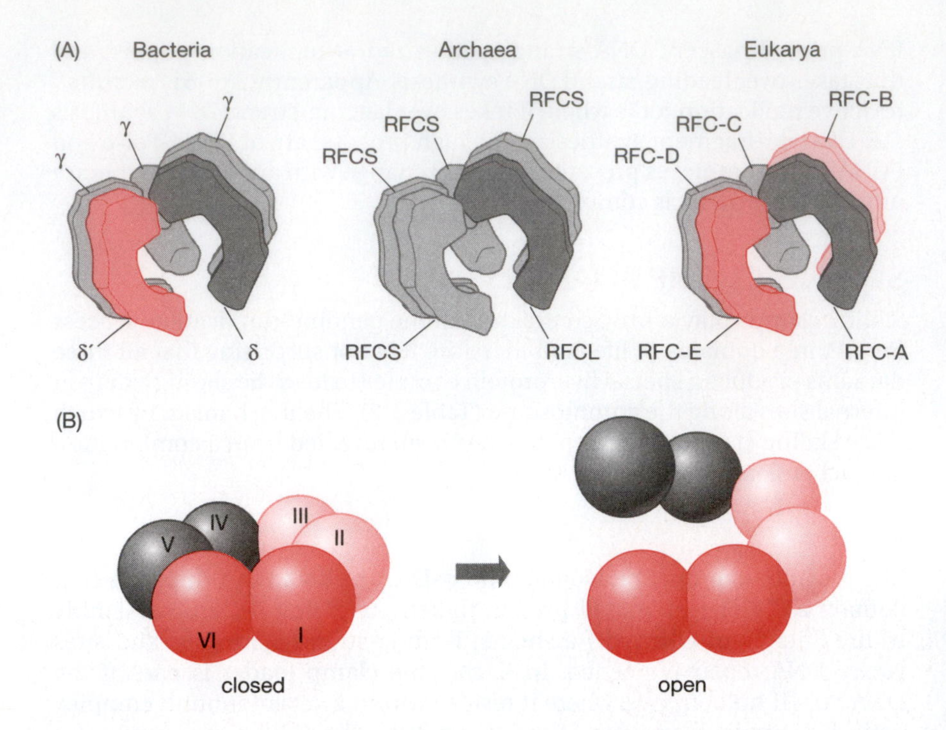

Figure 5-6. Architectural similarities in clamp loaders and sliding clamps. (A) Clamp loaders from bacteria, archaea, and eukarya. RFCS and RFCL refer to the small and large archaeal subunits, respectively. Color coding indicates orthologous subunits. A structure quite similar to the archaeal clamp loader is encoded by the bacteriophage T4 genome (T4 gp44 = RFCS, T4 gp62 = RFCL). (B) PCNA, the sliding clamp in archaea and eukarya, is a trimer in which each monomer contains two subdomains. Molecular dynamics simulations and electron microscopic analyses suggest that the six clamp subdomains are arranged as a planar ring in the closed form, and that when the clamp is opened, the subdomains adopt a right-handed spiral or 'lock washer' conformation. (Adapted from Jeruzalmi, D. (2005) *Proc. Natl. Acad. Sci. USA* 102: 14939–14940. With permission from the National Academy of Sciences.)

Archaea And Eukarya

The clamp loader is remarkably similar throughout the three domains of life (Figure 5-6A). Like the bacterial clamp loader, the archaeal and eukaryotic clamp loader, RFC, functions as a pentamer with all five subunits containing AAA+ domains. At the gross morphological level, RFC has one large subunit (RFC-A in eukarya, RFCL in archaea) complexed with four smaller subunits (RFC-B through RFC-E in eukarya, RFCS in archaea). In eukarya, RFC-B through RFC-E are distinct, but related, proteins. In archaea, the homologous small subunit complex is a homotetramer of a single gene product (RFCS). Structural studies have revealed that the subunits of RFC (Figure 5-6A), like the subunits of the sliding clamp (Figure 5-6B), form a rising right-handed spiral.

The crystal structure of a complex of yeast RFC and PCNA reveals that PCNA is a planar ring with the spiral of RFC rising above it (see Figure 5-7 in color plate section). In this structure, only three RFC subunits contact PCNA. However, if one invokes an out-of-plane opening of the PCNA ring to form a right-handed spiral or lock washer configuration (as depicted in Figure 5-6B), it could, in principle, contact all five subunits of RFC.

An elegant series of biochemical experiments have provided evidence that supports this out-of-plane mechanism for clamp loading by RFC. This is further supported by electron microscopy of the archaeal RFC•PCNA complex in the presence of a nonhydrolyzable analog of ATP. The arrangement of the ATPase domains within the complex suggests that these domains spiral around duplex primer:template DNA. The RFC-A subunit is positioned near the mouth of the PCNA ring, enabling it to bind readily to DNA emerging from the clamp. Assuming that this interaction involves the minor groove of the double helix, then the length of DNA associated with the RFC spiral (~11 bp) is such that the 3′-end of the double helix must terminate within the RFC complex. Hence, primer:template junctions are recognized specifically by the RFC complex. The PCNA ring is then cracked open by about 5 Å such that it parallels the helical rise of the RFC subunits. These data, combined with biochemical analyses, have led to a model similar to that described above for the β-clamp loading process. ATP binding by RFC stabilizes the interaction with PCNA. Hydrolysis of the bound ATP releases the clamp from the loader complex. Studies of the archaeal system

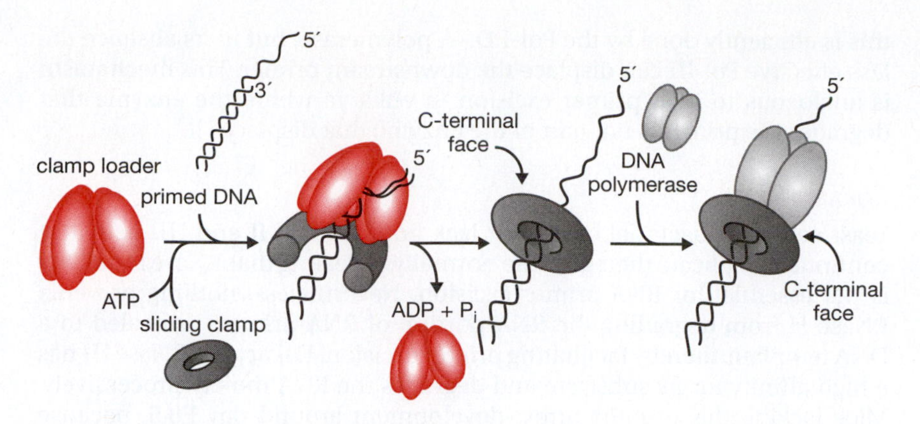

Figure 5-8. RFC-mediated PCNA clamp loading. The clamp loader uses the energy of ATP hydrolysis to assemble sliding clamps at primer:template sites. In the presence of ATP, the clamp loader binds to and opens a sliding clamp. The clamp loader selectively binds to primer:template junctions. This stimulates ATP hydrolysis and results in dissociation of the clamp loader from the clamp and closure of the sliding clamp around DNA. DNA polymerase is recruited to the primer:template junction by the sliding clamp assembled onto the RNA primer. (Adapted from Pomerantz, R.T. and O'Donnell, M. (2007) *Trends Microbiol.* 15: 156–164. With permission from Elsevier.)

reveal that while hydrolysis of ATP bound to the small subunits is needed for releasing the clamp, ATP hydrolysis by the large subunit is needed for recycling the RFC complex (Figure 5-8).

PRIMER EXCISION

Cellular genomes are made of DNA, not mosaics of DNA and RNA. Therefore, mechanisms must exist that remove all of the RNA primers during DNA replication. The mechanisms for RNA primer excision have been well documented in bacteria and eukarya. Archaea contain homologs of the same primer-excision proteins used in eukarya (Table 3-2), and therefore presumably employ the same mechanism as eukarya.

Bacteria

DNA elongation continues until the polymerase encounters the 5′-end of the nascent DNA strand downstream. At this point, the RNA primer on the downstream nascent DNA strand must be excised completely. In bacteria, this job is carried out by two enzymes, RNase HI and DNA Pol-I. RNase H is a nonspecific endoribonuclease that degrades RNA that is annealed to DNA in an RNA:DNA duplex analogous to A-form DNA (Table 1-2). RNase H cleaves the 3′-phosphoryl bond of RNA to generate 3′-OH, 5′-PO_4 ribonucleotide products. However, RNase H cannot cleave rN-p-dN linkages, and *in vitro* purified RNase H leaves from one to three NMPs at the 5′-ends of RNA-primed DNA chains. This would account for the fact that the 1000–2000 nt-long bacterial Okazaki fragments retain only one to three ribonucleotides at their 5′-ends. Neither RNase HI nor RNase HII, however, is essential for bacterial proliferation. The reason is that DNA Pol-I can not only fulfil the role of RNase H, it can also degrade rN-p-dN linkages.

DNA Pol-I is a unique DNA polymerase, because it has both 5′→3′ and 3′→5′ exonuclease activities in addition to DNA synthesis activity. Therefore, it can, in principle, displace Pol-III from the primer:template site and continue DNA synthesis while excising any RNA or DNA it encounters downstream with its 5′→3′ exonuclease activity. In fact, Pol-I is commonly used to radiolabel DNA by nick translation, the ability to degrade the 5′-end of one DNA strand while concomitantly extending the 3′-OH end of the other strand. Moreover, Pol-I could extend a DNA chain accurately by virtue of its 3′→5′ proofreading exonuclease activity. However, whereas genetic inactivation of the 5′→3′ exonuclease activity results in the accumulation of Okazaki fragments, mutations in the polymerase function have little, if any, effect on DNA replication. Thus, while the 5′→3′ exonuclease activity of Pol-I is essential for bacterial proliferation, its polymerase activity is not.

The explanation resides in the fact that the 5′→3′ exonuclease domain of Pol-I is a homolog of the mammalian FEN1 and the yeast RAD27 flap endonucleases (described below). Like FEN1, Pol-I 5′→3′ exonuclease requires that the 5′-end be partially displaced from the template. Normally, this is efficiently done by the Pol-I DNA polymerase, but in its absence the less effective Pol-III can displace the downstream primer. This mechanism is analogous to RNA primer excision in eukarya where the enzyme that degrades the primer is not part of the enzyme that displaces it.

Eukarya

Yeast cells, like bacterial cells, that lack both RNase HI and HII activities continue to replicate their genome normally, revealing that RNase H activity is not essential for RNA primer excision. Nevertheless, nothing prevents RNase H from degrading the RNA portion of RNA primers annealed to a DNA template, thereby facilitating primer excision. Eukaryotic RNase HI has a high affinity for its substrate and degrades the RNA moiety processively. Mice lacking this enzyme arrest development around day E8.5, because RNase HI is necessary during embryogenesis for mitochondrial DNA replication.

Although several of the eukaryotic DNA polymerases have a 3′→5′ exonuclease activity that increases the fidelity with which they copy the template, none of them have a 5′→3′ exonuclease to remove RNA primers. Instead, eukaryotes rely on the concerted efforts of three enzymes to excise RNA primers. One is PCNA•Pol-δ that not only extends the Okazaki fragment, but also displaces the 5′-end of the nascent strand downstream, thereby making it accessible to nuclease activity. Another is a structure-specific endonuclease called FEN1, and the third is DNA2, a helicase with endonuclease activity. Both PCNA and Pol-δ are essential for genome duplication in eukaryotes. Although FEN1 is not essential for survival in yeast, it is critical for normal cell growth and proliferation. DNA2 nuclease activity is essential for yeast proliferation. It ensures RNA primer excision occurs under conditions that are suboptimal for FEN1.

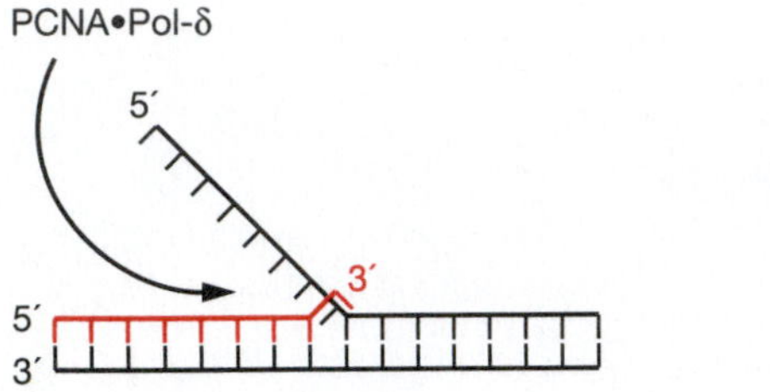

PCNA•Pol-δ
5′
3′
5′
3′
FEN1
5′
3′
N
H
OH
N
H
O-
P
5′
3′
DNA ligase I
5′
3′

Figure 5-9. A double-flap structure is the preferred substrate for FEN1. PCNA•Pol-δ displaces the 5′-end of the downstream strand as an eight-base flap (black) and leaves the upstream nascent DNA strand (red) with a single-nucleotide 3′-flap. FEN1 cuts one nucleotide into the duplex DNA below the 5′-flap, thereby releasing it for degradation by other nucleases and allowing the single-nucleotide 3′-flap to base-pair with the complementary strand (black). The result is a single phosphodiester bond interruption (nick) that can be sealed by DNA ligase I.

FEN1

Enzymatic activities that can digest either 5′-single-strand tails on duplex DNA, or flaps of 5′-ssDNA displaced from duplex molecules, are found in all three domains of life. One enzyme in particular, termed flap endonuclease 1 (FEN1), was first identified in mouse cell extracts from which it was isolated on the basis of its ability to cleave off the 5′-ssDNA tail on an otherwise dsDNA substrate. Once the gene sequence was identified, it became apparent that FEN1 had homologs in yeast, other eukaryotes, bacteriophages, and archaea.

FEN1 can cleave a variety of DNA substrates, including Y-shaped molecules containing a duplex region with both 3′- and 5′-extended single-stranded tails. However, the preferred substrate is duplex DNA from which an internal 5′-flap structure protrudes. In fact, the optimal substrate actually contains two flaps, a single-base flap at the 3′-end of the upstream DNA strand, and an eight-nucleotide flap on the 5′-end of the downstream DNA strand (Figure 5-9). Such a substrate could be generated *in vivo* by partial displacement of the downstream nascent DNA strand fragment by PCNA•Pol-δ in the process of elongating an Okazaki fragment. Simple competition for base pairing between the 3′-terminal nucleotide added by the polymerase and the displaced 5′-flap could lead to the generation of this sort of structure. FEN1 cleaves this structure in the duplex DNA portion one nucleotide downstream of the displaced 5′-flap. The reaction mechanism of FEN1 involves two metal ions that generate a hydroxyl group that then acts as a nucleophile to attack a phosphodiester bond and to stabilize the reaction intermediate. The end result is cleavage of the phosphodiester bond, leaving a 'nick' with a 3′-hydroxyl group and a 5′-phosphate, the only substrate upon which the replicative DNA ligase I will act.

FEN1 does not recognize a particular DNA sequence, but rather a particular DNA structure. It identifies the 5′-end of the flap then tracks along the single-strand flap until it reaches the junction between ssDNA and dsDNA. Structural studies have revealed that there is an extended and rather flexible α-helical loop structure in FEN1 that extends above the active site. This acts as a clamp that shuts down over the 5′-flap, essentially encircling it and allowing the enzyme to track along it. The single-nucleotide 3′-flap binds a pocket in FEN1, and in doing so, it triggers a series of conformational alterations that modulate the configuration of the α-helical clamp. One of the most dramatic features of FEN1's interaction with DNA is that it introduces a bend of about 90° at the flap junction point (Figure 5-10).

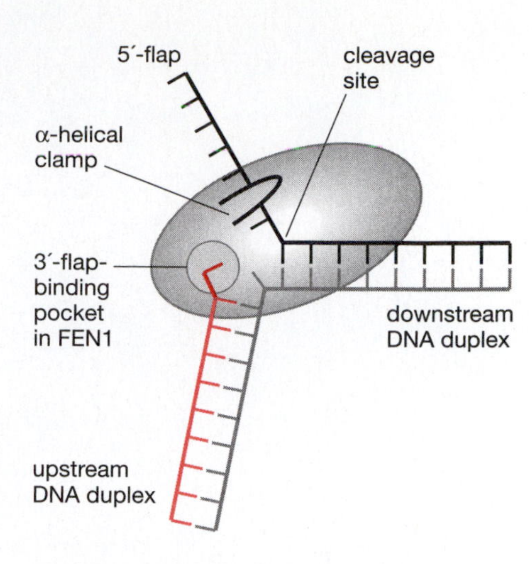

Figure 5-10. DNA–protein intermediate during cleavage of a 5′-flap by FEN1. (Data from Liu, Y., Kao, H.I. and Bambara, R.A. (2004) *Annu. Rev. Biochem.* 73: 589–615.)

DNA2

DNA2 was initially identified in a search for mutations that impair DNA replication. It possesses both an ATP-dependent 5′→3′ helicase activity and a ssDNA endonuclease activity. Significantly, DNA2 interacts with the ssDNA-binding protein RPA, and this interaction stimulates both its helicase and nuclease activities. This feature, together with its powerful exonuclease activity, makes DNA2 the perfect complement for FEN1. The longer the 5′-flap created by PCNA•Pol-δ, the greater the difficulty FEN1 has in removing it. Not only does FEN1 prefer short flaps of ssDNA, but flaps coated with RPA are extremely poor substrates. The longer the flap, the more likely it will bind RPA, and thus the less likely that FEN1 will release it. The RPA-coated flap then recruits DNA2 which progressively shortens the flap by tracking along it from the 5′-end, displacing RPA as it moves along. RPA prevents formation of secondary structures that would be inhibitory to flap removal. The end result is a short (~5 nt) 5′-flap. This length of flap is too short for RPA to bind and thus serves as an excellent substrate for FEN1 action (Figure 5-11).

DNA LIGATION

DNA ligases are encoded by the genomes in all three domains of life, as well as by a few viruses and bacteriophages, because they are essential for maintaining genome integrity during replication and repair of DNA. DNA ligases catalyze formation of a phosphodiester bond between adjacent 5′-phosphoryl and 3′-hydroxyl groups at a single-strand phosphodiester bond interruption (nick) in double-stranded DNA (Figure 5-12). They are categorized into two groups on the basis of the required cofactor for activity:

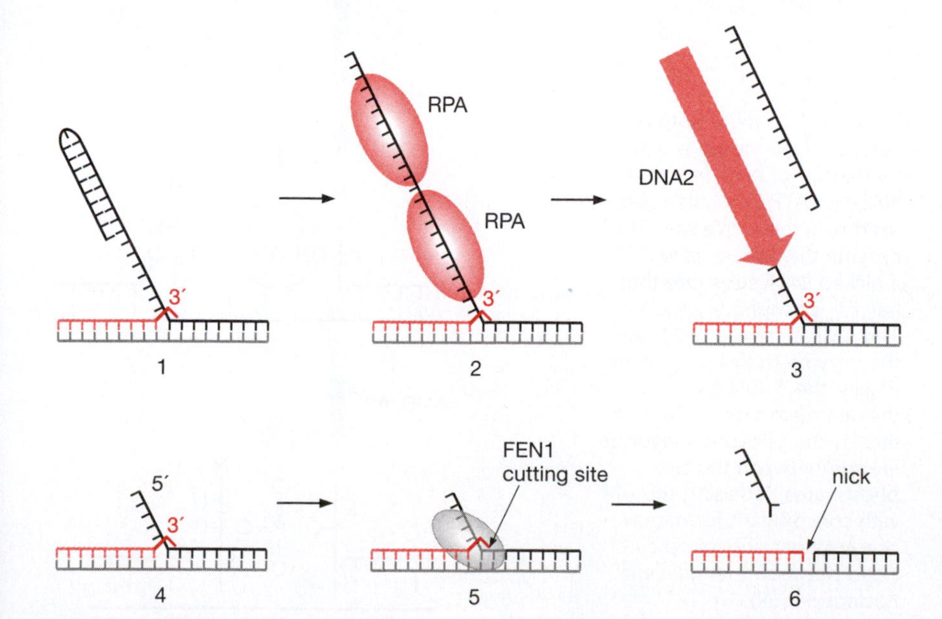

Figure 5-11. Six steps in the cleavage of long flaps by the combined actions of RPA, DNA2, and FEN1. RPA binds to the displaced ssDNA flap as it is displaced, thereby preventing it from forming secondary structures that would interfere with DNA2 activity.

one group requires ATP and the other requires NAD^+. ATP-dependent ligases are produced by viruses, bacteriophages, archaea, and eukarya, whereas NAD^+-dependent ligases are found exclusively in bacteria. *E. coli* encodes a single DNA ligase that is essential for DNA replication; mutations that inactivate this enzyme result in the accumulation of Okazaki fragments. Eukarya encode three different enzymes, but only DNA ligase I is required for lagging-strand template DNA synthesis; DNA ligases III and IV are involved in DNA repair and recombination. Nevertheless, all DNA ligases employ a common mechanism.

DNA ligases seal nicks in three sequential steps (Figure 5-12). First, the enzyme is activated by transfer of an AMP moiety from either ATP or NAD^+ to a conserved lysine residue in the enzyme's active site. This AMP is then transferred to the 5′-phosphate of the DNA substrate to produce an adenylated DNA intermediate. This generates a high-energy phosphoanhydride intermediate that is susceptible to nucleophilic attack by the 3′-OH group of a second DNA molecule, resulting in release of the AMP group and covalent joining of the two DNA molecules. While some DNA ligases such as the one encoded by bacteriophage T4 can efficiently join together two linear dsDNA molecules, most DNA ligases can only seal a nick in dsDNA, such as the nick generated by FEN1 (Figure 5-9).

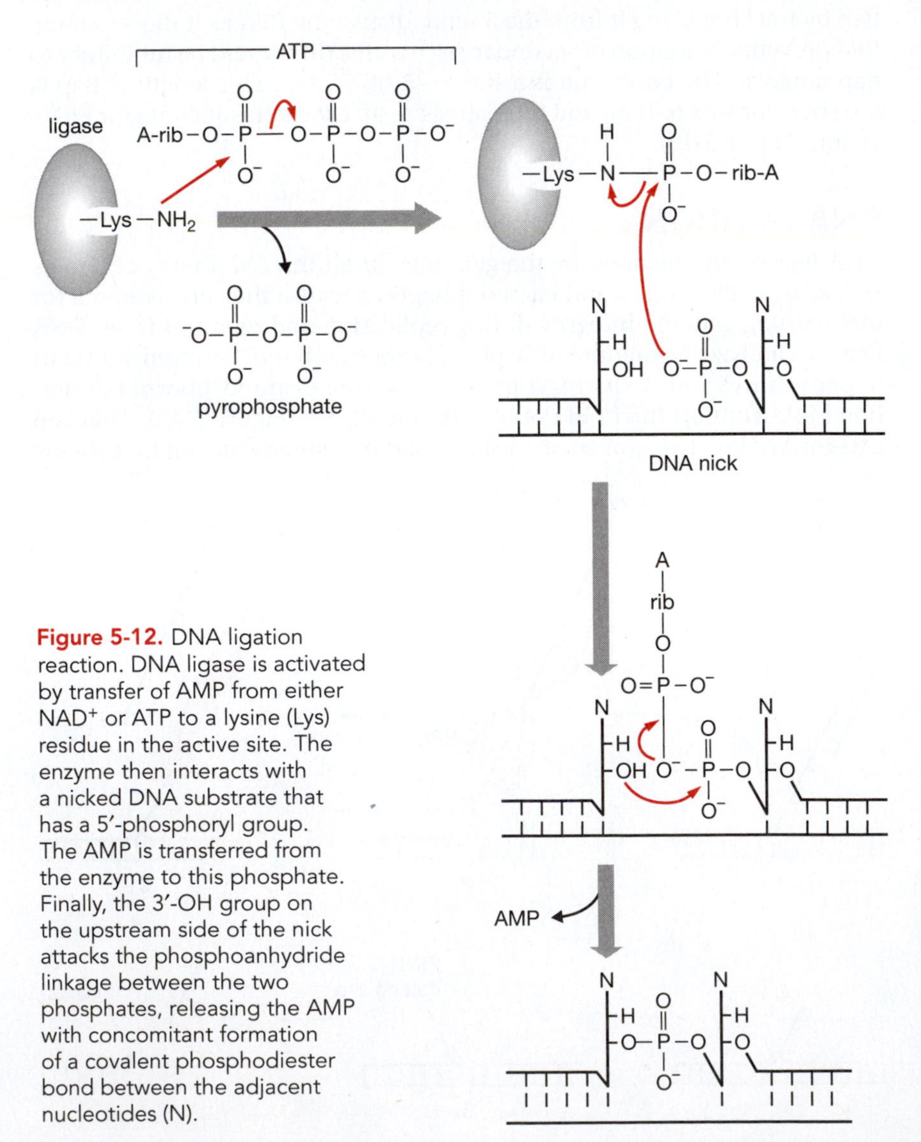

Figure 5-12. DNA ligation reaction. DNA ligase is activated by transfer of AMP from either NAD^+ or ATP to a lysine (Lys) residue in the active site. The enzyme then interacts with a nicked DNA substrate that has a 5′-phosphoryl group. The AMP is transferred from the enzyme to this phosphate. Finally, the 3′-OH group on the upstream side of the nick attacks the phosphoanhydride linkage between the two phosphates, releasing the AMP with concomitant formation of a covalent phosphodiester bond between the adjacent nucleotides (N).

Significant progress in the structural biology of DNA ligases has revealed a fascinating dynamism that is conserved between viral, bacterial, archaeal, and eukaryal DNA ligases. They are multidomain proteins with two domains shared by all ligases; a nucleotidyltransferase (NTase) domain and an oligonucleotide-/oligosaccharide-binding fold (OB fold) domain (see Figure 5-13 in color plate section). The catalytic activity resides in the NTase domain. This domain binds ATP or NAD^+ and contains the active-site residues including the lysine that is covalently attached to AMP during the reaction cycle of the ligase. The OB fold is located C-terminal to the NTase domain where it facilitates binding to the minor groove in dsDNA. Generally, DNA ligases possess additional DNA-binding domains, although their locations vary among the phyla. Eukaryotic and archaeal ligases have a DNA-binding domain that is N-terminal to the NTase domain. In bacteria, a helix-hinge-helix (HhH) domain is found C-terminal to the OB fold. Finally, in the Chlorella virus DNA ligase, there is a small insertion within the OB fold domain that constitutes an extra DNA-binding 'latch'. Despite the positional variation in the linear order of domains, all three ligases show a remarkably similar three-dimensional organization on DNA: viral, bacterial, and human DNA ligases all wrap themselves around their substrate, almost completely encircling it.

The remarkable similarity of DNA-bound ligases belies the conformational malleability of their domains. The three individual domains are connected by highly flexible linkers, which in the absence of DNA, allows the domains to adopt a range of configurations relative to one another. For example, crystal structures have been obtained for various intermediates in the Chlorella virus DNA ligase reaction cycle, allowing comparison of the structure of the enzyme as it goes through the various steps in DNA ligation. During adenylation, residues within the OB fold facilitate the transfer of AMP to the active-site lysine. Once this has been accomplished, the OB fold rotates and repositions to facilitate DNA binding. This is augmented by the latch-like insert present in the OB fold. This feature is disordered, and presumably highly flexible, in the absence of DNA. However, upon engagement of the ligase with DNA, as its name suggests, the latch becomes ordered and helps lock onto the DNA substrate.

Comparison of the structures of DNA-bound human ligase I and the orthologous *Sulfolobus solfataricus* DNA ligase I in the absence of DNA additionally provides evidence for extensive rearrangements in the archaeal/eukaryotic ATP-dependent DNA ligases. As can be seen in Figure 5-14 (in the color plate section), the *Sulfolobus* ligase has an extended conformation in the absence of DNA. In contrast, human ligase I on DNA has the archetypal wrap-around conformation, almost entirely surrounding the DNA substrate.

THE RINGMASTER

Completion of Okazaki-fragment synthesis in eukarya and archaea involves the concerted efforts of several proteins: PCNA•DNA Pol-δ, FEN1, DNA ligase I and, in eukaryotes, DNA2 (Figures 5-9 and 5-11). As in every circus, coordination of different events is accomplished by a ringmaster, which appropriately in this case is the sliding clamp, PCNA. Reconstituted systems have revealed that PCNA•Pol-δ has the ability to crash into the 5′-end of a downstream primer and lever it off the DNA, producing a 5′-tailed flap that is a substrate for FEN1. Not only does Pol-δ bind to PCNA, but FEN1 does as well (Figure 5-15). FEN1 contains a highly conserved C-terminal PCNA-interacting-protein (PIP) motif that binds to a groove in PCNA buried underneath the interdomain connector loop (Figure 4-23). This interaction greatly stimulates FEN1 activity. Like FEN1, Pol-δ and DNA ligase I also contain PIP motifs, and they also interact specifically with PCNA. In fact, more than 60 different proteins have been identified that bind PCNA, eight

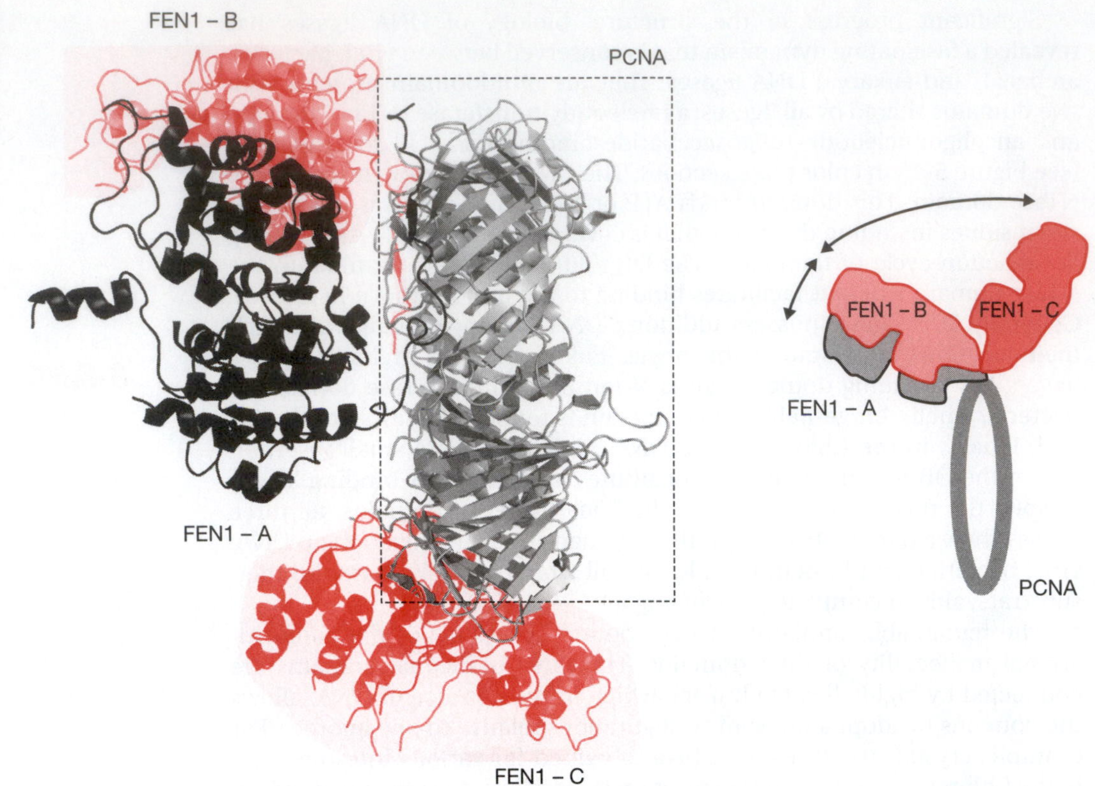

Figure 5-15. The structure of three molecules of FEN1 (red, black, and pink) in complex with PCNA (gray). The left-hand image shows the crystal structure. The cartoon indicates the relative positions of the three FEN1 molecules. Molecule A is in the correct position to act on an exposed flap. Molecules B and C are in carrier configurations, waiting their turn to act. The figure was prepared using Pymol with PDB file 1UL1. In the cartoon, the three distinct conformations of FEN1 are superimposed and modeled on a single PCNA subunit, to illustrate the motion that FEN1 undergoes in its transition from inactive carrier configurations to the active locked-down position.

of which are involved in DNA replication (DNA Pol-δ, DNA Pol-ε, FEN1, DNA ligase I, RFC, Mcm10, Topo IIa, and Cdt1).

How does PCNA coordinate the assembly and activity of its various client proteins? The answer may lie within recently elucidated crystal structures. For example, the structure of a complex of human FEN1 with PCNA revealed that three FEN1 proteins were bound to each PCNA ring and each FEN1 protein had a distinct conformation (Figure 5-15). In one conformation, FEN1 was locked forward in a position compatible with it engaging the DNA substrate. Another FEN1 protein was bent at a hinge region so that it was held back in an apparently inactive conformation. This important structure highlights two key points: first, a single PCNA ring can interact with multiple 'client' proteins and, second, clients can adopt a range of conformations, corresponding to 'carrier' and 'activated' configurations.

A second example of transit and active forms of PCNA partner proteins comes from structural studies of *Sulfolobus* DNA ligase I bound to PCNA. Here the ligase adopts an extended conformation, sticking out radially away from the PCNA ring. However, the structure of human ligase I bound to a nicked DNA substrate revealed that ligase almost completely encircled the DNA molecule, forming a ring-shaped structure of similar dimensions to PCNA. Thus, the nick-bound PCNA•ligase complex is proposed to be a stacked ring. This has the consequence that the interactions of other factors with PCNA would be sterically occluded. It is likely that the extended conformation represents the carrier conformation of the PCNA•ligase complex, whereas the stacked ring represents the active form. This is conceptually pleasing, as this stacked ring form could, in the process of catalyzing the final step in DNA synthesis, additionally displace polymerase and other factors from the PCNA ring, allowing their recycling to mediate synthesis of the next Okazaki fragment (Figure 5-16). Interestingly, however, *S. cerevisiae* Pol-δ releases spontaneously from the primer:template *in vitro* when it collides with the 5′-end of the nascent DNA strand, regardless of whether or not it encounters RNA or DNA. This suggests that both high

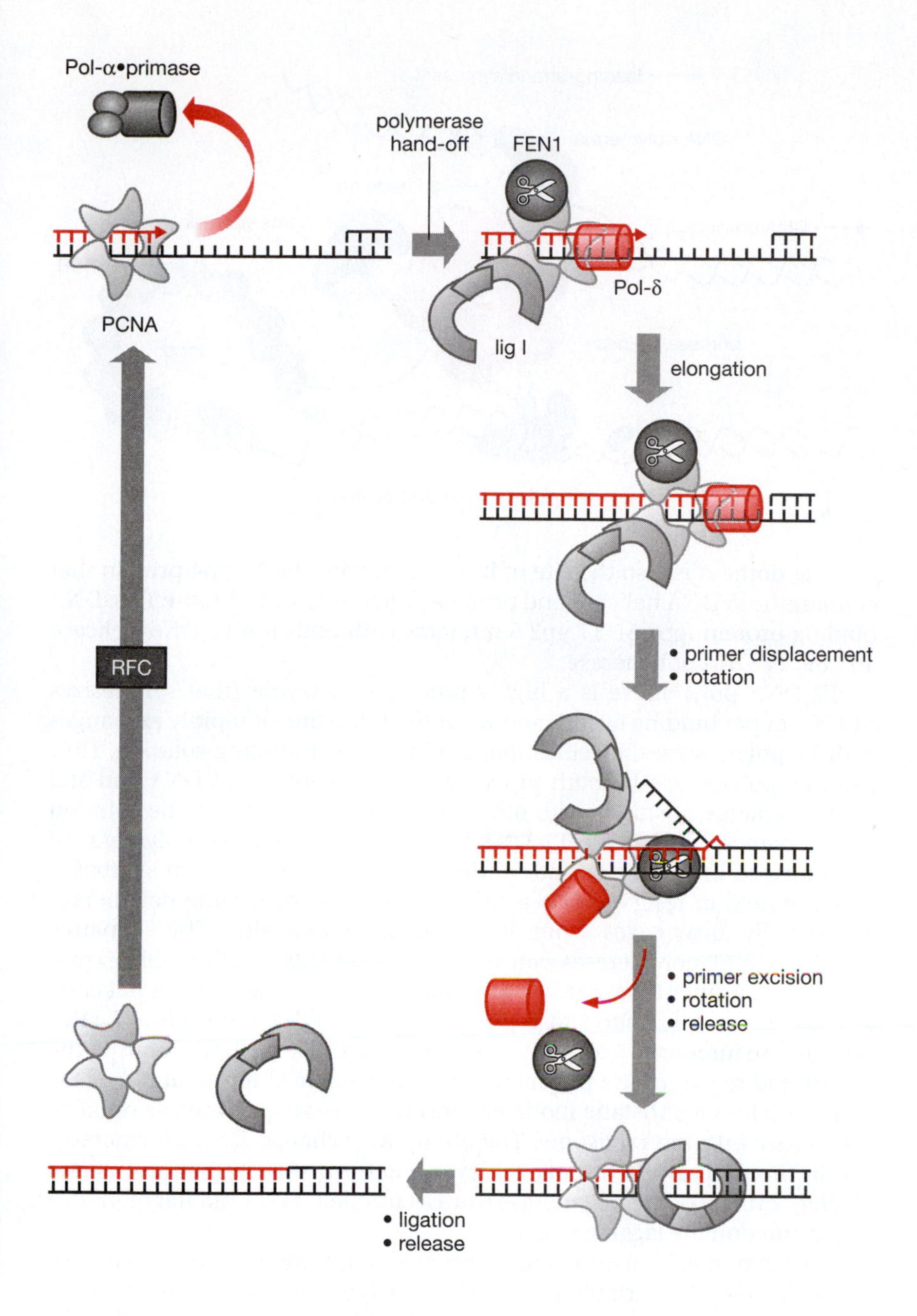

Figure 5-16. PCNA as a ringmaster. Loading of the PCNA sliding clamp on DNA requires the RFC clamp loader. The weakly processive Pol-α•primase is then prevented from rebinding to the primer:template site by the presence of the PCNA sliding clamp. Studies of the heterotrimeric PCNA of the archaeon *Sulfolobus* have demonstrated that PCNA can bridge between FEN1 and DNA polymerase and DNA ligase. It is, therefore, tempting to suggest that rotation of PCNA around the DNA results in a rotary hand-off of the DNA substrate from Pol-δ to FEN1 and finally to DNA ligase I, as their specific substrates become available. Although Pol-δ can be spontaneously released from PCNA when it collides with a downstream oligonucleotide duplex *in vitro*, these events may be coordinated *in vivo* by the action of DNA ligase.

processivity and the ability to recycle when its job is done are intrinsic to the PCNA•Pol-δ complex.

DYNAMIC PROCESSIVITY

The question remains as to how individual events at replication forks are coordinated. How is the synthesis of leading and lagging strands coordinated? How does the replication machine remain intact and on site until the job is done? The most definitive answers to these questions have come from studies on the replication machines produced by bacteriophages and bacteria.

The simplest replisome characterized so far is that of bacteriophage T7: it can be reconstituted with only four proteins (Figure 5-17). T7 DNA polymerase (T7 gp5) binds the *E. coli* host protein thioredoxin very tightly, causing the thioredoxin-binding domain in T7 gp5 to assume a claw-like structure that tethers the polymerase to its DNA substrate. The thioredoxin

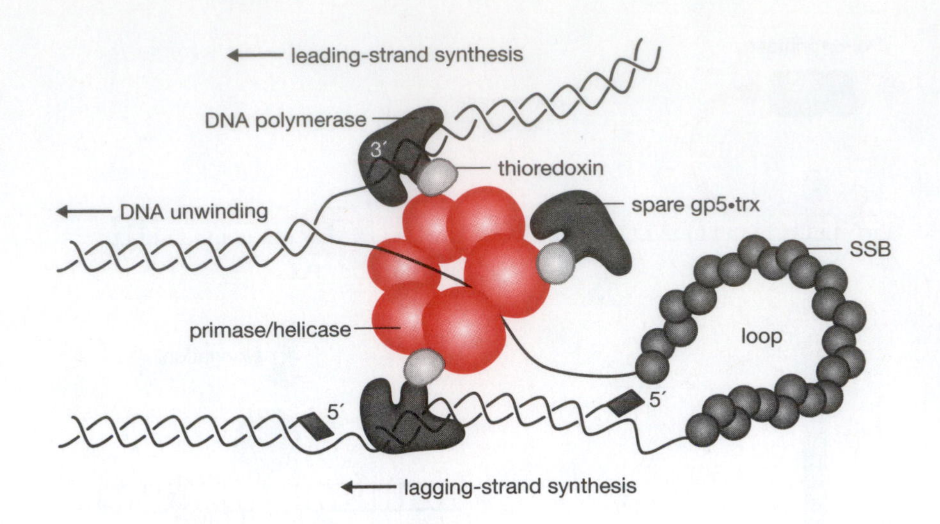

Figure 5-17. Bacteriophage T7 replication fork. The T7 replisome consists of two copies of the DNA polymerase (gp5), the processivity factor *E. coli* thioredoxin (trx), the homohexamer primase/helicase (gp4), and the ssDNA-binding protein (SSB/gp2.5). The gp5•trx complex synthesizes the leading strand continuously as the helicase unwinds the duplex. The lagging strand is synthesized as Okazaki fragments initiated from RNA primers (dark gray boxes) catalyzed by the primase domain of gp4. A DNA loop forms on the lagging strand to align both gp5•trx complexes. gp2.5 coats the ssDNA regions of the lagging strand generated as the helicase unwinds the DNA. The gp4 helicase/primase can bind multiple gp5 DNA polymerases. (Adapted from Johnson, D.E., Takahashi, M., Hamdan, S.M. *et al.* (2007) *Proc. Natl. Acad. Sci. USA* 104: 5312–5317. With permission from the National Academy of Sciences.)

binding domain is also the site of interaction with the T7 gp4 protein that contains both DNA helicase and primase functions, and with the T7 ssDNA binding protein (gp2.5). T7 gp2.5 interacts with both the T7 DNA helicase and the T7 DNA polymerase.

T7 DNA polymerase is a highly processive enzyme (that synthesizes >17 000 nt per binding event), and yet at the same time it rapidly exchanges with T7 polymerase molecules that are in the surrounding solution. How can this polymerase be both processive in its synthesis of DNA and still freely exchange positions with other polymerase molecules? The solution to this paradox is that the T7 DNA helicase binds multiple polymerases: one engaged in leading-strand synthesis, one in lagging-strand synthesis, and one held in reserve in case either the leading or lagging polymerase accidentally disengages from its primer:template site. This dynamic interchange of polymerases can occur because the T7 DNA polymerase bound to a primer:template site also binds tightly to the T7 DNA helicase, but when it is not bound to a primer:template DNA molecule, T7 DNA polymerase interacts weakly with its cognate helicase via basic loops in the C-terminal region of the polymerase and the acidic C-terminal tail of the helicase. This electrostatic mode ensures that a spare polymerase remains associated with the replisome. The ability to exchange DNA polymerases within the replisome not only achieves great processivity but also allows efficient cycling of the polymerase from a completed Okazaki fragment to a new primer on the lagging strand.

A similar mechanism occurs with the more complex bacteriophage T4 replisome. Bacteriophage T4 DNA polymerase, along with seven other T4-encoded proteins, constitutes the T4 replisome that carries out coordinated DNA synthesis on both templates. Among these, the T4 clamp loader (gp44 and gp62) and the sliding clamp protein (gp45) are polymerase accessory factors. These proteins greatly increase the processivity of T4 DNA polymerase during replication by forming a holoenzyme complex, one for each arm of the replication fork. Moving ahead of the holoenzyme is the helicase (gp41), primase (gp61), and helicase accessory protein (gp59) complex that rapidly unwinds the DNA and synthesizes the pentaribonucleotide primers for Okazaki-fragment synthesis. The ssDNA-binding protein gp32 stabilizes the ssDNA loop structure generated during lagging-strand synthesis. Electron microscopic analysis of bacteriophage T4 replication forks with biotin-labeled DNA polymerases suggests that about one-third of its replication forks carry three polymerases.

The composition of the *E. coli* clamp loader complex has undergone revision over the years, due to the biochemical complexities involved in determining stoichiometry of individual subunits within a large protein complex, and the fact that the *dnaX* gene encodes both the τ- and γ-subunits.

The three domains shared by τ and γ are sufficient to form a functional clamp loader complex in combination with δ and δ′. Thus, the consensus replisome-associated clamp loader is $\gamma\tau_2\delta\delta'$, which provides two τ-subunits to couple a leading- and lagging-strand Pol-III and a purpose for a single γ-subunit as a placeholder in the pentameric complex. The two τ-subunits present in each *E. coli* Pol-III holoenzyme complex bind to each of two core polymerases and serve to tether them together (Figure 4-21). In addition, τ binds to and stimulates the activity of the DnaB helicase. Thus, the τ-dimer can be viewed as the organizing center of the holoenzyme, providing a bridge between polymerases, clamp loader, and helicase.

Recent evidence suggests that a $\tau_3\delta\delta'$ clamp loader complex can indeed interact with a total of three DNA polymerase III core complexes and may be the natural replisome clamp loader (Figure 5-18). Several forms of *E. coli* clamp loaders have been reconstituted from purified subunits and found to be equally active at loading the β-clamp onto DNA. The $\tau_3\delta\delta'$-replisome, however, is 60% faster at DNA synthesis than the $\gamma\tau_2\delta\delta'$-replisome, and the size of Okazaki fragments at replication forks is smaller for the $\tau_3\delta\delta'$-replisome over the $\gamma\tau_2\delta\delta'$-replisome, revealing a higher efficiency of primer utilization on the lagging-strand template. The additional τ-subunit may either allow the replisome to carry a spare Pol-III, as proposed for the phage T7 replisomes, or allow three Pol-III enzymes to actively engage in DNA synthesis simultaneously, as proposed for T4 replisomes, one polymerase on the leading strand and two on the lagging strand simultaneously synthesizing two Okazaki fragments. When this concept is combined with the concept that DnaB helicase organizes DnaG primases at replication forks, the concept of dynamic processivity can be realized in full (Figure 5-18).

Whether or not such models are applicable to the more complex eukaryotic replisomes ploughing through chromatin remains to be

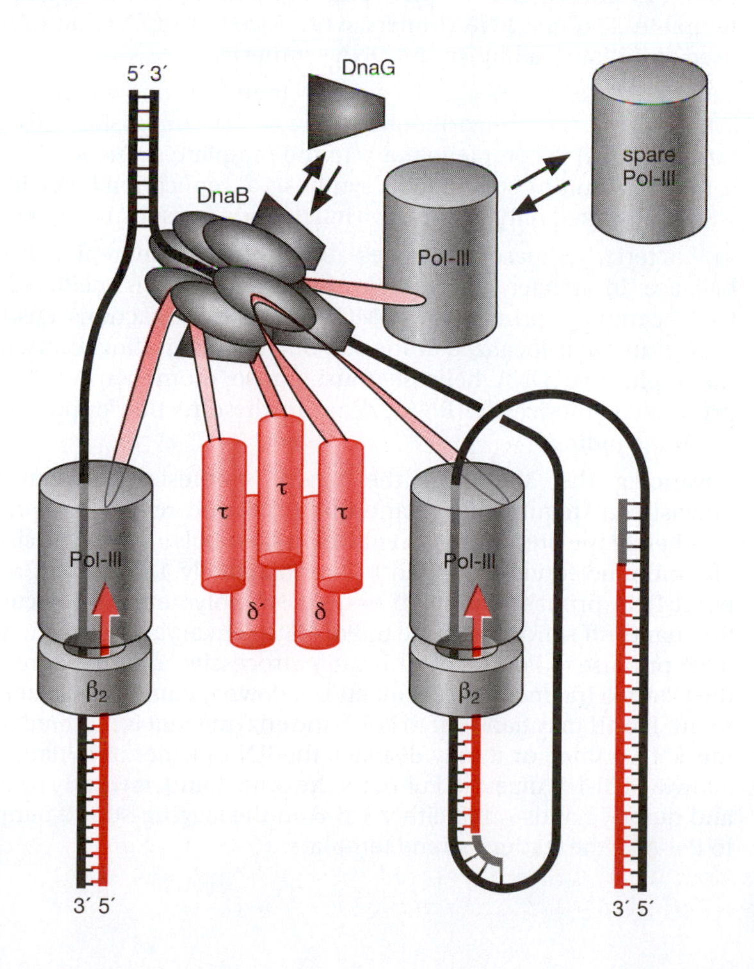

Figure 5-18. The dynamic processivity model for the *E. coli* replisome. The Pol-III holoenzyme $\tau_3\delta\delta'$ clamp loader complex links three Pol-III core enzymes to the DnaB helicase and to each other via three τ-subunits, but only two of the Pol-III enzymes are linked to a sliding clamp (β_2) and engaged in DNA synthesis. The third is held in reserve where it is free to exchange with Pol-III in the bacterial milieu. Similarly, the DnaB helicase is linked to three DnaG primase molecules (Figure 5-3), only one of which is actively engaged in synthesis of an RNA primer (gray line) on the lagging-strand DNA template (black line) to initiate synthesis of nascent DNA (red line). Helicase-bound DnaG can also exchange freely with DnaG in the bacterial milieu. Reserve primase and polymerase ensure that if unscheduled release of actively engaged enzymes from the template occurs, they are immediately replaced by new enzymes. (Data from Lovett, S.T. (2007) *Mol. Cell* 27: 523–526.)

determined. In particular, no clear homolog of the crucial τ-subunit has been identified. However, a growing body of work suggests that the Cdc45•MCM•GINS complex in eukarya (Figure 5-4) may play the role of the τ-subunits in *E. coli*. GINS can bind primase directly in both archaeal and mammalian systems, suggesting that GINS may serve to nucleate formation of a 'replisome progression complex'. Another candidate for this job is the eukaryotic clamp loader RFC, which bears close analogy with the bacterial clamp loader (Figure 5-6A).

SUMMARY

- Okazaki fragments are short, transient, RNA-primed nascent DNA chains. Their size distribution varies continuously about their mean length. In bacteria, the mean is 1000–2000 nt and the range is from 50 to 4000 nt. In archaea and eukarya, the mean length is 100–200 nt with a range from 30 to 300 nt. The frequency of Okazaki-fragment initiation is determined by the length of the lagging-strand template, and the frequency of RNA primer synthesis. The frequency of RNA primer synthesis is determined by the concentration of the replication factors involved. RNA primer synthesis is the rate-limiting step in Okazaki-fragment synthesis. On average, replication forks contain less than one Okazaki fragment per fork.
- DNA synthesis at replication forks begins with the synthesis of an RNA primer beginning with either 5′-ATP or GTP and proceeding, like all polymerases, in a $5' \rightarrow 3'$ direction. RNA primers are about 10 ribonucleotides in length, after which they are covalently linked to the 5′-end of a nascent DNA chain. This transition point is random with respect to DNA template sequence. RNA primers are synthesized once on the leading-strand template and many times on the lagging-strand template. The first RNA primers synthesized at replication origins are used to initiate leading-strand DNA synthesis.
- DNA primase requires only a DNA template on which to initiate synthesis of a short oligoribonucleotide whose composition (like that of nascent DNA) is complementary to the template sequence. This event is the rate-limiting step in DNA synthesis at replication forks. Initiation sites are marked only by a di- or trinucleotide consensus sequence.
- In bacteria, primase associates directly with the replicative DnaB helicase. In archaea and eukarya, this interaction is mediated by the GINS complex (primase•GINS•MCM). These interactions ensure that DNA primase is localized at the site of DNA unwinding. Consequently, the replicative DNA helicase must pause momentarily while DNA primase synthesizes an RNA primer in the direction opposite that of DNA unwinding.
- Advancing the 3′-end of the newly synthesized oligonucleotide downstream requires two hand-offs from one replication protein to another. In the presence of a replicative DNA polymerase, the size of the oligoribonucleotide is limited to approximately 10 residues, at which point DNA primase hands-off to the DNA polymerase. In bacteria, the first hand-off is from primase to Pol-III. In eukarya, the first hand-off is from primase to Pol-α. Pol-III is quite processive and therefore extends the Okazaki fragment until it meets the downstream RNA primer. At this point, Pol-III may hand-off to Pol-I, the enzyme that is required to excise the RNA primer, or it may displace the RNA primer as a 'flap', thereby allowing Pol-I to digest it. Pol-α, on the other hand, is weakly processive, and quickly hands-off to either Pol-δ on the lagging-strand template or to Pol-ε on the leading-strand template.

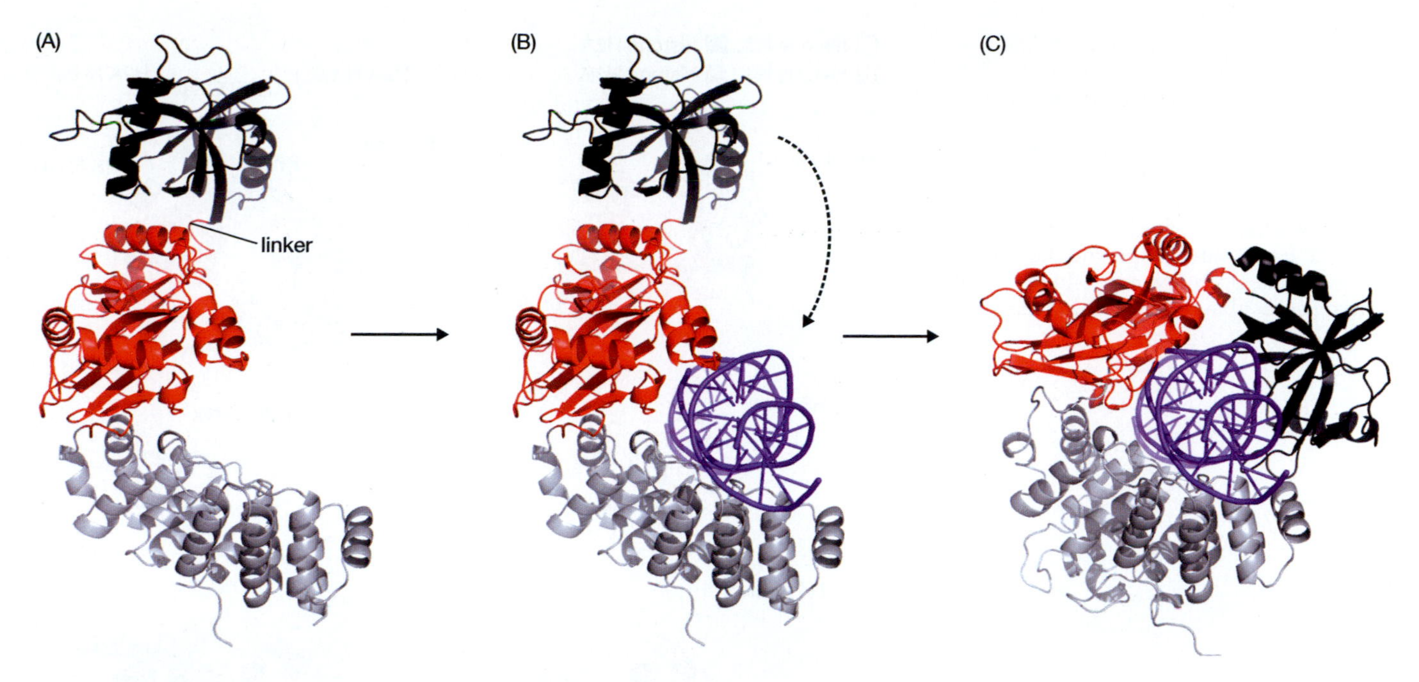

Figure 5-14. Model of the conformational shift in DNA ligase I. (A) Structure of DNA ligase I from *S. solfataricus*, determined in the absence of DNA, shows the ligase adopting an extended conformation. Domains are colored as in Figure 5-13. (C) Structure of human DNA ligase I bound to DNA, revealing the ligase wrapped entirely around its substrate. (B) Hypothetical intermediate that forms when ligase interacts with DNA, but the C-terminal OB fold has not yet moved to enfold the DNA molecule. Note that this view of the human ligase is rotated by 120° relative to the view in Figure 5-13.

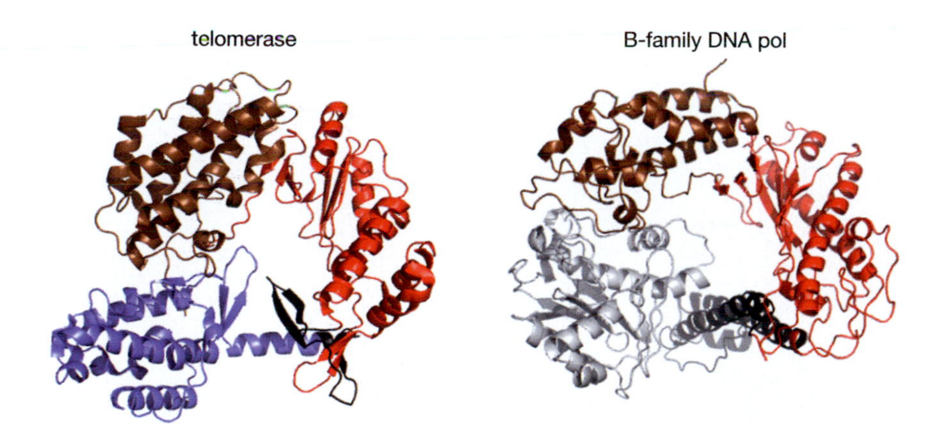

Figure 6-11. Structure of the TERT component of telomerase from the red flour beetle, *Tribolium castaneum*, and the B-family DNA polymerase encoded by bacteriophage RB69. Analogous domains are colored the same. Thumb domain (brown), fingers domain (black), and palm domain (red) represent common features of both proteins. The RNA-binding domain of TERT (blue) and the exonuclease proofreading domain of RB69 DNA polymerase (gray) fall into the same relative positions in each protein. Figures were prepared using Pymol with PDB files 3DU6 (TERT) and 1WAF (RB69 DNA polymerase). (Data from Gillis, A.J., Schuller, A.P. and Skordalakes, E. (2008) *Nature* 455: 633–637.)

Figure 7-6. Histone core particles. The H3•H4 tetramer, consisting of two molecules of H3 and two of H4, is a very stable protein complex that serves as the foundation for the histone octamer, consisting of the H3•H4 tetramer and two copies of histone H2A and two of H2B. Each of the four histones has both a central core that participates in forming the stable histone octamer, and N- and C-terminal tails that are accessible to protein modifications.

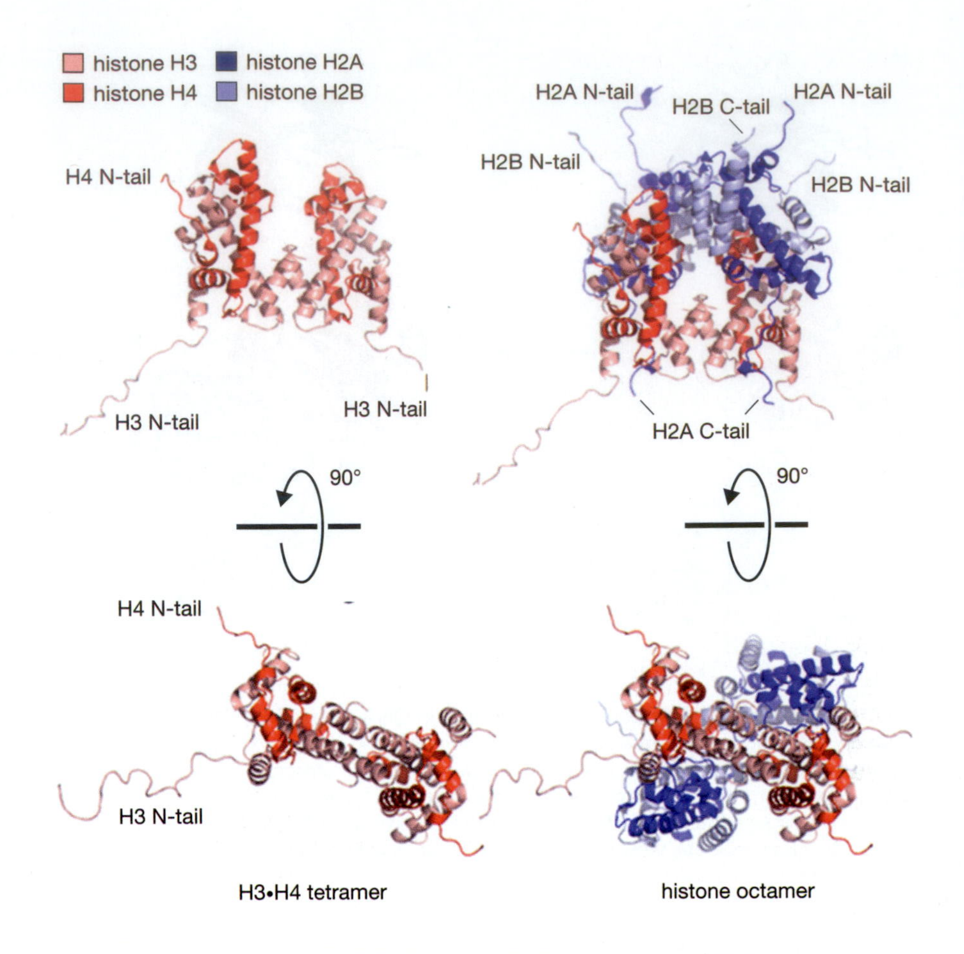

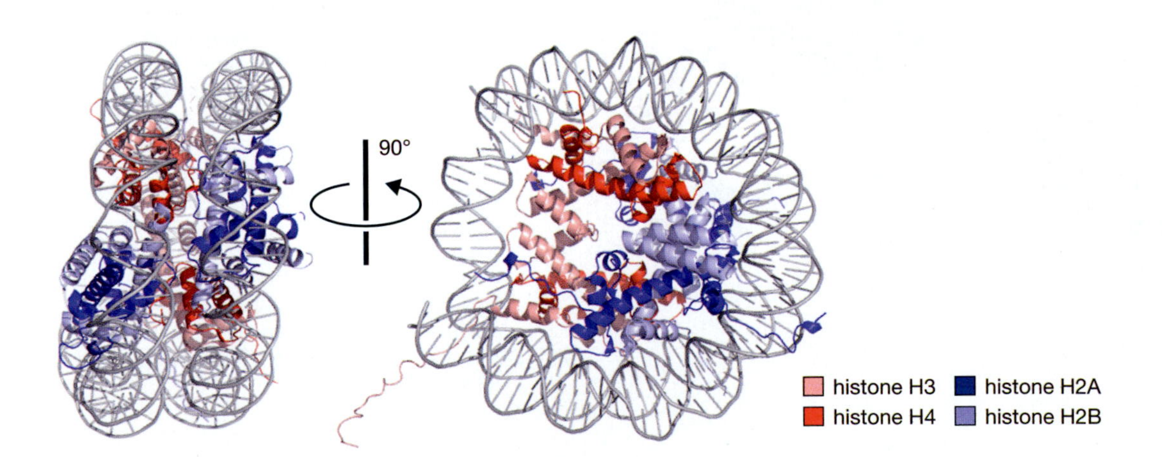

Figure 7-7. The structure of the eukaryotic nucleosome core particle (NCP). The NCP contains a histone octamer with two copies each of histones H3, H4, H2A, and H2B around which is wrapped 1.75 turns of dsDNA (146 bp) with a 20 Å diameter to form a 110 Å × 55 Å cylindrical structure.
10 Å = 1 nm.

- Replicative DNA polymerases depend upon their association with a sliding clamp to achieve a high degree of processivity. The sliding clamp is loaded onto dsDNA by a clamp loader, both of which are highly conserved in their structure and function throughout the three domains of life. Sliding clamps are circular molecules that are opened and resealed around dsDNA by the clamp loader. This action is driven, as with replicative DNA helicases, by ATP binding and hydrolysis. The mechanism of loading targets the 3′-end of a DNA primer:template site.
- The final stage in Okazaki-fragment metabolism is excision of the downstream RNA primer followed quickly by joining of the Okazaki fragment to the 5′-end of the newly synthesized DNA strand on the lagging-strand template. Since these events require DNA synthesis with concurrent excision of RNA and of an rN-p-dN linkage followed quickly by ligation of the 3′-OH end of the Okazaki fragment to the 5′-P end of the long DNA strand, several different enzymes are involved.
- In eukarya, the proteins required for Okazaki-fragment maturation are coordinated through association with PCNA (the 'ringmaster'). Pol-δ, FEN1, and DNA ligase I can each bind to PCNA. DNA Pol-δ extends the Okazaki fragment while RNase HII can nibble away the RNA primer. Pol-δ then displaces what is left of the RNA primer and its linkage to DNA as well as some of the DNA so that it becomes a substrate for either FEN1 or DNA2. Displacement of Pol-δ needs to be coordinated with the action of DNA ligase. In bacteria, the role of Pol-δ is played by Pol-III. The role of RNase HII is played by RNase HI. FEN1 and DNA2 are played by Pol-I, and DNA ligase I, respectively. Eukaryotic DNA ligase I and its cofactor ATP are played by bacterial DNA ligase and and its cofactor NAD^+.
- Events at replication forks are coordinated at multiple levels. The preceding chapters discussed formation of holoenzymes, replisomes, replication factories, and the 'trombone' configuration for replication forks. In addition, replication-fork assemblies can carry along additional copies of DNA primase and a replicative DNA polymerase. This concept of dynamic processivity ensures that, once started, replication forks can continue even if one of these components dissociates from its active site and must be replaced by another molecule that is kept as a 'spare part'.

REFERENCES

Additional Reading

Frick, D.N. and Richardson, C.C. (2001) DNA primases. *Annu. Rev. Biochem.* 70: 39–80.

Indiani, C. and O'Donnell, M. (2006) The replication clamp-loading machine at work in the three domains of life. *Nat. Rev. Mol. Cell Biol.* 7: 751–761.

Johnson, A. and O'Donnell, M. (2005) DNA ligase: getting a grip to seal the deal. *Curr. Biol.* 15: R90–R92.

Lao-Sirieix, S.H., Pellegrini, L. and Bell, S.D. (2005) The promiscuous primase. *Trends Genet.* 21: 568–572.

Liu, Y., Kao, H.I. and Bambara, R.A. (2004) Flap endonuclease 1: a central component of DNA metabolism. *Annu. Rev. Biochem.* 73: 589–615.

Chapter 6
TERMINATION

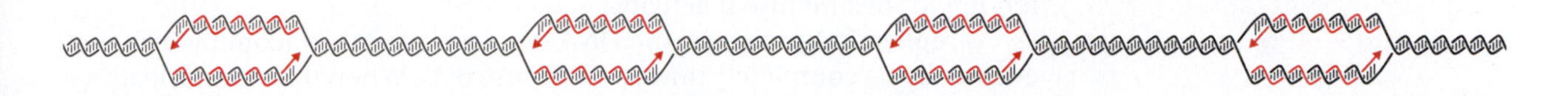

In previous chapters, we learned that organisms in all three domains of life duplicate their genomes by a semi-conservative mechanism termed the replication fork and that the replication fork is driven by a complex of proteins termed the replisome. Eventually, replication forks complete their task and terminate. Termination of DNA replication happens by three different mechanisms.

Termination occurs most often when two replication forks traveling in opposite directions collide with one another. When this happens, termination is a natural outcome of the structure and dynamics of replication forks: it does not require sequence-specific termination signals. Not only have such signals not been observed in nature, but, in principle, one would not expect them. If a sequence required for initiation of DNA replication is deleted, initiation events could still occur at other replication origins and the chromosome would still be duplicated, but if a sequence required for termination of DNA replication is deleted, then replication could not be completed, and the two sibling chromosomes could not separate during mitosis. That would be a lethal event for the cell.

The progress of replication forks is arrested when they encounter a **replication-fork barrier** (**RFB**). This is a specific ***cis*-acting** DNA sequence that binds a specific ***trans*-acting** protein that arrests replication forks arriving from one direction, but not from the other. RFBs commonly occur in the chromosomes of bacteria and bacterial plasmids where they ensure that genome duplication terminates at a specific locus (**termination site**) by arresting the progress of a fork arriving from one direction until the opposing fork arrives and the two forks collide. RFBs also appear in eukarya downstream of multicopy gene repeats, such as rRNA genes and tRNA genes, where they prevent DNA replication forks from colliding head-on with the DNA transcription machine. RFBs are distinct from naturally occurring impediments to replication-fork migration such as DNA-binding proteins, transcription units, unusual DNA structures, and sequences that retard the progress of DNA replication. Pausing of forks at natural impediments can lead to genomic instability, whereas arresting forks at RFBs does not.

Finally, replication forks come to an end: the end of a linear chromosome, that is. Herein resides a paradox. Replication forks cannot copy the 3′-terminus of the lagging-strand template at the ends of linear DNA molecules. Nature solved this problem by providing eukarya with a new tool, a unique DNA sequence attached to the ends of chromosomes, termed the **telomere**, and a special enzyme to maintain it, termed **telomerase**.

REPLICATION-FORK TERMINATION

When Forks Collide

Whenever two replication forks, traveling in opposite directions, collide with one another, DNA replication terminates and the two sibling molecules separate. This situation can arise in circular DNA molecules that contain one or more origins of bidirectional replication, and in linear DNA molecules that contain two or more origins of bidirectional replication (Figure 8-1). In these instances, termination is a natural consequence of DNA unwinding and DNA structure. It does not necessarily require specific DNA sequences. It does not occur at specific DNA sites. It does, however, require topoisomerase II activity.

As replication proceeds, the DNA between the two oncoming forks will eventually be completely unwound (Figure 6-1). When this occurs, leading-strand DNA synthesis at one fork will reach the 5′-end of the long nascent DNA strand synthesized by the other fork, separated only by a small gap of ssDNA. The RNA primer on the strand will be excised, the gap filled in, and the two long nascent DNA strands will be joined together as one. When this occurs in circular DNA molecules, the two sibling chromosomes are separated completely. In linear DNA molecules, such as the one represented in Figure 6-1, only the portion of the sibling chromosomes covered by the two replication bubbles involved will be separated.

Topological Problems Unique To Replication-Fork Termination

When DNA is unwound, one positive superhelical turn is produced to compensate for each negative helical turn removed (Figure 3-9). These superhelical turns can appear either in front of the fork as positive supercoils or behind the fork as negative precatenated intertwines. Thus, as two forks converge, this build-up of positive superhelicity becomes a significant barrier to unwinding the DNA between the two forks. Unless this topological stress is relieved, the intervening DNA strands will be melted but not unwound, and the two sibling molecules will become separated but catenated (Figure 6-2).

As long as there is sufficient DNA in front of a replication fork, type IB topoisomerases (Table 4-8) can relieve the positive supercoils in front of replication forks. However, as the distance between two oncoming forks decreases, it will become inaccessible to type I topoisomerases, thus forcing the torsional stress to be expressed in the form of precatenated intertwines behind the replication forks. Formation of precatenanes

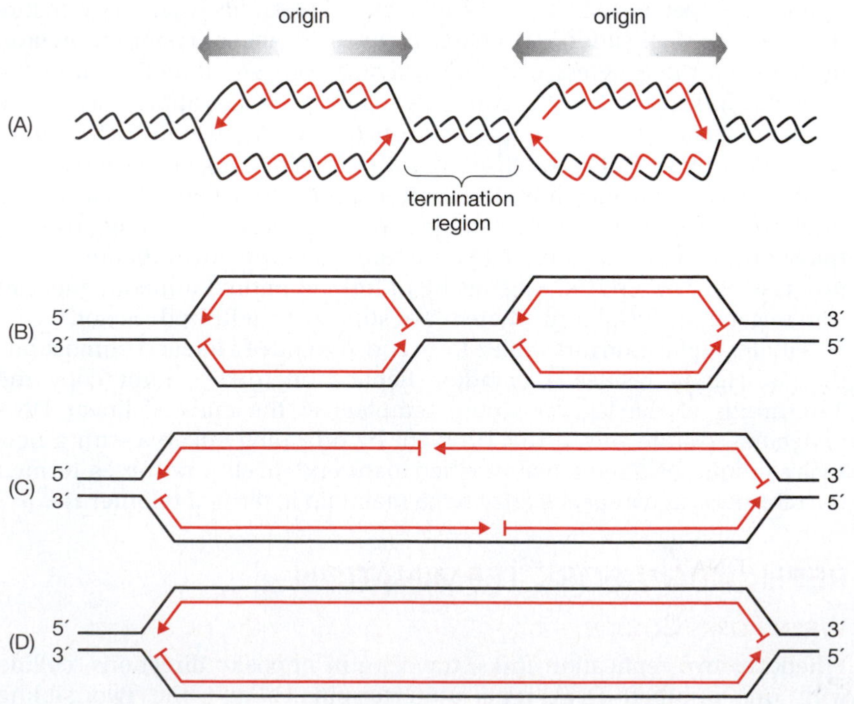

Figure 6-1. Head-on collision between two replication forks results in termination of DNA replication and separation of sibling DNA molecules. (A) Two adjacent replication bubbles are represented, each originating from a single origin of bidirectional replication (sites where two replication forks are assembled and then travel in opposite directions). (B) This is the same as (A) except DNA helical turns are omitted so that DNA strand polarities are clear. Nascent DNA strand (red) is extended at one end (arrowhead) and contains an RNA primer (vertical line) at the other end. (C) DNA unwinding continues until completed, resulting in a short gap in nascent DNA strand within the termination region of both sibling molecules. (D) DNA synthesis is completed and nascent strands are ligated together.

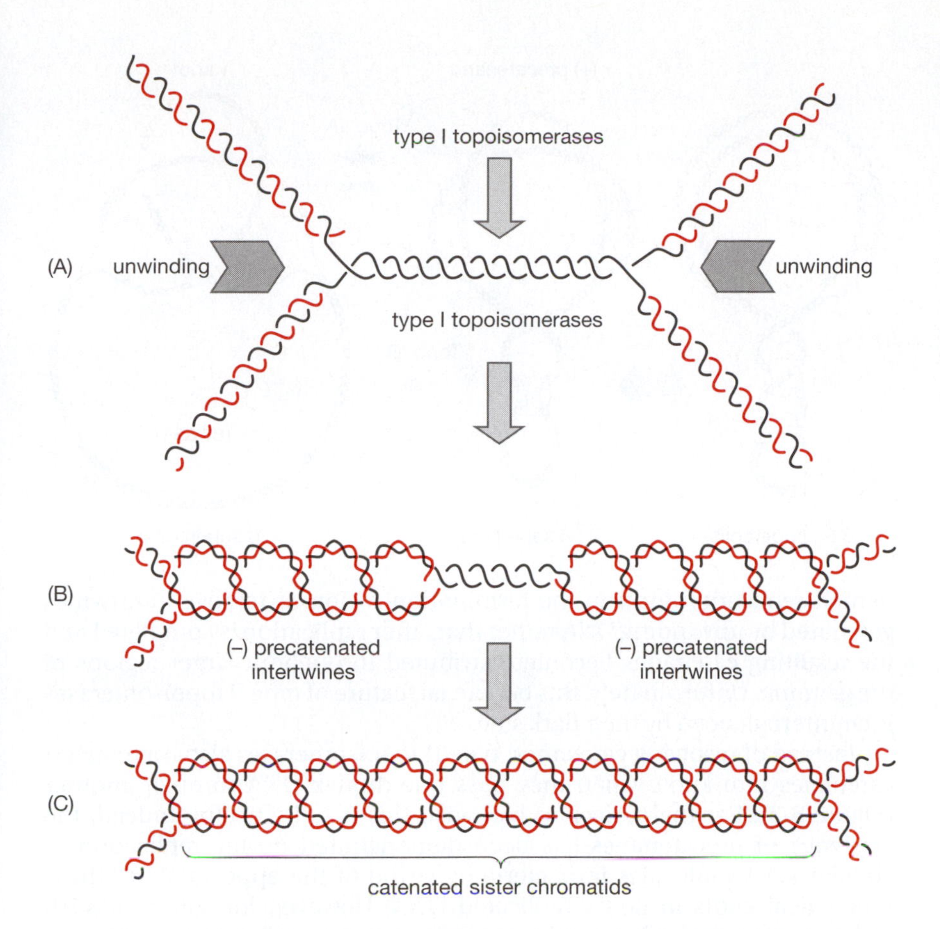

Figure 6-2. As two replication forks approach one another (A), the region between them will eventually become too small to allow type I topoisomerases to eliminate the positive superhelical turns generated in front of the fork by DNA unwinding (Figure 3-9). In that event, for each helical turn unwound, one negative precatenated intertwine will appear behind the fork (B). If these intertwines are not removed before the two sister chromatids separate, then the two chromatids will be catenated (C). (Adapted from Wang, J.C. (2002) *Nat. Rev. Mol. Cell Biol.* 3: 430–440. With permission from Macmillan Publishers Ltd.)

behind a replication fork creates two potential problems for the eventual separation of sister chromatids during mitosis. First, if the precatenanes are not removed prior to the collision of two replication forks (i.e. termination), then when the final helical turn is unwound and DNA synthesis is completed in that region, the two sister chromatids will be intertwined (i.e. catenated) (Figure 6-2). That is why a type II topoisomerase (Table 4-9) is required for termination of DNA replication and separation of sibling chromatids: they are the only enzymes that can separate precatenated intertwines behind the fork. These enzymes bind to two DNA duplexes, introduce a double-strand break in one duplex, pass the other duplex through the double-strand break, and then reseal the break, thereby potentially decatenating the two duplexes (Figure 4-25C). In bacteria, this role is played by Topo IV, and in eukaryotes by Topo II. Depletion of Topo II from budding yeast results in catenated DNA. Consequently, cells complete DNA replication and enter mitosis, but they suffer extensive chromosome missegregation.

These changes were first demonstrated with SV40 DNA replicating in the nuclei of monkey cells and with plasmid (episome) DNA replicating in the nuclei of the budding yeast *S. cerevisiae*. Following separation of sibling chromosomes, some parental DNA strands contain a gap in the nascent strand within the termination region (Figure 6-1C), and some contain catenated intertwines (Figure 6-2C). Under optimal conditions, both products are transient intermediates in DNA replication. However, conditions that inhibit Topo II activity also inhibit both DNA unwinding in the termination region and separation of catenated intertwines. Although specific DNA sequences are not required for these events to occur, unwinding parental DNA is more difficult at some genomic loci than at others. For example, the normal termination region in SV40 chromosomes as well as at *S. cerevisiae* centromeric sequences are unusual in that they promote formation of catenated intertwines when two converging replication forks enter to complete replication. This would occur if Topo II acts preferentially

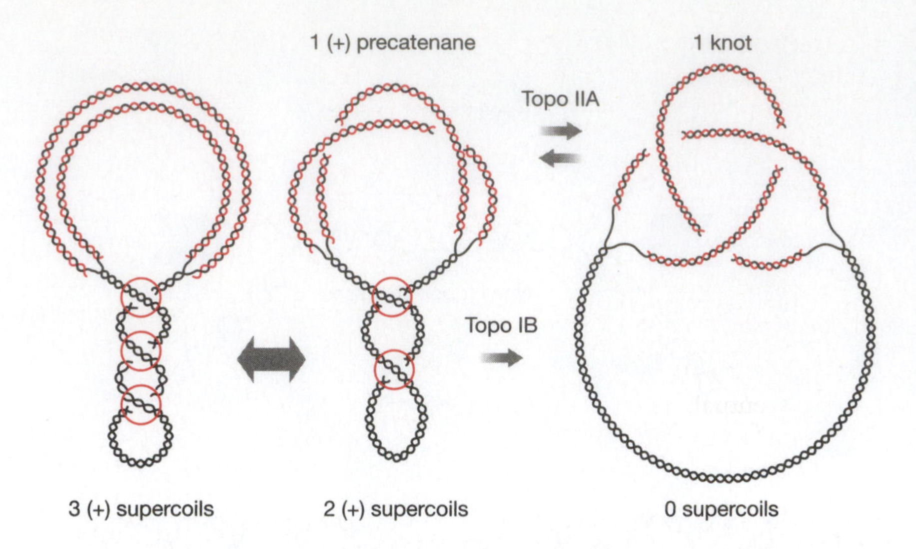

Figure 6-3. Precatenanes can become knots. DNA replication generates positive supercoils in front of the forks. A positive supercoil in front of the forks is topologically equivalent to a positive precatenane behind the forks. Precatenanes can be converted into knots by Topo II enzymes. Positive superhelical turns can be relaxed by Topo IB and Topo II enzymes (Tables 4-8 and 4-9). (Data from Schvartzman, J.B. and Stasiak, A. (2004) *EMBO Rep.* 5: 256–261.)

behind replication forks in the termination region to remove intertwines generated by unwinding DNA rather than after replication is completed and the resulting catenanes become distributed throughout larger regions of the genome. Unfortunately, this beneficial feature of type II topoisomerases is counterbalanced by their dark side.

Instead of resolving catenanes, type II topoisomerases also can convert catenanes into knots when they pass one duplex DNA through another (Figure 6-3). For the molecular biologist, this is quite useful. Indeed, the existence of precatenanes has been demonstrated during replication of circular DNA molecules in bacteria by virtue of the appearance of these topological knots in newly replicated DNA. However, formation of such knotted structures will severely impinge on the behavior and segregation of the DNA molecule. Thus, DNA sequences that promote formation of catenated intertwines in termination regions will also increase the chance that DNA knots occur. Hence, the activities of topoisomerases must be tightly regulated during the process of replication-fork convergence and termination.

REPLICATION-FORK BARRIERS

RFBs are specific DNA sequences that arrest the forward progress of a replication fork. In the circular genomes of bacteria and bacterial plasmids, RFBs ensure that genome duplication terminates on the opposite side of the genome from the origin of bidirectional replication. They do this by arresting the first fork that enters the termination region, forcing it to wait there for the inevitable collision with the fork coming from the opposite direction.

Replication-Fork Barriers In Bacteria

E. coli

In *E. coli*, replication begins at a single site termed oriC from which one replication fork proceeds in each direction. As both forks move at about the same rate, they will converge at the opposite side of the chromosome, a proposal confirmed by marker frequency mapping experiments in the early 1970s. At that time it was recognized that two potential mechanisms could exist for termination: either forks simply collide at some stochastically selected site within the termination region of the chromosome, or there exists a specific DNA site where termination occurs. Evidence that replication termination could occur at specific sites first came from studies of the *E. coli* plasmid R6K where electron microscopy revealed that fork progression occurred asymmetrically, with one replication fork terminating at a specific site within the plasmid. The first evidence that active termination mechanisms exist in the *E. coli* chromosome came from an elegant series of genetic studies that took advantage of a temperature-sensitive **allele** of the *dnaA* gene (encoding the replication initiator protein). Strains harboring this allele are incapable of initiating replication at elevated temperatures. However, chromosome replication could be restored by integration of a plasmid into the host chromosome. The origin of replication of the plasmid therefore serves as the **replicator** for the cellular genome. Plasmids were integrated at various positions around the chromosome and the pattern of replication was assayed. Remarkably, even though oriC was not functional, termination always occurred precisely across the chromosome from the chromosomal oriC, exactly as in wild-type cells. Attempts to delimit the termination region in the *E. coli* chromosome yielded initially puzzling results. Eventually, a number of specific termination sites were discovered within the overall termination region. Moreover, these termination sites were polar in nature: they blocked a replication fork coming from one side but not from the other. It is now known that there are a total of 10 RFBs (referred to as termination sites *TerA–J*) in the *E. coli* chromosome (Figure 6-4), five of which block the replication fork coming from one direction and five of which block the fork coming from the opposite direction.

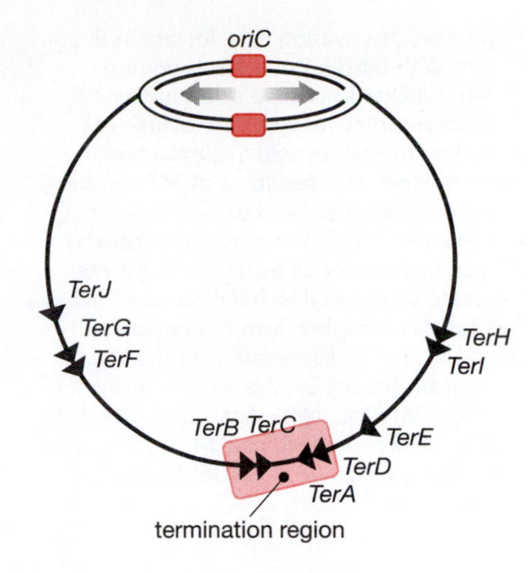

Figure 6-4. Replicating *E. coli* chromosome. Replication forks can proceed through termination sites (*Ter*) in the direction of the arrow, whereas forks arriving from the opposite direction are arrested. Thus, *TerJ, G, F, B,* and *C* permit passage of the counter-clockwise fork while *TerA, D, E, I,* and *H* permit passage of the clockwise fork. Therefore, replication usually terminates in the region defined by *TerB, C, A,* and *D* (termination region).

With the advent of genomic sequencing, it became apparent that *Ter* sites were 23 bp elements with closely related sequences based around a core consensus of 5′-AGNATGTTGTAACTAA-3′. The underlined GC base pair appears in all *Ter* sites and, as discussed below, plays a vital role in *Ter* function. Forks approaching from the 3′-end are blocked; forks coming from the 5′-end are not. The fact that there are 10 termination sites yet only two replication forks highlights the fact that the termination sites are not 100% effective: replication forks have a finite probability of getting through the barrier.

The ability of *Ter* sites to block replication forks depends upon the binding of the sites by a 36 kDa protein termed Tus (for termination-utilization substance; Figure 6-5). Tus binds specifically to *Ter* sites with an extremely strong affinity ($K_d = 3\times10^{-13}$ M), possibly the strongest recorded for a monomeric sequence-specific DNA-binding protein. Analysis of DNA replication *in vitro* has revealed that the Tus-*Ter* protein–DNA complex is unique in that it blocks the action of the *E. coli* replicative DnaB helicase. As with the block to replication-fork progression, inhibition of DnaB is dependent on the orientation of *Ter*. Thus, the polar nature of the fork block could be reconstituted with purified proteins and DNA. Now the mechanism becomes clear. DnaB translocates along the lagging-strand template in a 5′→3′ direction. DNA synthesis on the leading-strand template occurs to within four nucleotides of the absolutely conserved GC base pair in the *Ter* consensus sequence. This has been interpreted as indicating that DnaB dissociates from the replication fork when it is arrested by Tus. In support of this, lagging-strand synthesis continues up to 50 nucleotides from *Ter*. This can be attributed to the stalled DnaB positioning the DnaG primase to synthesize a primer prior to dissociation of DnaB from the stalled fork. It may be that, after dissociation of DnaB, DNA Pol-III holoenzyme is able to extend leading-strand synthesis up to the border of the *Ter* site, as well as completing synthesis of the last Okazaki fragment on the lagging-strand template. These observations also confirm that the 3′-end of nascent DNA on the leading-strand template is closer to the fork than the 5′-end of nascent DNA on the lagging-strand template, and suggests that the *E. coli* DNA primase•DNA helicase complex at replication forks requires at least approximately 50 nt of ssDNA template before it can initiate synthesis of an RNA primer.

This proposed dissociation of DnaB from the stalled fork appears to be a feature peculiar to the Tus-*Ter* complex. Studies of *E. coli* replication forks stalled *in vivo* by tandem arrays of binding sites for the tetracycline repressor (Tet-operator) revealed that these forks start moving again shortly

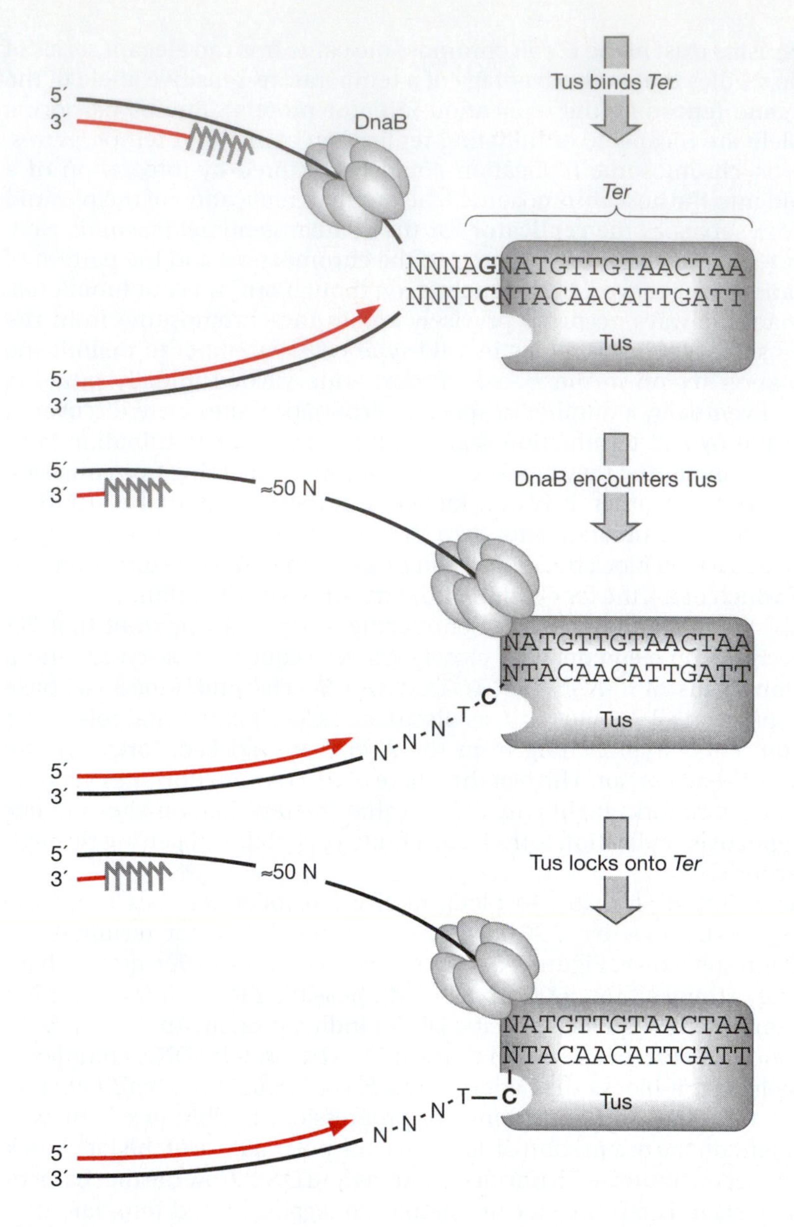

Figure 6-5. Activation of a *Ter* site in *E. coli* begins with binding of a Tus monomer. When the DnaB helicase approaches the 'nonpermissive' face of the Tus-*Ter* complex, the conserved GC base pair in *Ter* is melted. The resulting dCMP residue is then available to bind to Tus, locking the Tus-*Ter* complex in place and thereby preventing further DNA unwinding by the helicase. When a DNA helicase approaches the Tus-*Ter* complex from the opposite direction, Tus is displaced from the DNA. (Data from Bastia, D., Zzaman, S., Krings, G. *et al.* (2008) *Proc. Natl. Acad. Sci. USA* 105: 12831–12836, and Mulcair, M.D., Schaeffer, P.M., Oakley, A.J. *et al.* (2006) *Cell* 125: 1309–1319.)

after treatment of cells with anhydrotetracycline, an agent that causes the tetracycline repressor to dissociate from its DNA binding site. Therefore, in contrast to the dissociation of DnaB promoted by Tus-*Ter*, it appears that the entire replisome stalls at Tet-operator arrays.

The mechanism by which Tus arrests replication forks arriving from one direction but not from the other involves specific interactions between Tus, DnaB, and *Ter*. It appears that only one face of Tus interacts with DnaB, thereby defining the polarity of inhibition. In fact, the Tus-*Ter* complex can prevent DnaB helicase from sliding along the surface of dsDNA without unwinding it. As with DnaB helicase activity, the ability of Tus-*Ter* to block DnaB sliding activity exhibits a polarity; the block occurs only when the helicase encounters the nonpermissive face of the Tus-*Ter* complex. Moreover, a mutant form of Tus that reduces its interaction with DnaB but not its affinity for *Ter* DNA, is defective in arresting DnaB sliding. However, these data are difficult to reconcile with the observation that the Tus-*Ter* complex can block the helicase activity of a phylogenetically diverse range of replicative and repair helicases *in vitro* and does so from only one direction (i.e. polar effect).

A novel aspect of the Tus-*Ter* complex has been revealed by a series of biochemical and structural studies. An initial series of studies examined

Tus binding to artificial oligonucleotide substrates that contained varying lengths of single- and double-stranded DNA to mimic DNA unwound by a helicase. These analyses revealed that if the single-stranded regions come from the permissive site of *Ter* then progressively weaker binding of Tus is observed, in agreement with the notion that DnaB entering from the permissive side simply unwinds DNA and thus displaces Tus. In striking contrast, fraying from the nonpermissive end resulted in a 40-fold lower Tus-*Ter* dissociation rate, indicating that one role of the ssDNA was in stabilizing binding. Further experiments revealed that this stimulation is due to the accessibility of the cytosine in the invariant GC base pair of *Ter* (Figure 6-5).

These biochemical observations were substantiated by the crystal structure of Tus bound to a Y-shaped DNA in which the C was not base-paired (Figure 6-6). This revealed that the C residue flips out and away from the axis of the dsDNA, and that it is bound by a small pocket in the nonpermissive face of Tus. Thus, as the DnaB helicase approaches the nonpermissive face of the Tus-*Ter* complex it will melt the GC base pair. The C lies on the leading-strand template and so will be excluded from the inner channel of DnaB. Lying exposed on the outer surface of DnaB, it can bind the pocket on Tus, thereby locking down the Tus-*Ter* complex and consequently arresting DnaB helicase activity. Presumably, this eventually leads to dissociation of DnaB from the replication fork. How this happens remains unclear, but it isplausible that it involves direct contacts between DnaB and Tus.

Bacillus subtilis

Superficially, the termination system in *B. subtilis* is reminiscent of that in *E. coli*, with multiple *Ter* sites that act in a polar manner to arrest replication

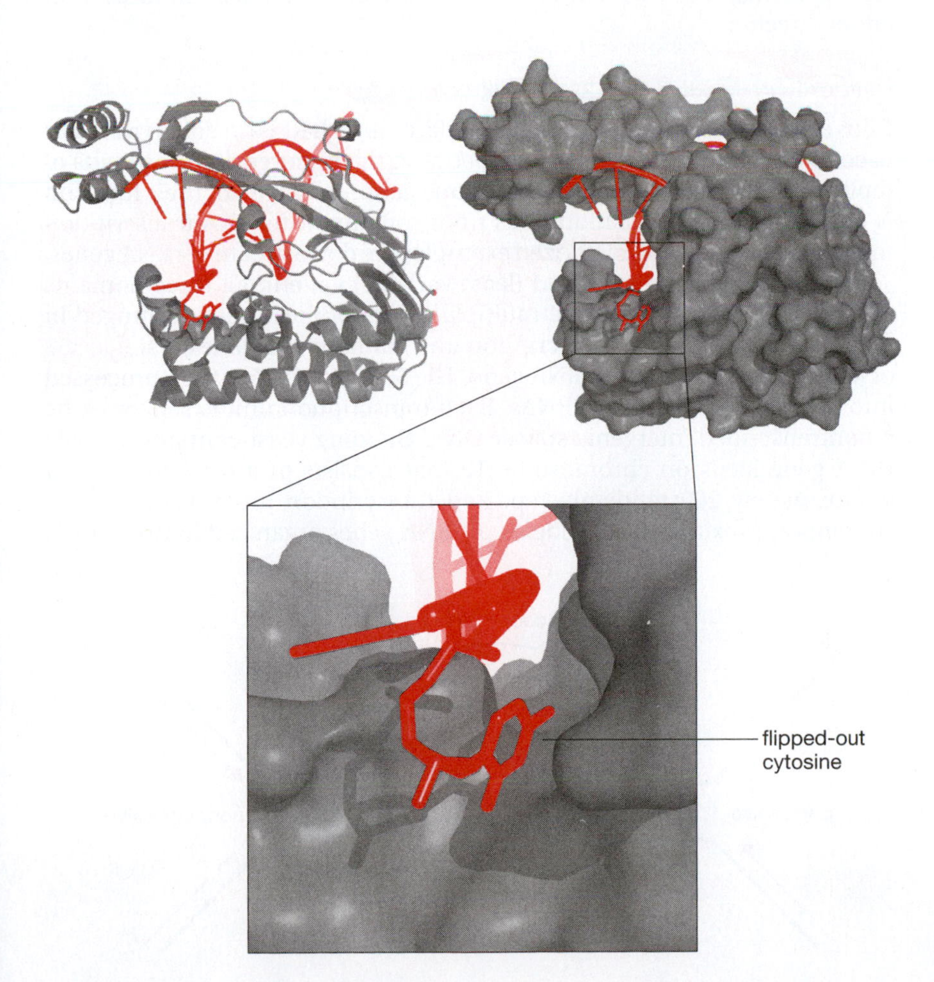

Figure 6-6. The crystal structure of Tus-*Ter* complex reveals a protein with a structural asymmetry consisting of two lobes and a DNA-binding domain consisting of a series of β-strands that invade the major groove of *Ter* DNA. The position of the pocket binding the flipped-out cytosine base (in stick form) is indicated (prepared with Pymol from PDB coordinates 2EWJ).

forks entering from either a clockwise or counter-clockwise direction. Remarkably, however, neither the *B. subtilis Ter* sites nor the protein that binds them (replication terminator protein, Rtp) are related to their counterparts in *E. coli*. The nine *B. subtilis* chromosomal *Ter* sites are 30 bp elements consisting of two binding sites for Rtp dimers. The A binding site shows considerable divergence between *Ter* sites and binds Rtp with low affinity, whereas the B binding site is conserved among *Ter* sites and contains an imperfect inverted repeat. Rtp shares no sequence or structural homology with the *E. coli* Tus protein.

The crystal structure of Rtp reveals a symmetric homodimer with each 14.5 kDa subunit possessing a winged helix-turn-helix DNA-binding fold. DNA-binding and structural studies reveal that two dimers of Rtp bind to a *Ter* site, one dimer contacts the A site, and the other contacts the B site. The B-site binding has high affinity ($K_d \approx 5\times10^{-11}$ M). Once a dimer is bound to the B site, a second dimer can bind to the much lower-affinity A site (Figure 6-7). This is an example of cooperative protein binding. It is mediated, at least in part, by direct protein•protein interactions between the two dimers, with contacts being mediated by the wing portion of the winged helix-turn-helix domain.

Like the *E. coli* Tus-*Ter* system, Rtp–*Ter* blocks the action of the replicative DNA helicase in a polar manner. However, the mechanism employed by Rtp–*Ter* is distinct from that employed by Tus–*Ter*. When *B. subtilis* DnaB helicase at replication forks encounters the A site first, it displaces the weakly bound Rtp dimer. Loss of the cooperative protein•protein interactions between the A and B sites weakens the B site-Rtp interaction, thereby releasing this Rtp as well. In contrast, forks approaching the tightly bound B site–Rtp complex are unable to displace the Rtp dimer and therefore are arrested. An atypical *Ter* site containing two B sites and no A sites arrests replication forks from either direction.

Replication-Fork Barriers In Eukarya

RFBs exist in eukarya as well as in bacteria, although RFBs in eukarya are not used to define termination regions between the hundreds to thousands of replication origins that exist in eukaryotic genomes. Instead, their function is to prevent DNA replication forks from colliding with DNA transcription machines. The best-characterized examples are downstream of rRNA genes.

rRNA is the most conserved (least variable) gene in all three domains of life, and eukaryotes encode multiple copies of these genes arranged in tandem arrays. A typical transcription unit contains two genes, a single 35S or 45S rRNA gene and a 5S rRNA gene. The 35S/45S rRNA is then processed into the 18S, 5.8S, and 28S rRNAs. Each transcription unit is separated by a nontranscribed intergenic spacer DNA. Budding yeast contains a single rRNA gene locus on chromosome 12 that consists of a tandem array of approximately 200 tandemly repeated transcription units. Fission yeast contains approximately 100 copies of rRNA genes organized in two arrays,

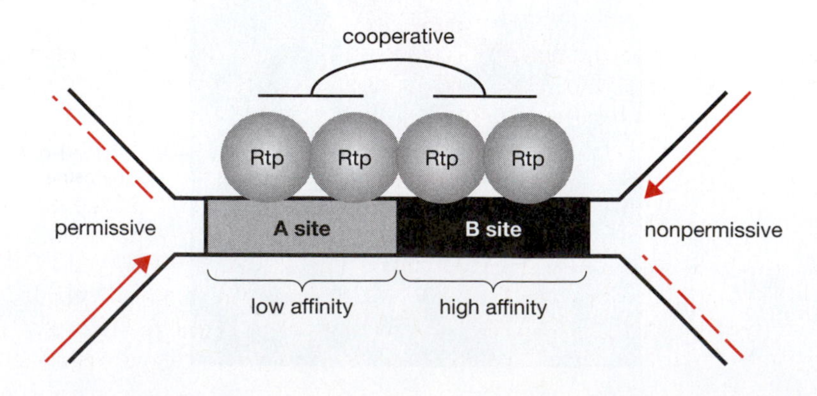

Figure 6-7. A dimer of the replication terminator protein (Rtp) binds to the high-affinity B site in the *B. subtilis Ter* sequence, which in turn facilitates binding of Rtp to the low-affinity A site through cooperative protein•protein interactions. Forks approaching the low-affinity A site can displace the bound Rtp, thereby reducing the affinity of Rtp for the B site and permitting the fork to pass. However, forks approaching the high-affinity B site cannot displace Rtp and therefore are arrested.

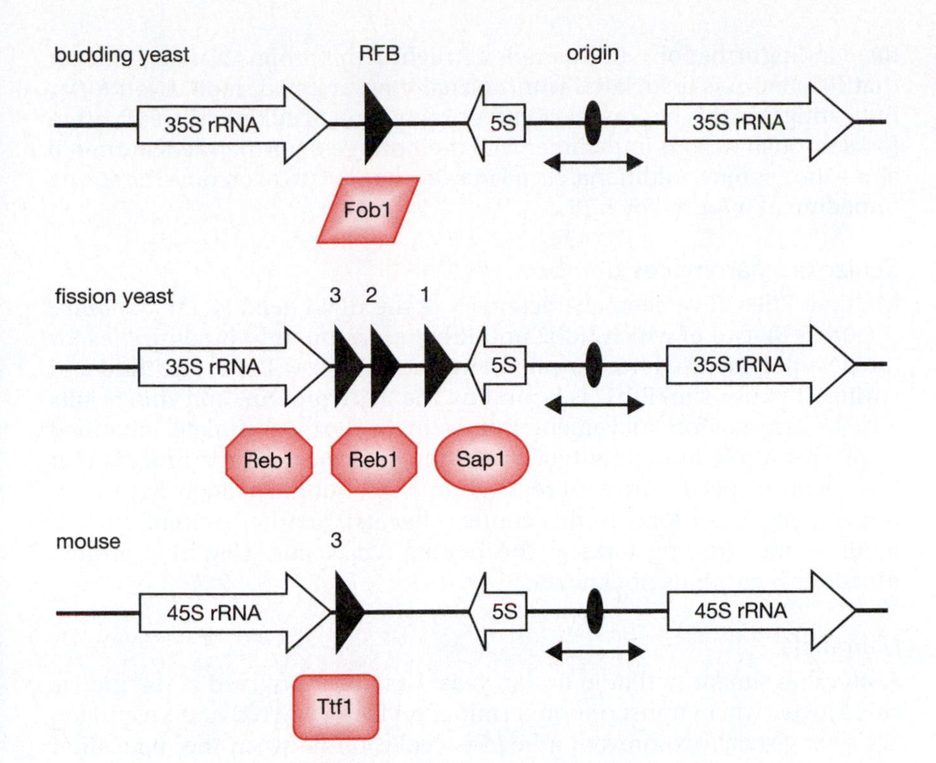

Figure 6-8. Replication-fork barriers (black triangles) in the rRNA gene repeats of eukarya, and the proteins that bind them. The directions of transcription of the 35S/45S rRNA gene cluster and the 5S rRNA gene are indicated by open arrows. An origin of bidirectional replication (black oval) is located between the 35S/45S rRNA gene cluster and the 5S rRNA gene. Forks originating here travel in the same direction as the transcription machine through the rRNA genes. RFBs located downstream of the 5S rRNA gene prevent the replication fork from entering the neighboring 35S/45S rRNA gene cluster.

one near each end of chromosome 3. Humans contain approximately 400 rRNA repeats organized into five clusters on chromosomes 13, 14, 15, 21, and 22. Each repeat contains an origin of bidirectional DNA replication (Figure 10-14) in which one fork travels in the same direction as the transcription machine through the 35S/45S rRNA gene cluster, and the other fork travels through the 5S rRNA gene (Figure 6-8). However, one of the forks would soon collide head-on with the transcription machine from the next 35S/45S precursor, if it were not for the presence of one or more RFBs.

Saccharomyces cerevisiae

One of the best-characterized blocks to eukaryotic replication forks is the RFB in the ribosomal DNA repeats of budding yeast. This RFB is bound by the Fob1 protein, although the mechanism by which this protein exerts its polar effects remains unclear. By analogy with the bacterial system, it might be anticipated that the Fob1-RFB complex will inhibit passage of the replicative MCM helicase. Genetic experiments have revealed that Fob1-RFB-mediated fork arrest results in the accumulation of a number of components of the replication-fork machinery at the arrest site, including MCM, GINS, Cdc45, DNA Pol-α, and DNA Pol-ε, suggesting that the replisome remains intact. Forks arrested at Fob1-RFB contain only a few nucleotides of ssDNA, and therefore do not activate the DNA damage-response pathway (Chapter 13). This pathway senses the accumulation of ssDNA-RPA complexes that occurs when replication forks stall because of poor replication conditions (e.g. low dNTP pools). Stalled yeast replication forks, for example, contain approximately 300 nt of ssDNA. Fob1-RFB functions normally even when the DNA damage-response pathway has been genetically inactivated.

Significantly, arrest of replication forks at this RFB is not absolute. Using two-dimensional gel electrophoresis to fractionate DNA replication intermediates (Figure 8-3), it was observed that the fork encountered the block, paused for an extended period of time, and then progressed through the barrier. Interestingly, the levels of the Rrm3 helicase increased at the arrest site during the course of the pause. Thus, the arrested fork may emit a distress signal that facilitates recruitment of Rrm3, and Rrm3 may help overcome the Fob1-induced barrier, perhaps by Rrm3-dependent dismantling of the Fob1-RFB complex. In light of this proposal, it is significant that Rrm3 also is associated with normal, non-arrested, replication forks, indicating that this helicase facilitates progression of eukaryotic replication forks through various impediments to their progress normally encountered along the genome. Additional Rrm3 may be required to overcome the strong impediment imposed by RFBs.

Schizosaccharomyces pombe

Multiple RFBs have been characterized at the rRNA gene loci in *S. pombe* (Figure 6-8), two of which (Rfb2 and Rfb3) correspond to binding sites for the RNA polymerase I transcription termination factor Reb1. The third and strongest pause site, Rfb1, is bound by the Sap1 protein, and this results in polar arrest of fork movement. Interestingly, Sap1 was initially identified as playing a role in mating-type switching in fission yeast, a process that also involves specific arrest of replication forks, but intriguingly Sap1 does not actually arrest forks in this context. Instead, another protein, Rtf1, is required for arresting forks at the mating-type locus. How this protein functions is currently not known.

Mammals

A situation similar to that in fission yeast has been observed at the mouse rRNA locus, where transcription termination factor 1 (Ttf1) and Ku antigen act synergistically to prevent a head-on collision between the replication and the transcription machinery (Figure 6-8).

OTHER IMPEDIMENTS TO REPLICATION FORKS

RFBs are not the only impediments to replication-fork movement. Forks stall (pause) when they encounter unusual DNA features that either impede DNA unwinding of duplex DNA or DNA synthesis on the template DNA. Such features include Z-DNA (Figure 1-9), hairpins in ssDNA or cruciforms in dsDNA that result from inverted repeat sequences (Figure 1-10), triple-helical (triplex) DNA (Figures 1-11 and 1-12), G-quadruplex DNA (Figures 1-13 and 1-14), and slipped-strand DNA (Figure 1-15). In the absence of a functional S-phase checkpoint, the replisome disengages, and the stalled replication forks collapse and give rise to chromosome breakage and rearrangements. Thus, DNA sequences that cause forks to slow down can appear as 'fragile sites'. These are genomic sites that are more sensitive to replication stress than other regions and thus require S-phase checkpoint proteins for fork progression. Replication-fork pause sites in mammalian genomes have been related to a number of human genetic diseases (Chapter 14).

DNA templates that contain inverted repeats generally arrest DNA synthesis when a DNA polymerase encounters the base of the presumptive hairpin structure. In fact, these 'hairpin-arrest sites' can be distinguished from nonhairpin-arrest sites by virtue of the fact that DNA polymerase will stop exactly at the base of the hairpin when arriving from either direction (either the $5' \rightarrow 3'$ or $3' \rightarrow 5'$ complementary template strand). In contrast, sequences also have been identified that arrest DNA polymerase coming from one direction, but not from the other. Similarly, G-quadruplex DNA and homopurine/homopyrimidine-rich DNA that can form triple-stranded structures also impede DNA polymerases. On a single-stranded template, the template strand can fold back on the newly synthesized strand to form a triplex DNA structure that could trap the polymerase. On a double-stranded (nicked circular) template, the nontemplate strand could fold onto the duplex in front of the polymerase as it is displaced by DNA synthesis. This would form a triplex structure in front of the polymerase through which the polymerase could not pass.

THE ROLE OF REPLICATION-FORK BARRIERS IN NATURE

With the exception of bacterial genomes, RFBs are infrequently used in nature. At a purely intuitive level, this makes sense. If defined termination sequences existed between all of the hundreds to thousands of replication origins in eukaryotic genomes, then every origin would need to be activated every time a cell divides. Otherwise, gaps of unreplicated DNA would appear and sibling chromosomes would fail to segregate during mitosis. Moreover, given the fact that DNA sequence signals are not required to terminate DNA replication-fork activity, simply a collision between two oncoming forks, there is no advantage for RFBs to define specific termination regions, unless of course their primary purpose is not to facilitate termination of replication, but to facilitate other cellular processes.

Many circular genomes in bacteria use RFBs to trap replication forks in the termination region, yet disruption of these termination systems does not cause an obvious defect in growth or viability. Nevertheless, replication-termination systems provide an evolutionary advantage by contributing to accurate chromosome partitioning. With a single circular bacterial chromosome, duplication of a region of the chromosome is followed closely by their segregation to opposite halves of the cell. Early replicated regions are partitioned before the bulk of the chromosome is duplicated. Once the two new chromosomes have been completed and partitioned to opposite sides of the cell, cell division occurs. Thus, chromosome partitioning involves three events. First, the region containing the replication origin is separated and repositioned. This is followed by chromosome compaction. Finally, the termination region separates. This includes chromosome decatenation, chromosome dimer-to-monomer resolution when necessary, and movement of the termini to either side of the cell before completion of medial septation. Partitioning of the terminus region requires that any chromosome dimers that may have arisen be resolved into monomers, and cleared from the region where the division septum forms. Inactivation of Rtp in *B. subtilis*, in combination with mutations in proteins required for movement of the chromosome away from the division septum, increases the production of anucleate cells.

A second advantage of RFBs that applies to both prokarya and eukarya is their ability to prevent replication forks from colliding with transcription machines coming from the opposite direction. Such collisions have been shown to cause fork stalling, a phenomenon known to increase the probability of homologous recombination at replication forks and chromosome breakage. Such problems would be greatest among closely packed arrays of genes. In the *E. coli* genome, for example, there is a strong correlation between the direction of DNA replication forks and the orientation of genes so that transcription machines and replication forks travel in the same direction from the replication origin to the termination region. Thus, RFBs in the termination region of bacterial genomes prevent forks from traveling counter to the general direction of bacterial gene transcription. Similarly, RFBs are distributed throughout the rRNA gene repeats in yeast and mammals such that they prevent replication forks from traveling counter to the direction of rRNA gene transcription. Although the Fob1 protein of budding yeast does not appear to be evolutionarily conserved, its analog in fission yeast (Reb1) has its counterpart in the Ttf1 protein of mice and humans.

THE TERMINATION PARADOX

Given the structure and dynamics of DNA replication forks, linear DNA genomes that initiate replication at one or more internal sites face a serious problem. Leading-strand DNA synthesis can continue until the replicative

DNA polymerase incorporates the nucleotide complementary to the 5′-terminal nucleotide in the leading-strand template, but lagging-strand DNA synthesis will inevitably leave the 3′-terminus of the lagging-strand template unreplicated. Note, for example, that replication forks arrested at RFBs contain at least 50 nt of single-stranded lagging-strand template. Even if DNA primase could initiate RNA primer synthesis at the 3′-terminal nucleotide in the lagging-strand template, no mechanism exists for completely removing the RNA primer and filling in the resulting gap. Thus, the replication-fork mechanism selected by nature to duplicate the genomes of all living things cannot accomplish this task by itself. Yet, it is axiomatic that the genome must be duplicated completely each time a cell divides.

Protein-Nucleotide Primers

The simplest solution to the termination paradox would be a mechanism that initiates replication at the ends of linear genomes. In fact, a mechanism exists that utilizes a protein-dNMP primer that binds to a unique sequence at the termini of linear DNA molecules and thereby initiates continuous DNA synthesis from the 3′-terminal nucleotide in the template to the 5′-terminal nucleotide. However, this mechanism is used only by moderately sized viral genomes such as ϕ29 bacteriophage and adenoviruses, because it generates a great deal of ssDNA and therefore is not suitable for the large genomes in eukaryotes (Figures 8-7 and 10-2).

Circularization

Another solution to the termination paradox is employed by linear genomes of bacteriophage and animal viruses such as phage λ, phage T4, and herpes simplex virus. These genomes are linear within the virus particle, but circularize as a result of their unusual termini (cohesive ends) when they enter their host cell's nucleus. They begin DNA replication by the θ-mode (Figure 8-1A), and then later switch to a rolling-circle or sigma (σ) mode (Figure 8-8) that allows a virus-encoded endonuclease to cut the long DNA **concatemers** into genome-length DNA monomers. The DNA monomers are then packaged into virus particles called **virions**.

Concatemer Formation

Still a third solution involves formation of linear concatemers containing multiple copies of the genome linked head-to-tail. This mechanism, as exemplified by phage T7, requires termini with complementary sequences. Phage T7 encodes several proteins that are involved in cutting the concatemer into genome-length segments that are then packaged into virions.

Telomeres And Telomerase

None of the above solutions are practical for the long linear chromosomes found in eukarya. Instead, eukaryotic chromosomes contain a unique DNA sequence at each end termed a telomere that is maintained by the enzyme telomerase. Telomeres provide a protective cap on chromosome ends by assembling a nucleoprotein structure on tandemly arrayed, short (5–26 bp), TG-rich sequences. Telomeres do not encode genetic information, rather they serve as binding sites for specific proteins. Some minimal length of the telomeric repeat appears necessary for maintaining the DNA–protein complex that prevents telomere loss. Loss of telomere function leads to defects in chromosome segregation and hence to genomic instability. A primary role for telomerase, therefore, is to maintain telomere length. In dividing cells with active telomerase, telomere length generally ranges from about 0.1 to 100 kbp, depending on the species, but the population of telomeres is maintained within well-defined limits specific to a particular cell type (Table 6-1). Loss of a critical number of TG repeats eventually

Table 6-1. Telomere Sequence And Length

Species	Telomere repeat	dsDNA portion	ssDNA portion
Ciliated protozoan, *Oxytricha nova*	5′-GGGGTTTT-3′	20 bp	16 nt
Ciliated protozoan, *Tetrahymena*	5′-TTGGGG-3′	250–300 bp	14–21 nt
Budding yeast, *S. cerevisiae*	5′-$TG_{(1-3)}$-3′	350 bp	12–14 nt
Fission yeast, *S. pombe*	5′-TTAC(A)GG-3′	300 bp	≤50 nt
Vertebrates, *H. sapiens*	5′-TTAGGG-3′	2–50 kb	35–600 nt

leads to activation of DNA damage checkpoint pathways and to cellular senescence or apoptosis, with important implications for human disease.

Telomere replication

The bulk of each telomere is replicated during **S phase** by the nearest replication fork. Telomere DNA terminates with a short 3′-end overhang (Figure 6-9). When the nearest replication fork passes through it, one chromatid will contain a blunt end, the result of leading-strand DNA synthesis, whereas the other chromatid will contain a recessed 5′-end on its C-rich telomeric strand, a consequence of lagging-strand DNA synthesis. This chromatid is the natural substrate for telomerase which requires a short region of ssDNA with which to align its RNA template (Figure 6-10). Since both ends of linear DNA molecules acquire G-rich overhangs during replication, their occurrence is not determined by which strand is the

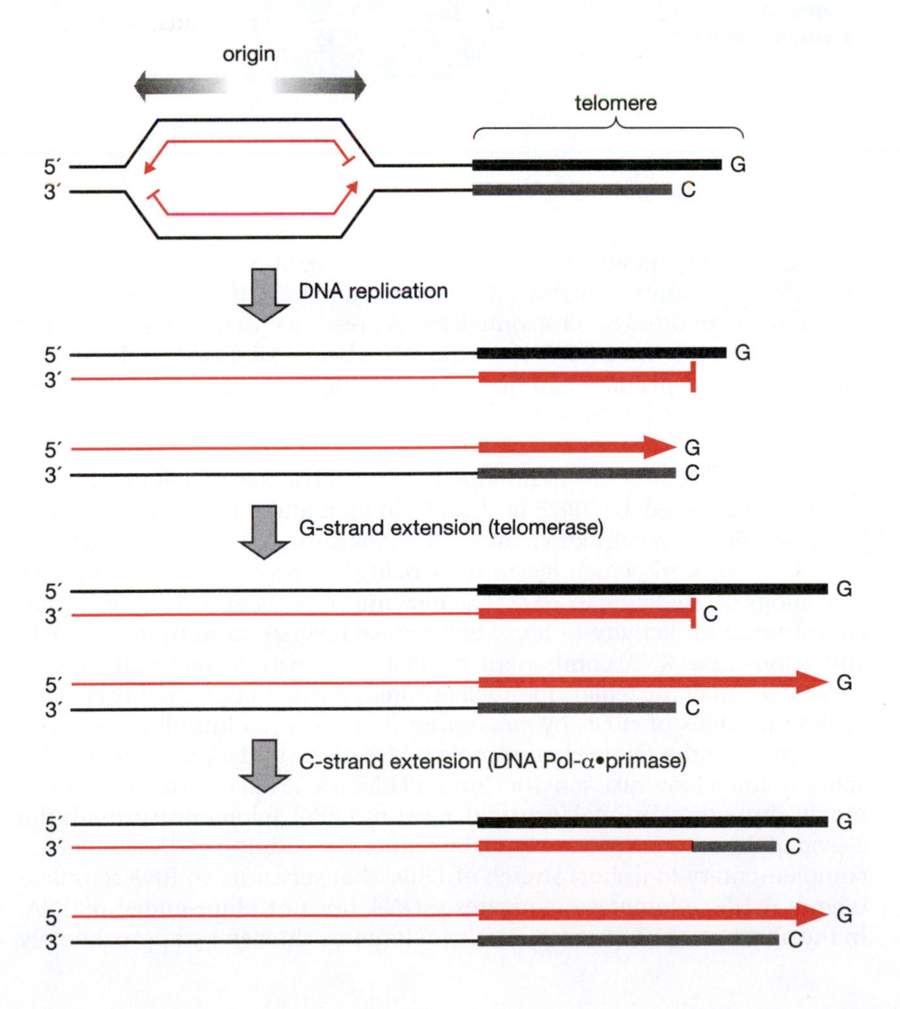

Figure 6-9. Telomere replication and telomerase-mediated extension. The G-rich telomere strand (dark bar labeled G) protrudes beyond the duplex. A DNA replication fork from a nearby origin of bidirectional replication enters the telomere. Leading-strand synthesis should completely replicate the C-rich template (gray bar labeled C), but lagging-strand template synthesis will leave the G-rich template terminus as a ssDNA overhang. Telomerase extends the G-rich strand (one C-rich end must be resected by a 5′→3′ exonuclease). The G-rich telomere extension product becomes the template for extending the C-rich strand using DNA Pol-α•primase to initiate RNA-primed DNA synthesis.

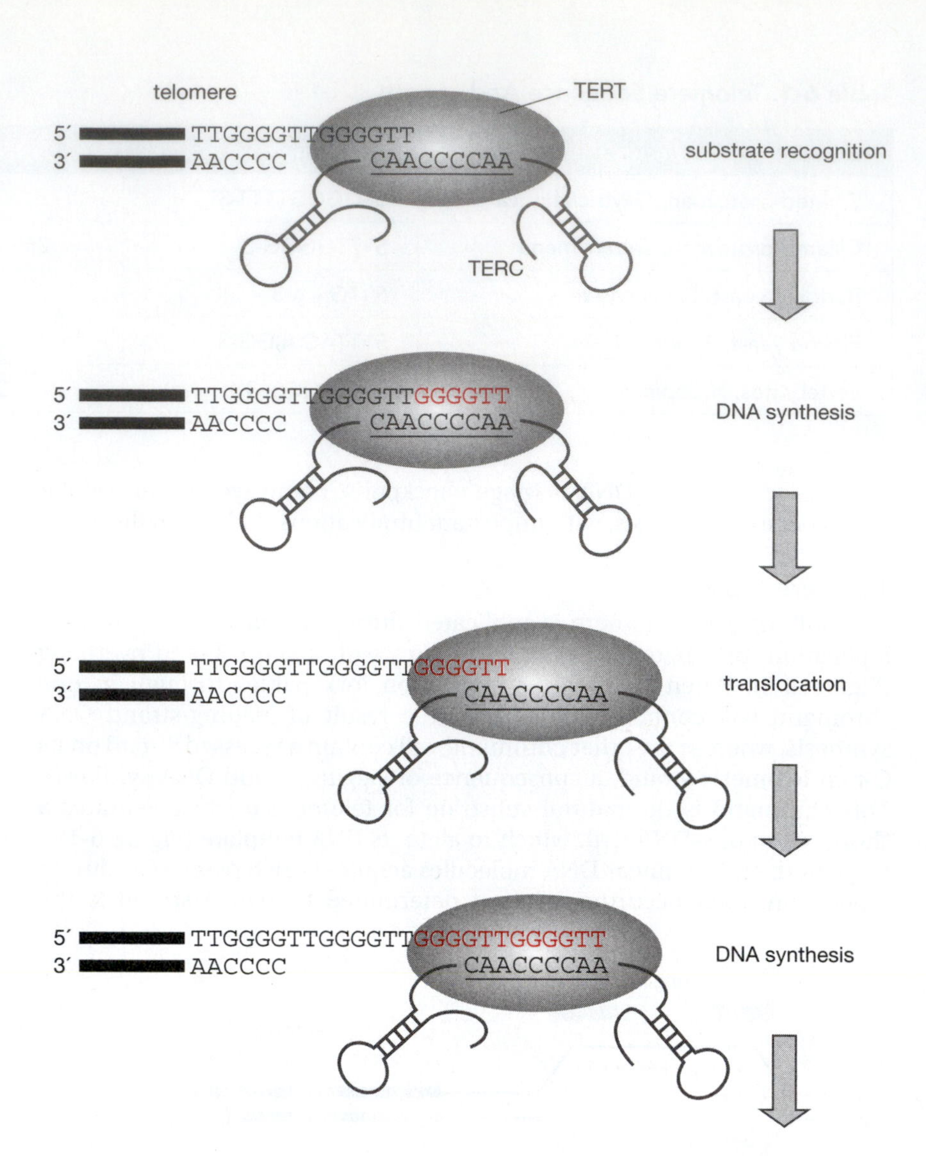

Figure 6-10. Processive extension of telomeres by telomerase. All telomerases contain an RNA template that recognizes the 3′-terminus of the G-rich strand of telomeres and incorporates the remaining nucleotides of the telomeric repeat. Thus, telomerase is a member of the reverse transcriptase family of enzymes that can synthesize DNA on an RNA template. In this example, *Tetrahymena thermophila* telomeric DNA is recognized by its cognate TERT protein (ellipsoid). TERT includes an RNA component that contains a hairpin structure at each end of the template sequence. These secondary structures serve to define the 3′- and 5′-boundaries of the template. The 3′-end of the telomere DNA anneals with the RNA template (underlined). Template-directed addition of nucleotides (red) to the 3′-end of the telomere occurs processively until the 5′-end of the template is reached. Telomerase translocates along the 3′-end of the DNA until it recognizes the 3′-template boundary whereupon another round of nucleotide addition begins. Repeat addition is regulated by interactions between TERT and the DNA template, as well as with telomerase-associated proteins. (Data from Osterhage, J.L. and Friedman, K.L. (2009) *J. Biol. Chem.* 284: 16061–16065.)

lagging-strand template and which is the leading-strand template. In fact, (3′)G-rich overhangs are detected in yeast and in mice that lack telomerase function. Therefore, a 5′→3′ exonuclease is presumed to resect the 5′-end of the C-rich strand. Once the G-rich strand has been extended by telomerase, then DNA Pol-α•primase can use it as a template to extend the 5′-end of the C-rich strand by the same mechanism used in lagging-strand DNA synthesis.

Telomeres cannot be maintained in the absence of telomerase, an enzyme discovered in 1985 by Carol Greider and Elizabeth Blackburn using *Tetrahymena* cell extract and substrates designed to mimic telomeric DNA termini, work which led to their being awarded the Nobel Prize for Physiology or Medicine in 2009. The enzyme consists of two subunits that are essential for activity *in vitro*: telomerase reverse transcriptase (TERT) and telomerase RNA component (TERC). Dyskerin, a nucleolar protein present in small nucleolar ribonucleoprotein particles that modify specific uridine residues of rRNA by converting them to pseudouridine, is also a component of the telomerase complex. Mutations in dyskerin affect TERC stability and telomerase activity. Human TERT is a 127 kDa protein that folds into its three-dimensional form with a 451 nt TERC. Telomerase extends the 3′-end of the telomeric DNA repeat by sequential addition of dNTPs that are complementary to a short stretch of TERC that serves as an RNA template (Figure 6-10). Telomerase elongates ssDNA but not blunt-ended dsDNA. In the absence of telomerase activity, telomeres shorten by approximately

3–5 bp per cell division in *S. cerevisiae*, and by 50–100 bp per cell division in cultured human cells.

The structure of the TERT protein component of telomerase has been solved. It consists of three highly conserved domains organized into a ring-like structure. It has an RNA-binding domain, a reverse transcriptase domain, and a domain analogous to the thumb domain of DNA polymerases (see Figure 6-11 in the color plate section). The RNA-binding domain mediates interactions with the RNA subunit of telomerase. The reverse transcriptase domain harbors the catalytic site of the TERT enzyme and is similar to the palm and finger domains of B-family DNA polymerases. The final domain occupies an analogous position to the thumb domain of DNA polymerases. Although the fold of this domain appears to be unrelated to any other proteins, its positioning in TERT relative to the fingers and palm domain suggests that it will play a role analogous to the thumb domain of DNA polymerases. The overall organization of these domains is similar to the gross architecture of B-family DNA polymerase, which includes the eukaryotic replicative polymerases α, δ, and ε (see Figure 6-11 in the color plate section).

DNA replication is a prerequisite for telomerase activity at the ends of telomeres. A functional origin of replication is necessary for activation of telomerase on the telomeres of a yeast episome, and elongation of shortened telomeres in yeast is dependent on origin firing. Presumably, the requirement for DNA replication has to do with the generation of ssDNA overhangs and the transient opening of the telomere-protein complex during S phase. The importance of the lagging-strand synthesis machinery in telomere maintenance has been demonstrated genetically in budding yeast and in mouse cells where telomere length increases when either DNA Pol-α or the clamp loader are made partially defective. Telomere elongation under these conditions depends on the presence of active telomerase, emphasizing the intimate relationship between the lagging-strand synthesis apparatus and telomerase. The emerging view is that C-strand synthesis is tightly coupled to telomerase activity. In yeast, this regulation is controlled by a specialized RPA-like ssDNA-binding complex (Cdc13, Stn1, and Ten1) that interacts with DNA Pol-α. Mutations in this complex lead to both overhang and telomere elongation. A second mechanism links telomerase activity to telomere length. In yeast, telomeres normally are replicated late during S phase, because the neighboring replication origin is activated late during S phase. However, shortened telomeres provoke the firing of nearby DNA replication origins early in S phase. Early telomere replication correlates with increased telomere length and telomerase activity. The mechanism is unclear, but it presumably reflects cell-cycle-dependent modulation of the composition of the telomere-protein complex.

Regulating telomere length

Genetic and biochemical analyses in budding yeast have made it clear that the presence of a catalytically active telomerase is not sufficient to guarantee telomerase activity at chromosome ends. Only 20–50 copies of telomerase are present in human cells. Instead, interaction between telomerase-inhibiting factors and telomerase-promoting factors that bind specifically to telomere DNA modulate telomerase activity (Table 6-2). For example, the TTAGGG repeats of mammalian telomeres associate with a six-protein complex termed shelterin (Figure 6-12). Shelterin enables cells to distinguish their natural chromosome ends from DNA breaks, represses DNA-repair reactions, and regulates telomerase-based telomere maintenance. The components of shelterin (Tin2, Trf1, Trf2, Rap1, Tpp1, Pot1) localize specifically to telomeres; they are abundant at telomeres throughout the cell cycle, and they do not function elsewhere in the nucleus. Three of these proteins are responsible for recognizing the TTAGGG. Telomeric repeat binding factors 1 (Trf1) and 2 (Trf2) bind the

Table 6-2. Telomere-Binding Proteins

Organism	Binds to dsDNA repeat	Recruited to dsDNA repeat	Binds to ssDNA repeat	Recruited to ssDNA repeat
Oxytricha			TEBPa, TEBPb	TERT (telomerase)
S. cerevisiae	Rap1	Rif1, Rif2	Cdc13	Stn1, Ten1, TERT (Est2•Est1)[a]
S. pombe	Taz1	Rap1, Poz1, Tpz1, Ccq1	Pot1	TERT
Vertebrates	Trf1, Trf2	Rap1, Tin2	Pot1	Tpp1, TERT

[a]Est2•Est1 are the catalytic and RNA-binding subunits, respectively, of *S. cerevisiae* telomerase.

duplex DNA portion of telomeres, whereas protection of telomeres 1 (Pot1) binds the single-strand TTAGGG repeats at the 3′-end of the telomere. Trf1 and Trf2 recruit the remaining four shelterin components to telomeres: Trf-interacting nuclear protein 2 (Tin2), Rap1 (the human ortholog of the yeast repressor/activator protein 1), Tpp1, and Pot1.

The current conundrum is to explain how the low-abundance telomerase is recruited to short telomeres containing few shelterin components and excluded from long telomeres containing many shelterin components. Trf1 and other shelterin components increase with the number of TTAGGG repeats, thus providing a mechanism by which the cell can measure the length of its telomeres. Increasing the amount of telomere-bound TRF1 through overexpression results in progressive telomere shortening, whereas a form of Trf1 that removes the endogenous Trf1 from telomeres induces telomere elongation. Tin2, Tpp1, and Pot1 behave as negative regulators of telomerase-mediated telomere elongation. At least one shelterin component, Tpp1, interacts directly with telomerase, and one shelterin component, Pot1, binds specifically to the 3′-overhang, the substrate targeted by telomerase. Pot1 plays a critical role in regulating telomerase activity, because diminished loading of Pot1 or replacement of the endogenous Pot1 with a mutant that lacks the DNA-binding domain leads to telomerase-dependent telomere elongation. Direct competition between Pot1 and telomerase for the 3′-end of the single-stranded overhang has been observed *in vitro*. When sufficient telomere–protein complex is present, the resulting higher-order structure prevents telomerase from

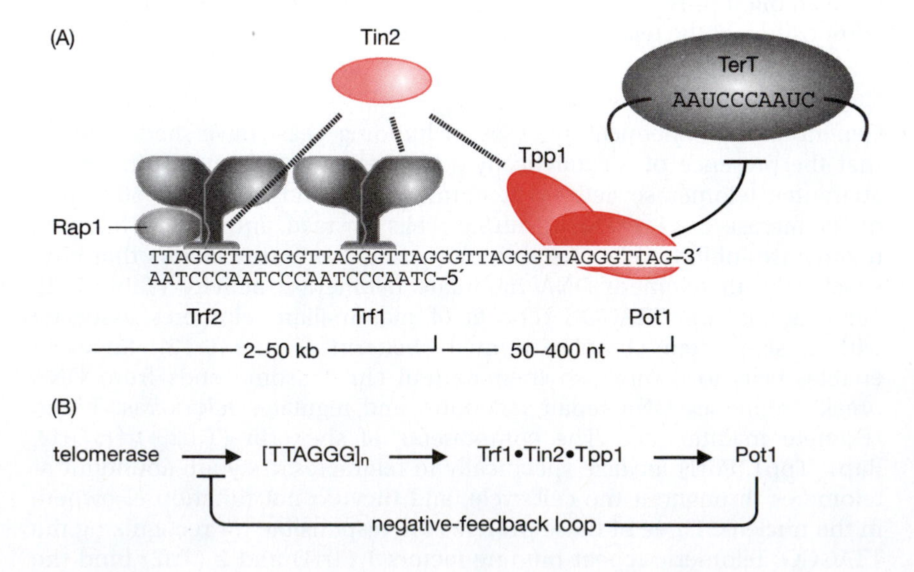

Figure 6-12. Regulation of telomerase activity by the shelterin complex. (A) Tpp1 interacts directly with telomerase and therefore may facilitate recruitment of the enzyme to telomeres in the absence of Pot1. Pot1 appears to be a negative regulator of telomerase by virtue of its ability to compete with the enzyme for the telomere's 3′-end. (B) Telomere length regulation is achieved through a negative-feedback loop in which a negative regulator of telomerase (Pot1) is loaded onto telomeres in a manner that is dependent on telomere length. (Data from Palm, W. and de Lange, T. (2008) *Annu. Rev. Genet.* 42: 301–334.)

extending the 3′-end of the G-rich strand. During subsequent cell divisions, telomeres will therefore become progressively shorter (the consequence of nuclease action) until at some point telomerase is again recruited to their 3′-termini.

The recruitment process has been investigated extensively in budding yeast. Telomeres associate with Cdc13, a ssDNA-binding protein with specificity for TG-rich telomeric repeats. In the absence of Cdc13, yeast telomeres get shorter. Cdc13 functions by recruiting the *S. cerevisiae* telomerase catalytic subunit (Est2) and a protein (Est1) that binds telomerase RNA. Est1 is required for telomerase activity *in vivo*. The ability of Cdc13 to recruit telomerase appears to be regulated by phosphorylation; when the telomere is sufficiently short, a protein kinase is recruited to the ssDNA terminus that phosphorylates Cdc13, thereby allowing it to recruit telomerase.

Some of these proteins also are required to facilitate passage of replication forks through telomeres. The ability of replication forks to move through the telomeres of *S. pombe* has been monitored using two-dimensional gel electrophoresis. In the absence of Taz1, forks stall as they enter the telomere, regardless of whether the telomeric G-rich strand is replicated by leading- or lagging-strand synthesis. In contrast, the Taz1-interacting protein Rap1 is dispensable for telomeric fork progression. This would account for the rapid loss of telomeres when Taz1 is inactivated. Presumably, the human orthologs of Taz1 (Trf1 and Trf2) facilitate fork passage through human telomeres.

Finally, the unusual GC-rich repeated sequence of telomeric DNA can give rise to unusual DNA structures called triple-helical (triplex) DNA (Figures 1-11 and 1-12D) and G-quadruplex DNA (Figures 1-13 and 1-14) that have been suggested as mechanisms for protecting the ends of telomeres and for regulating their length. The current data favor formation of t-loops in which the single-stranded telomere terminus loops back and invades the double-stranded telomeric DNA, creating a displacement loop (Figure 1-12E). This structure is stabilized by the shelterin complex.

SUMMARY

- Termination of DNA replication forks occurs naturally when two oncoming forks collide. No specific DNA sequences are required, only the action of topoisomerases that can decatenate intertwined duplex DNA molecules.
- Arrest of DNA replication forks occurs at specific DNA sequences that bind specific proteins termed replication-fork barriers (RFBs). Most, if not all, of these barriers arrest forks arriving from one direction, but not from the opposite direction. Bacteria use these barriers to create termination regions in their circular chromosomes that are about 180° from the replication origin.
- RFBs are found in eukarya as well as in bacteria where they are used to prevent DNA replication forks from colliding with DNA transcription machines, particularly in genomic regions that are very actively transcribed such as rRNA and tRNA gene clusters. Specialized helicases, such as Rrm3, may help replication forks overcome RFBs.
- The most important termination signals in eukaryotic cells are telomeres, a series of unique DNA repeat sequences that stabilize the ends of long linear chromosomes and thereby preserve the organism's genome. The length and sequence of telomeres are species-specific, and their maintenance is dependent on a special enzyme, telomerase, as well as a series of telomere-binding proteins that serve to stabilize the telomere and to regulate telomerase activity.

REFERENCES

Additional Reading

Autexier, C. and Lue, N.F. (2006) The structure and function of telomerase reverse transcriptase. *Annu. Rev. Biochem.* 75: 493–517.

Bastia, D. and Mohanty, B.K. (1996) Mechanisms for completing DNA replication. In *DNA Replication in Eukaryotic Cells*, DePamphilis, M.L. (ed.), pp. 177–216. Cold Spring Harbor Laboratory Press, Cold Spring Harbor, NY.

Bastia, D. and Mohanty, B.K. (2006) Termination of DNA replication. In *DNA Replication and Human Disease*, DePamphilis, M.L. (ed.), pp. 155–174. Cold Spring Harbor Laboratory Press, Cold Spring Harbor, NY.

Bianchi, A. and Shore, D. (2008) How telomerase reaches its end: mechanism of telomerase regulation by the telomeric complex. *Mol. Cell* 31: 153–165.

Mirkin, E.V. and Mirkin, S.M. (2007) Replication fork stalling at natural impediments. *Microbiol. Mol. Biol. Rev.* 71: 13–35.

Osterhage, J.L. and Friedman, K.L. (2009) Chromosome end maintenance by telomerase. *J. Biol. Chem.* 284: 16061–16065.

Palm, W. and de Lange, T. (2008) How shelterin protects mammalian telomeres. *Annu. Rev. Genet.* 42: 301–334.

Sealey, D. and Harrington, L. (2006) Telomere DNA replication, telomerase, and human disease. In *DNA Replication and Human Disease*, DePamphilis M.L. (ed.), pp. 561–592. Cold Spring Harbor Laboratory Press, Cold Spring Harbor, NY.

Chapter 7

CHROMATIN ASSEMBLY, COHESION, AND MODIFICATION

Genome duplication in all three domains of life involves not only replicating the DNA sequence, but assembling it into a protein–DNA complex termed **chromatin** that compacts it into the comparatively tiny space of a cell or nucleus, protects it from damage, restricts access to its genetic information, and ensures that sister chromatids remain together until cell division or mitosis. The problem is truly awesome. Each human cell contains a total of 2 m of DNA compacted into a nucleus that is only 5–10 μm in diameter. Thus, the length of DNA is reduced nearly one million fold! Clearly, mechanisms exist for folding DNA into a highly compacted form. Moreover, these mechanisms must be linked tightly to DNA replication. Otherwise, a cell would literally explode as it duplicates its genome.

DNA compaction is driven by its association with basic proteins, a process referred to as **chromatin assembly** that not only compacts the DNA but additionally shields it from access to most DNA-modifying enzymes. However, the mechanism for compaction must also allow for some genetic information to remain accessible for transcription, for other information to be repressed completely as cells differentiate into specialized functions, and for all of the genetic information to be duplicated each time a cell divides. Furthermore, while the replication fork is producing one identical copy of the genome, there must also be a mechanism that identifies sister chromatids (two **homologous chromosomes** in a diploid cell) so that daughter cells each receive one copy of each chromatid. It can be of little surprise that the investigation of mechanisms of DNA compaction is one of the most intensively studied and competitive fields in modern molecular biology.

In eukaryotes, chromatin assembly is accompanied by three other processes that are essential for cell division and for the organism's viability. The first is termed **sister chromatid cohesion**. The newly replicated copies of each chromosome, the sister chromatids, are attached to one another prior to their segregation in mitosis and meiosis. This association is critical for each sister chromatid to bind microtubules from opposite spindle poles and thus segregate away from each other during anaphase of either mitosis or meiosis II. The second is **histone modification** through phosphorylation, methylation, acetylation, or ubiquitination of specific amino acids. **Histones** are the proteins in eukarya that are responsible for compacting DNA. These changes affect both chromatin compaction and regulation of gene expression. The third process is **DNA methylation**. Both DNA methylation and histone modification are involved in establishing patterns of gene repression during development. Histone methylation can induce formation of localized heterochromatin, which is readily reversible, whereas DNA methylation leads to stable long-term repression.

DNA methylation is carried out by enzymes called **DNA methyltransferases**, because they transfer the methyl group from *S*-adenosylmethionine to specific bases in duplex DNA. In bacteria, most DNA methyltransferases act in concert with a **restriction endonuclease** that targets the same sequence. However, the restriction endonuclease can cleave its target only when the target is not methylated. Thus, foreign DNA from a bacteriophage infection or from another organism is cleaved, whereas the host's genomic DNA is protected by methylation at these sites. These **restriction-modification systems** constitute a barrier to horizontal gene transfer that plays a role in bacterial speciation. However, for this system to operate efficiently, DNA methylation also must be linked closely to DNA replication.

Some multicellular eukarya also methylate their DNA, but for an entirely different reason. DNA methylation in vertebrates and plants produces a particular pattern of base modifications that are inherited from one cell division to the next. These modifications suppress the expression of particular genes by recruiting **methyl-CpG-binding proteins** that prevent transcription factors from binding to DNA. This form of DNA modification is an example of an **epigenetic** change in the genome, a term that refers to heritable changes in DNA function that occur in the absence of changes to DNA sequence. More recently, histone modifications have been recognized as epigenetic changes in the genome that can regulate gene expression. In contrast to the tight link between the replication and methylation of DNA, the relationship between DNA replication and histone modification has yet to be defined.

Given the importance of chromatin structure to compacting the genome and regulating expression of its genetic information, the mechanisms by which chromatin is disrupted and then reassembled as replication forks pass through cannot be ignored if we are to understand how genomes are duplicated during cell division. Cells must not only replicate their DNA each time they divide, they must duplicate their genome, and must do so as rapidly as their replication forks travel.

CHROMATIN

Bacteria

Although defined by the absence of a nuclear membrane, prokaryotic organisms (bacteria and archaea) nevertheless compact their genomes into a discrete structure within the cell: the **nucleoid**. Nucleoids differ from eukaryotic nuclei not only in the absence of a membrane that separates it from the surrounding cytoplasm, but also in the absence of a stable ordered nucleosome structure. Bacterial nucleoids lack histones. However, they contain a set of abundant proteins that bind DNA and organize it into 40–200 topologically constrained loops, analogous to those described below for eukaryotic chromosomes. In the case of *E. coli,* the chromosome is compacted about 1000-fold compared to the length of a naked genome. This compaction is mediated by several abundant proteins, some of which are basic (isoelectric point ≥8) and some are not (Table 7-1). Of the six nucleoid-associated proteins identified so far, the most well characterized are IHF, Fis, H-NS, and HU. The primary feature of these proteins is to stabilize bends in DNA, thereby affecting DNA compaction and ability of replication origins and transcription promoters to unwind. In addition to nucleoid-binding proteins, DNA gyrase and topoisomerases I and IV (Tables 4-7 and 4-9) are important for changing the linking number of DNA as it is folded and twisted into a more compact form.

IHF protein

Of the four main nucleoid-associated proteins, integration host factor, or IHF, is the only one that binds to a clear high-affinity ($K_d \approx 0.3$ nM) consensus sequence. *E. coli* IHF is a heterodimer that binds 36 bp of dsDNA, and, in doing so, it introduces a bend of 160° (Figure 7-1). This remarkable degree of DNA deformation has topological consequences. Indeed, *in vitro*

Table 7-1. Bacterial Chromosomal Proteins

Protein	Molecular mass (kDa)	*p*I	Organization	DNA-binding specificity
IHFα IHFβ	11.4 10.7	9.3 9.3	Heterodimer	WATCARNNNNTTR
HUa HUb	9.5 9.2	9.6 9.7	Heterodimer	Nonspecific; prefers distorted DNA, binds dsDNA, dsRNA, ssDNA
Fis	11.2	9.3	Homodimer	GNrYAaWWWtTRaNC
H-NS	15.3	5.2	Dimer→oligomer	Bent DNA
StpA	15.3	8.0	Dimer→oligomer	Bent DNA
Dps	18.7	5.7	Dodecamer	Nonspecific

*p*I=isoelectric point; W=A or T, R=A or G, Y=C or T, and N=any base. Lower case letters indicate a preference rather than absolute conservation.

studies have revealed that binding of IHF to multiple sites on a DNA molecule leads to formation of a highly condensed structure. In addition, the sequence specificity of IHF means that it has the capacity to introduce highly localized architectural changes in DNA structure. These changes can have important consequences for gene expression, and as will be discussed below, also in the initiation of DNA replication. IHF belongs to a large superfamily of DNA-binding proteins, the DNABII family, that also includes the HU protein. IHF also binds DNA nonspecifically with an affinity similar to HU protein binding to linear DNA. IHF and HU together are estimated to cover 10–20% of the bacterial genome, depending on growth conditions.

HU protein

Histone-like protein from strain U93 (termed HU protein) is a paralog of IHF, but, in contrast to IHF, HU does not bind to specific DNA sequences. However, HU does bind preferentially to distorted forms of DNA such as bent, kinked, gapped, and cruciform structures with affinities as low as 2–20 nM. The K_d for linear dsDNA ranges from 0.4 to 30 μM, depending on salt conditions, but binding of HU to DNA is facilitated by the presence of negative supercoils. Thus, binding of HU to circular, covalently closed DNA

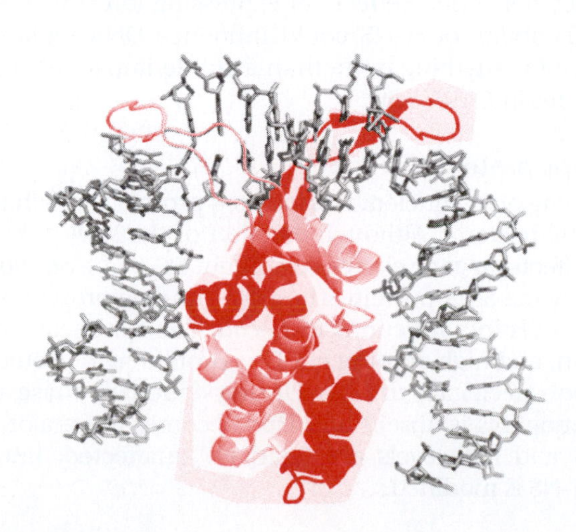

Figure 7-1. Structure of IHF bound to DNA. The gray DNA molecule is bent by 160°. The α and β subunits of the IHF heterodimer are shown in pink and red, respectively. Image generated from PDB file 2HT0, using Pymol.

is facilitated by gyrase (the bacterial topoisomerase that introduces negative superhelical turns into DNA), and discouraged by bacterial Topo I, which relaxes only negative superhelical turns. The degree of bending introduced by HU binding appears to vary depending on the nature of the substrate DNA molecule and suggests that more flexibility is observed than with IHF. Physiologically relevant levels of HU constrain negative supercoils, and make possible ring closure of short (~100 bp) linear DNA that would otherwise be prevented by the natural rigidity of dsDNA of this length.

Interestingly, the composition of HU varies during the cell cycle with both homodimeric (α_2 and β_2) and heterodimeric ($\alpha\beta$) forms being observed. During exponential growth, the α_2 dimer predominates, expression of the gene for the β form is elevated as cells enter stationary phase leading to preferential formation of the $\alpha\beta$ heterodimer. These different forms of HU have distinct DNA-binding affinities and so could potentially modulate the global architecture of the nucleoid during the distinct growth phases of a culture of *E. coli*. HU is the most important protein for compacting the DNA within the nucleoid.

Fis protein

Factor for inversion stimulation (Fis) protein, like HU protein, undergoes growth-phase-dependent regulation of its levels. It is the most abundant nucleoid-associated protein in exponentially growing cells but its levels drop rapidly at the onset of stationary phase. As with HU and IHF, Fis is a dimer, in this case a homodimer, that binds nonspecifically to 21–27 bp of DNA with a K_d of ≤20 nM, and in doing so, it introduces a bend of between 50° and 90°. It binds to specific sites with a K_d of approximately 2 nM. The main role of Fis in *E. coli* appears to be as a global regulator of transcription, but, as discussed below, it can also modulate DNA replication.

H-NS protein

Histone-like nucleoid structural (H-NS) protein lacks sequence specificity in its interaction with DNA, but it does prefer bent DNA 10–100 times over linear DNA ($K_d \approx 3$ nM) and occupies 8–10 bp. H-NS is not essential to maintain nucleoid structure. Like Fis, H-NS appears to act primarily as a global regulator of transcription. A second, related role has been recently proposed for H-NS. It has been observed that H-NS, upon binding, nucleates the formation of large assemblies of the protein that can repress or silence transcription. Remarkably, it was demonstrated that H-NS silencing happens preferentially on DNA with an AT-bias distinct from that of the host chromosome. Thus, foreign DNA introduced into the organism will be preferentially coated with H-NS and thus silenced. This xenogenic silencing ability of H-NS acts as a way of defending the host organism from the potentially deleterious effects of expressing foreign DNA. Whereas, in principle, this ability of H-NS could influence DNA replication, there is little evidence for anything more than a subtle impact of H-NS on overall replication rates in *E. coli* cells.

StpA and Dps proteins

StpA is a paralog of the nucleoid-associated protein H-NS that is conserved among enteric bacteria. Although deletion of the *stpA* gene in *E. coli* has only minor effects on gene expression, about 5% of the *Salmonella* genome is regulated by the StpA protein. The DNA-binding protein of starved cells (Dps) has two distinct functions in *E. coli*. The spherical Dps dodecamer can store iron, and high amounts of Dps protein help protect the genome by binding nonspecifically to DNA. During stationary phase, when cells are not proliferating, Fis is absent and Dps becomes the major DNA-binding protein. HU and IHF levels are relatively unaffected, but a significant fraction of H-NS is modified.

Nucleoid-associated proteins and DNA replication

DNA replication initiation in *E. coli* occurs at a single site termed oriC that is recognized by the DnaA protein (Figure 10-6). This interaction leads to DNA melting and recruitment of the replicative helicase, DnaB, in association with the helicase loader, DnaC (Figure 11-5). *In vitro* studies have revealed that Fis and IHF or HU play opposing roles in replication with Fis repressing oriC-dependent initiation of replication, and HU or IHF promoting it through their effects on either repressing or facilitating DNA unwinding. *In vivo* as well as *in vitro*, IHF can efficiently substitute for HU in activating oriC.

Using a reconstituted system, Hwang and Kornberg (1992) revealed that either HU or IHF could stimulate melting of oriC DNA by the initiator protein DnaA (Figure 7-2). The role of HU is likely due to its ability to bend and generally destabilize DNA. HU binds both dsDNA (36 bp/dimer) and ssDNA (24 nt/dimer), but in contrast to SSB, HU binds ssDNA non-cooperatively and does not destabilize dsDNA. *In vitro*, HU stimulates *de novo* initiation of oriC-dependent DNA replication when the ratio of HU to DNA is low (~5 HU dimers per oriC), but inhibits DNA replication when the ratio of HU to DNA is increased, suggesting that DNA unwinding is more difficult when DNA is coated with HU protein. Given fork velocities of approximately 1000 nt s^{-1} (Table 3-4), chromosomal proteins could dissociate and then reassociate so rapidly that the transition may not easily be detected. Therefore, whether or not these proteins are released from DNA as the replication fork passes through the chromosome, and then rebind as newly synthesized dsDNA appears, remains to be determined.

In contrast to HU, both Fis and IHF appear to be specifically involved in regulating this process, as both proteins bind to specific sites within the oriC region (Figure 7-3). The stimulation effected by IHF correlates with its ability to facilitate redistribution of the ATP-bound form of DnaA from high- to lower-affinity sites at the origin. Thus, presumably by virtue of its ability to deform DNA, IHF acts as a chaperone for the assembly of

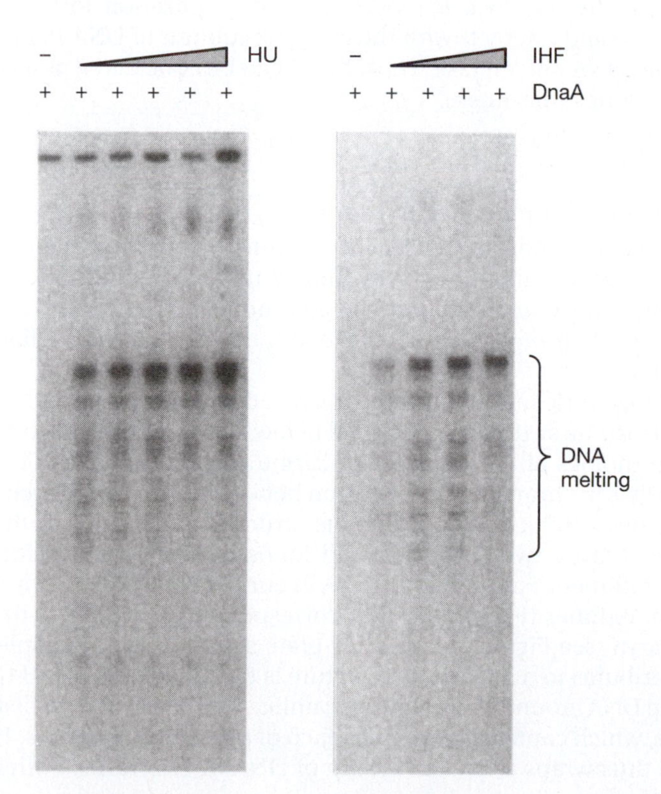

Figure 7-2. Melting of *E. coli* oriC is elevated by HU or IHF. Bands indicate sites cleaved by the ssDNA-specific nuclease P1, and thus reveal sites of DNA melting. (From Hwang, D.S. and Kornberg, A. (1992) *J. Biol. Chem.* 267: 23083–23086. With permission from The American Society for Biochemistry and Molecular Biology.)

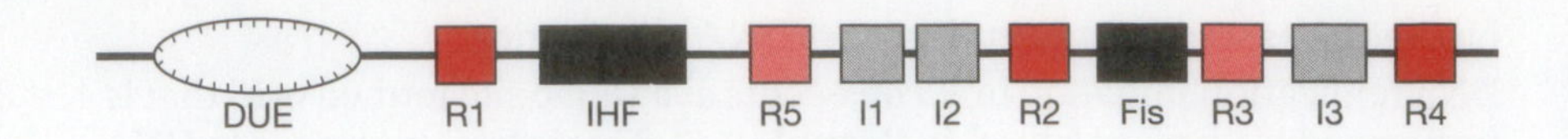

Figure 7-3. Cartoon of the *E. coli* oriC locus with high-affinity DnaA-binding sites (R1, R2, and R4) shown in bright red, intermediate affinity sites (R3 and R5) in pink, and non-consensus, low-affinity I-sites in gray. The binding sites of IHF and Fis are shown as white and black boxes, respectively. The site of unwinding, the DUE, is shown as an oval.

the full DnaA–ATP–oriC structure. In turn, this leads to origin melting. In contrast, Fis may act to occlude appropriate binding of DnaA to the I-sites and the R3 and R5 sites as well as inhibiting IHF binding. The combination of these events prevents origin melting. The opposing roles of Fis and IHF correlate well with their relative distributions during the cell cycle. More specifically, levels of Fis drop as cells approach the beginning of DNA synthesis and a concomitant elevated association of IHF with the origin is observed. Thus, the interplay of these bacterial chromatin proteins with the origin of replication is a component of the switch that ensures temporally appropriate firing of replication origins. In agreement with this, mutation of either Fis or IHF leads to initiation asynchrony in *E. coli*.

Chromatin disassembly and accessory DNA helicases

The mechanism by which DNA unwinding at replication forks deals with chromosomal proteins such as HU in bacteria and histones in eukarya is not clear, although accessory DNA helicases are likely involved. The paradigm for this group of enzymes is Dda, a monomeric helicase encoded by bacteriophage T4 that is able to displace bound proteins as it moves 5′→3′ along DNA templates. This ability to clear the tracks in front of the train appears to facilitate both the assembly and activity of the T4 replisome. Interestingly, Dda shares significant sequence homology with a small subset of accessory helicases that include the mammalian enzyme DNA helicase B. These helicases physically interact with DNA polymerase-α•primase and stimulate synthesis of short RNA primers on DNA templates coated with RPA, the eukaryotic ssDNA-binding protein. In yeast, the Rrm3 DNA helicase facilitates passage of replication forks through particularly stable non-nucleosomal protein complexes. Although Rrm3 affects replication only at discrete loci, it moves with the replication fork throughout the genome and interacts with the catalytic subunit of DNA Pol-ε, the leading-strand DNA polymerase. Thus, Rrm3 can be considered as a component of eukaryotic replisomes.

Archaea

Archaeal chromosomal proteins

Whereas a number of archaeal chromatin proteins have been identified, very little is known about how they impact upon DNA replication. Intriguingly, there is no chromatin protein common to all archaea; rather, distinct subsets of chromatin proteins are found in the various archaeal kingdoms (Figure 7-4).

Eukaryotic nuclear DNA is organized by histones into the now classical nucleosome structures described below. Some archaea encode from one to six orthologs of the eukaryotic histone proteins H3 and H4, suggestive of parallels in chromatin organization between eukarya and archaea. Archaeal histones exhibit a typical histone structure consisting of three α-helices separated by two short β-strand loops, and they form either homo- or heterodimers. For example, HMfA in euryarchaea and thaumarchaea forms a homodimer (Figure 7-5) that corresponds to the H3•H4 heterodimer in eukarya (see Figure 7-6, in color plate section). The principle feature that contributes to nucleosome structure is the ability of a $[H3{\bullet}H4]_2$ tetramer to wrap DNA around itself. However, unlike the full eukaryotic histone octamer core, which contains two copies each of two further histones, H2A and H2B, and thus wraps a total of 146 bp of DNA in 1.75 turns, both archaeal and

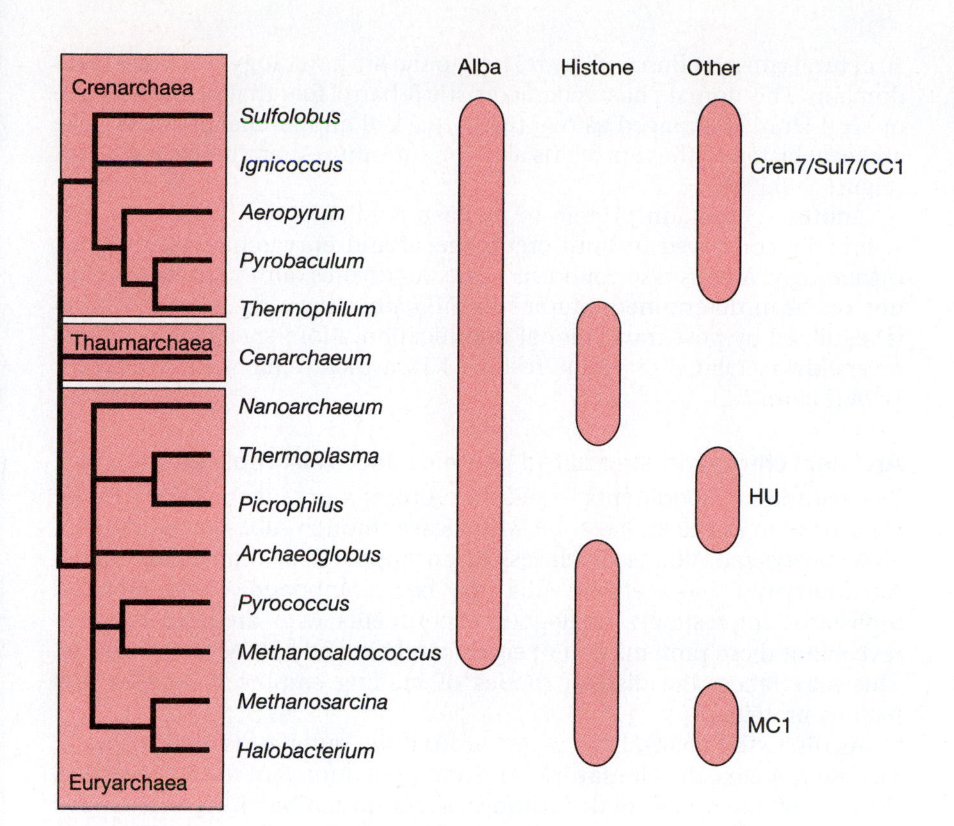

Figure 7-4. Distribution of chromatin proteins in the archaeal domain of life.

eukaryal histone $[H3{\cdot}H4]_2$ tetramers wrap approximately 90 bp in less than one circle. This results in a horseshoe-shaped structure that likely provides more flexibility at the expense of stability.

There are two exceptions to the ubiquity of histones in the euryarchaea. In *Thermoplasma* and *Picrophilus*, histones appear to have been replaced by a bacterial-like HU protein, presumably reflective of a gene replacement following lateral gene transfer. Histones also are absent from the hyperthermophilic crenarchaea where their place is taken by a diverse range of proteins such as Sul7, Cren7, and CC1 that show some degree of

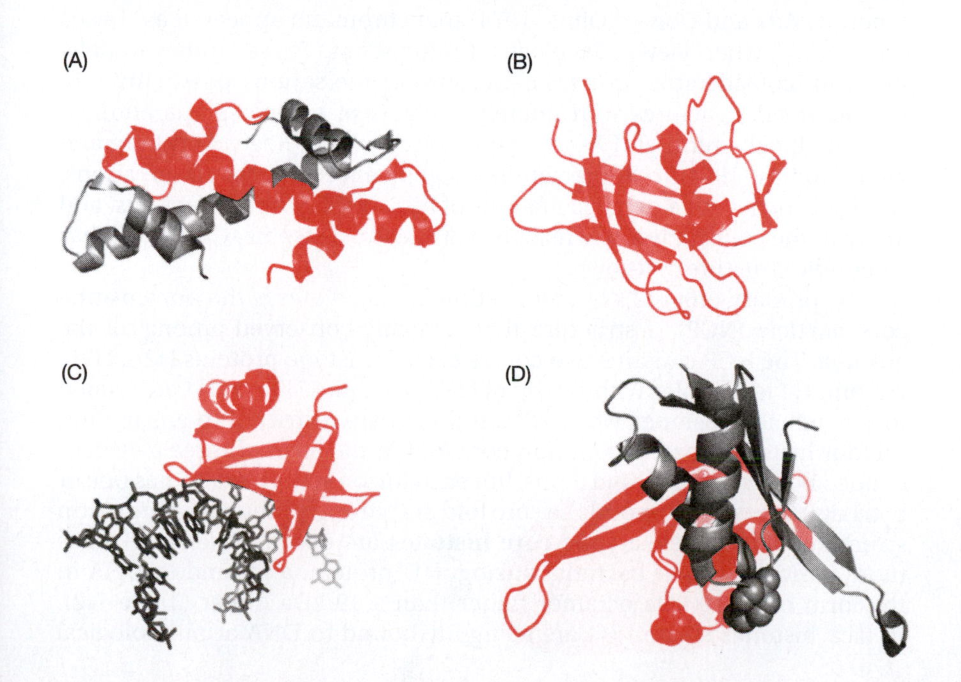

Figure 7-5. Structures of representative archaeal chromatin proteins. (A) Homodimer of histone HMfA. (B) MC1 protein. (C) Sul7d monomer bound to DNA introduces a sharp 70° bend in the DNA. (D) Homodimer of Alba from *Sulfolobus solfataricus* showing the positions of the lysine residues (spheres) that undergo reversible acetylation and impact upon DNA-binding affinity. Figures were prepared using Pymol with the following PDB coordinates: (A) 1B67 *Methanococcus fervidus* histone HMfA dimer; (B) 1T23 MC1; (C) 1WD0 *S. acidocaldarius* Sul7d; (D) 1H0X *S. solfataricus* Alba.

structural conservation and are related to the Src homology 3 domain (SH3 domain). This domain has a characteristic β-barrel fold that consists of five or six β-strands arranged as two tightly packed antiparallel β-sheets. Like archaeal histones, these proteins also can introduce sharp bends into DNA (Figure 7-5).

Another chromatin protein in archaea is Alba, a 10 kDa protein that is broadly conserved in both crenarchaeal and euryarchaeal kingdoms. Intriguingly, Alba is also found in some eukaryotes, although its role has not yet been determined. Studies in *Sulfolobus* have revealed that Alba is regulated by post-translational modification. More specifically, Alba is reversibly acetylated on lysine residue K16, which reduces its affinity for DNA (Figure 7-5).

Archaeal chromatin-associated proteins and DNA replication

Reconstitution experiments reveal that Alba is a potent roadblock to the translocation of the archaeal MCM helicase through Alba-coated dsDNA. Since acetylated Alba is much less of an impediment than unacetylated Alba, enzymes that acetylate Alba may be a component of the archaeal replisome. Interestingly, analogous experiments with archaeal histones reveal that these proteins do not significantly impede MCM translocation. This may reflect the distinct modes of binding employed by Alba and histone proteins.

As discussed below, DNA wraps around the outer surface of a core of histone proteins, thus it may be relatively straightforward for MCM to peel the DNA off the surface of the histones. In contrast, Alba forms an extended filament on DNA, effectively sealing the DNA in a tube of protein. However, beyond these preliminary observations, much remains to be learned about the interplay of archaeal chromatin with the replication process.

Eukarya

Throughout the 1960s it was largely believed that chromatin consisted of a uniform distribution of histones bound to DNA with no unique structure of its own. DNA was simply encased within the histone proteins. This view, however, was soon proven to be incorrect. In 1973, Dean Hewish and Leigh Burgoyne reported that rat liver nuclei contained an endonuclease that digested chromatin *in situ* into particles containing multiples of the smallest unit of DNA released, a discovery that correlated well with the report by Ada and Donald Olins (1974) that chromatin appeared as "beads on a string" when viewed by electron microscopy. These studies initially met with considerable resistance, because of the serious possibility that the observed structures were merely artifacts of sample preparation. It took the biochemical and X-ray diffraction results of Roger Kornberg and Jean Thomas (1974) to appreciate the significance of these observations, combine them with a detailed analysis of histone•histone interactions, and arrive at the concept we now recognize as the nucleosome. Chromatin has a periodic structure of its own.

The primary unit of DNA compaction in eukaryotes is the **nucleosome core particle** (**NCP**), a structure that is highly conserved among all the eukarya. The NCP contains two copies each of histone proteins H2A, H2B, H3, and H4 in complex with 146 bp of DNA. Histones H3 and H4 are related to the archaeal histones while H2A and H2B are restricted to eukaryotes. All four histones share a common core fold in the form of three α-helices connected by short β-strand loops, but each histone subtype also has one or two tails that extend beyond the core fold and strongly influence chromatin structure. Notably, eukaryotic **core histones** are approximately 10 times more basic than their bacterial analog, HU protein, and bind to DNA in the form of a 110 kDa octamer rather than a 19 kDa dimer (Table 7-2). In fact, histones H3 and H4 are so tightly bound to DNA at physiological

Table 7-2. Eukaryotic Chromosomal Proteins And Replication Assembly Factors

<table>
<tr><th>Protein</th><th>Molecular mass (kDa)</th><th>pI</th><th>Organization</th><th>DNA-binding specificity</th><th colspan="3">Structure</th></tr>
<tr><td>H3</td><td>15.5</td><td>11.1</td><td rowspan="2">Heterodimer</td><td rowspan="5">Nonspecific</td><td rowspan="4">Nucleosome core particle (NCP)</td><td rowspan="5">Euchromatin</td><td rowspan="7">Heterochromatin</td></tr>
<tr><td>H4</td><td>11.4</td><td>11.4</td></tr>
<tr><td>H2A</td><td>14.1</td><td>10.9</td><td rowspan="2">Heterodimer</td></tr>
<tr><td>H2B</td><td>14.1</td><td>10.9</td></tr>
<tr><td>H1</td><td>22.4</td><td>11.0</td><td>Monomer</td><td rowspan="3"></td></tr>
<tr><td>HP1</td><td>22.2</td><td>5.7</td><td rowspan="2">Homodimer</td><td rowspan="2">Methylated-H3</td><td rowspan="2"></td></tr>
<tr><td>Polycomb (PRC1)</td><td>71.7</td><td>6.2</td></tr>
</table>

Core histones are highly conserved among all eukarya, but histone H1 can vary significantly. For example, budding yeast have a recognizable H1, encoded by the *HHO1* gene, that differs markedly from vertebrate H1 in that it has two globular domains. Histone H1 participates in nucleosome spacing and formation of the higher-order chromatin structure, and it is actively involved in the regulation of gene expression. The heterochromatin protein-1 (HP1) family are highly conserved proteins that play a role in gene repression by heterochromatin formation, as well as in other chromatin functions. HP1 proteins are enriched at centromeres and telomeres of metazoan chromosomes, although they are not found in budding yeast. Polycomb-group proteins can remodel chromatin such that transcription factors cannot bind to promoter sequences in DNA. PRC1 (polycomb-group-repressive complex) is required for cytokinesis. Data are for human proteins.

salt concentrations that their dissociation cannot be measured. Thus, the eukaryotic nucleosome is a much more stable DNA•protein complex than either the bacterial chromatin complex, or the archaeal nucleosome structure. As such, eukaryotic nucleosomes present a serious impediment to DNA replication.

The nucleosome

A stunning step forward in our understanding of NCP organization came in 1997 with the determination of its high-resolution crystal structure by Karolin Luger, Tim Richmond, and colleagues. The core scaffold of proteins in the NCP is an octamer of histone proteins consisting of two copies of each of the four histones (see Figure 7-6 in color plate section). The octamer has a defined order of assembly. First, a H3•H4 dimer forms. Then this dimer forms a $[H3{\bullet}H4]_2$ tetramer with two copies each of H3 and H4. The $[H3{\bullet}H4]_2$ tetramer can interact with DNA to form the so-called **tetrasome**, an assembly that corresponds to the archaeal histone–DNA complex. The eukaryotic $[H3{\bullet}H4]_2$ tetramer then interacts with two $H2A{\bullet}H_2B$ dimers to form a [**histone**] **octamer** consisting of one $[H3{\bullet}H4]_2$ tetramer sandwiched by two H2A•H2B dimers. The octamer can then interact with 146 bp of DNA to form a nucleosome in which 1.75 turns of dsDNA is wrapped around the central histone octamer (see Figure 7-7 in color plate section). While this assembly can be recapitulated *in vitro* by stepwise reconstitution under carefully defined conditions, nucleosome assembly *in vivo* is carried out by a variety of **histone chaperones** that facilitate this process.

Organization of interphase chromatin

To understand the problem of replicating DNA in eukaryotic cells, one must understand the structure and organization of chromatin. In terms of the cell-division cycle (discussed in Chapter 12), replication complexes are established at replication origins during the transition from mitosis to $\mathbf{G_1}$ **phase**, and DNA synthesis occurs during S phase. G_1, S, and $\mathbf{G_2}$ **phases** are collectively known as interphase (the time between two sequential mitoses). The majority of the interphase genome is compacted by association with histones into NCPs. However, NCPs are simply the first step in a hierarchical series of higher-order assemblies (Figure 7-8).

The first step, the assembly of nucleosomes, results in the appearance of beads on a string when viewed by electron microscopy (Figures 7-8 and 7-9B). Individual NCPs are separated by linker DNA. Whereas all eukaryotic nucleosomes contain essentially the same number of base pairs of DNA, the distance between NCPs (the length of linker DNA) varies significantly from one organism to another, and from one chromosomal locus to another such that the average repeat distance between nucleosomes is 165 bp (budding yeast neurons) to 240 bp (sea urchin sperm) with most vertebrate chromatin between 175 and 195 bp. This variation is due primarily to the existence of histone H1 variants; cells with normal histone H1 exhibit a nucleosomal repeat distance of approximately 195 bp.

NCPs can associate with another histone protein termed H1. In doing so, H1 interacts with an additional 20 bp of DNA to form a particle referred to as the **chromatosome**, consisting of nine histone proteins with approximately 166 bp of DNA (Figure 7-10). Histone H1 stabilizes a higher-order structure termed the 30 nm fiber in which nucleosomes assemble to form a compacted solenoid-like structure (Figures 7-8 and 7-9C). The 30 nm fiber is the paradigm for the first order of folding in a nucleosome array. The basic architecture is a helical array of nucleosomes with a diameter of 30–32 nm that is essentially unaffected by lengths of linker DNA in the range of 30–60 bp. The 30 nm fiber contains approximately six to seven nucleosomes per 11 nm, a value consistent with that of unfolded euchromatin in nuclei.

The extent of chromatin compaction depends on coupling DNA wrapping around the octamer with coiling of the fiber. Increased wrapping that occurs in the presence of histone H1 induces more condensation of the

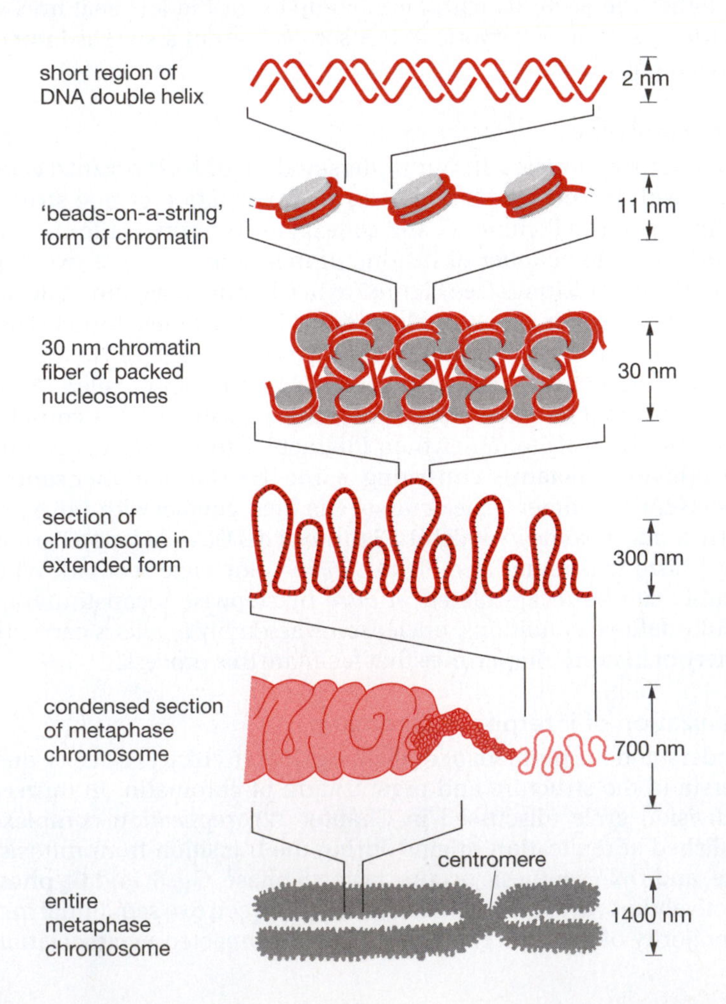

Figure 7-8. Model showing some of the many levels of chromatin packing that give rise to the highly condensed mitotic chromosome. The first three levels exist during interphase. (From Alberts, B., Johnson, A., Lewis, J. *et al.* (2007) *Molecular Biology of the Cell*, 5th ed. Garland Science/Taylor & Francis LLC.)

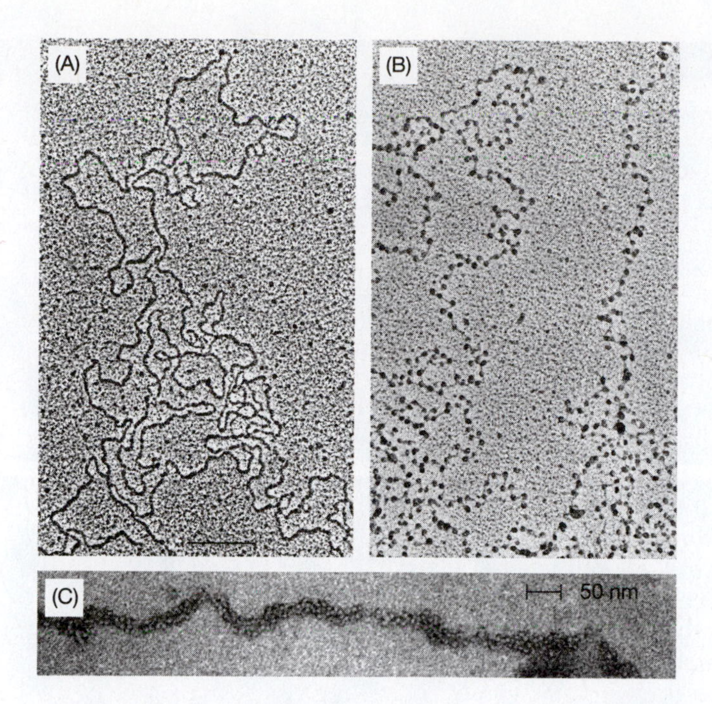

Figure 7-9. DNA and chromatin as seen in the electron microscope. (A) Yeast DNA. (B) Yeast chromatin stripped of histone H1 and nonhistone chromosomal proteins appears as a string of beads, each bead representing a single NCP. (C) Mouse chromatin isolated directly from interphase nuclei appears as a 30 nm-thick thread composed of beads. [(A) and (B) courtesy of José Sogo, ETH Zurich. (C) from Rattner, J.B. and Hamkalo, B.A. (1979) *J. Cell Biol.* 81: 453–457. With permission from The Rockefeller University Press.]

30 nm fiber. In this context, linker histones compact the fiber by neutralizing charge on linker DNA, thereby allowing it to bend. This linker-histone-dependent compaction is a highly cooperative process. Thus, chromatin structure is dynamic, not static. It responds to changes in environmental conditions such as ionic composition, pH, and DNA-binding proteins.

DNA loops

The 30 nm fiber is still not compact enough to account for interphase chromatin. For example, as a 30 nm fiber, the typical human chromosome would be 1 mm long, or approximately 100 times the diameter of a nucleus. Therefore, 30 nm fibers are apparently folded into loops and coils that compact them another 100-fold (Figure 7-8). Although the molecular basis for such folding is largely speculative, at least one clear example of such chromosome organization exists in the form of lampbrush chromosomes (Figure 7-11). These chromosomes are present in growing amphibian oocytes (immature eggs) and are highly active in gene expression. The decondensed loops evident in lampbrush chromosomes allow access to most regions of

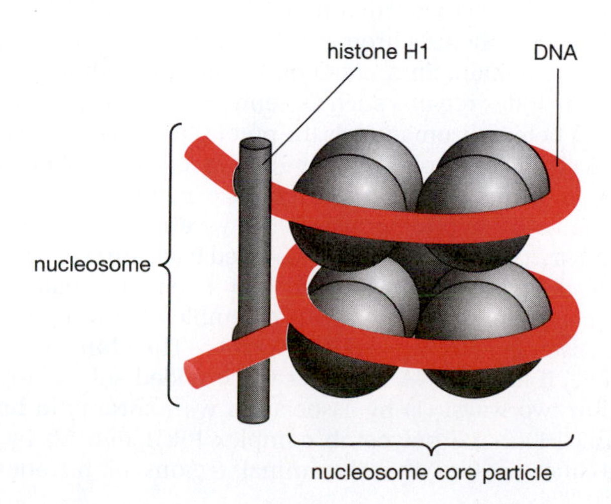

Figure 7-10. Chromatosomes consist of an NCP bound to one molecule of histone H1, as indicated. Chromatosomes contain approximately 166 bp of DNA protected from nuclease digestion. DNA is coiled around the histones in a left-handed direction. Thus, in a topologically-closed system, nucleosome disassembly produces one negative superhelical turn in the DNA (Figure 1-7).

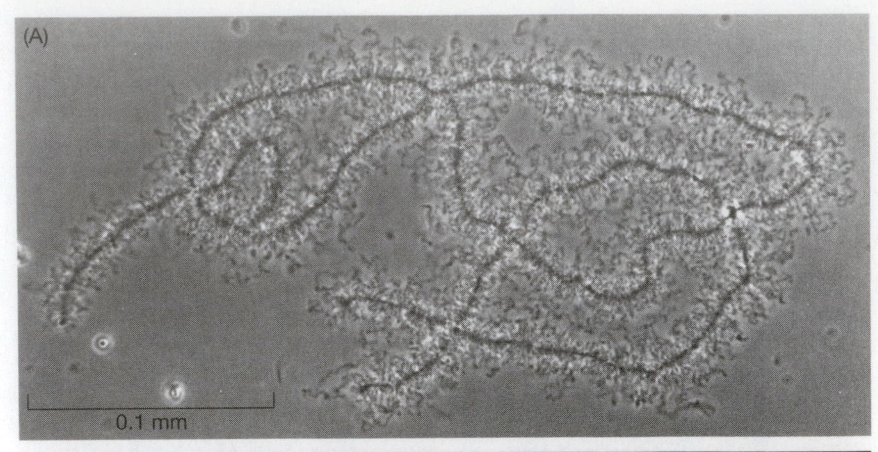

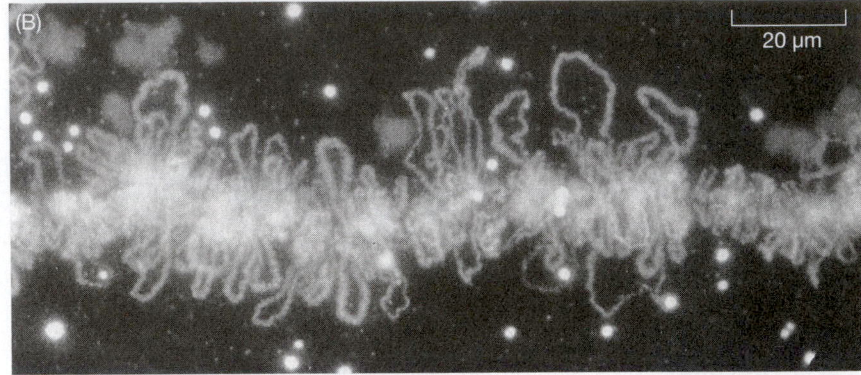

Figure 7-11. (A) Lampbrush chromosomes in an amphibian oocyte. Early in oocyte differentiation, each chromosome replicates to begin meiosis, and the homologous replicated chromosomes pair to form this highly extended structure containing a total of four replicated DNA molecules, or chromatids. The lampbrush chromosome stage persists for months or years, while the oocyte builds up a supply of materials required for its ultimate development into a new individual. (B) A lampbrush chromosome stained with antibodies against proteins that process RNA allows fluorescence light microscopy to reveal regions of the chromosome that are actively expressed. The round granules are large complexes of the RNA-splicing machinery. (Courtesy of Joseph G. Gall, Carnegie Institution.)

the chromosome while still maintaining a compact structure. Although lampbrush chromosomes are relatively rare among eukarya, virtually any DNA is packaged into lampbrush chromosomes when injected into amphibian oocytes. Therefore, these and other experiments have led to the concept that interphase chromosomes of all eukarya are arranged in loops, although most loops are normally too small or fragile to be observed easily.

Euchromatin and heterochromatin

Not all interphase chromatin is equally condensed. In 1929, Emil Heitz recognized two types of chromatin. Euchromatin and heterochromatin refer to the intensity with which different regions of chromatin stain; euchromatin stains with less intensity than heterochromatin. In fact, euchromatin has a more open conformation and exhibits high transcriptional activity, whereas heterochromatin is highly condensed and generally correlates with low transcriptional activity. Heterochromatin can be further subdivided into constitutive and facultative heterochromatin. Constitutive heterochromatin, as its name suggests, remains in a condensed state throughout the cell-division cycle and includes regions such as centromeres and telomeres. In contrast, facultative heterochromatin has the potential to reversibly convert to euchromatin. A classic example is the random inactivation of one of the two X-chromosomes during development of female mammals. The inactive X-chromosome stains as heterochromatin. However, if that inactivated copy is passed onto a male offspring, it is converted back into euchromatin.

The higher packing density observed with heterochromatin might reflect the fact that histone octamers can assemble into long arrays in which adjacent histone octamers stack together. Therefore, very tight packing is possible, if linker DNA regions can be folded sufficiently. This is accomplished in two ways: (1) by association with chromatin-binding proteins such as HP1 or the polycomb complex PRC1 and (2) by post-translational modifications of the N-terminal regions of histones that extend from the nucleosome core. HP1 interacts with histones H1, H3, and H4, as well as with histone methyltransferase and the methyl-CpG-binding protein MeCP2. Polycomb proteins are epigenetic gene repressors that are required in the maintenance of embryonic and adult stem cells. *In vitro*, PRC1 binds nucleosome arrays even if the N-terminal histone tails have been removed by tryptic digestion, and chromatin immunoprecipitation analyses indicate that specific polycomb-binding sites are depleted of nucleosomes. Nevertheless, the majority of polycomb target genes show extensive histone H3 methylated at lysine 27 (abbreviated to H3K27me) that typically extends well beyond the promoter region.

EUKARYOTIC CHROMATIN AND DNA REPLICATION

Given the remarkable structure and stability of eukaryotic chromatin, it should come as no surprise that it presents a formidable barrier to accessing DNA, whether for replication or for transcription. Nucleosomes have the potential to regulate access to replication origins, and to influence replication-fork translocation rates. Newly replicated DNA must be assembled rapidly into chromatin to prevent a 'nuclear explosion', but at the same time epigenetic marks in parental chromosomes need to be recapitulated in daughter chromosomes. To solve these problems, eukaryotic cells possess a bewildering range of proteins that dismantle chromatin and reassemble it at replication forks, and then modulate chromatin structure after the replication forks have moved on. These proteins are broadly classified into histone-modifying enzymes and ATP-dependent chromatin-remodeling enzymes.

Replication Origins

The activation of replication origins in eukarya, like the activation of transcription promoters, is profoundly influenced by chromatin architecture. For example, it has long been noted that replication of heterochromatin occurs later in S phase than DNA replication in euchromatin. One contributor to this phenomenon is histone modification in or around replication origins. Histone tails within heterochromatin are deacetylated, whereas those in the vicinity of actively transcribed genes in euchromatin are hyperacetylated, making these regions more accessible to transcription factors. In budding yeast, deletion of the *rpd3* gene encoding one of the cell's major histone deacetylases has the effect of bringing forward the timing of replication at 103 different origins (Figure 7-12). The locations of these advanced replication origins correlate well with the sites of increased histone acetylation in the mutant cells, suggesting that increased histone acetylation increases the efficiency of origin activation. Further studies revealed that artificially recruiting the Gcn5 histone acetyltransferase to the vicinity of a late-firing origin concomitantly increases histone acetylation at the origin and accelerates origin firing. Similar experiments performed in human and *Drosophila* cells have confirmed that histone deacetylase

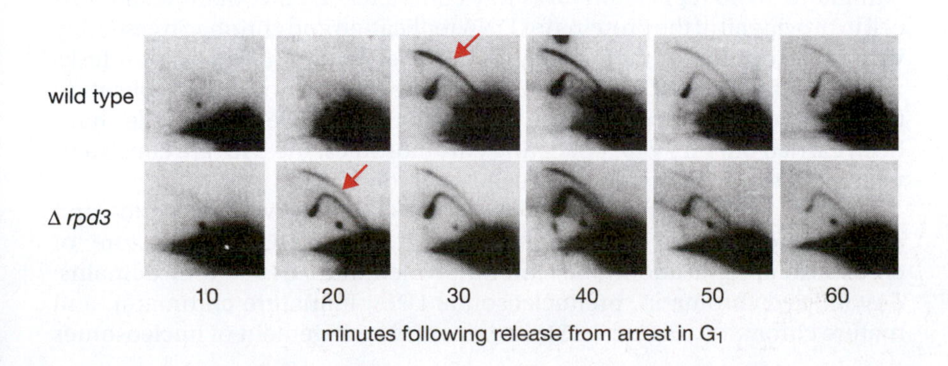

Figure 7-12. Two-dimensional gel analysis of the timing of replication initiation from budding yeast origin ARS609. The arrows indicate arcs on the gel that correspond to replication-initiation structures (Figure 8-3). Origin activity is detectable at least 10 min earlier in the absence of histone deacetylase Rpd3. (From Vogelauer, M., Rubbi, L., Lucas, I. *et al.* (2002) *Mol. Cell* 10: 1223–1233. With permission from Elsevier.)

activity reduces origin activity whereas histone acetyltransferase activity promotes origin activity. As with transcription promoters, the simplest mechanism to account for these results is that histone acetylation promotes an 'open' chromatin structure that makes replication origins more accessible to replication proteins.

As well as influencing both the timing of orign firing and use of origins, histone acetylation in higher eukaryotes may contribute directly to establishing the identity of replication origins. Experiments in *Xenopus* egg extracts reveal that introducing a plasmid with binding sites for transcription factors can localize replication initiation to the vicinity of these sites. This is in agreement with the observation that transcription factors often act to recruit histone-modifying factors, and nucleosomes in the vicinity of the new origin were found to contain hyperacetylated histones. Importantly, inhibition of transcription with α-amanitin (a potent inhibitor of RNA polymerase II) did not affect the ability of the transcription factor to specify the origin, indicating that origin specification was not a secondary effect of transcription. While a causal linkage has yet to be established, it seems likely that histone modification directly contributes to origin specification.

In addition to effects of post-translational modifications at origins of replication, chromatin-remodeling machines also may play a significant role. The yeast INO80 ATP-dependent chromatin-remodeling complex is recruited to origins at the beginning of S phase. INO80 also appears to progress with replication forks, presumably easing their passage through the chromatin template.

Replication Forks

Chromatin structure at replication forks

Following assembly of replication forks at replication origins, replication must proceed by disrupting chromatin structure in front of the fork and then reassembling it behind the fork. Since compact chromatin may well represent a barrier to fork progression, it is not surprising that several chromatin-altering activities are associated with the replisome. The passage of the replication machinery destabilizes the nucleosomal organization of the chromatin fiber over a distance of about 1 kb. In front of the fork, one or two nucleosomes are destabilized by dissociation of histone H1 and the advancing replisome. On daughter strands, the first nucleosome is detected at a distance of about 260 nucleotides from the first prefork nucleosome. The presence of histone H1 on newly replicated DNA is not detected until 450–650 bp has been replicated. Newly replicated chromatin is hypersensitive to endonucleases. It requires time to deacetylate newly incorporated histones and to condense the chromatin to form a 30 nm fiber.

The disposition of nucleosomes at replication forks in mammalian nuclei has been elucidated both by the sensitivity of replicating chromosomes to various nucleases, and by electron microscopy of replicating DNA molecules following psoralen-crosslinking of the nucleosome linker regions. These studies were carried out both on SV40 chromosomes and on cellular chromatin. SV40 replication forks rely completely on the mammalian host cell to provide all of the proteins for DNA replication and chromatin assembly with the exception of the replicative DNA helicase. SV40 replication forks use the SV40-encoded T-antigen DNA helicase whereas cellular replication forks use the MCM DNA helicase. Although the most detailed studies have been carried out on SV40 replicating chromosomes, the data are consistent with those studies carried out on cellular chromatin.

Based largely on the sensitivity of chromatin to various endo- and exonucleases and to DNA crosslinking by psoralen, the structure of chromatin at replication forks can be divided into at least four domains: destabilized chromatin, prenucleosomal DNA, immature chromatin, and mature chromatin (Figure 7-13). The size and arrangement of nucleosomes

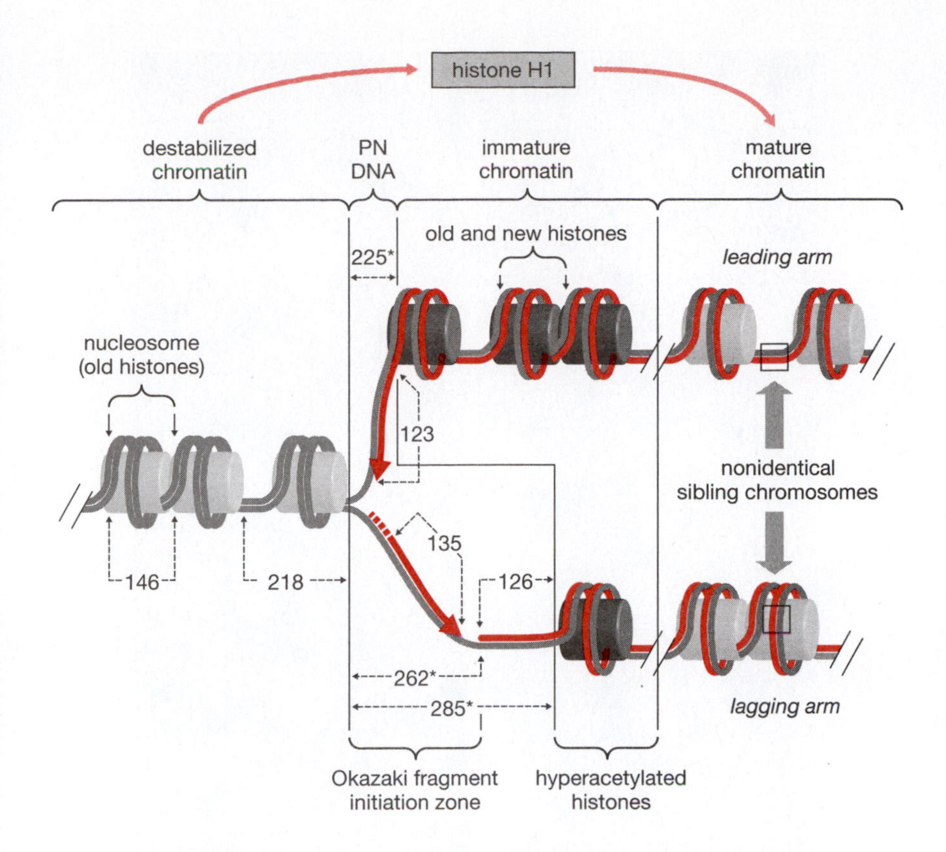

Figure 7-13. Nucleosome organization at SV40 replication forks. Nascent DNA is represented as a red ribbon with a single RNA-primed nascent DNA chain (Okazaki fragment) present on the lagging-strand template. Direction of synthesis is indicated by arrowheads. Numbers indicate the average distances in base pairs or nucleotides. Numbers with * are from electron microscopic analyses; remaining numbers are from biochemical analyses. Destabilized chromatin refers to the first one to three nucleosomes in front of replication forks that are more easily cross-linked with psoralen than mature chromatin, presumably by the loss of histone H1, histone modifications, or simply the act of DNA unwinding. Prenucleosomal DNA (PN DNA) refers to DNA that is fully accessible to psoralen-crosslinking and to digestion by both dsDNA- and ssDNA-specific endonucleases. Okazaki fragments can be selectively removed by exonucleases. Thus, PN DNA encompasses the sites where DNA synthesis occurs. The Okazaki initiation zone is the region wherein RNA-primed DNA synthesis is initiated at replication forks. Its length is taken from the length of ssDNA detected by electron microscopy on one arm of replication forks. Immature chromatin refers to newly synthesized DNA contained within nucleosomes that are enriched with acetylated histones. Mature chromatin refers to the 30 nm fiber compacted through the action of histone H1 and various modifications of histone tails. The same DNA sequence in the two sibling molecules (box) is not necessarily present in the same location relative to the nucleosomes. (Adapted from Cusick, M.E., Wassarman, P.M. and DePamphilis, M.L. (1989) *Methods Enzymol.* 170: 290–316. With permission from Elsevier. Additional data from Burhans, W.C., Vassilev, L.T., Wu, J. *et al.* (1991) *EMBO J.* 10: 4351–4360, Gasser, R., Koller, T. and Sogo, J.M. (1996) *J. Mol. Biol.* 258: 224–239, Sogo, J.M., Stahl, H., Koller, T. and Knippers, R. (1986) *J. Mol. Biol.* 189: 189–204, and Sogo, J.M., Lopes, M. and Foiani, M. (2002) *Science* 297: 599–602.)

in newly replicated chromatin are essentially the same as those in mature, nonreplicating chromatin. SV40 chromosomes, for example, contain 5232 bp of DNA organized into 24±1 nucleosomes, each containing 146 bp of DNA. Thus, the average distance from one SV40 nucleosome to the next is 218±9 bp, with the internucleosomal (linker) DNA varying from 0 to 60 bp. Both psoralen-crosslinking (Figure 7-14) and nuclease-digestion analyses find the same is true in newly replicated DNA.

The actual sites of DNA synthesis are free of nucleosomes and therefore referred to as **prenucleosomal DNA** (Figure 7-13). Based on exonuclease digestion of nascent DNA, prenucleosomal nascent DNA averages 123±20 nucleotides in length on the forward arm and up to one Okazaki fragment plus an average of 126±20 nucleotides on the retrograde arm. Based on electron microscopy of psoralen-crosslinked DNA, the distance from the branch point in replication forks to the first nucleosome is 225±145 nucleotides on the leading arm and 285±120 nucleotides on the lagging arm. These results reveal that the enzymes involved in DNA replication operate on DNA that is not assembled into nucleosomes, and that nucleosome assembly at replication forks occurs as rapidly as sufficient dsDNA becomes available to accommodate them. Similar analyses of the *S. cerevisiae* rDNA intergenic spacer region confirm that nucleosomes are present on newly replicated DNA within seconds after passage of the replication machinery.

Nucleosome assembly during *in vitro* DNA replication occurs preferentially at active replication forks, suggesting that single-stranded regions at replication forks allow the release of torsional strain that forms during nucleosome assembly. Nucleosomes on replicated chromatin are termed **immature chromatin**, because newly replicated nucleosomal DNA is hypersensitive to nonspecific endonucleases compared with nonreplicating chromatin, although it is resistant to exonucleases. Prenucleosomal DNA is hypersensitive to both endo- and exonuclease digestion. Thus, both prenucleosomal DNA and immature chromatin contribute to the commonly observed hypersensitivity of newly replicated chromatin to endonucleases. These regions are as hypersensitive to endonucleases

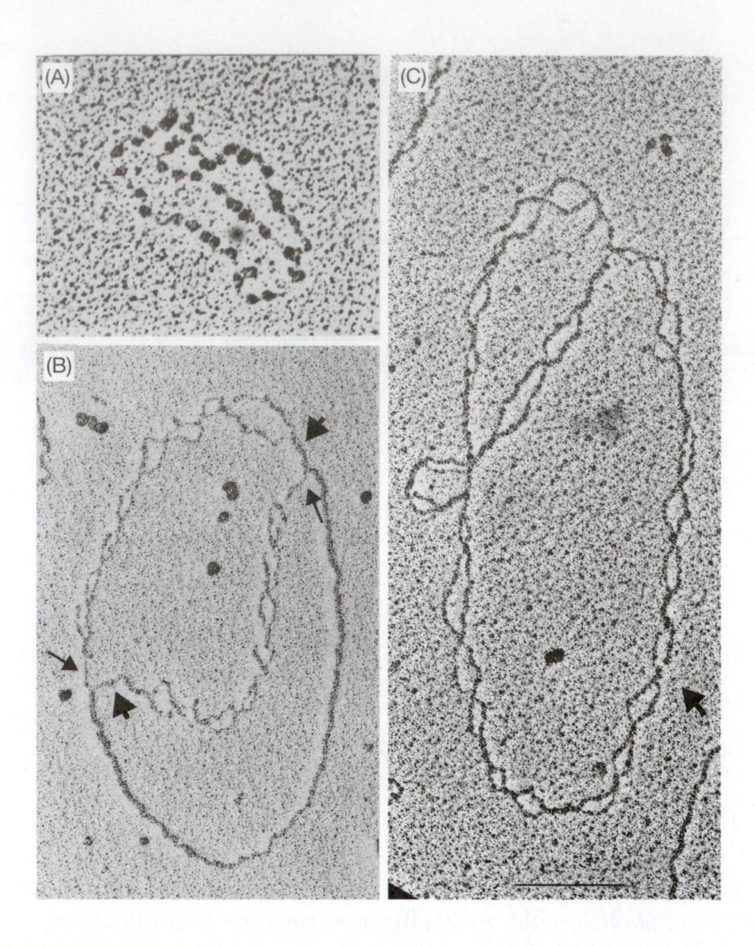

Figure 7-14. DNA from psoralen-crosslinked replicating SV40 chromosomes. With exception of the SV40 DNA helicase T-antigen, which unwinds DNA at replication forks assembled at the SV40 replication origin, this viral genome relies completely on its host cell to replicate its DNA and assemble it into chromatin. (A) SV40 replicating chromosome. Purified SV40 chromosomes were photoreacted with trimethyl-psoralen, a compound that intercalates into DNA and forms covalent adducts with pyrimidine bases when activated by ultraviolet light, thereby linking one strand to the other. DNA in NCPs is protected from psoralen. The DNA was then deproteinized and spread for electron microscopy under denaturing conditions to reveal the locations of nucleosomes (ssDNA bubbles) separated by linker DNA (cross-linked dsDNA) either before (B) or after (C) the DNA was lightly nicked with an endonuclease. (B) The unreplicated parental strand is covalently closed, which prevents separation of denatured DNA strands (i.e. bubbles) in the unreplicated region of a circular DNA molecule. Note the replication fork with a double-stranded leading strand (arrowheads) and a mostly single-stranded lagging strand (arrows). (C) In nicked molecules, bubbles are visible in front of, as well as behind, newly replicated DNA. The bar represents 500 nt. (From Sogo, J.M., Stahl, H., Koller, T. and Knippers, R. (1986) *J. Mol. Biol.* 189: 189–204. With permission from Elsevier.)

as is chromatin at actively expressed genes, typical of hyperacetylated nucleosomes and of chromatin lacking histone H1. Postreplicative, **mature chromatin** consists of newly assembled nucleosomes and a chromatin structure indistinguishable from nonreplicating chromatin. Chromatin maturation requires deacetylation and addition of linker histone H1.

Okazaki fragment initiation zones

One cannot ignore the striking correlation between the average length (135 nt) and size distribution (40–290 nt) of SV40 Okazaki fragments, the average size (262 nt) of the Okazaki fragment initiation zone, and the average distance (218 bp) between nucleosomes in SV40 chromosomes (Figure 7-13). Given the fact that RNA-primed DNA synthesis at replication forks (Okazaki fragments) in eukaryal as well as SV40 chromosomes occurs at stochastically selected sites within a region of ssDNA on the lagging-strand template (Figure 5-2 and Table 5-1), these data suggest that the size of this 'Okazaki fragment initiation zone' could be determined by the spacing of prefork nucleosomes randomly phased along the DNA.

If the rate-limiting step in DNA replication is unwinding DNA through the prefork nucleosome, then nucleosome spacing in front of replication forks would limit the size of the Okazaki fragment initiation zone behind replication forks. This model predicts that the average length of SV40 Okazaki fragments would be 115 nt with a maximum length of about 280 nt. These data are easily accommodated by the 262±103 nucleotides of ssDNA detected by electron microscopy of SV40 replicating DNA molecules.

Similarly, the mammalian nucleosomal repeat distance of about 195 bp should generate Okazaki fragments with an average length 15 nucleotides shorter than those of SV40. In fact, the typical length of mammalian Okazaki fragments is approximately 110 nucleotides. Moreover, the region of ssDNA on one arm of replication forks in yeast DNA of 215±126 nucleotides is proportionately shorter than in SV40, consistent with an average length of approximately 110 nucleotides for yeast Okazaki fragments.

Chromatin assembly at replication forks

Clearly, chromatin needs to form on the new daughter DNA molecules following passage of a replication fork. However, this process of **chromatin assembly** is not a spontaneous process but, rather, is actively directed by cellular factors. The primary forces governing nucleosome assembly are electrostatic. Thus, combining histones and DNA at physiological salt concentrations results in formation of insoluble aggregates because histones bind DNA irreversibly. At high salt concentrations histone aggregation on DNA is prevented by allowing DNA-bound histones to equilibrate with free histones. As the salt concentration is gradually reduced, $[H3{\bullet}H4]_2$ tetramers rapidly associate with DNA followed later by H2A•H2B dimers. The same sequential, two-step pathway occurs *in vivo* as well as *in vitro*. *In vivo*, however, interactions between histones and DNA are controlled by nucleosome assembly proteins (histone chaperones) that bind histones and deposit them onto DNA in an orderly manner.

In order to ensure an appropriate pool of new histone proteins, cells induce high levels of expression of the core histone genes and histone H1 during S phase. However, the deposition of histones onto DNA exhibits distinct patterns, according to the relative dynamics of the subassemblies; H2A•H2B dimers occupy an outer position on the nucleosome (Figure 7-15). In line with this observation, a range of studies have revealed that H2A•H2B dimers are far more dynamic than the H3•H4 components. Indeed, a wide body of work has established that newly synthesized H2A•H2B dimers are deposited on chromatin both ahead of and behind the replication fork, whereas new $[H3{\bullet}H4]_2$ tetramers are only found on newly synthesized DNA.

The old $[H3{\bullet}H4]_2$ tetramer is segregated as a stable entity. Parental nucleosomes are disrupted into two H2A•H2B dimers and one $[H3{\bullet}H4]_2$ tetramer. The latter is then transferred onto one of the two newly replicated arms of the fork to form a subnucleosomal particle onto which either old or newly synthesized H2A•H2B dimers are added to complete the nucleosome. This transfer is so rapid and efficient that only a large excess of naked DNA can compete with the daughter strands for parental histone binding during chromatin replication in cell-free systems.

Histones are highly basic proteins that are insoluble in the presence of DNA unless accompanied by a chaperone or organized into a histone octamer. Histone chaperones are proteins that associate with histones and stimulate a reaction involving histone transfer without being part of the final product. Remarkably, only six of the 19 different proteins or protein

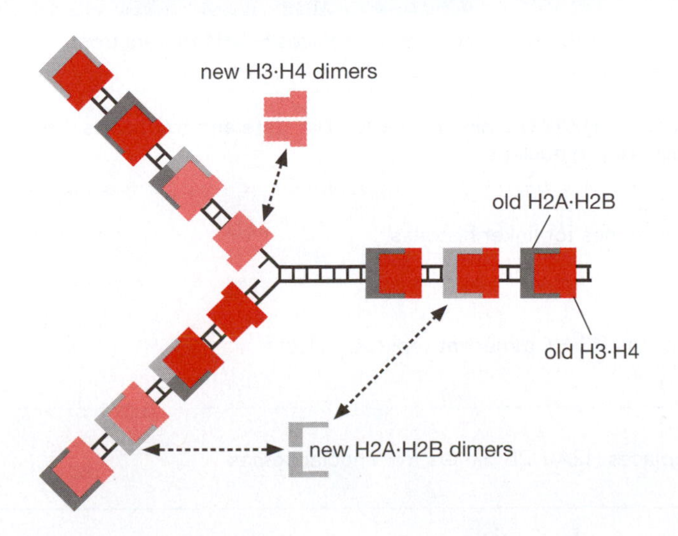

Figure 7-15. Dynamics of histone exchange and assembly at replication forks. Newly synthesized H2A and H2B associate to form dimers that can exchange with nucleosome-associated H2A•H2B dimers ahead of the fork and can additionally assemble onto $[H3{\bullet}H4]_2$ tetramers behind the fork. This appears to be a random process and individual nucleosomes can have two new, two old, or one new and one old H2A•H2B dimer. $[H3{\bullet}H4]_2$ tetramers assemble onto newly synthesized DNA only after fork passage. Whether or not they contain a mix of old and new $[H3{\bullet}H4]_2$ remains to be determined. Once the $[H3{\bullet}H4]_2$ tetrasome is assembled, it associates rapidly with H2A•H2B dimers to form the histone octamer.

complexes that have been identified as histone chaperones are involved directly in chromatin assembly at replication forks (Table 7-3); the others are involved primarily in regulating expression of genes in nonreplicating chromatin and in chromatin assembly linked to DNA repair.

With the development in 1986 of a cell-free system that recapitulated replication-coupled chromatin assembly *in vitro*, Bruce Stillman and colleagues were able to identify cellular factors that carried out the chromatin assembly reaction. This technical *tour de force* led to the identification of chromatin assembly factor-1 (CAF1), a highly conserved complex of three subunits. CAF1 chaperones individual H3•H4 dimers to replication forks for assembly into $[H3{\bullet}H4]_2$ tetramers. While CAF1 is essential for chromatin assembly in all cell-free systems tested thus far, remarkably, yeast mutants lacking CAF1 remain viable, suggesting the existence of additional redundant pathways. The way in which CAF1 couples H3•H4 deposition with the passage of the replication fork is both simple and elegant. CAF1 interacts directly with the sliding clamp PCNA at replication forks. Interestingly, of the four histone H3 variants, CAF1 interacts only with H3.1 and the canonical major H3 histone. H3.1 is the H3 variant most highly expressed during S phase. In addition, the H4 protein bound by CAF1 is enriched in acetylation at lysines 5 and 12, signatures of newly synthesized H4. Thus, CAF1 preferentially deposits new histones onto nascent DNA. CAF1 acts in concert with another histone chaperone termed antisilencing function-1 (Asf1). Asf1 interacts with both a H3•H4 dimer and with CAF1 via two interaction sites that lie on opposite faces of the Asf1 protein. It appears likely, therefore, that Asf1 simultaneously binds both CAF1 and H3•H4 dimer.

Once the CAF1•Asf1 system has assembled tetrasomes on newly replicated DNA, histone chaperones nucleosome assembly protein 1 (Nap1) and nucleoplasmin transfer H2A•H2B dimers from the cytoplasm to the nucleus where they are deposited directly onto tetrasomes. In contrast to CAF1, the role of Nap1 and nucleoplasmin is not limited to DNA replication, but can occur anywhere in chromatin where tetrasomes exist. In that sense, they are nucleosome maintenance proteins. Nuclear autoantigenic sperm protein (Nasp) and Nap1 are ubiquitous histone chaperones that bind the linker histone H1 (Nasp), and the maternally expressed linker histone B4 in *Xenopus* eggs (Nap1), suggesting that linker histones, like core histones, also require a chaperone to load onto chromatin.

Table 7-3. Histone Chaperones That Affect Chromatin Assembly At Replication Forks

Protein	Histone selectivity	Functions
Asf1	H3.1•H4, H3.3•H4	Provides histones to CAF1, and displaces H3•H4 dimers from nucleosomes
Nucleoplasmin	H2A•H2B	Maintains H2A•H2B dimers in a soluble state and transports them from cytoplasm to nucleus
Nap1		
Nasp	H1 (mammals)	Chaperones for linker histones
Nap1	B4 (amphibian eggs)	
CAF1 complex p150 p60 RbAp48	 H3.1•H4 H3.1•H4 H3•H4	Deposits H3•H4 dimers at replication forks
FACT complex Spt16 Ssrp1	 H2A•H2B H3•H4	Displaces H2A•H2B dimers from nucleosomes

Two models have been proposed to account for the synergistic relationship between Asf1 and CAF1 in mediating replication-coupled chromatin assembly. In the first model, Asf1 delivers H3•H4 dimers to CAF1, and CAF1 then deposits them onto the newly synthesized DNA duplex. In the second model, binding of an Asf1•H3•H4 complex to a CAF1•H3•H4 complex may trigger H3•H4 deposition by CAF1. The two models are by no means mutually exclusive. Both H3•H4 bound and unbound forms of Asf1 exist in the cell, and Asf1 can become saturated with histones when their levels are elevated. Thus, Asf1 can act as a buffer between the pool of newly synthesized histones and chromatin assembly. However, a second route exists whereby Asf1 can acquire its histone cargo. *In vitro* studies have revealed that Asf1 can actually disrupt $[H3 \cdot H4]_2$ tetramers by stripping one H3•H4 dimer out of the assembly. Thus, it is possible that Asf1 might also facilitate the disassembly of nucleosomes ahead of the replication fork before passing the H3•H4 dimer to the PCNA-bound CAF1 behind the fork. For this to be true, Asf1 should be associated with replication forks and, as discussed below, this is indeed the case. Furthermore, studies in a number of systems, including chicken, human, and *Drosophila* cells, have revealed that reduction in the levels of Asf1 results in defects at replication forks.

Chromatin disassembly at replication forks

The presence of stable NCPs presents a serious impediment for both DNA transcription and DNA replication. In the case of transcription, two distinct mechanisms allow progression of RNA polymerases through chromatin. Eukaryotic RNA polymerase III, bacteriophage RNA polymerases, and ATP-dependent chromatin-remodeling complexes can translocate through nucleosomes while concomitantly unwinding a small region of the DNA to allow transcription without releasing the histone octamer into solution. However, transcription by RNA polymerase II involves displacement of one H2A•H2B dimer. To accomplish this task, RNA polymerase II receives help from two histone chaperones. Nucleolin facilitates transcription of a number of genes by dramatically increasing the efficiency of chromatin-remodeling machines such as SWI/SNF and ACF. In addition, nucleolin is able to remove H2A•H2B dimers from nucleosomes already assembled on chromatin, thereby facilitating the passage of RNA polymerase II through nucleosomes. In this capacity, nucleolin is related to FACT (which stands for facilitates chromatin transcription), a highly conserved heterodimeric complex. Remarkably, the FACT complex associates with the replisome progression complex (Figure 13-16), a macromolecular assembly of proteins that includes the MCM helicase, the GINS complex, and Cdc45 (Figure 5-4), making it an excellent candidate for destabilizing nucleosomes in front of replication forks.

In order to completely unwind DNA bound to histone octamers, the histone octamer must disengage from both strands of the DNA. Current evidence suggests that this is accomplished by disrupting prefork nucleosomes into two parental H2A•H2B dimers and a $[H3 \cdot H4]_2$ tetramer. Furthermore, it is becoming increasingly appreciated that the FACT complex plays important roles in a range of chromatin-based transactions, including DNA replication. For example, if FACT is removed from *Xenopus* egg extracts by depletion with specific antibodies then the ability of the extracts to support DNA replication is severely impaired. Biochemical studies reveal that FACT acts to remove one or both H2A•H2B dimers from nucleosomes. The mechanistic basis of this remains unclear, but the interaction of FACT with the MCM complex strongly hints at a role for FACT in destabilizing histone octamers ahead of the replication fork. In agreement with this, FACT can stimulate unwinding of nucleosome-associated DNA *in vitro* by the MCM helicase subcomplex Mcm(4,6,7). Removal of H2A•H2B dimers will, of course, leave the $[H3 \cdot H4]_2$ tetramer associated with DNA. However, this tetramer is the eukaryotic equivalent of the archaeal histone–DNA

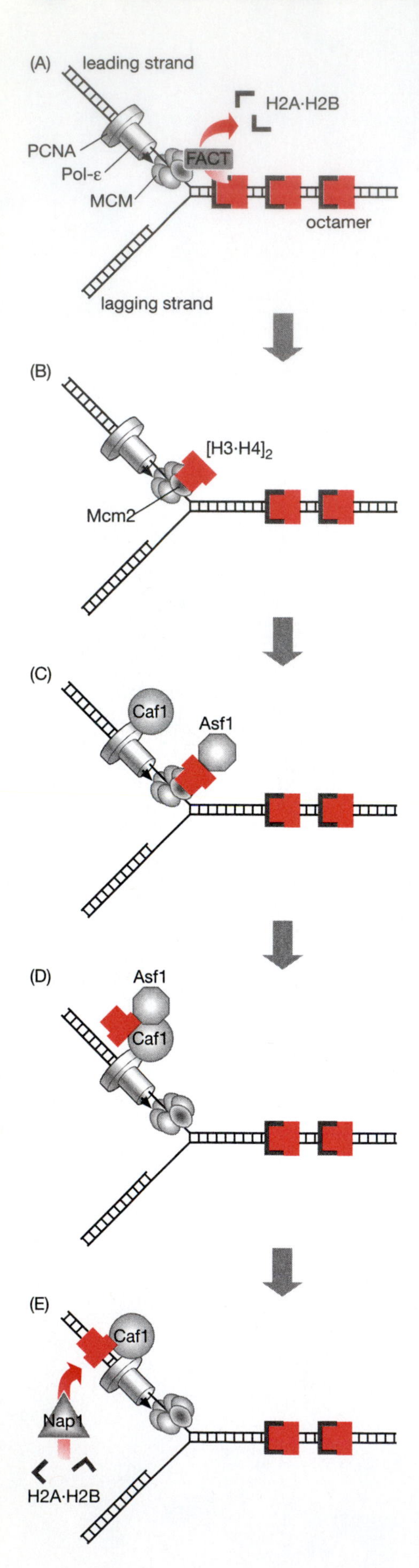

complex, and, as discussed above, the archaeal tetramer is not a barrier to MCM translocation. Thus, it is highly plausible that MCM can simply strip $[H3•H4]_2$ from the DNA. Since parental nucleosomes are known to re-associate with the newly synthesized daughter DNA molecules, $[H3•H4]_2$ doesn't simply diffuse away from replication forks.

How might $[H3•H4]_2$ tetramers be transferred from prefork nucleosomes to postfork nucleosomes? Two mechanisms are suggested by the data at hand. First, one of the MCM subunits, Mcm2, has been shown to bind the N-terminal tail of histone H3. Remarkably, histone H3 also binds the SV40 T-antigen helicase. Therefore, a $[H3•H4]_2$ tetramer that is displaced from DNA by either replicative helicase may be tethered to the helicase via interaction with its exposed H3 tail and thus remain at the replication fork for assembly onto newly replicated DNA. A second mechanism involves Asf1, which is found associated with the MCM helicase in nuclear extracts from human cells. This interaction is not direct, but rather mediated by a H3•H4 dimer that acts as a bridge between the MCM helicase and Asf1. In this manner, Asf1 links the supply of core histones to DNA unwinding at replication forks. As expected, either depletion of Asf1 by RNA interference, or overproduction of histone H3•H4, impedes DNA unwinding at replication sites.

These observations suggest a plausible model for histone dynamics at replication forks (Figure 7-16). As the replisome progression complex advances, FACT strips H2A•H2B dimers from the nucleosome directly ahead of the fork, thereby leaving the less stable $[H3•H4]_2$ tetramer. Prefork tetrasomes can then be physically peeled off DNA by the forward motion of the MCM helicase, releasing $[H3•H4]_2$ tetramer from the DNA; the released tetramer is then captured by Mcm2. Asf1 subsequently acts to split the tetramer into constituent H3•H4 dimers. Asf1 then acts in concert with the PCNA-associated CAF1 behind the fork to deposit $[H3•H4]_2$ tetramers onto nascent DNA duplexes. In addition, other Asf1 molecules are associated with newly synthesized H3•H4 dimers that also can be delivered to CAF1. Thus, both new and old $[H3•H4]_2$ tetramers will appear on newly replicated DNA. Finally, H2A•H2B dimers associate with the tetrasome to form NCPs. It is important to stress that although this model is compatible with the available data, it remains hypothetical.

SISTER CHROMATID COHESION

In eukaryotes, the two sister chromatids arising from DNA replication are held together following passage of the replication fork until the onset of anaphase in mitosis. Early models attempted to account for this **sister chromatid cohesion** either by invoking topological intertwining of the sisters, or by some inherent stickiness of chromatin that glues the two strands together. However, through a genetic analysis of fruit flies, Terry Orr-Weaver and her colleagues identified specific proteins that ensure sister chromatid cohesion. Subsequent work from a number of groups, studying a variety of eukaryotic organisms, has revealed that one of the principle architects of sister chromatid cohesion is an evolutionarily conserved complex of four proteins termed cohesin (Table 7-4) that embrace sister chromatids by forming a ring around them. For simplicity, the *S. cerevisiae* nomenclature is used hereafter.

Remarkably, most bacteria and archaea have a single SMC (see below) homolog that assembles into a homodimer that can associate with a kleisin subunit, a family of proteins that connect the two termini of an SMC protein.

Figure 7-16. FACT-finding model for chromatin assembly at eukaryotic DNA replication forks. Symbols are the same as in Figure 7-15).

Table 7-4. Cohesin Subunits In Various Organisms

Yeast		Flies	Frogs	Mammals
S. cerevisiae	*S. pombe*	*D. melanogaster*	*X. laevis*	*H. sapiens*
Smc1	Psm1	Smc1	Smc1	Smc1
Smc3	Psm3	Smc3	Smc3	Smc3
Scc1/Mcd1	Rad21	Rad21	Rad121	Rad21/Scc1
Scc3/Irr1	Psc3	SA	SA1, SA2	Stag1/SA1, Stag2/SAS2

Commonly used alternative names are separated by /. Adapted from Onn *et al.* (2008).

In bacteria, these proteins are involved in organizing DNA in the nucleoid and in chromosome segregation. The finding that SMC complexes exist in bacteria and archaea implies that these proteins existed before histones, and that the mechanism through which SMC proteins structure DNA is more ancient than nucleosomes.

Formation Of A Cohesin Ring Around Sister Chromatids

The key to understanding the cohesin complex lies in the structure of the Smc1 and Smc3 subunits, members of the structural maintenance of chromosomes (SMC) superfamily that is conserved in all three domains of life. These proteins possess N- and C-terminal globular domains that are separated by extremely long α-helical sections (Figure 7-17). The middle of the α-helical region contains an interruption in the form of a further globular domain that acts as a hinge. This hinge allows the two α-helical stretches to fold back upon each other to form a coiled-coil. A monomer of Smc1, thus folded, is then able to form a heterodimer with a similar coiled-coil monomer of Smc3. The initial dimerization occurs via the hinge region.

Interestingly, the formation of the coiled-coil also juxtaposes the N- and C-terminal domains of a given monomer. These domains interact and form a structure that corresponds to one half of an ATPase domain of the ABC transporter family. Thus, when the head domain of the Smc1 monomer in the Smc1•Smc3 heterodimer is brought into proximity of the head domain of the Smc3 monomer, two full ATP-binding sites are generated. Two molecules of ATP can therefore act as bridges between the SMC head groups, thereby forming a ring-shaped molecule. The fact that ring closure requires ATP allows the cohesin activity to be regulated by sequential ATP binding, hydrolysis, and then release of ADP. Once the two head groups have engaged ATP and dimerized, the kleisin subunit sister chromatin cohesion subunit-1 (Scc1) can form a bridge with the N-terminus of Scc1 bound to the Smc3 head domain and the C-terminal region of Scc1 bound to the Smc1 head domain. Finally, Scc1 interacts with the fourth cohesin subunit, Scc3. Considering that a single chromatin fiber composed of nucleosome core particles has a diameter of 10 nm, a single ring of cohesin could readily accommodate two such chromatin fibers. Note that the dimension would not accommodate two 30 nm fibers. This means that cohesin must be added to sister chromatids as they are born at the replication fork and before they have condensed into 30 nm fibers.

Cohesin localization

In budding yeast, cohesin loading onto chromatin begins during the $G_1 \rightarrow S$ transition. The main loading site is the region around the centromere. Additional loading is observed along the chromosome arms at a lower

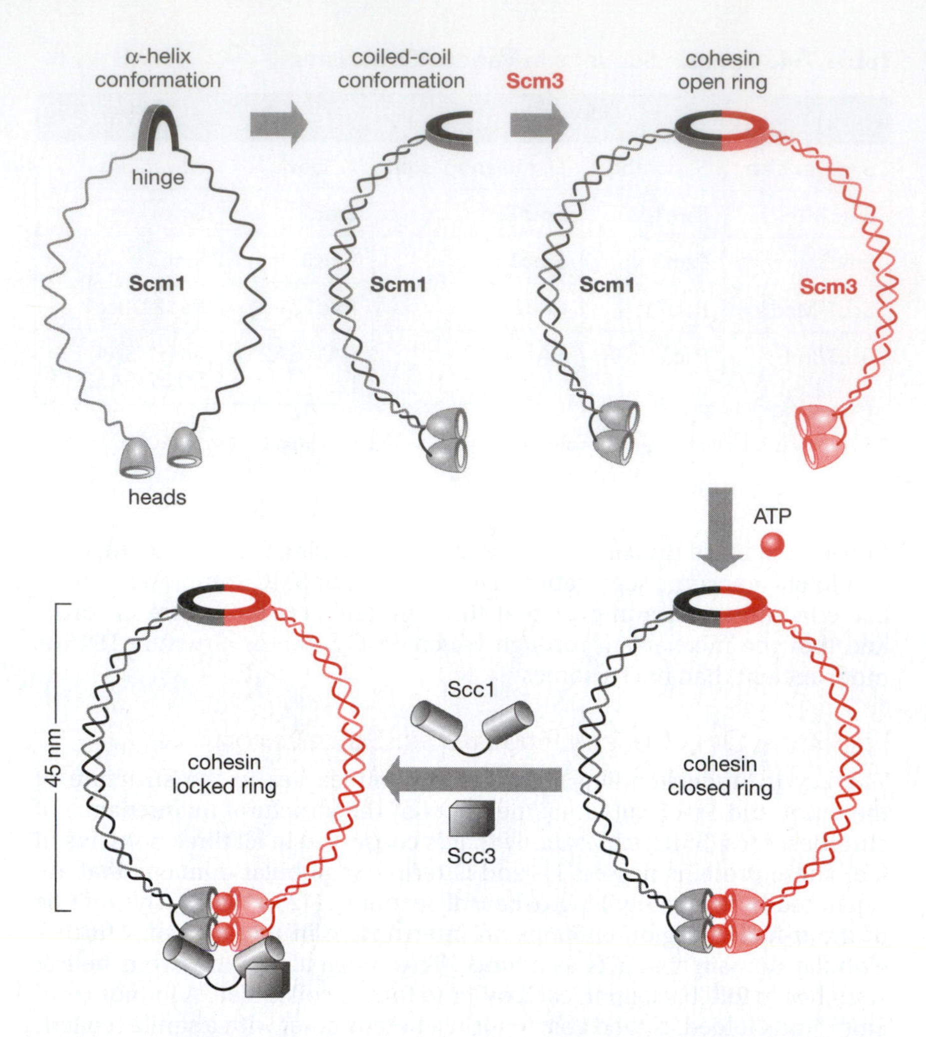

Figure 7-17. (A) Formation of cohesin rings from two SMC monomers, Smc1 (black) and Smc3 (red). The head and hinge domains are separated by about 45 nm. A single cohesin ring can encircle two 10 nm chromatin fibers, as modeled in Figure 7-18.

density, and, in budding yeast, this appears to occur periodically roughly once every 15 kb. Precisely what defines these cohesin-associated regions (CARs) is not fully understood. CARs are generally associated with AT-rich intergenic regions, and they often lie between convergently transcribed genes. In metazoa, cohesin also is enriched at centromeres, but other sites of cohesin enrichment are somewhat species-specific. In *Drosophila*, cohesin is found within transcribed regions, and in human cells where about 50% of cohesin is in intergenic regions, it is also found either in introns or the immediate environment of transcribed regions. There is an interesting correlation in human cells between the localization of cohesin and the binding sites for a site-specific DNA-binding protein, CTCF.

Cohesin loading

Without exception, cohesin loading is dependent on two additional proteins, a heterodimer of Scc2 and Scc4. Since Smc1 and Smc3 are required to bind and hydrolyze ATP, Scc2•Scc4 dimers are believed to stimulate the ATPase activity of cohesin, allowing the ring-shaped molecule to open and close repeatedly, encircling DNA as it does so. It should be emphasized at this point that loading of cohesin is temporally distinct from, and precedes the establishment of, sister chromatin cohesion. The interplay between cohesion and replication initiation is apparent in *Xenopus* where Scc2 and Scc4 are recruited to pre-replicative complexes (preRCs) via interaction with the Cdc7•Dbf4 complex. This interaction is conserved in fission yeast but not in budding yeast.

The precise timing of cohesin loading appears somewhat variable. In budding yeast the majority of loading takes place at the end of G_1 phase in the cell-division cycle, whereas in vertebrates, cohesin loading begins in telophase of mitosis. As will be discussed below, this likely reflects the distinct ways in which cohesin, and thus cohesion, are broken down in mitosis in different organisms.

Establishment Of Cohesion Between Sister Chromatids

Cohesin is loaded onto DNA prior to the start of S phase yet at the end of S phase is believed to encircle the two daughter molecules. How does it make the transition from encircling one DNA molecule to embracing both daughter duplexes? One model is that replication forks pass through the ring-shaped cohesin molecule, but the sheer size of the replisome may be incompatible with this appealing model (Figure 7-18). Alternatively, cohesin may open transiently as the replisome approaches and then close around the daughters immediately following its passage.

Although the molecular basis of cohesion establishment is not yet fully resolved, a number of establishment factors have been identified in genetic screens. Foremost among these is the acetyltransferase EcoI that is essential for budding yeast viability. However, EcoI is essential only during S phase; once cohesion is established, EcoI is dispensable. EcoI directly acetylates two conserved lysine residues, Lys-112 and Lys-113, in Smc3. Mutation of these residues to glutamine to mimic the structure of acetyl-lysine eliminates entirely the requirement for EcoI for viability. Since these two lysine residues are located near the ATP-binding site in the head group of Smc3, it has been suggested that acetylation of these lysines may influence ATP binding or hydrolysis by Smc3. Additional clues to the role of EcoI lie in the observation that it appears to associate with replication forks via association with PCNA. Also, EcoI is no longer essential if either Wpl1 or Pds5 is mutated. These two proteins form a complex that interacts with cohesin, and they have been implicated in destabilizing cohesin and promoting its dissociation from chromatin. Taken together, these observations suggest that the role of EcoI is to counteract the destabilizing role of the Wpl1•Pds5 complex. Clearly, much work remains to be done to elucidate the precise mechanisms of cohesion establishment. However, with the key players identified and significant insights gained into their mode of action, it is likely that progress in this field will be rapid.

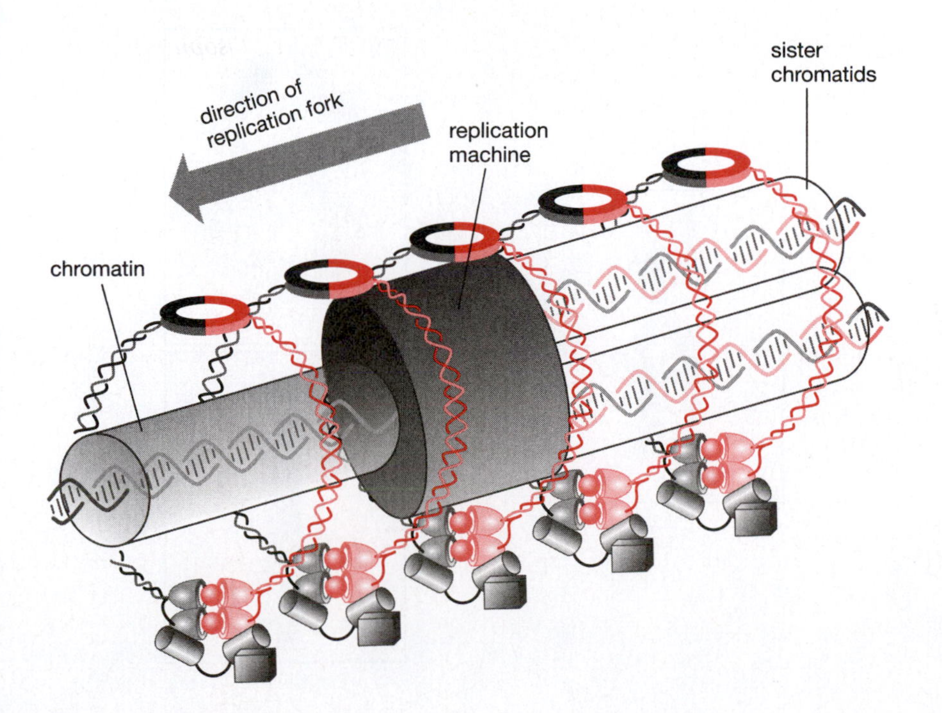

Figure 7-18. 'Embrace Model' for establishment of sister chromatid cohesion during S phase. In an elegant series of experiments, Haering, Nasmth, and colleagues demonstrated that single rings of cohesin could mediate cohesion between chromosomes *in vivo* by using site-directed mutagenesis to allow chemical cross-linking between cohesin subunits in order to covalently seal the cohesin ring. In the model shown the entire replication-fork machinery simply passes through the cohesin rings. Alternatively the cohesin rings open upon approach of the fork and then reseal behind the fork, thereby encircling the two daughter molecules.

Dissolution Of Cohesion Between Sister Chromatids

Sister chromatid cohesion is established in S phase and persists until anaphase in mitosis, at which point the sister chromatids are separated. This event requires dissolution of cohesin. In vertebrates, dissolution of cohesin occurs in two phases. First, cohesin is depleted along the chromosome arms during prophase, leaving only the cohesin in the vicinity of the centromere. This results in the classic X shape of prometaphase chromosomes (Figure 7-19A). The remaining centromeric and pericentromeric cohesion is then rapidly lost during the metaphase to anaphase transition. This pathway, which appears to be specific for vertebrates, depends both on the polo-like kinase (Plk1) that phosphorylates Scc1 and the vertebrate orthologs of Scc3, SA1, and SA2. However, the vertebrate Wpl1 ortholog may also play a role in a pathway that is distinct from the one involving the Plk1 kinase. What prevents dissociation of cohesin from centromeric and pericentromeric loci during prophase? It transpires that a centromeric associated protein called SgoI and an associated protein phosphatase termed PP2A specifically counteract the phosphorylation by Plk1, thereby inhibiting dissociation of cohesin in the vicinity of the centromere.

Precocious sister chromatid separation is observed after inactivation of a variety of cohesion regulators that are required at different stages of the cohesion cycle (Figure 7-19B). Unresolved sister chromatids are observed when proteins that regulate the removal of cohesin during prophase are inactivated (Figure 7-19C). Diplochromosomes, consisting of four chromatids lying side by side, instead of the normal two, are produced when cells go through two rounds of DNA replication without separation of chromatids. This phenotype is observed after inactivation of separase (see below) as well as at low frequency after expression of noncleavable Scc1.

When cells begin anaphase they must trigger the complete dissolution of all remaining cohesion in order to permit segregation of sister chromatids. This event is triggered by the ubiquitin ligase anaphase-promoting complex (APC) and regulated by the spindle-assembly checkpoint (Figure 13-19). The APC ubiquitinates a protein called securin thereby targeting it for degradation by the 26S proteasome. Prior to its degradation, securin is found in a complex with separase, a protease that targets Scc1, the principal subunit of cohesin. Separase cleaves Scc1 at two sites, resulting in the

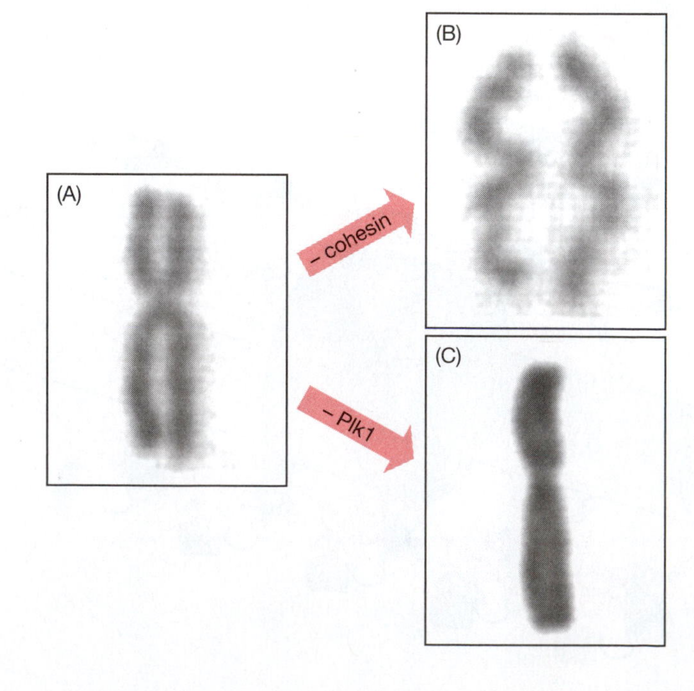

Figure 7-19. Giemsa-stained prometaphase chromosomes from human cells. (A) Sister chromosomes in wild-type cells have largely separated except at the centromere. (B) In contrast, sister chromatids are not associated in cells where cohesin, Scc2, or Scc4 have been inactivated. (C) Sister chromatids in cells deficient in Plk1, Aurora B, or condensin remain in close contact throughout their full length. (From Peters, J.M., Tedeschi, A. and Schmitz, J. (2008) *Genes Dev.* 22: 3089–3114. With permission from Cold Spring Harbor Laboratory Press.)

collapse of cohesin rings. When bound to securin, separase is held in an inactive form that cannot recognize its targets. Therefore, degradation of securin initiates release of cohesin from sister chromatids.

The function of separase and its regulation by securin are conserved and essential in all eukaryotic cells. However, vertebrates have an additional pathway for controlling separase activation. Cyclin-dependent kinase 1 (Cdk1) and its partner cyclin B regulate separase by two means. First, Cdk1•cyclin B phosphorylates separase resulting in inhibition of separase activity. Second, separase binds to, and is sequestered by, cyclin B. When the APC is activated at the beginning of anaphase, it targets both securin and cyclin B. Thus, in vertebrates, the action of the APC coordinately ablates three inhibitory mechanisms governing separase, and thus it leads to separase activation, consequent cleavage of Scc1, breakdown of the remaining cohesin rings and, ultimately, chromosome segregation.

The timing and mechanism of cohesin removal varies among species, just as the timing of loading varies, and it transpires that there is a causal linkage between these two phenomena. In vertebrates, most cohesin is removed from sister chromatids by the prophase pathway. This pathway does not result in the destruction of cohesin subunits, simply cohesin's displacement from chromatin. Only a small proportion of cohesin is degraded via the separase pathway. As a consequence a large pool of cohesin is available following completion of anaphase, and in vertebrates this cohesin is already re-associating with chromatin during telophase. However, in budding yeast the majority of cohesin is removed by the destructive action of separase and so cohesin accumulation on DNA is not observed until the end of the following G_1 phase.

DUPLICATION OF EPIGENETIC INFORMATION

As DNA is replicated and the sister chromatids are held together, only one problem remains: duplicating the epigenetic information encoded within the chromosomes. Such information comes in at least four forms: nucleosome phasing with respect to DNA sequence, histone modifications, histone variants, and DNA methylation. Although the precise mechanisms underpinning epigenetic inheritance remain poorly understood, progress in this field has been rapid and some insights into specific systems have been gleaned.

Nucleosome Segregation and Positioning

During the 1970s, the suggestion arose that asymmetric segregation of parental histones to one of the two sister chromatids could provide a vehicle for transmitting and segregating to daughter cells epigenetic information encoded within chromatin structure during the development of metazoan organisms. Given the structure of replication bubbles arising from bidirectional DNA replication and the possible formation of DNA replication loops and replication factories (Figure 3-11), 'old' NCPs in front of a replication fork could be distributed to newly replicated DNA in four possible modes (Figure 7-20). They could be distributed conservatively to only one of the two sister chromatids, or they could be distributed semi-conservatively to either the leading arm or the lagging arm of each fork. Unfortunately, this attractive hypothesis soon lost favor as it became apparent that histones in front of replication forks were distributed randomly to both arms of the fork. Analysis of cellular chromatin by Vaughn Jackson and Roger Chalkley and SV40 chromosomes by Paul Wassarman and Melvin DePamphilis in the 1970s and 1980s demonstrated that old H3 and H4 histones in front of replication forks are distributed with equal frequency to both arms of the fork, and that new nucleosomes are assembled on both arms of replication forks with equal frequency, either in the presence or

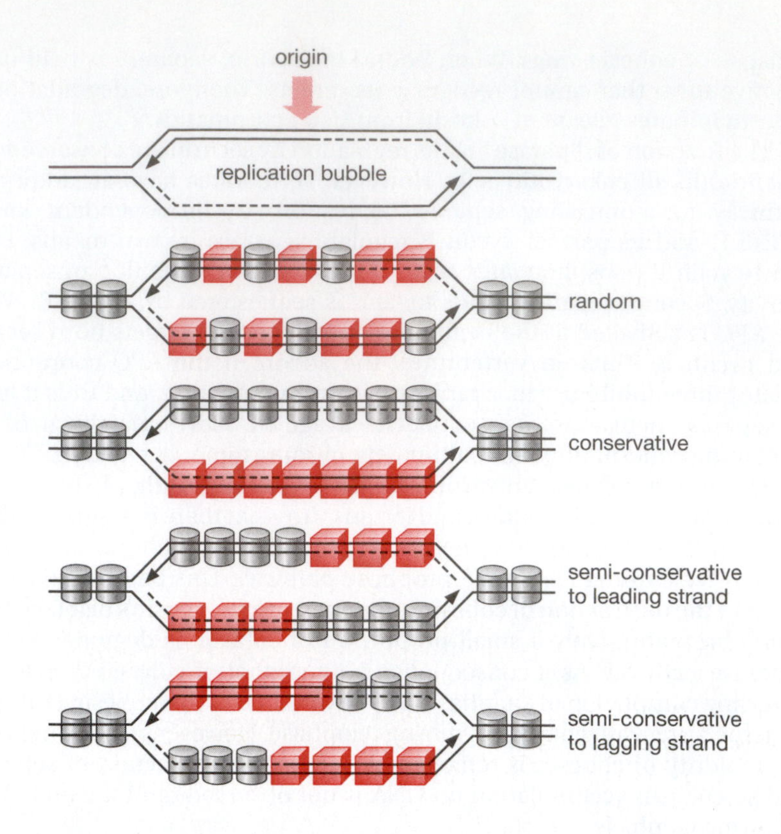

Figure 7-20. The four possible ways that old (gray cylinders) and new (red cubes) NCPs could distribute themselves in replication bubbles with two active forks in which DNA synthesis is continuous on one arm (solid arrow) and discontinuous on the other (dashed line).

absence of histone synthesis. Subsequent analysis of chromatin dynamics in several different eukaryotic organisms has since confirmed that, once formed, H3•H4 tetramers remain intact, and can form an NCP with either new or old H2A•H2B dimers (Figure 7-15).

Nevertheless, it remained possible that newly assembled nucleosomes on one arm of the fork might occupy positions on the DNA that were identical to those occupied by nucleosomes on the other arm of the fork. In other words, phasing of nucleosomes with respect to DNA sequence might be identical on two sister chromatids. However, nucleosome phasing at replication forks appears to be random. Nucleosomes can assemble *in vitro* on virtually any dsDNA as long as the core histones are maintained in a soluble state, and eukaryotic topoisomerase I is present to facilitate bending of DNA around the histone octamer. Sequences such as polydA:polydT, Z-form DNA, and triplex DNA discourage nucleosome assembly, whereas bent DNA and repeating CTG triplets facilitate nucleosome formation, but these sequences comprise a small fraction of the total genome. *In vivo*, nucleosomes are distributed virtually randomly throughout the bulk of the DNA such that the positions of nucleosomes on one sibling chromatid are not located at sites identical to those in its sister chromatid. Studies on SV40 and cellular replicating chromatin are consistent with the conclusion that parental histones are segregated randomly between the two daughter strands following fork passage, and 'new' histone octamers are assembled on both arms. For example, the accessibility of a restriction endonuclease cleavage site on one arm of a replication fork is unrelated to the accessibility of the same site on the other arm (Figure 7-14); the fraction of protected sites is that expected if site selection is a stochastic process. This suggests that chromatin assembly occurs independently on the two sibling molecules of a single replicating chromosome, and that nucleosome phasing with respect to DNA sequence is essentially random.

Despite the apparent absence of any conservation of nucleosome positioning at replication forks, analyses of nucleosome positions at specific gene loci suggest that cellular DNA contains information specifying the positions of nucleosomes, and that this code is used by the cell to facilitate nucleosome spacing by encoding information for multiple alternative overlapping nucleosomal arrays. How might these two concepts be reconciled? One possibility is that cellular replication forks, which employ the MCM DNA helicase to unwind DNA, differ from SV40 replication forks, which use the SV40-encoded T-antigen DNA helicase to unwind DNA. Therefore, cellular replication forks may distribute nucleosomes differently than viral forks by allowing a certain degree of nonrandom nucleosome assembly that would preserve epigenetic information. Alternatively, it would be advantageous for replication-linked nucleosome assembly to be as rapid and efficient as possible, and then for nucleosomes to be redistributed subsequently by replication-independent mechanisms.

Given the existence of ATP-dependent chromatin-remodeling machines in eukaryotic cells, the second hypothesis appears to be correct. Chromatin-remodeling enzymes use ATP hydrolysis as a fuel to push nucleosomes around on DNA. The consequence can be simple repositioning of the nucleosome, localized unwrapping of DNA from the surface of the nucleosome, or even the ejection of nucleosomes entirely from DNA. In addition some remodeling machines mediate exchange of histone dimers to allow selective deposition of histone variants. Other remodeling machines are large, multisubunit complexes that all include an SNF2 family ATPase that acts as the engine. Several remodeling machines contain recruitment domains that specifically recognize modified histones, for example the bromo-domain of the SWI/SNF remodeling machine recognizes acetylated lysine residues. Thus, the covalent modifications of histone tails described below can lead to recruitment of remodeling activities, thereby facilitating crosstalk between distinct classes of chromatin-modifying enzyme and adding to the huge complexity of nucleosome positioning at various sites.

Protein Modification

The regions surrounding centromeres are generally composed of constitutive heterochromatin, so-called pericentric heterochromatin. Nucleosomes within pericentric heterochromatin are enriched in histone H3 that is trimethylated on Lys-9 (H3K9me^3). This methylation mark allows binding of HP1, a protein that facilitates formation of heterochromatin. Following DNA replication, however, parental H3K9me^3 is randomly segregated to the two sister chromatids and interspersed with new, non-methylated H3-containing nucleosomes. Thus, the epigenetic information encoded by the methylation mark will disappear randomly from the two sister chromatids as cells divide unless a mechanism exists to replace it. It turns out that HP1 associates with SUV39H1, the methyltransferase responsible for converting histone H3 into H3K9me^3. Thus, when old H3K9me^3 recruits HP1, it also recruits SUV39H1, the very enzyme needed to methylate H3K9 in neighboring unmethylated nucleosomes, thereby creating a binding site for HP1 and allowing the methyl mark to spread (Figure 7-21). Unfortunately, while this paradigm explains how large regions of chromatin can be marked by a particular histone modification, little is known about how epigenetic marks at individual gene promoters are inherited.

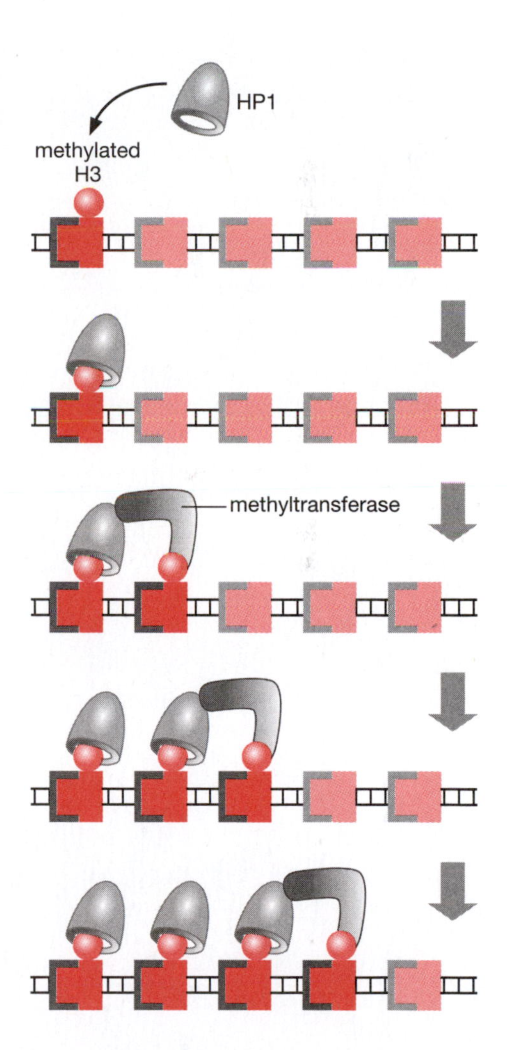

Figure 7-21. Model for extending histone H3K9 methylation throughout a genetic locus. The parental methylated histone H3 is recognized by HP1 which recruits the methyltransferase SUV39H1. SUV39H1 then methylates neighboring nucleosomes, thereby creating a new HP1-binding site and continuing to spread the methyl mark along the chromatin.

Histone-modifying enzymes

The crystal structure of a nucleosome reveals that the core folds of the histones form the inner heart of the NCP, but that some of the histones have extensive tails that extend out from the NCP (Figures 7-6 and 7-7). Not all of the histone tails were resolved in the initial crystal structure, but all four histones do possess conserved N- and C-terminal tails that extend beyond their core fold (Figure 7-22). These tails are rich in amino acid residues that are potential sites of post-translational modifications such as acetylation, phosphorylation, ubiquitylation, and methylation. In fact, many of them

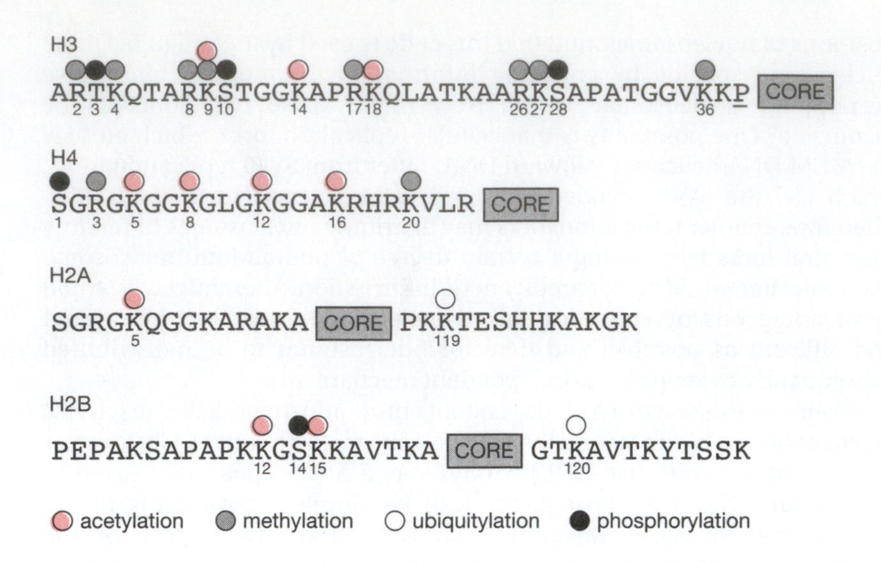

Figure 7-22. Sites of post-translational modification in the histone tails of NCPs. The N-terminal amino acid is number 1. CORE refers to that portion of the sequence engaged with other histones to form a NCP. In addition to the covalent modifications listed Pro-38 in H3 (underlined) is a target for isomerization. (Data from Kouzarides, T. (2007) *Cell* 128: 693–705.) Amino acids: A, alanine; R, arginine; E, glutamic acid; Q, glutamine; G, glycine; H, histidine; L, leucine; K, lysine; P, proline; S, serine; T, threonine; Y, tyrosine; V, valine.

are modified. For example, acetylation of Lys-14 of histone H3 (H3K14ac) or trimethylation of Lys-4 ($H3K4me^3$) are generally associated with open, actively transcribed loci, whereas others, such as $H3K9me^3$ or $H3K27me^3$, are associated with a repressed chromatin state. Newly assembled nucleosomes contain histone H4 acetylated on lysines 5 and 12. Because the tails extend beyond the NCP they have the potential both to contact neighboring NCPs and also to be recognized by additional factors. Modification, therefore, has the potential to both directly and indirectly modify nucleosome–nucleosome and nucleosome–DNA interactions, thereby influencing the accessibility of the DNA.

Based on the combinatorial nature of these tail modifications, it has been suggested that they constitute a code. This code is thought to dictate transitions between transcriptionally active and transcriptionally silent chromatin by either creating or destroying binding platforms for specific chromatin-remodeling factors. However, although it is well established that covalent histone modifications, alone or in combination, can influence nucleosome mobility and function as a scaffold for the recruitment of regulatory proteins, the universality of the histone code is controversial.

Histone variants

In addition to the panoply of post-translational modifications of histone tails, further complexity is introduced to eukaryotic chromatin with the realization that there are many variants of histone proteins, some of which have well-defined functions (Table 7-5). The only histone for which variants have not been discovered is histone H4. These variants show distinct temporal and spatial localization throughout the genome. One well-characterized example is CenpA, a histone H3 variant that replaces canonical H3 at centromeres. CenpA is critical to establishing the functional identity of the centromere. Thus, following DNA replication, it is necessary to ensure that the two daughter centromeres associate appropriately with CenpA to ensure propagation of their identity. This observation opens the door to the realization that the DNA sequence is not the sole information component of the genome. Rather, the nature of the packaging of the genome can influence both its form and function and even its inheritability: a chromosome that lacks a functional centromere will not be segregated into a daughter cell at mitosis. This is a manifestation of the phenomenon of epigenetics.

Some histone variants are substituted for canonical histones after DNA replication has occurred by 'histone-variant-specific exchange activities'. Two examples are the HIRA and Swr1 complexes that catalyze replacement

Table 7-5. Mammalian Histone Variants

Histone	Function
H3	Canonical
H3.1	?
H3.2	?
H3.3	Transcription activation
CenpA	Kinetochore assembly
H4	Canonical
H2A	Canonical
H2AX	DNA repair and recombination, roles played by canoical H2A histone in budding yeast
H2AZ	Gene expression, chromosome segregation
MacroH2A	X-chromosome inactivation, transcription repression
H2Abbd	Transcription activation
H2B	Canonical
TsH2B	Testis-specific H2B
H1.0	Terminal differentiation
H1.1	Canonical
H1.2	Cell proliferation, proper nucleosome spacing, expression of cell-cycle genes
H1.3	Repression of selected genes
H1.4	Cell proliferation
H1.5	Repression of selected genes
H1 testes	?
H1 oocyte	?

Adapted from Sancho *et al.* (2008) and Sarma and Reinberg (2005).

of H3.3 and H2AZ, respectively. In addition, several ATP-dependent chromatin-remodeling activities such as SWI/SNF have been shown to catalyze the displacement of H2A•H2B dimers.

DNA Methylation

Below the level of histone composition and modification, an even more fundamental level of epigenetic modification exists: differential methylation of specific bases in DNA. The methyl group is transferred from the cofactor *S*-adenosylmethionine by a class of enzymes called methyltransferases (MTases). In bacteria, DNA methylation occurs at the N6 position of adenine as well as the N4 and C5 position of cytosine (Figure 1-1). In metazoans only the latter type of modification is observed.

Bacteria

The only MTase directly involved with bacterial genome duplication is deoxyadenosine methyltransferase (Dam). In *E. coli,* Dam specifically methylates adenines at the N6 position when they are in the dsDNA sequence 5′-GATC:CTAG-5′. Since this target is a palindrome (sequence reads the same on both strands), adenosine is methylated on both strands. Thus, Dam can remethylate the hemimethylated GATC sites that result when DNA replication forks pass through, thereby restoring the original pattern of methylated Dam sites.

Regulation of origin activity

Studies in *E. coli* have revealed that an important aspect of the control of firing of the replication origin, oriC, is mediated at the level of DNA methylation. Work in the late 1980s revealed that if plasmid DNA containing an oriC sequence was prepared from cells that were methylation proficient (*dam*$^+$) and then introduced into a methylation deficient (*dam*$^-$) strain, it would undergo a single round of replication then fail to replicate any further. This suggested that the hemimethylated progeny DNA duplexes were somehow being sequestered from initiating further rounds of replication. This observation led to a conceptually simple genetic screen to look for mutants that would permit replication of methylated DNA in a *dam*$^-$ background. These screens bore fruit with the identification of the *seqA* gene.

The mechanistic basis of sequestration became apparent with the demonstration that SeqA protein binds preferentially to hemimethylated GATC sites in DNA. Significantly, there is an unusually high density of GATC sites at the origin of replication (Figure 7-23), leading to a high density of SeqA bound to the origin immediately following initiation of DNA replication. Tight binding of SeqA to DNA requires two hemimethylated GATC sequences oriented on the same face of the DNA helix and separated by up to three helical turns (31 bp).

Although normally a dimer, SeqA can form higher-order multimers, mediated by an N-terminal multimerization domain, that can extend beyond the SeqA DNA-binding sites. Consequently, juxtaposition of multiple binding sites at oriC leads to highly stable, cooperative binding of SeqA so that SeqA forms a stable filament structure at the origin, covering the entire oriC region and preventing access by the initiator protein DnaA. SeqA physically sequesters that origin away from the initiator protein. GATC sites are, of course, distributed throughout the chromosome, and studies using fluorescently tagged SeqA reveal that it associates with newly synthesized DNA as the replication fork progresses. Typically the association time of SeqA with a given locus is in the order of less than a sixth of a generation time of the *E. coli*. In contrast, the origin is sequestered for about a third of a generation time. This prolonged sequestration is probably in part due to the extremely high density of GATC sites at the origin and the consequential cooperative binding of SeqA and formation of an unusually stable complex. Release from sequestration is believed to be due to dissociation of SeqA from the hemimethylated GATC, allowing Dam to fully methylate the site and thus prevent re-association of SeqA.

The more stably bound SeqA is, the more difficult for Dam to gain access to the DNA. In agreement with this, overexpression of Dam reduces the sequestration time. In addition to being found in unusually high abundance at oriC, there are also a number of GATC sites in the promoter of the gene for the DnaA initiator protein. This gene is located close to the origin and so, as forks pass through it, it will be switched off by the binding of SeqA to its transcription control sequences. In doing so SeqA also acts to

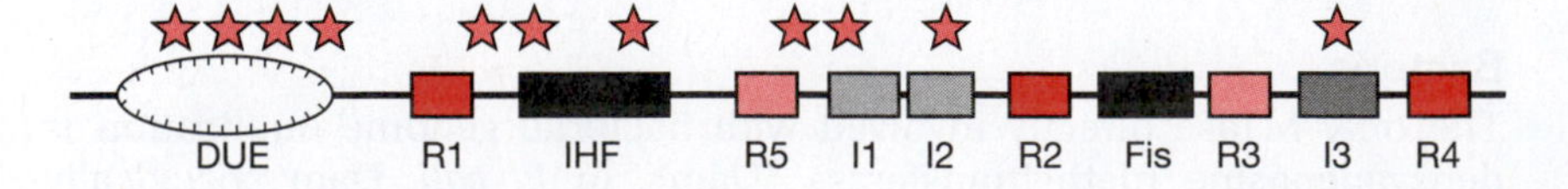

Figure 7-23. Sites of Dam-mediated methylation at GATC sites in the *E. coli* oriC are indicated by pink stars. There are 11 sites within the 245 bp minimal oriC locus. The binding sites for chromatin proteins and DnaA are shown in Figure 7-3.

reduce expression of DnaA following initiation of replication. Thus, SeqA by binding to hemimethylated DNA plays a significant role in ensuring the appropriate temporal control of origin activity both by modulating the production of DnaA and by physically preventing its association with origins. Nevertheless, as discussed in Chapter 12, additional mechanisms exist to impose tight control on origin firing in *E. coli*.

Mismatched base repair

The brief period during which methylation sites are hemimethylated after DNA replication allows the parental and daughter strand of DNA synthesis to be distinguished. In *E. coli*, this property enables repair of the replication errors by the MutHLS mismatch repair system. MutS recruits MutL and MutH to newly replicated DNA containing hemimethylated GATC sites. MutH cleaves the unmethylated strand, and the newly synthesized daughter strand containing the mismatched base is excised. DNA repair enzymes then fill in the ssDNA gap and ligate the nascent strand back together.

Eukarya

In mammalian cells cytosines within the dinucleotide 5′-CpG-3′ can be methylated. Because 5′-CpG is symmetric on the two DNA strands, the cytosines on both strands are methylated. The methyl groups protrude into the major groove and so can have a profound influence on the recognition of DNA. Indeed, differential methylation of DNA in gene promoters has a potent influence on gene expression. Generally, DNA methylation is associated with repression of genes and specific methyl-binding proteins have been identified that mediate this repression. For the reader of this book, it is important to know cytosine methylation occurs in the context of DNA, not in the dCTP precursor and so, following replication, the parental strand is methylated but the newly synthesized strand is not. Thus, following passage of a replication fork, DNA is transiently hemimethylated. Clearly, due to the semi-conservative nature of DNA replication, unless the new DNA is fully methylated, a further round of replication will entirely wipe the methyl mark from the DNA of one of the progeny molecules.

Mammals possess three active DNA methyltransferases (Dnmts), Dnmt1, Dnmt3a, and Dnmt3b. Dnmt3a and Dnmt3b function as *de novo* methyltransferases and set up novel methyl marks on unmethylated DNA. While fascinating from a gene expression perspective these proteins will not be discussed further here. From the standpoint of DNA replication, the most relevant enzyme is Dnmt1. This enzyme has a strong preference for hemimethylated DNA substrates and is known of as the maintenance methyltransferase. Since Dnmt1 has the capacity to fully methylate hemimethylated DNA, it might be expected to be linked to the passage of replication forks. Indeed, early studies revealed that Dnmt1 interacts with PCNA, providing a conceptually pleasing mechanism for delivering Dnmt1 to the replication fork. However, more recent work has demonstrated that it is possible to delete the mapped PCNA-interacting region of Dnmt1 without causing any severe impairment of its function. Further, the observed kinetics of methylation by Dnmt1 are much slower than the rate of fork movement. These data suggest that, while the PCNA interaction may facilitate Dnmt1 function, additional mechanisms exist for ensuring the targeting of Dnmt1 to sites of hemimethylation.

This conundrum was resolved recently with the characterization of Uhrf1, a protein that binds specifically to hemimethylated CpG-containing duplex DNA. Its mode of binding is reminiscent of some DNA repair enzymes in that it flips the methylated base out of the double helix and into a pocket on the protein. It is able to discriminate between hemimethylated and fully methylated DNA by having additional loops that read the other three bases of the CpG motif. Importantly, Uhrf1 interacts with Dnmt1 and mutations in

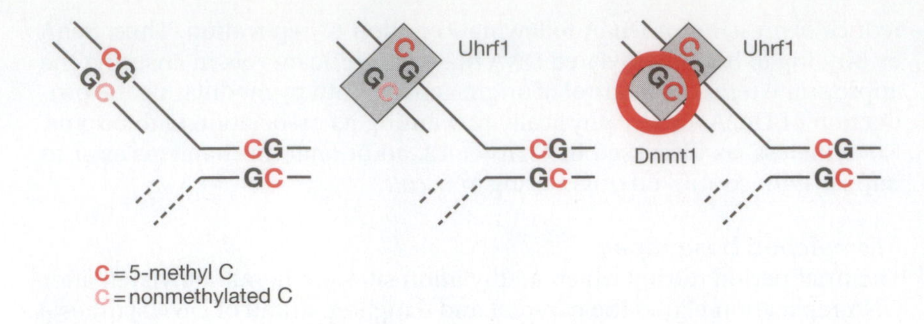

Figure 7-24. Model for the pathway of methylation maintenance in mammalian cells following the generation of hemimethylated DNA by passage of a replication fork. Uhrf1 recongnizes hemi-methylated DNA and recruits the methyltransferase Dnmt1. For simplicity only the leading-strand template arm is shown.

Uhrf1 result in severe deficiencies in the maintenance of DNA methylation. Thus, DNA methylation sites appeared to be maintained during genome duplication by three successive events. First, hemimethylated DNA is generated by the passage of a replication fork. Then Uhrf1 recognizes the hemimethylated CpGs and recruits Dnmt1, which in turn methylates the DNA (Figure 7-24).

SUMMARY

- DNA is associated with small basic proteins in cells, forming chromatin: this both compacts the DNA and protects it.
- Chromatin proteins can play vital architectural roles, for example at the bacterial origin of replication. In this case the Fis and IHF proteins play antagonistic roles with Fis acting to inhibit replication and IHF stimulating origin melting. Interestingly, these proteins show differential regulation during the cell cycle.
- By forming a compact structure on DNA, chromatin can be a barrier to processes such as replication and transcription that require access to DNA.
- Assembly of chromatin proteins in some archaea and all eukarya are regulated by post-translational modifications. These modifications include (among others) acetylation, phosphorylation, methylation, and ubiquitylation.
- In eukarya, the patterns of modified histones and histone variants regulate DNA replication and gene expression.
- The pattern of, and thus the control exerted by, modified histones can be passed to daughter molecules following DNA replication; this is a manifestation of epigenetics.
- Another manifestation of epigenetics is in the heritable site-specific methylation of DNA. In *E. coli*, DNA methylation serves to regulate origin firing; in eukarya, DNA methylation has fundamental impact on gene expression. In both cases hemi-methylated DNA, generated by passage of a replication fork, is recognised and the sites converted to fully methylated.
- In eukarya, chromatin structures ahead of the replication fork are disassembled and then reassembled behind the fork. This requires a complex series of transactions between a range of chromatin-modifying enzymes that are tethered to the replication fork machinery.
- Following the passage of the replication fork in eukaryotes, the daughter DNA molecules are held together by the ring-shaped cohesin complex. The ring-shaped cohesin is loaded prior to passage of the fork. Whether the entire replication fork passes through the ring, or the ring is transiently opened and re-sealed on fork encroachment and passage, is not yet clear.

- Sister chromatid separation at anaphase requires removal of the cohesin ring. Two pathways ensure this in vertebrates. First, a prophase pathway, dependent on the polo-like kinase (Plk1) removes the bulk of cohesin along chromosome arms. The remaining cohesin is removed at the metaphase to anaphase transition by the activation of separase, a proteinase that specifically degrades the cohesin component Scc1 and thus opens the cohesin ring.

REFERENCES

Additional Reading

Bassett, A., Cooper, S., Wu, C. and Travers, A. (2009) The folding and unfolding of eukaryotic chromatin. *Curr. Opin. Genet. Dev.* 19: 159–165.

Clark, D.J. (2010) Nucleosome positioning, nucleosome spacing and the nucleosome code. *J. Biomol. Struct. Dyn.* 27: 781–793.

Corpet, A. and Almouzni, G. (2009) Making copies of chromatin: the challenge of nucleosomal organization and epigenetic information. *Trends Cell Biol.* 19: 29–41.

Haering, C.H., Farcas, A.M., Arumugam, P., Metson, J. and Nasmyth, K. (2008) The cohesin ring concatenates sister DNA molecules. *Nature* 454, 297–301.

Henikoff, S. (2008) Nucleosome destabilization in the epigenetic regulation of gene expression. *Nat. Rev. Genet.* 9: 15–26.

Johnson, R.C., Johnson, L.M., Schmidt, J.W. and Gardner, J.F. (2005) Major nucleoid proteins in the structure and function of the *Escherichia coli* chromosome. In *The Bacterial Chromosome*, Higgins, N.P. (ed.), vol. 5, pp. 65–132. ASM Press, Washington DC.

Kaufman, P.D. and Almouzni, G. (2006) Chromatin assembly. In *DNA Replication and Human Disease*, DePamphilis, M.L. (ed.), vol. 6, pp. 121–140. Cold Spring Harbor Laboratory Press, Cold Spring Harbor, NY.

Literature Cited

Hwang, D.S. and Kornberg, A. (1992) Opening of the replication origin of *Escherichia coli* by DnaA protein with protein HU or IHF. *J. Biol. Chem.* 267: 23083–23086.

Kornberg, R.D. and Thomas, J.O. (1974) Chromatin structure: Oligomers of the histones. *Science* 184: 865–868.

Luger, K., Mäder, A.W., Richmond, R.K., Sargent, D.F. and Richmond, T.J. (1997) Crystal structure of the nucleosome core particle at 2.8 Å resolution. *Nature* 389: 251–260.

Olins A.L. and Olins D.E. (1974) Spheroid chromatin units (ν bodies). *Science* 183: 330–332.

Onn, I., Heidinger-Pauli, J.M., Guacci, V., Unal, E. and Koshland, D.E. (2008) Sister chromatid cohesion: a simple concept with a complex reality. *Ann. Rev. Cell Dev. Biol.* 24: 105–129.

Sancho, M., Diani, E., Beato, M. and Jordan, A. (2008) Depletion of human histone H1 variants uncovers specific roles in gene expression and cell growth. *PLoS Genet.* 4, e1000227.

Sarma, K. and Reinberg, D. (2005) Histone variants meet their match. *Nat. Rev. Mol. Cell Biol.* 6: 139–149.

Chapter 8
REPLICONS

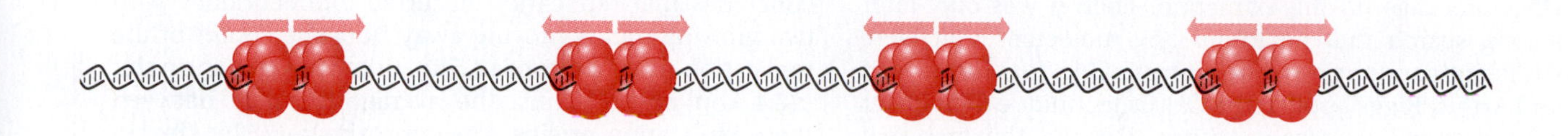

THE REPLICON MODEL

A major obstacle to the study of genome duplication is the enormous size and fragility of the DNA molecule. A combination of three different strategies finally solved the problem: development of novel techniques, analysis of DNA replication in the rapidly dividing embryos of frogs and flies, and the analysis of small (<10 kbp) chromosomes found in viruses and episomes. For example, it was not until 1963 that John Cairns demonstrated that the *E. coli* bacterium contains a single circular unbranched DNA molecule 700–900 μm in length (2–2.7 Mbp), and that this DNA was duplicated by forming a replication bubble with one fork at each end (Cairns, 1963) (see Box 8-1). Totally unexpected was the finding that the replicating form of the *E. coli* chromosome was an intact circle. Cairns recognized that some type of swivel mechanism would be needed to allow the parental DNA to unwind. The discovery of DNA topoisomerases was still 8 years in the future.

The Classical Replicon Model

In the same year that Cairns published his autoradiograms of *E. coli* chromosomes, Francois Jacob, Sydney Brenner, and Francois Cuzin proposed an elegantly simple model to explain how *E. coli* knows where and when to initiate DNA replication (Jacob *et al.*, 1963). They postulated that initiation of DNA replication, like initiation of DNA transcription, occurs at specific DNA sites termed replicators that bind a specific protein termed the **initiator** in response to environmental signals that regulate cell division (Figure 8-1A). The interaction between replicator and initiator triggers initiation of DNA synthesis at or near the replicator. The actual site where DNA synthesis begins is termed the **replication origin**. Note that the replicator and the replication origin are not necessarily coincident, and that the replicator will only affect the DNA molecule in which it resides. In contrast, the initiator gene affects all DNA molecules within reach of the initiator protein that it encodes. Thus, replicators are examples of *cis*-acting sequences and initiators are examples of *trans*-acting proteins. They further speculated that the initiator was either a DNA helicase that unwound the two strands of DNA to allow DNA polymerase to use them as templates for the synthesis of a complementary copy, or a DNA polymerase that could unwind DNA while concomitantly synthesizing a new DNA strand. Taken together, these three elements (replication origin, replicator, and initiator) comprised a single unit of DNA replication, which they termed a **replicon**.

The replicon model, as stated by Jacob, Brenner, and Cuzin, is correct for the genomes of bacteria and a variety of viruses and plasmids that contain a single chromosome with about 10 kbp of dsDNA or less. However, it is an oversimplification of genome duplication in more complex organisms.

Box 8-1. DNA Fiber Autoradiograph

John Cairns developed a technique for spreading ^{3}H-labeled DNA from *E. coli* lysates onto a flat surface and then visualizing the DNA fiber by autoradiography. Exposure times were over 2 months, and it took Cairns a couple of years to work out a technique for minimizing DNA breakage during extraction. Then it was only after a long search that he found any molecules that were sufficiently untangled to be interpretable. Joel Huberman and Arthur Riggs applied the Cairns technique to hamster cells cultured in plastic dishes. Despite the fact that mammalian chromosomal DNA molecules are invariably broken during preparation for fiber autoradiography, fragments as long as 1 mm in length (~3 Mbp) were found, which proved sufficient to detect multiple adjacent replication origins (panel A). Their results proved that the large DNA fibers in mammalian chromosomes are replicated at multiple sites, many of which are clustered in regions less than 500 μm long (~1.5 Mbp). Based on the change in grain density from dark in the center to light at the two ends of each ^{3}H-DNA section (panels B–E), one concludes that replication occurred bidirectionally with two growing points moving away from the center of the replication site at a rate of ≤2.5 μm min^{-1} per growing point (≤7.4 kbp min^{-1}). Thus, the average distance between these replication origins was approximately 300 kbp. In rare cases, the two sibling ^{3}H-labeled chromatids were visible (panels D and E). The work of Cairns together with that of Huberman and Riggs set the stage for the replicon model of DNA replication.

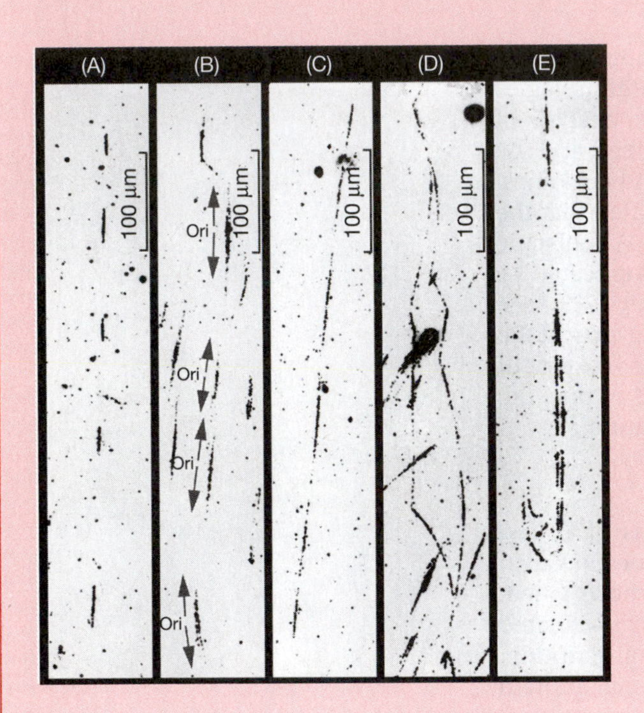

^{3}H-DNA fibers isolated from hamster cells. (A) Cells lysed 30 min after addition of ^{3}H-thymidine (^{3}H-TdR). Dark lines are ^{3}H-DNA. (B–E) Same as in (A) except that cells were then incubated an additional 45 min with nonradioactive TdR. The declining grain densities at the ends of the dark lines indicate the directions of fork movement (arrow) at the time when the specific activity of the intracellular dTTP pool was declining at the start of the incubation with nonradioactive TdR. (B, C) Tandem arrays indicative of bidirectional replication (two replication forks moving away from center of ^{3}H-DNA region) reveal the position of an origin of bidirectional replication (Ori). (D, E) Replication bubbles in which the two sibling ^{3}H-DNA molecules (chromatids) are visible. (E) Rare example of a single replication bubble created from the merging of a cluster of smaller bubbles. Emulsions were exposed for 4 months and then pictures were taken by dark-field microscopy. 100 μm is approximately 300 kbp (B-form DNA in solution contains 2941 bp μm^{-1}). ^{3}H-DNA fibers from *E. coli* appeared the same as those from hamster cells, except that tandem replication sites were never detected in *E. coli* chromosomes. (Adapted from Cairns, J.J. (1963) *J. Mol. Biol.* 6: 208–213, and Huberman, J.A. and Riggs, A.D. (1968) *J. Mol. Biol.* 32: 327–341. With permission from Elsevier.)

The Replicon Model Today

Many genomes, particularly the large linear ones in eukaryotes that are segmented into several separate chromosomes, contain multiple replication origins serviced either by a single initiator protein or by one set of several different initiator proteins (Figure 8-1C). Moreover, whereas most, if not all, replication origins contain one or more *cis*-acting sequences that are required for initiation of DNA replication (a replicator), some replication origins do not act as sequence-specific binding sites for initiator protein *in vitro*, although they do *in vivo*. These replicators are determined by both genetic and epigenetic information: that is, *cis*-acting DNA sequence information as well as information provided by DNA modifications and chromatin composition that can vary from one cell type to another. Therefore, replicators are defined by their function, not by their sequence. Accordingly, a replicon is simply a unit of DNA replication that is determined by the presence of a replication origin that functions only in the

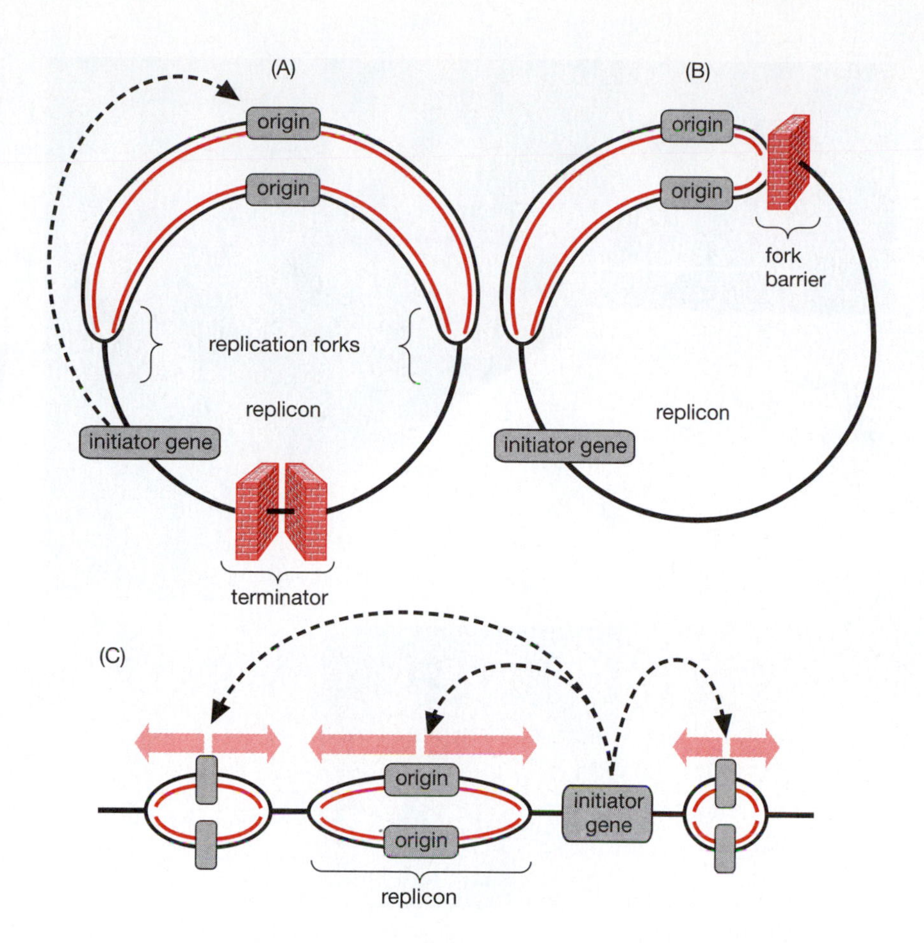

Figure 8-1. Internal replicons in circular and linear DNA molecules. (A) Most circular duplex DNA genomes exist as a single chromosome with a single replicator activated by a single initiator protein encoded by a gene within the replicon. Two replication forks are assembled at or close to the replicator and then proceed in opposite directions at about the same velocity (bidirectional replication). In some but not all cases, a termination signal is located 180° from the replicator. These replication intermediates are referred to either as Cairns or θ structures. (B) In some cases, a circular or linear chromosome contains a sequence close to the replicator that impedes passage of replication forks (fork barrier). This results in unidirectional replication. (C) With the exception of certain viruses, linear duplex DNA chromosomes have multiple internal replication origins (Figure 8-2), which in many cases have been shown to contain a replicator. These genomes invariably encode a single copy of one or more initiator proteins that act on all of the replicators. However, the replicators usually are not activated simultaneously, but in a temporal order. Parental DNA is black; nascent DNA is red.

presence of a specific initiator protein or group of proteins (Figure 8-1C). What constitutes a replication origin and what determines when it is active are discussed in Chapters 9–12.

Termination of DNA replication does not require specific *cis*-acting **terminator sequences** (Chapter 6), although clusters of specific *cis*-acting sequences termed replication-fork barriers are used in some circular chromosomes, such as those in *E. coli* and *B. subtilis*, to force replication to terminate at a specific chromosomal locus (Figures 6-4 and 8-1A). Terminators are simply collections of replication-fork barriers that prevent the passage of a replication fork coming from one direction but not from the other (Figures 6-4 and 8-1B). Such fork barriers are responsible for the appearance of unidirectional replication in some replicons. Both terminators and fork barriers serve to prevent interference of DNA replication with gene expression.

Visualizing Replicons

The structure of replication bubbles in eukaryotes was first visualized by electron microscopic analysis of the replicating chromosomal DNA in the early cleavage-stage embryos of the fruit fly, *Drosophila melanogaster*. Multiple tandemly arranged replication bubbles were observed in DNA fragments as long as 119 kbp (Figure 8-2A), many of which contained one ssDNA gap (Figure 8-2B). Notably, the two ssDNA gaps in the same bubble were positioned on opposite sides of the bubble, consistent with two replication forks traveling in opposite directions. The mean length for the ssDNA gaps in *D. melanogaster* replication bubbles was 200 bp, where the analogous ssDNA gaps in replication bubbles in bacteriophages λ and T7 were 500–1500 bp. This difference in gap sizes reflects the difference in the lengths of Okazaki fragments observed in flies and phages.

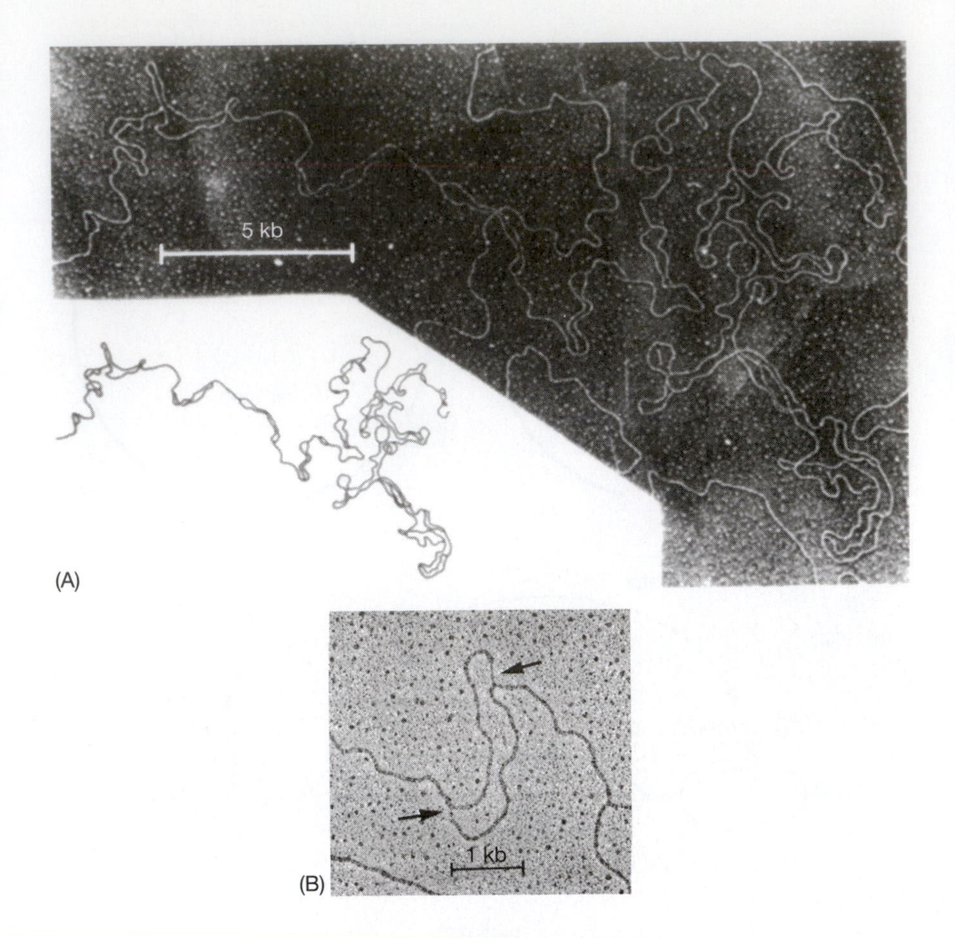

Figure 8-2. Fragment of replicating chromosomal DNA from *Drosophila* cleavage nuclei that contains 23 replication bubbles in a length of 119 kbp. Nuclei were isolated from cleavage-stage embryos and then lysed with a detergent. The DNA was isolated by density-gradient equilibrium centrifugation (Figure 3-2), dialyzed against a 100 mM salt solution at neutral pH, and then visualized by electron microscopy. The arrows in panel (B) indicate regions of single-stranded DNA. (From Kriegstein, H.J. and Hogness, D.S. (1974) *Proc. Natl. Acad. Sci. USA* 71: 135–139. With permission from The Rockefeller University Press.)

In 1987, Bonita Brewer, Walton Fangman, and Joel Huberman took advantage of the fact that replication forks, replication bubbles, and termination structures migrate with characteristic motilities during two-dimensional agarose gel electrophoresis (see Figure 8-3A in color plate section). DNA from proliferating cells is digested with one or more restriction endonucleases to release fragments of between 1 and 10 kbp that can be fractionated by gel electrophoresis. These fragments are first fractionated according to their size at neutral pH along the *x* axis of the gel, and then along the *y* axis of the gel according either to their shape at neutral pH, or to the lengths of their ssDNA at alkaline pH. The replication pattern of a specific genomic locus can then be visualized by transferring the DNA from the gel directly onto a membrane and then hybridizing it to a sequence-specific DNA probe that is labeled either with radioactivity or with a chromophore. In this way, the observer examines only those DNA fragments that originate from a particular site in the genome. The assignment of replication-fork and bubble structures to the migration of DNA molecules during gel electrophoresis is based on theoretical considerations, electron microscopy of DNA molecules recovered from two-dimensional gels, changes in the patterns when the locations and/or sizes of restriction fragments are altered, and agreement between the locations of initiation sites mapped by two-dimensional gel electrophoresis and the locations of autonomously replicating sequences (ARS elements) mapped genetically.

Two-dimensional gel electrophoresis allows identification of replication bubbles at specific genomic loci. It also allows one to determine the direction of replication-fork movement at specific genomic loci and thereby locate the sites from which replication forks originated, as well as sites where replication forks terminated. Brewer, Fangman, Huberman, and many others subsequently, have used this technique to identify the locations of replication origins in plasmid DNA molecules undergoing replication in budding yeast, as well as in the chromosomes of budding yeast, fission yeast, and mammalian cells. Results from such studies confirmed the long-standing assumption that *S. cerevisiae* ARS elements function as origins of replication and led to the adaptation of this technique to the identification of replication origins in a variety of eukaryotic and archaeal genomes. Moreover, data from two-dimensional gel analysis revealed that most, if not all, DNA replication intermediates in both single-cell and multicellular eukarya conform to the replicon model in which replication bubbles are formed at internal replicators and then expand bidirectionally using the replication-fork model (Figure 8-1). In fact, two-dimensional gel patterns from budding yeast are virtually superimposable on those from the distantly related fission yeast (Figure 8-3C), despite the fact the neither the replicators nor initiators in these organisms are interchangeable.

In 1997, Aaron Bensimon developed a method analogous to fiber autoradiography that could visualize replication forks and replication bubbles on single DNA molecules, but that could also map the position of these structures relative to unique genomic DNA sequences. As with two-dimensional gel electrophoresis, the technique of DNA combing, as it is popularly called, allows the direction and velocities of replication forks to be quantified and replication origins to be mapped, but with DNA combing one can determine if the results apply to every chromosome in a population of cells or simply reflect the statistical mean for the entire population of cells. DNA molecules from hundreds of cells are extended on silanized surfaces. Most DNA fragments are longer than several hundred kilobases. Hybridization of fluorescent DNA probes on combed DNA allows direct mapping of their respective positions along the fibers, and incorporation of fluorescent nucleotides allows mapping of replication origins and measurement of replication speed (Figure 8-4).

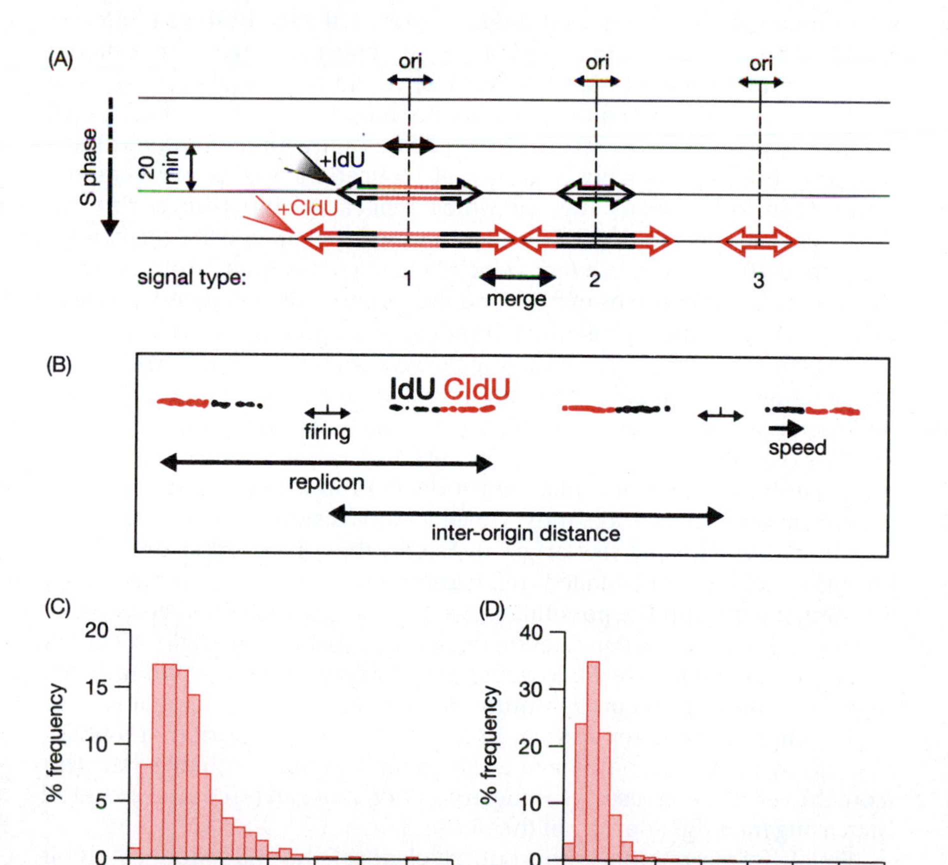

Figure 8-4. Measuring inter-origin distances and fork velocities by dynamic molecular combing. (A) Schematic representation of signals deriving from equal pulse-labeling first with iodo-deoxyuridine (IdU) and then with chloro-deoxyuridine (CldU). Replication forks progress bidirectionally at the same rate from the origin and incorporate the analogs forming a symmetrical replication bubble. Three types of signals may be obtained: a black and red signal with a gap between the black segments, corresponding to initiations that occurred before the beginning of the pulse (1); a dual-color signal with a continuous black segment corresponding to origins that fired during the first pulse (2); and a single isolated red signal corresponding to origins that fired during the second pulse (3). A continuous red signal flanked by two black ones is formed by the merging of two forks from adjacent origins. (B) Two adjacent replication bubbles are visualized on a combed DNA molecule (type 1 signal) from primary normal human keratinocytes. The replication origins are presumably located at the midpoints of the unlabeled segments. The distance between adjacent replication origins represents the inter-origin distance. Fork speed is calculated by dividing the length of each fluorescent signal by the time of the pulse. (C) The mean inter-origin distance for 606 molecules was 111 kbp. (D) The mean fork velocity for 5460 examples was 1.46 kbp min^{-1}. (Adapted from Conti, C., Saccà, B., Herrick, J. *et al.* (2007) *Mol. Biol. Cell* 18: 3059–3067. With permission from the American Society for Cell Biology.)

Why Replicons?

For a hypothesis to be useful, it must be biologically significant. In fact, at least five functions vital to genome duplication are associated with replicons: loading DNA helicases, separating replication from transcription, determining the length of S phase, maintaining genome stability, and establishing a temporal order for initiation events.

Replicons are defined by their central component, the replicator, and the primary function of the replicator is to load a DNA helicase onto duplex DNA in order to unwind the two template strands, thereby allowing DNA polymerase to synthesize a complementary copy of each template. But replicators serve other functions as well. For example, replicators prevent interference between genome duplication and gene expression. Since DNA and RNA polymerases remain associated with DNA as they synthesize their respective polynucleotides, it is easy to imagine ways in which head-on collisions between these molecules would result in stalled replication forks. Evolution avoided this problem in bacteria, bacteriophages, and bacterial plasmids by selecting for genomes in which the direction of transcription of most, if not all, genes lies in the same direction as replication forks traveling through these genes. This avoidance mechanism works well in small genomes, but in large genomes with multiple origins replication-fork barriers have evolved that prevent replication from colliding with transcription in high-traffic areas such as ribosomal RNA genes and tRNA genes.

Another problem facing cells is duplicating their genome within a reasonable period of time. Bacterial replication forks travel 20–60 times faster than eukaryotic replication forks (Table 3-4), and this allows most bacterial genomes (typically 4–8 Mbp) housed in a single chromosome driven by a single replicator to be duplicated in less than 1 h, depending on environmental conditions. In dramatic contrast, if each of the 24 different human chromosomes contained a single replicator, then the smallest human chromosome (47 Mbp; Table 1-3) would require about 5 days to duplicate, assuming, of course, that neither replication fork malfunctioned during this time! The time required to replicate large genomes is determined primarily by origin density (number of replication origins per genome) rather than fork velocity (rate at which replication forks travel through DNA). For example, during animal development, origin density can change more than 10-fold whereas fork velocity changes less than 2-fold. S phase in cleavage-stage embryos of frogs and flies is 8–10 min, whereas S phase in cells from adult animals is 6–8 h. In budding yeast, premature activation of S phase before all of the replication origins have been loaded with MCM DNA helicases results in a 25% reduction in active origins with a concomitant 30% increase in the length of S phase. Thus, one function for replicators is to regulate origin density.

Since the number of replication forks is proportional to the number of replication origins, forks must travel greater distances when the origin density is low. Thus, at low origin densities, there is a higher probability of forks stalling, and stalled replication forks increase chromosomal rearrangements and the possibility of apoptosis (programmed cell death). Conversely, high origin densities increase the probability of origin misfiring, an event that can also result in genome instability. Therefore, for each cell type, one would expect an optimum origin density, something that is most easily achieved using replicators. In fact, premature activation of S phase in budding yeast is accompanied by an increase in the loss of plasmid DNA from the cell (i.e. increased genome instability) that can be compensated by increasing the origin density of the plasmid.

Finally, for reasons that are not entirely clear, cells normally duplicate their genome according to a specific temporal pattern. In bacteria and archaea, the temporal order of replication is dictated by the simple fact that most of these organisms replicate their entire genome bidirectionally from a single replicator. Eukaryotic organisms, however, have multiple chromosomes, each with multiple replicators. Nevertheless, some sequences reproducibly replicate at the beginning of S phase while others replicate at later times. This program is believed to reflect a temporal program for activating the many replication origins in eukaryal genomes. The advantages to a temporal order of replication are speculative, but it does ensure that some genes are duplicated before others. The existence of early- and late-firing replicators would be one way to establish a temporal order of replication, but other factors, such as differences in chromatin structure and nuclear organization, appear to be more important. Interestingly, some archaea, although having single circular chromosomes, possess multiple origins of replication. For example, species of the genus *Sulfolobus* have three replication origins in a chromosome of between 2.2 and 3 Mb (depending on the species in question). All three origins fire in all cells and do so in a tightly coordinated fashion; however, the mechanisms that ensure this coordination are as yet unknown.

REPLICON TAXONOMY

DNA Structure And Replicator Location

The chromosomes of DNA genomes exist in only four forms: circular dsDNA, linear dsDNA, circular ssDNA, and linear ssDNA. When we consider that initiation of DNA replication can begin either at internal chromosomal sites or at the ends of a chromosome, then various types of replicon are possible, depending on the number and placement of replicators and the structure of the DNA. However, only five are found in nature: (1) circular dsDNA with one or more internal replicators, (2) linear dsDNA with one or more internal replicators, (3) linear dsDNA with terminal replicators, (4) circular ssDNA with one internal replicator, and (5) linear ssDNA with terminal replicators (Table 8-1). Of these, only two types of replicon are used to duplicate the

Table 8-1. The Five Types Of Replicon

Replicon	Chromosome	Examples
Type I Circular dsDNA with internal replicators	ori	Most bacteria, bacteriophage λ[a], plasmids, archaea, papillomaviruses, polyomaviruses, herpes simplex virus[a], Epstein–Barr virus[a], mitochondria
Type II Linear dsDNA with internal replicators	ori	Some bacteria, bacteriophages T7 and T4, budding and fission yeast, flies, vertebrates
Type III Linear dsDNA with terminal replicators	ori ori	Adenoviruses, bacteriophage ϕ29
Type IV Circular ssDNA with internal replicators	ori	Icosahedral bacteriophages, filamentous phages, geminiviruses, circoviruses
Type V Linear ssDNA with terminal replicators	ori ori	Parvoviruses

[a]These viral chromosomes are linear in their virions, but circular in their host cell.

genomes of cellular organisms: circular and linear dsDNA chromosomes with internal replicators.

Modes Of DNA Replication

There are three modes for DNA replication: conservative, semi-conservative, and dispersive (Figure 3-1). All dsDNA replicons (Type I, II, and III; Table 8-1) replicate DNA by the familiar semi-conservative mechanism in which the two strands of the double helix are unwound, and a complementary copy of each strand is synthesized (Table 8-2). The resulting duplex DNA molecules each contain one parental strand and one nascent strand. However, circular ssDNA replicons (Type IV) can be considered examples of conservative DNA replication in which the parental DNA molecule is duplicated without changing its structure or composition. As detailed below (Figure 8-8), the parental ssDNA is first converted into a transient dsDNA intermediate from which copies of the single-stranded parental DNA are produced. The first copy released in this mechanism is the original parental genome unchanged from the experience of duplication. Thus, the overall duplication of Type IV replicons (ssDNA→dsDNA→ssDNA) is conservative, although the DNA replication mode employed by the transient dsDNA intermediate, like all other dsDNA replication, is semi-conservative. The third mode of DNA replication, the dispersive mode, is employed by linear ssDNA replicons (Type V). Also as detailed below (Figure 8-9), parvoviruses follow a rather Byzantine route for genome duplication in which newly replicated sections of DNA are dispersed among old unreplicated DNA such that the newly replicated portions are covalently linked to the unreplicated portions.

Modes Of DNA Synthesis

Replicons not only select from three different modes for DNA replication, but also from four different modes for DNA synthesis. Given that all DNA polymerases can only add nucleotides to the 3′-end of a pre-existing primer (RNA, DNA, or protein-dNMP), and that all polynucleotide templates are

Table 8-2. Modes Of DNA Synthesis And Replication Among The Five Types Of Replicon

<table>
<tr><th>Replicon</th><th>Examples</th><th>DNA synthesis mode</th><th>DNA replication mode</th></tr>
<tr><td rowspan="2">Type I</td><td>Bacteria, bacteriophages, plasmids, archaea, animal viruses</td><td>Replication fork</td><td rowspan="4">Semi-conservative</td></tr>
<tr><td>Mitochondria</td><td>Single-strand displacement</td></tr>
<tr><td>Type II</td><td>Bacteria, bacteriophages, yeast, flies, vertebrates</td><td>Replication fork</td></tr>
<tr><td>Type III</td><td>Adenoviruses, bacteriophage ϕ29 family</td><td>Single-strand displacement and gap-filling</td></tr>
<tr><td>Type IV</td><td>Polyhedral and filamentous phages, geminiviruses, circoviruses</td><td rowspan="2">Gap-filling and single-strand displacement</td><td>Conservative</td></tr>
<tr><td>Type V</td><td>Parvoviruses</td><td>Dispersive</td></tr>
</table>

polar (5′-P-dNMP at one end, 3′-OH-dNMP at the other), then the ways in which replicative DNA synthesis can occur are limited to gap-filling, double-strand displacement, single-strand displacement, and replication-fork mechanisms. With the exception of double-strand displacement, each of these modes for DNA synthesis is employed by one of the five types of replicon (Table 8-2). Replicons of the same type, as defined by their DNA structure and the number and locations of their replicators, utilize the same replication mode and the same DNA synthesis mode. The only exception to this rule is mitochondrial DNA (mtDNA). It is a Type I replicon that replicates semi-conservatively by a single-strand displacement mechanism rather than a replication-fork mechanism.

Double-strand displacement

The simplest mechanism by which semi-conservative DNA replication could theoretically occur would be to completely unwind the DNA molecule and then synthesize a complementary strand on each template. However, this mechanism has never been observed in nature, presumably because it would expose the genome to extensive damage by chemical and enzymatic agents.

Single-strand displacement

The next simplest mechanism by which semi-conservative DNA replication could occur is to displace one of the two DNA strands with concomitant synthesis of the complementary copy of the other strand. Well-characterized examples of this mechanism are mtDNA replication (Figure 8-5), adenovirus DNA replication (Figure 8-7), and bacteriophage ϕ29 DNA replication (essentially the same as adenovirus DNA replication). In addition, replication of the dsDNA intermediate in the duplication of circular ssDNA chromosomes (see Figures 8-6 and 8-8) is a single-strand displacement reaction. One strand of the dsDNA form is nicked (a single phosphodiesterase bond disruption) and the 3′-OH end of this nick acts as a primer for DNA polymerase to extend as a rolling circle. The fact that the single-strand displacement mechanism has been observed only in cellular organelles and viruses further supports the supposition that nature avoids the production of extensive single-stranded DNA regions, because the probability of DNA damage is increased.

Replication fork

The mechanism by which nature replicates all of the genomes in bacteria and archaea, and all of the nuclear portion of the genome in eukarya, is the replication fork. Replication forks are essentially a single-strand displacement mechanism but with concomitant DNA synthesis on both template strands (Figure 3-4). Thus, the junction where the parental DNA molecule separates into two sibling molecules appears as a fork. The primary advantage of the replication-fork mechanism is that both template strands are rapidly converted into dsDNA. Not only is dsDNA more resistant to breakage and shear than ssDNA, but dsDNA is required for chromatin assembly. Therefore, the replication-fork mechanism allows rapid assembly of newly replicated DNA into chromatin. This further protects the genome and maintains it in a highly condensed form that can be stored within the nucleus.

Gap-filling

Gap-filling is a process in which DNA synthesis begins by adding nucleotides to the 3′-OH end of a DNA chain, extending that chain until it runs into the 5′-P end of another polynucleotide. Gap-filling commonly occurs during repair of damaged DNA where a segment of the damaged DNA strand is excised, and a DNA polymerase fills in the resulting gap by synthesizing a new strand whose sequence is complementary to the template strand. During genome duplication, gap-filling occurs on the lagging-strand template of replication forks where the gaps produced by excision of RNA primers must be filled in so that the Okazaki fragment can be ligated to the long growing nascent DNA strand (Figure 3-8 and Table 3-2). However, the fraction of DNA synthesized at replication forks by gap-filling is minor compared with the amount of DNA synthesized continuously on the leading-strand template and discontinuously as Okazaki fragments on the lagging-strand template.

The only time gap-filling plays a major role in DNA replication is with viral genomes. For example, circular ssDNA molecules are converted into circular dsDNA molecules by gap-filling (see Figure 8-8, below). Similarly, the 3′-OH end of the linear ssDNA genome of parvoviruses folds back upon itself to generate a DNA primer that allows gap-filling to convert the ssDNA form into a dsDNA form (Figure 8-9, below). The final stage in adenovirus DNA replication also utilizes a gap-filling reaction (Figure 8-7, below). Gap-filling is semi-conservative DNA replication (Figure 3-1) that produces a dsDNA molecule in which one strand is the parent and the other is a complementary copy.

SIX MANIFESTATIONS OF THE REPLICON

In principle, replicons can differ with regard to the structure of their DNA, the number and locations of their replicators, their mode of DNA replication, and their mode of DNA synthesis. Therefore, many manifestations of the replicon model could exist, but in actuality, all of the replicons found in nature fall into six groups (Table 8-3). Remarkably, only two of them are used to duplicate the genomes of cellular organisms. Genomes in bacteria and archaea are encoded in dsDNA that is replicated semi-conservatively from internal replicators using the replication-fork mode. The same is true for the nuclear portion of eukaryotic genomes. However, the mitochondrial portion of eukaryotic genomes is encoded in dsDNA that is replicated semi-conservatively using the single-strand displacement mode.

A. dsDNA With Internal Replicators That Use Replication Forks

The genomes of most bacteria, bacteriophages, bacterial plasmids, archaea, and small DNA animal viruses consist of a single circular dsDNA

Table 8-3. The Six Manifestations Of Replicons Found In Nature

<table>
<tr><th>Manifestation</th><th>DNA</th><th>Replicators</th><th>DNA replication mode</th><th>DNA synthesis mode</th><th>Examples</th></tr>
<tr><td>A</td><td rowspan="4">dsDNA</td><td rowspan="3">Internal</td><td rowspan="4">Semi-conservative</td><td>Replication fork</td><td>Bacteria, bacteriophages, plasmids, archaea, animal viruses, yeast, flies, vertebrates</td></tr>
<tr><td>B</td><td>Single-strand displacement</td><td>Mitochondria, chloroplasts, kinetoplasts</td></tr>
<tr><td>C</td><td>Rolling circle</td><td>Plasmids</td></tr>
<tr><td>D</td><td>Terminal</td><td>Single-strand displacement, then gap-filling</td><td>Adenoviruses, bacteriophage ϕ29 family</td></tr>
<tr><td>E</td><td rowspan="2">ssDNA</td><td>Internal</td><td>Conservative</td><td rowspan="2">Gap-filling then single-strand displacement</td><td>Polyhedral and filamentous phages, geminiviruses, circoviruses</td></tr>
<tr><td>F</td><td>Terminal and internal</td><td>Dispersive</td><td>Parvoviruses</td></tr>
</table>

molecule, several of which serve as paradigms for replicons (Table 8-3). Most of these contain only one replication origin although some archaea (e.g. *S. solfataricus*) contain as many as three. The chromosome in some bacteriophages, such as λ and T4, is linear in the phage particle, but it circularizes following injection into its host. Mechanisms unique to these genomes then restore them to their linear form during the final stage of the genome-duplication process so that they can be assembled into phage particles. All of these genomes are replicated predominately, if not exclusively, bidirectionally from a single replication origin using the replication-fork mechanism (Figure 8-1) described in Chapter 3.

Linear dsDNA chromosomes generally appear in viruses when their genome size reaches 30 kbp or more, but they are also found in a few bacteria and in the mitochondria of some fungi (Table 1-3). In these examples, the number of replication origins varies from two to four. However, the nuclear component of the genomes of all eukaryotes resides in large linear dsDNA chromosomes that contain hundreds to thousands of replication origins. The nuclear component of yeast genomes, for example, contains from 250 to 430 replication origins for an average of one origin every 30–50 kbp (with an origin density of 20–33 origins Mbp^{-1}). Mammalian cells, on the other hand, have genomes that are about 230 times larger than those in yeast, but contain less-dense replication origins. Human genomes, for example, contain from 22 000 to 44 000 replication origins for an average of one origin every 70–140 kbp (origin density 7–14 origins Mbp^{-1}). Despite the plethora of replication origins in eukarya, all of the origins within a single organism respond to the same initiator protein(s). Each virus, episome, bacterium, archaeon, and eukaryote produces a single initiator protein or a single set of initiator proteins that triggers initiation of DNA replication from all of its replication origins, although not necessarily at the same time.

In terms of the problem of unwinding the DNA templates and copying the DNA sequence, Type I and Type II replicons are the same, with the exception of mtDNA. Both are dsDNA with one or more internal replicators. Both suffer the same topological constraints during DNA unwinding, because rotation of long linear DNA molecules is as constrained at the structural level as covalently closed circular DNA molecules (Figure 3-9). Moreover, when one includes the fact that bare DNA does not exist within a cell, only chromatin, the restriction to rotation becomes absolute. As if that were not enough, eukaryotic chromosomes are attached to nuclear structure. Therefore, a unit of DNA replication in a circular dsDNA molecule (Figure 8-1A) is essentially the same as a unit of DNA replication in a linear dsDNA molecule (Figure 8-1C); DNA replication proceeds bidirectionally from the replicator locus using the replication-fork mode. The genes encoding the *trans*-acting factors that activate each replicator may be located either within or without the individual replicons upon which they act.

B. dsDNA With Internal Replicators That Use Single-Strand Displacement

Mitochondria, chloroplasts, and kinetoplasts (the mitochondria of flagellated protozoa) contain one or more copies of a DNA molecule that encodes some of the genes required for their function. The most extensively characterized is mtDNA from mammalian cells. mtDNA contains at least two internal replicators (oriH and oriL). mtDNA is replicated by a single-strand displacement mechanism in which only one of the two template strands is copied initially, and the displaced template strand is copied later (Figure 8-5). The unusual distribution of guanine and thymine between the two strands of mammalian mtDNA results in a significant difference in their density that allows them to be separated into a heavy (H) and a light (L) strand. DNA synthesis begins at the promoter for transcription of genes on the L-strand making this site the origin of replication for the H-strand (oriH).

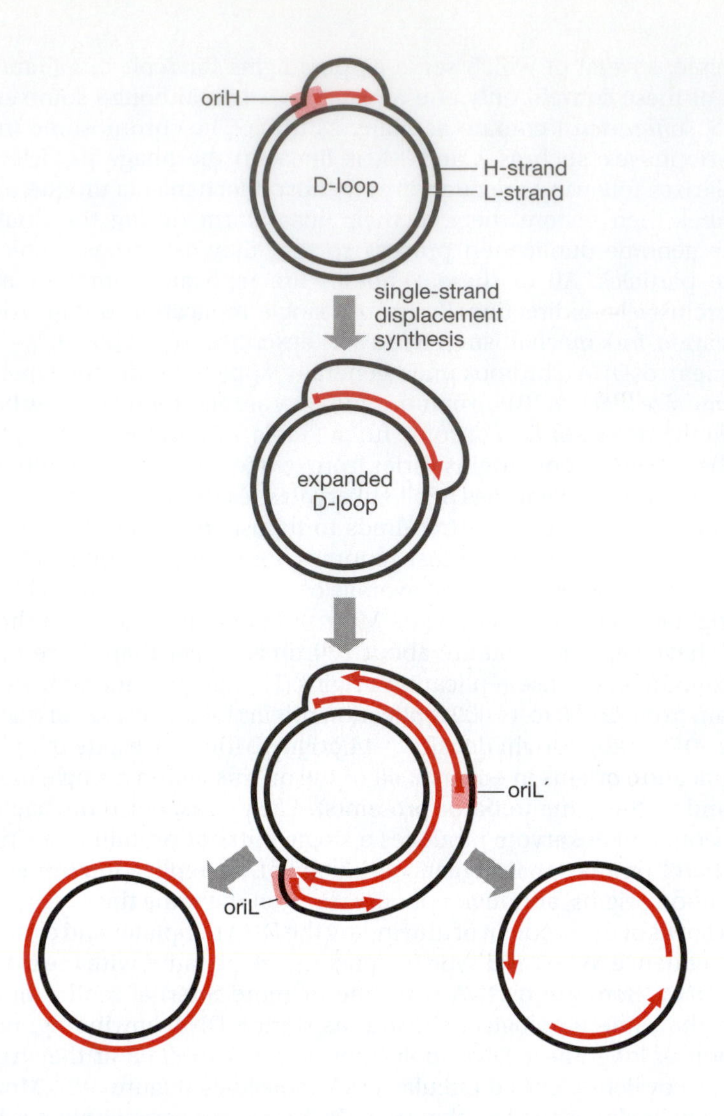

Figure 8-5. mtDNA replication. Black is parental DNA and red is nascent DNA. Arrowheads indicate the 3′-end of DNA. (Data from Clayton, D.A. and Larsson, N.-G. (2006) Mitochondrial DNA replication and human disease. In *DNA Replication and Human Disease* (DePamphilis, M.L., ed.), pp. 547–560.)

Synthesis of the nascent H-strand is then carried out by DNA polymerase-γ (the mtDNA polymerase) and proceeds unidirectionally with concomitant displacement of the parental H-strand as ssDNA.

In most cases, DNA replication pauses after about 1 kb of nascent H-strand DNA is synthesized to produce an easily identifiable replication intermediate termed a D-loop (Figure 8-5). The mechanism by which D-loop formation occurs is unknown, although the mtDNA DNA helicase ('Twinkle') is likely involved. The D-loop is then expanded until an origin of replication for the L-strand is encountered (oriL), usually at about two-thirds the way around the chromosome. These intermediates are termed expanded-D(*l*) structures. A natural consequence of such strand-asymmetric synthesis is that completion of nascent L-strand DNA synthesis is delayed relative to nascent H-strand synthesis, and separation of the two circular progeny molecules occurs prior to completing replication of the entire DNA molecule. Thus, one sibling molecule will be completely replicated, while the other will contain one or more gaps in the newly synthesized L-strand DNA, resulting in H-strand template regions that are still single-stranded. The number of gaps will reflect the number of initiation events that occurred during L-strand synthesis. Other light-strand origins (oriL′) may also be encountered before oriL is exposed, giving rise to replication intermediates that appear more like a θ-replication intermediate. If a significant number of L-strand initiation events occur within the broad area between the D-loop and oriL, then both strands would be synthesized bidirectionally as the double-stranded replication forks proceed through the length of the mtDNA. These intermediates would contain duplex DNA structures suggestive of **bidirectional replication** origins in θ structures and indeed such structures have been detected. Once the 3′-end of one nascent DNA strand meets the 5′-end of the nascent DNA strand downstream, the DNA ligase can form a covalent phosphodiester bond between the two (Figure 5-12).

Trypanosomatids, among the earliest branches of eukaryotes, contain a unique mtDNA network known as kinetoplast DNA that is composed of several thousand 1 kb minicircles interlocked like the links in medieval chain mail. Intertwined in this network are a few dozen 23 kb maxicircles. Maxicircles, like mtDNAs of higher organisms, encode rRNA and proteins such as subunits of respiratory complexes. Minicircles encode guide RNAs involved in editing of maxicircle transcripts. These catenated molecules appear to replicate by a θ-circle mechanism of the type used for bacterial genomes and their plasmids. Otherwise, it is assumed that all extranuclear genomic elements are replicated by the paradigm developed for vertebrate mtDNA. In contrast to kinetoplast DNA, mtDNA is not catenated, although individual mitochondria contain from one to three copies of mtDNA.

C. dsDNA With Internal Replicators That Use The Rolling Circle

A remarkable variation of the mtDNA replication mechanism appears in the duplication of plasmids in Gram-positive bacteria. Instead of simply displacing one DNA strand while copying the other, these genomes break one strand to create the classic **rolling-circle mechanism** first described in 1970 by Walter Gilbert and David Dressler. This mechanism was originally thought to be restricted to circular ssDNA genomes found in bacterial and plant viruses, but about 20 years ago small (<10 kbp) plasmids ubiquitous among Gram-positive bacteria, such as *Staphylococcus, Bacillus, Clostridium, Brevibacterium, Streptococcus, Lactococcus, Leuconostoc*, and *Streptomyces*, also were found to replicate by a rolling-circle mechanism. They have been reported in Gram-negative bacteria and in archaea as well. So far, more than 200 such plasmids have been identified and these can be divided into more than a dozen families based on sequence homologies in their initiator proteins and in their dsDNA replication origins. All of these plasmids encode initiators termed Rep proteins that bind specifically to their cognate dsDNA replicator. These replicators encode a Rep-specific nicking site. Initiation of plasmid DNA replication requires both the Rep-binding and nicking sites, while termination of plasmid replication can be promoted by a smaller sequence containing only the nick site. Each plasmid also encodes a ssDNA replicator that is required to convert the displaced strand into dsDNA.

The pT181-encoded Rep protein functions as a dimer. One monomer binds to a specific sequence within the dsDNA plasmid replicator. Rep binding to this replicator perturbs the DNA structure and exposes the nick site to cleavage by Rep, a process that results in Rep becoming covalently attached to the 5′-end of the nick (Figures 8-6 and 10-1). Rep also recruits a DNA helicase (PcrA in the case of Gram-positive bacteria) that unwinds the DNA. DNA unwinding is facilitated by association of the displaced strand with the host cell's single-stranded DNA-binding protein (SSB). DNA polymerase III initiates DNA synthesis on the free 3′-OH end of the nick and proceeds until the leading strand has been fully displaced. Rep, DNA Pol-III, and PcrA travel together around the circle. When DNA synthesis regenerates the dsDNA replicator and proceeds approximately 10 nucleotides beyond the Rep nick site, the second monomer of Rep then cleaves the displaced ssDNA. Following a series of cleavage/rejoining events, the circular, leading strand is released along with circular dsDNA containing the newly synthesized strand.

The dsDNA is then supercoiled by the host cell's DNA gyrase. The Rep protein in which the active tyrosine of one monomer is covalently attached

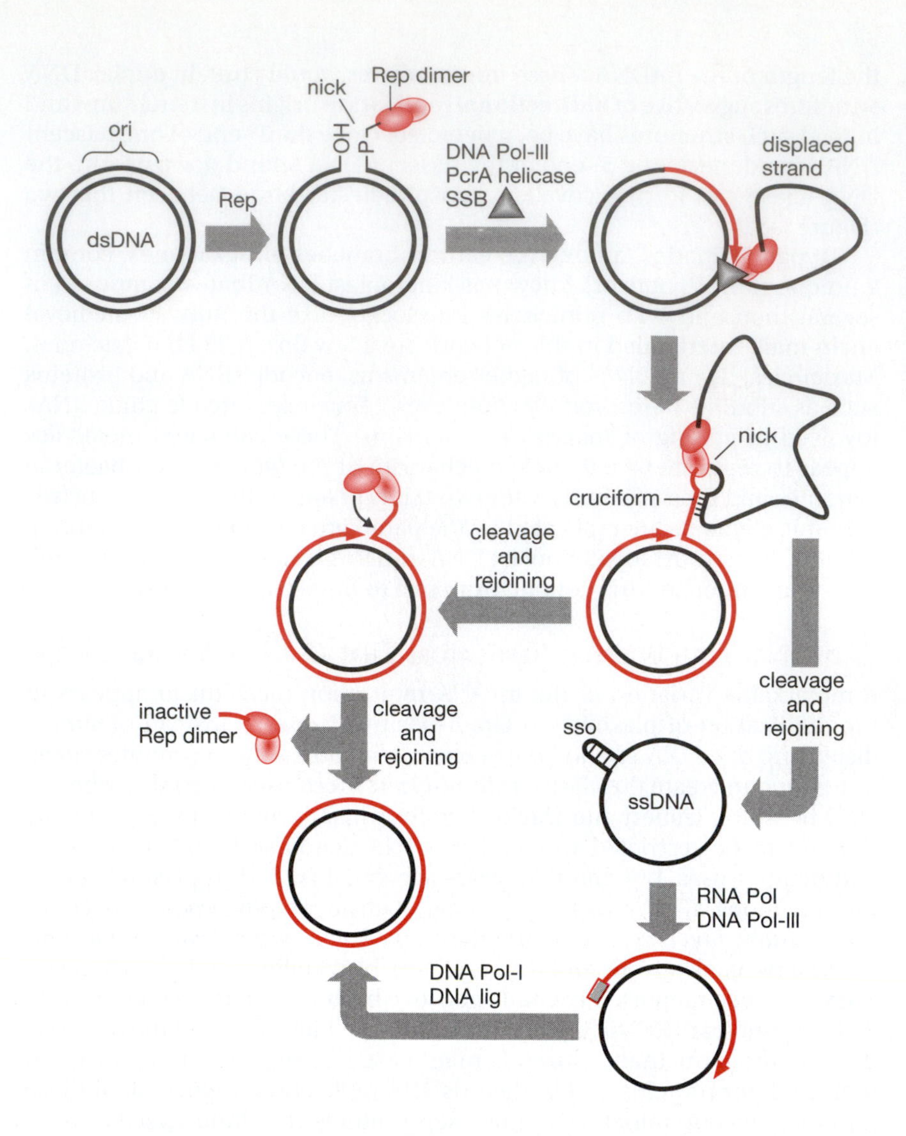

Figure 8-6. Plasmid pT181 replication in *Staphylococcus aureus*. A Rep dimer (two shaded ovals) binds to the dsDNA replicator (ori) and nicks one strand to generate a 3′-OH terminated DNA primer on which the host cell's DNA polymerase (red arrow) can initiate DNA synthesis (red line). *S. aureus* DNA helicase (PcrA) and single-stranded DNA-binding protein (SSB) displace one DNA strand while DNA polymerase copies the other strand. The pT181 replicator contains an inverted repeat that can form a palindrome in the displaced strand (Figure 10-1). This palindrome restores the nick site and thereby allows the other Rep monomer to release the displaced strand through a cleavage and rejoining reaction. *S. aureus* RNA polymerase uses a hairpin in the displaced ssDNA circle as a replication origin (sso) to synthesize an RNA primer (red box) that allows DNA polymerase III to initiate synthesis of the complementary strand and form a new dsDNA molecule. (Adapted from Khan, S.A. (2005) *Plasmid* 53: 126–136. With permission from Elsevier.)

to the 10-mer oligonucleotide is released and is inactive. The ssDNA released after leading-strand synthesis has been completed is converted to dsDNA utilizing the ssDNA replication origin and host-encoded proteins. First, an RNA primer is synthesized at the ssDNA origin by RNA polymerase. This primer is then extended by DNA polymerase III until it encounters the RNA primer. The RNA primer is excised by the 5′→3′ exonuclease activity of DNA polymerase I, and the gap is filled in by DNA polymerase I until the two DNA ends can be covalently joined by DNA ligase. The resultant dsDNA is converted into a supercoiled form by DNA gyrase.

D. dsDNA With Terminal Replicators That Use Single-Strand Displacement And Then Gap-Filling

Not all linear dsDNA chromosomes have internal replicators; some have replicators at their termini, but this type of replicon has been found only among the viruses. Human adenoviruses and *Bacillus subtilis* bacteriophage ϕ29 are examples of genomes encoded in a single linear dsDNA molecule with a replicator at each end. Adenovirus and ϕ29 DNA replication have been characterized extensively, and utilize the same mechanism for duplication of their respective genome (Figure 8-7). For example, the 5′-ends of adenovirus DNA are each covalently attached to a virally encoded protein termed adenovirus terminal protein (TP). Initiation of replication does not occur in the absence of either TP or the unique DNA sequence to which it is attached. Using another copy of TP to provide a protein-nucleotide primer,

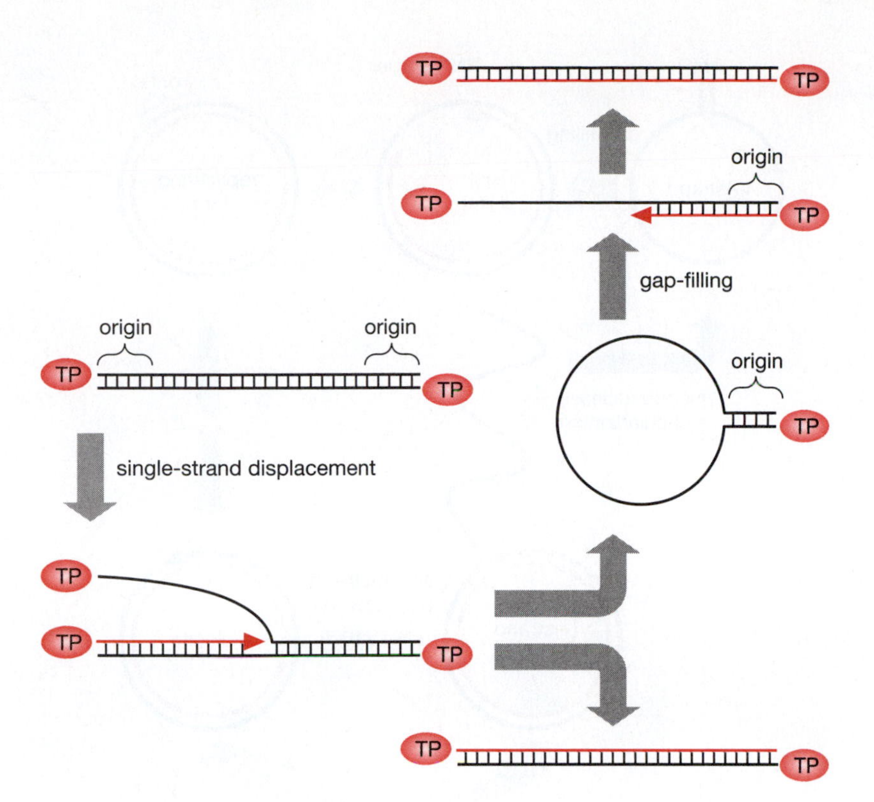

Figure 8-7. Adenovirus genome duplication. Black is parental DNA and red is nascent DNA. Arrowheads indicate the 3′-end of DNA. TP, terminal protein. (Data from Liu, H., Naismith, J.H. and Hay, R.T. (2003) *Curr. Top. Microbiol. Immunol.* 272: 131–164.)

adenovirus DNA polymerase initiates synthesis of one DNA strand while concomitantly displacing the other strand to produce a linear dsDNA and a linear ssDNA molecule. Since the adenovirus genome contains an inverted terminal repeat of approximately 100 bp at each end, the ssDNA molecule can circularize by annealing its 3′-end with its 5′-end to form a single-stranded circular molecule with a short stretch of duplex DNA, reminiscent of a panhandle. This panhandle now contains an adenovirus replication origin to which another initiator complex can bind. The displaced ssDNA strand is thereby converted into dsDNA by filling in the gap between the two TP proteins.

E. ssDNA With Internal Replicators That Use Gap-Filling And Then Single-Strand Displacement

Genomes consisting of a single circular ssDNA molecule ranging from 2 to 6 kb are found only in viruses, all of which employ replicon manifestation 'E' to duplicate their genome. Well-characterized examples include the lytic filamentous M13 bacteriophage and the lytic polyhedral ϕX174 bacteriophage that infect *E. coli*, as well as a group of more than 120 geminiviruses that infect both mono- and dicotyledonous plants, producing a variety of problems for the host. In addition, the smallest pathogenic DNA viruses that have been identified and characterized in animals, the circoviruses, utilize this mechanism of genome duplication. Circoviruses target the lymphoid tissue and cause immunosuppression in the host.

Small viral genomes are occupied primarily with producing proteins that are needed to assemble virions that allow it to survive after its host dies. Therefore, they rely entirely on their host to convert their ssDNA genome into a dsDNA replicative form. The viral genome encodes a single initiator protein that recognizes a unique sequence termed the (+)strand origin in the double-stranded form of the viral genome resulting in the initiation of viral DNA replication. Thus, with the exception of one viral protein, all of the proteins required for duplication of the viral genome are provided by the host cell.

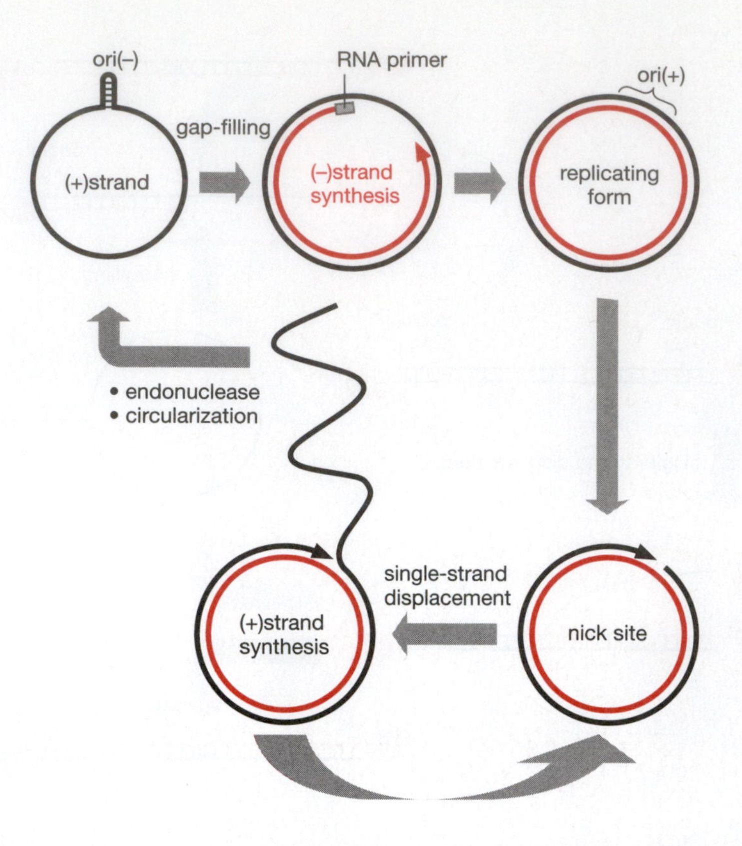

Figure 8-8. Bacteriophage ϕX174 and M13 genome duplication in *E. coli*. Black is parental DNA and red is nascent DNA. Arrowheads indicate the 3′-end of DNA. (Data from Weigel, C. and Seitz, H. (2006) *FEMS Microbiol. Rev.* 30: 321–381.)

Duplication of M13 and ϕX174 genomes occurs in two phases (Figure 8-8). First, the single-stranded viral genome [referred to as the (+)strand] is converted into a double-stranded replicating form by filling in what amounts to a large ssDNA gap. The second phase produces many copies of the (+)strand by a single-strand displacement mechanism referred to as rolling-circle or sigma (σ) replication, because replication intermediates at this stage appear in the electron microscope as dsDNA circles with attached ssDNA tails. Rolling-circle replication is also used by linear dsDNA phage genomes such as λ and T7 late during infection to produce large concatenated dsDNA molecules containing many copies of their genome tandemly linked head-to-tail. The genomes are then cut into genome-length molecules by phage-encoded enzymes, recircularized, and packaged into virions.

DNA replication begins when the viral (+)strand is released into its host cell. It is rapidly coated by SSB, a single-strand-specific DNA-binding protein encoded by the host. However, the (–)strand replication origin [ori(–), Figure 8-8] is an inverted repeat sequence that forms a hairpin structure, thereby preventing binding of SSB to this particular site. This duplex DNA hairpin is recognized in some bacteriophage by the host's RNA polymerase and in others by DnaG, the host-encoded DNA primase. Both enzymes synthesize a short (20 nt) RNA primer that is then extended by the host's DNA polymerase III. However, synthesis of an RNA with an exposed 3′-end annealed to the DNA template is not compatible with known conformations of the *E. coli* transcription complex. In fact, synthesis of the M13 replication primer is carried out by a novel form of the RNA polymerase transcription complex that contains an overextended RNA:DNA hybrid bound in the RNA polymerase trough normally occupied by downstream dsDNA. Thus, the 3′-end of the RNA is available for interaction with DNA polymerase. The existence of this novel transcription complex suggests that it may normally be used to initiate replication of bacterial mobile elements.

The RNA primer is excised by the 5′→3′ exonuclease activity of *E. coli* DNA polymerase I, with concomitant extension of the 3′-end of the nascent

DNA strand by the same enzyme. The 3′-end of the nascent DNA strand is then covalently joined to its 5′-end by DNA ligase. Conversion of the (+)strand DNA into a covalently closed, circular replicative-form (RF) DNA is completed upon introduction of negative superhelical turns by the host's DNA gyrase.

The phage-encoded initiator protein (ϕX174 gpA; M13 gpII) binds to the (+)strand origin [ori(+), Figure 8-8] in the replicating form of the viral genome. Ori(+) consists of two structural elements: a gpII-binding site and a nicking site where gpII breaks the phosphodiester backbone of the parental DNA strand. When gpII binds to ori(+) in supercoiled DNA, it induces unwinding of 7 bp that encompass the nick site. This partial melting of the DNA target is a prerequisite for the nicking reaction. Nicking occurs with concomitant formation of a transient covalent linkage between the 5′-end of the (+)strand and a specific tyrosine residue in gpII. The unwound region serves as an entry site for Rep, an *E. coli* DNA helicase that dimerizes upon DNA binding and unwinds the duplex by traveling along the DNA template in the 3′→5′ direction. The free 3′-OH end of the (+)strand serves as a primer on which the host's DNA Pol-III holoenzyme initiates DNA synthesis with concomitant displacement of the 5′-end of the (+)strand.

Geminivirus DNA replication follows the same basic mechanism outlined in Figure 8-8. Conversion of ssDNA into dsDNA involves RNA-primed DNA synthesis, but the proteins involved and the nature of ori(–) remain obscure. Presumably, it is coated with replication protein A (RPA), the eukaryotic version of *E. coli* SSB. The dsDNA to ssDNA phase, however, begins with activation of ori(+) by the virally encoded initiator protein Rep. Geminivirus ori(+) is an invariant 9 nt sequence found in the genomes of all geminiviruses sequenced to date. It is flanked by an inverted repeat with the potential to form a cruciform structure. Rep, the only viral protein essential for viral genome duplication, is a multifunctional protein with sequence-specific DNA binding, ssDNA endonucleolytic, and ATPase activities. Rep nicks ori(+) at a specific site, becoming in the process covalently linked to the 5′-end of the nick via a tyrosine residue, and recruiting RFC and PCNA to the 3′-OH end of the nick. Since PCNA is a cofactor of DNA polymerase-δ, this enzyme is presumably used to copy the (+)strand. In addition, mastrevirus Rep interacts with RepA, whereas Rep of the remaining geminiviruses interacts with REn. RepA and REn presumably facilitate site-specific initiation, and thus act as auxiliary initiation proteins. In addition, RepA binds the plant homolog of the mammalian retinoblastoma protein, and thereby may facilitate expression of genes required for DNA replication.

With each of these examples, the only virus-encoded DNA replication proteins are endonucleases that are required for nicking the double-stranded replicative form of the viral genome at the (+)strand origin. Each of these site-specific endonucleases remain bound to the 5′-end of the nicked DNA strand. All other DNA replication proteins are provided by the host cell.

F. ssDNA With Terminal Replicators That Use Gap-Filling And Then Single-Strand Displacement

Genomes encoded by linear ssDNA are unique to the *Parvoviridae*. These viruses infect a broad range of invertebrate and vertebrate hosts, from arthropods to humans, and are associated with certain human diseases. Minute virus of mice (MVM) and adeno-associated virus (AAV) are two well-characterized examples of parvovirus genome duplication.

Parvoviral genomes consist of a single ssDNA molecule with two replicators, one that provides a DNA primer on which DNA polymerase can initiate gap-filling, and one that forms as a consequence of DNA replication. Each replicator is embedded in a larger, hinge-like, hairpin terminus that can unfold to be copied, then refold to allow continuous amplification of the linear template. A hinge element exists in these termini that lowers the energy required for extending the hairpin, which together with an appropriately spaced duplex DNA-binding site for the viral initiator protein, allows both the double-stranded extended form and the hairpin form of the termini to be replicated. Accordingly, viral replication proceeds in two distinct phases. The first amplifies the number of copies of the viral genome by forming large palindromic concatemers. The second phase allows individual ssDNA genomes to be excised and displaced from these concatenated molecules so that they can be encapsulated into virus particles.

Phase 1: amplification

Parvoviral genomes consist of a single DNA molecule 4–6 kb in length that is terminated at each end by short (120–420 bases) imperfect inverted terminal repeats that can fold back on themselves to form duplex DNA structures referred to as hairpin termini (Figure 8-9A). The termini can vary significantly in sequence and secondary structure both within individual genomes and between species. The 3′-OH terminus is the parvovirus replication origin and acts as a primer for gap-filling DNA synthesis using the host DNA polymerase-δ•PCNA complex to initiate synthesis of the complementary DNA strand (Figure 8-9B). This reaction is facilitated by RPA, the single-stranded DNA-binding protein common to all eukarya. The result is a monomer-length, duplex DNA intermediate in which the two strands are covalently cross-linked at one end via a single copy of the viral 3′ inverted terminal repeat (Figure 8-9C).

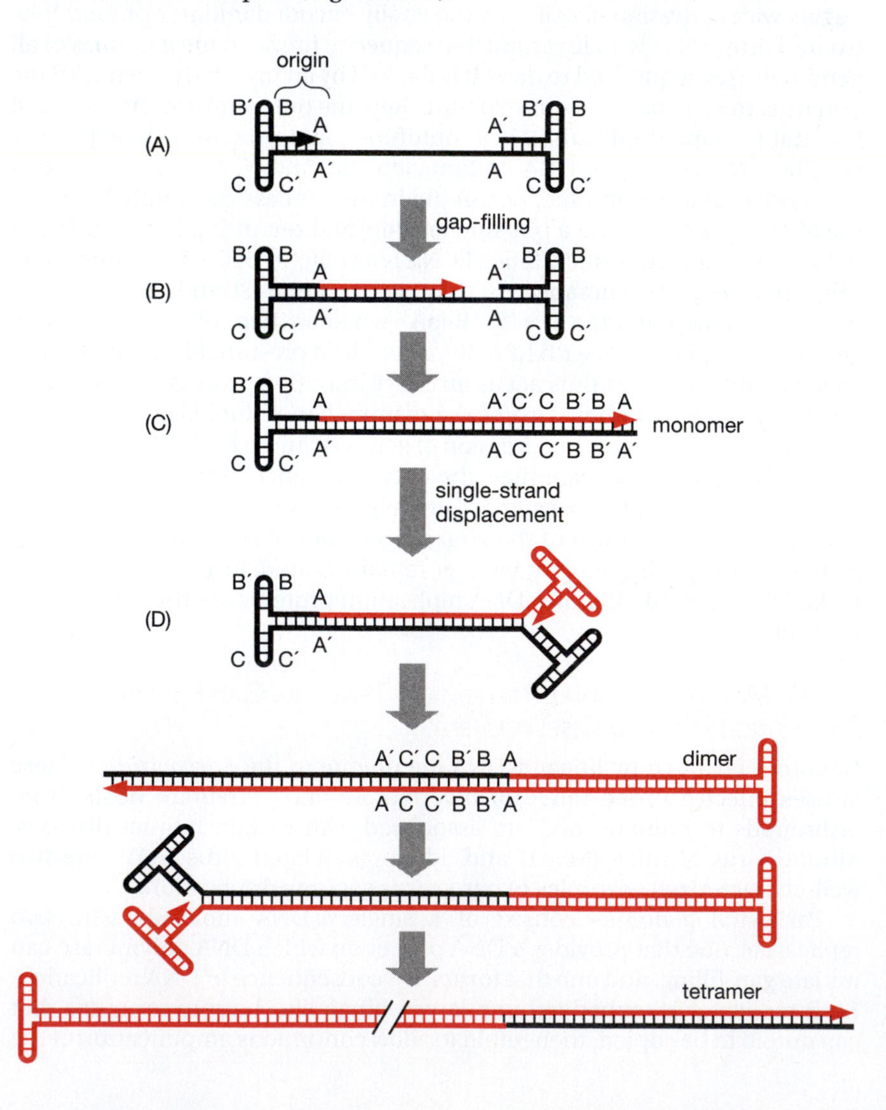

Figure 8-9. Parvovirus genome duplication: phase 1. The parental genome is converted into a duplex DNA molecule and then amplified by formation of palindromic concatemers in which the unit-length genomes are fused in left-end:left-end and right-end:right-end combinations. Black is parental DNA and red is nascent DNA. Arrowheads indicate the 3′-end of DNA. (Data from Cotmore, S.F. and Tattersall, P. (2006) Parovirus. In *DNA Replication and Human Disease* (DePamphilis, M.L., ed.), pp. 593–608.)

Parvovirus genomes are duplicated via a continuous, unidirectional, strand-displacement mechanism often described as rolling-hairpin replication (Figure 8-9D), an adaptation of the rolling-circle replication mechanism described in Figure 8-8. The replication process creates a series of palindromic, duplex, dimeric, and tetrameric concatemers. Successive rounds of replication are then initiated from these concatemers. The limited genetic capacity of small viruses dictates that they rely predominantly on the replication machinery of the host cell. Thus, DNA synthesis is carried out by the same enzymes and proteins that carry out DNA synthesis on the leading-strand DNA template of cellular replication forks.

Phase 2: resolution

Parvoviral ssDNA genomes are excised, and ultimately displaced, by site-specific single-strand nicks introduced into the duplex inverted terminal repeats by the viral initiator protein (Figure 8-10). Well-characterized examples are AAV Rep 78/68 and MVM NS1. This involves a *trans*-esterification reaction that liberates a base-paired 3′-nucleotide to act as a DNA primer on which DNA polymerase can initiate synthesis, leaving the initiator protein covalently attached to the 5′-nucleotide via a phosphotyrosine bond. Such

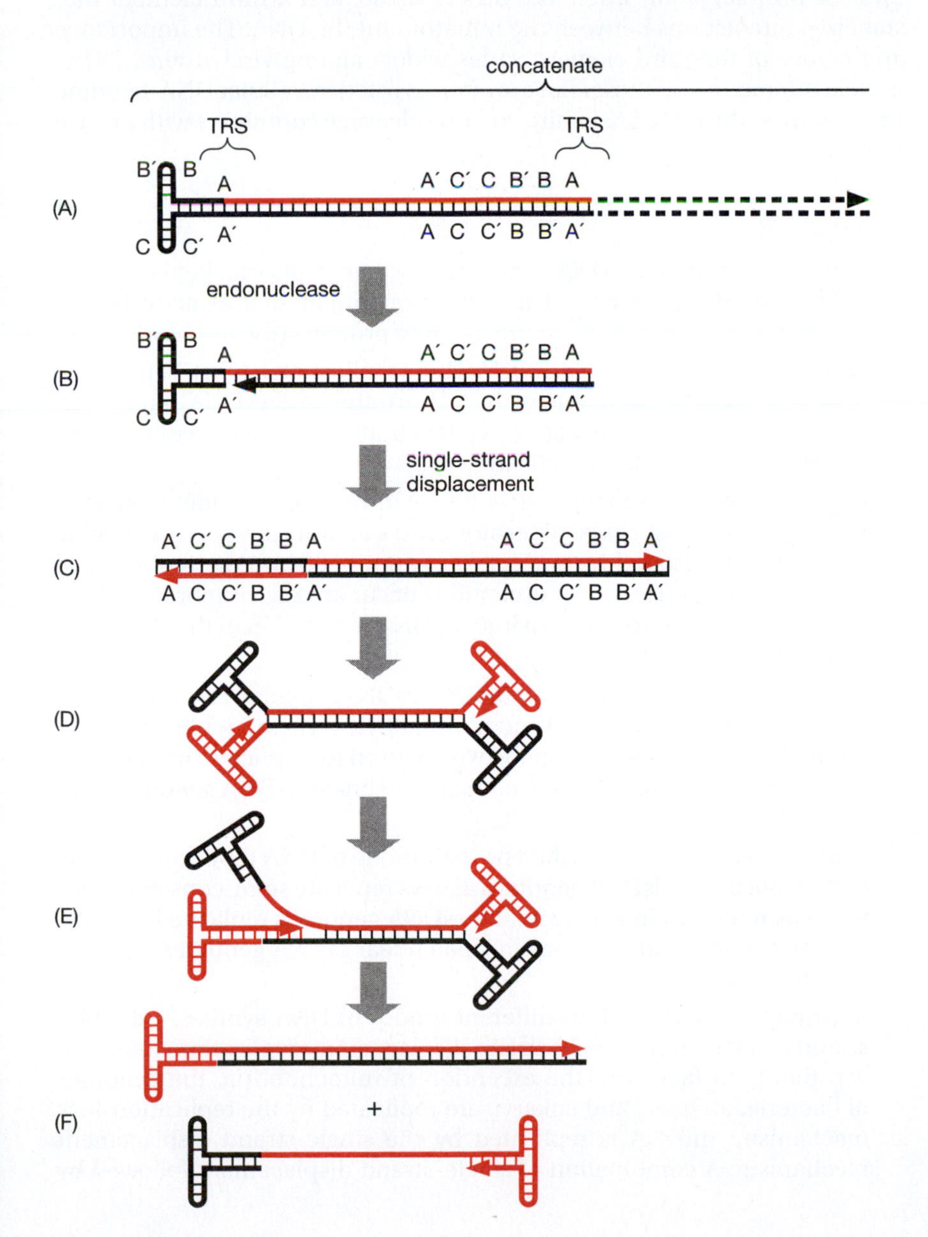

Figure 8-10. Parvovirus genome duplication: phase 2. Concatenates are resolved into monomers by site-specific nicking of one strand at the terminal resolution sequence (TRS) located in duplex DNA at the A:A′ junction in the replication origin. These nicks effectively separate one copy of the viral genome from the next. The ability to form terminal hairpin structures then allows formation of a DNA primer to initiate synthesis and displace one of the two strands.

a reaction is common to initiators of rolling-circle replication as well. However, whereas rolling-circle replication initiators carry out sequential nicking and rejoining reactions, parvovirus initiators only manage the nicking reaction, so that parvovirus initiator proteins remain covalently attached to the 5′-end of progeny genomes. Subsequent accumulation of displaced progeny single strands is entirely dependent upon the availability of preformed capsid protein, and packaging itself is driven by ongoing viral DNA synthesis.

Parvoviral replication origins are present as duplex DNA that must be melted before they can be cleaved by the viral initiator protein. This is accomplished by the helicase activity of the parvoviral initiator protein. All parvovirus initiator proteins have two separate enzymatic domains. The first is an N-terminal, site-specific, single-strand-specific nuclease domain that recognizes and nicks the replication origins. This domain is common to all rolling-circle initiators. The second is a C-terminal, superfamily III helicase domain that belongs to a group of viral 3′→5′ helicases that are related to the AAA+ family of cellular ATPases.

Although parvoviral origins vary in size and structure, they all contain three distinct DNA recognition elements: a duplex DNA site for binding the initiator protein, a site where the nick is made, and a third element that stabilizes interactions between the initiator and the DNA. The importance and nature of the third element varies widely among viral strains. DNA–protein interactions at these sites are often assisted by cellular DNA-bending proteins from the HMG1/2 family, creating cleavage complexes with precise ternary structure.

SUMMARY

- The simplest unit of DNA replication is the replicon. Replicons are defined by the presence of a replication origin that is activated by association with one or more *trans*-acting proteins (the initiator).
- Replication origins are sites in DNA where replication begins. Replicators are *cis*-acting DNA sequences that determine where DNA replication begins. Initiators are *trans*-acting factors that interact with replicators to initiate DNA synthesis at replication origins.
- Replication origins and replicators do not have to be coincident, but they usually are. In most cases, replicators and initiators are species-specific (the initiator encoded by one species cannot activate the replicator of another species). The major exceptions occur among vertebrates where initiators from one species can initiate DNA replication in the DNA from another species.
- Based solely on DNA structure (dsDNA, ssDNA, circular, and linear) and replicator locations (internal and terminal), five types of replicon exist in nature (Table 8-1). Of these, only two are used to duplicate the genomes of cellular organisms. They are circular and linear dsDNA genomes with internal replicators.
- Replicons can be distinguished by their mode of DNA replication (Table 8-2). Replicons in dsDNA genomes always replicate semi-conservatively, whereas replicons in small circular ssDNA genomes replicate by a conservative mode, and replicons in small linear ssDNA genomes replicate by a dispersive mode.
- In principle, there are four different modes of DNA synthesis: double-strand displacement, single-strand displacement, replication forks, and gap-filling. In fact, with the exception of mitochondria, the genomes of bacteria, archaea, and eukarya are replicated by the replication-fork mechanism; mtDNA is replicated by the single-strand displacement mechanism. A combination of single-strand displacement followed by

gap-filling is employed to replicate Type III replicons (linear dsDNA genomes with terminal replicators), and a combination of gap-filling followed by single-strand displacement is used to replicate Type IV and Type V replicons (circular and linear ssDNA genomes).

- If three parameters are considered (DNA structure, replicator location, and mode of DNA synthesis), many replicon manifestations are possible, but only six exist in nature (Table 8-3). Moreover, with the exception of mitochondria, the same replicon manifestation is used in the genomes of bacteria, archaea, and eukarya: dsDNA with one or more internal replicators that initiate bidirectional DNA replication using the replication-fork mechanism. mtDNA replicators initiate unidirectional replication by the single-strand displacement mechanism.

REFERENCES

Additional Reading

Reyes-Lamothe, R., Wang, X. and Sherratt, D. (2008) *Escherichia coli* and its chromosome. *Trends Microbiol.* 16: 238–245.

Herrick, J. (2010) The dynamic replicon: adapting to a changing cellular environment. *Bioessays* 32: 153–164.

Literature Cited

Cairns, J. (1963) The bacterial chromosome and its manner of replication as seen by autoradiography. *J. Mol. Biol.* 6: 208–213.

Jacob, F., Brenner, S. and Cuzin, F. (1963) On the regulation of DNA replication in bacteria. *Cold Spring Harb. Symp. Quant. Biol.* 28: 329–348.

Chapter 9
REPLICATION ORIGINS

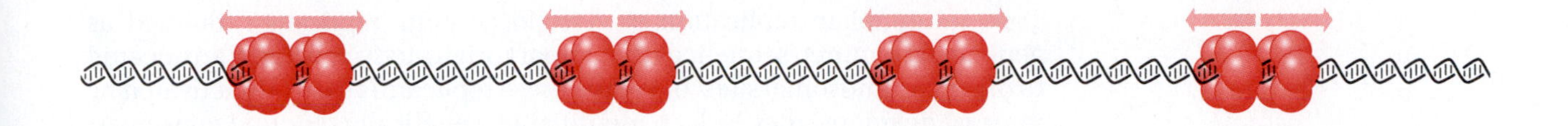

In previous chapters, we learned that the genomes of all cellular organisms (bacteria, archaea, and eukarya) are encoded by duplex DNA, and that they replicate semi-conservatively from internal replicons via the replication-fork mechanism. Therefore, every genome should have one or more specific sites where DNA replication begins (replication origins), and initiation of DNA replication at these sites should be dependent on *cis*-acting DNA sequences (replicators) that interact with one or more *trans*-acting factors (initiators). This chapter explores the nature of replication origins located within circular and linear dsDNA chromosomes, the replicons found in all cellular organisms (Table 8-1). The following chapter describes specific paradigms for replication origins in each of five types of replicon found in cells, viruses, and episomes.

The first replicators were reported in 1977–1978 by Yukinori Hirota in the bacterium *E. coli*, by Bill Dove and Fred Blattner in the bacteriophage λ, and by Dan Nathans and Tom Shenk in the mammalian simian virus 40 (SV40). Each of these paradigms for Type I replicons fulfills the requirements of the classical replicon model. They each contain a unique *cis*-acting sequence that is required for replication of their respective genome. Their function is sensitive to genetic alterations, and requires the presence of a cognate initiator protein to which they bind specifically. When transferred to other DNA molecules, they allow these molecules to replicate autonomously when provided with the required replication proteins. Moreover, they are origins of bidirectional replication. These features became less clear, however, as replication origins were elucidated in the chromosomes of eukaryotes.

Replication origins in cellular genomes are all determined by the interaction between initiator proteins and DNA, but the nature of the DNA sites and the manner in which they interact with replication proteins varies extensively among the bacteria, archaea, and eukarya, as well as among the viruses and episomes that inhabit these organisms. Nature appears to have experimented with every conceivable mechanism for genome duplication. The existence of replicators is now well established in single-cell organisms but the nature of replication origins is not clear when we move to multicellular organisms (metazoa) such as flies, frogs, and mammals where various methods for identifying replication origins have led to different, sometimes contradictory, conclusions.

In 1989–1990, three papers appeared in the journal *Cell*, each claiming to have mapped initiation sites for DNA replication in Chinese hamster cells within the intergenic region downstream of the DHFR gene. Remarkably, two of these studies concluded that initiation events originated from specific genomic sites, while the third concluded that initiation events were distributed uniformly throughout the intergenic region. Despite the development and application of novel origin-mapping methods, and the analysis of a variety of genomic sites in different organisms, a paradox emerged. Studies employing two-dimensional gel fractionation of total genomic DNA to detect replication bubbles or to map the polarity of replication forks generally concluded that initiation events were distributed uniformly over regions as large as 55 kb (initiation zones), whereas methods that mapped either the relative distribution or the relative abundance of nascent DNA strands along the genome invariably concluded that initiation events originated within specific loci of approximately 1 kb or less. Furthermore, site specificity was not detected in the early embryos of frogs, flies, and fish, and the question of whether or not these genomes, like those in yeast, contain replicator sequences remained controversial. True, mammalian replication origins do contain 'replicators' defined as genetically required sequences that impart origin activity when translocated to other chromosomal sites, but while these replicators can be inactivated by internal deletions, they lack an identifiable, genetically required consensus sequence, such as the ones found in budding yeast replicators. How might such disparate data be reconciled?

If one adopts the view that biological evolution does not invent new devices for solving old problems, but rather builds on what has gone before, then the simplest view is that replicators exist in the genomes of all organisms, but their function depends on the conditions under which they find themselves. Thus, some replicators are specified solely by a unique DNA sequence that is required for the initiator to carry out its function, while other replicators depend loosely on a certain DNA sequence or DNA composition that serves to direct initiation events like a policeman directs traffic but that are not required for initiator function. Still other replicators may lie dormant until they are recruited to serve when certain conditions occur. With this background in mind, consider how cells decide where to initiate genome duplication.

PATTERNS OF INITIATION

The goal of genome duplication is to produce at least one complete copy of the genome as quickly as possible, so as to avoid predators and minimize exposure of the genome to a hostile environment. So far, four different patterns of initiation events have been identified: site-specific, **distributive**, **endoreduplication**, and **gene amplification**.

The most common pattern occurs during normal cell division or virus proliferation wherein DNA replication generally, but not always, is site-specific; it begins at one or more specific sites within a genome. In bacteria, bacteriophages, bacterial plasmids (episomes), archaea, and animal and plant DNA viruses, initiation events usually occur at a single genomic site, although three to four specific sites have been identified in some cases. For example, the genome of the archaeon *S. solfataricus* (3 Mbp) contains three replication origins. The mammalian herpes simplex virus (HSV) genome (152 kbp) contains three replication origins, and the bacteriophage T4 genome (169 kbp) contains four replication origins. In the still larger genomes of eukarya, initiation events have been mapped to 250–430 specific sites in the 13 Mbp genomes of yeast, and based on the average inter-origin distances that are observed, the number of initiation sites in the 3000 Mbp genomes of mammalian cells has been estimated at 20 000–40 000 replication origins. Site-specific initiation in these examples is determined solely by DNA sequence.

Organisms such as flies, frogs, and fish whose fertilized eggs develop outside the mother face a unique problem not encountered by organisms whose embryos develop internally: predators. To counter this threat, externally developing embryos undergo a series of rapid cell cleavages. For example, in 2 days, a fertilized frog egg develops into a swimming tadpole, whereas a fertilized mouse egg produces only four cells! During these rapid cell cleavages, initiation events are distributed throughout the genome, and the distribution of initiation sites appears random with respect to DNA sequence. Remarkably, this distributive mode of initiation changes to a site-specific mode as these embryos develop into adult organisms and the time required for cell division lengthens. Thus, site-specificity also can be developmentally acquired through epigenetic changes, events that affect DNA function without affecting DNA sequence.

One of the cardinal rules in eukaryotic genome duplication is that genomes are duplicated once and only once during each cell-division cycle, but, like most rules, there are exceptions. During the development of vertebrates from a fertilized egg to an adult, specific terminally differentiated cell types arise in which the entire genome is duplicated multiple times without an intervening mitosis. This process, termed endoreduplication, produces cells with a single nucleus containing multiple copies of each chromosome. The mechanism by which this occurs results from changes in cell-cycle regulation (Chapter 12). In some instances, genomewide replication ceases and initiations become limited to replication origins at a few chromosomal loci, resulting in locus-specific gene amplification. Gene amplification occurs through multiple rounds of initiation at a small number of very specific replication origins. Because these origins reinitiate replication under conditions where the vast majority of replication origins in the cell do not, they are referred to as **amplification origins**, rather than replication origins.

So, what are 'origins' and how do they work?

FINDING REPLICATION ORIGINS

An arsenal of methods has been developed over the years to identify replication origins in every conceivable type of genome, but all of them are based on only five concepts: replicator activity assays, identifying origins of bidirectional DNA replication, quantifying nascent-strand abundance throughout a genomic locus, quantifying DNA sequence duplication, and detecting a sequence bias in leading-strand DNA synthesis. The following summaries provide the concepts involved; the literature cited at the end of this chapter provides the technical information.

Replicator Activity

There are three basic assays for replicator activity. The most widely used is **autonomously replicating sequence** (**ARS**) activity. ARS activity confers on a foreign DNA molecule the ability to replicate autonomously when provided with the appropriate replication proteins. This is usually done by transferring a DNA fragment to a small (3–12 kb) circular plasmid DNA molecule and then either transfecting the plasmid into an appropriate host cell, or incubating it *in vitro* together with the proteins and nucleotides required for DNA replication. If the plasmid replicates when the foreign DNA fragment is present, but not when it is absent, the fragment is considered to have replicator activity. Analysis of mutations, deletions, and insertions within the replicator fragment can then define the sequences required for DNA replication. Remarkably, the genomes of bacteria, bacterial plasmids, bacteriophages, yeast, and animal viruses all contain one or more ARS elements. These ARS elements are species-specific, because their activity depends on interaction with specific initiator proteins.

A variation of the ARS assay is the oriP replicator assay. oriP is a replicator found in the Epstein–Barr virus (EBV) genome that can drive plasmid DNA replication in human cells when provided with the EBV initiator protein EBNA1. oriP is unique in that it mimics replication origins in its host's chromosomes by restricting initiation events to once per cell-division cycle and by utilizing the host cell's initiator proteins. It contains a sequence that retains the plasmid with cellular chromosomes during mitosis and thereby prevents newly replicated DNA molecules from becoming lost or destroyed, and a binding site for the virus' initiator protein (Figure 10-18). Some human DNA sequences can substitute for the initiator protein binding site.

ARS elements have been difficult to identify in genomes from metazoa such as fruit flies, nematodes, and mammals. They appear to exist, but their activity is low in comparison with ARS elements from bacteria, plasmids, bacteriophages, animal viruses, and yeast. This reflects the fact that the complexity of replication origins increases as organisms become more complex. However, these sequences exhibit **ectopic site activity**, the ability of a DNA fragment to initiate replication when translocated to an ectopic chromosomal site. This assay has identified replicators in fruit flies and in mammalian cells.

Origins Of Bidirectional Replication

Origins of bidirectional replication lie midway between two replication forks, resulting in the appearance of 'replication bubbles' that are centered at a replication origin (Figure 8-1). With small genomes such as those in bacteria, archaea, plasmids, and viruses, replication bubbles can be detected either by fiber autoradiography (Box 8-1), electron microscopy (Figure 8-2), two-dimensional gel electrophoresis (Figure 8-3), or DNA combing (Figure 8-4). However, of these methods only two-dimensional gel electrophoresis and DNA combing allow mapping of replication origins to specific genomic sites, and only DNA combing reveals whether the results apply to the same chromosome in each cell within a population of cells, or simply reflect the statistical mean for the entire population of cells. A third method relies on identifying nascent DNA strands by the fact that they contain a short oligoribonucleotide covalently linked to their 5′-end and then mapping the 5′-ends of nascent leading-strand DNA to specific genomic sites (detailed below in Boxes 9-1 and 9-2). This method allows the actual initiation sites for DNA synthesis to be mapped with single-nucleotide resolution at unique genomic loci. The transition from discontinuous to continuous DNA synthesis at these loci marks an origin of bidirectional replication (Figure 9-1C).

Nascent-Strand Abundance

Replication origins can be identified from the enrichment of newly synthesized DNA at origin loci relative to non-origin loci. Three features of nascent DNA at replication origins distinguish it from either damaged DNA or newly repaired DNA. Nascent DNA at replication origins can be labeled by incorporation of deoxyribonucleotides during S phase. Nascent DNA strands synthesized at replication origins and at replication forks have RNA primers at their 5′-ends. Given the structure of bidirectional replication bubbles (Figure 9-1) and the size of Okazaki fragments (40–300 nucleotides), nascent DNA strands about 1 kb in size represent a fusion of the continuously synthesized nascent leading-strand DNA with the discontinuously synthesized nascent lagging-strand DNA (Okazaki fragments) from the same template strand. Therefore, quantifying the relative abundance of these 1 kb nascent DNA strands throughout a particular genomic locus by hybridization to a series of unique DNA fragments can locate a replication origin.

DNA Sequence Duplication

In an exponentially proliferating cell population, the copy number of a unique DNA sequence located close to a replication origin will be higher, on average, than that of another unique DNA sequence located near a termination site for DNA replication. Therefore, in combination with whole-genome DNA microarrays, this strategy has been used to monitor the progress of replication forks, and to map replication origins in bacteria, archaea, yeast, and flies.

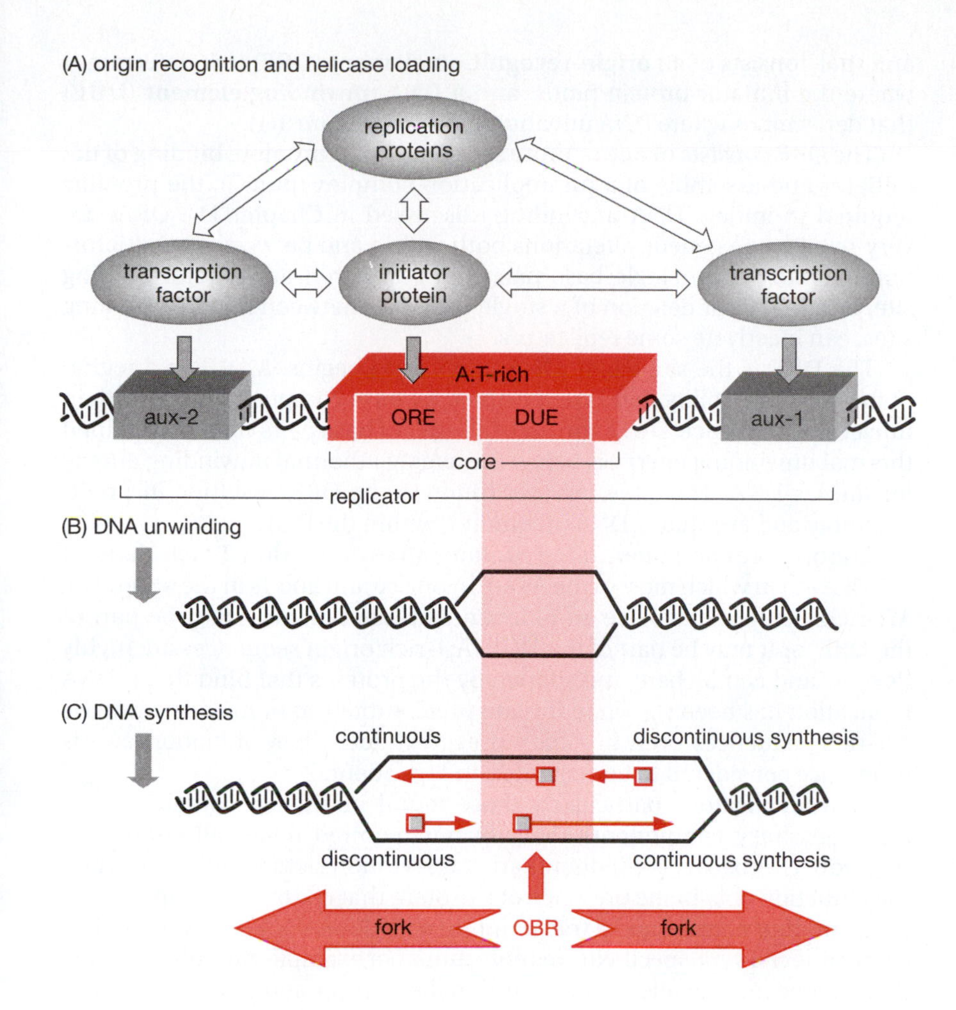

Figure 9-1. Initiation of DNA replication at internal replicators in dsDNA genomes. (A) Replicators consist of a core component that is required for DNA replication, and often contain one or more auxiliary components that facilitate replication under some, but not all, conditions. Auxiliary components are binding sites for transcription factors. The core component contains one or more binding sites for the initiator protein (origin-recognition element, ORE), a DNA-unwinding element (DUE), and an asymmetric A:T-rich element. (B) The first step in DNA replication is unwinding the DUE. (C) The second step is DNA synthesis via the replication-fork mechanism. RNA-primed (red square) DNA synthesis (red arrow) is continuous on the leading-strand template and discontinuous on the lagging-strand template. The first RNA primer is synthesized at replication origins either by DNA transcription or by DNA primase, after which RNA-primed DNA synthesis at replication forks occurs exclusively by DNA primase coupled with a replicative DNA polymerase. Nascent DNA strands are elongated continuously in the direction of DNA unwinding (fork movement) by a replicative DNA polymerase. The transition between continuous and discontinuous DNA synthesis marks the origin of bidirectional replication (OBR).

Leading-Strand Sequence Bias

In prokaryotes, prevalence of G over C and T over A is frequently observed in one arm of replication forks, the arm in which the direction of DNA synthesis is the same as the direction of fork movement (leading-strand synthesis). The sign of the resulting TA and GC skews changes abruptly when crossing replication-origin and replication-termination sites. The strength of the strand bias varies greatly among phyla and appears to correlate with growth rate. Much greater diversity of skew is observed among archaea than among bacteria. A computer program has been developed that locates origins of replication by measuring the bias between leading and lagging strand for all oligonucleotides up to 8 bp in length.

CHARACTERISTICS OF REPLICATION ORIGINS

Origin Anatomy

The dsDNA genomes of all bacteria, archaea, and single-cell eukaryotes, as well as the viruses, plasmids, and organelles that proliferate in them, all contain well-defined replicators that are required for efficient genome duplication. The same is true for ssDNA genomes of bacterial, animal, and plant viruses, although their mechanism of action differs from replicators in dsDNA genomes (described in Chapter 10). The essential function of a dsDNA replicator is to facilitate the assembly of a DNA replication machine at a DNA site that can be easily unwound so that DNA synthesis can initiate on both templates. To this end, replicators all contain an **origin-core component** (ori-core) that is required for origin activity under all conditions, and that consists of an **origin-recognition element** (**ORE**) that determines where the initiator protein binds, and a **DNA-unwinding element** (**DUE**) that determines where DNA unwinding begins (Figure 9-1).

The ORE consists of one or more sequences that promote binding of the initiator and assembly of a pre-replication complex (preRC), the proteins required to initiate DNA unwinding (discussed in Chapter 11). OREs are very sensitive to genetic alterations both within and between the initiator-binding sites. Even single base-pair changes within an initiator-binding site, or addition or deletion of a single base pair between initiator-binding sites, can inactivate some replicators.

The DUE is the site where DNA unwinding begins. A DUE is a region of low helical stability that is determined by base composition rather than nucleotide sequence. DUEs consist of a series of base pairs whose combined thermal unwinding energy is below the average thermal unwinding energy for duplex DNA. Thus, the initiator binds to the ORE, resulting in preRC assembly and eventually DNA unwinding within the DUE.

The ori-core component often contains an asymmetric **A:T-rich element** (>75% A+T) in which most of the As are in one strand and Ts in the other. The A:T-rich element may have an independent function, or it may be part of the ORE, or it may be part of the DUE. A:T-rich origin sequences are highly flexible, and can be bent and shaped by the proteins that bind them. DNA replication has been shown to initiate preferentially at or near asymmetric A:T-rich sequences even in *Xenopus* egg extracts where initiation events were once considered to be sequence-independent.

Many replicators, particularly those found in viruses, contain one or more **auxiliary components** that are not required under all conditions (Figure 9-1). Auxiliary components are transcription factor binding sites, and they function only in the presence of a protein that contains an appropriate DNA-binding domain and a transcription activation domain. As such, they determine cell-type specificity for replicators. For example, the polyomavirus (PyV) replicator includes a transcriptional enhancer and functions only in cells in which the enhancer is active. Auxiliary components are required under conditions either where the initiator protein concentration is low, or where the DNA is organized into a repressive chromatin structure. When DNA replication is allowed to take place *in vitro* (e.g. DNA plus a soluble cell extract or a collection of purified replication proteins), then auxiliary elements are dispensable. Dependence on auxiliary components varies considerably among replicators. For example, the PyV ori-core is less than 1% as active as the PyV origin *in vivo*, whereas the SV40 ori-core is about 20% as active as the SV40 origin *in vivo*. Auxiliary elements serve three different functions. First, they can facilitate binding of the initiator or other replication proteins [e.g. DNA polymerase-α•DNA primase, replication protein A (RPA)] to the replication origin. Second, they can facilitate DNA unwinding, and third, they can prevent chromatin structure from masking the replicator. In general, the dependence of ori-core on auxiliary components is inversely related to the affinity of the initiator protein for the ORE.

Bent DNA (Chapter 1) and inverted repeats (Figure 1-10), sequences that can facilitate binding of proteins to DNA, are also commonly associated with replication origins (e.g. *E. coli*, phage λ, SV40, *S. cerevisiae* ARS1). It has been suggested that inverted repeats fold into a cruciform conformation (Figure 1-10) during assembly of replication proteins at replication origins. Using anti-cruciform DNA antibodies to detect cruciform DNA, cruciform structures appear to accumulate at the G_1–S-phase boundary. These antibodies can bind to active replication origins in permeabilized mammalian cells and stimulate DNA replication, suggesting that stabilization of cruciform DNA facilitates assembly of initiation complexes. Several 14-3-3 protein isoforms have been found that bind cruciform DNA, and these proteins may transiently stabilize cruciform DNA structures *in vivo*.

Origin Specificity

Interaction between replicators and initiators is species-specific in bacteria, bacteriophages, plasmids, and animal viruses. For example, the bacterial initiator protein DnaA is found in all bacteria and DnaA from one bacterial species can bind to the replicator from another species, but DnaA can only initiate DNA unwinding when it binds to the replicator from its own species, its cognate replicator. The same is true among different members of the papovavirus family such as SV40 and PyV. Among eukarya, initiator proteins are interchangeable among species of the same class (e.g. mammals) and to a lesser degree the same phylum (e.g. vertebrates), but not necessarily among distantly related organisms within the same kingdom, such as budding yeast and fission yeast. ARS elements from the budding yeast, *S. cerevisiae*, do not function in the fission yeast, *S. pombe*, and vice versa, despite the fact that both ARS elements interact with homologs of the same initiator proteins, origin-recognition complex (ORC) and Cdc6.

Interaction between viral replicators and host cell proteins can also be species-specific, because of specific interactions between viral initiator proteins and host proteins. For example, the SV40 replicator and initiator can initiate DNA replication in monkey and human cells, but not in mouse and hamster cells. The PyV replicator and initiator function in mouse cells, but not in human cells. Host-range specificity in these cases results from the need for the viral initiator protein to bind to the host cell DNA polymerase-α•DNA primase and recruit it to the viral replicator.

Origin Function

The first step in DNA replication is binding the initiator protein to the ORE, sometimes bringing with it additional proteins necessary for DNA unwinding and DNA synthesis (Figure 9-1A). As detailed in Chapter 10, there are five different mechanisms by which initiator proteins interact with replicators to begin the process of DNA replication (Table 10-1). Of these, two involve initiators that associate with other proteins before binding to DNA and transport them to the replication origin. Protein priming initiators such as adenovirus terminal protein are associated with the DNA polymerase responsible for replicating the adenovirus genome. DNA helicase initiators such as the SV40 large T-antigen (T-ag) are associated with RPA, the cell's single-stranded DNA-binding protein, and DNA polymerase-α•primase, the only cellular enzyme that can initiate RNA-primed DNA synthesis. RPA facilitates both DNA unwinding and DNA synthesis (Chapter 4). Papillomavirus E1 protein associates with the papillomavirus transcription factor E2 which then facilitates binding of E1 to the ORE by binding of E2 to an adjacent auxiliary component. Other replicators contain auxiliary components, but there is not evidence that the corresponding transcription factors are recruited to the replicator by their association with the initiator. T-ag binds multiple transcription factors including TBP, TFIIB, TEF1, p300, CBP, and tumor-suppressor gene products p53 and pRB. These proteins help T-ag to drive quiescent cells into S phase and regulate gene expression, but they do not facilitate T-ag's role as an initiator protein. Other initiators such as helicase loaders (DnaA in bacteria, ORC+Cdc6 in eukarya) recruit replication proteins to the replicator, but only after the initiator is bound to DNA.

The second step in DNA replication is DNA unwinding (Figure 9-1B). In those replicators where this event has been characterized, DNA melting begins within the DUE, thereby creating two ssDNA templates upon which DNA polymerase can initiate synthesis of the complementary strand. DNA unwinding in active replicators begins when the initiator binds to the replicator, and is then continued by the assembly of an active DNA helicase at the initial site of DNA unwinding. The process of DNA unwinding is either facilitated or opposed by localized superhelical turns in the DNA.

Negative superhelical turns facilitate DNA unwinding, whereas positive superhelical turns inhibit DNA unwinding (Figure 3-9). Superhelicity at replication origins is determined by topoisomerases, chromatin structure, and transcription (Chapter 1).

The third step in DNA replication is DNA synthesis (Figure 9-1C). The properties of DNA polymerases together with the structure of DNA dictate that every **origin of bidirectional replication (OBR)** must exhibit a transition from discontinuous to continuous DNA synthesis on each template. All DNA polymerases synthesize DNA only in one direction (5′→3′), and none can initiate synthesis *de novo*; they simply extend the 3′-OH end of an RNA or DNA strand. Therefore, given that the two DNA templates are antiparallel (Figure 1-3), DNA synthesis can be continuous on one arm of a replication fork, but not on the other arm. One of the two DNA template strands must be synthesized discontinuously through the repeated synthesis and joining of Okazaki fragments (Figure 3-8). Assuming that most, if not all, DNA synthesis occurs continuously on one template strand of a replication fork and discontinuously on the other, the OBR is marked by a transition from continuous to discontinuous DNA synthesis on each DNA template. Thus, an OBR is the site where DNA synthesis begins, whereas the replicator is the site that is genetically required for initiation of DNA replication. In principle, the OBR and the replicator can occur at separate sites in the genome. In practice, they are coincident in all of the replication origins characterized so far, although they appear to be separated widely in amplification origins.

The original paradigm for replication origins that function in the nuclei of mammalian cells was that of SV40, a small tumor virus with a circular dsDNA genome of only 5.2 kbp. Its genome was sequenced completely by 1978 in the laboratories of Walter Fiers and Sherman Weissman. This feat allowed the first genetic and molecular analysis of DNA replication and gene expression in mammalian cells at the level of single-nucleotide resolution. Unraveling the relationship between DNA sequence and replicator function began in 1982 when Hay and DePamphilis mapped a clear transition between discontinuous and continuous DNA synthesis that identified the origin of bidirectional replication at the resolution of a single nucleotide (Box 9-1) (Hay and DePamphilis, 1982). Soon after, similar results were

Box 9-1. Mapping Initiation Sites for *de novo* DNA Synthesis in Simian Virus 40 Chromosomes

Initiation sites for nascent RNA-primed DNA chains in replicating SV40 DNA molecules were mapped with single-nucleotide resolution by radiolabeling the 5′-termini of Okazaki fragments isolated from viral DNA undergoing replication in monkey cells. These ^{32}P-DNA fragments were then annealed to cloned single-stranded DNA sequences, cut at a unique restriction site, and sequenced in order to identify the nucleotide position in the DNA template that corresponded to the ^{32}P-labeled nucleotide in the Okazaki fragment. The methods used allowed them to map the locations of both rN-p-dN covalent linkages (see figure, panel B) and 5′-ends of RNA primers that were located six to nine nucleotides upstream of the 5′-end of the nascent DNA chains. The results demonstrated that discontinuous synthesis occurred predominantly, if not exclusively, on only one arm of replication forks. In fact, initiation events within the replication origin were indistinguishable from initiation of Okazaki fragment synthesis at replication forks throughout the genome. Moreover, leading-strand DNA synthesis began only on one template strand within the replicator core, usually at T-antigen (T-ag)-binding site-2 (the ORE), despite the fact that T-ag-binding site-2 is a perfect palindrome, and therefore the same potential initiation signal exists on both template strands. Thus, only one copy of DNA polymerase-α•DNA primase is loaded at the replicator, and that copy has a unique orientation relative to the replicator.

The initiator protein encoded by SV40 is a DNA helicase called large tumor antigen (T-ag) that binds to specific DNA sequences in the presence of ATP and then unwinds DNA with concomitant ATP hydrolysis. Borowiec and Hurwitz identified structural changes in the replication origin induced by binding of T-ag to SV40 origin DNA *in vitro* by treating the protein–DNA complex with chemicals that detect modified forms of DNA and with enzymes that digest ssDNA. They detected an 8 bp region of melted DNA within the early palindrome region of the SV40 replication origin that was not dependent on either ATP hydrolysis or binding of T-ag to site-1. Melting was not detected in the

A:T-rich element, although a conformational change in the dsDNA was indicated. Thus, both the early palindrome and A:T-rich domains exhibited helical instability, despite their dissimilar A+T compositions. However, Lin and Kowalski showed that origin activity correlated inversely with the helical stability of mutations within the early palindrome domain but not within the A:T-rich domain or the T-ag-binding domain. A specific subdomain of the early palindrome coincided with the site identified as melted after T-ag binds to the replicator *in vitro* and with the OBR identified *in vivo*. Mutations in this subdomain indicated that its helical instability was required to facilitate T-ag-induced melting of the replication origin and initiation of DNA replication. Therefore, this site was the DNA-unwinding element (DUE).

Taken together, these results suggested a sequence of events. T-ag binds to site-2 within the replicator core sequence and begins melting the replication origin at the DUE. When the amount of ssDNA is sufficient to accommodate DNA polymerase-α•DNA primase, RNA-primed DNA synthesis begins on only one of the two template strands within the replicator core. Although DNA synthesis can begin at several preferred sites within the core region, most of the time it begins within T-ag-binding site-2. The second initiation event, the one for leading-strand synthesis in the opposite direction on the complementary template, does not occur until DNA unwinding proceeds downstream into the area of T-ag-binding site-1.

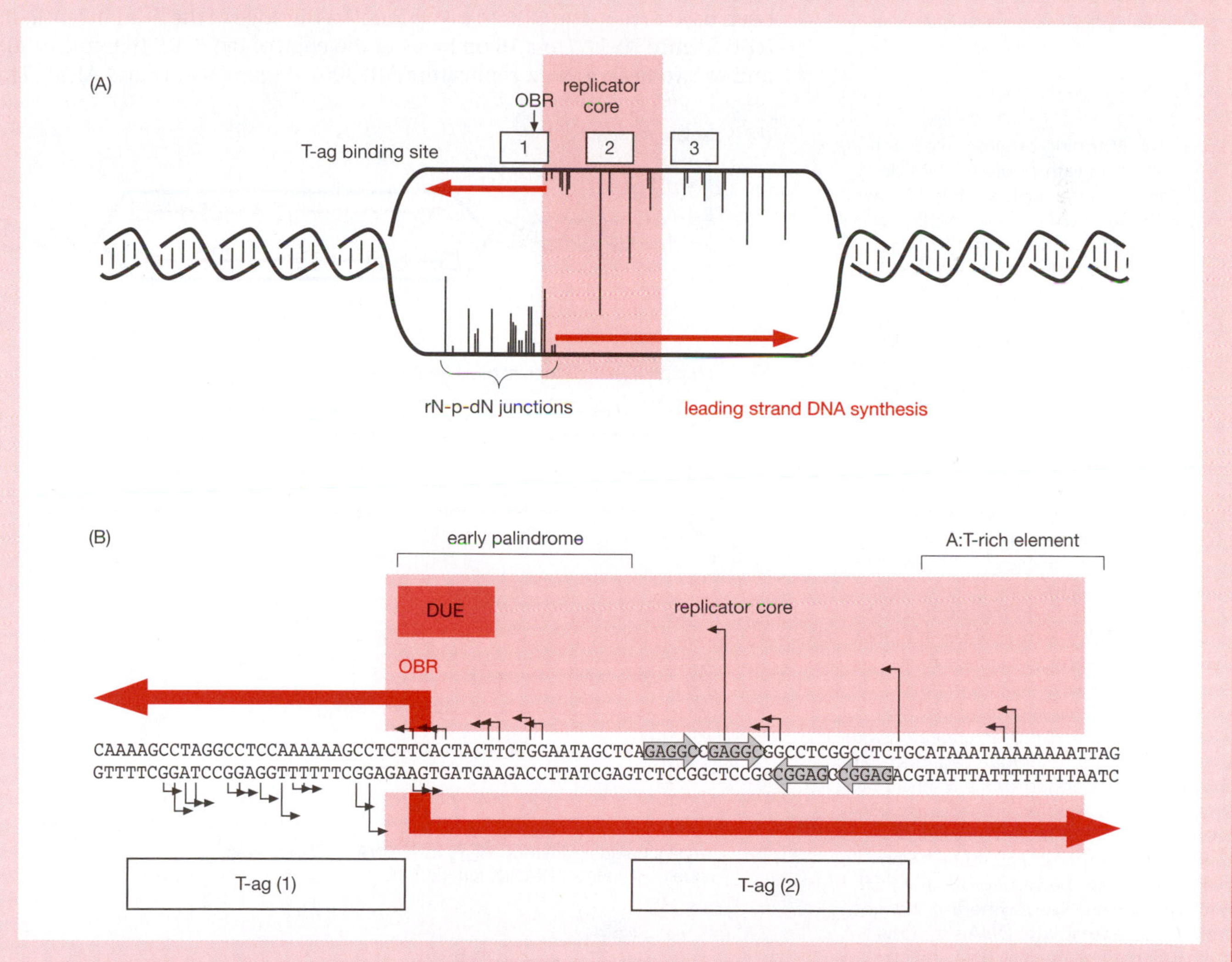

The SV40 replicator. (A) The transition between the presence of rN-p-dN covalent linkages and their absence marks the transition between discontinuous DNA synthesis (Okazaki fragments) and continuous DNA synthesis (leading strand). The replicator core sequence and the three T-ag binding sites are indicated. The origin of bidirectional replication (OBR) is defined as the transition between discontinuous and continuous DNA synthesis that occurs on each strand of the origin of DNA replication. (B) In the SV40 genome, the OBR maps to a two-nucleotide site at position 5210–5211 in the DUE. The DUE corresponds to the 8 bp site that melts when T-ag binds to site-2 in the presence of ATP. The nucleotide locations of the dN in rN-p-dN junctions in each DNA template strand are indicated as bars whose length reflects their relative abundance. Arrowheads indicate their direction of synthesis. (Data from Borowiec, J.A. and Hurwitz, J. (1988) *Proc. Natl. Acad. Sci.* USA 85: 64–68, Hay, R.T. and DePamphilis, M.L. (1982) Cell 28: 767–779, and Lin, S. and Kowalski, D. (1994) *J. Mol. Biol.* 235: 496–507.)

reported for the *E. coli* replicator, suggesting that the model for replicators in Figure 9-1 applied to cells as well as to viruses.

However, given the paradoxical results that arose in studies of eukaryotic replication origins, the question as to whether or not this model applied to eukaryotic replications remained to be addressed. In 1998–1999, Bielinsky and Gerbi developed a new method for mapping RNA-p-DNA chains (Figure 9-2) that was sensitive enough to detect OBRs in eukaryotic genomes (Bielinsky and Gerbi, 1999). Two critical innovations were the use of λ-exonuclease to destroy all nonnascent DNA fragments, and the use of a heat-resistant DNA polymerase to allow repeated cycles of primer annealing and extension to amplify the signal. λ-Exonuclease digests ssDNA from 5′→3′ termini, but it does not digest RNA. Therefore, λ-exonuclease does not digest RNA-primed nascent DNA strands. Using this method, they detected the same OBR at the SV40 replication origin described in Box 9-1, and then went on to identify a strikingly similar OBR at one of the many replicators in the budding yeast genome (Box 9-2). Similar strategies have been used to map the OBR to a 4 bp locus at the edge of the human lamin B2 ORE (Figure 10-17), to a 16 bp locus at the edge of PyV ORE (Figure 10-5B), and to two fission yeast replicators ARS3001 (Figure 10-11) and ARS1. The

(A) isolate replicating DNA

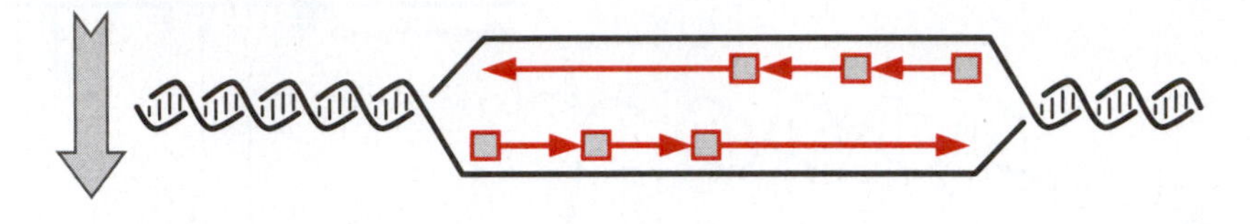

(B) phosphorylate 5′-termini of all DNA chains

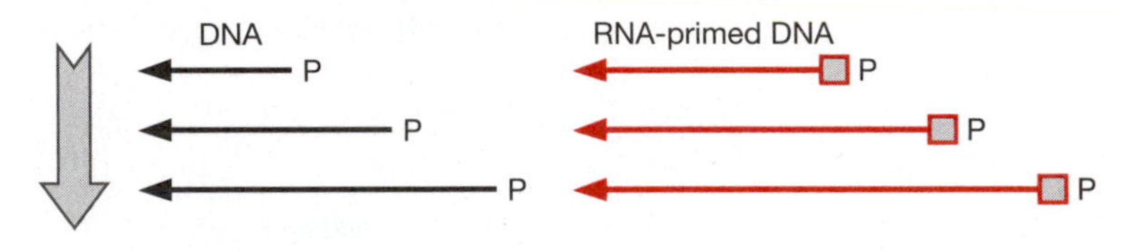

(C) digest 5′-P-DNA

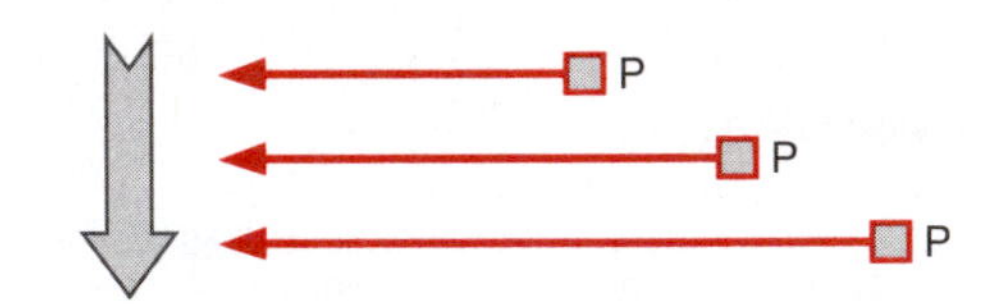

(D) anneal unique primer (^{32}P-□) to 5′-RNA-p-DNA chains
extend primer to dN-p-rN junction

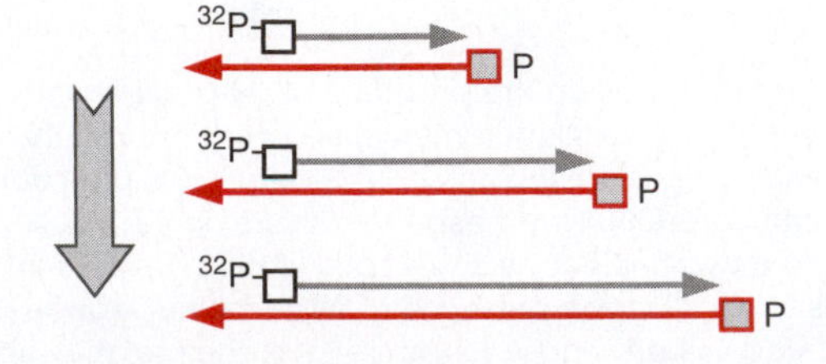

(E) determine sequence
of ^{32}P-labeled DNA chains (^{32}P-□——▶)

Figure 9-2. Mapping origins of bidirectional DNA replication with single-nucleotide resolution using the replication initiation point (RIP) method. (A) Newly replicated DNA is partially purified from the huge background of unreplicated DNA by virtue of its affinity for benzoylated naphthoylated DEAE cellulose, a resin that binds ssDNA more tightly than dsDNA and therefore selectively binds replication forks. Red lines are nascent DNA. Black lines are old DNA, some of which is broken or nicked during purification. Arrows indicate the direction of DNA synthesis (5′→3′). Red squares indicate RNA primer. (B) The partially purified replicating intermediates are then treated with phage T4 polynucleotide kinase (T4 PK) and ATP to ensure that 5′-ends of all DNA chains are phosphorylated and therefore sensitive to digestion with λ-exonuclease. (C) DNA chains without RNA primers are then removed by digestion with λ-exonuclease. (D) The λ-exonuclease-resistant ssDNA (RNA-p-DNA chains) is then annealed with a ^{32}P-DNA primer of unique sequence, and this primer is extended to the 5′-end of the DNA template using a heat-resistant DNA polymerase so that multiple cycles of primer annealing, extending, and release from the template can be carried out in a PCR machine. Since DNA polymerases cannot use RNA as a template, DNA synthesis stops when it reaches the dN-p-rN junction on the 5′-end of the nascent DNA strand. (E) The nucleotide location of the 3′-end of the ^{32}P-PCR product is determined by comparing its length to the lengths of ^{32}P-PCR products arrested randomly by ddNTPs when the same primer is used to sequence this region of DNA by the Sanger ddNTP method. (Data from Gerbi, S.A. and Bielinsky, A.K. (1997) *Methods* 13: 271–280.)

Box 9-2. Mapping Initiation Sites for *de novo* DNA Synthesis in Budding Yeast Chromosomes

Replication origins in yeast genomes have been mapped with single-nucleotide resolution using the replication initiation point (RIP) method developed by Bielinsky and Gerbi (Figure 9-2). The improved sensitivity of the RIP method then allowed them to identify the OBR in ARS1, a replication origin in budding yeast genomes. In the context of the yeast chromosome, ARS1 exhibited an OBR spanning a 2 bp locus between the ORE and the DUE (see figure, panel B). In the context of a circular episome, the same OBR appears but it spans an 18 bp locus at the intersection of the ORE and DUE (panel C). Thus, initiation-site selection in yeast is determined by DNA sequence, but other factors, perhaps chromosome structure, affect the sites where DNA synthesis begins. Moreover, the yeast used in these experiments were deficient in DNA ligase I activity, the enzyme required to join the 3′-end of the Okazaki fragment to the 5′-end of the leading strand. These cells replicated their genome, but did not join Okazaki fragments to the leading strand. Leading-strand synthesis in ARS1 initiated at only a single site, the OBR. Thus, DNA replication in *S. cerevisiae* begins at precisely defined sites within the genome.

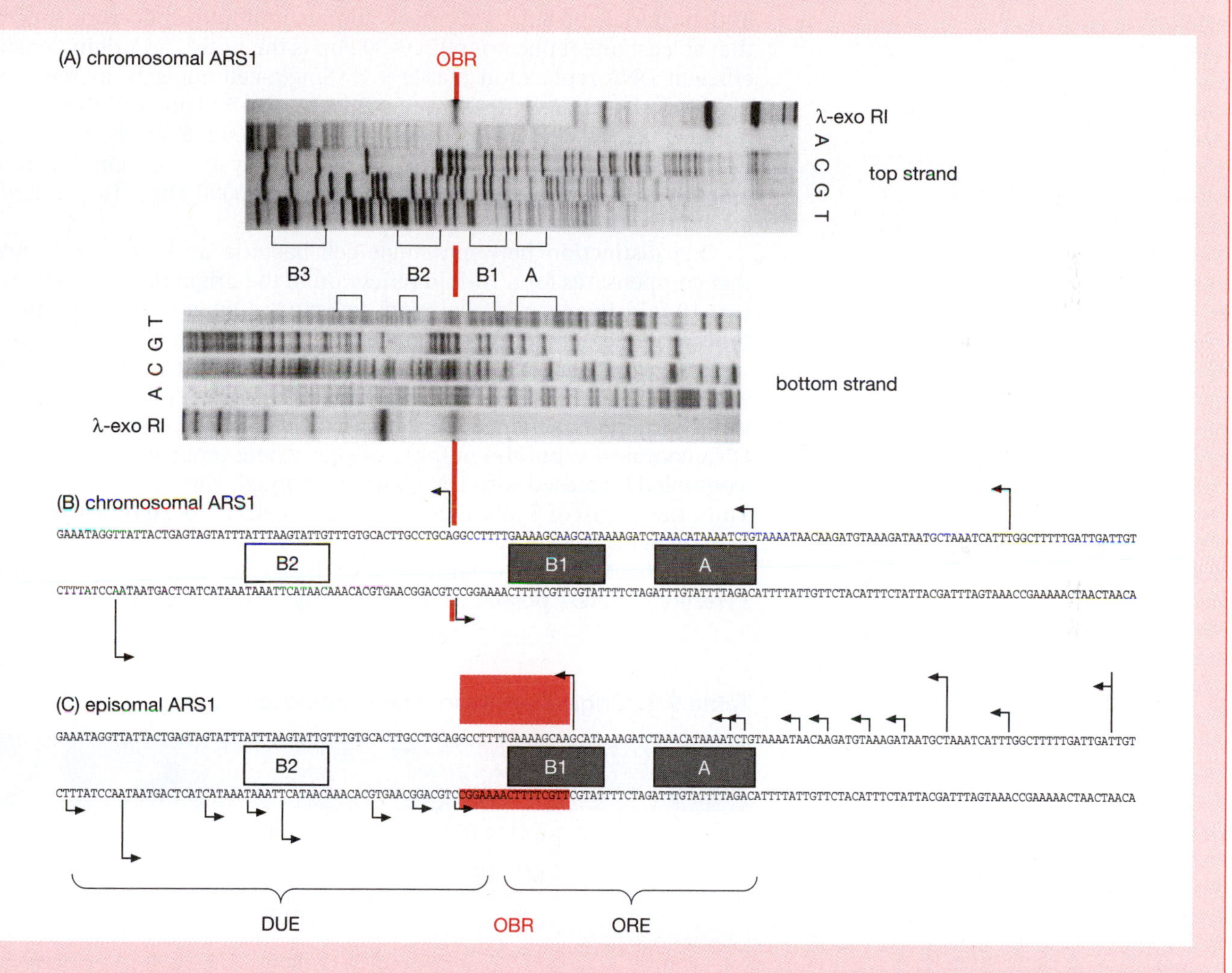

Mapping the OBR in ARS1, a replicator in the genome of the budding yeast *S. cerevisiae*, using the RIP method (Figure 9-2). (A) The ^{32}P-labeled primer extension products were fractionated by polyacrylamide gel electrophoresis in parallel with Sanger sequencing reactions carried out on another sample of ARS1 DNA using the same primers [ddATP (A), ddCTP (C), ddGTP (G), or ddTTP (T)]. Each ^{32}P-labeled band therefore denotes an individual DNA start site that can be determined from the adjacent DNA sequence information. The long blank region beneath the ladder of bands in the λ-exonuclease (λ-exo)-treated replicating intermediate (RI) DNA revealed the region of continuous DNA synthesis. The transition between continuous and discontinuous DNA synthesis (OBR) was marked by the smallest detectable ^{32}P-band on the top and bottom strands of the replication origin (red line). ARS1 elements A, B1, B2, and B3 are indicated for orientation. Elements A+B1 constitute the ORE, and element B2 lies in the center of the DUE. (B) Start sites for DNA synthesis (rN-p-dN junctions) within the chromosomal ARS1 replication origin. (C) Start sites for DNA synthesis with ARS1 in a stable episome (plasmid). The sequences flanking those shown here differed between chromosome and episome. Arrowheads indicate the direction of DNA synthesis. The length of each bar is proportional to the intensity of its band in the gel. (Adapted from Bielinsky, A.K. and Gerbi, S.A. (1999) *Mol. Cell* 3: 477–486. With permission from Elsevier.)

striking similarity of these results suggests that the replicon model in Figure 8-1 and the replicator model in Figure 9-1 applies to all dsDNA genomes with internal replicators.

Origin Density

The length of time required to duplicate a single chromosome is determined by the number of functioning replication origins it contains (**origin density**) and the velocity of their replication forks. Is there an optimal origin density, and is it different for viruses and the cells they infect? All viral genomes require a minimum of one replicator per chromosome to compete with the much larger genomes of their host cells for DNA replication factors. Some bacteriophage genomes as large as 50 kbp are housed in a single chromosome with a single replication origin (Table 9-1). Larger genomes in both bacterial and eukaryal viruses contain multiple replicators, suggesting that at least one replicator per 40–50 kbp is the optimum origin density for efficient DNA replication (Table 9-1). Single-cell eukarya, as represented by yeast, also contain about one origin per 40–50 kbp, and the density in multicellular eukarya, as represented by mammalian cells, is about one per 100 kbp (Table 9-2). Remarkably, the density of origins in bacteria and archaea is 10 times less (one origin per 1000–5000 kbp). How might one account for the variation?

One distinction between single-cell bacteria and single-cell eukarya that compensates for a 10-fold reduction in the origin densities of bacteria is a 10-fold increase in their fork velocities (Table 9-2). Thus, the total time required for genome duplication (S phase) in these organisms is remarkably similar. As the lengths of DNA molecules continued to increase with evolutionary time, the velocity at which their replication forks could travel remained unchanged or even reduced as a consequence of increased DNA complexity, but the number of sites where replication forks could be assembled increased with DNA length, although clearly not in proportion. Thus, the length of S phase in mammalian cells is 10–20 times longer than S phase in bacteria or yeast.

In archaea, there appear to be two distinct modes of replication used. *Pyrococcus abyssi* possesses a single origin of replication in its 1.8 Mb

Table 9-1. Origin Density In Viral Chromosomes

Virus	Genome	Genome size (kbp)	Origins	Origins kbp^{-1}
Bacteriophages	ϕX174 RF[a]	5.4	1	1/5
	M13 RF	6.4	1	1/6
	ϕ29	19.3	2	1/10
	λ	48.5	1	1/49
	T7	40	1	1/40
	T4	169	4	1/42
Animal viruses	SV40, PyV	5.2	1	1/5
	Papillomavirus	7.9	1	1/8
	Adenovirus	30–38	2	1/15–19
	HSV	152	3	1/51
Plant viruses	Geminivirus	2.5–3.1	1	1/2.5–3.1

[a]RF is 'replicative form'.

Table 9-2. Origin Density In Cellular Chromosomes

Organism	Genome	Genome size (Mbp)	Origins	Origin density	Fork velocity (kb min^{-1})	S phase (min)	Origins required per S phase[a]
Bacteria	*E. coli*	4.6	1	1/4.6 Mbp	60	40	1
	B. subtilis	4.2	1	1/4.2 Mbp	40	40	1
Archaea	*Aeropyrum pernix*	1.7	2	1/0.8 Mbp			2
	S. solfataricus	3.0	3	1/1 Mbp	4.8	90	3
	Pyrococcus abyssi	1.8	1	1/1.8 Mbp	20	45	1
Yeast[b]	*S. cerevisiae*	12.2	247–429	1/35 kbp 1/49 kbp	2.9	36	60
	S. pombe	13.8	272–426	1/20 kbp 1/32 kbp	2.8	24	103
Mammals[c]	*H. sapiens*	3080	~29 000	1/106 kbp	1.5	480	2138

[a]The number of bidirectional replication origins required to account for the observed S phase using the mean replication-fork velocity.

[b]The same replication origins are used during both mitotic and meiotic cell cycles (Feng *et al.*, 2006; Heichinger *et al.*, 2006).

[c]Replication origins have been mapped throughout a 30 Mbp region in the human genome by measuring the relative abundance of RNA-primed DNA chains using microarrays containing overlapping DNA sequences. The results reveal 283 replication origins distributed in clusters for an average of one origin every 106 kbp (Berezney *et al.*, 2000; Cadoret *et al.*, 2008).

chromosome and its replication forks move at approximately 20 kbp min^{-1}. In contrast, the larger genomes of *Sulfolobus* species (between 2.2 and 3 Mbp) are replicated from three origins. However, the *Sulfolobus* forks are considerably slower, moving at 4.8 kbp min^{-1}.

Taken together, these data suggest that the product of the density of bidirectional replication origins (O_D) and the replication-fork velocity (F_V) is a replication constant (Table 9-3). They support the hypothesis that large genomes require some minimum number of replicators per chromosome to efficiently duplicate the genome within a reasonable period of time. Too few replicators increases the probability that replication forks will fail as the distance over which they operate increases, thereby resulting in DNA damage and possibly cell death. The so-called MCM paradox (discussed below) supports the notion that too many replicators also pose a threat to genome stability by increasing the probability that failures will occur during the initiation process. Nature appears to select for an optimum origin density.

Table 9-3. Origin Density × Fork Velocity = Replication Constant

	O_D (origins kbp^{-1})	F_V (kbp min^{-1})	Replication constant (origins min^{-1})
Bacteria (*E. coli*)	1/4400	50	~0.01
Archaea (*Sulfolobus*)	1/1400	12.4	~0.01
Eukarya (yeast)	1/34	2.85	~0.08
Eukarya (mammals)	1/106	1.5	~0.01

F_V, replication-fork velocity; O_D, density of bidirectional replication origins.

Origin Usage

The time required for genome duplication (S phase) depends on the number of replication origins per genome and replication-fork velocity (Mbp per min per fork). Thus, the number of active replication origins can be calculated from the length of S phase, the replication-fork velocity, and the assumption that all replication origins act bidirectionally.

$$\text{S phase (min)} = \left[\frac{\text{No. of Mbp}}{\text{Genome}}\right] \times \left[\frac{\text{Genome}}{\text{No. of origins}}\right] \times \left[\frac{\text{1 Origin}}{\text{2 Forks}}\right] \times \left[\frac{\text{(Fork) (min)}}{\text{No. of Mbp}}\right]$$

Cellular organisms exhibit an inverse relationship between origin density and fork velocity (Table 9-2), suggesting that origin density increased as genomes became larger and chromatin structure made DNA unwinding more difficult. The fraction of origins that are active during S phase is 100% in bacteria, 100% in archaea, 15–25% in *S. cerevisiae*, 25–40% in *S. pombe*, and approximately 7% in mammals (Table 9-2). For eukarya, origin usage is generally estimated from the ratio of bubble arcs to fork arcs in two-dimensional gel electrophoresis patterns (Figure 8-3). Although clusters of origins may fire together – the equivalent of a single strong origin – there clearly is an excess of replication origins in the genomes of eukarya. One explanation is that when two or more yeast replicators are placed in the same plasmid DNA molecule, only one is active during each cell-division cycle, demonstrating that the rate-limiting step in genome duplication is initiation; once replication forks are assembled and activated, DNA replication is rapid and the forks can suppress initiation events from neighboring origins.

Active origins in budding yeast fire on average once every two cell cycles, but active origins in fission yeast fire less than once in 10 cell cycles, and the frequency in mammals is even less. Some replicators in yeast chromosomes are active each time a cell divides whereas others are active in fewer than 10% of cell divisions. Similarly, most of the 29 000 or so origins in the human genome are not used when a single cell divides. These results simply reflect the fact that replicators within a single species of eukarya are not identical; those in *S. cerevisiae*, for example, share only approximately 15% of their sequences in common, and those either in *S. pombe* or in human cells lack a required consensus sequence. What is observed by the experimenter is origin usage within a population of cells: some cells use one origin, whereas other cells choose another origin. Thus, eukarya contain a higher density of origins than necessary for genome duplication, and both their sequence and efficiency varies considerably. Analysis of *S. pombe* replication origins suggests that differential recruitment of ORC to origins during mitosis followed by competition among origins for limiting replication factors determines the frequency of origin usage.

Budding yeast chromosomes are very stable, with loss rates in the range of 10^{-5} loss per cell division. To determine the number of replicators required to maintain chromosomes during cell division, Carol Newlon and colleagues systematically deleted the 19 ARS elements in chromosome III of *S. cerevisiae*. Chromosome III consists of 315 kbp of DNA that contains seven efficient replicators, three inefficient replicators, and nine ARS elements that function as replicators in plasmids but not in their native habitat. Deletion of single ARS elements, as well as combinations of double, triple, and quadruple ARS elements, had no detectable effect on chromosome stability. Not until the five efficient replicators were deleted was the loss rate of this chromosome increased 2 to 3-fold, revealing that yeast chromosomes contain more replicators than are needed to ensure genome duplication. The data suggest that as few as one replicator per 225 kbp is sufficient to replicate a yeast chromosome, consistent with the measured fork rate of 2.9 kb min^{-1} during an S phase of approximately 55 min. Intriguingly, the nine silent ARS elements were required to efficiently duplicate the chromosome when all of the efficient replicators were deleted, revealing the presence of a hierarchy of replicators.

These results reveal two remarkable features of replicons. First, a single active replicator in 315 kbp of DNA is sufficient to maintain a eukaryotic chromosome. Therefore, the density of replication origins in Table 9-2 represents the optimal origin density, not the minimal origin density. Second, some replicators are primary targets of initiator proteins whereas others lie dormant until conditions exist in which initiators seek them out for activation. When the primary replicators are unavailable, the less efficient 'secondary' replicators are utilized. In other words, eukaryotic genomes contain many more replicators (or potential replicators) than utilized during normal cell-division cycles.

Origin Timing

Metazoan genomes, such as those of flies and mammals, are regulated at the level of large chromosomal domains (0.5–5 Mbp in mammals) within which clusters of replicons are activated coordinately. These **replicon clusters** are activated in a specific temporal order during S phase that is a stable property of specific cell types. The fact that the same chromosomal segments are replicated at the same time during S phase each time the cell divides defines these replicon clusters as a second, higher-order replication unit. Replicon clusters that are active at the beginning of S phase tend to reside in the interior of the nucleus, whereas replicon clusters that are activated later during S phase are enriched in perinucleolar regions and the nuclear periphery. Furthermore, chromosomal loci that replicate early in either flies or mammals tend to be GC-rich regions and contain actively transcribed genes, whereas those that replicate late tend to be A:T-rich regions and contain silent genes.

The temporal order of DNA replication appears to be established by the distribution of heterochromatic and euchromatic chromatin in interphase nuclei which, in turn, are established during the transition from mitosis (M phase) to G_1 phase of the cell-division cycle. Actively transcribed regions of the genome generally, but not always, correspond to euchromatic regions, whereas regions that are transcriptionally inactive generally, but not always, correspond to heterochromatic regions. More direct evidence comes from the fact that one of the two X chromosomes is inactivated during cell differentiation by changes in its chromatin structure, a process that silences most of its genes and causes the inactivated chromosome to replicate late during S phase. Selective silencing of one of the two alleles of a specific gene in mammalian chromosomes (termed imprinting) results in the expressed allele replicating earlier than the silent allele. Overall, the DNA replication-timing profiles of the two different human cell types are remarkably similar with differences reflecting the differential expression of specific genes. Thus, chromatin structure and nuclear organization appear to be the dominant parameters in determining the temporal order of genome duplication. Whether a temporal order of genome duplication is a critical component of metazoan genome duplication that ensures that some genes are duplicated before other genes remains to be determined. It may simply be a consequence of genome organization. Some chromosomal loci may be more accessible to one or more replication factors than other loci, and the process of DNA replication may simply alter the microenvironment of inaccessible loci, thereby making them more accessible.

Short *cis*-acting sequences have been identified in yeast that can cause replication origins to activate late during S phase when they reside in close proximity to the origin. For example, origins close to telomeres replicate late. These sequences appear to inhibit loading of replication proteins, since analysis of *S. pombe* replication origins revealed that inefficiently used origins also replicate late in S phase. However, in contrast with metazoa, yeast do not exhibit a correlation between replication timing and gene expression. In fact, analysis of replication of specific chromosomes in either budding yeast or fission yeast using DNA combing and fluorescence microscopy at approximately 1 kbp resolution (Figure 8-4) reveals that no two molecules undergo the same pattern of DNA replication. Each segment of a chromosome replicates independently of all other segments of the same chromosome. Therefore, despite the fact that yeast replication origins are determined by specific DNA sequences (replicators), the temporal order in which these replicators are activated is a stochastic process rather than an ordered one. This simply means that each replicator has a certain probability of initiating replication at a certain time during S phase, but the order in which replication origins are activated is not predetermined. Thus, some origins are generally activated early during S phase and some late, but not every cell fires the same origin at the same time during each cell-division cycle.

DEVELOPMENTAL ACQUISITION OF SITE-SPECIFICITY

In 1974, David Hogness and Harold Callan reported a remarkable discovery. Using electron microscopy (Figure 8-2) and fiber autoradiography (Chapter 8, Box 8-1) they found that the density of replication origins throughout the chromosomes of flies, newts, and frogs decreased significantly as rapidly cleaving nuclei in the fertilized eggs of these organisms developed into adults. At the beginning, replication bubbles were closely spaced, but as cells differentiated the number of active replication origins in the genome decreased dramatically. With the development of origin-mapping techniques, it soon became clear that site-specific DNA replication in metazoa is developmentally acquired.

Fertilized *Xenopus* eggs undergo 12 rapid cleavage events before DNA transcription begins. During this time, DNA replication exhibits little, if any, dependence on DNA sequence or composition. Virtually any dsDNA molecule injected into unfertilized eggs or incubated in an extract from artificially activated eggs will initiate replication with little dependence on DNA sequence. Replication bubbles in these molecules are distributed uniformly throughout the DNA with a periodic spacing of about 10 kb. As development proceeds, however, some initiation sites disappear while others are enhanced, resulting in a clear pattern of preferred initiation sites at specific genomic loci. For example, the endogenous ribosomal RNA genes are not transcribed during the rapid cleavage stages, and DNA replication in ribosomal gene loci initiates and terminates at 9–12 kbp intervals with no obvious dependence on specific DNA sequences. Transcription of these genes, which begins during the late blastula stage, is accompanied by specific repression of replication initiation within transcription units; the frequency of initiation within intergenic spacers remains as high as in early blastula. Thus, circumscribed zones of replication initiation emerge in intergenic DNA regions during the time in metazoan development when the chromatin is remodeled to allow gene transcription. The relative levels of nascent-strand abundance throughout ribosomal RNA gene loci in mammalian cells reveal that initiation events exhibit a strong preference for the promoter region.

A similar story exists in *Drosophila* where blastoderm embryos undergo 13 rapid nuclear divisions within a syncytium before transcription begins. In the early stage of *Drosophila* embryogenesis, DNA replication initiates at unspecified sites in the chromosome. In the DNA polymerase-α gene locus, repression of origin activity in the coding region is detected soon after formation of cellular blastoderms. By 5 h after fertilization, only four initiation sites are detected in the 65 kb locus between the DNA polymerase-α and E2f genes, and only two of these are detected in cultured *Drosophila* cells.

Developmental changes in origin specification also occur during mammalian development. A region of 450 kb in the IgH locus is devoid of initiation events prior to B-cell development, but initiation events appear in both the expressed and nonexpressed allele soon after B-cell development begins. As with flies, initiation site-specificity increases with cell differentiation. For example, a 200 kbp region at the human β-globin gene and a 500 kbp region at the mouse IgH gene are both replicated from a single initiation locus.

These studies reveal that initiation of DNA replication can change from stochastic to site-specific as fertilized eggs develop into adult organisms. Therefore, the selection of initiation sites in metazoan organisms depends on epigenetic as well as genetic factors.

PARADOXES AND SOLUTIONS

The Replicator Paradox

The genomes of all bacteria, bacteriophages, bacterial plasmids, archaea, single-cell eukarya, mitochondria, animal viruses, and plant viruses examined so far contain replicators that are required for efficient genome duplication. These replicators impart replication activity when transferred to a foreign DNA molecule, and their activity is altered or abolished by small deletions or insertions or even single-base-pair changes within them. Then, imagine the surprise when, in the early 1980s, Ron Laskey and coworkers demonstrated that most, presumably all, DNA molecules could replicate when either injected into frog eggs, or incubated in frog egg extracts! The only sequence requirement identified so far is a preference for asymmetric A:T-rich regions (adenines on one strand bonded to thymines on the other). Similar results were subsequently obtained with eggs from fish, sea urchins, and flies, suggesting that replicators were not required for genome duplication in metazoan cells.

In apparent contradiction to these results, DePamphilis and coworkers demonstrated that DNA plasmids injected into mouse eggs or into the nuclei of mouse preimplantation embryos do not replicate unless they contain a viral replicator that functions in mouse cells and are provided with the cognate viral initiator protein. Furthermore, origin-mapping experiments carried out in a number of different laboratories revealed that replication does not begin randomly throughout the genomes of mammalian cells, but at specific chromosomal loci. The nature of these loci, however, has remained elusive. Some are as small as 0.5 kbp, while others are as large as 50 kbp. Some can be translocated to other chromosomal sites where they function as replication origins, and their activity is sensitive to internal genetic alterations. Others, however, are insensitive to internal genetic alterations, but strongly influenced by external or even distal genetic alterations. These results suggest that replicators are determined by **sequence context** as well as by sequence content.

Attempts to isolate autonomously replicating sequences from mammalian cells have yielded mixed results. Despite some notable successes, most attempts either failed or yielded replicator sequences whose activity was only marginally above the background. One promising approach introduced by Michele Calos was the oriP replicator assay described above. This assay revealed that many human DNA fragments could support plasmid replication as long as they were 10–15 kbp in length. Bacterial DNA sequences were significantly less effective, suggesting that autonomous DNA replication in human cells is stimulated by multiple sequence features. Subsequently, Hans Lipps and coworkers constructed a plasmid that mimics the oriP replicator assay, but without the need for either viral replicator components or viral initiator proteins. pEPI (Figure 10-19) requires only a nuclear retention sequence that prevents plasmid loss when cells undergo mitosis, and an active promoter that drives transcription through the nuclear retention sequence. DNA replication of pEPI is entirely dependent on the cell's replication machinery. Initiation occurs at many sites throughout the plasmid. These results suggest that replication origins in metazoan cells are determined primarily by epigenetic factors.

Thus, the question remains: do the genomes of multicellular organisms contain replicators? The answer is yes, but only if the definition of a replicator is expanded to include a passive as well as an active role, and to include epigenetic as well as genetic factors. Replicators can be classified into three groups: active replicators that are required both for initiator binding and for initiator function, passive replicators that facilitate initiator binding but are not required for initiator function, and conditional replicators or passive replicators that are strongly dependent on epigenetic factors.

Active replicators

Active replicators are specific *cis*-acting DNA sequences that are required both for the binding and function of its cognate initiator protein. For example, six molecules of the SV40 initiator protein large T-ag must be assembled into a hexameric complex directly on the SV40 replicator before T-ag can function as a DNA helicase and unwind the origin DNA (Figures 11-2 and 11-3). The same is true for the closely related PyV genome (Figure 10-5). However, SV40 T-ag cannot activate the PyV replicator, and PyV T-ag cannot activate the SV40 replicator. In other words, T-ag cannot initiate DNA unwinding at internal DNA sites in the absence of its cognate replicator. The singular advantage of active replicators is their rigid specificity: they ensure that replication begins at a specific genomic locus (the replicator itself). This advantage accounts for their popularity among the genomes of bacteria, bacterial plasmids, bacteriophages, viruses, and mitochondria (Table 8-1). The replicator in bacteria determines the direction of fork movement through bacterial genes and where replication termination will occur, an event that is linked with cell division. The replicator in plasmids, viruses, and cellular organelles allows these comparatively small DNA molecules to compete effectively for replication proteins encoded by the large genomes of their cellular host.

Passive replicators

Passive replicators are DNA binding sites for initiator proteins, but they are not required for initiator protein function; the initiator simply must be attached to an internal DNA site to initiate DNA replication. Passive replicators contain one or more specific DNA sequence motifs that are required for initiation of DNA replication, and that allow them to be identified on the basis of DNA sequence alone. However, passive replicators lack the rigid DNA sequence requirements of active replicators. Examples of passive replicators are the hairpin structures in ssDNA genomes of some bacteriophages and viruses (Figures 8-8 and 8-9), and ssDNA replication intermediates of some plasmids (Figure 8-6). In eukarya, passive replicators are represented by oriL, oriL′, oriL″, etc. of mtDNA (Figure 8-5), by the replication origins in yeast chromosomes (Figures 10-10 and 10-11), and by gene amplification origins that are active only in specific tissues at specific times during development (Figures 10-20 and 10-21). Passive replicators provide several advantages. They recruit initiator proteins to the DNA substrate. They can do so at specific sites distributed throughout the genome. If low concentrations of initiator protein were allowed to initiate replication randomly throughout the genome, then some regions would be replicated efficiently, while other regions would not. They allow separation of initiation sites for DNA replication from sites for DNA transcription. This helps to prevent collisions between replication forks and transcription forks. They do not require a unique DNA sequence. Thus, they allow evolution greater flexibility in selecting initiator proteins.

The conclusion that passive replicators are not required for initiator protein function arises from several observations. Yeast replicators, for example, are not required for chromosome replication. When all of the ARS elements are deleted from a 142 kbp budding yeast chromosome fragment, it is lost from the population about 200 times more frequently than its wild-type counterpart. Nevertheless, it continues to replicate and segregate correctly in 97% of cell divisions, as long as its ends are attached to telomeres and the cell provides the initiator proteins ORC and Cdc6. Thus, yeast replicators are clearly advantageous in promoting efficient chromosome duplication, but they are not required for assembly and activation of a yeast replication complex. Similarly, artificial attachment of human initiator proteins (either ORC or Cdc6) to DNA in human cells is sufficient to create an origin of replication. Presumably, the same would be true for the yeast orthologs. Initiator proteins that recognize passive replicators can initiate replication at a wide variety of DNA sequences, although the efficiency at which they do so is greatest at asymmetrical A:T-rich sequences. This is true for dsDNA molecules in frog egg extracts, and it is true for binding of ORC to DNA *in vitro*. Consequently, initiator proteins from mammals or flies can substitute for those from frogs to initiate DNA replication in frog egg extracts.

Active and passive replicators are distinguished by their anatomy, the effect of sequence context on their activity, and their sensitivity to regulation. Active replicators have a rigid modular anatomy, whereas passive replicators have a flexible modular anatomy. For example, both viral and yeast replicators are composed of several identifiable sequence elements, but the spacing and orientation of sequence elements in viral origins is far more critical than in yeast origins. Moreover, the same functional modules from different origins within the same yeast are interchangeable, even though they do not share a common sequence.

Passive replicators are sensitive to the nature of the DNA sequences that surround them, whereas active replicators are insensitive. This phenomenon, referred to as sequence context, is well documented but poorly understood. It may be that the sequences on either side of a replicator affect chromatin structure and that chromatin structure can either suppress or stimulate origin activity. Alternatively, differences in the composition or sequence of flanking DNA regions may affect the ability of passive replicators to bind initiators. Thus, not all DNA sequences that exhibit ARS activity in plasmids function as replicators within the context of real cellular chromosomes.

Active replicators can initiate replication many times within a single cell-division cycle, whereas the activity of passive replicators is restricted to no more than one initiation event per cell division. This regulation occurs primarily, if not exclusively, through cell-cycle-dependent modifications of various DNA replication proteins described in Chapter 12. There are, however, two notable exceptions to this rule. Epstein–Barr virus oriP is an active replicator that binds specifically the Epstein–Barr virus initiator protein EBNA1. However, the ability of the oriP–EBNA1 complex to initiate replication is dependent on its association with its human host's initiator proteins (ORC+Cdc6) and thereby becomes subject to the same regulatory controls that govern the replication of its host's genome. The *E. coli* replicator, oriC, also differs to some degree from those in bacteriophages and animal viruses in that oriC is restrained, but not restricted, to initiating replication once per cell cycle. DNA methylation at oriC serves to delay a second initiation event from occurring until the replication forks from the first event can terminate. However, this control is overcome by rich growth conditions, and the replicators in other bacterial genomes are not methylated.

Conditional replicators

Conditional replicators are passive replicators whose activity depends on epigenetic factors, such as chromatin structure, as well as on the sequence or composition of their DNA. Thus, conditional replicators are dormant in some cells and active in others. Like passive replicators, conditional replicators are sensitive to genetic alterations, exhibit replicator activity when translocated to other chromosomal sites, and, in some cases, exhibit ARS activity. Moreover, conditional replicators such as those at the human Lamin B2, TOP1, MYC, MCM4, and FMR1 loci bind ORC preferentially relative to adjacent sequences. Unlike passive replicators, conditional replicators are developmentally acquired and cannot be identified on the basis of DNA sequence alone.

Conditional replicators are of two types. Most serve as S-phase specific origins during mitotic cell cycles (genome duplication followed by cell division) and presumably endocycles (genome duplication in the absence of mitosis) as well. Others serve as amplification origins, unique DNA sequences that amplify specific genes in certain terminally differentiated cells that do not proliferate. Amplification origins differ from S-phase replication origins in that amplification origins can be activated multiple times within a single S phase. Gene amplification occurs only in specific cell types during specific times during the organism's life cycle.

Conditional replicators provide a major advantage to the evolution of species in that the number and locations of replication origins can change as a single fertilized egg develops into an animal consisting of hundreds, perhaps thousands, of different cell types. Thus, changes in genome composition that are necessary for speciation, and developmentally regulated changes in gene expression that are necessary for cell differentiation, can be accomplished without diminishing the organism's ability to initiate DNA replication throughout its genome. Furthermore, the availability of a large number of potential replication origins provides the organism with the opportunity to license additional origins under conditions of environmental stress that cause replication forks to stall. In human cells, for example, excess chromatin-bound MCM helicase (the eukaryotic enzyme required to unwind DNA at replicators and at replication forks) licenses many replication origins that remain dormant during normal DNA replication, but that are activated when replication-fork progression is inhibited.

Genetic factors that determine eukaryotic replicators

DNA sequence

Indirect evidence that mammalian replicators are determined, at least in part, by DNA sequence comes from the simple fact that they map to specific sites in both single-copy and multicopy loci. The same OBR identified in cells containing one copy of a particular genomic locus per genome copy also appears in cells containing many tandem copies of the same locus, and the same OBR is used by different cell lines derived from the same species. Therefore, each copy of the amplified region that initiates replication must use the same OBR. Otherwise, initiation would appear to occur at many different sites within the same DNA locus. Further evidence comes from the existence of specific protein–DNA interactions at replication origins adjacent to the lamin B2, β-globin, and Myc genes in human cells, downstream of the DHFR gene in hamster cells, and at amplification origins in *Drosophila*, *Sciara*, and *Tetrahymena*.

Direct evidence that metazoan replication origins are genetically determined comes from identification of autonomously replicating sequences and from genetic analyses of origin activity. Several reports of ARS elements that function in mammalian cells and cell extracts have been documented in detail and shown to correspond to sites where replication occurs in mammalian chromosomes. The most extensively characterized ARS activity in mammalian cells lies in the promoter region of the human Myc gene. Why all mammalian replicators do not exhibit ARS activity is unknown, but as with yeast, many sequences may exhibit ARS activity in the context of a plasmid that do not exhibit origin activity in the context of a real chromosome. This concern has been addressed by measuring the ability of origins to function at ectopic chromosomal sites. For example, neither the lamin B2, β-globin, nor DHFR replication origins exhibit ARS activity, but

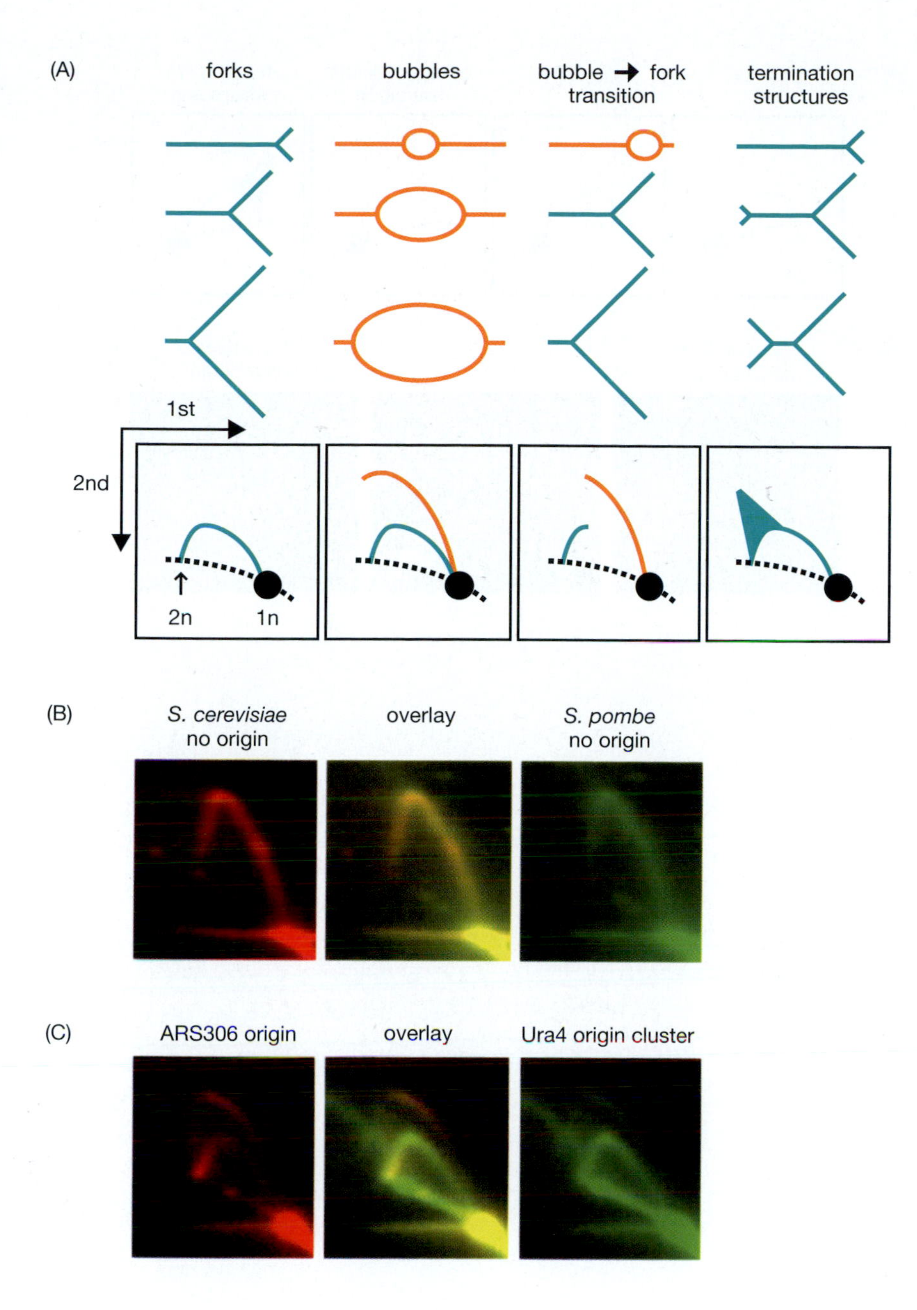

Figure 8-3. Comparisons of DNA replication intermediates in the chromosomes of budding and fission yeast were obtained by mixing the two DNA restriction endonuclease digests before fractionation by neutral-neutral two-dimensional gel electrophoresis. (A) The pattern of migration expected for various dsDNA replication intermediates is illustrated. (B) Comparison of two unique 5.7 kbp DNA fragments that lack a replication origin, one from *S. cerevisiae* and one from *S. pombe*. Only replication forks are detected, and the pattern from *S. pombe* is indistinguishable from the pattern in *S. cerevisiae*. (C) Comparison of a 4.3 kbp site in *S. cerevisiae* that contains a single replication origin (ARS306) with a 4.3 kbp site in *S. pombe* (Ura4) that contains two replication origins called ARS3002 and ARS3004. Replication bubbles are detected at these sites in both organisms, and their structures are indistinguishable. Replication forks also are present. They occur when one of the two forks in a bubble travels past one of the restriction sites (bubble to fork transition). They also occur when forks from other replication origins enter the fragment under examination. Thus, the ratio of bubbles to forks is proportional to the frequency at which the origin analyzed is used during each cell-division cycle. The significant increase in the fraction of forks present in the Ura4 origin cluster is due to the presence of another origin (ARS3003) in the adjacent DNA fragment. The presence of termination structures in the Ura4 cluster results from collision of forks originating from the fact that ARS3002, 3003, and 3004 are all located within a 6 kbp locus. (Adapted from Brun, C., Dijkwel, P.A., Little, R.D. *et al.* (1995) *Chromosoma* 104: 92–102. With permission from Springer.)

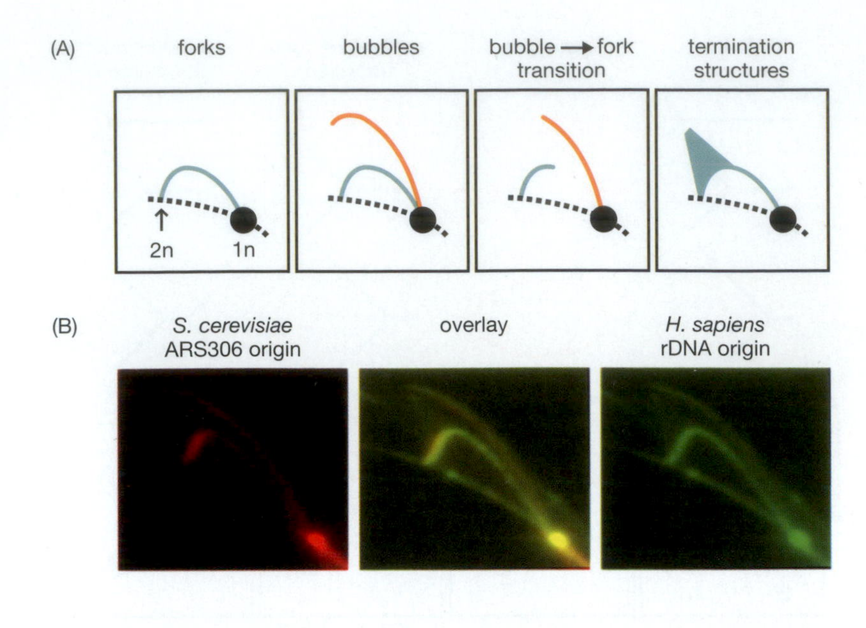

Figure 10-15. Comparisons of DNA replication intermediates in the chromosomes of budding yeast with rDNA in human cells were obtained by mixing the two DNA restriction endonuclease digests before fractionation by two-dimensional gel electrophoresis (Figure 8-3). (A) The pattern of migration expected for various replication intermediates is illustrated. (B) Two unique 8.5 kbp DNA fragments, one from *S. cerevisiae* that contains ARS306 replication origin and one from *H. sapiens* rDNA that contains the primary initiation site (see Figure 10-14A), were analyzed for replication intermediates. Replication bubbles were detected in both organisms, and their structures were indistinguishable. Replication forks were also present. The higher ratio of forks to bubbles at the rDNA replication origin reflects the fact that only a fraction of the rDNA origins are active in each cell, and therefore this region is passively replicated by forks that originated elsewhere. Termination structures are also detected in the rDNA origin region, but their low abundance requires longer exposures of the autoradiogram. (From Brun, C., Dijkwel, P.A., Little, R.D. *et al.* (1995) *Chromosoma* 104: 92–102. With permission from Springer.)

they each exhibit origin activity when DNA fragments from 2 to 16 kbp in size are translocated to other sites in their host's genome. DNA fragments that lack origin activity *in situ* also lack origin activity at ectopic sites, and small deletions within the replicator can eliminate its activity. Such experiments have been done with both S-phase replication origins and amplification origins, leading to the conclusion that sequences as small as those required for fission yeast replicators (~0.5 kbp) are required for metazoan replicators.

Origin activity is also influenced by sequences distal to the site where DNA synthesis begins. Some of these sequences activate replicators and some of them suppress replicators. Examples include the amplification control element 3 in the *Drosophila* chorion gene locus (Figure 10-21), the chick olfactory receptor and insulator sequences (Figure 10-13), locus control regions about 30 kbp distal to the human and the murine β-globin origins, the DHFR gene promoter (Figure 10-16), and ecdysone-response elements (Figure 10-20). The effect of these sequences is epigenetic, because they are active only at certain times during the organism's development or during the cell-division cycle.

Initiator proteins

Unraveling the biochemistry surrounding eukaryotic replication origins began in 1992 with the discovery by Steve P. Bell and Bruce Stillman of the budding yeast ORC, a complex of six different proteins that bind specifically to budding yeast ARS elements. Since then, ORC homologs have been discovered in most, if not all, eukaryotic genomes analyzed to date, endowing them with the role of initiator in eukaryotic genome duplication. Each species has only one set of ORC genes, eliminating the possibility that different initiator proteins arise in different cell types. Surprisingly, however, ORC from human, frog, or fly cells shows little, if any, sequence specificity when binding either to purified DNA or DNA in cell extracts. The only consistent preference is for asymmetric A:T-rich DNA of the type targeted by *S. pombe* ORC. Such sequences are commonly found at eukaryotic replication origins. One explanation is that additional proteins, such as Cdc6, are required for site specificity. All of the Cdc6 proteins characterized from archaea and eukarya contain a DNA-binding domain called the winged helix domain in addition to an ATP-binding site known as the AAA+ domain (Figure 11-7). In fact, site specificity in budding yeast requires both ORC and Cdc6, and formation of stable ORC•Cdc6–DNA complexes requires ATP as well as suppression of ATPase activity in both ORC and Cdc6. Thus, the initiator for DNA replication in budding yeast, and presumably in other eukaryotes as well, actually consists of ORC and Cdc6 together. Intriguingly, *Xenopus* has two Cdc6 genes, one of which is expressed in eggs where initiation-site selection is nonspecific, and then this Cdc6 isoform is replaced by another isoform during gastrulation when site-specific initiation events appear.

Epigenetic factors that determine eukaryotic replicators

Nuclear structure

Site-specific initiation events have been observed *in vitro* only with nuclei whose membrane remains intact and impermeable to large molecules such as proteins. Initiation of DNA replication in permeabilized nuclei, chromatin, or DNA substrates from yeast or mammalian cells invariably occurs at 'randomly' selected sites throughout the genome. Moreover, nuclei assembled in the absence of lamin B3, one of the fibrous proteins that line the inner surface of the nuclear membrane, cannot initiate DNA replication. This sensitivity of initiation-site selection to nuclear structure may reflect the existence of physical attachments between chromatin and nuclear components referred to as matrix attachment sites or nuclear scaffold attachment sites that result in formation of chromatin loops. It may also reflect the need to localize and concentrate specific proteins involved in the initiation process.

Chromatin structure

Chromatin structure that is accessible to transcription factors and therefore allows genes to be expressed also allows replication origins to be active, whereas chromatin structure that represses gene expression also represses origin activity. Thus, increases in histone acetylation loosens chromatin compaction and increases the activities of both transcription promoters and replication origins. In vertebrates, Cdt1 (a protein essential for loading the replicative DNA helicase) assists binding of HBO1 (the primary histone acetyltransferse) to replication origins. This results in localized acetylation of histone H4 which facilitates loading the DNA helicase. In contrast, increases in histone H1 increases chromatin compaction and represses initiation of both transcription and replication. Nucleosomes located close to yeast replicators reduce their ability to initiate replication, and in mammalian cells an ATP-dependent chromatin-remodeling complex is required for efficient replication of heterochromatin, a tightly packed form of chromatin that tends to be transcriptionally silent. The appearance of repressive chromatin structures during frog and fly development correlates with the transition from 'random' initiation events throughout their genomes to locus-specific initiation events. Genomic replication origins have not been mapped in preimplantation mammalian embryos, but an analogous transition occurs during formation of two- and four-cell mouse embryos in which the requirement for enhancers to drive promoters or viral replicators first appears with the loss of histone acetylation and the expression of histone H1.

Transcription factors

Transcription factors can facilitate initiation of DNA replication, either by acting directly to facilitate binding of initiator proteins or indirectly by preventing chromatin structure from masking the ori-core. The clearest examples are animal virus replicators such as SV40, PyV, and papillomavirus, although transcription factor Abf1 facilitates several budding yeast replicators such as ARS1. Transcription factors Myb, E2f, and Rb facilitate amplification origins in flies, and transcription factors Myc and Hox bind to human replication origins where they interact with pre-replication complexes and stimulate DNA synthesis. Whether these proteins facilitate origin specification or origin activation remains to be determined. However, initiation can be made to occur preferentially at a specific DNA site by tethering the transcriptional co-activator VP16 to the site and then incubating the DNA in a *Xenopus* egg extract. Since active transcription does not occur in these extracts, site specificity is induced by modifying the way in which the DNA is organized into chromatin and then packaged into a nucleus. This may account for the observation that 85% of the replication initiation sites in mouse embryonic stem cells are associated with transcription units. The strong coincidence of initiation sites for DNA replication with those for transcription suggests that replication origins and transcription promoters evolved together.

Transcription

Transcription through replication origins suppresses initiation events, presumably either by preventing binding of initiator proteins or by creating superhelical turns in the DNA. Unwinding DNA either by transcription or by replication generates positive superhelical turns downstream that suppresses DNA unwinding, and negative superhelical turns upstream that facilitates DNA unwinding (Figures 1-8 and 3-9). Thus, transcription toward a replicator should suppress it whereas transcription away from a replicator should activate it. In bacteria, *E. coli* oriC (Figure 10-6) and bacteriophage λ (Figure 10-7) are examples of transcriptionally activated replicators. Transcription may also create regions of DNA that are accessible to initiator proteins. In eukarya, replication origins located among actively transcribed regions of metazoan genomes replicate earlier during S phase than those located among nontranscribed genes. In addition, intergenic regions are likely to be more accessible to replication proteins than actively transcribed genes. In this manner, active transcription can affect both when and where DNA replication initiates.

DNA topology

Binding of ORC to DNA is stimulated approximately 30-fold by negative superhelical turns in the DNA. Negative supercoils could promote wrapping of DNA around the ORC analogous to the way DNA is coiled around histones in a nucleosome (Figure 1-7). Alternatively, negative supercoils could facilitate partial unwinding of DNA, thereby transforming an initial weak electrostatic interaction between ORC and DNA into a stable ORC–DNA complex.

Initiator protein concentration

Discrimination among potential DNA binding sites depends on their relative affinity for the protein involved and whether or not the interaction is binary (between two molecules) or greater. Thus, at low concentrations, a protein will bind preferentially to DNA sequences with the greatest affinity for that protein, whereas at high concentrations a protein will bind to weak as well as to strong DNA binding sites. If multiple proteins must interact with a DNA binding site to produce a stable protein–DNA complex, then the effect of protein concentration is exacerbated. In metazoa, the cellular levels of initiator proteins can change markedly during animal development. For example, the amount of ORC in frog eggs is about 100 000 fold greater than in cells from adult animals, a fact that would likely eliminate any preferences that may exist for specific DNA sequences. Conversely, a low ratio of initiator to DNA in proliferating cells in adult organisms would favor some DNA sites over others. In fact, titration of *Xenopus* egg extract with either sperm chromatin or mammalian nuclei reduces the frequency of initiation without affecting the rate of fork migration. Thus, changes in the ratio of initiation proteins to DNA can occur during animal development that would favor assembly of replication complexes at some sites over others.

Nucleotide pools

Changes in the cellular concentrations of dNTPs can affect both replication-fork velocity and the density and utilization of replication origins. At normal cellular concentration of dNTPs, DNA synthesis in hamster cells proceeds at about 1.3 kb min^{-1} $fork^{-2}$, and DNA replication in the 128 kbp AMPD2 locus begins predominantly at oriGNA13. In cell lines with reduced dNTP pools, forks travel at about 0.6 kb min^{-1} and initiation events are distributed equally among oriGNA13 and five other (secondary) initiation sites, revealing an average origin density of approximately one origin every 21 kb. Analysis of this system using the DNA-combing assay (Figure 8-4) has revealed that replication forks emanating from an efficient (primary) replicator suppress initiation events at inefficient (secondary) replicators.

DNA methylation

DNA methylation can interfere with binding of ORC to DNA *in vitro*, thereby preventing initiation events at potential replication origins. This is analogous to DNA methylation at *E. coli* oriC, preventing binding of the *E. coli* initiator protein, DnaA.

The Site-Selection Paradox

Initiation of DNA replication in metazoa does not occur randomly throughout the genomes of proliferating cells in adult organisms, but the specificity of initiation site selection is not clear. One view is that most initiation events in cells from adult organisms occur predominantly at specific sites within the genome as it exists *in vivo*, and that these sites generally occupy about 1 kbp or less (Figure 9-3A). Another view is that most initiation events in these cells are distributed stochastically throughout a larger region (10–60 kbp) with boundaries such as actively transcribed DNA that define an 'initiation zone' (Figure 9-3B). In either case, bidirectional replication occurs from a particular genomic locus. What differs is the size of that locus and the nature of the factors that determine its location. Thus, the OBR for site-specific initiation is narrow, whereas the OBR for an initiation zone is comparatively broad. Moreover, to the extent that assembly of replication complexes is easier at sites distal to the boundaries, initiation events within an initiation zone will follow a Gaussian distribution about the center of the zone.

The Jesuit model

The simplest explanation for the observations described above is that metazoan genomes contain many potential replication origins, but that as multicellular organisms develop epigenetic changes mask some origins and activate others. This concept was referred to as the Jesuit model, because it is the Jesuits who remind us that "many are called, but few are chosen" (Matthew 22:14). Many of these sites appear to recruit MCM helicase, but remain dormant unless the cells are subjected to physiological stress. Thus,

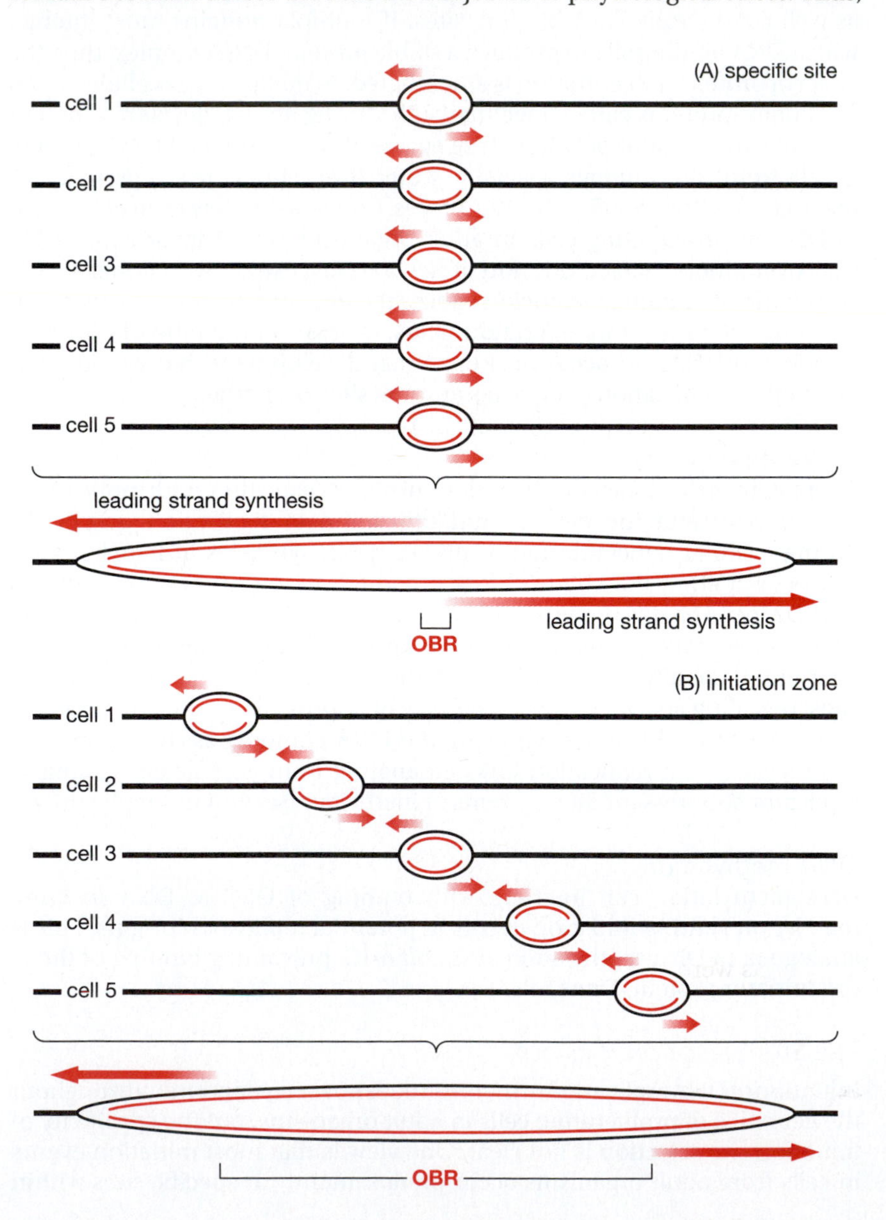

Figure 9-3. Two views of replication initiation sites in metazoan genomes. (A) If each cell within a population selects the same specific site within the same chromosome at which to initiate replication, then a single site-specific OBR (transition from discontinuous to continuous DNA synthesis, marked by red arrows) will appear at this locus. (B) If each cell selects a different site from many potential sites within a defined initiation zone, then bidirectional replication will originate from a broad initiation zone. Assuming that site selection within the zone is stochastic, more initiation events will occur toward the center of the zone than toward its peripheries. The distance between leading-strand initiation sites (the OBR) will appear much greater.

replication origins in metazoan chromosomes appear to consist of primary initiation sites that are used at a high frequency and secondary initiation sites that are used at a low frequency. Therefore, initiation zones may consist of one or more strongly preferred primary origins within a field of weakly preferred secondary initiation sites, a situation that reflects the variation in origin usage in yeast.

In some cases, the difference between site-specific initiation and initiation zones may simply reflect differences in resolution between various origin-mapping strategies. Origin-mapping methods based on quantifying the relative amounts of nascent DNA strands at various locations reveal the origins most frequently used (i.e. primary replication origins), whereas methods-based analysis of DNA structures by gel electrophoresis or an analysis of single DNA fibers reveal both efficient (primary) and inefficient (secondary) replication origins in a format that is difficult to quantify. Taken together, however, the results suggest that single-cell and multicellular eukaryotes select initiation sites based on the same principles. The similarity between two-dimensional gel analyses of replication origins in budding yeast and in human cells is striking (Figure 10-15). Most replication origins in either fission yeast or budding yeast consist of a single replicator and appear as a specific initiation site. However, two-dimensional gel analyses of the Ura4 region in *S. pombe* suggested that initiation events were distributed throughout an approximately 5 kbp initiation zone, but genetic analyses demonstrated that the replication bubbles were coincident with three separate replicators (ARS elements). Thus, what at first appeared to be a continuous initiation zone is actually a cluster of three replication origins. Similarly, ARS101 and ARS310 in *S. cerevisiae* contain multiple, redundant binding sites for the ORC, and each binding site must be altered to abolish origin function. Such redundant replicator organization also appears to exist in human cells where the 8 kbp β-globin replication origin consists of at least two genetically distinguishable replicators.

The MCM Paradox

In prokaryotes, initiator proteins bind to DNA and recruit the replicative helicases. In bacteria, two helicases per origin (one for each fork) are sufficient to allow bidirectional DNA replication. In single-cell eukaryotes, MCM helicase has been shown to travel with replication forks, but eukaryotic cells such as budding yeast, frog eggs, and mammalian cells contain from 10 to 40 times more MCM helicase than active replication origins, and a large fraction of chromatin-bound MCMs in animal cells are not localized to sites of DNA replication in S-phase cells. Thus, we have a paradox: how can MCM proteins function as the replicative helicase yet fail to coincide spatially with sites of DNA replication?

One explanation is that many more replication origins are 'licensed' (assembled into pre-replication complexes, as described in Chapter 11) than necessary to duplicate the genome, but the extra origins remain dormant unless needed, for example if environmental stress causes replication forks to stall. In that case, dormant origins downstream of the stalled forks are activated so that genome duplication can be completed in a reasonable time period. In other words, $O_D \times F_V$ = replication constant (Table 9-3). This hypothesis is supported by several studies in which the density of active replication origins significantly increased when dNTP pool levels were reduced or DNA polymerase activity was inhibited either in mammalian cells or in *Xenopus* egg extract. Dormant origins appear to contain the replicative MCM helicase, but for some unknown reason they are slow to activate the helicase and initiate DNA synthesis. Thus, under normal conditions, replication forks from more efficient origins apparently disrupt any dormant replication complexes they encounter so that genome duplication is restricted to once and only once during a single cell-division cycle.

How do cells use a minimum number of initiator proteins to load a maximum number of putative replicators with MCM helicase? The answer may well reside in the simple fact that each initiator–DNA complex can load multiple MCM helicases onto chromatin. This fact allows the possibility that MCM helicases might unwind DNA by binding to chromatin at a distance from the site where they were loaded (the initiator–DNA complex). The observation that the fluorescence signal from immunostained MCM helicases located at replication forks is weak compared with the signal emanating from MCM helicases bound elsewhere simply means that the number of MCMs at each fork (presumably one) is less than the number of MCMs clustered at sites elsewhere (potential replication origins).

In principle, there are three ways in which MCM helicases could be distributed to sites distal to the place where they were loaded. (1) They could migrate along the chromatin until they find a DNA sequence that is easily unwound. (2) They could be translocated to distal sites by the mechanism used to load helicases onto chromatin. (3) They could operate independently of DNA synthesis and simply unwind large regions of DNA, allowing DNA synthesis to begin at many sites along the exposed DNA templates. Model one seems unlikely, because eukaryotic genomes consist of DNA wrapped around an octamer of histones (the nucleosome), and nucleosomes are 11 nm×11 nm×5.5 nm in size, far too large to fit inside the MCM helicase. One could postulate that the helicase migrates along the surface of the DNA, but DNA is bound tightly to the histones, preventing MCMs from tracking the double helix without the assistance of chromatin-remodeling complexes to release DNA from histones. Therefore, models two and three are most plausible. These models are relevant to resolving the site-selection paradox.

Loading DNA helicases at distal DUEs

The MCM paradox leads to the conclusion that cells may load MCM helicases onto chromatin at sites distal to the helicase-loading mechanism (the initiator–replicator complex). Given the requirements for helicase loading and action (Chapters 10 and 11), these sites would likely be DUEs where DNA unwinding could begin and replication forks be assembled. This would result in site-specific initiation events giving rise to initiation zones. In fact, such a mechanism has been demonstrated in bacteria.

Bacterial plasmid R6K provides a model for how initiator proteins can bind to a single replicator but initiate replication at sites quite distant from the replicator by forming DNA loops. The bacterial plasmid R6K has three replication origins (oriα, oriβ, and oriγ) distributed over a distance of 5.2 kb (Figure 9-4), but only one of the three origins is active on a given molecule at any given time. oriγ is the primary binding site for the plasmid-encoded initiator protein π (a member of the Rep family of plasmid-encoded initiator proteins). oriγ contains seven tandem π protein DNA-binding sites. In addition, oriγ contains two binding sites for the *E. coli* initiator protein (DnaA) and one for the *E. coli* IHF protein. All three of these DNA sequence motifs, as well as the A:T-rich region, are essential for oriγ activity *in vivo* and *in vitro*. In contrast, oriα and oriβ each contain only one π-binding site. In fact, neither oriα nor oriβ bind π protein by themselves. Thus, it is remarkable that oriα and oriβ are the most frequently utilized origins in R6K, and that oriγ is generally silent!

When oriγ is present in the same DNA molecule as oriα and oriβ, then π protein forms a bridge between oriγ and either oriα or oriβ, resulting in formation of a DNA loop (Figure 9-5). Binding of IHF at the ihf locus causes bending of the DNA that presumably brings the DnaA protein bound to the dnaA1 site into contact with π protein bound to its seven binding sites and with DnaA bound to the dnaA2 site (Figure 11-4). DNA bending also promotes contact between the π–DNA complex and the A:T-rich region. Mutants of π protein that are defective in DNA looping fail to activate either oriα or oriβ, but they do allow initiation of DNA replication to occur at oriγ.

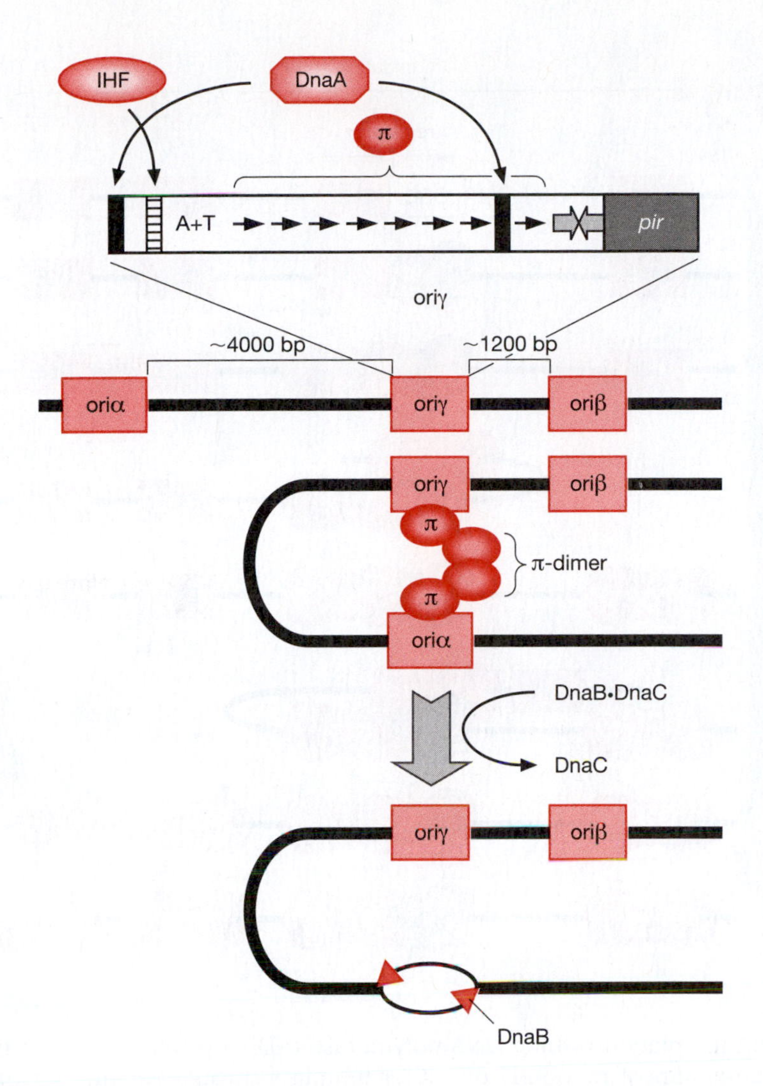

Figure 9-4. Action at a distance: the R6K paradigm. *E. coli* plasmid R6K contains three replication origins (oriα, oriβ, and oriγ). The initiator protein π mediates formation of DNA loops between π DNA-binding sites. DNA looping occurs either by two DNA-bound π-monomers bridged by a π-dimer (shown), or by two interacting DNA-bound π-dimers (not shown). DnaB is the *E. coli* replicative DNA helicase. DnaC is an *E. coli* protein required for loading DnaB onto ssDNA generated through the action of π, DnaA and IHF bending of the DNA as demonstrated for oriγ in Figure 11-4. (Adapted from Zzaman, S. and Bastia, D. (2005) *Mol. Cell* 20: 833–843. With permission from Elsevier.)

The dimeric form of π is biologically inert and has to be activated by the DnaK chaperone system into a form that forms monomers upon contact with DNA. Thus, oriγ functions as a replicator, and oriα and oriβ function as DUEs. The ability to form a DNA loop allows initiation of DNA replication at some distance to either side of the oriγ replicator.

The R6K paradigm could be applied to replicators in metazoan genomes by allowing a single initiator–replicator complex to load MCM helicases at several different DUEs that lie distal to the replicator but within the same DNA molecule (Figure 9-5). Mammalian MCM helicase and ATPase activities are stimulated specifically by the presence of thymine-rich ssDNA sequences commonly found at DUEs, and MCM is a prime target of the DDK protein kinase, which is essential for activation of replication origins. Thus, a combination of T-rich template DNA, protein phosphorylation, and interaction with other replicative proteins may contribute to selection of replication initiation sites in higher eukaryotes. The net effect would be that some cells would initiate DNA unwinding at the replicator, but other cells would initiate DNA unwinding at one or more sites distal to the replicator. Replication forks would then be assembled at the unwound regions and DNA synthesis would originate from multiple sites within the vicinity of the replicator. The result would be an initiation zone of the type found between the DHFR and 2BE2121 genes in Chinese hamster cells (Figure 10-16).

Uncoupling DNA unwinding from DNA synthesis

A second mechanism for generating initiation zones involves uncoupling DNA unwinding from DNA synthesis, thus exposing large regions of

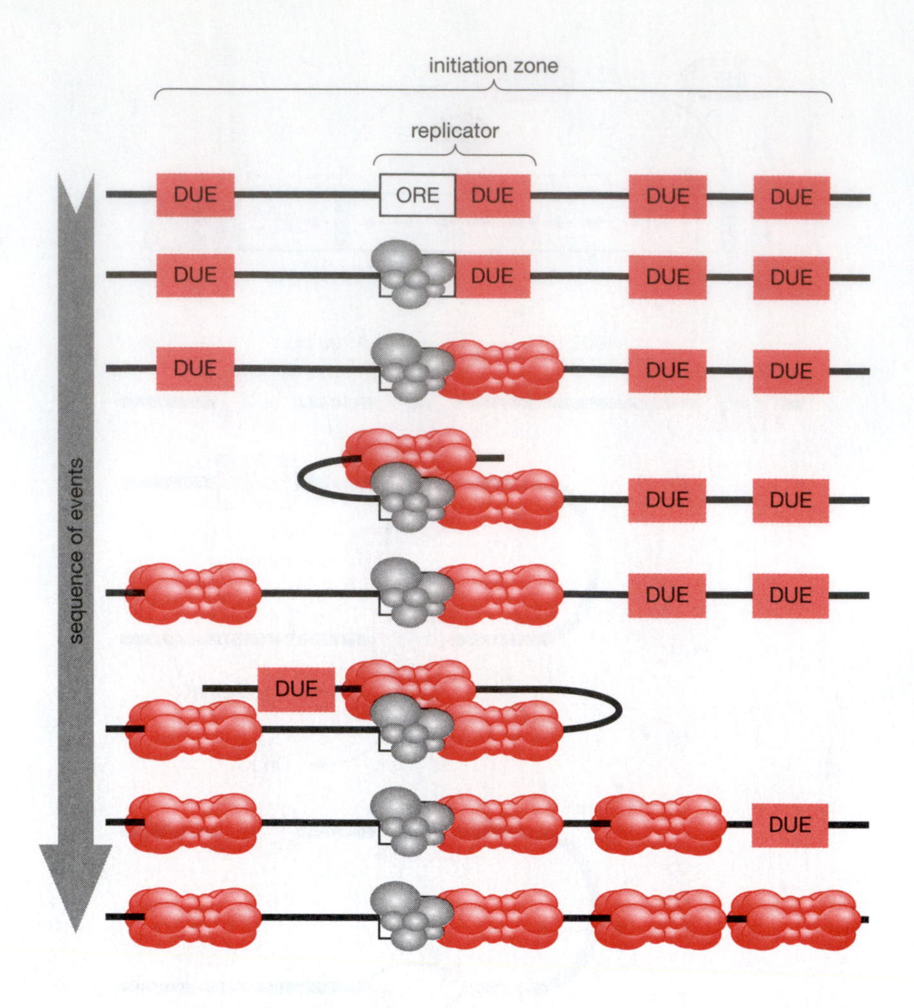

Figure 9-5. Initiation zones could result from a single replicator loading MCM helicases onto distal DUEs. Eukaryotic initiator proteins origin-recognition complex (ORC) and Cdc6 (gray spheres) bind to the replicator's ORE. MCM helicase (pink asymmetrical dumbbells) is loaded onto the replicator's DUE through the action of proteins Cdt1 and Mcm9. This would constitute site-specific initiation. However, additional MCMs could be loaded onto distal DUEs by a DNA-looping mechanism analogous to the one used by bacterial plasmid R6K (Figure 9-4). Either each cell within a population could load MCMs onto all of the available DUEs in the vicinity of a replicator, or different cells within a population could load different DUEs. The end result would be the same; multiple initiation events within an 'initiation zone'. (Data from Lutzmann, M. and Méchali, M. (2008) *Mol. Cell* 31: 190–200, Masai, H., You, Z. and Arai, K. (2005) *IUBMB Life* 57: 323–335, and Takahashi, T.S., Wigley, D.B. and Walter, J.C. (2005) *Trends Biochem. Sci.* 30: 437–444.)

ssDNA template on which DNA polymerase-α•DNA primase could initiate synthesis almost randomly on either template throughout the unwound DNA region (Figure 9-6).

One explanation for this phenomenon is that preRCs are assembled at many potential replication origins, but that replication forks traveling from the primary origin disrupt the preRCs at secondary origins before they have a chance to initiate DNA synthesis. Reducing fork velocity would prevent this.

Alternatively, reducing the rate of DNA synthesis frequently results in uncoupling DNA synthesis from DNA unwinding, resulting in large stretches of ssDNA in the vicinity of a replicator. RNA-primed DNA synthesis could then begin at many sites within the unwound DNA region to produce an initiation zone. Thus, synchronizing cells at their G_1/S interphase either by reducing nucleotide pools with drugs such as hydroxyurea or mimosine, or by inhibiting DNA polymerase activity with drugs such as aphidicolin, can uncouple DNA unwinding from DNA synthesis and thereby produce large regions of ssDNA. This accounts for the accumulation of RPA at replication foci (Figure 3-11A) with concomitant activation of the DNA damage response (discussed in Chapter 13). Loading multiple MCM helicases at a single replicator will also increase the probability that a DNA helicase will proceed with DNA unwinding before DNA synthesis can begin.

SUMMARY

- Replication origins in dsDNA genomes all exhibit two or more of the following characteristics: (1) the ability to initiate replication when translocated to a foreign DNA molecule or to an ectopic chromosomal

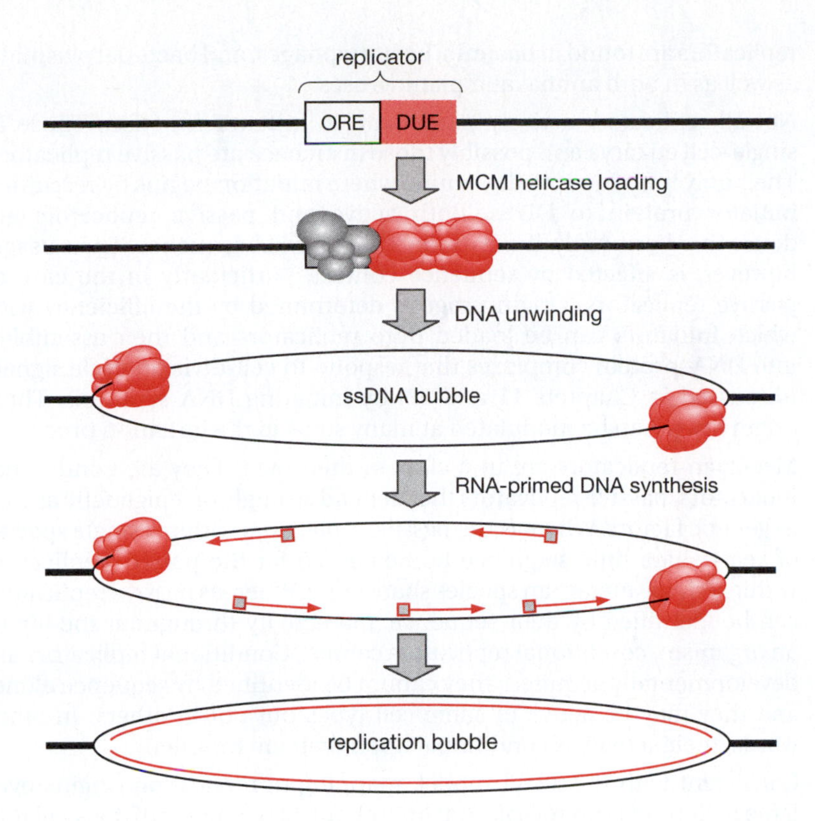

Figure 9-6. Initiation zones could result from uncoupling DNA unwinding from DNA synthesis. Two MCM helicases are loaded at a single DUE, one for each replication fork. Newly synthesized DNA is red. Other symbols are as in Figure 9-5.(Adapted from Benbow, R.M., Zhao, J. and Larson, D.D. (1992) *Bioessays* 14: 661–670. With permission from John Wiley and Sons, Inc.)

locus, (2) sites where DNA synthesis first occurs, (3) a transition between discontinuous and continuous DNA synthesis on each of the two DNA strands, (4) DNA bubbles appear, and (5) sites from which replication forks originate.

- Replicators in dsDNA genomes of bacteria, archaea, single-cell eukarya, and the viruses and plasmids that inhabit these organisms consist of two essential components: an origin-recognition element to which the initiator protein binds and a DNA-unwinding element that is the site where DNA unwinding begins. They may also have auxiliary components that bind specific transcription factors. Since the transcription factors are expressed in some cells but not in others, they can impose cell-type specificity upon replicator activity.
- The function of replicators is to determine where DNA replication begins. The function of initiator proteins is to begin the process of DNA unwinding. Thus, the sequence of events at a replicator is (1) bind initiator, (2) initiate DNA unwinding, (3) initiate leading-strand DNA synthesis, and (4) initiate lagging-strand DNA synthesis.
- With the exception of embryos undergoing rapid cell cleavage events, initiation of genome duplication does not occur randomly throughout the genome, but at specific sites. The complexity of these sites, however, increases as a function of biological evolution (viruses $\leq$ bacteria $\approx$ archaea $\leq$ single-cell eukarya $<$ multicellular eukarya).
- All genomes contain replication origins, but not all replication origins contain active replicators. Active replicators are required to bind the initiator protein in such a manner that the protein can carry out its function. Since the function of the initiator protein is to initiate DNA unwinding, an active replicator either facilitates the ability of the initiator to organize itself into a DNA helicase, or facilitates the ability of the initiator to load the replicative DNA helicase onto DNA. Active

replicators are found in bacteria, bacteriophages, and bacterial plasmids, as well as in both animal and plant viruses.

- Not all replicators actively participate in initiator function. Those in single-cell eukarya and possibly those in archaea are passive replicators. Their only function is to determine where initiation begins by recruiting initiator protein to DUEs. Both active and passive replicators are determined primarily, if not exclusively, by DNA sequence. Their usage, however, is affected by sequence context, particularly in the case of passive replicators. Origin usage is determined by the efficiency with which initiators can be loaded onto replicators and then assembled into DNA–protein complexes that respond to cell-division cycle signals (discussed in Chapters 11 and 12) by initiating DNA synthesis. Thus, origin usage can be modulated at many steps in the initiation process.
- Metazoan replicators are in a class of their own. They are conditional replicators, passive replicators that depend strongly on epigenetic as well as genetic factors. Whereas the passive replicators within a single species of yeast share little sequence homology, so far the passive replicators within a single metazoan species share none. Whereas passive replicators can be identified by their sequence and activity throughout the life of an organism, conditional replicators cannot. Conditional replicators are developmentally acquired; they cannot be identified by sequence alone, and they may be active in some cell types but not in others. In other words, their activity is conditional upon their environment.
- Consistent with the Jesuit model, mapping of replication origins over large regions of DNA reveals that only about 30% or fewer of the available replication origins are used regularly on individual chromosomes in the cells of adult animals; the remaining sites are generally dormant. The mean origin-to-origin distance on individual human chromosomes is 90–120 kb, and many of the same origins appear in different human cell lines, suggesting a genetic determinant. However, less frequently used replication origins can be detected as well, reducing the mean origin-to-origin distance to approximately 30 kb in human cells, and approximately 15 kb for DNA replicating in frog egg extracts.
- The product of origin density and replication-fork velocity appears to be constant in bacteria, archaea, and eukarya (Table 9-3). This allows changes in one of these parameters to be compensated by changes in the other.
- In genomes with multiple origins, there can be great variation in the frequency at which individual origins are used. Some origins remain dormant throughout many cell divisions, despite the fact that they appear to contain one or more of the proteins required to produce the pre-replication complexes described in Chapter 11.
- Origin timing is imposed on replicators primarily by chromatin structure and the various factors, such as transcription and histone modification, that influence chromatin structure.
- Both site-specific (origins of bidirectional replication) and nonspecific (initiation zones) initiation events occur in metazoan genomes, apparently through the existence of efficient (primary) and inefficient (secondary) replicators. A similar, although less dramatic, situation exists in yeast. Two other mechanisms can generate initiation zones. A single eukaryotic replicator can load multiple DNA helicases onto chromatin, perhaps at locations distal to the initiator binding site. This could generate a zone of initiation events surrounding the replicator. Uncoupling DNA unwinding from DNA synthesis generates large regions of ssDNA that could allow DNA synthesis to initiate at randomly chosen sites surrounding the replicator.

- Large genomes contain more replication origins than they actually use during cell division. This ensures that loss of origins from genetic mutations or rearrangements does not hamper genome duplication.
- Metazoa contain excessive levels of MCM helicases relative to the number of active replication origins. This contributes to the presence of MCM helicases loaded onto dormant replicators that can be activated under stressful conditions.

REFERENCES

Additional Reading

Finding Replication Origins

DePamphilis, M.L. (1997) Identification and analysis of replication origins in eukaryotic cells. *Methods* 13: 209–324.

Gerbi, S.A. (2005) Mapping origins of DNA replication in eukaryotes. *Methods Mol. Biol.* 296: 167–180.

Gerhardt, J., Jafar, S., Spindler, M.P., Ott, E. and Schepers, A. (2006) Identification of new human origins of DNA replication by an origin-trapping assay. *Mol. Cell. Biol.* 26: 7731–7746.

Heichinger, C., Penkett, C.J., Bahler, J. and Nurse, P. (2006) Genome-wide characterization of fission yeast DNA replication origins. *EMBO J.* 25: 5171–5179.

Lebofsky, R., Heilig, R., Sonnleitner, M., Weissenbach, J. and Bensimon, A. (2006) DNA replication origin interference increases the spacing between initiation events in human cells. *Mol. Biol. Cell* 17: 5337–5345.

MacAlpine, D.M. and Bell, S.P. (2005) A genomic view of eukaryotic DNA replication. *Chromosome Res.* 13: 309–326.

Newlon, C.S. and Theis, J.F. (2002) DNA replication joins the revolution: whole-genome views of DNA replication in budding yeast. *Bioessays* 24: 300–304.

Todorovic, V., Giadrossi, S., Pelizon, C., Mendoza-Maldonado, R., Masai, H. and Giacca, M. (2005) Human origins of DNA replication selected from a library of nascent DNA. *Mol. Cell* 19: 567–575.

Worning, P., Jensen, L.J., Hallin, P.F., Staerfeldt, H.H. and Ussery, D.W. (2006) Origin of replication in circular prokaryotic chromosomes. *Environ. Microbiol.* 8: 353–361.

Acquisition Of Site Specificity

Anglana, M., Apiou, F., Bensimon, A. and Debatisse, M. (2003) Dynamics of DNA replication in mammalian somatic cells: nucleotide pool modulates origin choice and interorigin spacing. *Cell* 114: 385–394.

Hyrien, O., Maric, C. and Mechali, M. (1995) Transition in specification of embryonic metazoan DNA replication origins. *Science* 270: 994–997.

Norio, P., Kosiyatrakul, S., Yang, Q., Guan, Z., Brown, N.M., Thomas, S., Riblet, R. and Schildkraut, C.L. (2005) Progressive activation of DNA replication initiation in large domains of the immunoglobulin heavy chain locus during B cell development. *Mol. Cell* 20: 575–587.

Sasaki, T., Sawado, T., Yamaguchi, M. and Shinomiya, T. (1999) Specification of regions of DNA replication initiation during embryogenesis in the 65-kilobase DNA-polα-dE2F locus of *Drosophila melanogaster*. *Mol. Cell. Biol.* 19: 547–555.

Replication Origins That Function In Bacteria

Cunningham, E.L. and Berger, J.M. (2005) Unraveling the early steps of prokaryotic replication. *Curr. Opin. Struct. Biol.* 15: 68–76.

Kaguni, J.M. (2006) DnaA: controlling the initiation of bacterial DNA replication and more. *Annu. Rev. Microbiol.* 60: 351–375.

Leonard, A.C. and Grimwade, J.E. (2005) Building a bacterial orisome: emergence of new regulatory features for replication origin unwinding. *Mol. Microbiol.* 55: 978–985.

Weigel, C. and Seitz, H. (2006) Bacteriophage replication modules. *FEMS Microbiol. Rev.* 30: 321–381.

Replication Origins That Function In Archaea

Robinson, N.P. and Bell, S.D. (2005) Origins of DNA replication in the three domains of life. *FEBS J.* 272: 3757–3766.

Replication Origins That Function In Eukarya

Aladjem, M.I. (2007) Replication in context: dynamic regulation of DNA replication patterns in metazoans. *Nat. Rev. Genet.* 8: 588–600.

Aladjem, M.I., Falaschi, A. and Kowalski, D. (2006) Eukaryotic DNA replication origins. In *DNA Replication and Human Disease*, DePamphilis, M.L. (ed.), pp. 31–62. Cold Spring Harbor Laboratory Press, Cold Spring Harbor, NY.

Bell, S.P. and Dutta, A. (2002) DNA replication in eukaryotic cells. *Annu. Rev. Biochem.* 71: 333–374.

Blow, J.J. and Ge, X.Q. (2008) Replication forks, chromatin loops and dormant replication origins. *Genome Biol.* 9: 244.

Cvetic, C. and Walter, J.C. (2005) Eukaryotic origins of DNA replication: could you please be more specific? *Semin. Cell Dev. Biol.* 16: 343–353.

DePamphilis, M.L. (ed.) (1996) *DNA Replication in Eukaryotic Cells.* Cold Spring Harbor Laboratory Press, Plainview, NY.

DePamphilis, M.L. (1999) Replication origins in metazoan chromosomes: fact or fiction? *Bioessays* 21: 5–16.

Dubey, D.D., Zhu, J., Carlson, D.L., Sharma, K. and Huberman, J.A. (1994) Three ARS elements contribute to the ura4 replication origin region in the fission yeast, Schizosaccharomyces pombe. *EMBO J.* 13: 3638–3647.

Hiratani, I. and Gilbert, D.M. (2009) Replication timing as an epigenetic mark. *Epigenetics* 4: 93–97.

Stillman, B. (2005) Origin recognition and the chromosome cycle. *FEBS Lett.* 579: 877–884.

Theis, J.F. and Newlon, C.S. (2001) Two compound replication origins in *Saccharomyces cerevisiae* contain redundant origin recognition complex binding sites. *Mol. Cell. Biol.* 21: 2790–2801.

Literature Cited

Berezney, R., Dubey, D.D. and Huberman, J.A. (2000) Heterogeneity of eukaryotic replicons, replicon clusters, and replication foci. *Chromosoma* 108: 471–484.

Bielinsky, A.K. and Gerbi, S.A. (1999) Chromosomal ARS1 has a single leading strand start site. *Mol. Cell* 3: 477–486.

Cadoret, J.C., Meisch, F., Hassan-Zadeh, V., Luyten, I., Guillet, C., Duret, L., Quesneville, H. and Prioleau, M.N. (2008) Genome-wide studies highlight indirect links between human replication origins and gene regulation. *Proc. Natl. Acad. Sci. USA* 105: 15837–15842.

Feng, W., Collingwood, D., Boeck, M.E., Fox, L.A., Alvino, G.M., Fangman, W.L., Raghuraman, M.K. and Brewer, B.J. (2006) Genomic mapping of single-stranded DNA in hydroxyurea-challenged yeasts identifies origins of replication. *Nat. Cell Biol.* 8: 148–155.

Hay, R.T. and DePamphilis, M.L. (1982) Initiation of SV40 DNA replication in vivo: location and structure of 5′ ends of DNA synthesized in the ori region. *Cell* 28: 767–779.

Heichinger, C., Penkett, C.J., Bahler, J. and Nurse, P. (2006) Genome-wide characterization of fission yeast DNA replication origins. *EMBO J.* 25: 5171–5179.

Chapter 10
ORIGIN PARADIGMS

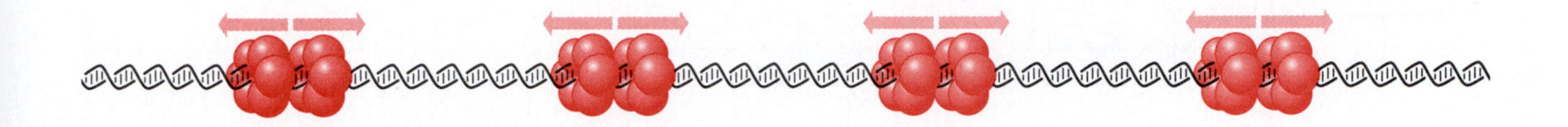

An enormous amount of information has accumulated over the past 50 years on the mechanisms by which proteins interact with replication origins to initiate genome duplication in living organisms as well as in the viruses and plasmids that inhabit them. Remarkably, all of the data can be grouped into five mechanisms (Table 10-1). The first four mechanisms use **active replicators** (sequence-specific sites in the genome with which the **initiator** protein must bind in order for it to function). Only the DNA helicase loader mechanism employs either **passive replicators** (DNA sites that are required for initiator binding but not for initiator function) or **conditional replicators** (DNA sites that serve as replicators only under certain conditions). Note that bacteria and archaea employ the DNA helicase loader mechanism exclusively to initiate duplication of all of their genome, and that eukarya employ the same mechanism to initiate replication of the bulk of their genome (the nuclear portion). Only the mitochondrial portion of the eukaryal genome is replicated by a different mechanism: transcriptional activation. Some genomes employ more than one of these mechanisms. For example, mtDNA replication begins at an active replicator (oriH) but then uses passive replicators or nonspecific initiation events to complete its replication. Circular ssDNA genomes follow the opposite pattern. They begin with a passive replicator or nonspecific initiation events to form a dsDNA replicative form, and then switch to an active replicator to produce additional copies of the dsDNA form.

The essential characteristics of paradigms for each of the five mechanisms are described below. Several paradigms are described for the same mechanism when important variations on the same theme appear, or when striking similarities exist among diverse species or virus groups. A brief rationale is also offered for why each biological system was selected.

DNA-NICKING MECHANISM

Initiators known as Rep proteins introduce site-specific breaks into one of the two DNA strands (referred to as a **nick**) to generate a 3′-OH-terminated DNA primer that can then be extended by a DNA polymerase in the company of a DNA helicase (Figure 10-1). The result is single-strand DNA displacement synthesis, sometimes referred to as rolling-circle synthesis (Figures 8-6 and 8-8). In this case, the initiator protein is an origin-recognition protein with an endonuclease activity. However, Rep cannot nick the DNA unless it binds to its specific target sequence; mutations in either Rep or the Rep protein DNA-binding site that prevent binding also prevent nicking. Therefore, Rep binding must perturb the DNA-nicking site so as to expose one strand to the endonuclease. For example, Rep may convert the inverted repeat component [equivalent to the **DNA-unwinding element** (**DUE**) in Figure 9-1] into its cruciform structure, thereby exposing a ssDNA loop. This is an example of how an active replicator and its cognate initiator interact to begin DNA synthesis.

Table 10-1. The Five Mechanisms For Initiating Replicative DNA Synthesis

Mechanism	Genome	Examples
DNA nicking DNA pol 5′ OH P 3′ 3′ 5′	Circular ssDNA viruses, circular dsDNA plasmids, internal active replicator for dsDNA replicating forms	Icosahedral phage ϕX174, filamentous phage M13, geminiviruses, circoviruses, parvoviruses, some bacterial plasmids (pT181)
Protein priming DNA pol 5′ -dN-OH 3′ 3′ 5′	Linear dsDNA, terminal active replicators	Adenoviruses, phage ϕ29 family
DNA transcription 5′ 3′ 3′ 5′ OH DNA pol	Linear and circular dsDNA, internal active replicators	Bacteriophage λ, T7, and T4, plasmid ColE1, mitochondria
DNA helicase DNA pol–primase 5′ 3′ 3′ 5′	Circular dsDNA, internal active replicators	SV40, polyomavirus, papillomavirus, HSV
DNA helicase loader DNA helicase 5′ 3′ 3′ 5′	Circular and linear dsDNA, internal passive and conditional replicators, as well as nonspecific initiation events	Bacteria, some bacterial plasmids (R6K), archaea, eukaryal nuclear DNA, Epstein–Barr virus

Cartoons illustrate the physical state of the replicator after the initiator has acted; arrow indicates the next event.

The DNA-nicking mechanism is restricted to the replicating forms of ssDNA genomes in bacteriophages, plant viruses, parvoviruses, and small multicopy dsDNA plasmids. So far no parallels exist between the DNA-nicking mechanism used in these bacteriophages and bacterial plasmids to initiate single-strand displacement synthesis and events that occur in the genomes of living things.

PROTEIN-PRIMING MECHANISM

Some initiators contain a single dNMP residue that serves as a primer on which a DNA polymerase can initiate DNA synthesis. In principle, such initiator proteins could be used to prime DNA synthesis in virtually any genome, but nature has restricted their use to linear dsDNA chromosomes from 19 to 38 kbp in length. Examples include the human adenoviruses, bacteriophage PRD1 that infects **Gram-negative bacteria** such as *Escherichia coli*, Cp-1 that infects **Gram-positive bacteria** such as *Streptococcus pneumoniae*, and ϕ29, PZA, phi15, BS32, B103, Nf, M2Y, and GA-1 that infect the Gram-positive bacterium *Bacillus subtilis*. These genomes have two active replicators, one at each end, which are unique

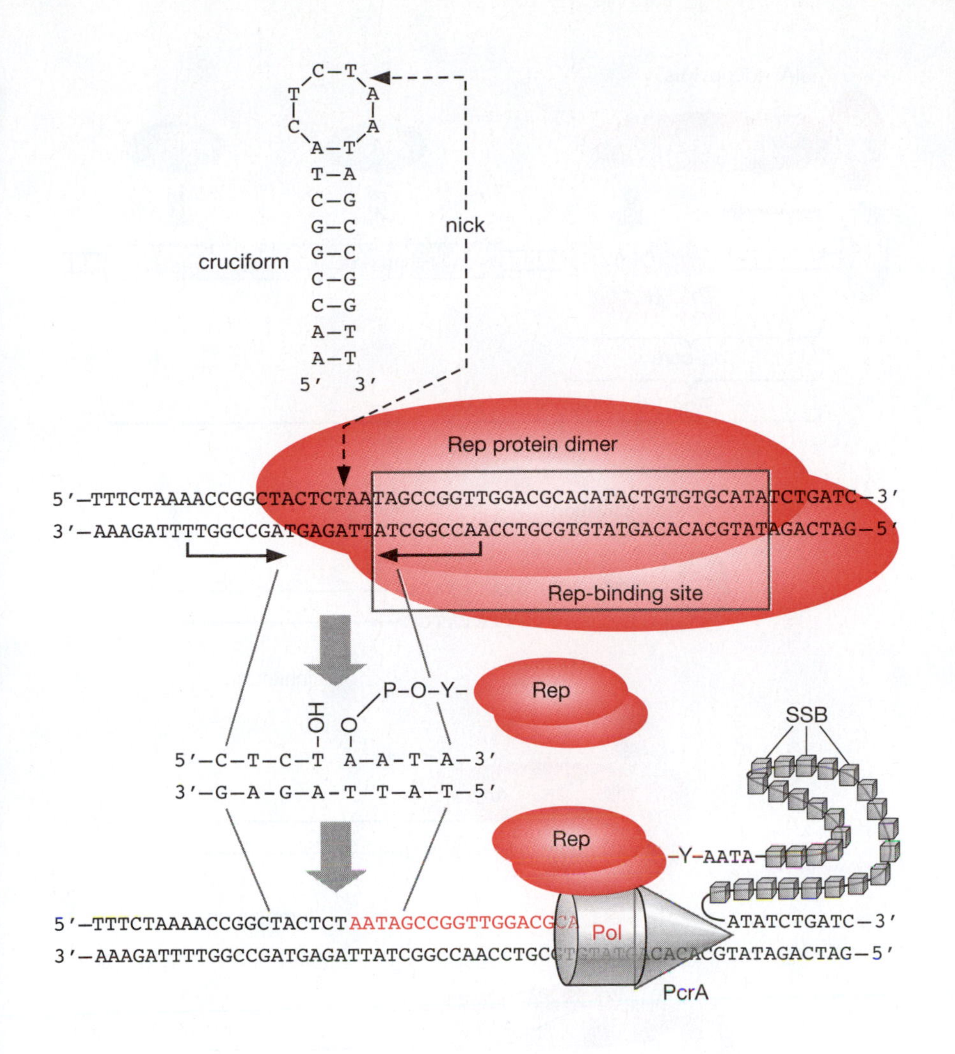

Figure 10-1. Replication origins in plasmids from Gram-positive bacteria and in the dsDNA replicative forms of circular ssDNA bacteriophages and plant viruses. For example, the plasmid pT181 contains a 46 bp replicator and encodes an initiator protein (Rep) that binds to a unique 31 bp sequence within this replicator (Rep-binding site). Rep then cleaves a phosphodiester bond in one strand of an inverted repeat upstream of this site (Rep-nicking site) to create a 3′-OH terminus upon which the host cell's DNA polymerase (Pol) can initiate DNA synthesis (red). During this reaction, a tyrosine (Y) residue in Rep becomes covalently linked to the 5′-phosphoryl of the nicked DNA strand. DNA synthesis occurs concomitantly with DNA unwinding by the host cell's replicative DNA helicase PcrA, resulting in displacement of the DNA strand covalently attached to the Rep initiator protein. This displacement reaction is facilitated by the host cell's single-stranded DNA-binding protein SSB. When DNA synthesis comes full circle and replicates through this region, the inverted repeat at the nicking site allows the displaced strand to form a palindrome that recreates the nicking site. Rep cuts this site again to release the displaced strand in a cleavage and rejoining reaction (Figure 8-6). It is also possible that Rep binding to the dsDNA replicator induces formation of a cruciform that exposes the site to endonuclease action by Rep. (Adapted from Khan, S.A. (2005) *Plasmid* 53: 126–136. With permission from Elsevier.)

among replicators in that they contain a protein covalently attached to the 5′-end of one of the two DNA strands. Such replicators follow the general model described in Figure 9-1, except that one end of the **essential origin component** (**ori-core**) is the end of a linear dsDNA genome. Therefore, DNA replication can proceed only in one direction. In addition, these viral genomes are replicated by single-strand displacement synthesis (Figure 8-7) instead of replication-fork synthesis (Figure 8-1).

Two extensively characterized paradigms exist – adenovirus and bacteriophage ϕ29 – but since they are virtually identical mechanisms, only adenovirus is described here. Adenovirus (or 'Ad') DNA replication occurs within its host cell's nucleus as soon as three adenovirus-encoded proteins are expressed in virus-infected cells: DNA polymerase (Adpol), pre-terminal protein (pTP), and **DNA-binding protein** (**DBP**). pTP and Adpol form a stable heterodimer. The adenovirus ori-core begins with the repetitive sequence 5′-CATCAT-3′ whose 5′-end is covalently linked to adenovirus terminal protein (TP), a 55 kDa C-terminal fragment of the adenovirus pTP that remains covalently linked to the 5′-end of each DNA strand (Figures 8-7 and 10-2A). pTP is processed to TP by an adenovirus-encoded protease prior to encapsulation of the viral genome. Adpol catalyzes formation of a phosphodiester bond between the α-phosphoryl of dCTP (the first nucleotide incorporated into the nascent DNA strand) and the β-OH on serine 580 in pTP. DBP increases the efficiency of this reaction by decreasing the K_m of Adpol for dCTP. The pTP(dCMP)•Adpol complex then binds to the **origin-recognition element** (**ORE**). The interaction between the Adpo•pTP heterodimer and the ORE is stabilized by the presence of

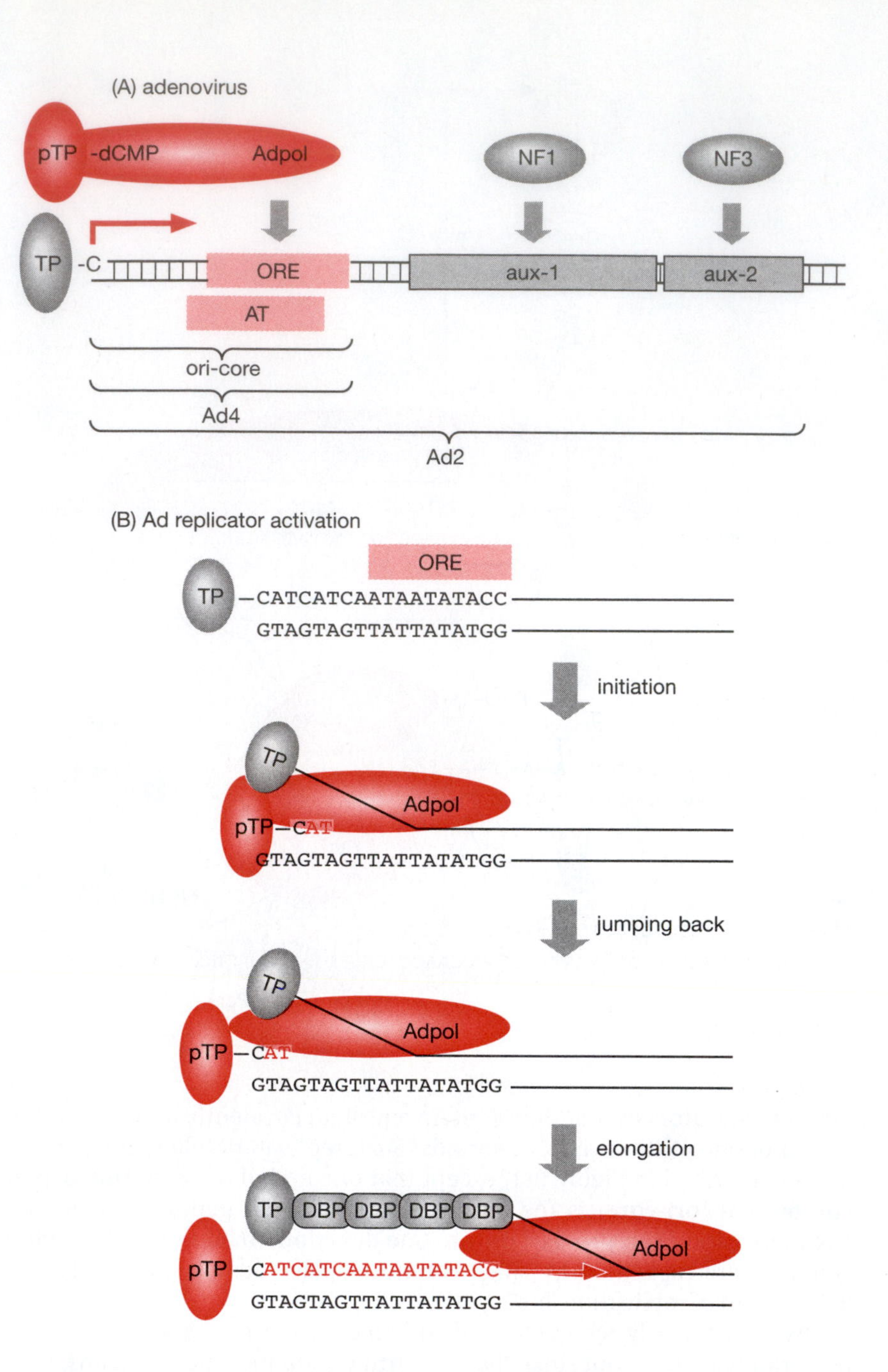

Figure 10-2. Adenovirus (Ad) replication origin. The replicator in Ad2 is 49 bp and consists of an 18 bp ori-core that is used in all adenovirus replicators and two auxiliary sequences that bind either nuclear factor 1 or 2 (NF1/CTF1 or NF3/Oct-1). Not all adenovirus replicators include these auxiliary sequences. Activation of adenovirus DNA replication begins with the binding of the adenovirus DNA polymerase (Adpol)–pre-terminal protein (pTP) complex and template-dependent addition of two nucleotides to the dCMP residue in the pTP protein primer to the ORE. The CAT triplet then 'jumps back' to the terminal position where it is elongated by Adpol in a single-strand displacement reaction that is facilitated by the adenovirus DNA-binding protein (DBP). AT is an A:T-rich region. (Adapted from de Jong, R.N., van der Vliet, P.C. and Brenkman, A.B. (2003) *Curr. Top. Microbiol. Immunol.* 272: 187–211, and Liu, H., Naismith, J.H. and Hay, R.T. (2003) *Curr. Top. Microbiol. Immunol.* 272: 131–164. With permission from Springer.)

cellular transcription factors NF1/CTF1 and NF3/Oct-1 bound to auxiliary sequences 1 and 2 (aux-1 and aux-2), respectively, and by the TP attached to the 5′-end of the adenovirus replicator to form a **pre-replication complex** (**preRC**).

Since DBP binds dsDNA as well as ssDNA, it actually coats the entire genome, thereby increasing the affinity of NF1 for aux-1. The phage ϕ29-encoded analog for DBP (protein 6) produces a remarkable conformational change in the DNA reminiscent of nucleosome assembly in eukaryotes, suggesting that DBP may do likewise. Interestingly, some adenoviruses, such as Ad4, use only ori-core, whereas others, such as Ad2, use the complete origin. The reason for this difference is unknown. Since both Ad2 pTP•Adpol and Ad4 pTP•Adpol have essentially the same affinity for ori-core, one would have expected aux-1 and aux-2 to facilitate both of their activities.

The dCMP residue that is covalently attached to pTP serves as the primer for elongation of the nascent DNA strand by Adpol (Figure 10-2B).

Adpol displaces the nontemplate strand, and positions the serine-dCMP in pTP opposite the G residue in the second GTA triplet within ori-core. Adpol then synthesizes a pTP-CAT trinucleotide intermediate. Remarkably, the newly synthesized pTP-trinucleotide intermediate then jumps backward and pairs with the first GTA triplet within ori-core. This event is followed by continuous extension of the pTP-CAT trinucleotide primer by Adpol with concomitant displacement of the nontemplate strand. Unwinding of adenovirus DNA is unusual in that it does not require a specialized DNA helicase. Rather, it is carried out by the highly processive Adpol in a reaction facilitated by DBP to prevent reannealing of the displaced strand.

The **jumping-back mechanism** provides several advantages to terminal replicators. First, it enables the polymerase to correct mistakes at the beginning of replication. Like all replicative DNA polymerases, Adpol has an intrinsic 3′→5′ exonuclease. In the event that Adpol accidentally inserts the wrong nucleotide, its 3′→5′ exonuclease can remove it and allow the polymerase to try again. However, the presence of a terminal protein covalently attached to the 5′-end of the nascent DNA strand interferes with this proofreading mechanism. By starting replication internally, a mismatched nucleotide that was not corrected by the impaired exonuclease activity will be corrected in the next round of replication, because this trinucleotide will not serve as a template. In addition, small deletions of the genome termini can be repaired using the jumping-back mechanism. Jumping back also ensures that shortening of this genome does not occur during replication. Finally, jumping back allows the polymerase and its protein primer to align themselves over a larger region of template, and therefore ensure greater fidelity for the initiation reaction.

DNA TRANSCRIPTION MECHANISM

Some initiator proteins are **RNA polymerases**; they transcribe one strand of the replicator to generate an RNA primer that allows a DNA polymerase to initiate either bidirectional DNA replication using the replication-fork mechanism (e.g. bacteriophages T7 and T4, and bacterial plasmid ColE1), or single-strand displacement synthesis (e.g. mitochondrial genomes). In both cases, a transcript is processed by the action of an RNase H-like activity that cuts the nascent RNA strand in the RNA:DNA hybrid duplex at one or more specific sites. In addition, transcription through the replicator partially melts the dsDNA, thereby providing the ssDNA necessary to load the replicative DNA helicase. The helicase unwinds DNA in front of the DNA polymerase so that one or both templates can be copied.

The DNA transcription mechanism provides an unusual opportunity to select initiation sites for DNA replication from any one of the many gene promoters inhabiting the genome. The fact is, however, that in each of the examples that have been studied in detail, nature has selected only one site (four in the case of the larger T4 genome) to act as a replicator. Again, these are active replicators, because the initiator, guided by various auxiliary transcription factors, produces an RNA primer only at specific DNA sequences.

Bacteriophage T7

Bacteriophage T7 contains a single linear DNA chromosome 40 kbp in length. The primary DNA replication origin in phage T7 (oriT7) consists of two 23 bp T7 RNA polymerase promoters (ϕ1.1A and ϕ1.1B) followed by a 61 bp AT-rich (79% A+T) region that functions as a DUE (Figure 10-3). Transcription initiated at either of these two promoters by the T7 RNA polymerase unwinds the DUE. T7 RNA polymerase is rapidly displaced by the T7 DNA polymerase which uses the 3′-end of the short RNA transcript as a primer on which to initiate DNA synthesis on one template while

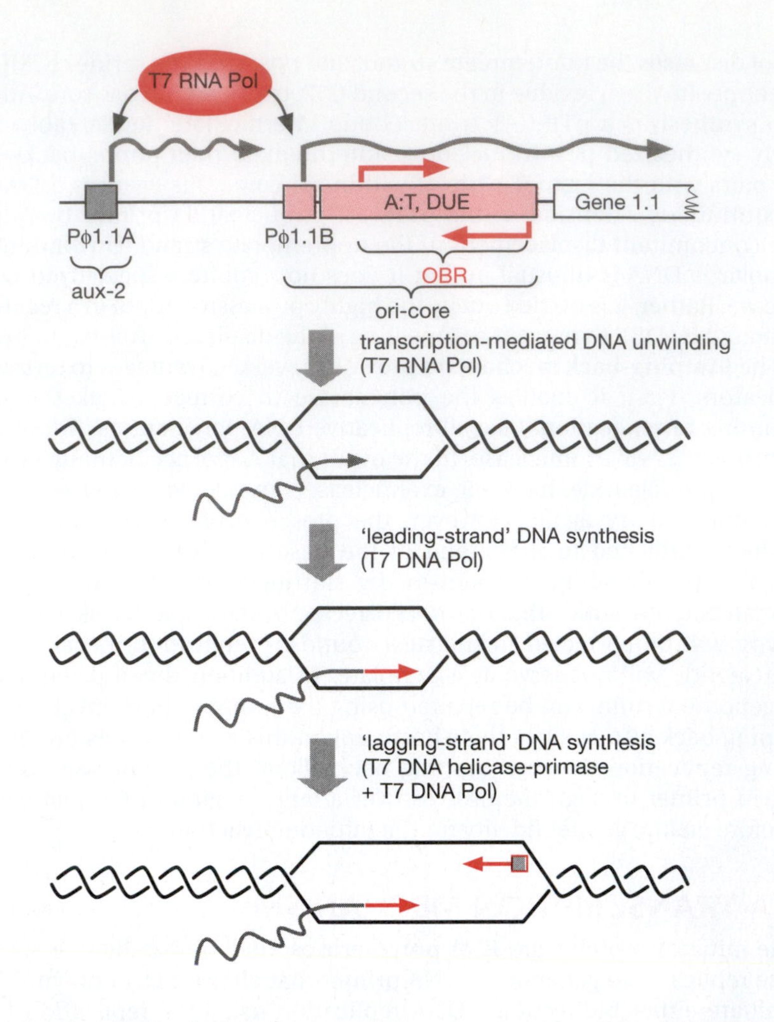

Figure 10-3. Bacteriophage T7 replication origin. Phage T7 RNA polymerase initiates transcription of mRNA at either of two upstream promoters (solid boxes show origin-recognition elements), and the transcripts are used as RNA primers by T7 DNA polymerase to initiate leading-strand synthesis (red arrow) at one or more sites within the DUE. DNA synthesis in the opposite direction (lagging-strand synthesis) is initiated by a T7-encoded DNA helicase-primase. The T7 ori-core is 159 bp. (Data from Benkovic, S.J., Valentine, A.M. and Salinas, F. (2001) *Annu. Rev. Biochem.* 70: 181–208.)

concomitantly displacing the complementary template. T7 **ssDNA-binding protein** facilitates this event by inhibiting the displaced ssDNA from reannealing to its complementary sequence. The transition from RNA to DNA synthesis occurs at several sites within the DUE, resulting in nascent DNA strands with 10–60 ribonucleotides covalently linked to their 5′-termini. oriT7 is not essential for duplication of the T7 genome; if the primary origin is inactivated, then DNA replication begins at one of several secondary origins using the same transcriptional activation mechanism.

DNA unwinding requires the T7 DNA helicase-primase, a single protein with two separate functions. T7 DNA polymerase alone cannot displace one template strand from the other, but in the presence of the T7 helicase-primase T7 DNA polymerase can synthesize thousands of nucleotides of DNA at a rate of approximately 300 nucleotides per second. This remarkable feat requires an unusual cofactor, *E. coli* thioredoxin. In the absence of thioredoxin, T7 DNA polymerase incorporates only one to 15 nucleotides before dissociating from the primer:template. Thus, thioredoxin functions like the multimeric clamp proteins β-clamp and PCNA. T7 DNA polymerase also interacts stoichiometrically with T7 ssDNA-binding protein which facilitates its progress by eliminating hairpins and twists from the DNA template.

The T7 helicase component interacts asymmetrically with both ssDNA strands at replication forks, binding about 25 nucleotides of the 5′-strand through the central hole in the hexameric helicase. The helicase can then move unidirectionally 5′→3′ on the lagging-strand template fueled by NTP hydrolysis at a rate of approximately 260 bp s^{-1} (30°C). Most helicases prefer ATP for this purpose, but the T7 helicase uses either ATP or dTTP. Unwinding probably occurs by exclusion of the 3′-tail from entering the hole in the helicase. T7 DNA polymerase and T7 DNA helicase clearly interact with one another, because T7 helicase-primase does not catalyze strand-displacement synthesis by DNA polymerases from either *E. coli* or phage T4. It also interacts stoichiometrically with T7 ssDNA-binding protein which facilitates DNA unwinding by preventing the two strands from reannealing.

DNA synthesis in the opposite direction (lagging-strand synthesis) is initiated by T7 DNA helicase-primase, a protein that binds preferentially to four specific sequences (5′-GGGTC-3′, TGGTC, GTGTC, TTGTC), one or more of which are found at oriT7 as well as at secondary origins in the genome. However, primase activity is not dependent on binding to a specific DNA sequence (Table 5-1), thereby allowing flexibility in selection of initiation sites for initiation of Okazaki-fragment synthesis. T7 DNA primase alone binds template DNA weakly; tight DNA binding occurs via interactions between DNA templates and the helicase domain, but the binding of either ATP or CTP significantly increases the affinity of the primase for its DNA template. DNA lacking a primase-recognition site does not inhibit oligoribonucleotide synthesis; Okazaki fragments can vary in length from 0.5 to 6 kb with the concentration of T7 DNA helicase-primase as the primary determinant of the frequency of Okazaki-fragment initiation events. T7 ssDNA-binding protein stimulates the frequency of primase initiation approximately 10-fold.

Bacteriophage T4

The bacteriophage T4 genome consists of a single linear 169 kbp DNA molecule. It contains four replicators that have been mapped to specific sequences called oriA, oriF, oriG, and oriE. Like oriT7, these replicators are promoters that are recognized by a unique RNA polymerase. In this case, however, T4 does not encode a new RNA polymerase, but instead modifies the host RNA polymerase. *E. coli* RNA polymerase initiates transcription at T4 replicators only after its σ70 subunit is modified by the T4-encoded AsiA factor and T4 transcriptional activators MotA and DbpC have associated with one of the four replicators. The RNA polymerase is then displaced by T4 DNA polymerase (gene 43), which uses the RNA transcript as a primer to initiate DNA synthesis. T4 ssDNA-binding protein (gene 32) facilitates this transition by binding to the displaced ssDNA template. T4 helicase-loading protein (gene 59) then loads T4 DNA helicase (gene 41) onto the displaced ssDNA template, simultaneously arresting progress of the T4 DNA polymerase. The helicase recruits T4 DNA primase (gene 61) which initiates synthesis of the first RNA primer on the displaced template. This primer is used by another T4 DNA polymerase to initiate DNA synthesis in the opposite direction. Thus, T4-modified *E. coli* RNA polymerase, T4 helicase, and T4 primase serve as initiator proteins for T4 replicators, whereas two T4-encoded transcription factors, T4 ssDNA-binding protein (SSB), and the T4 helicase loader facilitate the initiation reaction.

Bacterial Plasmid ColE1

Most bacterial plasmids encode their own initiator protein, but a few rely entirely on their host cell to provide the machinery for duplicating their genome. The best characterized of these is the *E. coli* plasmid ColE1. ColE1 can replicate even when bacterial protein synthesis is inhibited in the presence of chloramphenicol, thus allowing selective amplification of the plasmid under conditions that prevent bacterial proliferation. Such plasmids utilize a mechanism for initiation analogous to that used by phage T7.

ColE1 DNA replication begins when *E. coli* RNA polymerase initiates synthesis of an RNA primer (RNA II) 555 bp upstream of the initiation site for leading-strand DNA synthesis on the 'continuous' arm of one replication fork (Figure 9-1). This process is dependent on the secondary structure at the 5′-end of RNA II. RNase H, a nonspecific ribonuclease that cleaves the 3′-phosphoryl bond of RNA in DNA:RNA duplexes to produce 3′-OH-terminated RNA, cleaves RNA II at the replication origin. *E. coli* DNA polymerase I extends the 3′-end of this RNA primer, and the plasmid replicates unidirectionally from a unique origin (the RNA-p-DNA transition site). Two primosome assembly sites (Chapter 13, replication restart) exist downstream of the ColE1 origin, one is responsible for initiation of discontinuous DNA synthesis on the lagging-strand template, the other for switching from DNA Pol-I to DNA Pol-III on the leading-strand template.

Remarkably, ColE1 can replicate in bacteria lacking either RNase H or both RNase H and DNA Pol-I, revealing the existence of alternative pathways. In the absence of RNase H, RNA II transcripts can still be recognized as an RNA primer by Pol-I without cleavage of RNA II at the replication origin. In the absence of both DNA Pol-I and RNase H, DNA synthesis can occur at various sites on the nontranscribed (i.e. displaced) DNA strand by assembly of replisomes, but this reaction is not efficient and always terminates at a specific DNA fork barrier called terH. Replication can be extended beyond terH by DNA helicase.

The number of copies of ColE1 per cell is limited by synthesis of a 108-nucleotide RNA I molecule that is the complementary (antisense) sequence of RNA II. Thus, RNA I inhibits binding of RNA II with its DNA template. The pairing of RNA I and RNA II is regulated by a small protein dimer called Rom or Rop that binds to specific sites on the RNA I:RNA II hybrid. The net result is that the number of plasmids is determined by the concentrations of RNA I and Rom.

Mitochondrial DNA

Part of the eukaryotic genome resides in the cell's mitochondria, chloroplasts, and kinetoplasts (Table 1-3). Each of these organelles contains one or more copies of a unique DNA molecule that is duplicated independently of the nuclear DNA portion of the genome. The replicators from mouse and human **mitochondrial DNA** (**mtDNA**) initiate DNA synthesis at a specific site called oriH (Figures 8-5 and 10-4). oriH appears to be required for initiation of the **D-loop** intermediate, although this has never been demonstrated by genetic analysis. The origins of nascent DNA strands with RNA covalently attached to their 5′-ends have been mapped to specific nucleotide positions within oriH. The 5′-end of the RNA maps to the transcription promoter on this strand, and the RNA-p-DNA covalent linkages map to several sites slightly downstream of the transcription start site. Three **conserved sequence blocks** (**CSBs**) reside between the light-strand transcription promoter and the sites of transition from RNA to nascent DNA. These conserved sequences may determine where the transition from RNA synthesis to DNA synthesis begins. Thus, initiation of DNA replication at oriH is analogous to the phage T7 replicator in that both require a specific RNA polymerase to initiate transcription at a specific promoter, and both require a mechanism by which DNA polymerase can use these transcripts to initiate DNA synthesis.

mtDNA replication begins with initiation of transcription at the light-strand promoter (Figure 10-4). This involves the core mtRNA polymerase, the HMG-box transcription factor mtTFA, and transcription factor mtTFB2. As transcription proceeds through oriH, an RNA:DNA hybrid forms that involves some subset of the conserved sequence blocks (CSBs 1, 2, and 3). The precise nucleotide sequence requirements for RNA:DNA hybrid formation appear to be flexible, which could explain the absence of CSB 2 and CSB 3 in certain vertebrates. The RNA:DNA hybrid creates the substrate for RNA processing by RNase MRP that leads to the formation of RNA primers. The cleavage pattern generated on synthetic substrates suggests that RNase MRP is an important mtRNA primer processing activity in mitochondria. Alternatively, one or more of the CSBs might promote

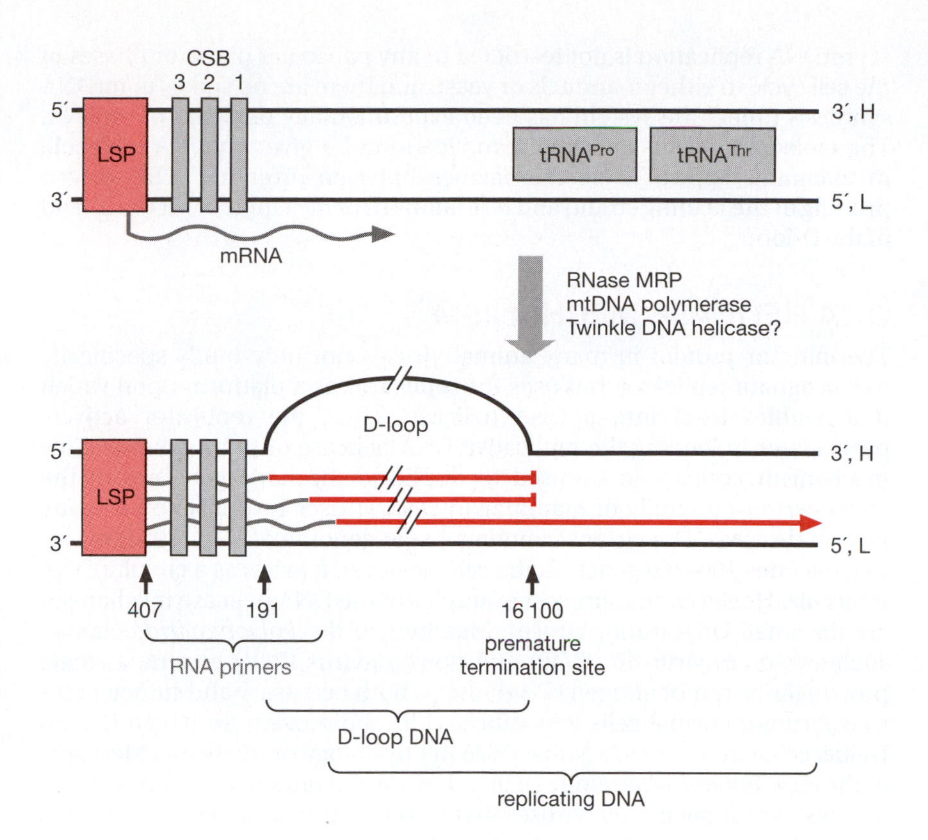

Figure 10-4. Human mtDNA heavy-strand replication origin (oriH). DNA replication at oriH begins with transcription from the 97 bp light (L)-strand promoter (LSP) at map position 407 through the downstream tRNA genes. Some transcripts either are terminated prematurely, or they are cleaved by the mitochondrial RNA-processing endonuclease (RNase MRP). In either case, the transition from RNA to DNA synthesis appears to be determined by three conserved sequence blocks (CSBs) located between map positions 407 and 191. These RNA primers are used by the mtDNA polymerase to initiate heavy (H)-strand DNA synthesis. However, some of these RNA-p-DNA chains terminate prematurely to produce D-loop structures (map positions 191 to 16 100) containing approximately 660 nucleotides of nascent DNA, whereas others continue to replicate the mtDNA genome, perhaps with the help of the mtDNA helicase, Twinkle. Human mtDNA contains 16 569 bp. (Data from Clayton, D.A. and Larsson, N.-G. (2006) Mitochondrial DNA replication and human disease. In *DNA Replication and Human Disease* (DePamphilis, M.L., ed.), pp. 547–560, Fish, J., Raule, N. and Attardi, G. (2004) *Science* 306: 2098–2101, and Pham, X.H., Farge, G., Shi, Y. *et al.* (2006) *J. Biol. Chem.* 281: 24647–24652.)

premature termination of the LSP transcript. In either case, synthesis of the heavy (H) DNA strand occurs through extension of an RNA primer by mtDNA polymerase (also called DNA polymerase-γ).

Two types of initiation events occur at oriH, one that is productive and one that is not. Most DNA synthesis events at oriH terminate prematurely, resulting in a D-loop DNA structure that contains a nascent H-strand about 660 nucleotides long (Figures 8-5 and 10-4). In human mtDNA, nascent DNA chains that start at map positions 191 and 151 terminate prematurely at map position 16 100. However, nascent DNA chains starting at map position 57 proceed well beyond this control point, behaving as 'true' replicating intermediates. This origin appears to play a dominant role in maintaining the number of copies of mtDNA in each organelle. The formation of D-loops appears to play a major role in the initial recovery of the normal mtDNA complement after mitochondrial depletion has occurred, for example in accelerating mtDNA replication to satisfy the demands of developmental, physiological, or age-related changes.

DNA unwinding appears to involve the human mtDNA helicase, Twinkle. Twinkle is encoded by a nuclear gene that is essential for mtDNA maintenance and regulation of mtDNA copy number. Twinkle bears amino acid sequence homology with phage T7 gene 4 protein, a bifunctional DNA primase/DNA helicase. Recombinant Twinkle forms a hexamer that prefers UTP as an energy source while it translocates along the DNA template in a 5′→3′ direction and unwinds duplex DNA (Table 4-1).

Based on a combination of two-dimensional gel electrophoresis analysis of replication intermediates and mapping of the 5′-ends of RNA-primed DNA chains, **light (L)-strand DNA synthesis** occurs in the major noncoding region of vertebrate mtDNA and is effectively unidirectional. In some cases, conversion of nascent RNA strands to DNA starts at defined loci, the most prominent of which maps, in mammalian mtDNA, in the vicinity of the site known as oriL (Figure 8-5). The mechanism for RNA-primed DNA synthesis of the L-strand is unknown.

mtDNA replication is not restricted to any particular phase or phases of the cell cycle in either mammals or yeast, and there are no spikes in mtDNA synthesis unless the system has been experimentally depleted of mtDNA. The molecular events controlling replication of a given mtDNA molecule in mammals appear to be the balance between promoter (LSP)-driven priming of the leading strand and termination (or lack thereof) at the 3′-end of the D-loop.

DNA HELICASE MECHANISM

The initiator protein in many animal viruses not only binds specifically to its cognate replicator, but uses the replicator as a platform upon which it assembles itself into a DNA helicase. Thus, the replicator actively participates in loading the replicative DNA helicase onto its genome. This mechanism appears to be used to duplicate the large genomes of the *Herpesviridae*, a family of mammalian DNA viruses responsible for many human diseases. The virions contain a linear genome of 150–170 kb in size that encodes 100–200 genes. In the cell, however, it exists as a circular DNA molecule. However, the simplest examples of the DNA helicase mechanism are the small DNA tumor viruses. Members of the *Polyomaviridae* family such as simian virus 40 (SV40) and polyomavirus (PyV) encode a single protein [large tumor antigen (T-ag)] that is both necessary and sufficient for transforming normal cells into cancer cells, a discovery for which Renato Delbecco shared the 1975 Nobel Prize in Physiology or Medicine. Members of the *Papillomaviridae* family such as bovine and human papillomaviruses (or 'PVs') are small dsDNA viruses that infect mammals and cause diseases ranging from warts to cervical cancer. *Polyomaviridae* and *Papillomaviridae* contain 5–8 kbp circular genomes that encode only five to eight genes. Except for their initiator protein, they rely entirely on their host cell to provide the DNA replication machinery.

Each viral genome encodes a single initiator protein that must assemble into a complex of six identical proteins (homohexamer) in order for it to exhibit ATP-dependent unwinding of long stretches of duplex DNA (DNA helicase activity). Only the replicators in *Polyomaviridae* and *Papillomaviridae* allow assembly of their cognate initiator protein (T-ag and E1, respectively) into a DNA helicase that can unwind DNA at an internal site. Initiator protein that assembles into hexamers in the absence of replicator can unwind DNA only by threading its way onto linear dsDNA molecules with a terminal ssDNA extension (Figures 4-1 and 4-3). Thus, they are inactive as initiator proteins. Once loaded onto a DNA molecule, however, these viral initiators can proceed to unwind dsDNA and thereby create replication forks. Thus, the initiator not only melts the replicator's DUE, but it then becomes part of the replication machinery that drives a replication fork.

Simian Virus 40

The SV40 replicator is the paradigm for small dsDNA viral genomes that replicate in the nuclei of mammalian cells. The 64 bp SV40 ori-core is required for DNA replication both *in vivo* and *in vitro*. It consists of three functional domains: a 15 bp inverted repeat that is part of the DUE, an ORE consisting of a 27 bp inverted repeat that contains four GAGGC pentanucleotide binding sites for the SV40 initiator protein T-ag, and a 17 bp asymmetrical A:T-rich sequence in which most of the As are on one strand and the Ts are on the other (Figure 10-5A). T-ag binds to each GAGGC pentanucleotide with a 10-fold greater affinity than to random DNA sequences in either dsDNA or ssDNA. In fact, a single pentanucleotide is sufficient to mediate assembly of a T-ag hexamer, and the presence of two sites is sufficient for assembly of two T-ag hexamers (one for each replication fork) referred to as a double hexamer. However, the intact ORE is required to assemble a double

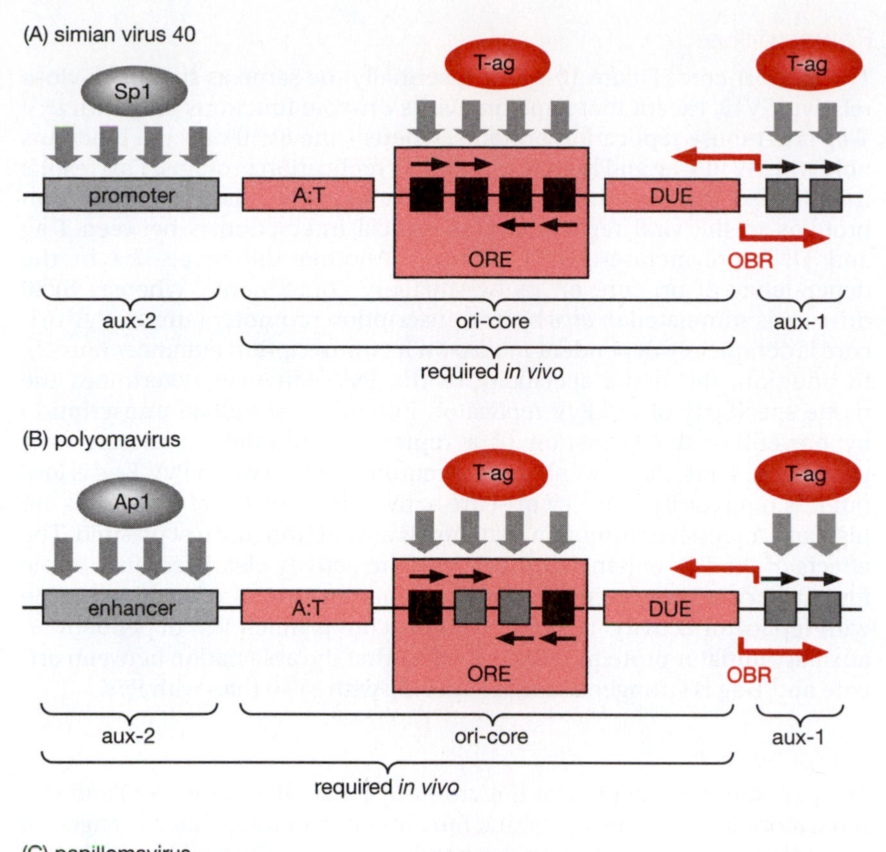

Figure 10-5. Animal virus replication origins. (A) SV40 replicator consists of an ORE containing four strong T-antigen (T-ag)-binding sites (GAGGC, solid boxes), a DUE, and an asymmetric A:T-rich sequence (A:T). Black arrows indicate relative sequence orientations. Red arows indicate start sites for leading-strand nascent DNA. (B) The PyV replicator is driven by PyV T-ag. Shaded T-ag-binding sites are weaker than solid GAGGC T-ag-binding sites. (C) The papillomavirus replicator is driven by papillomavirus E1 protein and facilitated by papillomavirus E2 protein. Shaded region and one E2-binding site are found in the human origin (HPV), but not in the bovine origin. (Data from DePamphilis, M.L. (1996) Origins of DNA replication. In *DNA Replication in Eukaryotic Cells* (DePamphilis, M.L., ed.), pp. 45–86, and Schuck, S. and Stenlund, A. (2007) *J. Virol.* 81: 3293–3302.)

hexamer of T-ag in a manner that allows it to initiate DNA unwinding in the DUE. The orientation and spacing of the four pentanucleotides are critical for replicator function. Thus, SV40 provides another example of an active replicator, a unique DNA sequence that is required for its cognate initiator protein to assemble into a DNA helicase that can initiate DNA unwinding at an internal site.

The SV40 ori-core is flanked by two ori-auxiliary sequences that facilitate origin activity *in vivo*, but that are dispensable *in vitro*. aux-1 binds a T-ag dimer, and aux-2 binds the transcription factor Sp1. aux-2 can be replaced with binding sites for other transcription factors. Some, such as AP1 and NF1, will also stimulate ori-core activity, but many others will not. Both the transcription factor activation domain and its DNA-binding domain are required, but stimulation of ori-core does not involve stimulation of transcription (i.e. RNA polymerase activity). Instead, it appears that transcription factor binding to auxiliary sequences facilitates ori-core activity by preventing chromatin structure from interfering either with T-ag binding to the ORE, or with T-ag unwinding origin DNA.

Polyomavirus

The PyV ori-core (Figure 10-5B) is essentially the same as that of its close relative, SV40, except that papillomavirus ori-core functions only with PyV T-ag and mouse replication proteins, whereas the SV40 ori-core functions only with SV40 T-ag and human or monkey replication proteins. This results from the fact that the initiator protein must recruit cellular DNA replication proteins to the viral replicator. One critical interaction is between T-ag and DNA polymerase-α•DNA primase. Another difference lies in the dependency of ori-core on its ori-auxiliary components. Whereas SV40 ori-core is stimulated *in vivo* by its transcription promoter (aux-2), PyV ori-core is completely dependent *in vivo* on its transcription enhancer (aux-2). In addition, the tissue specificity of the PyV enhancer determines the tissue specificity of the PyV replicator. Enhancers stimulate transcription by preventing the formation of a repressive chromatin structure over promoters. Thus, the PyV enhancer is required *in vivo* when PyV T-ag is low, but it is not required for PyV ori-core activity *in vitro* where T-ag levels are high and repressive chromatin structure is absent from the test plasmid. The effects of the PyV enhancer on PyV ori-core activity clearly confirm a role for auxiliary sequences in preventing chromatin structure from interfering with replicator activity. That the SV40 ori-core is much less dependent on auxiliary initiator proteins reflects the fact that the association between ori-core and T-ag is stronger and more specific with SV40 than with PyV.

Papillomavirus

The papillomavirus replicator (Figure 10-5C) is similar to the SV40 and PyV replicators in that it also contains four initiator-binding sites arranged in two pairs, and it is bound by the papillomavirus initiator protein E1. E1 is the replicative DNA helicase of papillomavirus, and it shares extensive homology with SV40 and PyV T-antigens. As with T-ag, the affinity of E1 monomers for DNA depends on the number and arrangement of its DNA binding sites in the papillomavirus ORE. Assembly of the E1 double hexamer proceeds by an ordered pathway. First a double monomer appears, then a double dimer, a double trimer, and finally a double hexamer. As with T-ag, E1 binds to specific sequence elements in the papillomavirus ori-core, forms a hexameric helicase that unwinds the origin DNA, and then travels with the replication fork, unwinding the DNA ahead of DNA synthesis. DNA unwinding begins within the approximately 12 bp asymmetric A:T-rich element. In this case, the A:T-rich element and the DUE are coincident.

At low E1 concentrations, binding to ori-core is facilitated by interaction with E2, a dimeric transcription-replication factor that binds with high affinity to aux-1 and aux-2. E2 is not required *in vitro*, where high levels of E1 alone can initiate DNA replication. The preferred binding site for the E1•E2 complex is 5′-ATTGTT-3′; four to six related sites are present in the ORE. The E1•E2–DNA complex consists of an E2 dimer and an E1 dimer. This serves as a starting point for the assembly of larger oligomeric E1 complexes. *In vitro*, E1 binds to DNA with little sequence specificity in absence of E2. Nevertheless, E1 can still assemble into replication-competent oligomers at high protein concentrations. Thus, E2 imparts origin specificity onto E1 while facilitating assembly of larger E1 complexes.

Herpes Simplex Virus

The herpes simplex virus (HSV) genome contains three replicators (oriL and two copies of oriS) and encodes one initiator protein (UL9) that binds specifically to each of the three HSV replicators. This interaction is required for HSV DNA replication. UL9 is similar to the initiators encoded by SV40, PyV, and papillomavirus viral genomes in that all four initiator proteins exhibit ATPase and DNA helicase activities *in vitro*. UL9 is able to unwind

DNA oligonucleotides up to 100–200 bp on its own, and together with ICP8/UL29 (an HSV-encoded ssDNA-binding protein) UL9 can unwind up to 2 kb of DNA. However, in contrast to T-ag and E1 initiator proteins, UL9 cannot unwind blunt-ended linear or circular dsDNA containing an HSV replicator, even in the presence of ICP8. Intriguingly, UL9 together with ICP8 and in the presence of ATP can convert oriS into a cruciform structure that stably binds UL9. This structure, in turn, appears to recruit a replisome composed of the HSV DNA polymerase, a trimeric helicase/primase enzyme that is required to unwind DNA at HSV replication forks (Table 4-1), and ICP8/UL29. However, as yet, HSV replicator-dependent DNA replication has not been observed using purified proteins.

DNA HELICASE LOADER MECHANISM

The DNA genomes of bacteria, archaea, and the nuclear portion of eukaryotic genomes as well as some bacteriophages and bacterial plasmids all utilize initiator proteins that function as DNA helicase loaders (Table 10-2). Remarkably, evolution has selected a single initiator mechanism for the genomes of all cellular organisms. More remarkable still is the fact that the interaction between replicator and initiator appears to vary among these genomes. Some of these genomes employ active replicators (bacteria, bacteriophages, bacterial plasmids), some employ passive replicators (yeast and perhaps archaea), and some employ conditional replicators (metazoa). When a DNA helicase loader binds to an active replicator it distorts the DNA and forces it to partially unwind. The unwound DNA region appears within the DUE component of the replicator and serves as an entry site for the replicative DNA helicase. So far, the replicators in archaea, yeast, and metazoan genomes do not appear to unwind in response to initiator binding. Nevertheless, the requirements for helicase loading described in Chapters 4 and 11, and the effects of DNA superhelicity on the initiation process described below, suggest that they do.

The replicative DNA helicase associates with a chaperone (Table 10-2) that brings it to the initiator–replicator complex. This event is followed closely by the loading of a DNA primase and a DNA polymerase that will initiate DNA synthesis on one of the two templates. The net result is that DNA unwinding is coupled with DNA synthesis in the form of two replication forks traveling away from the replicator in opposite directions.

One notable difference is that the DNA helicase loaders in bacteria, archaea, and eukarya all bind and hydrolyze ATP, whereas those encoded by bacteriophages and bacterial plasmids do not. Therefore, the energy for the initial distortion of the DNA is provided by the free-energy change of initiator–DNA interaction; energy from ATP hydrolysis is required only to

Table 10-2. Initiator Proteins As DNA Helicase Loaders

Genome	Paradigm	Helicase loader	Helicase chaperone	Helicase
Some plasmids	R6K	π+DnaA+IHF	DnaC	DnaB
Some bacteriophages	λ	O	P	DnaB
Bacteria	*E. coli*	DnaA	DnaC	DnaB
Archaea	*S. solfataricus*	Cdc6/Orc1?	Cdc6/Orc1?	Mcm
Yeast	*S. cerevisiae*, *S. pombe*	ORC(1-6)+Cdc6	Cdt1	Mcm(2-7)
Metazoa	*H. sapiens*	ORC(1-6)+Cdc6	Cdt1/RLF-B	Mcm(2-7)/RLF-A

Different names for the same protein are separated by a '/'.

translocate the replicative DNA helicase along the DNA. Moreover, in the examples of bacteria, temperate bacteriophages, and plasmids that encode their own initiator proteins, the replicators all contain multiple tandem initiator DNA-binding sites within their ORE, and in each case, binding of the cognate initiator protein results in localized DNA bending and partial melting of the DUE (Figures 11-4 and 11-5), despite the fact that initiators encoded by these phage and plasmids do not require ATP either for binding to the replicator or for initiating DNA unwinding, whereas the initiator encoded by bacteria does require ATP for these functions. Thus, evolution has imposed an additional level of regulation on the activation of replicators in cellular organisms.

Bacteria

Three well-characterized examples of DNA helicase loaders that function in bacteria are the *E. coli* DnaA protein, the *E. coli* bacteriophage λ O protein, and the *E. coli* bacterial plasmid R6K π protein. In each case, the initiator binds to its cognate replicator, distorts the DNA, and induces localized DNA melting. This, in turn, allows loading of the replicative DNA helicase. In one sense, these initiator proteins are static DNA helicases; proteins that can melt the DNA at or near their binding site, but that cannot continue to unwind DNA at the replication fork.

Escherichia coli

Two well-characterized examples of bacterial genome duplication are *E. coli*, a bacterium that lives in our intestines and can cause human illness, and *B. subtilis*, a soil bacterium that undergoes a primitive form of cell differentiation called sporulation. They contain a single, circular duplex DNA genome of 4.6 and 4.2 Mbp, respectively, with a single chromosome replicator (oriC) that binds the initiator protein DnaA. DnaA binds specifically to a sequence called the DnaA box that is found in all bacterial genomes sequenced to date. Therefore, the *E. coli* replicator has become the paradigm for all bacteria, although some differences exist.

E. coli oriC is a 245 bp sequence that is essential for duplicating the bacterial genome (Figure 10-6). The efficiency of initiation, however, is influenced by flanking sequences (i.e. sequence context) and by the position and orientation of oriC within the DNA genome. oriC contains several functional domains. Adjacent to the array of DnaA boxes and I-sites is an AT-rich DUE composed of three 13-mer repeats (5′-GATCTaTTtaTTT-3′). This region contains a third class of DnaA-binding sequences, the ATP–DnaA boxes, that are similar to I-sites in that they are selectively bound by DnaA only in the presence of ATP. Initial strand separation takes place in

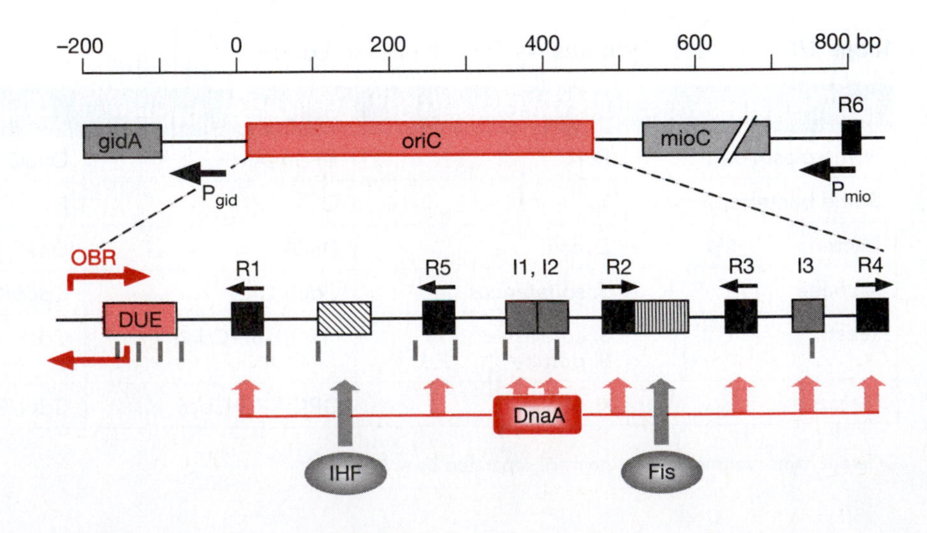

Figure 10-6. *E. coli* oriC. The minimal oriC replicator is flanked by two genes, gidA and mioC, whose promoter (P) and direction of transcription (arrow) are indicated. DnaA-binding sites [5′-TT(A/T)TNCACA-3′] are indicated as black boxes (R1–R6) and shaded boxes (I1–I3). Arrows above boxes indicate their orientation relative to one another. The sites from which leading-strand synthesis begins [origin of bidirectional replication (OBR)] are indicated by red arrows. GATC sites are indicated as red vertical bars. (Data from Leonard, A.C. and Grimwade, J.E. (2005) *Mol. Microbiol.* 55: 978–985, and Messer, W. (2002) *FEMS Microbiol. Rev.* 26: 355–374.)

the DUE. Binding of DnaA to the ssDNA 13-mer repeats might maintain the DUE in a melted conformation, ready for loading the replicative DNA helicase, DnaB.

The bacterial initiator protein, DnaA, is a 52.6 kDa protein that binds to each of the five R-boxes and three I-sites with a K_d between 0.6 and 50 nM. DnaA binds with highest affinity to the duplex DNA with the sequence 5′-TTATCCAC-3′. DnaA is an ATPase, and the nucleotide-binding status of DnaA influences its association with DNA. I-sites differ in several bases from R-boxes and bind the active form of DnaA (DnaA–ATP) threefold more tightly than the inactive form (DnaA–ADP). In contrast, R-boxes bind both DnaA–ATP and DnaA–ADP with equal affinity. Mutations that reduce the affinity of DnaA to either R-boxes or I-sites diminish oriC activity *in vivo*. Mutations that permit I-sites to bind DnaA–ADP and DnaA–ATP with equal affinity result in reinitiation of DNA replication prior to completion of S phase.

Located in the left and right half of oriC are binding sites for IHF and Fis proteins, respectively. These are origin auxiliary sequences; they facilitate oriC activity, but they are not required. Strains lacking Fis or IHF DNA-binding sites in oriC have altered initiation timing, but otherwise appear normal. Thus, in concert with negatively supercoiled DNA and the architectural factors IHF and Fis, DnaA melts the DUE, generating the ssDNA necessary for loading the replicative DNA helicase and for assembly of the replication machinery.

The nucleotide sequence GATC is repeated at 11 separate locations within oriC. Seven are located at weak DnaA-binding sites in the 13-mer region and positioned either within or adjacent to R5, I1, I2, and I3. GATC is the recognition site for deoxyadenosine methyltransferase, and also for the interaction of SeqA protein with DNA. Immediate reactivation of oriC is prevented by sequestering it to the cell membrane, a mechanism dependent both on SeqA and methylation of the GATC sites in oriC.

Transcription of DNA by RNA polymerase in a region less than 500 bp from oriC can facilitate the ability of DnaA to open the DNA duplex. However, in contrast to phages T7, T4, and mtDNA, transcriptional activation of oriC does not provide a primer for DNA synthesis, but a mechanism for altering the local DNA structure. Transcription stimulates melting at the DUE only under conditions that inhibit oriC melting, such as organization of the DNA into chromatin with HU protein, low temperature, or low negative superhelicity in the vicinity of oriC. HU protein is the bacterial equivalent of the histones that organize eukaryotic DNA into nucleosomes, and when HU binds to DNA it inhibits DNA unwinding. A similar phenomenon occurs at the bacteriophage λ replicator. As with the auxiliary components of the SV40 and polyomavirus replicators, transcription may prevent chromosomal proteins from otherwise interfering with initiator protein binding or activity.

Transcription can also affect the ability of DnaA to induce unwinding of the DUE by changing the superhelical density in the oriC locus. Transcription creates positive superhelical turns downstream and negative superhelical turns upstream of the site of transcription (Figure 1-8). Therefore, the effect of transcription on the ability of initiator to unwind replicator DNA depends on the direction of transcription relative to the replicator. Two genes, gidA and mioC, flank oriC (Figure 10-6), but neither gidA nor mioC transcription is required in cells deficient in the histone-like proteins Fis or IHF. Transcription from the mioC promoter in an oriC-driven plasmid correlates with an increase in plasmid copy number, revealing that mioC transcription can stimulate oriC activity. However, deletion of the mioC promoter, including the DnaA box (R6), does not affect the synchrony of initiation events at oriC. However, oriC that is impaired by deletion of DnaA box R4 requires transcription of at least one of these two genes. Thus, transcription stimulates oriC only under suboptimal conditions.

Other bacterial replicators mimic *E. coli* oriC in that they contain a cluster of DnaA-binding sites, but differences do occur in their organization and regulation. For example, *B. subtilis* oriC contains two regions with multiple DnaA-binding sites that are separated by the *dnaA* gene. Three AT-rich 16-mers constitute a DUE at the upstream end of *B. subtilis* oriC and another AT-rich region exists at the downstream end. However, neither of these DnaA-binding regions alone exhibits autonomously replicating sequence (ARS) activity, the ability to impart DNA replication activity when present in a plasmid. In addition, the *dnaA* gene coding region can be removed without affecting ARS activity of the *B. subtilis* oriC-containing fragment, and the minimum distance between the two DnaA-box regions that is required for oriC activity is 274 bp. Thus, although the organization of *B. subtilis* oriC at first appears to differ significantly from that of *E. coli* oriC, the two replicators are functionally the same. They do differ, however, in that GATC sites are absent from the *B. subtilis* replicator.

Bacteriophage λ

Lysogenic or temperate bacteriophage are those that can either enter a lytic cycle in which they multiply and kill their host cell, or enter a quiescent state in which transcription of the phage genome is repressed and therefore does not replicate. In the quiescent state, the phage chromosome integrates into the host chromosome where it is replicated along with the host chromosome and passed on to the daughter cells. *E. coli* phage λ and *Salmonella* phage P22 are two well-characterized examples of lysogenic bacteriophages. Both employ the same mechanism to initiate genome duplication, and comparisons of bacteriophage protein sequences suggest that other lysogenic phage use this mechanism as well.

The λ chromosome is linear within the phage, but it circularizes when it enters the host cell as a consequence of its cohesive termini (the 5′-ends are 12 nucleotides longer than the 3′-ends and complementary to one another). The λ chromosome contains a single replicator (oriλ) and encodes two initiator proteins (products of the O and P genes). The O protein is a small 33.7 kDa protein that is very unstable ($t_{1/2} \approx 1.5$ min) unless it is bound to DNA. Upon binding to DNA, however, O protein begins unwinding DNA at oriλ, a process facilitated by the host cell's ssDNA-binding protein (SSB) binding to the exposed ssDNA bubble and thereby preventing it from reannealing. P protein captures the host cell's DNA helicase (DnaB) and brings it to the oriλ–O-protein complex. Thus, the O and P proteins work together to load the *E. coli* replicative DNA helicase (DnaB) onto the λ genome. Phage λ DNA replication, like that of its *E. coli* host cell, is strongly dependent on a negatively supercoiled DNA template to facilitate DNA unwinding at the replicator. Other than the O and P proteins encoded by the λ genome, the phage relies entirely upon the host cell's replication-fork machinery (Figure 5-8) to duplicate its genome.

oriλ is located within the structural gene for the O protein and consists of three components: the promoter (P_R), a series of four 19 bp O-binding sites, and a 40 bp asymmetrical A:T-rich region (Figure 10-7). Transcription is required for activation of oriλ *in vivo*, although it does not have to pass through the replicator. *In vitro*, transcription is not required for oriλ activity unless high levels of HU protein are present. Thus, transcription serves to disrupt repressive chromatin structure at oriλ. This allows O protein to bind to four tandem sequences in oriλ, thereby bending the DNA in such a way that melts part of the DUE region.

Plasmids that encode their own initiator

Bacterial plasmids are invaluable tools for the analysis of replication origins, termination sites, and regulation of DNA replication, as well as for the production of large quantities of recombinant proteins. The number of

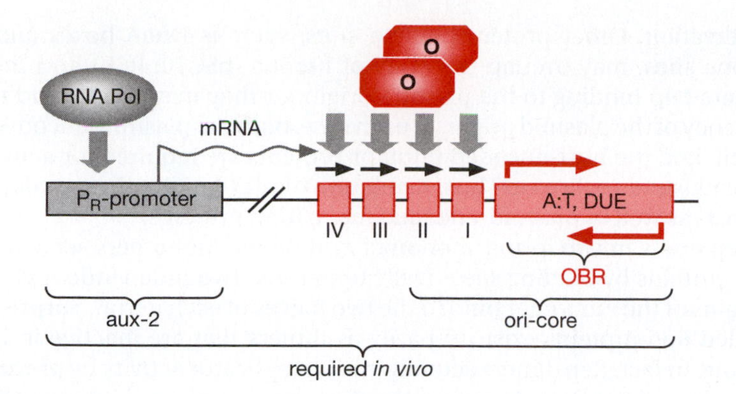

Figure 10-7. Bacteriophage λ replication origin. O-protein dimers bind specific DNA sites (5′-ATCCCTCAAAWYRRGGGRW-3′) indicated as black boxes I–IV each oriented in the same direction. The 5′-end of the λ P_R transcript is 1011 bp from the binding site IV. Leading-strand synthesis (red arrows) begins in the AT-rich DUE. (Data from Dodson, M., McMacken, R. and Echols, H. (1989) *J. Biol. Chem.* 264: 10719–10725, and Yoda, K., Yasuda, H., Jiang, X.W. and Okazaki, T. (1988) *Nucleic Acids Res.* 16: 6531–6546.)

copies of a plasmid that exist in a single bacterium (generally between one and 20) must be regulated to prevent death of the host cell. In most cases, the number of plasmid genomes is limited by the availability of initiator protein, either ones encoded by the plasmid or ones encoded by its host.

Some plasmids replicate bidirectionally from a single replicator, as does their host's genome. Others, such as ColE2, R6K, pSC101, and R1, replicate unidirectionally because the progress of one of the replication forks is held back by a replication-fork barrier (Figure 8-1). For example, replication from all three R6K origins is unidirectional until the first fork arrives at the replication terminus where its progress is arrested. Eventually a fork arrives from the opposite direction and the two sibling genomes separate. Thus, the replication mode is 'sequentially bidirectional'. Some plasmids encode their own initiator protein (Rep) that binds to specific sequences of less than 20 bp called **iterons** that reside in close proximity to the *rep* gene (Figure 10-8). The sequence and number of iterons are characteristic for each plasmid, and they bind specifically their cognate Rep protein. Thus, iterons belonging to the same plasmid show a high degree of sequence

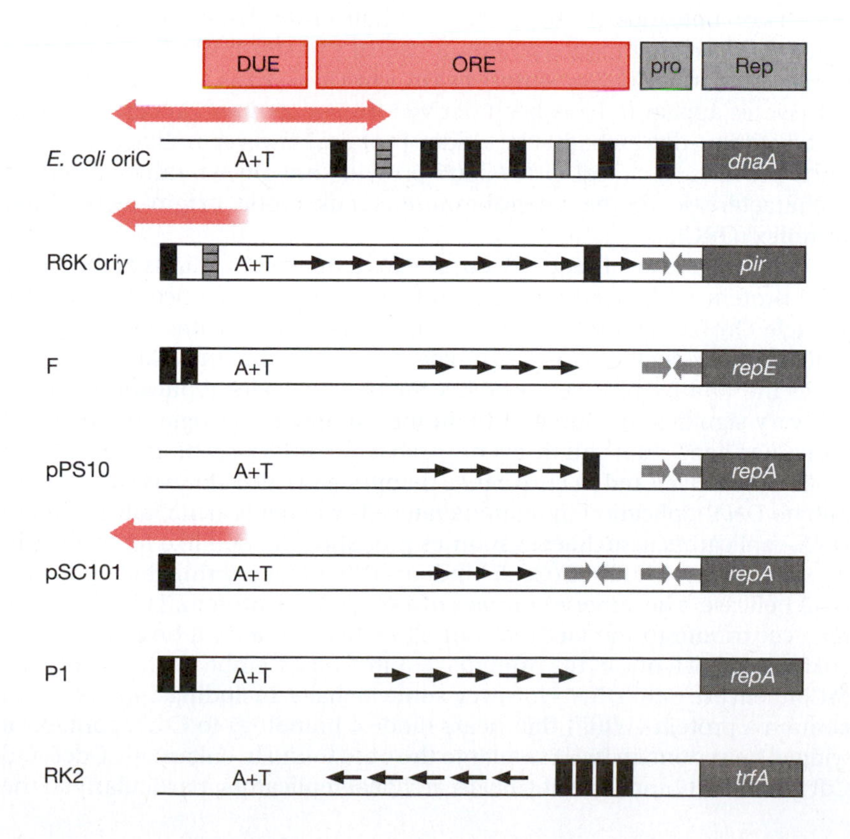

Figure 10-8. Replication origins in Gram-negative bacterial plasmids that encode their own initiator protein. Plasmid replicators (not drawn to scale) are located predominantly, if not exclusively, just upstream of the gene that encodes their initiator protein (Rep) and consist of several sites (iterons, →) that bind specifically their cognate Rep protein. This region accounts for about half of the replicator. DnaA-binding sites (▮) are also present, and in some cases Fis- (light-gray bar) and IHF- (striped bars) binding sites. Inverted repeat sequences (→ ←) frequently lie outside the replicator that both bind Rep and act as promoters for the initiator gene. In some cases, such as P1, one or more iterons serve as promoters (pro) for the *rep* gene. For historical reasons *rep* genes have various names. (Adapted from Chattoraj, D.K. (2000) *Mol. Microbiol.* 37: 467–476. With permission from John Wiley and Sons, Inc.)

conservation. Other protein-binding sites, such as DnaA boxes and Fis-binding sites, may overlap the array of iterons (pSC10, R6K) and thereby regulate Rep binding to the plasmid origin, or they may simply add to the efficiency of the plasmid origin. In each case, both the plasmid-encoded Rep protein and the host-encoded initiator proteins are required for activation of the plasmid replicator. Thus, regulation of plasmid duplication depends on host- as well as plasmid-encoded replication proteins.

Rep exists in two forms, monomer and dimer. Monomers serve as initiator proteins by binding specifically to iterons. Two independent globular domains of the monomer bind to the two halves of each iteron. Surprisingly, purified Rep proteins exist primarily as dimers that are inactive in iteron binding. In fact, Rep dimers actually inhibit replicator activity by preventing unwinding of the DUE. However, the dimers are converted into monomers by a conformational change mediated by chaperone proteins such as *E. coli* proteins DnaK, DnaJ, and GrpE.

Rep dimers also bind DNA sequences that specifically serve as promoters for *rep* genes in some plasmids. In these operators, half-iteron sequences are present as an inverted repeat (Figure 10-8). In P1, the iterons also act as the promoter for RepA. Thus, Rep dimers stimulate expression of the *rep* gene, but inhibit replicator activity. Rep monomers have the opposite effects, they inhibit expression of the *rep* gene, but activate replicator activity. In this manner, the steady-state equilibrium between Rep monomer and Rep dimer regulates replication of these plasmids.

ARCHAEA

Archaeal chromosomes range in size from 0.5 to 6 Mb and contain from one to three replication origins. Compared to bacteria and single-cell eukarya, little is known about the nature and function of their replication origins. What is known, however, reveals an intriguing blend of features associated with bacterial and eukaryal origins. Unfortunately, this has led to nomenclature that is borrowed from both bacterial and eukaryal systems. Archaea contain homologs of the eukaryotic initiator protein Cdc6. Since Cdc6 is homologous to the C-terminal half of the largest subunit, Orc1, of the origin-recognition complex, archaeal initiator proteins are referred to either as Orc1/Cdc6 or simply as Orc1 or Cdc6. Thus, Orc1 and Cdc6 in eukaryotes appear to have been derived from a common ancestor similar to the present-day archaeal Orc1/Cdc6 protein. However, unlike eukaryotic Cdc6 proteins, archaeal Orc1/Cdc6 proteins are DNA-binding proteins, a characteristic of the heterohexameric eukaryotic **origin-recognition complex** (**ORC**).

Almost all archaeal genomes encode from one to 10 variants of the Orc1/Cdc6 protein, with most species encoding two or three. *Pyrococcus* encodes a single Orc1/Cdc6 protein, and *Methanothermobacter* encodes two, but other archaea, such as *Haloferax*, have as many as 14 Orc1/Cdc6 proteins. Since the affinity of these proteins for their cognate replication origins can vary significantly, some of them may simply be vestigial remnants of an earlier gene-duplication event. Archaeal Orc1/Cdc6 binds to archaea replicators *in vivo* and *in vitro*. Since mammalian Cdc6 alone is sufficient to initiate DNA replication in mammalian cells when it is artificially bound to DNA, replicators in archaeal genomes probably function like replicators in yeast genomes, with Orc1/Cdc6 binding DNA and recruiting the replicative DNA helicase, a hexameric complex of a single Mcm protein. This difference may contribute to the fact that not all archaea contain a homolog of the eukaryotic Cdt1, one of the proteins required for assembly of the eukaryotic MCM helicase onto DNA. However, some archaea, including *Sulfolobus*, do contain a protein (WhiP) that bears modest homology to Cdt1, contains a winged helix domain (wH) similar to the ones found in eukaryotic Cdc6 and Cdt1, and that binds specifically to archaeal replicators, particularly in the presence of archaeal Orc1/Cdc6 proteins. Thus, archaeal replicators appear similar to eukaryal replicators, except that they employ a simpler system for loading the MCM DNA helicase onto chromatin.

Although the amino acid sequences of all archaeal Orc1/Cdc6 proteins are similar, they cluster into two main families that may represent two different modes of DNA binding. Nevertheless, there exists a general pattern in which one or more Orc1/Cdc6 proteins bind specifically to a DNA consensus sequence [(t/c)TNCANNNGAA(a/c)] that is represented multiple times within a locus of 1 kbp, and that is commonly found in archaeal replication origins.

Some archaea, such as *Pyrococcus*, contain a single replicator, called oriC, immediately upstream of a Orc1/Cdc6 gene. This is reminiscent of bacteria wherein the *dnaA* gene lies within or near the bacterial oriC replicator. Like bacterial replicators, *Pyrococcus* oriC possesses an asymmetrical A:T-rich sequence as well as repeated sequence motifs, and an **origin-recognition box** (**ORB**), equivalent to the ORE in Figure 9-1. Some archaea, such as *Methanothermobacter* (Mth), contain as many as 10 Orc1/Cdc6 DNA-binding sites, centered around an asymmetric A:T-rich region that may act as a DUE (Figure 10-9A). Of the two Orc1/Cdc6 variants encoded by this species, only Cdc6-1 binds tightly to Mth oriC. Some archaea, such as the hyperthermophilic *S. solfataricus* (Sso), have a single chromosome

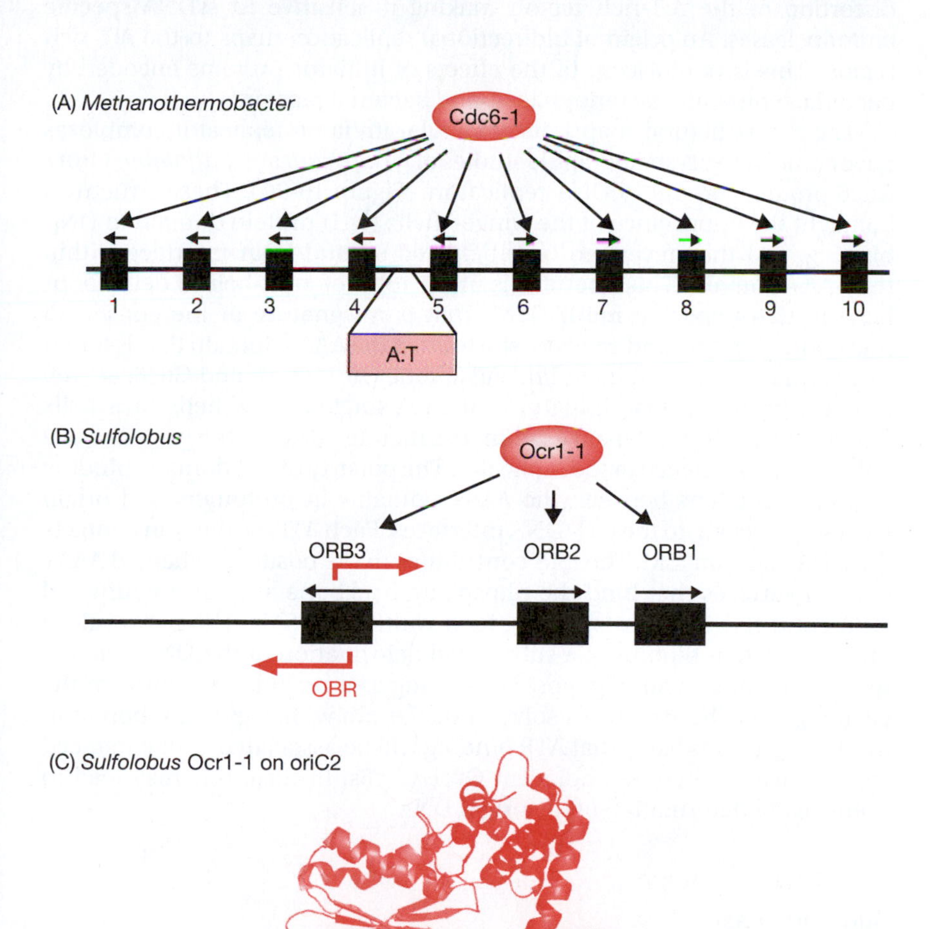

Figure 10-9. Archaeal replicators. (A) oriC in *Methanothermobacter thermoautotrophicus* genome contains 10 binding sites for the single initiator protein Orc1/Cdc6-1. Arrows indicate relative sequence orientation. (B) *Sulfolobus solfataricus* oriC1 contains three binding sites for Orc1-1. (C) Crystal structure of *S. solfataricus* Orc1-1 on oriC2. Positions of the winged helix (wH-DNA) and α-helical helix-loop-helix initiator-specific motif (ISM-DNA) DNA-binding modules are indicated. (Adapted from Capaldi, S.A. and Berger, J.M. (2004) *Nucleic Acids Res.* 32: 4821–4832, with permission from Oxford University Press. Additional data from Dueber, E.L., Corn, J.E., Bell, S.D. and Berger, J.M. (2007) *Science* 317: 1210–1213, and Robinson, N.P. and Bell, S.D. (2007) *Proc. Natl. Acad. Sci. USA* 104: 5806–5811.)

encoding three Orc1/Cdc6 homologs and three replication origins, two of which are located upstream of Orc1/Cdc6 genes.

The replicon organization of *Sulfolobus* may represent an evolutionary link between the single-origin systems of bacteria and some archaea and the multiorigin systems in eukarya. The three Sso Orc1/Cdc6 proteins exhibit a complex pattern of binding to their three cognate replication origins. Sso Orc1-1 and Orc1-2 bind to all three origins *in vivo* and *in vitro*, but Sso Orc1-3 binds only to oriC2 and oriC3. Sso Orc1-1 binds to three 36 bp ORBs within oriC1 (Figure 10-9B). These ORBs possess a region of dyad symmetry with an adjacent, asymmetrically positioned, G-rich sequence. These are the actual binding sites for Orc1-1, and various arrangements of this dyad element may constitute replication origins among the various species of archaea. However, the affinity of Orc1-1 is sharply reduced for individual dyad elements in comparison with the complete ORB.

The ORB consensus sequence identified in *Sulfolobus* also appears in *Aeropyrum* at a 685 bp locus that contains four canonical ORBs with an asymmetrical 67 bp A:T-rich region in the center. This organization is analogous to the Mth origin (Figure 10-9A). The *Aeropyrum* genome encodes two Orc1/Cdc6 genes; only Orc1-1 binds cooperatively to this putative replication origin. Each ORB binds Orc1-1 protein which then induces an ATP-mediated structural change in the DNA that results in distortion of the A:T-rich region making it sensitive to ssDNA-specific endonucleases. An origin of bidirectional replication maps to the A:T-rich region. This is reminiscent of the effects of initiator proteins encoded by bacteria, temperate bacteriophages, and bacterial plasmids.

The first structural snapshots of cellular initiator–replicator complexes have come from crystallographic studies of *Aeropyrum* and *Sulfolobus* Orc1/Cdc6 proteins bound to DNA replicators (Figure 10-9C). These structures highlight the importance of the winged helix (wH) protein domain in DNA binding, and they reveal an unanticipated second DNA interface within the **AAA+ domain**. This interface is in the form of an α-helical helix-loop-helix **initiator-specific motif** (**ISM**) that is a signature of the conserved nucleotide-binding and catalytic site termed the AAA+ domain that is found in all initiator proteins, including eukaryotic Orc1, Orc4, and Cdc6, as well as in the bacterial DnaA initiator protein. A single winged helix-turn-helix domain binds the ORB in an asymmetric manner, despite the presence of a palindromic sequence in the replicator. The polarity of wH domain-binding facilitates contacts between the AAA+ domains in protomers and origin DNA sequences 3′ to the wH–DNA interfaces. Each ATPase domain contacts the DNA using an ISM. The ISM contributes to the positively charged AAA+ domain surfaces that bind the phosphate backbone and form additional DNA contacts via the intervening loop within the ISMs. The net result of initiator-protein binding is a substantial deformation of the DNA, leading to localized underwinding, possibly serving as a prelude to melting of the DUE regions. The structures solved thus far are with Mg^{2+}-ADP bound in the AAA+ site. It is likely that ATP binding will be associated with enhanced communication between adjacent Orc1/Cdc6s. In turn, this may lead to even greater deformation of the origin DNA.

Single-Cell Eukarya

Budding yeast

The paradigm for replicators in the genomes of single-cell eukaryotes is the budding yeast *S. cerevisiae*. These replicators are determined primarily, if not exclusively, by specific DNA sequences, because at least 70% of the replicators in the *S. cerevisiae* genome can be identified by sequence analysis alone. Nevertheless, only about 15% of the entire replicator sequence is common to all replicators in this genome. Moreover, the activity of replicators in budding yeast is strongly influenced by sequence context.

Budding yeast replicators are composed of AT-rich sequences of approximately 150 bp (Figure 10-10A). These replicators were discovered by their ability to confer autonomous replication on DNA plasmids (ARS assay) and to allow these plasmids to be maintained in yeast over many generations (plasmid maintenance assay). Hence, they are commonly referred to as ARS elements. The same sequence elements that are required in plasmid-based assays also have been shown to be required at chromosomal origins by altering the DNA sequence of the natural replicator and analyzing the appearance of replication bubbles at the altered locus using two-dimensional gel electrophoresis. Estimates of the number of replicators in the 10.2 Mb *S. cerevisiae* genome range from 247 to 429. They contain two essential components: an ORE to which the six-subunit ORC binds preferentially, and a DUE. Historically, these two components have been called the A and B domains, respectively. Of the five sequence elements identified in budding yeast replicators, only the **A-domain** and the B1-element are conserved among all of them. In general, each of the five elements can be interchanged with the corresponding element from another budding yeast replicator, despite the fact that, with the exception of the A-domain, there is little recognizable sequence conservation among them.

The A-domain contains a 17 bp sequence that is essential for replicator function and conforms to an extended ARS consensus sequence (WWWWTTTAYRTTTWGTT) where W is A or T, Y is C or T, and R is A or G. The **B-domain** consists of several genetically identifiable elements within a broad, nonconserved sequence 3′ to the T-rich strand of the A-domain. If the orientation of the A-domain relative to the B-domain is inverted, replicator activity is lost. Functional elements within the B-domain vary in number, type, and position among yeast replicators. Collectively, B-domain

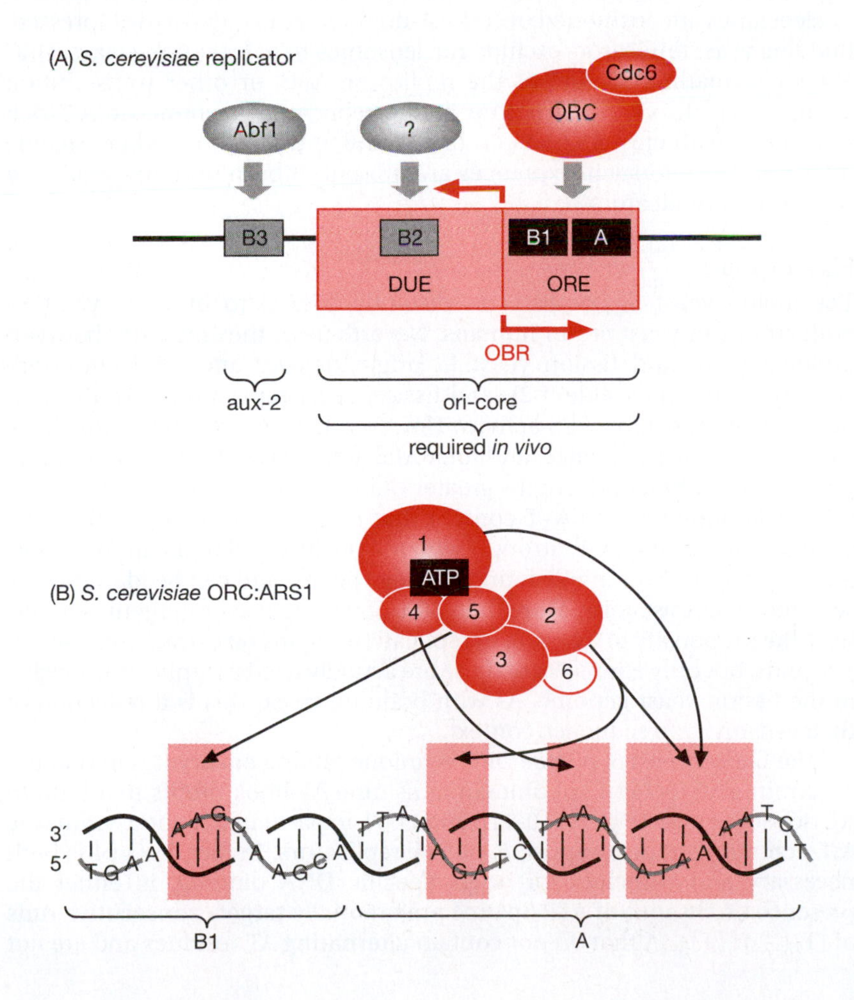

Figure 10-10. Budding yeast replicators. (A) *S. cerevisiae* ARS1. Genetically defined elements are indicated by boxes (solid for strongly required, shaded for weakly required). The origin of bidirectional DNA replication is indicated by red arrows. The ORE was originally identified as the ORC-binding site, although subsequent studies suggest that ORC and Cdc6 proteins together constitute the *S. cerevisiae* initiator protein complex. (B) ATP-dependent association of ScORC with ARS1 *in vitro*. ScORC exists as a stable complex of six different subunits. ATP binds to ScOrc1, ScOrc4, and ScOrc5. In the presence of ATP, ScORC binds specifically to the A and B1 elements in *S. cerevisiae* replicators. Interactions (arrows) are indicated between specific ScORC subunits and specific sequences (gray boxes) within the ScARS1 replicator. Only ScOrc6 is not required for DNA binding. (Data from Bell, S.P. (2002) *Genes Dev.* 16: 659–672.)

elements are essential for origin activity. However, individual B-elements are not essential, although they contribute significantly to origin efficiency.

The A-domain and the B1-element together comprise the approximately 43 bp ORE. *In vitro*, ORC binding is mediated by interactions between the major groove in DNA and Orc1, Orc2, and Orc4, whereas Orc5 appears to contact B1 at a specific position (Figure 10-10B). The B1-element is not essential for origin function, but it is important for efficient initiation. Orc6 is dispensable for specific ORC–DNA binding *in vitro*. *In vivo* ORC requires Cdc6 to target initiation to specific DNA sequences in chromosomes, and together with Cdt1, to load the MCM DNA helicase onto DNA.

B2, B3, and B4 are auxiliary components. The B2-element is part of the DUE, and the DUE must reside on the B1 side of the ori-core. Although B2-elements are functionally interchangeable, their sequences are not conserved among ARS elements. Mutations that raise the helical stability of the origin proportionately reduce replication activity and confer cold-sensitive replication of ARS plasmids. The cold sensitivity results from an elevation of the threshold temperature below which the initiator protein can still bind to origin DNA, but cannot unwind it. Genetic analysis of B2-elements suggests that some of them also interact with specific replication proteins.

The B3-element is the only element other than A and B1 known to directly bind a protein, in this case transcription factor Abf1. As with auxiliary components in viral replicators, other transcription factor-binding sites can replace B3. Some replicators contain a B4-element consisting of runs of As and Ts, similar to sequences in B2-elements. Since B4 can substitute for B3, B4 may be a binding site for an unidentified protein in addition to being easily unwound and part of the DUE.

S. cerevisiae replication origins all appear to be located in intergenic regions, and these regions tend to be free of nucleosomes. In fact, when nucleosomes are positioned over the A-domain, origin activity is repressed. Budding yeast replicators exclude nucleosomes in at least three ways. ORC alters chromatin structure at the replicator. Abf1 or other transcription factors alter chromatin structure in the replicator. Asymmetric A:T-rich sequences, with clusters of As on one strand and Ts on the other, exclude nucleosomes, and such sequences are present in both budding yeast and fission yeast replicators.

Fission yeast

The fission yeast *S. pombe* is as distantly related to budding yeast as both strains of yeast are to humans. Nevertheless, the similarity between budding yeast and fission yeast in origin density and replication-fork velocity is striking (Table 9-2), and fission yeast replicators, like those in budding yeast, exhibit ARS activity. However, fission yeast replicators lack a clear consensus sequence, and their ARS activity is determined primarily by high AT content and lengths greater than 0.5 kbp. Nevertheless, regions up to 1 kb long with an A+T content that is significantly higher than the genome average coincide strongly with the positions of replication origins. Such regions are commonly found between genes, and can be identified by sequence analysis alone. More than 50% of fission yeast intergenic regions have the propensity to function as replication origins on extrachromosomal plasmids, but only about 400 of these are actually used as replication origins in the fission yeast genome. As with budding yeast, this is a reflection of their sensitivity to sequence context.

The fission yeast *S. pombe* ORC is unique among eukaryotes in that the N-terminus of its Orc4 subunit contains nine AT-hook motifs that bind to AT-rich sequences (Figures 10-11B and 11-7), in particular to the asymmetric A:T-rich motifs that comprise *S. pombe* replicators. In fact, SpOrc4 is both necessary and sufficient for origin-specific DNA binding, in either the presence or absence of ATP. SpOrc4 preferentially targets consecutive runs of $(T)_{3-7}$ or $(T)_{3-4}A$ that do not contain alternating AT residues and are not interspersed with G or C residues. Therefore, the role of the other SpORC subunits appears to be to recruit SpCdc18 (the *S. pombe* equivalent of Cdc6) and to load the SpMCM helicase onto the *S. pombe* genome.

S. pombe replicators vary from 500 to 1000 bp in size. Genetic analyses of several replicators have revealed the presence of two or three required regions. These regions consist of asymmetric A:T-rich sequences with clusters of As on one strand and Ts on the other, but otherwise lack a clearly defined sequence. Nevertheless, they exhibit distinct genetic and biochemical properties. For example, ARS3001 (Figure 10-11A) contains four such elements of 50–70 bp, but only two of them bind SpORC. Elements Δ3 and Δ9 are required for replicator activity, whereas Δ2 and Δ6 facilitate replicator activity. Remarkably, SpOrc4 binds tightly to Δ3, but not to Δ9. Therefore, SpOrc4 can distinguish among various asymmetric A:T-rich sequences, suggesting that sequence specificity depends on the number and arrangement of AT-hook motifs in this protein. Moreover, Δ3 can substitute for Δ9, but Δ9 cannot substitute for Δ3, revealing that these two required elements have different functions. Δ3 is orientation-dependent relative to Δ2, and together they serve as the ori-core component of this replicator, analogous to the ori-core component of budding yeast replicators. Δ9 cannot be moved further away from Δ2+Δ3 (ori-core) or it loses activity, suggesting that Δ9 facilitates the action of ori-core.

Thus, fission yeast, like budding yeast, contains replicators, but the interaction between replicator and initiator in fission yeast is simpler than in budding yeast. A single subunit (Orc4) of the fission yeast ORC targets asymmetric A:T-rich sequences that lack a specific consensus sequence common to all fission yeast replicators, whereas in budding yeast ORC subunits 1–5 are required to target a specific A:T-rich sequence that is common to all budding yeast replicators. Since fission yeast intergenic sequences are exceptionally AT-rich, this mechanism prevents the demands of DNA replication from interfering with the demands of gene expression without increasing the genetic load with the need to maintain hundreds of unique sequences as replicators.

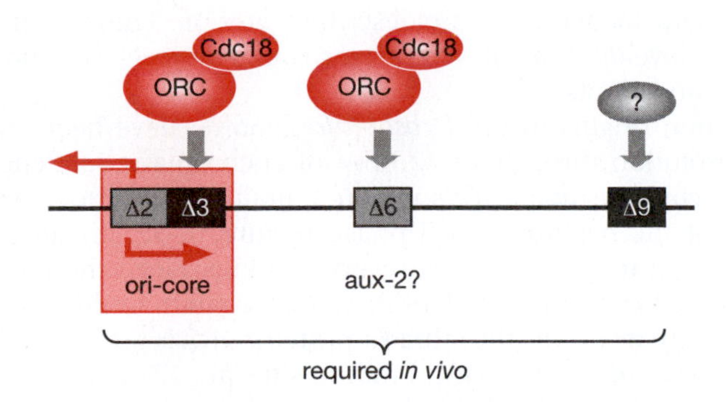

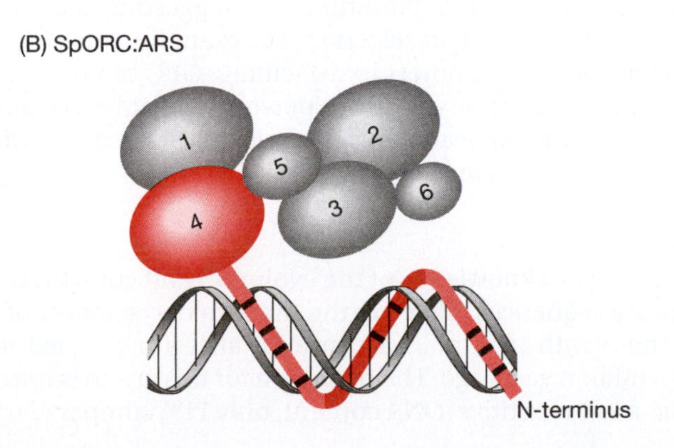

Figure 10-11. Fission yeast replicators. (A) *S. pombe* ARS3001. Cdc18 is the fission yeast equivalent of Cdc6 in budding yeast. SpORC is 444 kDa. (B) *S. pombe* Orc4 is solely responsible for binding to Sp replicators by wrapping its N-terminal half, containing nine AT-hook motifs, around the minor groove of asymmetric A:T-rich sequences. Based on atomic-force microscopy, the length of the DNA fragment complexed with either SpORC or SpOrc4 is shortened by approximately 140 bp, suggesting that two turns of the DNA are wrapped around the protein. [(A) data from Kong, D. and DePamphilis, M.L. (2002) *EMBO J.* 21: 5567–5576. (B) data from Gaczynska, M., Osmulski, P.A., Jiang, Y. *et al.* (2004) *Proc. Natl. Acad. Sci. USA* 101: 17952–17957.)

Despite the differences in the sequence composition of their replicators and the way in which their initiator proteins bind, replication origins in budding and fission yeast use the same basic mechanism to begin genome duplication, produce the same DNA replication intermediates (Figure 8-3), and use the same replicators during both mitotic and meiotic cell-division cycles.

Multicellular Eukarya

Replication origins in cells from adult metazoan animals are found, on average, once every 75–300 kbp, or about eight origins per Mbp. Multicellular eukarya encode initiator proteins (ORC and Cdc6) that are homologs of those found in the budding yeast, *S. cerevisiae*, and they produce replication intermediates whose structures are indistinguishable from those in yeast cells (see Figure 10-15). Therefore, metazoan replication origins presumably bear some resemblance to those in budding yeast. What exactly that resemblance is, however, is not clear. What is clear is that site-specific replication origins in metazoa are developmentally acquired and determined by both genetic and epigenetic factors (see Chapter 9). Moreover, both site-specific and nonspecific initiation events can be detected within various replication origins or 'zones'. Thus, metazoan genomes contain replicators that are detected under some, but not all, conditions. The presence of nonspecific initiation events in the genomes of flies, frogs, and mammals may simply reflect the fact that they are 14–240 times larger than those in yeast, an increase that reflects the larger nontranscribed intergenic regions in more complex organisms (the C-value paradox, Chapter 1).

Flies

The fruit fly *D. melanogaster* provides a convenient system for genetic analysis of genome duplication and gene expression as an insect develops from a fertilized egg into an adult. It is easy to maintain, inexpensive to house in large numbers, and has a life cycle of only 2 weeks. Thus, numerous generations can be studied within a single year. Flies also offer a unique insight into the process of endoreduplication (multiple S phases without an intervening mitosis) and gene amplification (selective re-replication of a single genetic locus) which are described later on. These events are rare during the development of vertebrates, but relatively common among arthropods and plants.

Replication origins in the *Drosophila* genome have been mapped at 1.5 kbp resolution throughout 22 Mbp of euchromatic sequences in the left arm of chromosome 2. This region contains 62 distinct sites that are replicated at the beginning of S phase (Figure 10-12), or an average of one early-firing replication origin every 354 kbp. Since the mean size of these sites is 32 kbp, some or all of them may contain multiple replicators. Consistent with its role as the initiator protein, ORC is preferentially bound to early-replicating origins. ORC-binding sites are AT-rich and associated with a subset of RNA Pol-II-binding sites, suggesting that transcription acts locally to influence origin selection. For example, transcription factors associated with active promoters may facilitate ORC binding to DNA. Note that early-replicating regions are interspersed with late-replicating regions, suggesting that some replication origins are activated early during S phase whereas others are activated late.

Birds

Birds fill a gap in our knowledge of the evolution and conservation of genes and regulatory sequences. They are the modern descendants of dinosaurs, and they share with mammals a common ancestor that existed approximately 310 million years ago. The paradigm for this group is the chicken. Its genome has a low repetitive-DNA content, only 11% compared with 40–50%

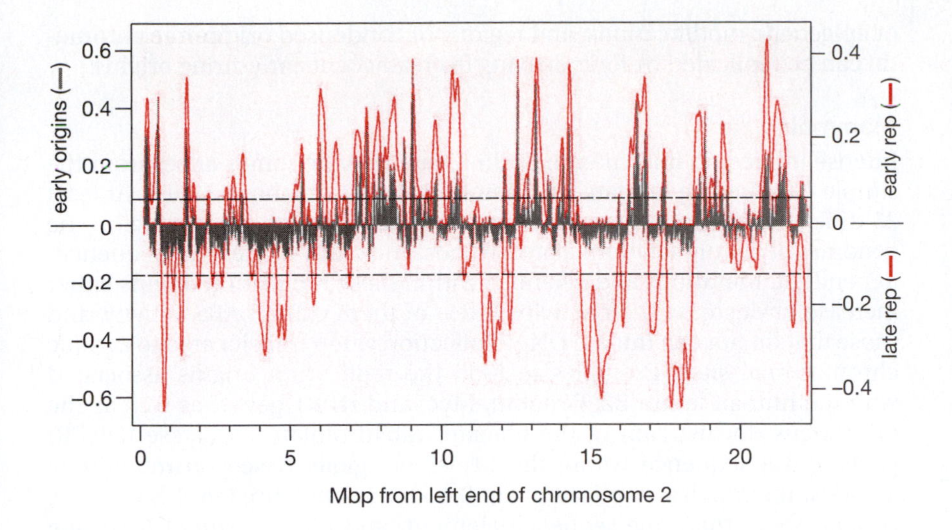

Figure 10-12. *Drosophila* replication origins on the left arm of chromosome 2. *Drosophila* Kc cells were synchronized by arresting them in G_2 phase with the molting hormone ecdysone. Early-firing replication origins (gray) were identified by releasing the cells from the ecdysone block into medium containing hydroxyurea to inhibit DNA synthesis by reducing dNTP pools and BrdU to label nascent DNA. Cells accumulated as they entered S phase, and DNA synthesis was restricted to those replication origins that were activated during this time. Alternatively, cells were released from hydroxyurea to proceed synchronously through S phase over the next 7 h. Early-replicating regions of the genome were labeled with BrdU for 1 h after releasing cells from hydroxyurea. Late-replicating (rep) regions were labeled for 1 h near the end of S phase. The two BrdU-containing DNA fractions were purified, labeled with either Cy5- or Cy3-conjugated dUTP to distinguish one from the other, and then hybridized to a single genomic microarray. The ratio of early- to late-replicating DNA was plotted (red). Peaks above or below the dashed black lines ($P<0.001$) are enriched significantly for either early- or late-replicating sequences (scale is logarithmic). (Adapted from MacAlpine, D.M., Rodríguez, H.K. and Bell, S.P. (2004) *Genes Dev.* 18: 3094–3105. With permission from Cold Spring Harbor Laboratory Press.)

found in mammals, making it possible to compare the frequency of replication origins in genomes with very different overall sequence composition.

Analysis of the relative abundance of RNA-primed DNA chains within a 52 kbp region in the chicken genome between the folate receptor gene and the olfactory receptor gene reveals the presence of four or five site-specific replication origins (Figure 10-13). One origin is located at the 5′-insulator sequence that separates condensed chromatin from accessible chromatin, two are near the ρ-globin gene, and two are near the β^A-globin gene. This entire region, including the condensed chromatin, replicates early in S phase, and these sites remain during differentiation of primary erythroid cells into erythrocytes. Three origins consist of GC-rich sequences enriched in CpG dinucleotides, typical of promoters and other regulatory sequences. The fourth origin is AT-rich. The insulator origin has unmethylated CpGs, hyperacetylated histones H3 and H4, and histone H3 methylated at lysine-4, consistent with an accessible chromatin structure. However, opposite modifications are observed at the other GC-rich origins. Thus, different early-firing origins within the same locus can have seemingly opposing patterns

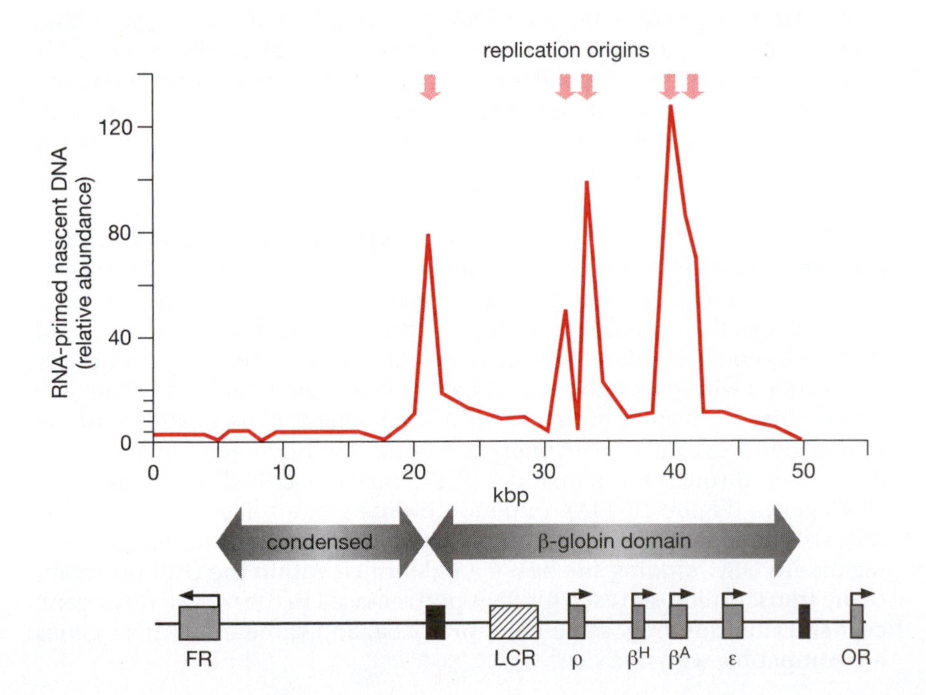

Figure 10-13. Chicken replication origins. The chicken β-globin domain in erythroid progenitor cells contains at least four and probably five site-specific replication origins. The relative abundance of short RNA-primed DNA chains 2–3.5 kb in length was measured at 42 sites distributed throughout a 52 kbp locus using real-time PCR (red line). The β-globin domain contains the embryonic (ρ, ε), hatching (β^H), and adult (β^A) globin genes whose expression is regulated by a locus control region (LCR) upstream of ρ and a strong enhancer between the β^A and ε. The β-globin domain is bound by an insulator element (solid box) at each end. This locus is bounded at one end by the folate receptor (FR) gene and at the other end by an olfactory receptor (OR) gene. The FR gene is expressed in these cells, but not the globin genes. A region of condensed chromatin is recognized by its resistance to micrococcal nuclease. (Adapted from Dazy, S., Gandrillon, O., Hyrien, O. and Prioleau, M.N. (2006) *EMBO Rep.* 7: 806–811. With permission from Elsevier.)

of epigenetic modifications, and regions of condensed origin-free chromatin can be replicated by forks arriving from adjacent early-firing origins.

Mammals

Intense interest in how mammals duplicate their genomes arises from the simple fact that we humans are members of this taxonomic class. At least 25 different replication origins have been identified at specific loci in the genomes of cultured mammalian cell lines, and many of these are sequence-dependent. Mutations and deletions within these replication origins either increase or decrease their activity. A few of them exhibit ARS activity, and those that do not can initiate DNA replication when translocated to ectopic chromosomal sites. Examples include the replication origins associated with the human lamin B2, β-globin, Myc, and HPRT genes, as well as the ori-β locus downstream of the hamster dihydrofolate reductase (DHFR) gene, and a sequence within the 14-3-3 eta gene. Based on mutational analysis, mammalian replicators are 0.4 kb to approximately 1 kb in size, contain more than one sequence element, and bind *in situ* at least one copy of the mammalian initiator proteins, ORC and Cdc6. As with yeast, only a fraction of the available metazoan replication origins in each cell are required to duplicate the genome each time a cell divides (Table 9-2). In fact, most eukaryotic origins are not active during every cell cycle. Nevertheless, mammalian replicators exhibit neither a recognizable consensus sequence nor sequence composition and therefore cannot be identified through sequence analysis alone.

Studies on mammalian cultured cell lines have led to conflicting conclusions as to the nature of their replication origins. On the one hand, methods that quantify the relative number of newly synthesized DNA strands at various DNA loci (Chapter 9, Finding replication origins) reveal the presence of discrete (0.4–1 kb) **origins of bidirectional replication** (**OBRs**; Figure 9-1) at specific genomic locations. Since these methods quantify the relative activity of initiation events, they reveal the most active replication origins. On the other hand, methods that rely upon detecting the presence of replication bubbles or replication forks by two-dimensional gel electrophoresis or DNA combing detect large (30–55 kb) **initiation zones** in which initiation events are distributed virtually randomly throughout the locus. These methods reveal a broader range of initiation sites whose relative efficiencies are difficult to determine. Subsequent analysis of single DNA molecules by the DNA-combing technique suggests that replication origins are spaced on average once every 20–25 kbp, remarkably similar to the spacing of replicators in yeast. As with yeast, only a fraction of the mammalian origins are active during a single S phase. Two well-characterized examples of this phenomenon are the replication origins associated with the rRNA and the DHFR genes.

rRNA genes: a paradigm for tandemly repeated replication origins

Analyses of replication origins associated with rRNA genes in single-cell and multicellular eukarya reveal a remarkably conserved genomic organization that includes the presence of a single, primary replicator just upstream of the rRNA gene promoter. In human, mouse, and rat genomes, rRNA genes are encoded within a 44 kbp sequence that is repeated tandemly about 400 times. Most replication events occur a few kilobase pairs upstream of the transcribed region along with many additional low-frequency initiation sites distributed through the remainder of the nontranscribed region between rRNA genes (Figure 10-14A). Initiation events are not detected within the regions encoding genes, the promoter, or the transcription-termination region. An ORC-binding site has been identified within the OBR upstream of the transcription start site for RNA polymerase I in the mouse rRNA gene cluster. Data from frog, sea urchin, protozoa, and slime mold rRNA genes are compatible with this view.

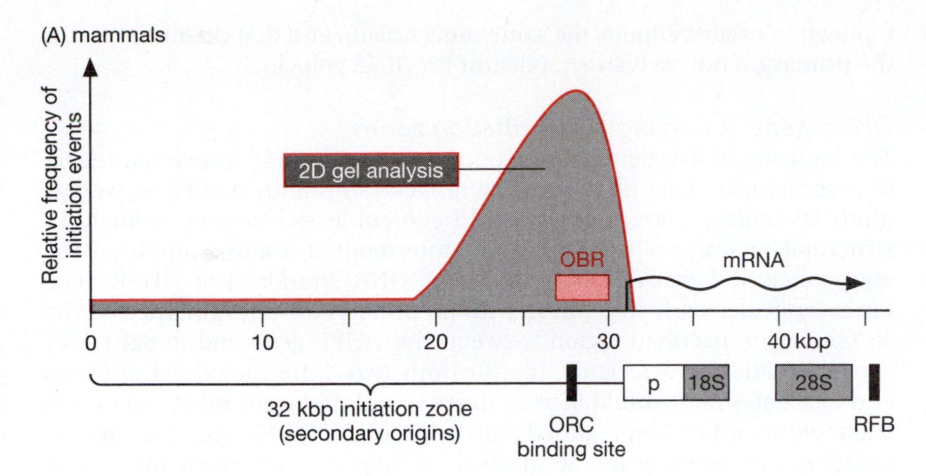

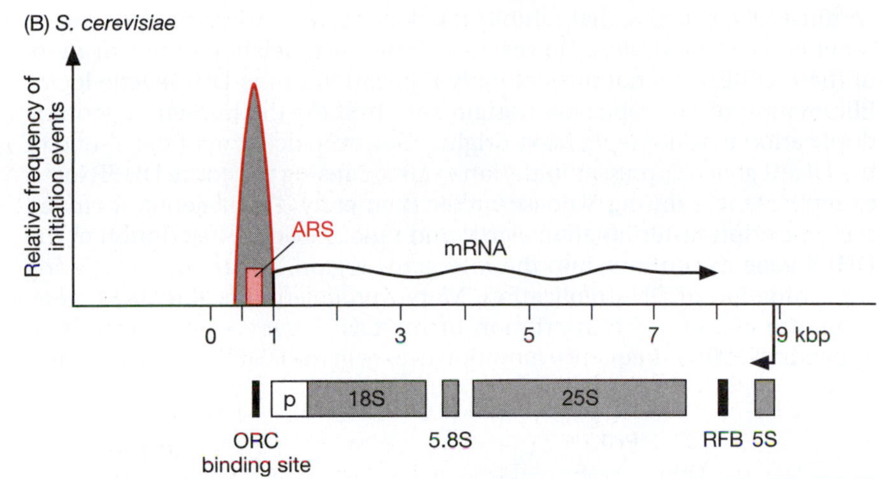

Figure 10-14. rRNA genes in mammals and yeast. (A) Human, mouse, and rat cells contain a 44 kbp rRNA gene repeat. (B) The budding yeast *S. cerevisiae* contains a 9.1 kbp rRNA gene repeat. In each case, a replication-fork barrier (RFB) lies just downstream of the rRNA transcript that prevents replication forks from traveling upstream and colliding with the transcription machine, but does not prevent forks from traveling downstream in the same direction as the transcription machine. Nascent-strand abundance analysis reveals an origin of bidirectional DNA replication (OBR) just upstream of the promoter (p) for mammalian rRNA gene expression, and genetic analysis reveals an ARS element at the equivalent site in *S. cerevisiae*. 2D, two-dimensional. (Data from Gencheva, M., Anachkova, B. and Russev, G. (1996) *J. Biol. Chem.* 271: 2608–2614, Linskens, M.H. and Huberman, J.A. (1988) *Mol. Cell. Biol.* 8: 4927–4935, and Zellner, E., Herrmann, T., Schulz, C. and Grummt, F. (2007) *Nucleic Acids Res.* 35: 6701–6713.)

The *S. cerevisiae* rRNA gene locus also consists of a tandem array of 100–200 repeated transcription units, but these are only 9.1 kb in length and contain a single replicator (227 bp ARS element) upstream of the transcription unit from which DNA replication proceeds bidirectionally (Figure 10-14B). However, due to the inefficiency of the rDNA ARS element, fewer than one-third of them are actually used as replication origins during a single S phase. In both mammalian and yeast rRNA gene clusters, the leftward-moving fork is arrested at a replication-fork barrier (Figure 6-8) located at the 3′-end of the transcription unit, while the rightward-moving fork proceeds on average through about five repeats until it terminates at a stalled leftward-moving fork. Thus, replication within rDNA regions begins bidirectionally, but soon appears unidirectional. These replication-fork barriers prevent replication machines from colliding with transcription machines, thereby ensuring that rRNA gene transcription can proceed even during S phase.

Taken together, the data from yeast and metazoa are consistent with a single site-specific replication origin just upstream of the rRNA gene promoter that functions as the primary initiation site, and multiple nonspecific initiation events throughout the nontranscribed intergenic region. The fact that these nonspecific sites are not detected at yeast rDNA loci may simply reflect the relative sizes of the nontranscribed region. In mammals, this region is at least 30 kbp long, whereas in budding yeast the same region is only 2 kbp long, thus reducing the probability that nonspecific initiation events will occur in this region. The striking similarities between the structures of replicating intermediates from the human rRNA gene replication origin and from a single yeast replication origin (see Figure 10-15 in color plate section) further supports the conclusion that both replication origins employ the same mechanism, and that both function as the primary, if not exclusive, replicator for rRNA gene loci.

DHFR gene: a paradigm for initiation zones

The hamster DHFR gene locus is an example of a large intergenic region that contains at least three primary replication origins (OBRs) as well as many secondary ones, and that has been subjected to genetic analysis, structural analysis of the replication intermediates, and analysis of the abundance and distribution of nascent DNA strands. The DHFR gene locus replicates early in S phase with all initiation events confined to the 55 kbp nontranscribed region between the DHFR gene and the 2BE2121 gene (initiation zone; Figure 10-16). Both two-dimensional gel analyses and nascent-strand abundance analyses reveal that most initiation events occur within a 12.5 kbp region downstream of the DHFR gene. This region contains two primary initiation sites, at least one of which (ori-β) lies within a 5.8 kbp region that exhibits replicator activity when translocated to other chromosomal sites. However, as with yeast, deletion of one or more of these OBRs does not prevent early replication of the DHFR gene locus. Elimination of one replication origin simply shifts the burden of genome duplication to other replication origins. However, deletion of the 3′-end of the DHFR gene suppresses initiation events, causing the entire DHFR locus to replicate late during S phase rather than early. This deletion included the transcription-termination signal and thus allowed transcription of the DHFR gene to continue into the intergenic region, apparently interfering with initiation of DNA replication. Moreover, deletion of the DHFR gene promoter eliminated transcription of the DHFR gene with concomitant appearance of low-frequency initiation events in the DHFR gene itself. Thus,

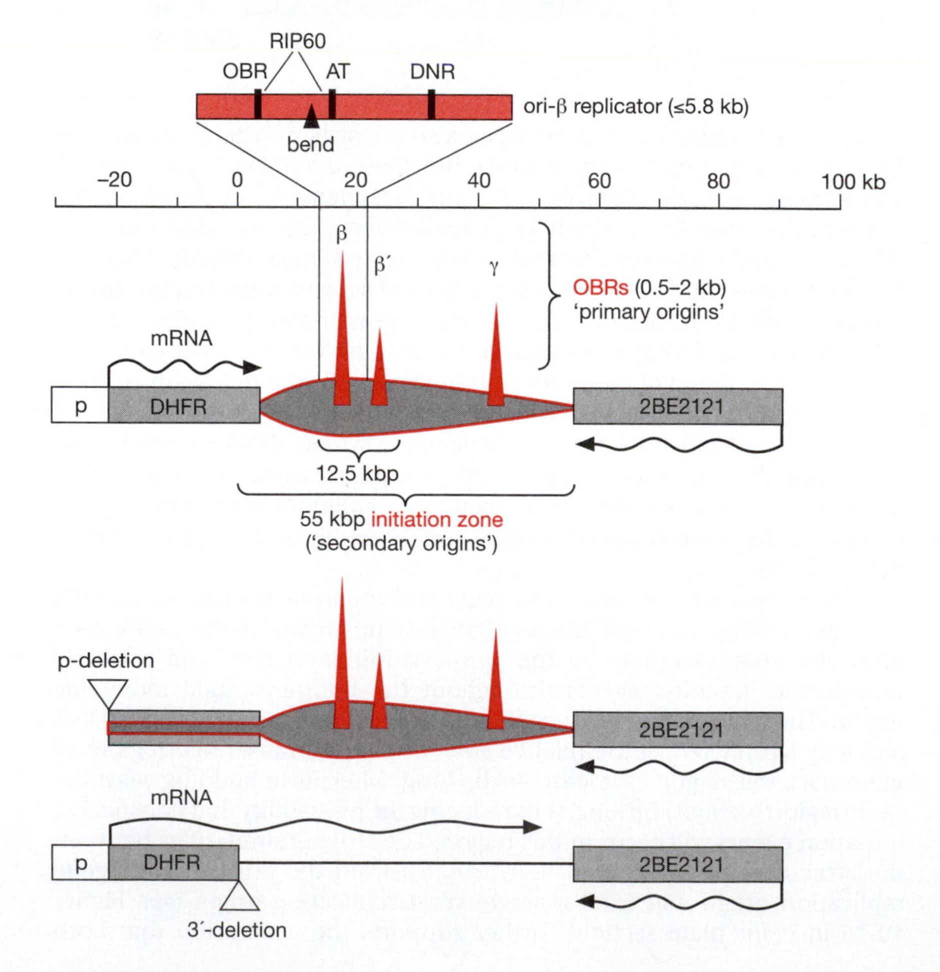

Figure 10-16. Replication origins and replicators in hamster cells. The 55 kbp region between the DHFR and 2BE2121 genes in Chinese hamster cells provides a paradigm for the relationship between initiation zones and OBRs in amplified gene regions. The ori-β replicator is contained within a 5.8 kb region encompassing the ori-β OBR, two binding sites for the protein RIP60, a sequence-induced stable bend in the DNA (bend), an AT-rich region (AT), and a GA+CA dinucleotide repeat element (DNR), all of which have been implicated in replicator activity. Deletion of the DHFR promoter (p) results in loss of DHFR gene transcription and the appearance of initiation events within the 25 kbp DHFR gene locus. Deletion of the 3′-end of the DHFR transcription unit eliminates the transcriptional termination signal, resulting in extended DHFR transcription into the intergenic region with concomitant loss of initiation events. (Adapted from Gray, S.J., Liu, G., Altman, A.L., Small, L.E. *et al.* (2007) *Exp. Cell Res.* 313: 109–120. With permission from Elsevier. Additional data from Kobayashi, T., Rein, T. and DePamphilis, M.L. (1998) *Mol. Cell. Biol.* 18: 3266–3277, and Mesner, L.D. and Hamlin, J.L. (2005) *Genes Dev.* 19: 1053–1066.)

the absence of transcription permits nonspecific, low-frequency initiation events within this initiation zone. At other genomic loci, however, such as the mouse IgH locus, low-frequency initiation events were not detected throughout a 450 kbp intergenic region. Therefore, while transcription can suppress or activate replication origins, the absence of transcription does not automatically result in the appearance of replication origins.

Lamin B2 gene: a paradigm for site-specific replication origins

The paradigm for site-specific initiation events in mammalian genomes is the replication origin that lies between the human lamin B2 gene and the TIMM 13 gene (Figure 10-17). Lamin B2 replicator activity, as measured by the ability of this locus to initiate replication when translocated to other chromosomal sites in the genome, requires about 1.2 kbp of sequence information. This region includes a **CpG island**, a region that contains a higher frequency of CG dinucleotides than would be statistically expected. Deletions within the lamin B2 replicator identify a 290 bp ori-core region that is necessary but not sufficient for replicator activity. The ori-core resides within the 0.6 kbp nontranscribed region between the two genes and contains both the OBR and ORE. Leading-strand DNA synthesis begins at specific nucleotides on each template to define the OBR. Orc1, Orc2, and Cdc6 contact specific nucleotides within the ORE in G_1-phase cells. As with SV40, polyomavirus, and *S. cerevisiae* replicators, the lamin B2 OBR is adjacent to one side of the ORE. Furthermore, topoisomerases I and II interact with replication origins prior to assembly of a preRC and initiation of DNA synthesis, and inhibition of topoisomerase I activity prevents origin activation. Orc2 competes for the same origin DNA sites that are bound by

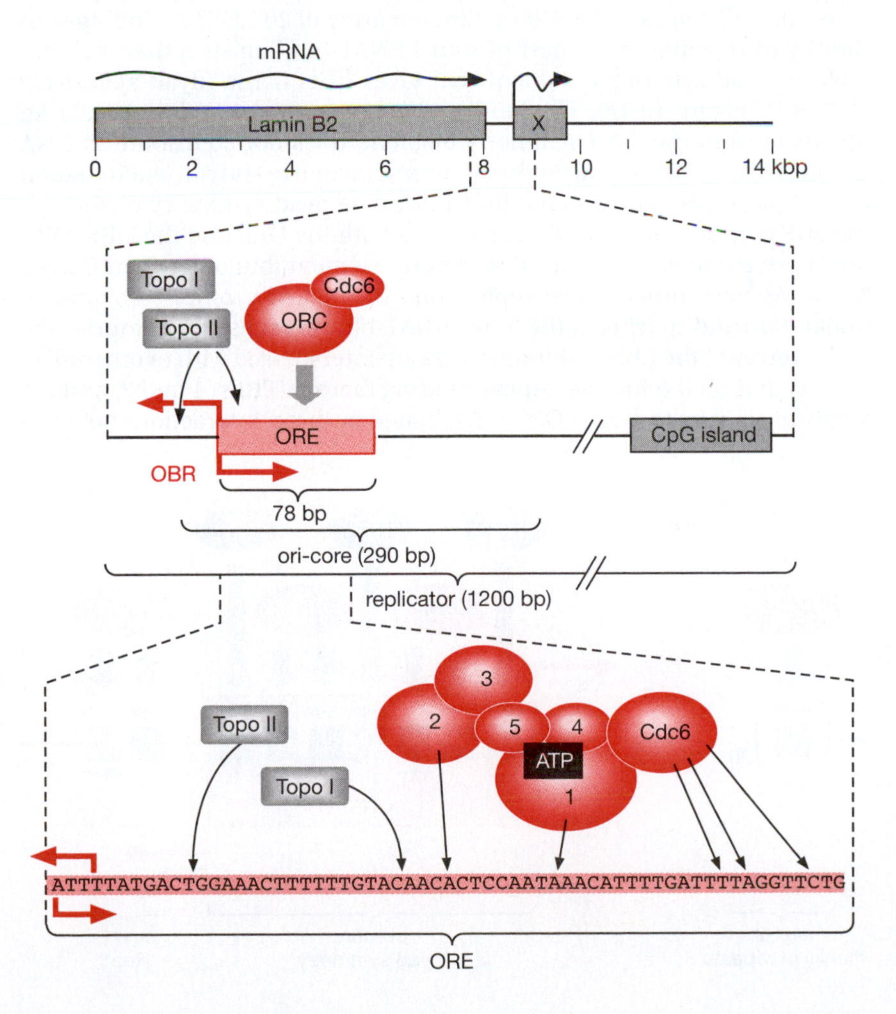

Figure 10-17. Interaction of initiator and replicators in human cells. The human lamin B2 replicator is located between the lamin B2 gene and a gene of unidentified function (X) called TIMM 13. This replicator lies within a 1.2 kbp DNA fragment consisting of at least three identifiable components: a 290 bp core that is required but not sufficient for origin activity, a 78 bp origin-recognition element (ORE) that interacts with ORC and Cdc6 proteins, and an origin of bidirectional replication (OBR) that marks the beginning of leading-strand DNA synthesis in both directions. Orc1, Orc2, and Cdc6 bind to specific sites within ori-core, interacting with the indicated nucleotides, at specific times during cell division. Similarly, sequence-specific cleavage by Topo I and Topo II occurs both within and without the ORE. (Data from Abdurashidova, G., Radulescu, S., Sandoval, O. et *al.* (2007) *EMBO J.* 26: 998–1009.)

either topoisomerase at different times during cell division, suggesting that DNA topology plays an essential role in origin activation. ORC also has been found bound preferentially to OBR sites in the human *Lamin B2, TOP1, MYC, MCM4*, and *FMR1* gene loci. ORC and topoisomerase II colocalize on chromatin, and both have been identified at the MCM4 replication origin.

Epstein–Barr virus

Epstein–Barr virus (EBV), a member of the *Herpesviridae* family, is a human tumor virus that causes both lymphomas and carcinomas. Remarkably, it is the only virus known to mimic the replication pattern of its host's nuclear chromosomes. EBV is a 165 kb dsDNA genome that is linear in the virion but circularizes in the host cell's nucleus. EBV utilizes its host's replication machinery to maintain itself as an episome in mammalian cells. The EBV genome is duplicated only during S phase and only once each time a cell divides. Viral progeny are not synthesized under these conditions, and the cells are said to be latently infected. EBV also has an additional mode of replication in which rare cells in an infected population change to support EBV's lytic cycle. Under these conditions, viral DNA replicates 100-fold or more without cell division, viral progeny are synthesized, and the host cell dies. This lytic mode of replication utilizes a unique viral replication origin and many viral proteins. In contrast, viral genome duplication during the latent period requires only a single virus-encoded protein called Epstein-Barr nuclear antigen 1 (EBNA1) and a single viral replicator termed oriP.

oriP was discovered by screening cloned fragments of EBV DNA for ARS activity in EBV-positive cells. Subsequent studies revealed that oriP-dependent plasmid DNA replication requires only EBNA1 and the human replication machinery in order to duplicate itself once and only once per cell division. oriP consists of a 650 bp tandem array of 20 EBNA1-binding sites (**family of repeats**) and a nest of four EBNA1-binding sites that includes a 65 bp dyad symmetry element that gives it its name (**dyad symmetry element**) (Figure 10-18). The family of repeats, which resides about 1 kb upstream from the dyad symmetry element, does not contribute to DNA synthesis, but it is required for the stable maintenance and nuclear retention of oriP plasmids and genomic EBV DNA. The dyad symmetry element is the oriP core component, which contains both the ORE and the OBR. ORC binds preferentially to the dyad symmetry element, but only when EBNA1 binds. As with other active replicators, oriP activity requires a precise orientation and spacing of the four EBNA1-binding sites that comprise the ORE. However, the EBNA1-binding sites are interspersed with telomere-like repeats that bind **telomere-repeat binding factors** (**TRFs**) 1 and 2, and the ability of ori-core to recruit ORC is facilitated by direct interactions between

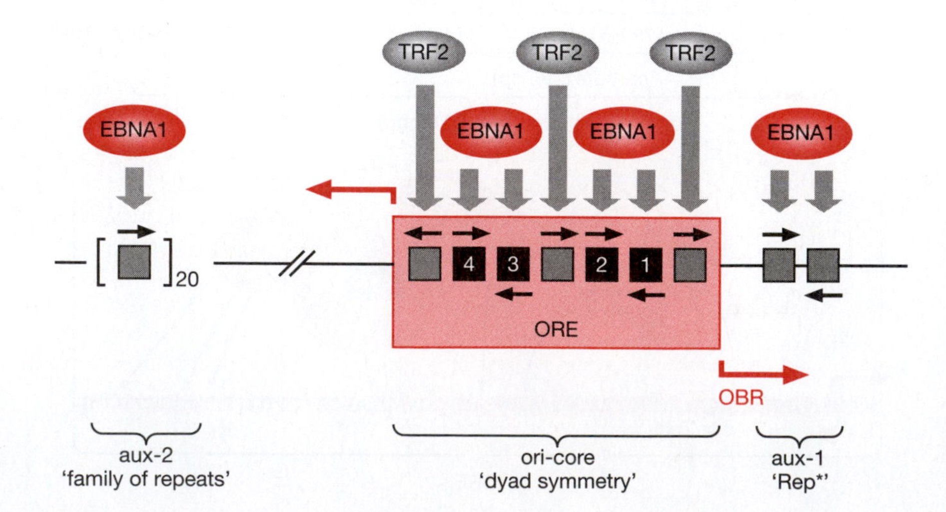

Figure 10-18. Epstein–Barr virus (EBV) oriP replicator. oriP is driven by the EBV-encoded EBNA1 initiator protein, but uses its host cell's ORC, Cdc6, Cdt1, and Mcm(2-7) preRC proteins to initiate DNA replication. A family of 20 tandemly repeated EBNA1 DNA-binding sites (aux-2) are about 1 kb from the origin-recognition element (ORE). The ORE contains an inverted repeat (dyad symmetry element) that includes four EBNA1-binding sites. Inserting or removing a single base pair between the EBNA1 sites of a functional pair (sites 1 and 2, or sites 3 and 4) abolishes oriP activity, because cooperative binding of EBNA1 is lost. Proper spacing is also required to recruit ORC. aux-2 is not required for DNA replication, but it is required for plasmid maintenance. aux-1 (also known as Rep*) stimulates oriP activity. (Data from Hammerschmidt, W. and Sugden, B. (2006) Epstein-Barr virus. In *DNA Replication and Human Disease* (DePamphilis, M.L., ed.), pp. 687–706.)

TRF2 and ORC1. However, since these telomere repeats are not essential for replicator activity, they are auxiliary components. Two additional downstream EBNA1 sites (termed Rep*) that also facilitate replicator activity, and the family of repeats upstream of ori-core constitutes an aux-2 component, as represented in the idealized origin. (Figure 9-1).

Surprisingly, oriP can be deleted from EBV and the genome will still replicate as an extrachromosomal element, suggesting that DNA synthesis can originate throughout much or all of the EBV genome. In fact, two-dimensional gel electrophoresis and DNA combing have revealed a broad initiation zone located about 25 kb from oriP. Nevertheless, the dyad symmetry element in oriP is required to establish an episomal state in mammalian cells; although the EBV initiation zone can support DNA synthesis for short periods of time, oriP is required to support DNA replication for long periods of time. The EBV paradigm suggests that genomes in multicellular organisms contain both site-specific and nonspecific replication origins or initiation zones, and that these two types of replication origin serve different biological roles.

Episomes

Episomes are DNA molecules that can replicate autonomously outside of the host cell's genome. They are common in bacteria where they are generally referred to as plasmids. However, establishing an episome in mammalian cells is a challenging task. In either single-cell or multicellular eukaryotes, the episome must be either a covalently closed, circular dsDNA molecule, or it must be a linear dsDNA molecule with telomeres at each end. Otherwise, it will be degraded. Telomeres are DNA sequences that protect the ends of linear chromosomes from degradation and prevent the chromosome from inserting itself into another chromosome (Chapter 6). The DNA must also contain at least one sequence that functions as a replication origin in the host's nucleus. In yeast, this sequence is an ARS element, but in metazoa virtually any DNA sequence appears to work, although the efficiency of replication increases markedly with DNA length. Finally, the DNA must contain a sequence that allows it to be retained within the nucleus and segregated into each of the two daughter cells during cell division. In budding yeast, this is a centromere. Centromeres are DNA sequences that bind two chromatids together during the prophase and metaphase stages of mitosis and to which the microtubules attach that pull chromatids apart during anaphase. Well-characterized centromeres range in size from the 125 bp centromere of budding yeast to those of several megabases in human chromosomes. Since mammalian centromeres are too large to be of practical application, viral or cellular nuclear attachment sites have been used in their place. DNA attachment sites within the nucleus, referred to as either **matrix-attachment regions** (**MARs**) or **scaffold-attachment regions** (**SARs**), are believed to play a crucial role in chromatin organization and function (Chapter 7).

Two episomes have been developed that can be maintained in mammalian cells without continual selection for expression of an episome-encoded gene product (Figure 10-19). Both are circular DNA molecules between 6 and 12 kbp. One of them utilizes the EBV oriP, and one of them functions without the help either of viral DNA sequences, or viral proteins. oriP activity requires the EBV EBNA1 protein, which is provided either by cells that have been engineered to produce the protein, or by one or more copies of the EBNA1 gene encoded by the episome itself. EBV-based vectors function primarily, if not exclusively, in human cells. EBNA1 bound to the family of repeats tethers the episome to condensed mitotic chromosomes as they segregate during mitosis, thus ensuring that the plasmid will be inside the nucleus when the nuclear membrane re-forms during telophase, and that the episome cosegregates with the host cell's chromosomes during cell division. The chromosome-binding domains in EBNA1 are AT-hook

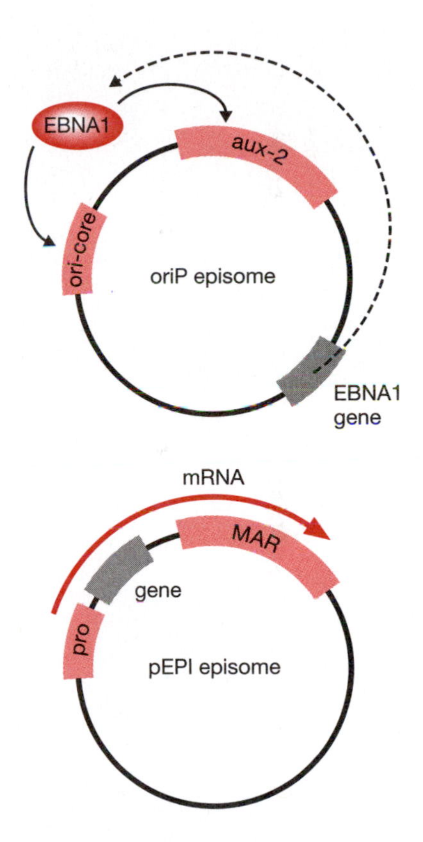

Figure 10-19. Episomes that function in mammalian cells. EBV oriP-dependent episomes replicate once per cell division in human cell lines, but they require EBNA1 for both DNA replication during S phase and segregation into daughter cells during mitosis. pEPI episomes depend only on transcription through a nuclear matrix-attachment region (MAR). This allows virtually any gene to be expressed from a strong promoter (pro). (Data from Stehle, I.M., Postberg, J., Rupprecht, S. *et al.* (2007) *BMC Cell Biol.* 8: 33, and Stoll, S.M. and Calos, M.P. (2002) *Curr. Opin. Mol. Ther.* 4: 299–305.)

motifs, typical of proteins that bind the scaffold-associated regions on metaphase chromosomes. Large (about 12 kbp) segments of human DNA can be substituted for the oriP dyad symmetry element with some success, but EBNA1 and the family of repeats are still required for maintaining the episome within the nucleus during cell division.

The autonomously replicating vector pEPI does not rely on either viral sequences or proteins. pEPI can be established in a wide variety of proliferating cells where it is maintained at a low copy number (2–15 per cell). It replicates once per cell cycle during early S phase and is stable during mitosis over hundreds of generations in the absence of selection for a pEPI-encoded gene product. Replication of pEPI relies on binding of ORC to one of several stochastically selected sites in the plasmid followed by assembly of a normal cellular replication complex. The establishment of pEPI in mammalian cells requires the presence of a MAR in the plasmid as well as a transcription unit upstream that drives transcription into the MAR sequence. pEPI interacts with the prominent nuclear matrix protein SAF-A and with metaphase chromosomes. The episome's host range appears to be limited only by the ability of the cell to utilize the promoter driving transcription through the MAR. The mechanism is not yet understood, but transcription through the MAR may prevent the episome from binding to nuclear structure in interphase cells, thus allowing plasmid replication to occur, whereas the absence of transcription that normally occurs during mitosis may allow the MAR to tether the plasmid to chromosomes.

WHY DNA HELICASE LOADERS?

If the object is simply to load a DNA helicase at a particular genomic site, one might think that initiators in bacteria, archaea, and eukarya would all be DNA helicases, as indeed they are in many animal viruses. So, why then did nature not select powerful DNA helicases as initiators for DNA replication in cellular organisms? The answer revolves around two principles. The first is to minimize the genetic burden. Direct helicase loading, as exemplified by several animal viruses, requires a sequence-specific loading site (active replicator), and such sites would have to be distributed at least once every 50 kbp or less, based on known origin densities (Tables 9-1 and 9-2). While this is not a problem for small genomes such as those found in bacteria and archaea, it would increase the genetic burden of the larger more complex genomes in metazoa. The second principle is regulation. All cells need to produce at least one complete genome each time they divide; complex cells such as eukarya need to restrict genome duplication to one complete copy each time a cell divides. To accomplish this, cells must link genome duplication to cell division, a subject discussed in Chapters 12 and 13. Thus, while direct loading of a DNA helicase provides a rapid, efficient way to initiate DNA replication in viral genomes, it is not one that is easily regulated.

Why did evolution select the helicase loader mechanism? Activation of replicators by transcription would lend itself to regulation, but it is difficult to apply to cellular genomes, because DNA replication may interfere with the temporal and spatial requirements for gene expression. Moreover, it cannot be applied to large genomes in multicellular organisms where structural genes and RNA-encoding genes account for only a small fraction of the genome. Less than 2% of the human genome is used to encode genes! If organisms with large genomes were to evolve, then a mechanism was needed that could distribute DNA helicases uniformly throughout the entire genome.

The protein-priming mechanism used at terminal replicators is untenable for large genomes, because it generates extensive amounts of ssDNA. Since ssDNA cannot be assembled into nucleosomes, it is much more vulnerable to chemical and enzymatic modification than is dsDNA. Evolution might have selected protein priming as a mechanism for initiation of Okazaki-fragment synthesis, but it did not. Perhaps excision of a protein primer required a novel mechanism that does not exist in nature's repertoire, whereas excision of an RNA or DNA primer was easily accomplished using one or more of the endo- and exonucleases that already exist. Note, for example, that the adenovirus and ϕ29 terminal proteins remain covalently attached to the ends of their respective genomes.

The DNA-nicking mechanism, like the DNA helicase mechanism, requires sequence-specific active replicators distributed throughout the genome. In addition, it introduces breaks in the DNA backbone, a potentially dangerous mechanism when applied to many sites in large genomes. Moreover, all of the initiators that use the DNA-nicking mechanism produce a single-strand displacement mechanism (the rolling-circle mechanism) that works well for small circular genomes, but is not suitable for large circular genomes. mtDNA, for example, uses a strand-displacement mechanism, not a rolling-circle mechanism.

Instead, evolution appears to have selected a protein with very limited helicase activity to begin the unwinding of DNA at replication origins in simple organisms such as bacteria and perhaps archaea, and then loaded a much more efficient helicase to continue DNA unwinding at replication forks. By separating the task of initiation from elongation, the problem of regulating DNA replication could focus more easily on the initiation events.

AMPLIFICATION ORIGINS

Amplification origins are specialized replicators that respond to development signals present only in specialized cells that are terminally differentiated (they never divide again). Once activated, amplification origins recruit the same initiator proteins that activate replication origins during normal mitotic cell cycles, but in contrast to normal S phase replication origins, amplification origins continue to initiate DNA replication without cell division. The result is re-replication of a particular DNA locus multiple times. This process, referred to as gene amplification, differs from genome endoreduplication (Chapter 12), in that gene amplification uses specialized replication origins to amplify specific genomic loci, whereas endoreduplication modifies the cell-cycle controls that restrict genome duplication to once per cell division to produce multiple copies of the entire genome in the absence of cell division.

Sciara coprophila

The giant polytene chromosomes found in larval tissues of dipteran insects arise from pairing of homologous chromosomes followed by rounds of endoreduplication. The result is a parallel array of up to 1024 chromatids in *D. melanogaster* and 8192 chromatids in the dark-winged fungus gnat *Sciara coprophila*. Specific loci within these chromosomes then undergo additional amplification to produce puffs of DNA (Figure 10-20A).

Site-specific gene amplification at the end of larval life in *Sciara* has been studied extensively at locus II/9A. Prior to a burst of transcription, this locus is specifically amplified about 17-fold over the rest of the genome. Gene expression at this locus is triggered by expression of the hormone ecdysone, the master regulator of insect development. Injection of ecdysone into pre-amplification stage larvae induces amplification, and expression of the genes at II/9A. Location of an ecdysone-response element (DNA-binding site) adjacent to the ORE (ORC-binding site) suggests that ecdysone may regulate binding of ORC to this site (Figure 10-20B).

Drosophila melanogaster

The most extensively studied example of developmentally regulated gene amplification occurs during formation of oocytes (female gametes)

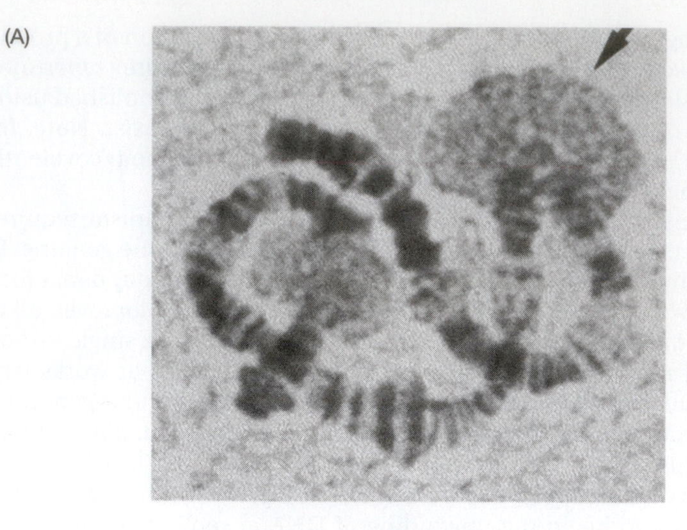

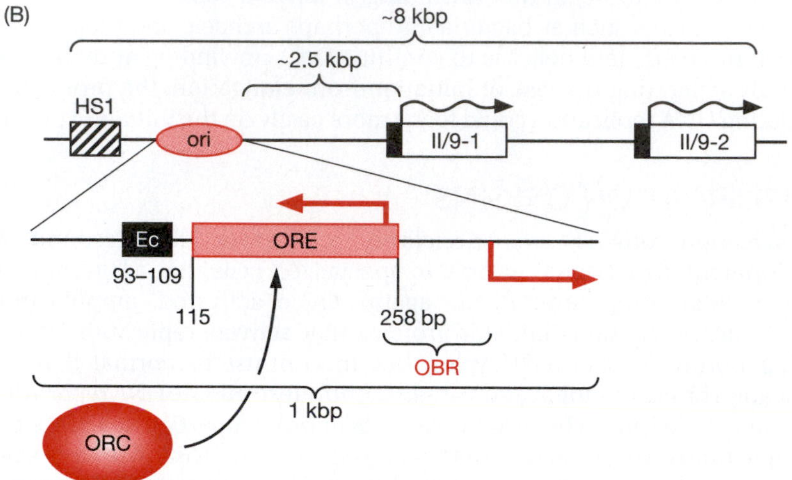

Figure 10-20. *Sciara* amplification origin and replicator. DNA puffs in *Sciara* polytene chromosomes are sites of DNA amplification. (A) The largest DNA puff in the *Sciara* salivary glands is found in late larval life at locus 9A on polytene chromosome II (arrow). Other puffs are also apparent. Chromosomes were stained with Giemsa, a dye that is specific for the phosphate groups of DNA and binds preferentially to AT-rich regions. (B) Puff II/9A consists of a replication origin and two genes (II/9-1, II/9-2) that are transcribed away from ori (wavy arrows). The 1 kbp amplification origin (ori) is located approximately 2.5 kb upstream of gene II/9-1. One ecdysone-response element (Ec) is located within the 1 kb origin, adjacent to the ORC-binding site (ORE). One start site for leading-strand DNA synthesis (leftward red arrow) occurs at the edge of the ORC-binding site and a second start site (rightward red arrow) outside the ORE but within ori define the boundaries of the OBR. A strong DNase-hypersensitive site (HS1) exists at the left boundary of ori. Such sites occur commonly at DNA binding sites for transcription factors or other proteins that regulate genomic functions. [(A) from Gerbi, S.A., Strezoska, Z. and Waggener, J.M. (2002) *J. Struct. Biol.* 140: 17–30. With permission from Elsevier. (B) adapted from Foulk, M.S., Liang, C., Wu, N. *et al.* (2006) *Dev. Biol.* 299: 151–163. With permission from Elsevier.)

in *Drosophila.* The chorion gene products produced by the follicle cells that envelop each oocyte form a shell that surrounds the oocyte. About 20 chorion proteins altogether are present in this eggshell, and the genes encoding the major chorion proteins are localized in two clusters, one on the X chromosome and one on the third chromosome. Near the end of oogenesis, the chorion gene locus at position 66D on chromosome 3 is amplified about 64-fold. This is followed by a period of robust chorion gene expression and then apoptosis of the follicle cells.

Chorion gene amplification occurs by repeated initiation of DNA synthesis from defined sites within each chorion gene cluster. Following amplification there is a gradient of DNA copy number surrounding the chorion clusters. This gradient has a peak in the vicinity of the chorion genes and extends out to about 40 kbp on either side. The center of this peak corresponds to a region in which DNA replication is initiated repeatedly. The gradient arises from progressive movement of replication forks to either side (Figure 10-21A). This image was confirmed by electron microscopy of chromatin isolated from egg chambers. Clusters of replication forks within replication forks were seen in egg chambers undergoing chorion gene amplification, but not in early, non-amplifying egg chambers.

This chorion locus contains four genes and five partially redundant amplification-enhancing regions (AERs, Figure 10-21B). Of the five AERs, only amplification control element-3 (ACE3) and ori-β together are

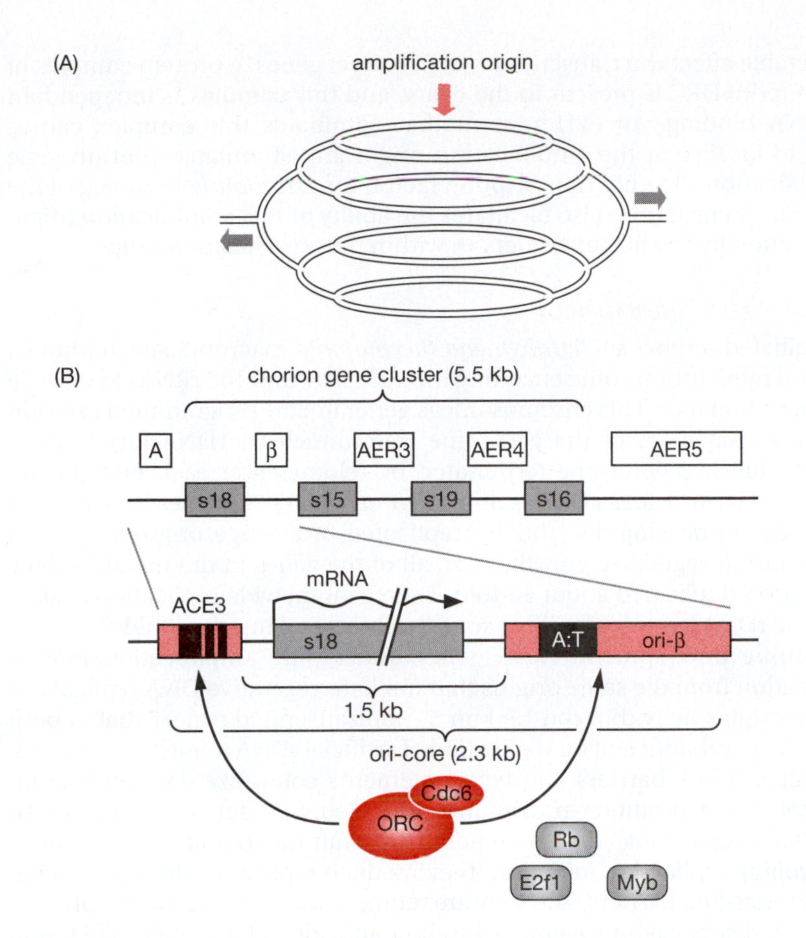

Figure 10-21. *Drosophila* amplification origin and replicator. (A) The onion skin model for specific gene amplification reflects the result of repeated initiation events at an amplification origin with replication forks eventually stalling as they travel further away from their origin. (B) Chorion locus III/66D consists of four chorion genes (s18, s15, s19, s16) interspersed with five amplification-enhancing regions (A, β, AER3, AER4, and AER5). A is ACE3 (320 bp) and β is ori-β (840 bp). They are necessary and sufficient for DNA amplification. They are separated by the s18 chorion gene (1.2 kb). Deletion mapping of ACE3 revealed an evolutionarily conserved 142 bp core sequence and two smaller core elements (black bars), suggesting that ACE3 consists of multiple, partially redundant sequence elements. Ori-β contains two required sequence elements, a 140 bp 5′-element and a 226 bp A:T-rich 3′-element. The A:T-rich element has significant homology to ACE3. [(A) data from Osheim, Y.N., Miller, O.L., Jr and Beyer, A.L. (1988) *Mol. Cell. Biol.* 8: 2811–2821. (B) data from Calvi B.R. (2006) Developmental gene amplification. In *DNA Replication and Human Disease* (DePamphilis, M.L., ed.), pp 233–255.)

necessary and sufficient for DNA amplification. Most replication forks (~80%) originate within ori-β. The sequences of ACE3 and ori-β are reminiscent of fission yeast replicators. They are 60–70% AT-rich, but otherwise have only limited similarity to one another, and small deletions of the most conserved sequences do not affect amplification. Nevertheless, *S. cerevisiae* ARS1 cannot substitute for ori-β, just as it cannot substitute for fission yeast origins, thereby confirming the sequence specificity of ori-β. The level of amplification is context-sensitive; it depends on the genomic site wherein the amplification locus resides. However, this context sensitivity can be eliminated by placing chromatin insulator sequences at each end of the amplification region. Insulators are sequences that prevent a transcription enhancer from stimulating a promoter on the opposite side of the insulator. Under these conditions, essential sequences within ACE3 and ori-β elements can be identified. In contrast, placing an insulator between ACE3 and ori-β cripples the origin, revealing that the proteins that bind to these two essential components must interact with one another.

What regulates the activity of this amplification origin? The retinoblastoma tumor-suppressor protein (Rb) binds to the E2f transcription factor thereby preventing it from interacting with the cell's transcription machinery. In the absence of Rb, E2f and its binding partner DP mediate the *trans*-activation of E2f target genes that facilitate the G_0-to-S-phase transition. In *Drosophila,* E2f1 function during S phase to regulate initiation of DNA replication at replication origins. The current model is that an E2f1•Dp•Rb complex binds near ORC at the *Drosophila* chorion gene-amplification origin and form a complex with ORC. Although E2f1 does not direct ORC binding, it restricts its activity through Rb. Reduced levels of Rb increase gene amplification and genomic replication without detectable effects on transcription of E2f target genes. A protein complex of E2f1•Dp•Rb•ORC is present in the ovary, and this complex is independent of DNA binding. An E2f1 mutant that eliminates this complex causes ORC to localize at the amplification origin and stimulates chorion gene amplification. Another transcription factor, the *Drosophila* homolog of the Myb oncogene family, also facilitates the ability of this amplification origin to function by binding to sequences within the amplification origin.

Tetrahymena thermophila

The ciliated protozoan *Tetrahymena thermophila* macronucleus harbors a natural minichromosome encoding the 26S, 5.8S, and 16S rRNAs as a single transcription unit. This chromosome is generated by programmed excision and rearrangement of the germ line (**micronuclear**) rDNA locus into a 21 kbp linear palindrome terminated by telomeres at each end (Figure 10-22). **Macronuclear rDNA** is amplified up to 10 000 copies in a single S phase during development, but it is replicated, on average, only once per cell cycle during vegetative growth. First, all of the genes in the macronucleus are endoreduplicated about 45-fold. Then genomewide replication ceases, and the rDNA locus is amplified specifically several hundred fold.

During development, rRNA genes undergoing amplification initiate replication from the same origins that mediate vegetative DNA replication. These origins lie within the 1.9 kbp 5′-nontranscribed spacer that is both necessary and sufficient for replication of artificial rDNA minichromosomes. Replication-fork barriers and type I elements colocalize with replication origins, while promoter-proximal type I elements act at a distance to regulate origin firing. Type I elements are multifunctional. In addition to controlling replication initiation, they mediate replication-fork pausing at replication-fork barriers, and they are required for rDNA transcription.

Four distinct type I element-binding activities (TIF1, TIF2, TIF3, and ORC) have been identified. In contrast to previously described replication-initiation factors, these proteins bind exclusively to single-stranded DNA, indicating that these regions are predominantly single-stranded in native chromosomes. TIF2 and TIF3 bind specifically to the A-rich strand of type I elements. TIF1 binds specifically to the A-rich strand at the rDNA origins and to the T-rich strand at the rDNA promoter. *Tetrahymena* ORC contains an integral RNA subunit that spans the terminal 282 nucleotides of the 26S rRNA and participates in rDNA origin recognition by base pairing with the type I element T-rich strand. Since ORC and non-ORC proteins are targeted to opposite strands at the rDNA origin, the interplay between these *trans*-acting factors may regulate origin activity.

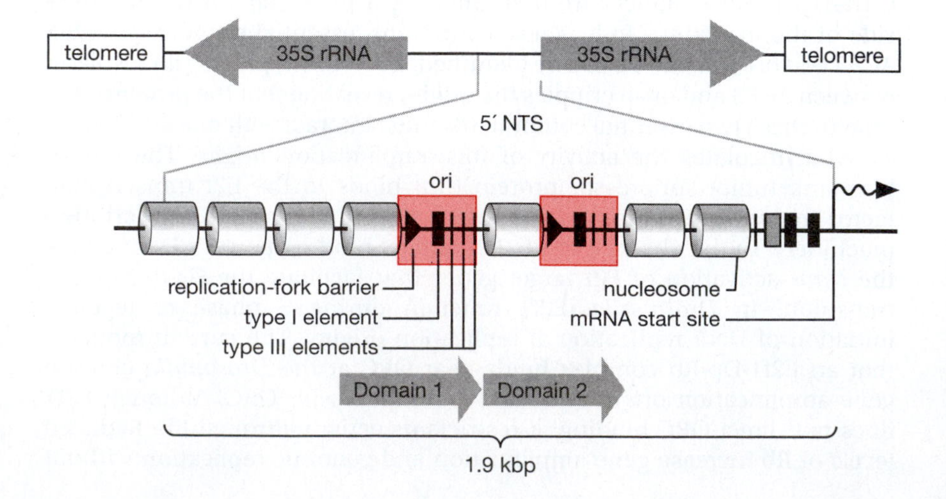

Figure 10-22. *Tetrahymena* rDNA minichromosome is a 21 kbp sequence containing two inverted copies of the rRNA coding region (35S) separated by approximately 3.8 kbp of noncoding sequence bound by 11 kbp telomeres. The 1.9 kb 5′-nontranscribed spacer (5′ NTS) contains positioned nucleosomes that flank two nucleosome-free regions that contain all of the initiation sites for cell-cycle-regulated DNA replication and gene amplification (ori). Dispersed, repeated type I elements and pause site elements are required for both replication initiation and the transient pausing of replication forks. Domains 1 and 2 are 430 bp tandemly duplicated segments that have undergone subsequent sequence divergence. (Adapted from Mohammad, M.M., Donti, T.R., Sebastian Yakisich, J. *et al.* (2007) *EMBO J.* 26: 5048–5060. With permission from Elsevier.)

SUMMARY

- Five different mechanisms are used to initiate DNA replication at replicators in all cells, viruses, and episomes (Table 10-1). Replication in circular and linear ssDNA genomes of bacteriophages, animal and plant viruses, and in some dsDNA bacterial plasmids is initiated by DNA nicking. Replication in linear dsDNA viral and bacteriophage genomes is initiated by protein priming. Replication in circular dsDNA chromosomes of most animal viruses is initiated by a DNA helicase. Replication in bacteria, archaea, and in the nuclei of eukarya is initiated by a DNA helicase loader. The same mechanism also is used to initiate replication in several bacterial plasmids and in the EBV chromosome. Replication of the tiny portion of the genome encoded by mtDNA is initiated by DNA transcription, the same mechanism used in temperate bacteriophage, some lytic bacteriophages, and some bacterial plasmids. Thus, mtDNA replication differs from nuclear DNA replication both in the way replication is initiated and extended (nuclear DNA uses replication forks and mtDNA uses single-strand displacement). These differences support the hypothesis that the mitochondrial genome originated from a bacterial ancestor.
- The complexity of replicators increases with biological complexity. Replication origins that function in bacteria (cellular, phage, and plasmid origins) consist of unique DNA sequences that bind a cognate initiator protein that then partially unwinds the DNA-unwinding element. These origins contain an active replicator. Active replicators are also found in the dsDNA genomes of animal viruses and in mtDNA. However, replication origins in single-cell eukarya appear to be passive. They bind the origin-recognition complex (ORC), but ORC does not appear to modify the replication origin. Instead, ORC recruits other proteins (Cdc6, Cdt1, and MCM helicase) that unwind the DNA-unwinding element. Nevertheless, yeast replicators rely strongly enough on DNA sequence that most of them can be identified by sequence analysis alone. With the evolution of metazoa, replicators no longer exhibit a species-specific consensus sequence. Although DNA sequence is still a determinant, they rely heavily on epigenetic factors to the extent that their activity is conditional.
- The binding sites for initiator proteins range in size from 5 to 20 bp and vary extensively in DNA sequence and nucleotide composition. Binding sites for transcription factors are variable in size. Replicators from viruses and plasmids as well as from bacteria, archaea, and eukarya, range in size from 50 to 1200 bp. Their origin-recognition elements (which often contain multiple protein-binding sites) range in size from 18 to 280 bp.
- Eukarya encode seven different initiator proteins [ORC(1-6)+Cdc6] that act in concert to activate their cognate replicators. So far, three species-specific ORC–replicator DNA interactions have been reported. In budding yeast, ORC(1-5) binds to the ORE. In fission yeast, the Orc4 subunit targets the ORE bringing with it the remaining ORC subunits. In *Tetrahymena*, ORC contains an RNA component that interacts specifically with the rDNA origin by base pairing with an essential *cis*-acting replication determinant termed the type I element.
- Replication origins in single-cell eukarya rely more heavily on specific DNA sequences than do replication origins in multicellular eukarya, thus making replicators easier to identify in single-cell organisms. Nevertheless, the density of their replication origins, the structure of their replication bubbles and forks, and their flexibility in the numbers and locations of replication origins required for survival appear to be conserved throughout the eukarya.

- As previously observed for proteins that operate at replication forks (Figure 4-21), proteins encoded by bacterial and bacteriophage genomes that initiate DNA replication commonly form macromolecular complexes that carry out DNA replication. The same appears to be true for the analogous proteins produced by eukaryotes and their viruses.
- The remarkable diversity of sequence composition and organization in the replication origins found in the genomes of cells, viruses, and episomes (plasmids) is testimony to the vigorous experimentation that occurred during biological evolution. Essentially every possible mechanism that can be imagined for duplicating dsDNA genomes is utilized by some living or nonliving thing to duplicate its genome.
- Remarkably, only two of these mechanisms were selected by nature to duplicate the genomes of living things. The primary mechanism is a DNA helicase loader, a protein or protein complex that determines where on the genome the replicative DNA helicase will bind and begin DNA unwinding. This protein is called DnaA in bacteria and Orc1/Cdc6 in archaea. Eukarya employ seven different proteins to accomplish the same task. They are called Orc1 through Orc6 and Cdc6. All of these proteins are initiator proteins, because they determine where in the genome replication will begin. Once they bind to DNA, the replicative DNA helicase is bound at or near the initiator–DNA complex. However, eukarya also contain cytoplasmic organelles called mitochondria or chloroplasts that contain a small portion of their genome. A different mechanism, involving transcriptional activation, is used to replicate this DNA.

Chapter 11
INITIATION

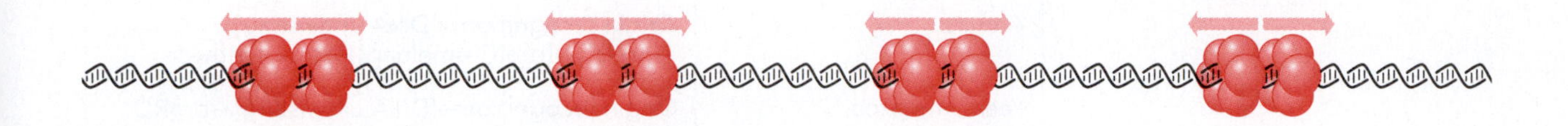

DNA synthesis during genome duplication in bacteria, archaea, and the nuclei of eukarya begins with the construction of replication forks at replication origins. Therefore, to understand how replication origins in cellular organisms function is to understand how replication forks are constructed, a process that involves a sequence of events that can be illustrated as in Figures 3-8 and 9-1 and summarized as eight events in Table 11-1. The anatomy of replication origins is detailed in Chapters 9 and 10, and events 5–8 at replication forks are detailed in Chapters 3–5. In this chapter, events 1–4 are detailed, and their similarities and differences compared among the three domains of life.

While each event in the construction of a replication fork is readily identifiable in all of the cells, viruses, and episomes known to use this mechanism, three differences are apparent. The first is that origin recognition in bacteria, bacterial plasmids, bacteriophages, and mammalian DNA tumor viruses results in localized melting of the DNA-unwinding element (DUE), whereas there is no evidence that origin recognition in archaea, yeast,

Table 11-1. Sequence Of Events In The Construction Of A Replication Fork

Step	Event	Description
1	Origin recognition	Binding initiator protein to replicator.
2	preRC assembly	Loading the replicative DNA helicase[a] at replication origin. Also referred to as origin licensing, because it permits replication origin to initiate DNA synthesis.
3	Origin melting	Localized unwinding of the DNA-unwinding element (DUE) by the replicative DNA helicase to expose the DNA templates.
4	preIC assembly	Loading the initiator DNA polymerase[b] at replication origin. Eukaryotic preIC assembly includes activation of the DNA helicase and origin melting.
5	Leading-strand synthesis	RNA-primed DNA synthesis by the initiator DNA polymerase[b] on the leading-strand DNA template. In eukarya, the initiator DNA polymerase (Pol-α•primase) is quickly replaced by the more processive Pol-ε•PCNA complex.
6	Lagging-strand synthesis	RNA-primed DNA synthesis by the initiator DNA polymerase on the lagging-strand DNA template. In eukarya, Pol-α•primase is quickly replaced by the more processive Pol-δ•PCNA complex.
7	Primer excision and gap-filling	Replacing RNA in nascent DNA strands with DNA.
8	DNA ligation	Joining Okazaki fragments to the 5′-ends of nascent DNA strands.

preIC, pre-initiation complex; preRC, pre-replication complex.
[a]DNA helicase that unwinds DNA specifically at replication forks (see Table 4-1).
[b]DNA polymerase that initiates DNA synthesis at replication origins (see Table 4-3).

Table 11-2. Comparison Of Events In The Formation Of DNA Replication Forks Among The Three Domains Of Life

Genome	Events
Bacteria, bacterial plasmids, bacteriophages, DNA tumor viruses	Origin recognition→DNA melting→preRC→preIC→DNA synthesis
Archaea	Origin recognition→[DNA untwisting] →preRC→[preIC→melting]→DNA synthesis
Yeast, metazoa, Epstein–Barr virus	Origin recognition→[DNA untwisting]→preRC →preIC→melting→DNA synthesis

Events in brackets are supported by data, but not yet firmly established. The order of events in archaea and eukaryotes remains somewhat speculative.

metazoa, and Epstein–Barr virus directly causes DNA melting (Table 11-2). At most, some DNA untwisting is imposed by the initiator on its replicator, or perhaps formation of a DNA loop (Chapter 7). In eukarya, DNA unwinding may be part of the preIC assembly process wherein the inactive replicative DNA helicase is converted into an active one. The second difference is that preIC assembly in bacteria is essentially a continuation of preRC assembly, whereas preRC and preIC assembly in eukarya are clearly separate events, because cell proliferation can pause after completion of preRC assembly. The third is that the number of different proteins required to form replication forks (preRC+preIC) varies from as few as 12 in some viruses to as many as 37 in eukarya. Although increasing the number of events in replication-fork assembly increases the risk of mistakes, it also increases the ability of cells to regulate when and where replication forks are constructed, and to link their construction to cell division and differentiation.

THE ROAD TO REPLICATION FORKS

The first event in creating replication forks at any replication origin is origin recognition (Table 11-1). This occurs when one or more initiator proteins binds to a DNA sequence that can function as an origin of replication. All genomes encode at least one protein that functions as an initiator. With the genomes of noncellular entities such as simian virus 40 (SV40), the initiator is a single protein (large tumor antigen, T-ag) that binds to its cognate active replicator and then rapidly, and without intervention by regulatory pathways, triggers DNA replication. With cellular organisms such as bacteria the initiator is also a single protein (DnaA) that binds to an active replicator (oriC). However, the productivity of this event is regulated so that DNA synthesis does not begin before the cell reaches a critical mass per origin ratio. Archaea can encode more than one initiator protein, each of which may recognize a different replicator. With eukarya, however, origin recognition reached its pinnacle of complexity. The eukaryotic initiator consists of seven different proteins [ORC(1-6)+Cdc6] that bind to a wide assortment of passive replicators, some of which are sensitive to epigenetic parameters (conditional replicators) and some that remain silent until induced by physiological stress (dormant replicators).

At some point in the initiation process, localized melting of the DNA must occur to engage the replicative DNA helicase. This is necessary because these enzymes are composed of six subunits organized into a ring structure that surrounds a single strand of DNA. Therefore, the hexameric structure must either be threaded onto the end of a DNA molecule, or assembled around a ssDNA molecule (Figure 4-3). This initial DNA-unwinding event is carried out at the DUE either by the replicative DNA helicase itself (e.g. SV40 T-ag) or by a helicase loader (e.g. *E. coli* DnaA). This means that initiator proteins must either be DNA helicases that assemble onto active replication origins, or be DNA helicase loaders that assemble the replicative DNA helicase onto a DUE. Clear evidence for localized DNA melting by initiator proteins has been obtained only with replicators from bacteria, bacteriophages, bacterial plasmids, and small DNA tumor viruses. The initiator proteins from archaea, fission yeast, fruit flies, and humans can alter the pitch of a DNA helix by untwisting it so as to increase the number of base pairs per turn (Figure 1-4 and Table 1-2), but localized DNA melting (disruption of base pairs) may occur only during or after loading of the replicative helicase.

The initiator proteins of small dsDNA viruses, such as SV40 T-ag, are assembled into a functional hexameric DNA helicase through specific binding of T-ag monomers to their cognate replicator sequence, a process that requires ATP. With cellular organisms, however, loading the replicative DNA helicase is accomplished by a group of proteins that includes the initiator and is referred to as a helicase loader. The resulting protein•DNA complex that contains the replicative DNA helicase, but not the DNA polymerase and DNA primase required to initiate DNA synthesis, is referred to in eukaryotes as a pre-replication complex (preRC). However, the term applies equally well to the analogous situation in bacteria, archaea, many bacteriophages (e.g. λ), viruses (e.g. herpes), and plasmids (e.g. R6K). preRCs differ not in their function, but in the number of proteins they contain. They contain as few as one gene product in small viruses such as SV40 to five in bacteria and 15 in eukarya (Table 11-3).

Cellular genomes encode many DNA helicases, but it appears that only one serves as the replicative DNA helicase (Table 4-1). Since DNA helicases travel in only one direction along ssDNA while concomitantly displacing (unwinding) the complementary strand in the dsDNA downstream, a minimum of two helicase hexamers must be loaded at each replication origin to achieve bidirectional replication. Their mechanism of action is described in Chapter 4. Replicative DNA helicases fall into two categories: those that can assemble themselves into an active hexameric DNA helicase at specific DNA sequences (e.g. SV40 T-ag), and those that require an independent loading mechanism. The latter include the replicative helicases of all cellular organisms (Table 11-3).

The final event in constructing a replication fork is loading the initiator DNA polymerase. Cellular genomes encode many DNA polymerases (Table 4-4), but only one that interacts with DNA primase (Table 5-1), the enzyme

Table 11-3. preRC Proteins

Viruses (SV40)	Bacteria (*E. coli*)	Archaea (*S. solfataricus*)	Single-cell eukarya (*S. cerevisiae*)	Multicellular eukarya (*H. sapiens*)	Function
T-ag monomer	DnaA	[Orc1/Cdc6]-1, -2 and -3	ORC	ORC	Initiator (origin recognition)
	IHF		Cdc6	Cdc6	Co-initiator
	Fis			Mcm9	Geminin antagonist
				HBO1	Histone H4 acetylation
	DnaC	WhiP	Cdt1	Cdt1	Helicase loader
T-ag hexamer	DnaB	Mcm	MCM	MCM	Replicative DNA helicase

The origin-recognition complex (ORC) and the minichromosome maintenance (MCM) complex each consist of six different proteins. Nonspecific dsDNA-binding proteins such as HU in bacteria, histone-like proteins in archaea, and histones in eukarya can either facilitate or interfere with assembly of pre-RCs, depending on the location and type of replicator. The archaeal WhiP protein is related to eukaryal Cdt1 but its precise role remains unclear.

that synthesizes a short oligoribonucleotide on the DNA template in order to provide the 3′-terminal polynucleotide primer required by all DNA polymerases. In fact, some viruses and episomes encode their own DNA polymerase DNA-priming mechanism so that duplication of their own genome competes successfully with that of their host. Historically, loading the initiator DNA polymerase is termed replisome assembly in bacteria and pre-initiation complex (preIC) assembly in eukarya, but here the term preIC is used to describe equivalent events in bacteria, archaea, and eukarya.

preIC assembly describes those events that follow preRC assembly and precede RNA-primed DNA synthesis. Whereas preRC assembly determines where DNA replication *can* begin, preIC assembly determines where DNA replication *will* begin. All of the replication origins in a genome are not activated each time a cell divides (see Origin usage, Chapter 9), and the frequency at which a particular origin is activated depends primarily on whether or not a preRC is converted into a preIC. preIC assembly depends on the strength of the replicator and its genomic location (i.e. sequence context). The major differences between preIC assembly in small genomes of viruses such as SV40 and those of bacteria, archaea, and eukarya are the number of proteins involved and the imposition of regulatory events. For example, the number of additional proteins required to produce a preIC range from 11 in SV40 to 15 in bacteria and 22 in eukarya (Table 11-4). This means that, prior to initiation of DNA synthesis during genome duplication in human cells, at least 37 different proteins must be assembled at a replication origin! The pathways by which this occurs are described below.

Replication-Fork Assembly And Cell Proliferation

The transition from preRC to preIC to initiation of DNA synthesis is uninterrupted during the duplication of viral and episomal genomes. Thus, they appear as a single event. Whether or not preRC assembly can

Table 11-4. preIC Proteins

<table>
<tr><th>Viruses (SV40)</th><th>Bacteria (E. coli)</th><th>Archaea (S. solfataricus)</th><th>Eukarya (S. cerevisiae)</th><th>Function</th></tr>
<tr><td>Pol-α (2 subunits)</td><td>Pol-III core (3 subunits)</td><td>Pol-B</td><td>Pol-α (2 subunits)</td><td>Initiator DNA polymerase</td></tr>
<tr><td rowspan="3">DNA primase (2 subunits)</td><td rowspan="3">DnaG</td><td rowspan="3">DNA primase (2 subunits)</td><td>DNA primase (2 subunits)</td><td>DNA primase</td></tr>
<tr><td>Mcm10</td><td>Load initiator DNA polymerase•primase</td></tr>
<tr><td>DDK (2 subunits)</td><td>'Dbf4-dependent protein kinase' needed for preIC assembly</td></tr>
<tr><td rowspan="5">T-ag</td><td rowspan="5"></td><td rowspan="2"></td><td>Dpb11</td><td>Load DNA polymerase</td></tr>
<tr><td>Sld2</td><td>Load DNA polymerase</td></tr>
<tr><td rowspan="3">GINS (2 subunits)</td><td>GINS (4 subunits)</td><td>DNA polymerase loading complex</td></tr>
<tr><td>Cdc45</td><td>Load DNA polymerase</td></tr>
<tr><td>Sld3</td><td>Load DNA polymerase</td></tr>
<tr><td>Pol-ε (4 subunits)</td><td>Pol-III holoenzyme (10 subunits)</td><td>Pol-B</td><td>Pol-ε (4 subunits)</td><td>Leading-strand polymerase</td></tr>
<tr><td>RPA (3 subunits)</td><td>SSB</td><td>SSB</td><td>RPA (3 subunits)</td><td>ssDNA-binding protein</td></tr>
</table>

In addition, changes in DNA topology require the intervention of topoisomerases, histones, or histone-like proteins.

be distinguished from preIC assembly in bacteria depends on the rate of proliferation. Bacterial cells undergo a regulated sequence of events that coordinate cell growth and division (Figure 11-1). Growth of bacteria on rich medium speeds up the process and eliminates the B period. Thus, in slowly proliferating bacteria one can distinguish binding of initiator protein (DnaA) to replicator DNA (oriC) in newly formed cells (D period) from preIC assembly as cells begin chromosome duplication (C period). Bacteria lack a 'G_2 phase' because separation of sibling chromosomes occurs concomitantly with DNA replication, although cytokinesis (separation into two cells) is delayed until separation of the two nucleoids (equivalent to eukaryotic nuclei) has been completed. The transition from preRC to preIC in bacteria is triggered by increasing levels of ATP.

preRC assembly in eukarya only occurs when proliferating cells transit from M to G_1 phase of the cell cycle and when quiescent cells enter the cell cycle by transiting from G_0 to G_1 phase (Figure 11-1). Thus, preRC assembly marks a commitment to cell proliferation; it occurs spontaneously in the absence of cyclin-dependent protein kinase (CDK) activity, and it can be recapitulated *in vitro* with purified proteins and DNA. preIC assembly, on the other hand, marks the beginning of DNA replication by creating DNA replication forks, a process not yet reconstituted with purified proteins.

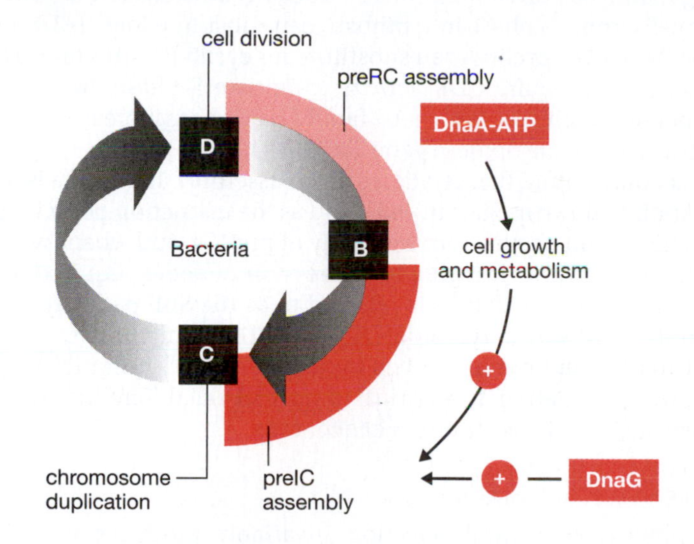

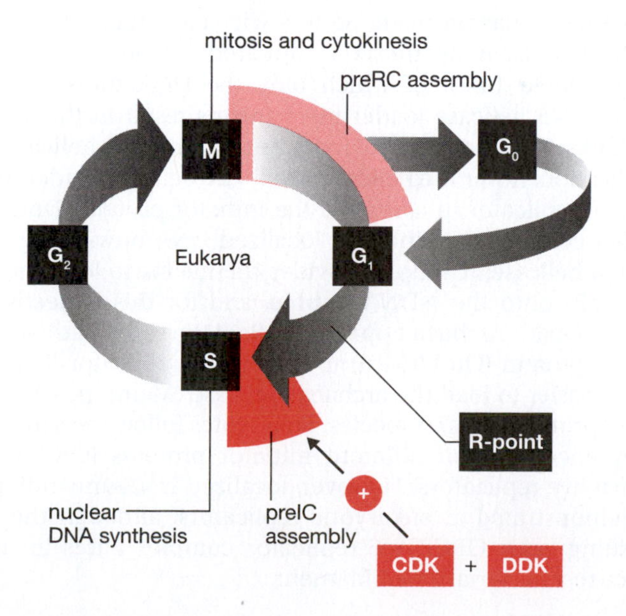

Figure 11-1. Cell cycles and their relationship to preRC and preIC assembly at DNA replication origins. The bacterial cell cycle consists of three periods: between termination of chromosome duplication and the end of cell division (D), the interval between cell division and initiation of DNA synthesis (B), and the period of DNA synthesis or chromosome duplication (C). preRC assembly occurs when the ATP-activated form of initiator protein (DnaA) has titrated all the binding sites in the replicator oriC. ATP also is required for loading the replicative DNA helicase, DnaB. preIC assembly begins with addition of DNA primase (DnaG). The eukaryotic cell cycle consists of four phases: the time between cell division (cytokinesis) and nuclear DNA replication in which cells increase in size and synthesize the proteins needed for DNA replication and chromatin assembly (G_1), the time in which nuclear DNA synthesis occurs (S), the time between completion of nuclear DNA synthesis and mitosis (G_2), and mitosis (M) which normally includes cytokinesis. G_0 is a quiescent state from which cells can re-enter the cell cycle through G_1 phase. preRC assembly occurs during the transition from M to G_1 phase, and preIC assembly marks the transition from G_1 to S phase. preIC assembly in eukarya is activated by two protein kinases, CDK and DDK. The restriction point (R) marks a critical event when mammalian cells commit to proliferation and become independent of growth stimulatory factors.

In eukarya, preIC assembly is distinguished from preRC assembly by the fact that preRC assembly is inhibited by CDK activity, whereas preIC assembly is triggered by CDK activity. The number of different CDKs and their cyclin activators varies among species (Table 12-3). Yeast contain a single CDK called Cdc28 in budding yeast and Cdc2 in fission yeast. Cdc28 is activated by two different S-phase cyclins (Clb5 and Clb6) for entry into S phase. Clb5 appears to be more central to the initiation of DNA replication since cells lacking Clb5 exhibit a slow S phase, whereas cells lacking Clb6 show no defects in cell-cycle progression. Remarkably, cells lacking both Clb5 and Clb6 can still replicate their DNA, because mitotic cyclins can substitute for S-phase cyclins. Similarly, fission yeast DNA replication normally is triggered by Cdc2 and the S-phase cyclin Cig2, but it can also be triggered by Cdc2 and the mitotic cyclin Cdc13.

Among the metazoa, both Cdk2•cyclin E and the protein kinase Cdc7•Dbf4 (termed DDK for Dbf4-dependent protein kinase) are required to trigger S phase (Figure 11-1). For many years, it was believed that Cdk2 was essential for initiation of genome duplication in metazoa. In fact, Cdk2•cyclin E is essential for initiation of DNA replication in *Xenopus* egg extracts and *Drosophila* embryos, but mice lacking Cdk2 are viable, revealing that Cdk2 is dispensable for DNA replication in mammals. Mammals have at least 10 CDKs, only four of which (Cdk1, Cdk2, Cdk4, and Cdk6) are involved in regulating cell division. In this case, Cdk1, an enzyme that is required to drive cells from G_2 phase into mitosis, can substitute for Cdk2 in initiating S phase. Moreover, cyclin A can substitute for cyclin E in this task. Thus, while all eukaryotes require CDK activity to initiate S phase, by increasing the number of proteins within a gene family, nature has increased the chance of survival should one of these gene activities be lost.

A second feature that regulates preIC assembly in eukarya is a G_1-phase checkpoint known as Start in yeast and as the restriction point in mammals. This checkpoint occurs after assembly of preRCs, and whenever cells enter and leave G_0 phase. It regulates expression of genes required for S phase using analogous mechanisms referred to as the SBF pathway in yeast and the Rb•E2f pathway in mammals (Chapter 13). In mammals, the restriction point marks the time in the cell cycle when a cell becomes independent of growth stimulation. It is critical to normal cell differentiation and tissue homeostasis, and it is absent in cancer cells.

Where The Road Splits

The process of natural selection invariably produces more than one solution to a particular problem. So it is with construction of replication forks. Replicative DNA synthesis is initiated by one of five different mechanisms (Table 10-1), of which only the DNA transcription, DNA helicase, and DNA helicase loader mechanisms result in the assembly of replication forks. Of these, all cellular organisms employ a helicase loader to initiate replication, never the helicase itself. The helicase loader in bacteria uses an active replicator to assemble the initiator protein (DnaA) into an initiator•DNA complex that induces localized DNA unwinding. However, DnaA is not a helicase; it simply provides the means to load the bacterial helicase (DnaB) onto the ssDNA bubble, and for this it needs a helper protein called DnaC. Archaea appear to follow the same pathway by using their initiator protein (Orc1/Cdc6) to distort the replicator in such a way as to make it easier to load the archaeal helicase (Mcm), possibly with the help of WhiP protein in some species. Eukaryotes follow the same pathway, except they encode seven different initiator proteins (ORC+Cdc6) that recognize passive replicators. However, localized DNA unwinding has not yet been demonstrated at eukaryotic replicators, although the structure of the budding yeast ORC•Cdc6-replicator complex bares an intriguing resemblance to a bacterial DnaA filament.

With the evolution of viruses and episomes, these nonliving entities plagiarized their host's genes to develop simple, efficient mechanisms that would ensure their own survival. In bacteria, a pathway evolved for lysogenic phage and plasmids in which a single initiator protein mimics the role of the bacterial initiator (DnaA) by bending their cognate replicator in such a way as to induce DNA unwinding, but without the need for ATP and without intervening levels of regulation. Lytic viruses and episomes (including mitochondrial DNA) with dsDNA genomes and internal replicators took a different road. They rely either on an RNA polymerase or a DNA helicase to unwind DNA and then assemble the replication-fork machinery at each end of the unwound region. Of these, the simplest introduction to understanding DNA helicase loaders is to understand how a DNA helicase is used to initiate replication.

VIRAL GENOMES

SV40 and papillomavirus contain small, circular, dsDNA genomes with a single replicator (Figure 10-5) that replicate in the nucleus of their mammalian host. With the exception of their cognate initiator protein, all subsequent events in the duplication of these genomes rely solely on proteins that are produced by the host cell during S phase. Since the goal of these viruses is simply to produce as many progeny as possible before the host cell dies, they need not concern themselves with regulatory pathways designed to couple genome duplication with cell division. They do, however, need to drive their host cell into S phase, and this they accomplish by expression of an initiator protein (T-ag) that activates cellular gene expression.

SV40 T-ag is the prototype for replicative DNA helicases. Although it can assemble spontaneously into a hexamer in the absence of DNA, preformed T-ag hexamers can initiate DNA unwinding only at the ends of duplex DNA molecules, not at internal sites. To initiate DNA replication at internal sites, T-ag must be assembled into a functional DNA helicase by binding to the SV40 replicator. Electron microscopy of negatively stained specimens, together with two-dimensional digital image processing and analysis of single particles, has taken microscopy of proteins and protein•DNA complexes to a new dimension. This technique allowed visualization of the SV40 initiator complex (two T-ag hexamers) assembled at the SV40 replicator (Figure 11-2). The resulting structure is approximately 24 nm long with a width that varies along the length of the particle from 9 to 12 nm. The large structure appears to consist of two identical smaller structures placed head-to-head. These smaller structures (~12 nm long) each consist of a wide region at the distal end of the particle and a narrow region located at the center of the larger structure. Taking into account the available biochemical data, along with the results of the image processing, the larger structures are side-view projection images of double hexamers of T-Ag formed at the SV40 replication origin, arising from two single T-Ag hexamers positioned at the replication core in a head-to-head orientation.

SV40 DNA replication begins when SV40 T-ag assembles onto its cognate replicator to produce two functional DNA helicases (the SV40 preRC). This event requires ATP and the SV40 origin-recognition element (ORE; Figure 11-3). Sequence-specific contacts occur between the origin-binding domain in the N-terminal portion of T-ag and the pentanucleotide boxes in the ORE. Binding of successive T-ag monomers to the viral origin is a concerted process; once the first monomer binds, a hexameric complex assembles rapidly. Moreover, assembly of a single hexamer on origin DNA is less stable than assembly of a double hexamer. The pentanucleotides in the ORE form a palindromic sequence that orients the two helicase hexamers in opposite directions. The two hexamers face each other, offering a docking surface constructed by the same origin-binding domains that interact with the ORE. Since the spatial arrangement of pentanucleotides is essential in aligning

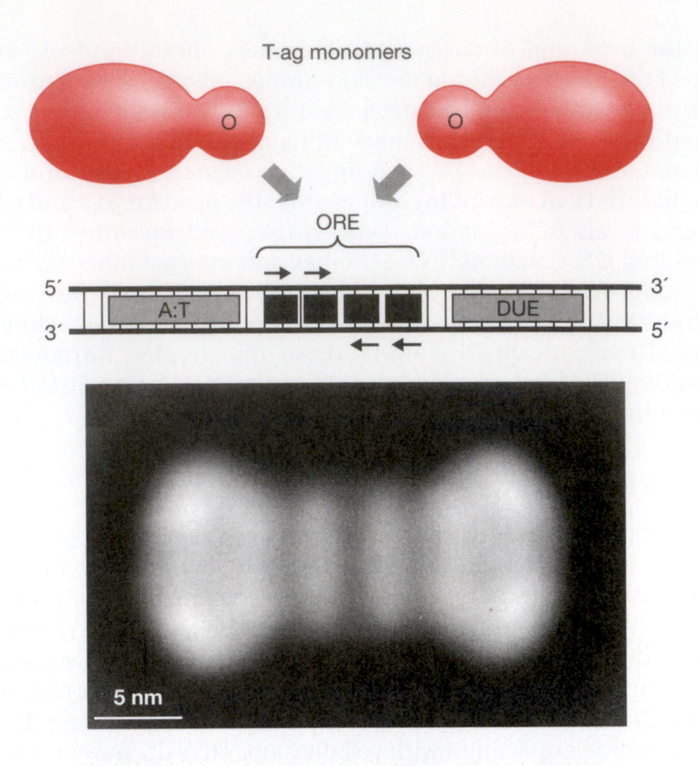

Figure 11-2. Image of two SV40 T-ag double hexamers on the SV40 replicator. Purified T-ag monomers (pink double spheres) were incubated for 1 h at 37°C in the presence of 2 mM ADP with an 80 bp DNA fragment containing the SV40 origin core component. ADP was present instead of ATP, because the latter triggers helicase activity of the T-ag hexamers, resulting in the dissembling of the dodecameric complexes. T-ag monomers assemble face-to-face on an SV40 replicator to form two T-ag hexamers. Their origin-binding domains (O) are aligned on the four pentanucleotides (solid boxes with arrow indicating sequence polarity) that comprise the origin-recognition element (ORE). The image shown below the DNA is the average of a total of 1010 individual electron micrographs of negatively stained protein•DNA complexes filtered to the calculated resolution of 2.8 nm. The two hexamers contact each other head-to-head, in part through the origin-binding domains. The helicase domains resemble propellers on each end. The DNA in the interior is not visible. A simplified diagram of the SV40 ori-core (Figure 10-5A) is included for orientation. (From Valle, M., Gruss, C., Halmer, L. *et al.* (2000) *Mol. Cell. Biol.* 20: 34–41. With permission from the American Society for Microbiology.)

the two hexamers on the replicator, it is easy to understand why single base-pair changes, insertions, or deletions within the ORE dramatically block initiation of DNA replication. Assembly of the double hexamer is accompanied by remodeling of contacts between T-ag subunits that distorts the origin DNA. An eight-nucleotide bubble forms in the DUE region, and DNA untwisting occurs in the A:T-rich region. The single-stranded bubble thus represents an early intermediate in the unwinding reaction catalyzed by the double hexamer and marks the origin of bidirectional replication (OBR) from which the two leading strands initiate and diverge (Figure 9-1).

The next step is DNA unwinding by the newly assembled T-ag helicase, a process that requires ATP hydrolysis (ATP→ADP+P) to provide the energy for separating base pairs and RPA to prevent reannealing of the two DNA templates, and formation of secondary DNA structures such as palindromic hairpins (Chapter 4). T-ag recruits RPA to the replicator through a specific association between the T-ag origin-binding domain and the 70 kDa subunit of RPA. This interaction ensures that RPA binds to its minimal target site of 8 to 10 nucleotides of ssDNA, thereby coupling activation of origin unwinding with RPA loading onto ssDNA. As T-ag helicase unwinds more DNA (15–30 nucleotides), RPA binds cooperatively to the newly exposed ssDNA template, releasing T-ag in the process so that it can continue to unwind DNA. Once the ORE has been converted from dsDNA to ssDNA it no longer binds T-ag. This sequence of events converts a sequence-specific DNA-binding protein (T-ag monomer) into a sequence-independent DNA helicase (T-ag hexamer) that can unwind the entire viral genome at a rate of approximately 200 nucleotides min^{-1}.

Genome duplication requires both DNA unwinding and DNA synthesis. To this end, the helicase domain of T-ag also binds specifically DNA polymerase-α•primase, the enzyme used by eukaryotes to initiate DNA synthesis on the leading-strand template at replication origins and to initiate synthesis of Okazaki fragments on the lagging-strand template at replication forks. Pol-α is a tetrameric protein (Table 4-5), and each T-ag hexamer binds one Pol-α tetramer, primarily through the T-ag helicase domain. In this way, T-ag recruits Pol-α•primase to the replication fork to form a preIC.

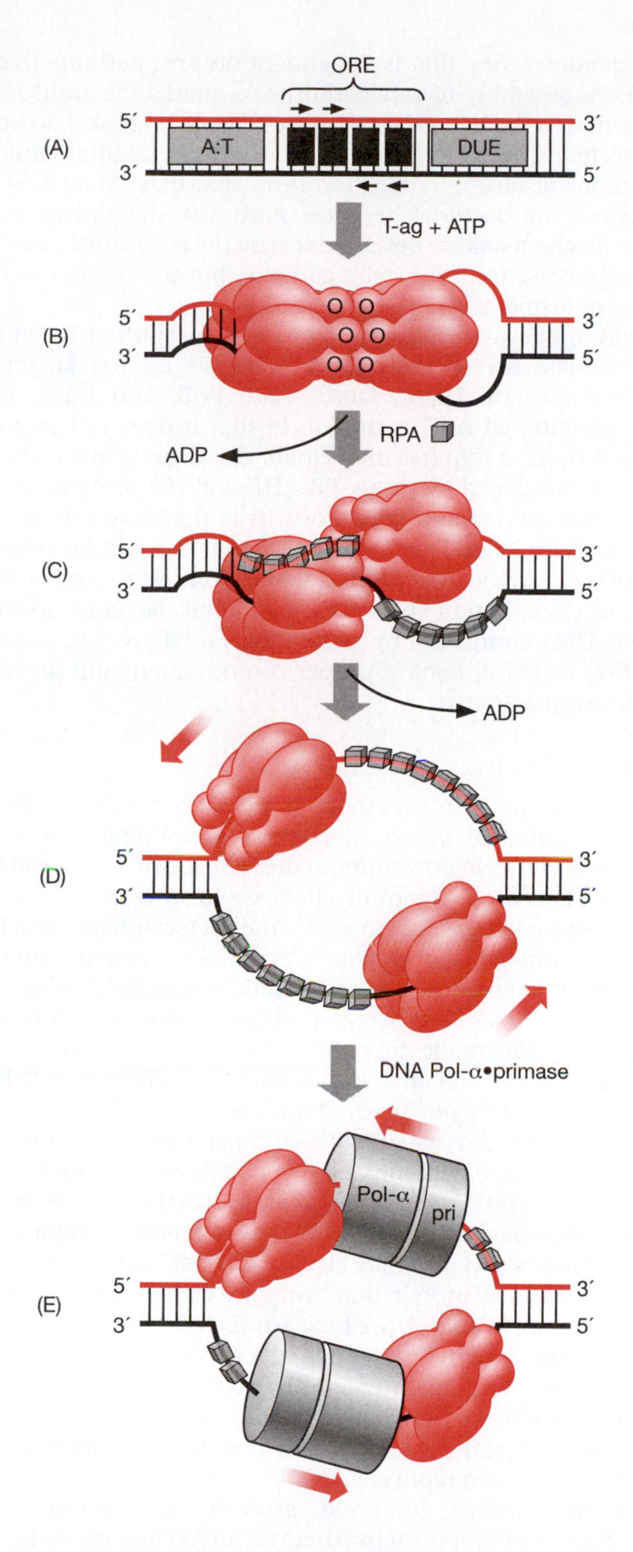

Figure 11-3. Creation of replication forks at the SV40 replicator. Two SV40 T-ag hexamers are assembled face-to-face (B) on an SV40 replicator (A, and Figure 10-5A). The origin-binding domain (O) of each T-ag monomer is aligned on the four pentanucleotides (solid boxes with arrow indicating sequence polarity) that comprise the origin-recognition element (ORE). Their helicase domains (larger pink spheres) induce deformation of the A:T-rich element (A:T) and partial melting of the DNA-unwinding element (DUE). ATP hydrolysis then allows DNA helicase activity to unwind the ORE (C). RPA (small cubes) binds to ssDNA and prevents reannealing. As each T-ag hexamer travels 3′→5′ along the leading-strand template (D), ssDNA coated with RPA becomes long enough so that DNA polymerase-α•primase can bind (E) and then initiate leading-strand DNA synthesis (Table 3-1). (Data from Jiang, X., Klimovich, V., Arunkumar, A.I. *et al.* (2006) *EMBO J.* 25: 5516–5526, Meinke, G., Bullock, P.A. and Bohm, A. (2006) *J. Virol.* 80: 4304–4312, and Valle, M., Chen, X.S., Donate, L.E. *et al.* (2006) *J. Mol. Biol.* 357: 1295–1305.)

One byproduct of DNA unwinding is formation of superhelical DNA (Figures 3-9 and 6-2). To remedy this problem, T-ag also binds to and thereby recruits to the replicator topoisomerase I, an enzyme that relaxes both positive and negative superhelical turns (Table 4-8).

BACTERIAL GENOMES

In contrast to small animal-virus genomes, the initiator protein in bacteria (DnaA) functions as a DNA helicase loader rather than a DNA helicase. Nevertheless, there are two mechanisms for creating replication forks in bacterial genomes, one that is dependent on oriC and one that is not. oriC-dependent assembly of replication forks marks the transition from the B to C period in the bacterial cell cycle (Figure 11-1) and, as such, it is subject to mechanisms that regulate the cell cycle. In addition, there is also a mechanism for restarting replication forks that have 'stalled' at various sites throughout the bacterial genome. Both oriC-dependent and oriC-independent mechanisms are designed to load the replicative DNA helicase and DNA polymerase onto the cell's genome, but the difference between them can be confusing.

Historically, the term primosome refers to a complex of seven proteins required to initiate RNA-primed DNA synthesis on the lagging-strand template (DnaG, DnaB, DnaC, DnaT, PriA, PriB, and PriC). However, primosome assembly at oriC is unique in that it does not require PriA, PriB, PriC, and DnaT; it requires only DnaB, DnaC, and DnaG. The protein complex at oriC produced by DnaA, Fis, IHF, and HU proteins creates the correct DNA helicase loading conditions. It is therefore referred to as a preRC. PriA, PriB, PriC, and DnaT are needed only to assemble primosomes at sites outside of oriC. This is termed replication restart and is discussed in Chapter 13. In addition, heat-shock proteins facilitate assembly of these protein–DNA complexes by facilitating protein•protein interactions such as folding, establishment of proper conformation, and prevention of inappropriate aggregation.

preRC Assembly On Bacterial Genomes

In the 1980s it became clear that most, if not all, bacteria encode a single initiator protein called DnaA that binds to an active replicator called oriC and induces localized DNA unwinding at the DUE in oriC. Thus, the bacterial initiator is not a helicase, but part of a helicase-loading mechanism, a concept that also appears in the genomes of some bacteriophages and bacterial plasmids. For example, bacteriophage λ encodes a single initiator protein called O and plasmid R6K encodes a single initiator protein called π, both of which mimic the ability of the bacterial initiator DnaA to induce localized melting at their cognate replicator's DUE (Figure 11-4). R6K makes use of its host's initiator protein DnaA and co-initiator protein IHF, as well. However, in contrast to the assembly of DnaA at oriC, assembly of O at oriλ and π at oriγ does not require ATP. As with SV40 T-ag, once O and π bind their cognate replicators the process of genome duplication follows without interruption or regulation. In contrast, bacterial replicators, like those in single-cell and multicellular eukaryotes, use ATP as a means to regulate origin activation. Some bacteriophages and plasmids also encode additional preRC proteins, such as bacteriophage λP protein that competes with its host's DnaC protein to load the host's replicative DNA helicase (DnaB) onto oriλ rather than oriC. Otherwise, the remaining events that result in preIC assembly at oriλ and oriγ use the same proteins employed by the host. Larger bacteriophages such as T4 and animal viruses such as herpes produce many of their own preIC and replication-fork proteins which give them complete dominance over their host and lead to rapid cell lysis.

E. coli is the paradigm for preRC assembly at bacterial replication origins. The *E. coli* initiator protein (DnaA) is an ATPase, the N-terminus of which binds specifically to its cognate replicative DNA helicase (DnaB) and the C-terminus binds specifically to the OREs within oriC. DnaA exists in two forms, an active form that is bound to ATP and an inactive form that is bound to ADP (Figure 11-5). The five R-boxes in oriC bind both the active and inactive forms of DnaA equally well, whereas the three I-boxes and the three AGATCT sites within the DUE preferentially bind the active form of DnaA three- to fourfold more tightly. The affinity of DnaA for these sites varies such that R4≥R1>R2>(R5, I2, I3)>I1>R3>>DUE. Thus, DnaA is bound to R1, R2, and R4 throughout the cell cycle as a stable replicator–initiator complex. In addition, Fis, a protein that facilitates oriC activity *in vivo*, is bound to

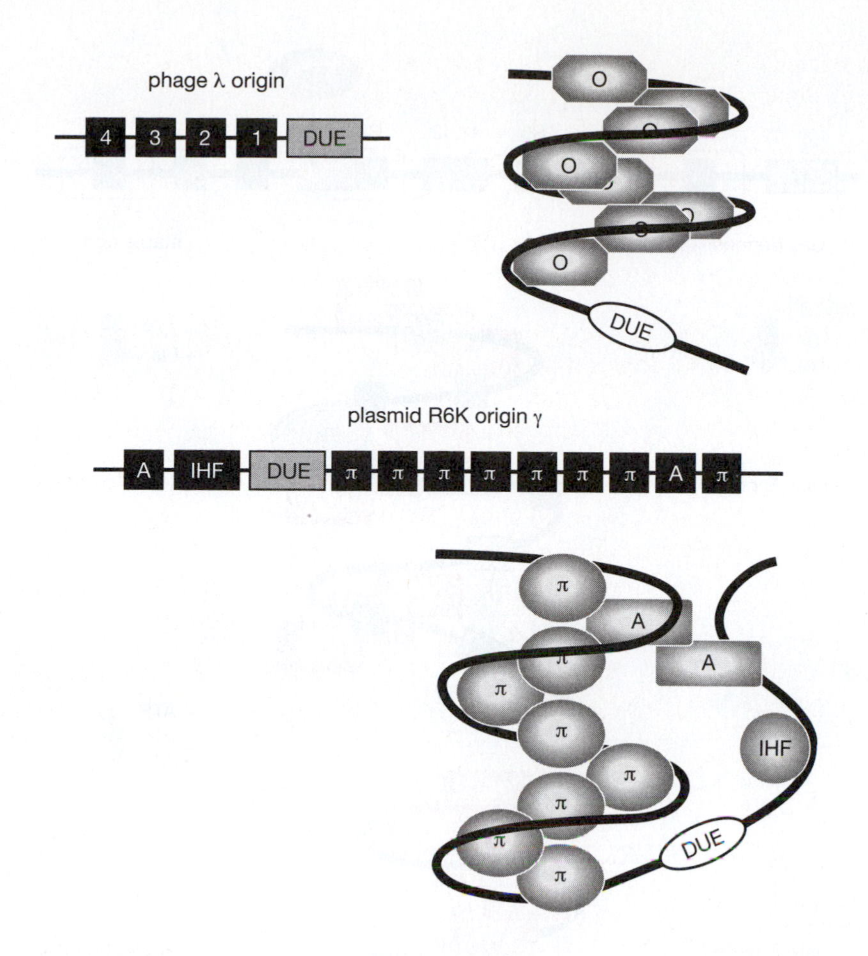

Figure 11-4. *E. coli* bacteriophage λ and plasmid R6K are well-characterized examples of genomes that encode initiator proteins (λO and R6Kπ) that function as part of a helicase-loading mechanism in the genomes of viruses and episomes, respectively. The phage λ origin-protein complex consists of four dimers of O protein (a 270 kDa protein complex) around which DNA is wrapped in a toroidal form, the topological equivalent of superhelical turns (Figure 1-7). The DNA is wrapped in a right-handed helix around the initiator proteins, as determined from crystallographic analysis of *E. coli* DnaA–oriC complexes (Figure 11-5). Neither the λO initiator protein nor the R6Kπ initiator protein binds ATP. As with *E. coli* oriC, concomitant binding of π-protein monomers, DnaA, and IHF proteins to R6K oriγ results in melting of the oriγ DUE. This hypothetical initiation complex is also based on a model for *E. coli* DnaA-oriC interactions. Solid black boxes indicate essential DNA binding sites for the indicated protein. Unwound DUE is indicated as gray box. Melted DUE is indicated as white oval. Proteins (gray geometrics) interact with both DNA and with one another, as indicated. (Data from Krüger, R. and Filutowicz, M. (2003) *Nucleic Acids Res.* 31: 5993–6003, Schnos, M., Zahn, K., Blattner, F.R. and Inman, R.B. (1989) *Virology* 168: 370–377, and Zzaman, S. and Bastia, D. (2005) *Mol. Cell* 20: 833–843.)

an oriC auxiliary site (Fis). Fis appears to facilitate the linkage between cell growth and initiation of DNA replication by suppressing IHF binding and further DnaA binding at oriC. Fis (factor for inversion stimulation; 12 kDa monomer), IHF (integration host factor; 20 kDa heterodimer), and HU (small heat stable protein; 18 kDa homodimer) are three histone-like proteins in bacteria that alter DNA architecture by wrapping and bending DNA. However, Fis and IHF bind to specific DNA sequences whereas HU binds DNA nonspecifically.

Oligomerization of DnaA is required for oriC activation, with the result that about 10 DnaA monomers are eventually bound to each activated oriC (Figure 10-6). This transition is also marked by release of Fis, and binding of IHF to oriC, a change that appears to facilitate wrapping oriC DNA around the DnaA protein complex. The net result is to partially unwind DNA at the DUE.

Total DnaA levels are constant during cell division, but as intracellular levels of DnaA–ATP rise during the B period (equivalent to the transition from G_1 to S phase in eukarya), Fis is released and IHF binds to oriC along with DnaA•ATP binding to the R3, R5, and I-sites in oriC. This change facilitates wrapping oriC DNA around the DnaA protein complex and induces unwinding of the DUE. Crystallographic studies reveal that DnaA•ADP binds oriC as a monomer, whereas ATP facilitates packing of adjacent DnaA molecules through extensive DnaA oligomerization. Fis remains bound to oriC throughout most of the cell cycle, but it is released at the time DNA synthesis begins. IHF binds specific asymmetric DNA sites as a heterodimer, bending DNA by 160°. IHF binds to oriC at the time of initiation and presumably facilitates DUE melting. The opposing activities of Fis and IHF ensure an abrupt transition from a repressed complex with unfilled weak-affinity DnaA-binding sites to a completely loaded complex with underwound DNA, increasing both the precision of DNA replication

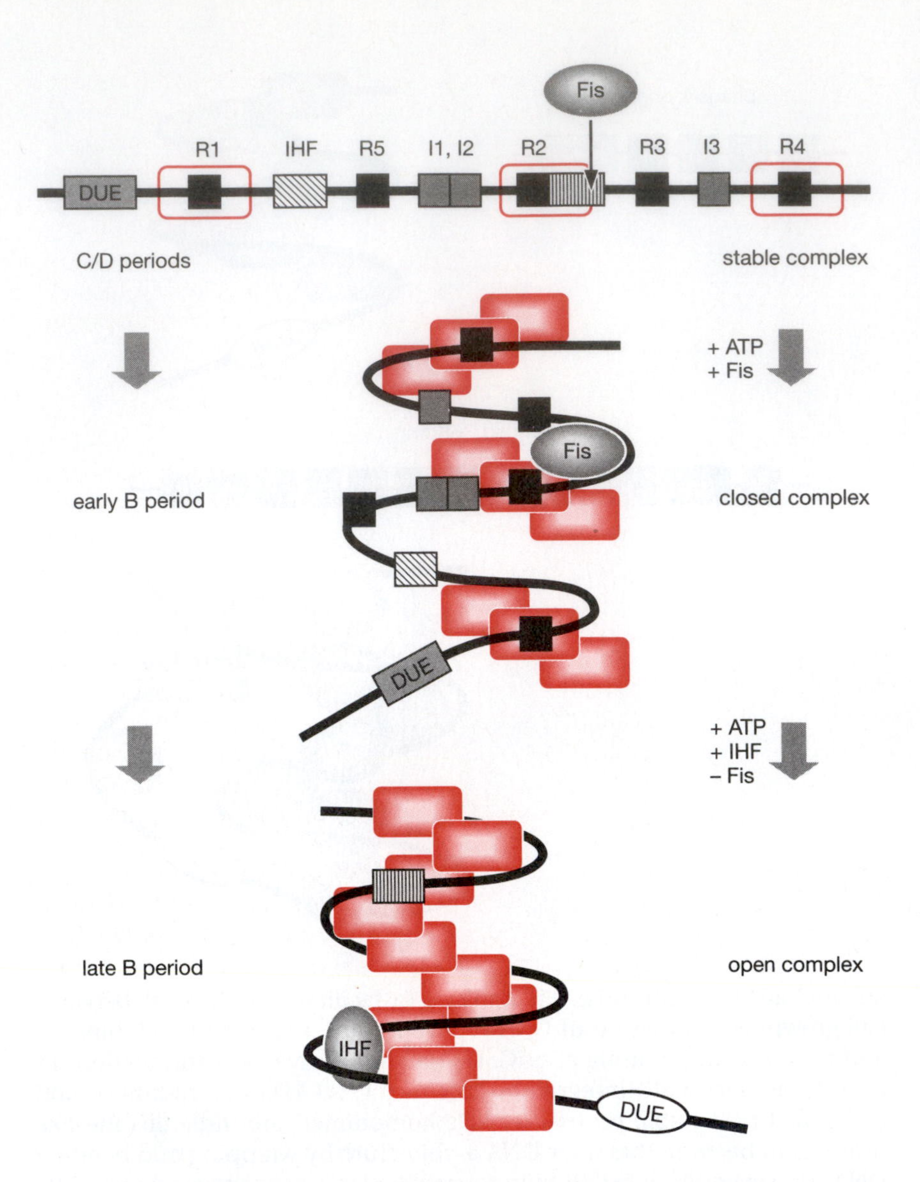

Figure 11-5. ATP-dependent activation of *E. coli* oriC by DnaA protein. C/D period bacterial cells contain inactive DnaA protein (red-bordered rectangle) bound to specific sites in oriC (R1, R2, R4). ATP increases the ability of DnaA to oligomerize (red filled rectangle) and to bind extensively to oriC (closed complex), resulting in partial unwinding of the DUE (open complex). Formation of an open complex requires sequence-specific DNA-binding proteins DnaA, Fis, and IHF, and the sequence-nonspecific histone-like DNA-binding protein HU (not shown). Binding of DnaA protein to oriC initiates unwinding of the DUE by introduction of negative superhelical turns (toroidal coil, Figure 1-7). Five DnaA dimers and one IHF dimer represent a protein–DNA complex of about 550 kDa. In the presence of ATP, DnaA alone forms protein filaments with a right-handed organization. Such filaments would impose a directionality to the DNA wrapped around it, stabilizing positive DNA supercoils around a central protein core. In a closed topological system, compensatory negative writhe would be expected to arise in response to a positive toroidal wrap. This negative superhelical stress would facilitate unwinding of the DUE. (Data from Leonard, A.C. and Grimwade, J.E. (2005) *Mol. Microbiol.* 55: 978–985, and Mott, M.L. and Berger, J.M. (2007) *Nat. Rev. Microbiol.* 5: 343–354.)

timing and initiation synchrony. Since negative supercoils are a source of energy for unwinding DNA (Figure 1-6) and for wrapping DNA around protein complexes (Figure 1-7), initiation of DNA replication at oriC is temperature sensitive. Slight increases in temperature stimulate initiation, whereas slight decreases inhibit initiation.

The next step in preRC assembly is loading the DnaB replicative DNA helicase (see Figure 11-6 in color plate section). DnaB is an ATP-dependent DNA helicase consisting of six identical subunits (Figure 4-2). When DnaB is not associated with DNA, it is complexed with six molecules of DnaC, a small AAA+ protein that, when bound to ATP, inhibits DnaB helicase activity. The central channel in the DnaB•DnaC(ATP) complex is closed until DnaB interacts specifically with DnaA bound to oriC. This leads to opening of the helicase ring, dissociation of DnaC, and loading of two DnaB hexamers within the ssDNA portion of the DUE, one DnaB hexamer on each template strand (Figure 4-1). Thus, DnaC not only facilitates loading the replicative DNA helicase onto oriC, but it prevents loading DnaB onto non-origin DNA.

preIC Assembly On Bacterial Genomes

Assembly of the *E. coli* equivalent of a preIC begins with loading its DNA primase (DnaG) onto the DnaB•ssDNA template complexes at oriC (Figure 11-6). DnaG synthesizes an oligoribonucleotide about 10 residues in length that serves as a primer for initiation of DNA synthesis on the leading-strand template (Table 5-1 and Figure 5-3). This DNA helicase•DNA primase complex is equivalent to the single helicase-primase protein produced by gene 4 of the phage T7 genome (Chapter 5). Subsequent association of DnaB with the Tau (τ) subunit of the DNA polymerase III holoenzyme (Table 4-6 and Figure 4-21) links DNA unwinding to DNA synthesis and thereby ensures that long stretches of ssDNA will not be generated by a runaway DNA helicase. Excess ssDNA makes the genome unstable, and, in the case of eukarya, triggers checkpoint controls that rapidly arrest DNA replication. Once DNA synthesis begins, regulatory mechanisms inactivate DnaA by stimulating ATP hydrolysis.

ARCHAEAL GENOMES

The initiator proteins in archaea (Orc1/Cdc6) are architecturally related to those in bacteria (DnaA) and bear clear sequence homology to two initiator proteins in eukarya (Orc1 and Cdc6). What is striking about these initiator proteins is the similarity between their C-terminal regions (Figure 11-7); they all belong to the AAA+ family of ATPases. These are proteins that can bind ATP and cleave it to ADP, using both the act of binding this small molecule as well as the energy released upon its hydrolysis to change their conformation, to carry out specific tasks such as unwinding DNA, and to assemble multisubunit protein complexes that can act as molecular machines. Moreover, DnaA, Orc1/Cdc6, and Orc1 proteins also contain either a helix-turn-helix domain (HTH) or a winged helix (wH) domain that allows them to bind DNA. Orc1 is unique among ORC proteins in all species in that it contains a bromo-adjacent homology (BAH) domain that mediates protein–protein interactions. Orc4 in fission yeast is unique among ORC proteins in that it binds DNA through its multiple AT-hook domains. Therefore, it is not surprising that it lacks a wH domain. The ability of DnaA, Orc1, and Cdc6 to bind and hydrolyze ATP is critical to their function in DNA replication, and the same is likely to be true for archaeal Orc1/Cdc6 as well.

X-ray analyses of an archaeal Orc1/Cdc6 heterodimer bound to origin DNA demonstrate that, in addition to conventional DNA-binding elements, initiators use their AAA+ ATPase domains to recognize origin DNA. In fact,

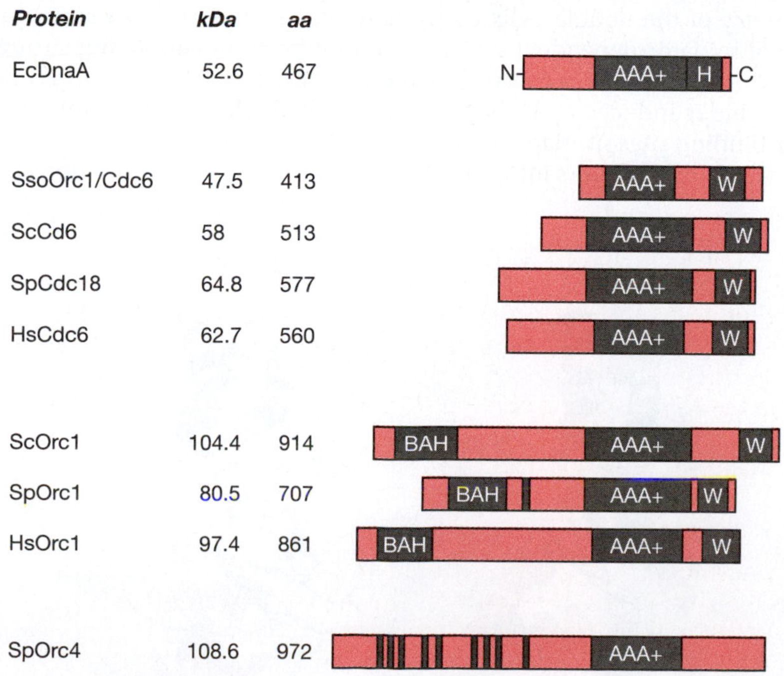

Protein	*kDa*	*aa*
EcDnaA	52.6	467
SsoOrc1/Cdc6	47.5	413
ScCd6	58	513
SpCdc18	64.8	577
HsCdc6	62.7	560
ScOrc1	104.4	914
SpOrc1	80.5	707
HsOrc1	97.4	861
SpOrc4	108.6	972

Figure 11-7. Initiator proteins in bacteria, archaea, and eukarya. The AAA+ domain binds ATP and provides ATPase activity. Helix-turn-helix (H) and winged helix (W) domains are putative DNA-binding sites, a feature common among transcription factors. The bromo-adjacent domain (BAH) interacts with chromatin-binding proteins. SpOrc4 binds DNA via its nine AT-hook domains (gray bars). Sequence domains were located using the program Pham from the Sanger Institute, and structure-guided sequence analysis. Sequences (pink bars) were obtained from the NCBI website, plotted in proportion to their length in amino acids, and aligned at the centers of their AAA+ domain. HsCdc6 has striking amino acid sequence similarity with HsOrc1, largely because both proteins contain the AAA+ and wHTH domains. Abbreviations: aa, amino acids; Ec, *E. coli*; Hs, *H. sapiens*; Sc, *S. cerevisiae*; Sp, *S. pombe*; Sso, *S. solfataricus*. (Data from Neuwald, A.F., Aravind, L., Spouge, J.L. and Koonin, E.V. (1999) *Genome Res.* 9: 27–43, and Aravind, L., Anantharaman, V., Balaji, S. *et al.* (2005) *FEMS Microbiol. Rev.* 29: 231–262.)

the N-terminal domains of both *E. coli* DnaA and *S. solfataricus* Orc1/Cdc6 proteins contain AAA+ domains, and both have C-terminal domains that are required for replicator-specific DNA binding (HTH in DnaA, wHTH in Orc1/Cdc6). The wH domain in archaeal Orc1/Cdc6 is also present in eukaryal Cdc6 and Orc1 proteins (Figure 11-7). However, whereas binding of the bacterial and eukaryal initiator proteins to their cognate replicators is dependent on ATP, this is not true for the archaeal Orc1/Cdc6 initiator. The AAA+ domain in archaeal Orc1/Cdc6 can bind DNA in the absence of ATP. However, it may still utilize its ATPase function to assemble a preRC, since that seems to be the primary function of the AAA+ domain in eukaryal initiator proteins.

For comparison, note that the SV40, polyomavirus, and papillomavirus initiators (Figure 10-5), as well as the bacteriophage λ (Figure 10-7) and some Gram-negative bacterial plasmids that encode their own initiator protein (Figure 10-8), lack HTH and wH domains. Moreover, although the small viral initiators contain AAA+ domains characteristic of DNA helicases, these domains are absent from bacterial phage and plasmid initiators. In other words, the initiator proteins of cellular organisms utilize DNA-binding domains not found in the genomes of viruses and episomes, and *vice versa*. Moreover, initiators of cellular organisms all encode AAA+ domains characteristic of DNA helicases, even though they lack the ability to unwind long stretches of DNA, as well as DNA-binding domains that allow much greater flexibility in DNA-site selection. These would be excellent characteristics for helicase loaders.

Cellular initiators form higher-order assemblies on replication origins, by using ATP to locally remodel duplex DNA and facilitate proper loading of synthetic replisomal components. The 3.4 Å resolution structure of an archaeal Orc1/Cdc6 heterodimer bound to origin DNA demonstrates that, in addition to conventional DNA-binding elements, initiators use their AAA+ ATPase domains to recognize origin DNA (Figure 11-8). Together these interactions establish the polarity of initiator assembly on the origin and induce substantial distortions into origin DNA strands (Figure 11-9).

Remarkably, *S. solfataricus* Orc1-1 and Orc1-3 achieve selective recognition of their respective target sites, including the asymmetry of initiator binding, with nominal sequence-specific interactions. They both substantially unroll the double helix while preserving base pairing. Footprinting studies using agents that are sensitive to distortion of the geometry of the double helix corroborate these distortions *in vitro*, with sites of initiator-dependent increases in modification correlating strongly with the bend points observed in the crystal structure. This Orc1/Cdc6 heterodimer induces a 20° bend in the DNA duplex at the point where their binding sites overlap (see Figure 11-10 in color plate section). Thus, these archaeal initiators introduce local and global DNA deformations that

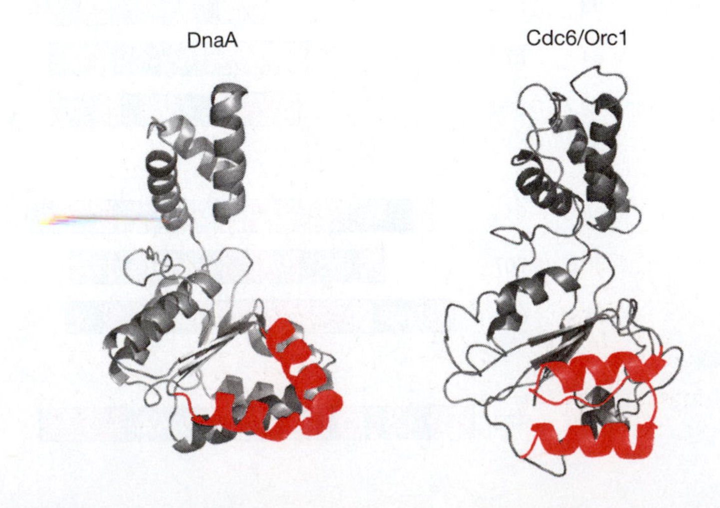

Figure 11-8. Comparison of *Aquifex aeolicus* DnaA and *S. solfataricus* Orc1-1 initiator proteins. The Orc1/Cdc6 initiator-specific motif (ISM) (red) contains a large loop that allows two α-helices to pack together in a parallel manner. The much shorter loop between the two α-helices in the DnaA ISM forces them to form a V-shaped wedge that is critical for binding replication-origin DNA. This figure was prepared using Pymol with PDB files 1L8Q (DnaA) and 2V1U (Orc1/Cdc6). (Data from Dueber, E.L., Corn, J.E., Bell, S.D. and Berger, J.M. (2007) *Science* 317: 1210–1213.)

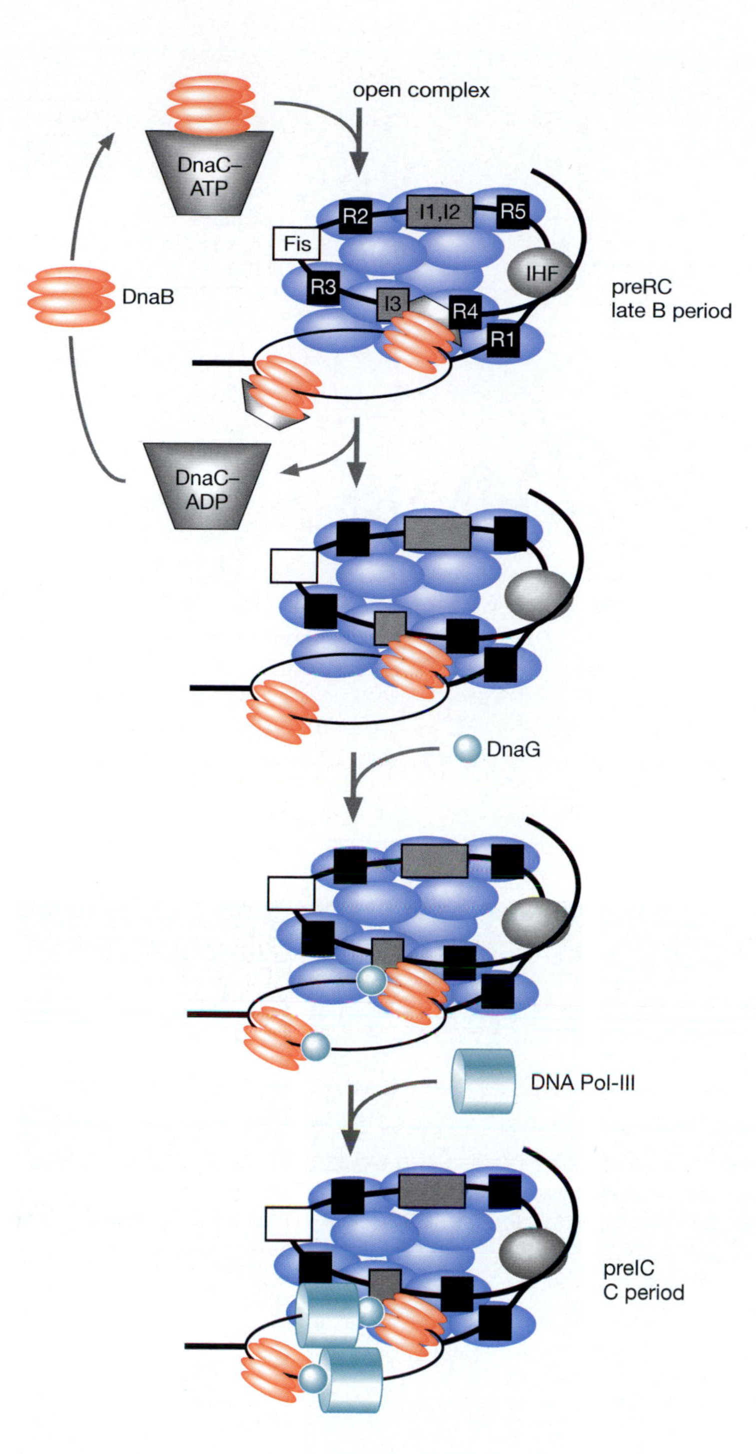

Figure 11-6. preRC and preIC assembly in *E. coli*. Assembly of a preRC at *E. coli* oriC requires formation of an open complex (Figure 11-5). DnaC–ATP chaperones DnaB onto the partially melted DUE. ATP hydrolysis is accompanied by release of DnaC•ADP to produce a preRC. The *E. coli* DNA primase DnaG can then bind to DnaB and recruit the DNA Pol-III holoenzyme to form a preIC. SSB is ubiquitous; it binds to all ssDNA and it facilitates both DNA helicase and DNA polymerase activities. Formation of the preIC leads directly to DNA synthesis.

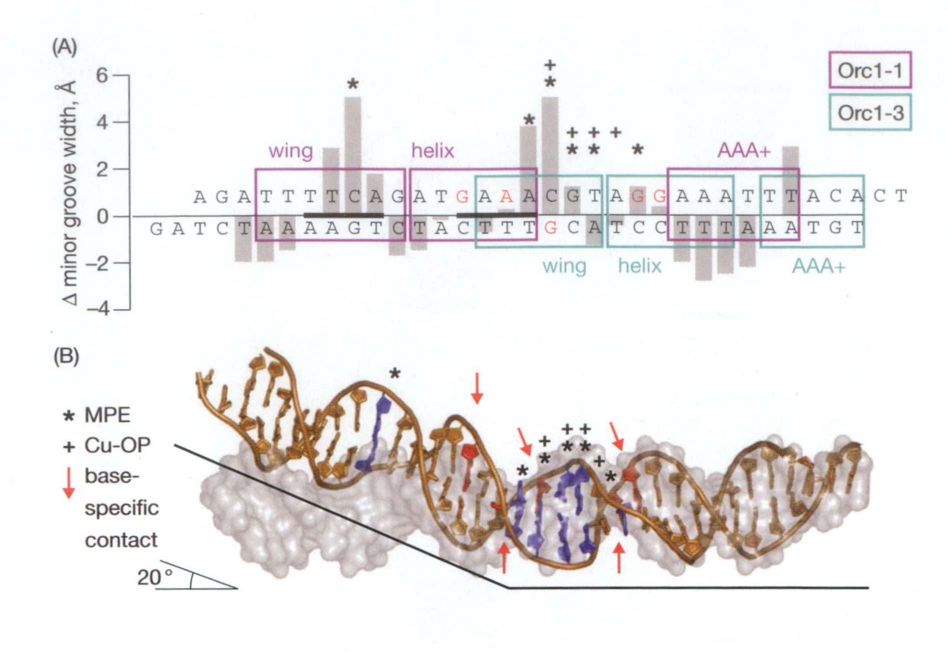

Figure 11-10. Archaeal initiators deform origin DNA. (A) Overview of DNA distortions. Highly conserved origin-binding sequence positions are underlined. Bases specifically recognized by side chains are in red. Distortion of the minor groove width from B-form DNA is plotted in gray along the DNA sequence. Positions sensitive to methidiumpropyl-EDTA (MPE; asterisks) and copper phenanthroline (Cu-OP; crosses) modification in the presence of Orc1-1 and Orc1-3 proteins correspond with deformations seen in the structure. (B) Comparison between initiator-bound DNA (orange cartoon) and B form DNA (gray surface). Base pairs sensitive to modification are in blue and distinguished as in (A). Bases specifically recognized by side chains are highlighted in red. (From Dueber, E.L., Corn, J.E., Bell, S.D. and Berger, J.M. (2007) *Science* 317: 1210–1213. With permission from the American Association for the Advancement of Science.)

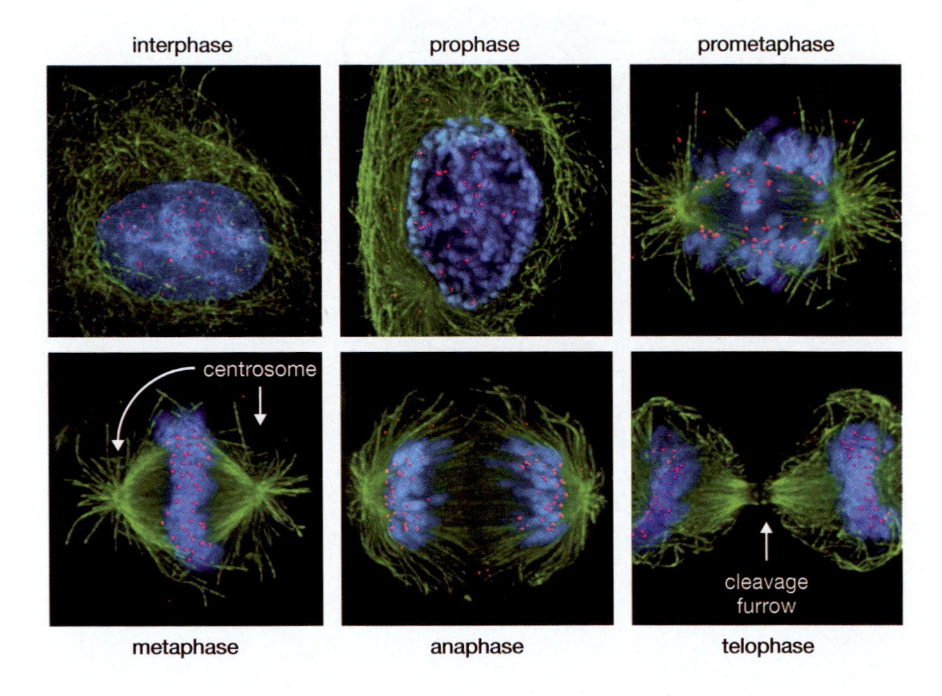

Figure 12-4. Anatomy of the human mitotic cell cycle. DNA is stained blue, microtubules are stained green, and the kinetochore is stained red. Nuclear and cellular membranes as well as nucleoli are not visible in these images. Mitosis occurs exclusively in eukaryotic cells, but details can be species-specific. For example, animals undergo an 'open mitosis' where the nuclear envelope breaks down before the chromosomes separate, whereas yeast and fungi undergo a 'closed mitosis' where the nuclear membrane remains intact. (From Cheeseman, I.M. and Desai, A. (2008) *Nat. Rev. Mol. Cell Biol.* 1: 33–46. With permission from Macmillan Publishers Ltd.)

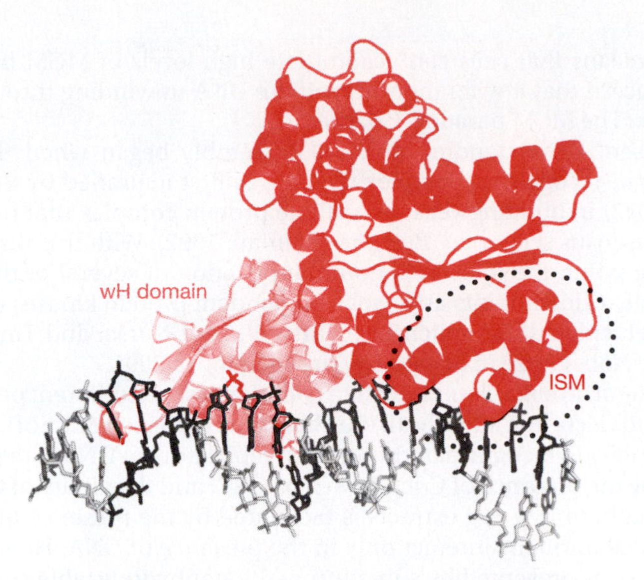

TCTCCACAGGAAACGGAGGGGT
AGAGGTGTCCTTTGCCTCCCCA

Figure 11-9. Orc1/Cdc6 binding to an origin-recognition box (ORB) element. The AAA+ domain is in red and the ISM indicated, the wH domain is in pink. The sequence of the ORB element is shown below and residues contacted by the respective domains of the protein are indicated with appropriate colors. Generated from PDB file 2GXA. (Data from Gaudier, M., Schuwirth, B.S., Westcott, S.L. and Wigley, D.B. (2007) *Science* 317: 1213–1216.)

substantially underwind the origin and thereby may help prepare the way to load the archaeal MCM helicase (Table 4-1).

EUKARYAL GENOMES

The sequence of events by which replication forks are formed at replication origins in eukaryotic cells has been characterized extensively in budding yeast (*S. cerevisiae*), fission yeast (*S. pombe*), flies (*D. melanogaster* embryos), frogs (*X. laevis* eggs), and mammals (*H. sapiens* tumor cell lines). Remarkably, despite a 200-fold increase in genome size and complexity, the sequence of events involved in assembling replication forks in humans is virtually identical to that in yeast. Moreover, of the 67 different proteins that are required to assemble replication forks in the human genome, homologs of 61 of these proteins carry out the same functions in yeast. The six additional proteins in metazoa are Mcm9, HBO1, geminin, Cdk2, p21, and p57. Mcm9 binds to ORC•Cdc6•chromatin sites and facilitates the ability of these sites to recruit Cdt1. Cdt1 then recruits the eukaryal MCM helicase (Table 4-1). In yeast, Cdt1 binds the MCM helicase even in the absence of chromatin. HBO1, an H4-specific histone acetylase, is a coactivator of Cdt1. DNA in acetylated chromatin is more accessible to DNA-binding proteins. Geminin is a specific inhibitor of Cdt1 activity that is critical to regulation of preRC assembly in multicellular animals, but absent from single-cell eukaryotes and from plants. Geminin binds Cdt1 and prevents it from binding the MCM helicase. Geminin also inhibits Cdt1-associated HBO1 acetylase activity. Mcm9 is a geminin antagonist. Cdk2 is a cyclin-dependent protein kinase that specifically activates the transition from G_1 to S phase during cell division in metazoa and regulates assembly of preRCs. p21 and p57 are CDK-specific inhibitors that regulate the transitions from G_1 to S and G_2 to M in mammals (Chapters 12 and 13).

preRC Assembly (Origin Licensing) On Eukaryotic Genomes

The enzyme required for unwinding DNA at replication origins and replication forks in eukaryotic cells is the MCM helicase (Table 4-1), but, in contrast to the SV40 T-ag helicase, the MCM helicase cannot assemble itself onto DNA without the help of a DNA helicase-loading mechanism. In eukarya, this mechanism is called either preRC assembly or origin licensing. preRC assembly restricts DNA unwinding by the MCM helicase to those chromosomal sites where initiator proteins form a complex with

DNA. This means that cells can accumulate high levels of MCM helicase without concern that it will randomly initiate DNA unwinding throughout the genome (The MCM paradox, Chapter 9).

Our present understanding of preRC assembly began when the six-subunit origin-recognition complex (ORC) was first identified by Bell and Stillman (1992) in budding yeast as a stable protein complex that binds to the ARS consensus sequence (Bell and Stillman, 1992). With this discovery as a starting point, the roles in genome duplication of several of the cell-division cycle (Cdc) mutants and cyclin-dependent protein kinases (CDKs) were soon elucidated, for which Lee Hartwell, Paul Nurse, and Tim Hunt received the Nobel Prize in Physiology or Medicine in 2001.

The 'initiator' in eukaryotes is really a complex of seven different proteins, ORC(1-6) and Cdc6, that affect both the affinity and site specificity of ORC for DNA. The ability to recognize budding yeast replicators in DNA is markedly enhanced by the presence of Cdc6 as well as ORC, and the ability of ORC to bind chromatin in frog egg extracts is facilitated by the presence of Cdc6. ORC and Cdc6 normally interact only in the presence of DNA. However, if ATP hydrolysis is prevented by substituting the nonhydrolyzable substrate ATP-γ-S for ATP, then a stable ORC•Cdc6 complex can form in the absence of DNA. Thus, Cdc6•ATP binds to and changes the structure of ORC, and together they bind cooperatively to origin DNA with higher DNA sequence specificity than ORC binding alone (Figure 11-11). On DNA that lacks origin activity, Cdc6 ATPase activity promotes dissociation of Cdc6•ADP, whereas DNA that contains origin activity suppresses Cdc6 ATPase activity, resulting in a stable ORC•Cdc6•DNA complex that can then promote MCM loading. Thus, Cdc6 ATPase activity contributes to origin DNA sequence specificity by promoting dissociation of Cdc6 from nonreplicator DNA. The dissociation rate ($t_{1/2}$) for ORC•Cdc6•DNA complexes is less than 0.125 min. In fission yeast, Cdc6/Cdc18 and Cdt1 both contribute significantly to the stability of the initial SpORC•DNA complex and enhance SpORC-dependent topological changes in origin DNA.

preRC assembly begins when initiator binds replicator. ORC can bind DNA in the absence of Cdc6, but it cannot load the MCM helicase. Conversely, Cdc6 can initiate DNA replication in the absence of ORC, if it is artificially bound to DNA by fusing it to a transcription factor DNA-binding domain. However, wild-type Cdc6 cannot bind to DNA in the absence of ORC. Therefore, it is the ORC•Cdc6•chromatin complex that determines where the MCM helicase can assemble onto chromatin (Figure 11-11). The ORC•Cdc6•DNA complex, however, cannot load the MCM helicase in the absence of Cdt1, a protein that binds to the MCM helicase through its carboxyl terminus, and to the ORC•Cdc6•DNA complex through its N-terminus. Once the MCM helicase is loaded, Cdc6, Cdt1, and one or more of the ORC subunits become dispensable for DNA replication; they can be eluted from chromatin in cell lysates without preventing the completion of S phase. The Mcm(2-7) complex can now slide along dsDNA without unwinding it. Conversion of an MCM double hexamer into an active helicase occurs during preIC assembly.

As with bacteria, preRC assembly in eukaryotes is highly dependent on ATP and ATP hydrolysis. Assembly of human or fly ORC requires ATP binding to both Orc4 and Orc5, and binding of ORC to DNA requires ATP binding to Orc1 (Figures 11-11 and 11-12). Moreover, ATP hydrolysis by both ORC and Cdc6 is essential for cell viability, and for assembly of multiple MCMs at yeast replicators. Thus, ORC•Cdc6 is part of a molecular machine that can repeatedly load MCM helicases onto DNA. At least two MCM helicases must be loaded onto each replicator, one for each of the two replication forks that must be constructed in order to produce bidirectional replication, but the ability to load multiple MCMs at a single ORC•Cdc6-DNA site allows for the possibility that one replicator could produce multiple origins of bidirectional replication (Figure 9-5).

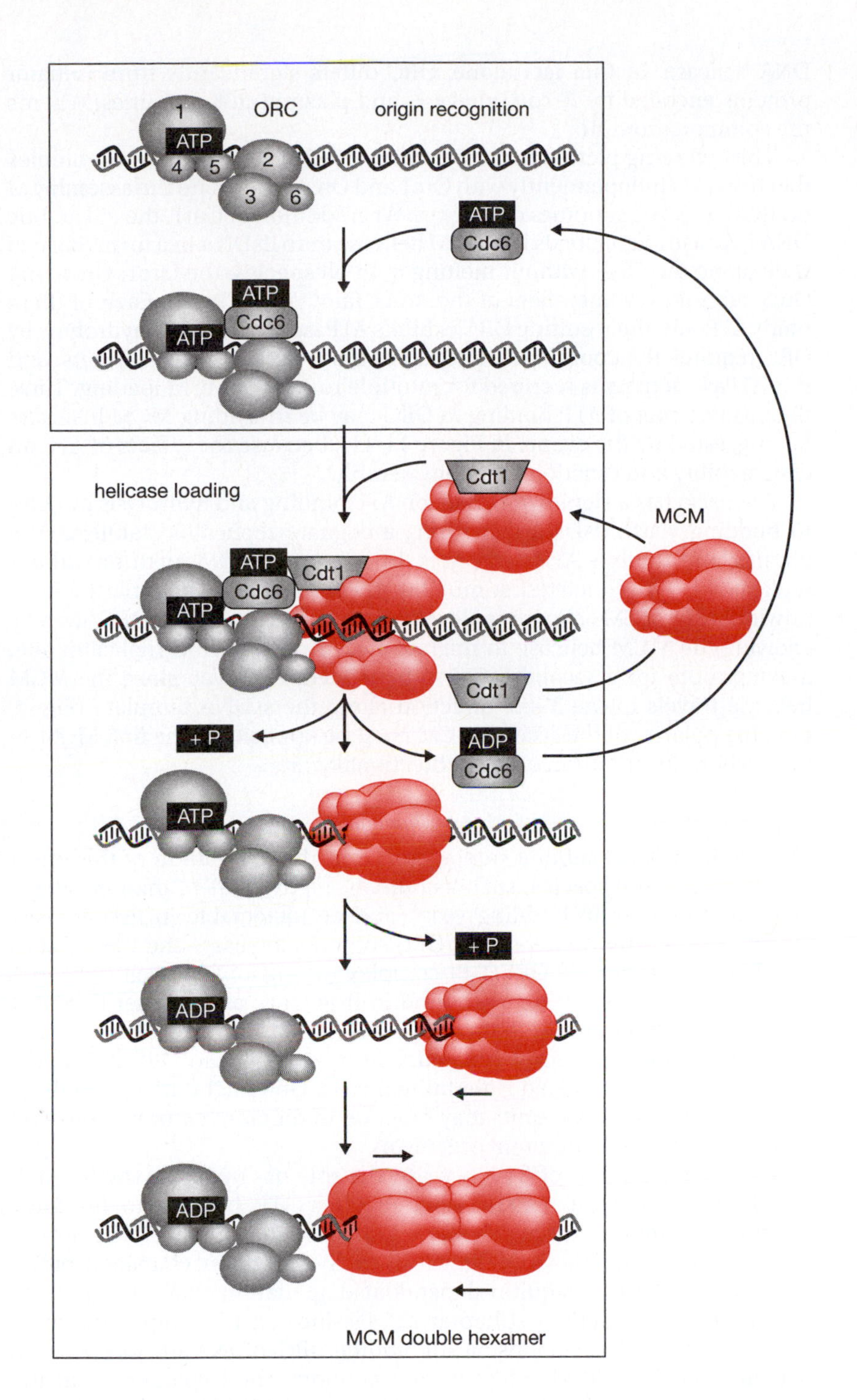

Figure 11-11. Assembly of preRCs in single-cell and multicellular eukarya, and the role of ATP. ORC•ATP binds to DNA and thereby allows Cdc6•ATP to also form a stable complex with the ORC•DNA site. This initiator–replicator complex is then targeted by a Cdt1•MCM helicase complex, and the MCM helicase is loaded onto the DNA. Loading the MCM helicase requires ATP hydrolysis by Cdc6, because inhibiting Cdc6 ATPase activity stabilizes Cdt1 on origin DNA and prevents MCM loading. By analogy with sliding DNA clamp loaders (Figure 5-7), association of the Cdt1•MCM complex with the DNA-ORC•ATP•Cdc6•ATP complex opens the MCM hexamer, exposing a DNA-binding site within its central channel (shown for DnaB in Figure 4-2). This event, in turn, allows adjacent DNA to associate with the Mcm subunits (shown for E1 in Figure 4-7), and activates the Cdc6 ATPase. Hydrolysis of the Cdc6-bound ATP releases Cdt1 and Cdc6•ADP to be recycled into Cdc6•ATP and Cdt1•MCM, with concomitant loading of the MCM helicase onto replicator DNA. Once loaded, the Mcm(2-7) complex can slide along dsDNA without unwinding the DNA; its helicase activity is not activated until preIC assembly. Loading of the MCM helicase, in turn, stimulates hydrolysis of ORC-bound ATP and thereby releases the loaded MCM helicase. Exchange of ADP for ATP resets ORC for a new round of preRC assembly. In multicellular eukarya, this event may also trigger release of Orc1 from ORC•chromatin sites and its subsequent ubiquitin-dependent degradation (Chapter 12). Note that the second helicase must be loaded in opposite orientation to the first helicase in order to achieve bidirectional DNA replication. (Data from Randell, J.C., Bowers, J.L., Rodríguez, H.K. and Bell, S.P. (2006) *Mol. Cell* 21: 29–39, Sivaprasad, U., Dutta, A. and Bell, S.P. (2006) Assembly of pre-replication complexes. In *DNA Replication and Human Disease* (DePamphilis, M.L., ed.), pp. 63–88, and Speck, C., Chen, Z., Li, H. and Stillman, B. (2005) *Nat. Struct. Mol. Biol.* 12: 965–971.)

The ORC paradox

A single set of ORC subunits is encoded by the genomes of all eukaryotes examined so far, and the individual subunits are sufficiently conserved throughout evolution that, with the exception of the smallest subunit (Orc6), homologs are readily identifiable. Moreover, genetic and biochemical analyses of ORC proteins in yeast, flies, and frog eggs confirm that all six of the ORC subunits are required for genome duplication in most, if not all, organisms. Thus, it seems paradoxical that interactions between subunits vary among different organisms, that subunits appear to have different functions in different organisms, that most ORCs exhibit weak DNA site specificity, and most of all, that ORC does not appear to induce the localized DNA melting that is believed to be necessary for loading the replicative

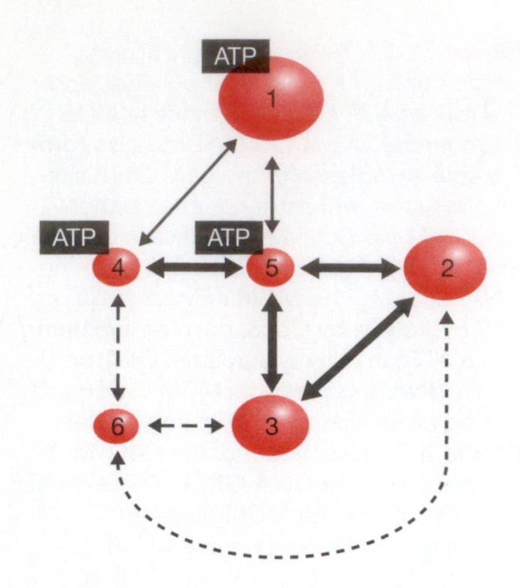

Figure 11-12. Organization of the mammalian ORC. Strong interactions (thick arrows) exist between the stable ORC(2-5) core subunits. Orc1 in yeasts, flies, and frogs is tightly bound to the other ORC subunits, but in mammals weak interactions (thin arrows) exist between Orc1 and ORC(2-5). Barely detectable interactions (dashed arrows) exist between Orc6 and ORC(1-5). Orc1, Orc4, and Orc5 bind ATP. (Data from DePamphilis, M.L. (2005) *Cell Cycle* 4: 70–79, Ranjan, A. and Gossen, M. (2006) *Proc. Natl. Acad. Sci. USA* 103: 4864–4869, and Siddiqui, K. and Stillman, B. (2007) *J. Biol. Chem.* 282: 32370–32383.)

DNA helicase. In that fact alone, ORC differs significantly from initiator proteins encoded by *E. coli*, phage λ, and plasmid R6K genomes. Where's the common ground?

The emerging picture is one in which ORC(2-5) forms a 'core' complex that interacts independently with Orc1 and Orc6 to regulate the assembly of preRCs along the genomes of eukarya. With addition of Cdt1, the eukaryotic DNA helicase loader loads the MCM helicase onto dsDNA in a form that can slide along the DNA without melting it. In all species, the Orc1, Orc4, and Orc5 subunits are members of the AAA+ family of ATPases. Each of them binds ATP, and the resulting ORC exhibits ATPase activity. ATP hydrolysis by ORC requires the coordinate function of the Orc1 and Orc4 subunits, and this ATPase activity is required for multiple rounds of MCM loading. Thus, the primary role of ATP binding to ORC may be in loading MCM helicases (as suggested by the events in Figure 11-11), because the effects of ATP on ORC stability and function vary considerably.

DNA also has a significant effect on ATP binding and hydrolysis by ORC. In budding yeast, dsDNA containing a cognate replicator stabilizes ATP binding and inhibits ATP hydrolysis. In contrast, ssDNA, with or without replicator DNA sequences, stimulates ATPase activity. Thus, partial DNA unwinding by MCM helicase could stimulate ATP hydrolysis by ORC, thereby allowing the MCM helicase to migrate away from the ORC•replicator site, making room for a second MCM to be loaded. However, since the MCM helicase travels in the 3′→5′ direction along the ssDNA template (Figure 4-1), the polarity of the second MCM must be opposite to the first MCM so that DNA replication can occur bidirectionally.

ORC structure

ATP stabilizes ORC subunit interactions, but the magnitude of this effect is greater in some species, such as human (Figure 11-12), than in others. The effect is weak in budding yeast, and nondetectable in fission yeast (Table 11-5). Furthermore, the six ORC subunits in yeasts and flies exist as a stable, stoichiometric ORC(1-6) complex throughout the cell cycle, but the Orc6 subunit is not tightly bound in frog ORC, and neither Orc6 nor Orc1 are tightly bound in mammalian ORC (Figure 11-12). Thus, ORC(1-5) is the predominant form isolated from frog eggs, and ORC(2-5) is the predominant form isolated from human cells. One might infer from these differences that ORC subunits may function in different combinations and play different roles in different organisms.

With the exception of fission yeast, the role of Orc1 is to stabilize the ORC•DNA complex, a reaction that requires ATP binding to the Orc1 subunit. The importance of this role is underscored by the fact that Orc1 in mammals, frogs, and flies undergoes cell-cycle-dependent changes in either phosphorylation or ubiquitin-dependent degradation that contribute to regulation of ORC activity (Chapter 12). Fission yeast is unique in that its Orc4 subunit is solely responsible for binding ORC to DNA and this reaction is unaffected by ATP. This fact further supports the hypothesis that the primary role of ATP in eukaryal ORCs is in loading MCM helicases.

Specific roles for Orc2, Orc3, Orc4, and Orc5 are as yet unknown, but associations between subunits 2, 3, 4, and 5 appear quite stable in all species, suggesting that ORC(2-5) plays the role of a stable core complex with which other proteins can interact. For example, in humans, neither ORC(2-5) nor Orc6 nor Orc1 alone binds DNA either *in vitro* or *in vivo*. But in the presence of ATP, Orc1 causes ORC(1-5) to bind to DNA both *in vitro* and *in vivo*. ATP binding by Orc4 and Orc5 is required for stable association between Orc4 and ORC(2,3,5) (Figure 11-12). Thus, mutations in the Orc1, Orc4, or Orc5 ATP-binding motifs can each inhibit the ability of human ORC to initiate DNA replication. The role of Orc6 in metazoa is less well defined. In flies, Orc6 is required for binding ORC to DNA, although it does not appear to play this role in humans. Depletion of Orc6 from either

Table 11-5. Origin-Recognition Complexes (ORCs)

Species	Stable form	ATP dependence	DNA binding	ATP dependence
Fission yeast (*S. pombe*)	ORC(1-6)	No	Yes, only Orc4 required	No
Budding yeast (*S. cerevisiae*)	ORC(1-6)	Yes	Yes, Orc6 not required	Yes
Flies (*D. melanogaster*)	ORC(1-6)	No	Yes, all six subunits required	Yes
Frogs (*X. laevis*)	ORC(1-5)	No	Yes	Yes
Mammals (*H. sapiens*)	ORC(2-5)	Yes	No	No
	Orc1+ORC(2-5)⇌ORC(1-5)	No	Yes, role of Orc6 unclear	Yes
	Orc6+ORC(2-5)⇌ORC(2-6)	?	No	No

Drosophila embryos or cultured human cells appears to affect cytokinesis, as well as DNA replication, suggesting the possibility that Orc6 in metazoa may help to coordinate genome duplication with cell division.

Fortunately, the role of Orc6 in budding yeast has been clearly elucidated. Orc6 is dispensable for ORC binding to DNA, for mitosis, and for cytokinesis, but it is required for maintaining association of the MCM helicase with chromatin and therefore for initiation of DNA replication. Without Orc6, the remaining ORC(1-5) complex fails to interact with Cdt1 and fails to load the MCM helicase onto origin DNA. Moreover, replacing Orc6 with a fusion protein consisting of Cdt1 linked to the Orc6 'ORC assembly domain' allows ORC to load MCM helicase, but only once. Therefore, multiple rounds of helicase loading appear to require a dynamic association between Cdt1 and ORC *vis-à-vis* the Orc6 subunit.

The differences between yeast and metazoan Orc6 may reside in the simple fact that there is no obvious evolutionary relationship between fungal Orc6 and human-type Orc6 proteins (Iyer and Aravind, 2006). Fungal Orc6 is not found outside the ascomycete group of fungi and seems to be a late addition to ORC. It is likely that fungal Orc6 displaced the ancestral and more widespread Orc6 of the type found in humans.

Electron microscopic analysis of the budding yeast ORC bound to DNA has revealed the overall shape of this structure (Figure 11-13). Cooperative binding of *S. cerevisiae* Cdc6 to ORC on DNA in an ATP-dependent manner induces a change in the pattern of origin binding that requires the Orc1 ATPase. The ORC•Cdc6 complex forms a ring-shaped structure with dimensions similar to those of the ring-shaped MCM helicase. The ORC•Cdc6 structure is predicted to contain six AAA+ subunits, analogous to other ATP-dependent protein machines, and suggests that Cdc6 and origin DNA activate a molecular switch in ORC that contributes to preRC assembly.

Whether or not eukaryotic initiator proteins ORC and Cdc6, like bacterial initiator protein DnaA, melt the DUE prior to loading of the replicative DNA helicase is not clear. Binding of fission yeast, fly, or human ORC to DNA is strongly facilitated by the presence of negative superhelical turns in the DNA, consistent either with wrapping the DNA around ORC as a left-handed superhelix (Figure 1-7), or with facilitating melting or untwisting of the DNA (Figure 1-8).

ORC assembly

In human cells, ORC binding to DNA appears to follow a sequential pathway in which the ORC(2-5) core is assembled in nuclei but does not bind to DNA in the absence of Orc1 (Figure 11-14). Orc1 is transported into the nucleus independently of ORC(2-5) and Orc6. Orc1 has its own nuclear localization signal that functions independently of its ORC-binding

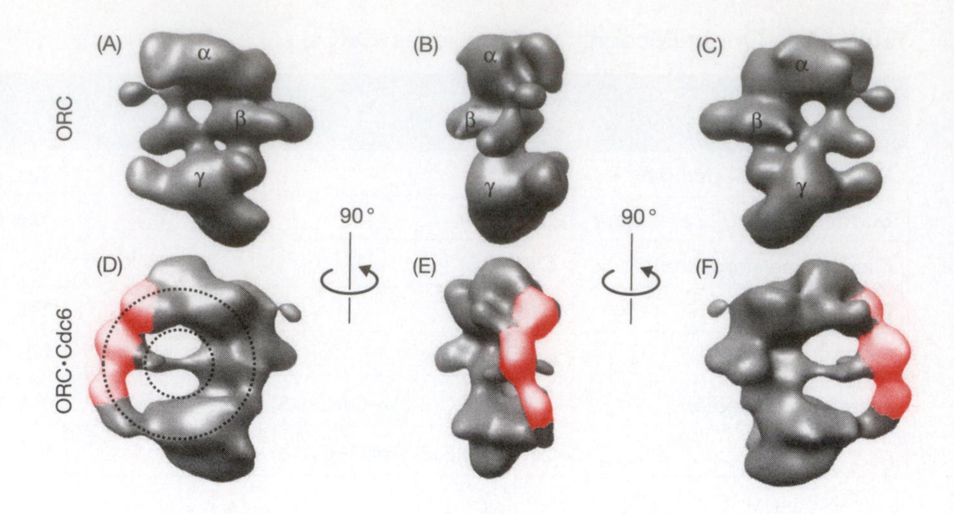

Figure 11-13. Three-dimensional image of ORC and Cdc6 bound to DNA. The stable binding of *S. cerevisiae* ORC and Cdc6 in the presence of ATP-γ-S allowed the structures of ORC and ORC•Cdc6 complexes to be visualized to a resolution of approximately 25 Å by single-particle reconstruction from thousands of electron micrographs. ORC (gray) appeared as an elongated structure with three physical domains (α, β, and γ). Although changes in ORC were noted when it was bound to Cdc6 (red), ORC still maintained its basic three-domain structure. The atomic structure of an archaeal Orc1/Cdc6 ortholog can be fitted into the space occupied by the presumed yeast Cdc6 protein. Cdc6 (red) binding to ORC gave rise to a pronounced ring-like feature (dashed black circles in ORC•Cdc6 complex) with inner and outer diameters of approximately 60 and 120 Å, respectively. These match the collar and external diameters of an archaeal MCM structure. This similarity suggests that the ring-like feature of the ORC•Cdc6 complex might facilitate loading the MCM helicase onto origin DNA. (From Speck, C., Chen, Z., Li, H. and Stillman, B. (2005) *Nat. Struct. Mol. Biol.* 12: 965–971. With permission from Macmillan Publishers Ltd.)

domain. Orc6 is the only one of the ORC subunits that binds karyopherin-α, a nuclear transport protein. This interaction is required for Orc6 nuclear localization. ORC(2-6) complexes in human cells are not bound tightly to chromatin and their function is unknown.

Only ORC from single-cell eukaryotes (yeasts) exhibits sequence-dependent DNA binding. ORC from flies, frogs, and humans does not bind to specific DNA sequences in cell extracts or as purified proteins. However, all three ORCs bind preferentially to asymmetric A:T-rich sequences of the type favored by *S. pombe*. In fact, DNA replication in *Xenopus* egg extracts and in *Drosophila* cells begins preferentially at asymmetric A:T-rich sequences. Characteristic features of A-tract DNA include an unusually

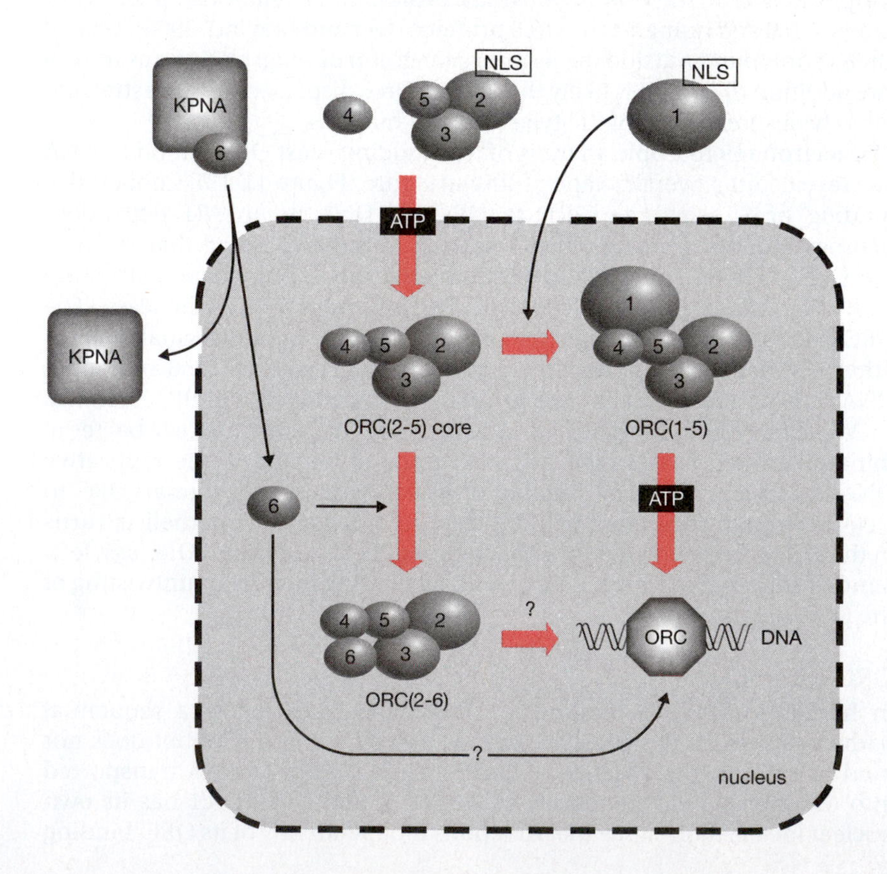

Figure 11-14. ORC assembly in mammalian cells. A stable ORC(2-5) core complex forms in the presence of ATP and is transported into the nucleus. Orc1 enters the nucleus and binds to ORC(2-5) to form a complex that, in the presence of ATP, can bind to DNA. Karyopherin-α (KPNA) releases Orc6 into the nucleus where it is assembled into an ORC(2-6) complex. ORC(1-6) has not been detected.

narrow minor groove and a high base twist that gives A-tract sequences their propensity to bend the helical axis when incorporated into otherwise non-A-tract DNA (Bent DNA, Chapter 1). This feature could stimulate DNA–protein interactions. In addition, ORC may associate with other proteins that increase its affinity for specific DNA sequences. For example, Cdc6/Cdc18 and Cdt1 facilitate the binding specificity of yeast ORC, and the Orc6 subunit facilitates the binding specificity of *D. melanogaster* ORC. The Orc1 BAH domain, which mediates protein–protein interactions, facilitates Orc1 activity in human cells. This domain allows budding yeast Orc1 to bind the transcription repressor Sir1, and it may account for the reported interactions between human Orc1 and transcription repressor AIF-C1, telomere-binding protein TRF2, chromatin-binding protein HP1, and the Epstein–Barr virus initiator protein EBNA1.

preIC Assembly On Eukaryotic Genomes

preIC assembly in eukaryotic cells involves a further 24 proteins in addition to those that comprise the preRC. Two of these proteins (CDK and cyclin) are not part of the preIC–chromatin complex (Table 11-4). preIC assembly involves two critical transitions. The first is changing an inactive Mcm(2-7) complex into an active MCM helicase with the 3′→5′ template strand passing through its center opening (Figure 4-1). The MCM helicase can then begin unwinding DNA at replication origins so that the second transition can occur, loading the replicative DNA polymerases onto the exposed DNA templates.

Activating the MCM helicase at a preRC requires the action of two protein kinases (DDK and CDK) and the participation of nine additional proteins (Mcm10, Cdc45, Dpb11, GINS (four subunits), Sld2, and Sld3). GINS and Cdc45 remain associated with the MCM helicase as it travels with the replication fork (Figure 5-4). Apparently these proteins are required continuously to maintain the proper organization of the replication machine. For example, the four-subunit GINS complex (Figure 5-4) may play the role of the bacterial $\tau_3\delta\delta'$ clamp loader complex that coordinates the action of several different polymerases at a single replication fork (Figure 5-18). Of the eight DNA polymerases encoded by the budding yeast genome and the 14 DNA polymerases encoded by the human genome, only three (Pol-α, Pol-δ, and Pol-ε) are required for duplication of the nuclear genome (Figure 4-19 and Table 4-4); duplication of the mitochondrial portion of the genome is carried out solely by Pol-γ. Therefore, preIC assembly must discriminate among a large number of DNA polymerases to load the correct enzyme at the correct time. This is accomplished by a series of events (Figure 11-15) that culminates with the loading of DNA Pol-α•primase and the initiation of RNA-primed DNA synthesis (Figure 11-16) on what will become the leading-strand DNA template (Table 3-1).

Activating the MCM helicase

The first step in preIC assembly is marked by association of Mcm10 protein and DDK with the completed preRC and concomitant displacement of Cdc6 and Cdt1, the two proteins that originally recruited Mcm(2-7) to the initiator–chromatin complex. Mcm10 also can bind ssDNA, and it is required for efficient phosphorylation of Mcm(2-7) in the preRC, suggesting that Mcm10 facilitates DNA unwinding and the conversion of Mcm(2-7) into the MCM helicase. Electron microscopic analysis suggests that Mcm10 can form a ring-shaped hexamer that would allow it to encircle a displaced DNA strand. DDK consists of the serine/threonine protein kinase Cdc7 and its regulatory subunit Dbf4. DDK is not required either for preRC assembly or for loading Mcm10, but it is required for binding the preIC protein Cdc45 to replication origins. The primary target for DDK-dependent phosphorylation appears to be the MCM helicase. Budding yeast Mcm2, Mcm4, and Mcm6

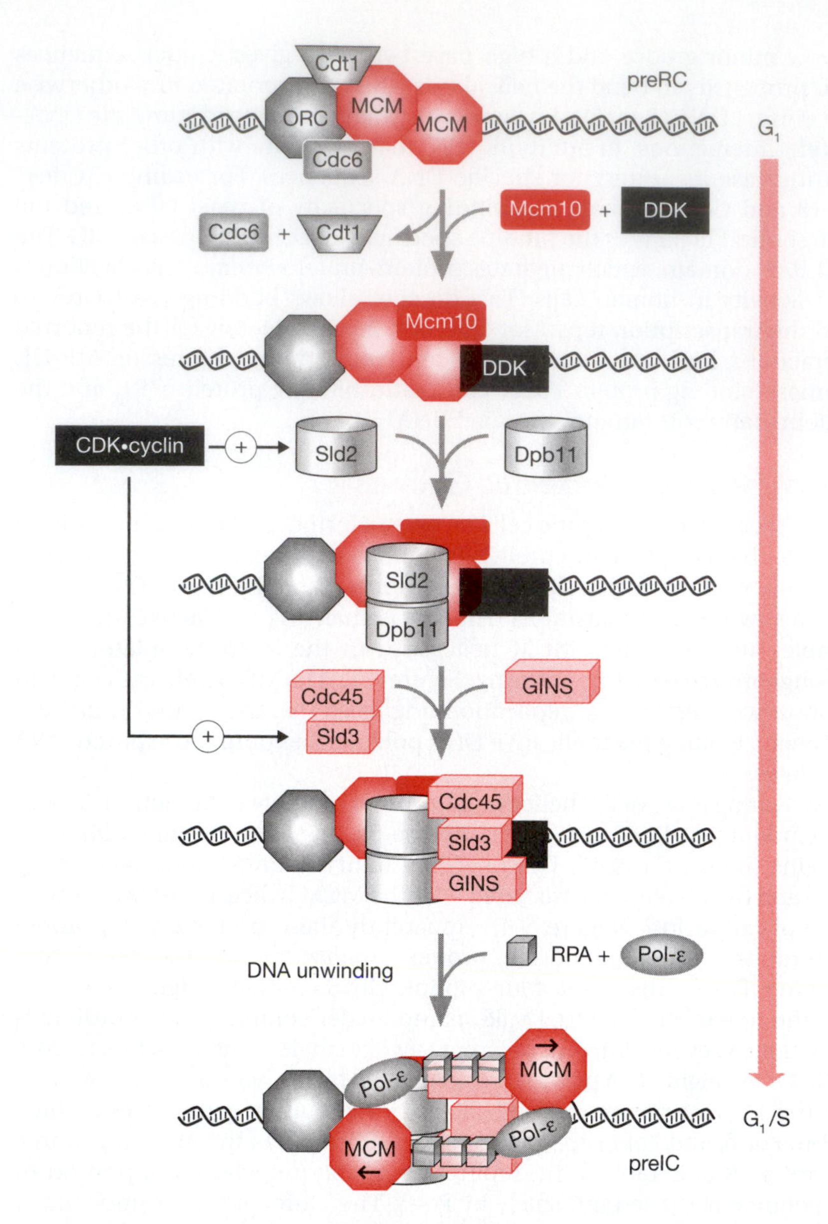

Figure 11-15. Conversion of preRC into preIC in budding yeast. preRCs appear in G_1 phase of cell division. preICs appear at the transition from G_1 to S phase. CDK•cyclin is the only direct participant that is not bound to chromatin. Except for the fact that at least two copies of the MCM helicase must bind at each ORC site that gives rise to bidirectional replication, the stoichiometry of other proteins is not indicated, only their order of addition.

are all known to be substrates for DDK, but only mutations in Mcm5 that appear to affect this protein's flexibility can bypass the requirement for DDK activity. Since similar mutations in Mcm2 and Mcm4 cannot bypass the role of DDK, Mcm5 may be the final target of DDK regulation. Thus, despite the fact that Mcm5 is not phosphorylated by DDK, phosphorylation of one or more of the Mcm(2-7) subunits by DDK appears to alter the shape of this complex and thereby convert it into an active DNA helicase.

Among the metazoa, MCM subunits 2, 3, and 4 are phosphoproteins. Mcm2 can be phosphorylated *in vitro* by either Cdc7, Cdk1, or Cdk2, but *in vivo* Mcm2 is phosphorylated by DDK specifically at Ser-5, an event required for assembly of Mcm2 with the remaining MCM subunits. Similarly, Mcm3 can be phosphorylated *in vitro* by either Cdk1 or Cdk2, but *in vivo* Mcm3 is phosphorylated by Cdk1•CcnB specifically at Ser-112, an event that promotes assembly of Mcm3 with the remaining MCM subunits followed by loading of the MCM helicase onto chromatin. Thus, phosphorylation of Mcm3 during mitosis and Mcm2 during the subsequent G_1 phase facilitates preRC assembly. Mcm4 also is targeted *in vivo* by DDK, but this event promotes stable binding between Mcm4 and Cdc45. Thus, phosphorylation of Mcm4 serves to stabilize the Cdc45•MCM•GINS complex that is required for efficient DNA unwinding at replication forks (Figures 5-4 and 13-16).

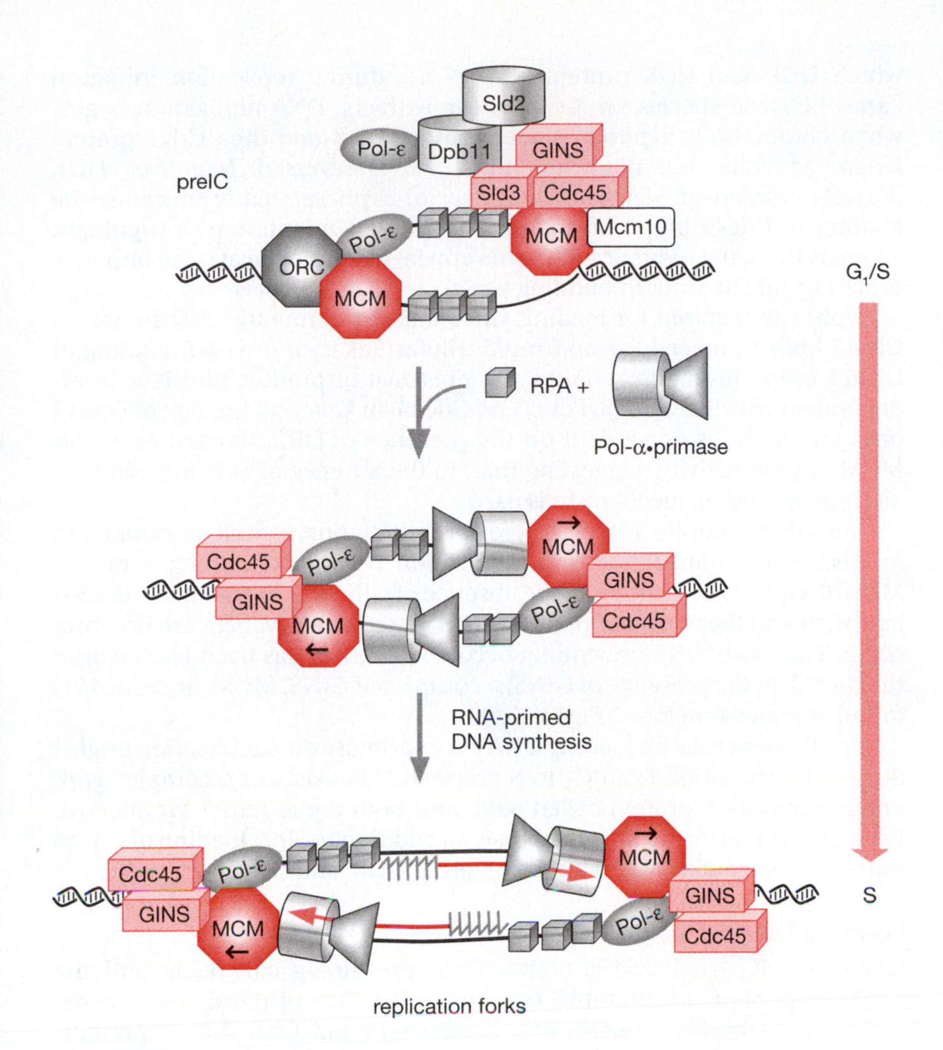

Figure 11-16. Conversion of a preIC into two replication forks in budding yeast. The preIC has been expanded to indicate the putative interactions between different proteins. These interactions are based on coprecipitation of proteins using antibodies directed against a single protein. Two GINS complexes and two Cdc45 proteins must be present in each preIC that gives rise to bidirectional replication. Otherwise, the stoichiometry is unknown. The critical step that distinguishes G_1 from S phase is RNA-primed (gray nucleotides) DNA synthesis (red arrow). This event is carried out by DNA Pol-α•primase to produce nascent DNA on the leading-strand DNA template where the direction of DNA synthesis is the same as the direction of DNA unwinding. Note that replication forks must either pass through or displace ORC from the chromosome in order to replicate the DNA.

Whereas all eukarya encode Dbf4, vertebrates contain an additional regulator of Cdc7 called Drf1 (Dbf4-related factor 1). Surprisingly, Dbf4 is dispensable for DNA replication in *Xenopus* egg extracts, but Drf1 is essential. Similar to the function of Cdc7•Dbf4 in yeast, Cdc7•Drf1 is required for Mcm phosphorylation and Cdc45 loading. Drf1 expression declines after gastrulation whereas Dbf4 increases, indicating that Cdc7•Drf1 functions specifically during the early embryonic cell cycles.

The next step requires CDK phosphorylation. In budding yeast, Sld2 and Sld3 must be phosphorylated by Cdk1•Clb5 or Cdk1•Clb6 before they can then bind to Dpb11 and promote loading of other factors, such as GINS and Cdc45. CDK-dependent phosphorylation of Sld2 is required for its interaction with Dbp11, and the Sld2•Dbp11 complex must interact with MCM that is bound to DNA as well as GINS, Pol-ε, and Cdc45 in order to assemble a preIC. Sld3 appears to be a chaperone for Cdc45, which in turn is required to recruit DNA Pol-α•primase. Thus, CDK phosphorylation of Sld3 is required for binding of Cdc45 to the preRC. The roles of Sld2 and Sld3 in preIC assembly have been characterized in yeast, but clear homologs have not been identified in metazoa. However, limited homology has been observed between the N-terminal 210 amino acids of Sld2 and the N-terminal domains of the metazoan RecQL4 DNA helicase. Furthermore, it has been demonstrated that RecQL4 is required for initiation of DNA replication in frog egg extracts. Further, it is clear that CDK phosphorylation is required for initiation of DNA synthesis (S phase) in multicellular as well as unicellular eukarya, but the metazoan targets remain to be identified.

The existence of different phosphorylation targets in unicellular and multicellular eukarya may account for the observation that the order in which DDK and CDK protein kinases act during replication initiation varies between species. In *Xenopus* egg extracts, DNA replication begins when chromatin is exposed sequentially to DDK and then Cdk2 protein kinase activities, but not when the order is reversed. Moreover, DDK phosphorylation of Mcm4 at the onset of S phase clearly precedes the loading of Cdc45 onto chromatin, a Cdk2-dependent step. Intriguingly, whereas the same sequence of events appears in fission yeast, the opposite order appears to occur in budding yeast.

Dpb11 is required for loading GINS and for chromatin-association of Cdc45 both in unicellular and multicellular eukarya. In yeast, loading of Dpb11 onto chromatin requires the presence of preRCs, and it is interdependent with Sld2, GINS, Pol-ε, and Cdc45. In *Xenopus*, loading of Dpb11 onto chromatin is dependent on the presence of ORC, but not on either MCM or CDK activity, suggesting that, in the absence of Sld2 and Sld3, an alternative loading mechanism is used.

The GINS complex is required for both initiation as well as elongation of DNA replication. It binds to replication origins containing a preRC, Mcm10, Dpb11, and Cdc45. In the absence of GINS, recruitment of Cdc45 is inhibited and the conversion of Mcm(2-7) into the MCM helicase does not occur. Therefore, DNA unwinding occurs after GINS has been loaded onto the preRC. In the presence of GINS, a complex of GINS, MCM, and Cdc45 is found at replication forks (Figure 5-4).

Cdc45 is essential for loading DNA Pol-α•primase onto replication origins during the transition from G_1 to S phase. Cdc45 exists as a complex with another initiation protein called Sld3, and both are required for effective DNA unwinding by the MCM helicase. In addition to Sld3, loading of Cdc45 onto preRCs requires Mcm10, DDK, Dpb11, CDK, and GINS.

Loading DNA Pol-α•primase

Once the MCM helicase is active, DNA unwinding can begin and the final components of the preIC can be loaded. One of these components is the ssDNA-binding protein RPA (Figure 4-13 and Table 4-2). It protects ssDNA from nucleolytic damage, prevents hairpin formation, and blocks reannealing of the complementary templates until nascent-DNA-strand synthesis is completed. More to the point, DNA unwinding at replication origins either in bacterial or eukaryal cells is ineffective in the absence of a single-stranded DNA-binding protein. Moreover, in the case of eukaryotes, RPA interacts with the replicative DNA helicases and DNA polymerases. For example, bacterial SSB cannot substitute for eukaryal RPA during initiation of DNA replication at the SV40 replicator or in human nuclei.

DNA Pol-ε, the enzyme responsible for DNA synthesis on the leading-strand template (discussed in Chapter 4), is recruited to the preIC by Dpb11 and GINS. Since Dpb11 is required to load GINS onto chromatin, and GINS is required to load Cdc45, and Cdc45 is required for DNA unwinding to occur, Pol-ε appears to play some role prior to origin unwinding that does not require DNA synthesis. In *Xenopus* and perhaps other metazoans, DNA Pol-ε does not appear to be required for preIC assembly, because helicase activation, DNA unwinding, and recruitment of DNA Pol-α•primase can occur in its absence.

Once sufficient DNA template is exposed, DNA Pol-α•primase is recruited to the preIC. Since Pol-α•primase is the only DNA polymerase that can initiate DNA synthesis *de novo*, S phase (i.e. nuclear DNA synthesis) cannot begin until it is loaded onto the activated preIC. Mcm10 and Cdc45 associate specifically with Pol-α•primase. Interaction between Mcm10 and DNA Pol-α depends on a third protein called And1/Ctf4/Mcl1. Since Mcm10 was one of the first proteins to bind to the preRC, it appears that Cdc45 and And1 chaperone Pol-α•primase to the unwound origin, thus ensuring that the right enzyme for the right job is localized at the right places on the genome. Mcm10 physically interacts with DNA Pol-α and with Cdc45.

In doing so, Mcm10 not only facilitates loading of DNA Pol-α•primase, it stabilizes the preIC–DNA Pol-α•primase complex. Once Pol-α•primase has initiated RNA-primed DNA synthesis, RFC loads the sliding clamp PCNA (Figure 4-23) onto the RNA–DNA primer and the resulting complex recruits DNA Pol-ε to extend the 3′-end of the nascent leading strand while concomitantly displacing Pol-α•primase (Figure 5-8). PCNA greatly increases the processivity of Pol-ε (Table 4-5).

SUMMARY

- Construction of replication forks can be considered as eight consecutive events beginning with origin recognition (Table 11-1).
- All genomes encode at least one protein that functions as the initiator whose role is origin recognition and initiation of preRC assembly. Bacteria encode DnaA. Archaea encode as many as 14 different Orc1/Cdc6 proteins. Eukarya encode seven different proteins [ORC(1-6) and Cdc6] and perhaps Cdt1 as well. DnaA, Orc1/Cdc6, Orc1, and Cdc6 all share in common homologous ATP-binding sites as well as orthologous DNA-binding domains. One advantage to increased complexity is an increased ability to regulate when and where genome duplication occurs (Chapter 12).
- Initiator proteins that produce replication forks are either DNA helicases or DNA helicase loaders.
- To engage the DNA helicase, localized melting must occur at the replicator's DUE. In principle, this event is required for loading any hexameric DNA helicase at internal sites in duplex DNA genomes. Proof of this principle has been obtained with viral, plasmid, bacterial, and archaeal replicators, but it has not yet been demonstrated for the passive replicators in eukaryotic cells. Localized DNA unwinding does not always require ATP binding and subsequent hydrolysis (e.g. phage λ and plasmid R6K), but they are involved in bacteria and eukarya where they contribute to regulating the transition from G_1 to S phase. Although the DNA is distorted in ORC•Cdc6•DNA complexes, localized DNA melting has not been detected.
- Only in the case of small viral genomes is the initiator protein and the helicase protein one and the same; initiator protein is organized by the replicator into a helicase. Otherwise, all initiator proteins are part of a helicase loader mechanism that involves a minimum of three different proteins in bacteria to as many as eight proteins in eukarya to load the helicase at the correct sites. Replicative DNA helicases such as DnaB and MCM are capable of sliding along dsDNA without unwinding it. However, whereas the DnaB appears to be loaded directly onto ssDNA generated by the interaction of DnaA with oriC, the MCM helicase is first loaded onto dsDNA during preRC assembly and then converted into an active DNA helicase during preIC assembly. The fact that the loaded MCM can slide along dsDNA strongly supports the concept that dsDNA is threaded through the center hole in the MCM hexamer.
- Loading the initiator DNA polymerase and DNA primase is the final step in preIC assembly. In eukaryotes, these two enzyme activities form a single stable multiprotein complex called DNA Pol-α•primase. The loading process can be as simple as association of DNA Pol-α•primase with the initiator/replicative DNA helicase in small animal viral genomes, or as complex as assembly of an independent preIC in which at least 11 other proteins are required to ensure that the DNA Pol-α•primase is loaded at the correct sites.
- ATP and ATP hydrolysis are required for all DNA helicase activities

encoded by bacteria, archaea, or eukarya, but they are not necessarily required for proteins to distort DNA sufficiently to load a DNA helicase at internal genomic sites. Lysogenic bacteriophage, Rep-dependent bacterial plasmids, and archaea all encode initiator proteins that can distort their cognate replicators without the help of ATP or ATP hydrolysis. Nevertheless, all of the initiators encoded by living organisms have the ability to bind ATP and hydrolyze it, and the primary role of these activities appears to be to regulate preRC assembly. Although ATP affects ORC, Cdc6 association and binding to DNA in most eukaryotes, it is not required for initiator binding to replicators in either fission yeast or archaea.

REFERENCES

Additional Reading

Barry, E.R. and Bell, S.D. (2006) DNA replication in the Archaea. *Microbiol. Mol. Biol. Rev.* 70: 876–887.

Fanning, E. and Zhao, K. (2009) SV40 DNA replication: from the A gene to a nanomachine. *Virology* 384: 352–359.

Labib, K. and Gambus, A. (2007) A key role for the GINS complex at DNA replication forks. *Trends Cell Biol.* 17: 271–278.

Mott, M.L. and Berger, J.M. (2007) DNA replication initiation: mechanisms and regulation in bacteria. *Nat. Rev. Microbiol.* 5: 343–354.

Sclafani, R.A. and Holzen, T.M. (2007) Cell cycle regulation of DNA replication. *Annu. Rev. Genet.* 41: 237–280.

Stillman, B. (2005) Origin recognition and the chromosome cycle. *FEBS Lett.* 579: 877–884.

Tada, S., Kundu, L.R. and Enomoto, T. (2008) Insight into initiator-DNA interactions: a lesson from the archaeal ORC. *BioEssays* 30: 208–211.

Takeda, D.Y. and Dutta, A. (2005) DNA replication and progression through S-phase. *Oncogene* 24: 2827–2843.

Tanaka, S., Tak, Y.S. and Araki, H. (2007) The role of CDK in the initiation step of DNA replication in eukaryotes. *Cell Div.* 2: 16.

Wigley, D.B. (2009) ORC proteins: marking the start. *Curr. Opin. Struct. Biol.* 19: 72–78.

Zakrzewska-Czerwinska, J., Jakimowicz, D., Zawilak-Pawlik, A. and Messer, W. (2007) Regulation of the initiation of chromosomal replication in bacteria. *FEMS Microbiol. Rev.* 31: 378–387.

Zegerman, P. and Diffley, J.F.X. (2007) Phosphorylation of Sld2 and Sld3 by cyclin-dependent kinases promotes DNA replication in budding yeast. *Nature* 445: 281–285.

Literature Cited

Aravind, L., Anantharaman, V., Balaji, S., Babu, M.M. and Iyer, L.M. (2005) The many faces of the helix-turn-helix domain: transcription regulation and beyond. *FEMS Microbiol. Rev.* 29: 231–262.

Bell, S.P. and Stillman, B. (1992) ATP-dependent recognition of eukaryotic origins of DNA replication by a multiprotein complex. *Nature* 357: 128–134.

Iyer, L.M. and Aravind, L. (2006) The evolutionary history of proteins involved in pre-replication complex assembly. In *DNA Replication and Human Disease* (M.L. DePamphilis, ed.), pp. 751–760. Cold Spring Harbor Laboratory Press, Cold Spring Harbor, NY.

Chapter 12
CELL CYCLES

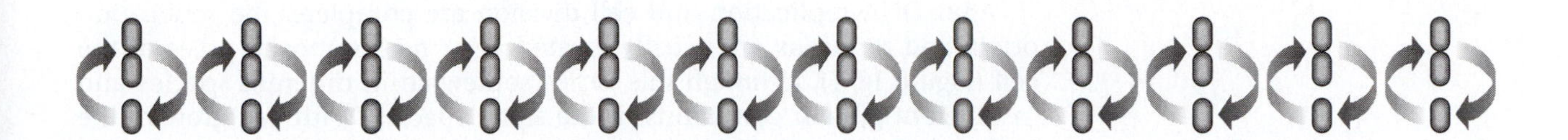

The cell-division cycle, or simply **cell cycle**, describes a series of events that drive the cell from growth to genome duplication to separation of sibling genomes to cell division. The purpose of cell cycles is to ensure that genome duplication precedes cell division, and to ensure that each daughter cell receives one complete copy of the genome. All cell cycles can be divided into two periods: interphase, during which the cell grows, accumulates nutrients needed for cell division, and duplicates its DNA, and then cell division. Bacteria and archaea divide by **binary fission**, the process wherein the cell duplicates its single chromosome and segregates the two sibling chromosomes so that when the cell pulls apart, each daughter cell receives one copy of the cell's genome. Eukarya divide by **mitosis**, the process by which a cell with multiple nuclear chromosomes separates them into two identical sets, each contained within a nucleus, and then undergoes **cytokinesis**, the process by which the nuclei, cytoplasm, organelles, and cell membrane are divided into two separate daughter cells. Animal cells undergo an **open mitosis** wherein the nuclear envelope breaks down before the chromosomes separate. Fungi such as yeast undergo a **closed mitosis** wherein the cell nucleus remains intact throughout the cell cycle.

THE BACTERIAL CELL CYCLE

One striking difference between genome duplication in bacteria and in eukarya is the ability of bacteria to initiate a second round of DNA replication before the first round is completed. While bacteria generally produce only one copy of their genome prior to cell division, they are capable of producing more than one copy when they find themselves in rich growth medium, and they can undergo cell division even though the second or third set of replication forks has not yet terminated. In eukarya, reinitiation of DNA replication prior to cell division results either in unique terminally differentiated polyploid cells (discussed below) or in 'DNA re-replication' (discussed in Chapter 13), an aberrant form of DNA replication that triggers apoptosis (programmed cell death).

Nutrient availability is one of the strongest determinants of bacterial cell size. When grown in rich media where generation times are short, bacterial cells can be up to twice the size obtained when cultured in nutrient-poor medium with longer generation times. Under slow-growth conditions, the cell cycle is divided into periods of cell growth (B period), chromosome duplication and separation of sister chromatids (C period), and termination of DNA replication to cell division (D period) (Figure 11-1). In *E. coli*, B+C+D normally require about 1 hour to complete (Table 12-1). However, the B period is nonexistent in cells grown in a rich culture medium where generation times are equal to or even less than the time required for C+D. Remarkably, bacteria are capable of sustaining generation times shorter than the time required to complete chromosome replication. This requires cells to initiate new rounds of DNA replication prior to completion of the previous round, thereby sustaining four to eight sets of replication forks simultaneously. Once established, all of the replication origins within a single cell are activated simultaneously each time a cell divides. By increasing cell size with growth rate, bacteria can maintain a constant ratio of DNA to cell mass, even at faster growth rates, and therefore the 'initiation mass' (discussed below) remains constant.

After DNA replication and cell division are complete, the replication origin and terminus are usually located at or near opposite poles of the cell (Figure 12-1), although this varies somewhat in different species and in different growth conditions of the same species. With cell growth, the bacterial replicator oriC localizes to the middle of the cell where it remains during the initiation of DNA replication. Once genome duplication begins, however, the two copies of oriC migrate toward opposite poles of the cell, and the terminus moves toward the center of the cell where cell division will occur. As replication progresses, the remaining regions of the chromosome separate from one another and follow their respective origins toward opposite poles. The 'cytokinetic ring' marking the site where cell division will occur assembles sometime after initiation of DNA replication, but cell division does not occur until a complete nucleoid (the bacterial equivalent of a nucleus, but without a nuclear membrane) is present in each newly forming daughter cell (nucleoid segregation). The terminus migrates from the pole of a newborn cell to midcell at the end of replication. The duplicated termini flank the invaginating septum during cytokinesis, and are thus positioned at the poles of the new cell prior to the next round of the cell cycle. Although the complex of replication proteins at each replication fork (replisome) is more mobile than originally thought, it is believed that the DNA passes through the replisome rather than the replisome sliding along the DNA (Figure 3-11).

The force and directionality for nucleoid separation and positioning are believed to come from the process of DNA replication itself and from chromosome condensation. By tagging DNA polymerase subunits with green fluorescent protein, the clusters of replisomes that form foci (replication factories, Figure 3-11) could be monitored in living cells.

Table 12-1. Typical Times Required For Transit Through Each Phase Of Cell Division

Organism	G_1(B)	S(C)	G_2	M(D)	Total
E. coli (bacteria)	25 min	40 min		1 min	66 min
Sulfolobus acidocaldarius (archaea)	0.1 h	1.5 h	1.5 h	0.2 h	~3.3 h
S. cerevisiae (budding yeast)	30 min	60 min	30 min	5 min	~2 h
S. pombe (fission yeast)	10 min	10 min	90 min	10 min	~2 h
Xenopus (frog)					
Cleavage embryo		20 min		10 min	~30 min
Cultured cells[a]					
Drosophila (fly)					
Syncytial embryo		0.5 min		0.5 min	~1 min
Cultured S2 cells	10 h	8 h	10 h	1 h	~29 h
Mus musculus (mammal)					
Embryonic fibroblasts	16 h	6 h	2 h	1 h	~25 h
Embryonic stem cells	2 h	6 h	1 h	1 h	~10 h

[a]Cultured *Xenopus* cells appear similar to cultured mammalian cells, although specific data are not available. Cell-cycle phases are identified as G_1, S, G_2, and M in eukarya, and as B, C, and D in bacteria (Figure 11-1). Times are given either in hours (h) or minutes (min). Data are typical for optimal laboratory conditions, but the time required for cell division depends greatly on temperature, nutrients, and the proximity of other cells.

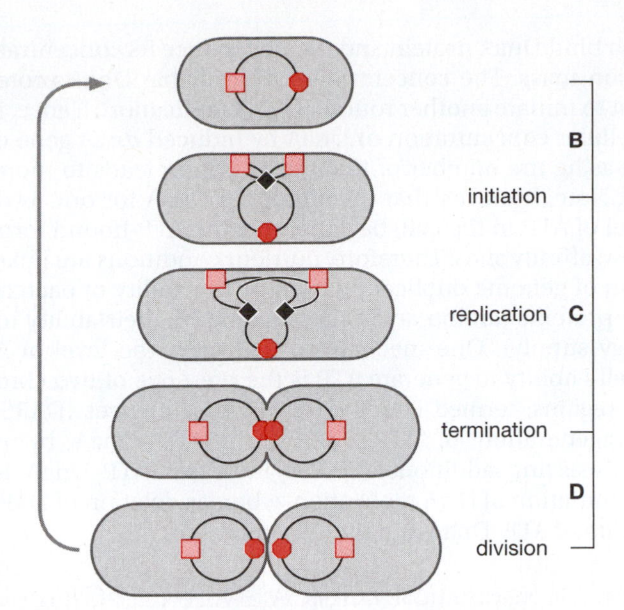

Figure 12-1. Bacterial cell-division cycle. Duplex DNA chromosome (black line) begins replication at oriC (light squares) and terminates at Ter (dark circles). The two replisomes (black diamonds) are not always linked together.

During the C period, the replisome localizes to midcell (Figure 12-1), and DNA is spooled through a replication factory that is either stationary or passively moving. Daughter chromosomes are extruded to opposite halves of the cell. Thus, the force expended by DNA polymerase to move the template may be used to power chromosome segregation. The effects of mutations in various chromosomal proteins upon the general disposition of the nucleoid and the distribution of specific DNA sequences suggest that chromosome segregation is facilitated by compacting and organizing daughter chromosomes as they are extruded from the replisome into the two halves of the predivisional cell.

The Initiator/Replicator Ratio Triggers Genome Duplication

In 1968, William Donachie enunciated the principle that "initiation of a round of DNA replication always takes place at a time when the cell mass/chromosome origin reaches a particular critical value". Although his work was done in the enterobacterium *E. coli*, it has since been found to be true for other species as well. Initiation of chromosome duplication is triggered when bacteria reach a critical cell mass, and this **initiation mass** is largely independent of their rate of growth. By 1986, Erik Boye and colleagues identified the critical factor in the initiation mass as DnaA, the bacterial initiator protein that activates the bacterial replicator oriC. Artificially increasing DnaA expression induces DNA replication in smaller cells, whereas reducing DnaA expression delays DNA replication until cells reach a larger mass. In fact, the number of initiation events increases in proportion to the concentration of DnaA up to almost twice the normal level of DnaA (Figure 12-2). In other words, the initiation mass halves when the DnaA concentration doubles. At higher levels of DnaA, other proteins required for DNA replication become rate limiting and the velocity of replication forks decreases dramatically. The basic conclusion from these studies is that bacterial chromosome duplication is regulated primarily, if not exclusively, by regulating the ability of the bacterial initiator protein to bind to its cognate replicator.

An attractive model is that DnaA accumulates during the B period and then triggers initiation when it reaches a critical level called the initiation mass. This level allows DnaA to bind to low-affinity sites within oriC after it has bound to the high-affinity sites (Figure 11-5). Following initiation, duplication of the chromosome increases the number of high-affinity

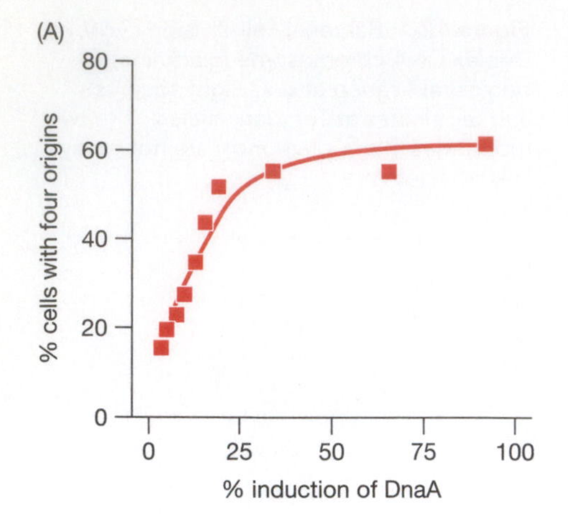

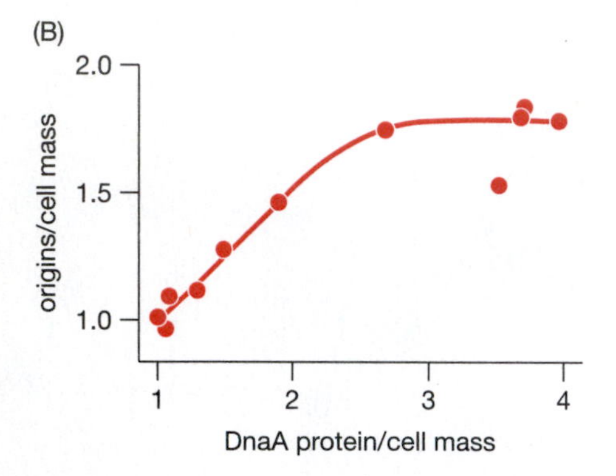

Figure 12-2. Relationship between number of origins per cell mass and DnaA protein concentration. The DnaA protein concentration in *E. coli* was increased above the wild-type level by expression of a *dnaA* gene located on a plasmid. The ectopic *dnaA* gene was driven by a *lacP*-controlled promoter whose activity was dependent on the concentration of isopropyl-*p*-D-thiogalactopyranoside added to the culture medium. (A) Increasing the DnaA protein concentration above endogenous levels (0% induction) rapidly increases the number of replication origins activated per cell. (B) The average number of replication origins per cell was proportional to the concentration of DnaA protein until levels exceeded an approximately twofold increase over the normal cellular level (ratio=1). At that point, other proteins required for DNA replication became limiting, preventing accumulation of additional newly replicated chromosomes. (Data from Atlung, T. and Hansen, F.G. (1993) *J. Bacteriol.* 175: 6537–6545, and Løbner-Olesen, A., Skarstad, K., Hansen, F.G. *et al.* (1989) *Cell* 57: 881–889.)

sites, which bind DnaA protein and thereby reduce its concentration below the initiation mass. The concentration of available DnaA protein is now insufficient to initiate another round of DNA replication. Hence, increasing the intracellular concentration of DnaA by induced *dnaA* gene expression or by increasing the number of *dnaA* gene copies leads to more frequent initiations. Note, however, that the affinity of DnaA for oriC is dependent on the level of ATP in the cell, because only the ATP-bound form of DnaA binds to low-affinity sites. Therefore, nutrient conditions are linked directly to initiation of genome duplication both by the ability of bacterial cells to synthesize proteins (amino acids supply), and by their ability to generate ATP (energy supply). One mechanism that links the level of ATP–DnaA with the cell's ability to generate ATP is the presence of two chromosomal intergenic regions, termed DnaA-reactivating sequences (DARSs). DARSs promote regeneration of ATP–DnaA from ADP–DnaA by nucleotide exchange. Inserting additional DARSs increases ATP–DnaA levels and stimulates initiation of DNA replication, whereas deletion of DARSs retards accumulation of ATP–DnaA and inhibits initiation.

Suppression Of Reinitiation Within A Single Cell Cycle

At least three mechanisms repress extra initiation events in bacteria, although they are not necessarily represented in all species. They are sequestration of newly replicated oriC to prevent its interaction with DnaA, regulating the amount of DnaA that is bound to ATP, and sequestration of excess DnaA protein to sites in the bacterial chromosome that do not contain oriC (Figure 12-3). The fact that oriC binds from 10 to 20 copies of DnaA allows fine-tuning of this initiator titration mechanism in regulating the transition from the B to C period.

Sequestration of oriC

Following initiation of DNA replication at oriC in enterobacteria, cells enter an eclipse period in which oriC is refractory to new initiation events. This mechanism prevents immediate reinitiation at oriC in the presence of high levels of DnaA activity. Moreover, sequestration of oriC as well as synchronous firing of all oriCs at the onset of the C period ensure that initiation of DNA replication occurs only once per cell cycle. Disrupting either regulatory mechanism leads to extra initiation events.

The eclipse period depends on the sequestration of newly replicated oriC through the binding of SeqA protein to hemimethylated GATC sequences. *E. coli* oriC contains 11 GATC sites that are methylated by deoxyadenosine methyltransferase (Dam) (Figure 7-23). Eight of these GATC sequences are phylogenetically conserved. When the replication fork passes through them, they become hemimethylated (Figures 7-24 and 12-3), because neither bacteria nor eukarya can incorporate methylated dNTPs into DNA. GATC sequences within oriC remain hemimethylated for about one-third of the cell cycle, after which remethylation rapidly occurs. SeqA prevents DnaA from binding to newly replicated oriC replicators and unwinding the DNA strands. SeqA binds to the hemimethylated weak-affinity DnaA-binding sites I2, I3, and R5M in oriC, but not to the high-affinity DnaA-binding sites R1, R2, and R4. Thus, SeqA allows the bacterial initiator protein to re-establish the closed DnaA-oriC complex, but not the open DnaA-oriC complex (Figure 11-5). This is equivalent to the transition in eukaryotic cells from a preRC, in which the MCM helicase binds to DNA without unwinding it, to a preIC, in which the MCM helicase begins to unwind the DNA. Such mechanisms allow cells to re-establish initiator-replicator complexes in preparation for the next cell cycle, while at the same time preventing reinitiation of DNA replication at the same replication origins within a single cell cycle.

This sequestration of oriC by SeqA results in the association of hemimethylated oriC with the bacterial cell membrane. Since proliferating *E. coli* cells contain about eightfold more SeqA than Dam enzyme, and the

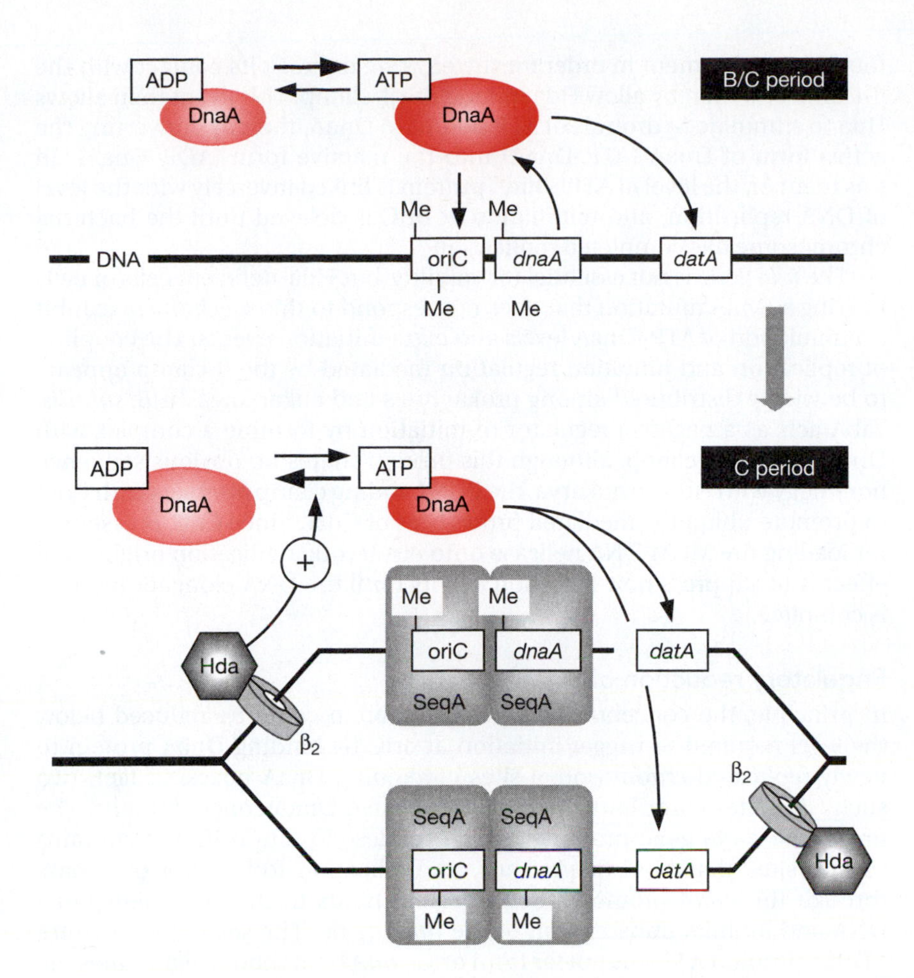

Figure 12-3. Three pathways act in concert to prevent reinitiation of DNA replication during *E. coli* chromosome replication by interfering with the interaction between ATP–DnaA and oriC. (1) Sequestration of hemimethylated oriC by SeqA protein prevents newly replicated copies of oriC from responding to activation by DnaA. (2) The concentration of DnaA protein is reduced by SeqA suppressing expression of the newly replicated *dnaA* gene, and by binding to the *datA* locus. (3) ATP–DnaA (the form that triggers initiation of DNA replication) is converted into ADP–DnaA by Hda•β-clamp-mediated hydrolysis of the bound ATP. While all three pathways are active in a variety of bacteria, the primary pathway appears to be the Hda•β-clamp. Sizes of red ovals indicate relative concentrations. Me indicates methylated DNA.

affinity of SeqA for hemimethylated oriC exceeds that of Dam, SeqA prevents Dam from immediately restoring oriC to its fully methylated status. The equilibrium shifts in favor of Dam each time it successfully remethylates the GATC sites, an event that cannot be reversed by SeqA. Although SeqA is not essential for bacterial growth and proliferation, its absence results in an increasein asynchronous initiation events and rapid methylation of oriC. The *seqA* gene together with the *dam* gene is found only in a subset of Gram-negative bacteria. However, other bacteria may utilize analogous DNA methylation mechanisms to regulate origin utilization.

Regulatory inactivation of DnaA

Since oriC activation requires binding of DnaA protein that is bound to ATP (Figure 11-5), initiation of DNA replication is strongly affected by regulating the level of the ATP-bound form of DnaA protein. *E. coli* cells contain from 1 to 3 mM ATP, and DnaA-dependent unwinding of oriC requires ATP at millimolar concentrations ($K_d \approx 1$ mM). If the available ATP concentration is near the K_m for ATP in DnaA-dependent unwinding of oriC, then cell-cycle-dependent fluctuations in ATP levels will strongly influence initiation frequency. In fact, Hda (a protein homologous to DnaA domains II and III) and the β-subunit of the DNA Pol-III holoenzyme (the bacterial sliding clamp; Figure 4-22) stimulate hydrolysis of ATP bound to DnaA protein in a process termed regulatory inactivation of DnaA. DnaA and Hda can each interact with the β-clamp when it is bound to DNA. During Okazaki-fragment synthesis, the β-clamp is mounted on duplex DNA and is in direct contact with the Pol-III holoenzyme. If only one region of the β-clamp interacts either with the α-subunit of the Pol-III core enzyme or with Hda, then Pol-III must first complete the synthesis of the Okazaki fragment in order for the enzyme to break its contact with the β-clamp and thereby allow Hda to bind the β-clamp. This event then allows Hda to stimulate hydrolysis of ATP bound to DnaA, thereby converting the active form of DnaA (ATP–DnaA) into the inactive form (ADP–DnaA). In this manner, the level of ATP–DnaA protein is linked inversely with the level of DNA replication, and reinitiation at oriC is delayed until the bacterial chromosome has completed replication.

The *hda* gene is not essential for viability, but Hda-deficient cells or cells bearing a DnaA mutation that does not respond to this mechanism exhibit accumulation of ATP–DnaA levels and extra initiation events. The coupling of replication and initiation regulation mediated by the β-clamp appears to be widely distributed among prokaryotes and eukaryotes. In *B. subtilis*, YabA acts as a negative regulator of initiation by forming a complex with DnaA and the β-clamp, although this protein shares no obvious sequence homology with Hda. In eukarya, the PCNA sliding clamp interacts with Cdt1 to promote ubiquitin-mediated proteolysis of Cdt1. Since Cdt1 is essential for loading the MCM DNA helicase onto eukaryotic replication origins, the effect is to suppress new initiation events until the DNA elongation phase is complete.

Regulatory reduction of DnaA

In principle, the concentration of DnaA protein could be reduced below the level required to trigger initiation at oriC by binding DnaA protein to newly replicated chromosomal sites containing DnaA boxes. In fact, two such DNA sites contribute to reducing elevated DnaA concentrations. The first is the *dnaA* gene promoter which resides close to oriC and contains several sites that bind SeqA. Thus, as replication forks from oriC pass through the *dnaA* promoter, SeqA protein binds to the hemimethylated DNA and inhibits transcription of the *dnaA* gene. The second is the *datA* ('DnaA titration A') locus not far from oriC. *datA* binds about eight times the number of DnaA protein monomers that are bound to oriC. The *datA* locus is fairly close to oriC and should be replicated during the eclipse period. Therefore, as replication forks pass through the *datA* locus, the additional *datA* loci should reduce the number of DnaA protein molecules that would otherwise be available to reactivate oriC.

The *E. coli* chromosome contains about 300 DnaA boxes, but the *datA* locus (which contains four or five DnaA boxes) appears to be solely dedicated to initiator titration, inasmuch as the simultaneous deletion of seven other strong DnaA-binding loci does not affect replication. The unusual affinity of *datA* for DnaA protein is dependent on two central, properly spaced DnaA boxes flanking a site that binds IHF protein. Although *datA* is not an essential regulatory element, deletion of the *datA* site does lead to asynchronous initiations. However, these asynchronous initiations arise from initiation events that do not occur during normal exponential growth. A range in the number of extra copies of *datA* was effective in maintaining synchronous initiations, but surplus *datA* hindered growth, presumably by titrating DnaA away from oriC. As replication forks move away from oriC, duplication of *datA* then titrates the excess DnaA protein remaining in the cytoplasm.

Deletion of *datA* results in additional initiation events, but only under non-physiological conditions. Deregulation occurs either when wild-type cells are treated with rifampicin (a specific inhibitor of bacterial RNA polymerase), or when they are engineered to overexpress DnaA protein. Activation of oriC is sensitive to transcription in its vicinity, because transcription can affect the ability of DnaA to unwind oriC, and the transcription affecting oriC is modulated by DnaA protein binding to the DnaA-box sequences within or around the oriC region (Figure 10-6). Apparently, in the absence of transcription, oriC becomes more sensitive to increased levels of DnaA.

THE ARCHAEAL CELL CYCLE

Little is known about archaeal cell cycles, but it is anticipated that their regulatory mechanisms are more similar to bacteria than to eukarya, because none of the proteins that regulate eukaryotic cell cycles (see Table 12-3, below) are found either in bacteria or in archaea. On the other hand, some archaeal chromosomes, such as that of *Sulfolobus*, contain multiple replication origins, analogous to eukaryotic chromosomes. Whether or not archaea, like bacteria, can tolerate the presence of replication forks during cell division remains to be seen.

What is known about archaeal cell cycles is reminiscent of other single-cell organisms. The cell cycle of *Sulfolobus* occupies about 3 h (Table 12-1). It has a short G_1 period in its cell cycle where the cells have a single copy of the chromosome. There is then a prolonged period where cells have two complete chromosomes prior to nucleoid segregation and cell division. Analyses of both the frequency of various genes throughout the genome in synchronous and exponentially proliferating cultures of *Sulfolobus acidocaldarius*, revealed that all three replication origins fire within a narrow time window and are utilized in every cell, with DNA replication forks moving bidirectionally away from each origin at approximately 88 bp s^{-1}. This coordination of origin firing is reminiscent of bacteria growing in rich media, but dissimilar to eukaryotic chromosomes where the frequency of activation of replication origins varies markedly. It should be noted, however, that archaeal replication forks travel more slowly than those in bacteria, at a rate similar to eukaryotic forks (Table 3-4). The three *Sulfolobus* Orc1/Cdc6 initiator proteins exhibit a uniformity of cellular abundance and origin binding throughout the cell cycle. This is analogous to yeast, but not to metazoa. The cellular level of MCM helicase also is constant during the cell cycle, but its origin localization is regulated. During early S phase, MCM is strongly enriched at all three replication origins. This is analogous to both bacteria and eukarya. How the regulation of origin firing is effected in archaea is currently unknown.

THE EUKARYOTIC MITOTIC CELL CYCLE

The prototypic cell-division cycle in eukarya is the **mitotic cell cycle**, a sequence of events that can be grouped into four phases: DNA synthesis (S phase), mitosis (M phase), and two intervening gaps of time called the G_1 and G_2 phases (Figure 11-1). As with bacteria, cell growth occurs primarily after mitosis and before DNA replication, the period referred to as G_1 phase. In addition, eukaryotic cells can exit their mitotic cell cycle and enter a quiescent state called G_0 in which the ability to maintain the living state is maintained in the absence of either cell growth or proliferation. Among the single-cell eukarya, cell cycles have been characterized most completely in the budding yeast *Saccharomyces cerevisiae* and the fission yeast *Schizosaccharomyces pombe*. Among the metazoa, the systems of choice have been cultured mammalian cells and eggs from the frog *Xenopus laevis* to dissect the biochemistry, and the fruit fly *Drosophila melanogaster* and mice to dissect the genetics. The conservation among these systems is striking. Of the 37 genes required to produce a pre-initiation complex (preIC), only Mcm9 is unique to metazoa (Tables 11-3 and 11-4). Of the 70 gene products that regulate cell division during cell proliferation (see Table 12-3; and geminin), only five are unique to metazoa (Cdk2, Emi1, Cdc25B, Cdc25C, and geminin). This evolutionary conservation allows one to construct a fairly unified picture of the events that drive eukaryotic cell cycles forward and that restrict genome duplication to once and only once per cell division.

Common Features

Eukaryotic mitotic cell cycles share several features in common. First, events move only in one direction, namely $M \rightarrow G_1 \rightarrow S \rightarrow G_2 \rightarrow M$. When cells exit M phase they have a choice of either continuing to proliferate ($G_1 \rightarrow S \rightarrow G_2 \rightarrow M$) or resting ($G_0$). Second, progress through the mitotic cell cycle is regulated at two critical transitions: $G_1 \rightarrow S$ and $G_2 \rightarrow M$. Third, once a cell is committed to genome duplication (S phase) it must complete cell division or die. Thus, at the optimal temperature for proliferation in a particular species, the time required to transit through S, G_2, and M phases is fairly constant among different cell types from the same organism. Changes in nutrient conditions affect primarily the length of G_1. Finally, eukaryotic cells cannot divide unless they contain two complete copies of their genome, thus giving rise to two daughters, each with one complete copy of the genome.

At least 90% of all mitotic cell cycles is occupied by **interphase** (G_1, S, and G_2; Figure 11-1), that portion of the cell cycle in which most cell functions are performed, including preparation for cell division. Interphase cells can be recognized visually, because the nuclear membrane is present, the chromatin has not yet condensed, and therefore individual chromosomes are not visible (see Figure 12-4 in color plate section), although the nucleolus may be visible as an enlarged dark spot. G_1-phase cells obtain nutrients, grow, and carry out functions required to maintain them in a living state. S-phase cells duplicate the genome, most of which resides in the nucleus. Mitochondrial DNA replication, however, occurs throughout interphase. The period of time that elapses between completion of S phase and the onset of mitosis is called G_2 phase.

The beginning of mitosis (**prophase**) is marked by the appearance of chromosomes (condensed chromatin fibers) that look like a bowl of tangled noodles within the cell's nucleus (Figure 12-4). Each of the 'noodles' actually consists of two homologous chromatids (also called sister chromatids) that are identical copies of one of the mother cell's chromosomes. The homologs are held together tightly by a protein complex called **cohesin** that binds all along the length of each chromatid (Figure 7-18). Cohesin binds particularly strongly at the centromere, a region near the center of each chromatid. As mitosis progresses, most of the cohesin along the chromatids' arms is released, but the cohesin complex at the centromere remains, thus holding the two sister chromatids together (Figure 7-19).

As mitosis continues, the membrane around the nucleus in metazoa dissolves (but not in yeast and fungi) and the cell enters **prometaphase**. Thin, strong filaments of protein called microtubules attach to the DNA. The microtubules are the mechanism by which sister chromatids are pulled apart. In humans, one end of each microtubule fiber is anchored at either end of the cell to a structure called the **centrosome**, and the other end attaches to a protein complex called the **kinetochore** that is attached to the centromere of each chromatid. If a microtubule misses a kinetochore, it retracts and starts over. This process continues until each kinetochore grips a microtubule. Only then does the next stage of mitosis begin.

Metaphase occurs when the microtubule fibers shorten, pulling the chromatids toward the centromere. At the same time, the microtubule attached to a chromatid's sister begins to pull toward the opposite side of the cell. Because the sister chromatids are still firmly attached by the cohesin complex, all of the chromatids end up aligned in the middle of the cell. If only one of the two sister chromatids is attached to a microtubule, there is no tension on the pair and both are dragged toward one anchor point. This lack of tension signals that something is wrong, and mitosis stops until the problem is fixed (spindle-assembly checkpoint; Chapter 13).

With the chromatids lined up and each sister chromatid attached to opposite anchor points, the cell is ready to separate sister chromatids. The cohesin complex holding the centromeres together is cut, and the sister chromatids are pulled apart from one another (**anaphase**). In this way, duplicate copies of the parent cell's nuclear genome are divided between the two daughter cells. The remaining portion of the genome, the mitochondria, are distributed randomly between the two sibling cells.

The cell continues to divide, forming membranes around the two areas of DNA to create a nucleus for each daughter cell (**telophase**). Once inside the nucleus, the DNA decompresses from its highly compacted state. Mitosis is now complete, but the job of cell division is not.

Cytokinesis, the process of separating the cytoplasm of a single eukaryotic cell into two daughter cells, and nuclear envelope assembly usually occur concurrently, although they are distinct processes. In animal cells, a cleavage furrow develops where the metaphase plate used to be, pinching off the separated nuclei. A contractile ring, made of non-muscle myosin II and actin filaments, assembles in the middle of the cell adjacent to the cell membrane. The resulting actin-myosin fibers constrict the cell membrane to form a cleavage furrow. This process continues until a so-called midbody structure is formed, the process of abscission then physically cleaves this midbody into two. Abscission depends on septin filaments beneath the cleavage furrow that provide a structural basis to ensure completion of cytokinesis. After cytokinesis, nonkinetochore microtubules reorganize and disappear into a new cytoskeleton as the cell cycle returns to interphase.

Abbreviated Mitotic Cell Cycles

The time required for mitotic cell cycles can vary from minutes to hours, and the time spent at each stage of the cycle can vary extensively (Table 12-1). While budding yeast and mammalian cells have long, nutrient-dependent G_1 phases, fission yeast has a long G_2 phase. Mammalian embryonic stem cells and human cancer cells typically have a much shorter G_1 phase than normal cells from adult animals. In some cases, G phases are not evident.

Fertilized eggs of frogs, fish, and echinoderms undergo a series of cleavages that subdivide the maternally inherited cytoplasm into smaller and smaller cells, each of which contains a single nucleus (see Table 12-6, below). These **cleavage cell cycles** are so rapid they consist simply of a series of S and M phases, reminiscent of bacterial cells proliferating in rich growth medium. Similarly, the fertilized eggs of flies begin with a **syncytial cell cycle** in which the first 10 nuclear divisions occur as a series of S→M transitions in the absence of cytokinesis to produce a single cell (syncytial blastoderm) containing 1024 nuclei. Such rapid, abbreviated mitotic cell cycles can occur, because the egg is a storehouse of the proteins needed for genome duplication in the absence of transcription. Nevertheless, in contrast to bacteria, proliferating eukaryotic cells do not reinitiate DNA replication prior to mitosis, regardless of whether the cell cycle takes several minutes or several hours. Invariably, pre-replication complexes (preRCs) assembled during the $M \rightarrow G_1$ transition are inactivated during S phase, and assembly of new preRCs is prevented until mitosis is complete and a nuclear membrane is present.

Cell-Cycle Analysis

Analysis of cell cycles in metazoa relies on three strategies for identifying specific stages in cell division: differences in cell morphology, fractionating cells according to their size and DNA content, and arresting cell proliferation with inhibitors that target specific metabolic events, addition of pheromones, and use of temperature-sensitive mutants (Table 12-2). Cells isolated or arrested at specific stages in their cell cycle can then be allowed to continue proliferating as a synchronized population. However, cell synchrony never lasts beyond two or three generations.

Without metabolic inhibitors

Detecting differences in morphology relies on staining cells with dyes that detect DNA and with antibodies targeted to specific proteins, and the expression of those proteins can then be related to their subcellular localization and to the morphological state of the nucleus and chromatin.

Table 12-2. Methods For Characterizing Eukaryotic Cells At Specific Stages In Their Division Cycle

Method	Cells staged at			
	G_0 or G_1	S	G_2	Mitosis
Morphology	Nucleus No S-phase-specific proteins	Nucleus DNA synthesis S-phase-specific proteins	Nucleus No DNA synthesis S-phase-specific proteins remain	No nucleus Prophase Prometaphase Metaphase Anaphase Telophase
Centrifugal elutriation	Small cells 2N DNA	Larger cells 2N→4N DNA	Largest cells, 4N DNA	
Inhibitor				
α-Factor (*S. cerevisiae*)	Late G_1 at 'R-point'			
Serum starvation Amino acid deprivation	Late G_1 at 'R-point'			
Lovastatin	Late G_1			
Hydroxyurea Mimosine		S		
Thymidine		S		
Methotrexate		S		
Aphidicolin		S		
RO3306			G_2	
Nocodazole				Metaphase
Cdc15 temperature-sensitive mutant (*S. cerevisiae*)				Anaphase/telophase

The primary advantages are the simplicity of the technology, its application to small samples, and the ability to relate the presence or absence of a protein to cellular morphology and to its subcellular location. For example, one can distinguish the various steps in mitosis (Figure 12-4) without the need for metabolic inhibitors. Limitations of this strategy include its reliance on antibody specificity, accessibility of the target protein, artifacts of cellular fixation, and difficulty in quantifying data.

Centrifugal elutriation is a method for fractionating cells according to their size and density. Since cell density is strongly affected by its DNA content, this method separates small G_1-phase cells with 2N DNA content from cells at later stages in their growth and genomic duplication. A cell suspension is forced into the bottom of a specially designed horizontal centrifuge chamber while the rotor is spinning. Thus, two forces act on the cells. Centrifugal force drives cells away from the spinning axis, while the counter-flow pushes them toward the spinning axis. The net result is that the smallest particles with the least density exit the top of the spinning chamber first and the largest particles with the greatest density exit the top of the chamber last. In practical terms, G_1-phase cells exit first, followed by S-phase cells (earliest first; latest last), followed by G_2- and M-phase cells as a single group. In this manner, one can obtain large populations of cells at specific stages in their cell cycle. The purity of each population can be determined either by their cell morphology or by an analytical technique called fluorescence-activated cell sorting (FACS) that can separate cells according to their DNA content. The limitations of centrifugal elutriation include the expense of the instrument, the need for large quantities of a pure cell population, and the lack of resolution among cells with 4N DNA content (G_2 and mitosis). The best synchrony is achieved in G_1 and in early and mid S phase.

The **baby machine** is a method that enriches for newborn ('baby') cells. An asynchronous population of cells is attached to a membrane that is then inverted and clamped into a double funnel chamber. Medium is applied to the top of the chamber and it slowly flows through the membrane, allowing the attached cells to grow and divide. Newborn cells are therefore washed off the membrane and can be collected. This technique has the advantage that the cell cycle is not artificially perturbed by drug treatments but results in relatively small sample sizes. The baby machine has been successfully employed with a range of bacterial, archaeal, and eukaryotic cells.

With metabolic inhibitors

Inhibition of specific metabolic pathways is commonly used to synchronize populations of eukaryotic cells. To be effective, the inhibitor must be easily reversible and generally non-toxic. Still, they must be used with caution. Inhibiting a single event during cell proliferation may yield cells that are morphologically arrested in G_1, S, G_2, or M phase and that are no longer synthesizing DNA, but it does not mean that all other metabolic pathways have been arrested, as well. On the contrary, once cells have passed their restriction point (Figure 11-1), arresting their progress with metabolic inhibitors inevitably results in DNA damage and eventually cell death. When the inhibitor is withdrawn, the fraction of cells that recover and re-enter their cell cycle diminishes with the length of time they were exposed to the inhibitor. Moreover, the tolerance of cells to metabolic inhibitors varies markedly among species and cell types. With these caveats in mind, inhibitors provide simple, effective, inexpensive ways to obtain large quantities of synchronized cells.

Serum starvation, alone or coupled with deprivation of an amino acid that is not synthesized *de novo* by the cell, arrests mammalian cells after they have assembled pre-replication complexes but before they have passed the restriction point (Figure 11-1). With time, preRC proteins Orc1, Cdc6, and MCM disappear from arrested cells as they adjust to a quiescent G_0 state.

Lovastatin is a reversible competitive inhibitor of 3-hydroxy-3-methylglutaryl-coenzyme A reductase, an enzyme required for cholesterol biosynthesis. It appears to prevent the onset of S phase by inducing accumulation of CDK-specific inhibitors. Hydroxyurea and mimosine inhibit ribonucleotide reductase, an enzyme required for *de novo* synthesis of dNTPs (Figure 3-12). These drugs rapidly reduce both the dATP and dGTP pools in eukaryotic cells, thereby arresting cells as they enter S phase as well as cells already within S phase. A similar effect is seen with excess thymidine, which inhibits thymidylate synthase, an enzyme required for dTTP synthesis (Figure 3-12). A more general inhibitor of nucleotide metabolism is methotrexate, a powerful inhibitor of dihydrofolate reductase, an enzyme required for the synthesis of tetrahydrofolate. Since tetrahydrofolate is an intermediate in the synthesis of purines, thymidine, and certain amino acids, methotrexate inhibits synthesis of RNA and proteins as well as DNA. Aphidicolin inhibits specifically the replicative DNA polymerases Pol-α, Pol-δ, and Pol-ε, thereby arresting DNA synthesis at replication forks. However, aphidicolin does not inhibit DNA primase. In the presence of aphidicolin, DNA synthesis is initiated at replication origins, but it is limited to RNA–DNA primers 20–40 nucleotides long. RO3306 selectively inhibits the cyclin-dependent protein kinase Cdk1, which is 10–1000 times more sensitive to this drug than other protein kinases. Since Cdk1 activity is essential for entrance into mitosis, cells are arrested in G_2 phase. Nocodazole prevents assembly of microtubule filaments by binding to tubulin, thereby arresting cells in prometaphase and metaphase.

Two methods are unique to synchronization of yeast cell cycles. α-Factor is a peptide pheromone produced by *S. cerevisiae* cells with mating type-α. α-Factor arrests cells with mating type-a at their restriction point (Figure 11-1) in late G_1 phase by inhibiting the Cdk1/Cdc28•Cln activity that is required for preIC assembly. α-Factor is particularly useful because cells recover rapidly and progress synchronously through two or three cell cycles. The second method uses a temperature sensitive mutation in Cdc15, a protein kinase required for cells to exit mitosis. At the restrictive temperature (37°C), cells arrest in late anaphase/telophase with a characteristic dumbbell morphology.

Artifacts

Cell synchronization is not without its artifacts, particularly when inhibiting a specific step in cell metabolism. Reducing the activity of a protein required for replication-fork activity (e.g. Dpb11) can cause chromosome breaks that result in gross chromosomal rearrangements. Similarly, either forcing cells to enter S phase prematurely before preRC assembly is completed (e.g. stimulating CDK activity) or inhibiting preRC assembly during G_1 phase (e.g. overexpression of G_1 cyclins) results in a paucity of initiation events and therefore a longer S phase. Consequently, dsDNA breaks occur that lead to gross chromosomal rearrangements. Prolonged incubation with aphidicolin also induces DNA damage at replication forks, apparently by exposing ssDNA regions to the action of endonucleases. Moreover, arresting DNA synthesis with aphidicolin, hydroxyurea, or mimosine does not prevent DNA unwinding from continuing. Molecules accumulate with long stretches of ssDNA on both arms of replication forks. Thus, arresting DNA synthesis can not only cause DNA damage, but it can alter the number and locations of sites where DNA synthesis begins (i.e. replication origins). Similarly, arresting protein biosynthesis with inhibitors such as emetine in order to block histone synthesis to study chromatin assembly also rapidly inhibits synthesis of Okazaki fragments and creates forks with ssDNA on one arm and dsDNA on the other. Finally, blocking the metaphase to anaphase transition with nocodazole does not necessarily prevent other events leading to G_1 phase. For example, inhibiting Cdk1 activity in metaphase-arrested cells triggers preRC assembly and subsequent DNA replication despite the fact that the cells remain morphologically in metaphase.

Accelerators And Brakes

During the last quarter of the twentieth century, extraordinary advances were made in understanding, at the molecular level, the mechanisms that regulate eukaryotic cell proliferation. In recognition of this progress, the 2001 Nobel Prize in Physiology or Medicine was awarded to Leland Hartwell, Paul Nurse, and Tim Hunt for pioneering the discovery of specific genes that regulate the cell cycle, for identification and characterization of Cdk1, a cyclin-dependent protein kinase (CDK) that is essential for the $G_2 \rightarrow M$ transition, and for the discovery of cyclins, a group of proteins that regulate the activities of CDKs. The pace of discovery has not slackened, and we now know that the mitotic cell cycle is driven by two interacting accelerators that drive the cell cycle forward, and two brakes that prevent premature transitions from one phase of the cell cycle to the next.

The two accelerators are CDKs and **ubiquitin ligases** (more formally termed **E3** ubiquitin protein ligases). The two brakes are cyclin-dependent kinase inhibitors (CKIs) that specifically inhibit CDKs, and early mitosis inhibitor (Emi) that specifically inhibits the anaphase-promoting complex (APC), one of the two ubiquitin ligases that drive cell cycles. The other ubiquitin ligase is the SCF complex (named for its three core subunits, Skp1, Cullin, and F-box). Together, these four components drive the mitotic cell cycle in one direction, $M \rightarrow G_1 \rightarrow S \rightarrow G_2 \rightarrow M$. All of these proteins are unique

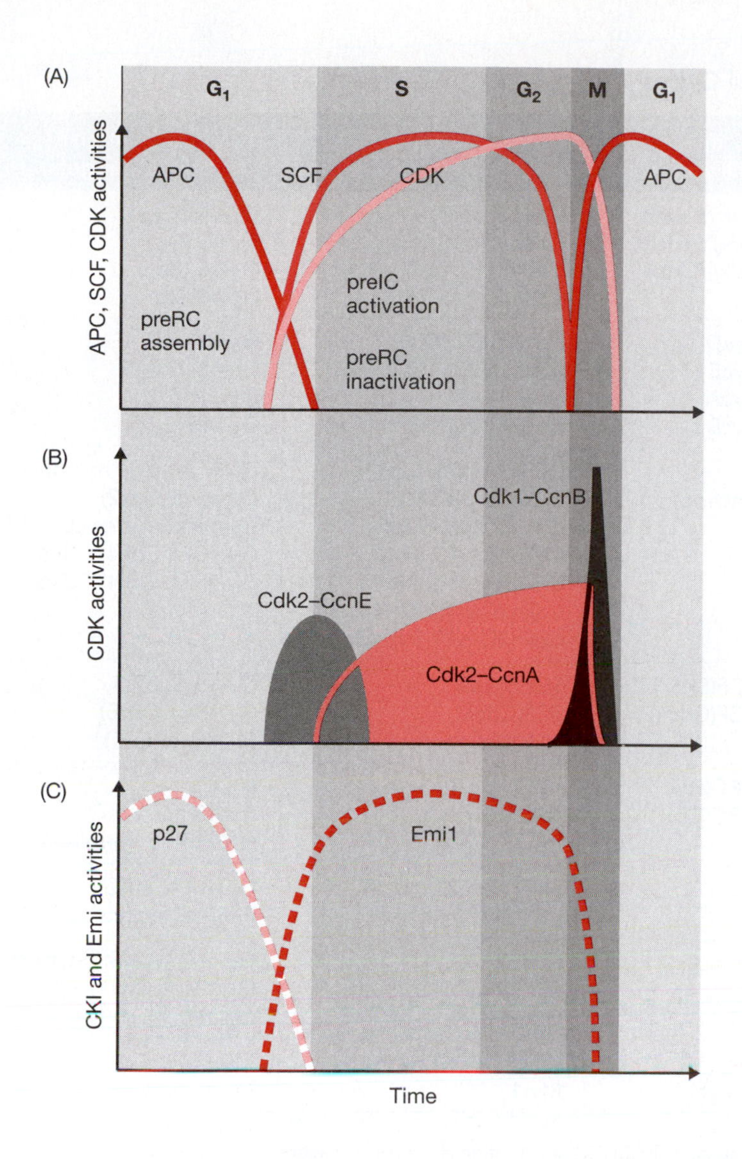

Figure 12-5. Time course for the appearance of activities critical to the regulation of eukaryotic cell cycles. (A) The APC (red) and SCF (dark orange) ubiquitin ligase activities and the total cellular CDK activity (light orange) rise and fall as eukaryotic cells grow (G_1 phase), replicate their nuclear DNA (S phase), and divide [G_2 phase and mitosis (M)]. (B) Mammals employ two CDKs (Cdk1 and Cdk2) and three cyclins (CcnA, CcnB, and CcnE) to produce three distinct CDK•cyclin enzymatic activities [Cdk2•CcnE (gray), Cdk2•CcnA (orange), and Cdk1•CcnB (black)] that appear at specific times during cell division. Cdk1•CcnA activity (not indicated) closely mirrors Cdk1•CcnB activity. Yeast employ a single CDK and three to six B-type cyclins, and flies employ two CDKs and three cyclins to accomplish the same tasks (Table 12-3). (C) Eukaryotic cells express a single CKI (p27 in mammals) during G_1 that inhibits CDK activities, and a single Emi (Emi1) during S and G_2 phases that inhibits APC activity during S, G_2, and prophase. Human proteins are indicated, most of which have analogs in other eukarya (Table 12-3).

to eukarya, and their *activities* during cell proliferation appear only during specific stages of the cell cycle (Figure 12-5). Some of the protein components are expressed constitutively. For example, proteins such as the core subunits of ubiquitin ligases and CDKs are generally expressed throughout the cell cycle. Their activities are regulated by changes in the cellular levels of regulatory subunits such as F-box proteins and cyclins, respectively.

All eukarya encode at least one CDK as well as several different cyclins (Table 12-3). Cyclins are required for CDK activation and substrate specificity. Single-cell organisms such as yeast encode a single CDK whose specificity is determined by which of the 3 to 6 different cyclin proteins it binds. In contrast, complex multicellular organisms such as humans encode at least 10 different CDKs and at least 24 different cyclins! However, only CDKs 1, 2, 4, and 6 regulate cell division directly, and the activities of these CDKs are determined by 10 cyclins whose sequence homologies group them into four different families termed A, B, D, and E. Note that the cellular concentrations of CDK proteins are approximately 30-fold greater than the maximum concentrations observed for cyclins, and that CDK proteins are synthesized throughout cell proliferation, whereas cyclin proteins are present only during specific periods of the cell cycle. Therefore, cell-cycle-dependent changes in CDK activities (Figure 12-5B) result from cell-cycle-dependent changes in cyclin protein levels. Cyclin levels, in turn, are regulated by ubiquitin-dependent proteolysis.

Table 12-3. Proteins That Regulate Eukaryotic Cell Cycles

Protein	Budding yeast (*S. cerevisiae*)	Fission yeast (*S. pombe*)	Insects (*Drosophila*)	Mammals (human)	Regulation (primary targets)
CDKs	Cdk1/Cdc28 Cdk1/Cdc28	Cdk1/Cdc2 Cdk1/Cdc2	Cdk1/Cdc2 Cdk2/Cdc2c Cdk4/Cdk6	Cdk1 Cdk2 Cdk4 Cdk6	$G_2 \rightarrow M$; $G_1 \rightarrow S$ $G_1 \rightarrow S$ $G_0 \rightarrow S$ (Rb) $G_0 \rightarrow S$ (Rb)
Cyclins	Cln3 Cln1, Cln2 Clb5, Clb6[a] Clb1 to Clb4	Puc1 Puc1 Cig2[a] Cig1, Cdc13	CycD CycE CycA CycB	CcnD1, D2, D3 CcnE1, E2 CcnA1, A2[a] CcnB1[a], B2, B3	$G_0 \rightarrow S$ $G_0 \rightarrow S$; $G_1 \rightarrow S$ Restricts S $G_2 \rightarrow M$
CKIs	Sic1	Rum1	Dacapo	Cdkn1A/p21/Cip1 Cdkn1B/p27/Kip1 Cdkn1C/p57/Kip2 Cdkn2B/p15/INK4b Cdkn2A/p16/INK4a Cdkn2C/p18/INK4c Cdkn2D/p19/INK4d	DNA damage response $G_1 \rightarrow S$ (⊥ Cdk2) $G_2 \rightarrow$endocycles (⊥ Cdk1) $G_0 \rightarrow S$ (⊥ Cdk4, Cdk6) $G_0 \rightarrow S$ (⊥ Cdk4, Cdk6)
SCFs	SCF(Cdc4) SCF(Grr1)	SCF(Pop1)	SCF(Skp2) SCF(Cdc4)	SCF(Skp2) SCF(Cdc4) SCF(βTrCP)	$G_1 \rightarrow S$ (⊥ p27) $G_0 \rightarrow S$ (⊥ CcnE) $M \rightarrow G_1$ (⊥ Emi1)
APCs	APC(Cdc20) APC(Cdh1)	APC(Slp1) APC(Ste9)	APC(Fzy) APC(Fzr)	APC(Cdc20) APC(Cdh1)	$M \rightarrow G_1$ $G_1 \rightarrow S$ (⊥ CcnA, B; geminin)
Emi			Rca1	Emi1, Emi2	$G_2 \rightarrow M$ (⊥ APC) Emi2 is "oocyte specific"
Cdc25s	Mih1	Cdc25	String	Cdc25A Cdc25B Cdc25C	$G_2 \rightarrow M$; $G_1 \rightarrow S$ (↑ Cdk2) $G_2 \rightarrow M$ (↑ Cdk1) $G_2 \rightarrow M$ (↑ Cdk1)
Wee and Mik	Swe1	Wee1 Mik1	Dwee1 Dmyt1	Wee1 Myt1	$M \rightarrow G_1$ (⊥ Cdk1) $M \rightarrow G_1$ (⊥ Cdk1)

Genes aligned horizontally are functional analogs of the human protein. Inhibition (⊥) or activation (↑) of specific activities is indicated.

[a]Clb5, Clb6, and Cig2 are yeast B-type cyclins that drive S phase rather than mitosis. Of the cyclin B family, only CcnB1 is required for mitosis. Of the cyclin A family, only CcnA2 is required for S phase. Multiple names for the same gene are separated by /. The first name is the official name, although those who work with these genes often prefer one of the earlier names.

Polyubiquitinated proteins are substrates for large (~2000 kDa) protein complexes called **proteasomes** that have a sedimentation constant of 26S. They are located in both the nucleus and cytoplasm of all eukarya. Proteasomes degrade unneeded or damaged proteins into peptides of seven to eight amino acids that can then be further degraded into amino acids by various proteases and reused in the synthesis of new proteins. Proteins targeted for degradation are covalently linked at specific lysine residues to a small (~8.5 kDa) highly conserved protein called **ubiquitin**. Addition of a single ubiquitin (**monoubiquitination**) can alter a protein's localization from nucleus to cytoplasm, but addition of a polymer of four or more ubiquitin molecules (**polyubiquitination**) generally (but not always) directs the protein to the 26S proteasome. **Ubiquitination** occurs in three steps: activation of ubiquitin by an E1 ubiquitin activating enzyme, transfer of ubiquitin from the E1 enzyme to the active site of an E2 ubiquitin conjugating enzyme, and finally formation of an isopeptide bond between a lysine residue in the target protein and the C-terminal glycine of ubiquitin. This final step requires the activity of one of the hundreds of E3 ubiquitin ligases (or simply ubiquitin ligase). Ubiquitin ligases function as the substrate recognition component of the ubiquitination pathway; they interact both with E2 enzymes and with ubiquitin. Two ubiquitin ligases regulate cell division. SCF complexes regulate the $G_1 \rightarrow S$ transition, and APC regulate the $M \rightarrow G_1$ transition. They do so by regulating the levels of cyclins and CDK-specific inhibitors (and therefore CDK activities), preRC proteins (and therefore initiation of DNA replication), and proteins involved in chromatid cohesion (and therefore the onset of anaphase).

SCF ubiquitin ligases are about 170 kDa in size and contain four subunits (F-box protein•Skp1•Cul1•Rbx1). The F-box protein, of which about 20 have been identified in yeast and hundreds in higher eukaryotes, determines substrate specificity. For example, SCF(Skp2) targets the Cdk1-specific inhibitor p27 (and its orthologs; Table 12-3), whereas SCF(Cdc4) targets CcnE (Table 12-3). SCF(βTrCP) targets the APC-specific inhibitor Emi1. In each case, ubiquitination by SCF requires prior phosphorylation of the target protein that is then recognized by a unique F-box protein.

APC ubiquitin ligases regulate the metaphase to anaphase transition in mitosis (Figure 12-4). They are much larger (1.5 MDa) than SCFs, and they contain 12 subunits. APC and SCF complexes share in common a Cullin subunit and a RING subunit that binds the E2 ubiquitin conjugate. However, in contrast to SCFs, APC substrate specificity is determined by only two subunits, Cdc20 and Cdh1, neither of which requires phosphorylation of the target for substrate recognition. However, APC(Cdc20) assembly requires CDK-dependent phosphorylation of Cdc20 as well as several core APC subunits, whereas APC(Cdh1) assembly occurs only in the absence of CDK activity. This difference allows APC(Cdc20) to be assembled during early M phase while CDK activity is high, and APC(Cdh1) to be assembled during the metaphase to G_1-phase transition when CDK activity is low. Both APCs target essentially the same proteins, thereby allowing APC functions to continue as cells transit from mitosis to G_1.

Cyclin-dependent kinase inhibitors (CKIs) are proteins that bind specifically to CDKs and, in most cases, inhibit their activities. CKIs are divided into two groups. The INK4 (inhibitors of Cdk4) family binds specifically to Cdk4 and Cdk6 and inhibits their activities. They regulate the ability of quiescent cells to re-enter the cell cycle ($G_0 \rightarrow S$ transition, Figure 13-4). The Cip/Kip family binds specifically to Cdk1•cyclin and Cdk2•cyclin complexes by first anchoring to the cyclin and then inserting itself into and distorting the kinase active site. Their primary function is to prevent premature DNA synthesis. Mammals contain three Cip/Kip proteins (p21/Cip1, p27/Kip1, and p57/Kip2). p27 regulates entry into S phase. p21 is part of the DNA damage response discussed in Chapter 13. p57 is involved in various types of cell differentiation, among which is triggering endoreduplication in cells that are developmentally programmed for this process. Neither p21 nor p27 are essential for cell proliferation, since mice have been constructed that lack one or the other gene. However, inactivation of both genes prevents embryonic development, suggesting that Cip/Kip gene functions are, to some extent, redundant (discussed below).

Early mitosis inhibitors (Emis) are proteins that bind specifically to one or more subunits in the APC and thereby prevent the premature appearance of APC activity. Only one (Emi1/Rca1) has been reported so far in somatic cells, and it is essential for cell proliferation. In the absence of Emi1, cyclin proteins do not accumulate during interphase. Premature activation of the APC by suppression of Emi1 induces DNA re-replication by degrading cyclin A, a key regulator of preRC assembly. A second example, Emi2, is found only in oocytes where it is part of the spindle assembly checkpoint in meiotic cells (Figure 13-20).

Yin And Yang

In Chinese philosophy, the concept of yin and yang is used to describe how seemingly opposing forces are interconnected and interdependent in the natural world, giving rise to each other in turn. Yin and yang are

Table 12-4. The Yin And Yang Of Eukaryotic Cell-Division Cycles

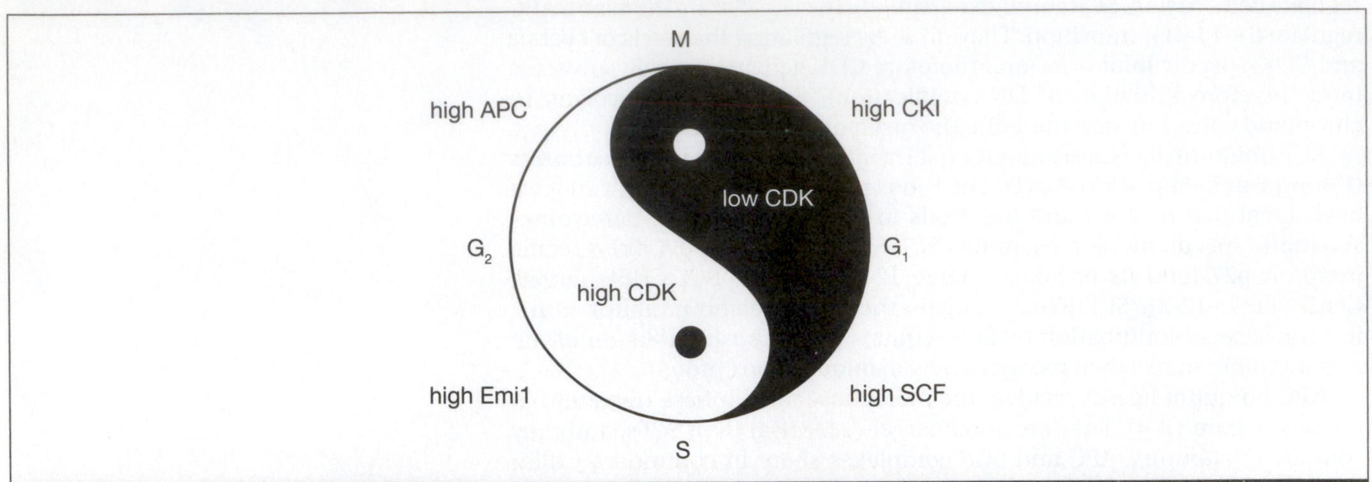

Cell cycle	preRC assembly	preIC activation	CDK	CKI	SCF	APC	Emi1
Anaphase→late G_1	Yes	No	Low	High	Low	High	Low
Late G_1→S	No	Yes	High	Low	High	Low	High
S	No	Yes	High	Low	High	Low	High
G_2→metaphase	No	Yes	High	Low	High	Low	Low

complementary opposites within a greater whole. Everything has both yin and yang aspects, which constantly interact, never existing in absolute stasis. So it is with the regulation of genome duplication (Table 12-4).

preRC assembly (Figure 11-11) requires low CDK activity, whereas preIC assembly (Figure 11-15) requires high CDK activity. Changes in the level of CDK activity during cell division are determined by changes in the levels of CKIs and cyclins. The net result is that preRC assembly occurs when the CKI to cyclin ratio is high, and preIC assembly occurs when the CKI to cyclin ratio is low. The CKI to cyclin ratio, in turn, is regulated by their selective degradation. Cyclins A and B are targeted for ubiquitin-dependent protein degradation by the APC ubiquitin ligase, whereas cyclin E and CKIs are targeted by the SCF ubiquitin ligase. The net result is that preRC assembly occurs when the APC to SCF ratio is high, and preIC assembly occurs when the APC to SCF ratio is low. Finally, the ubiquitin ligase activities themselves are regulated by specific inhibitors such as Emi1 as well as by feedback loops (described below) in which each enzyme targets for destruction a critical subunit in the other enzyme's composition. Taken together, these events drive cell division from M to G_1 to S to G_2 and then back to M.

GENOME DUPLICATION PRECEDES CELL DIVISION

It is axiomatic that genome duplication must precede cell division, but how this feat is accomplished is one of the marvels of evolution. It is based on a single principle: preRC assembly in eukarya occurs only in the absence of CDK activity. In other words, several of the proteins that comprise a preRC are subject to CDK-dependent phosphorylation events that prevent preRC assembly. Moreover, once the MCM DNA helicase has been loaded onto the genome, preRC assembly is complete, and the eight ORC, Cdc6, and Cdt1 proteins are no longer required until the next cell cycle. This allows the cell to inactivate these proteins as soon as DNA replication begins, thereby preventing any new initiation events from taking place before cell division can occur.

To accomplish this, the cell cycle is driven in one direction by a series of **feedback loops** in which the product of one biochemical reaction feeds back to either enhance or repress its own activity. Among the metazoa, these loops are centered about two enzymes, Cdk1 and Cdk2. Cdk1 regulates the $G_2 \rightarrow M$ transition, and Cdk2 regulates the $G_1 \rightarrow S$ transition. In single-cell eukarya, one CDK does both jobs by partnering with different cyclins (Table 12-3). Feedback loops occur when a change in the concentration or activity of a target protein is reinforced by the activity of the target protein itself. To illustrate this concept, consider a protein A that inhibits either the activity or production of protein B, which itself inhibits either the activity or production of A (Figure 12-6). The action of B on A constitutes a negative-feedback loop. Consider a protein D that stimulates either the activity or production of protein B, which itself stimulates either the activity or production of D. The action of B on D constitutes a positive-feedback loop. Thus, when protein D appears on the scene and increases either the amount or activity of B, B reinforces the action of D by stimulating D and by inhibiting its nemesis A, thereby further increasing the amount or activity of B while concomitantly shutting down the role of A. The net result is that the downstream targets or product of B will increase dramatically. Examples of feedback loops are seen during the transition from G_2 phase to M phase.

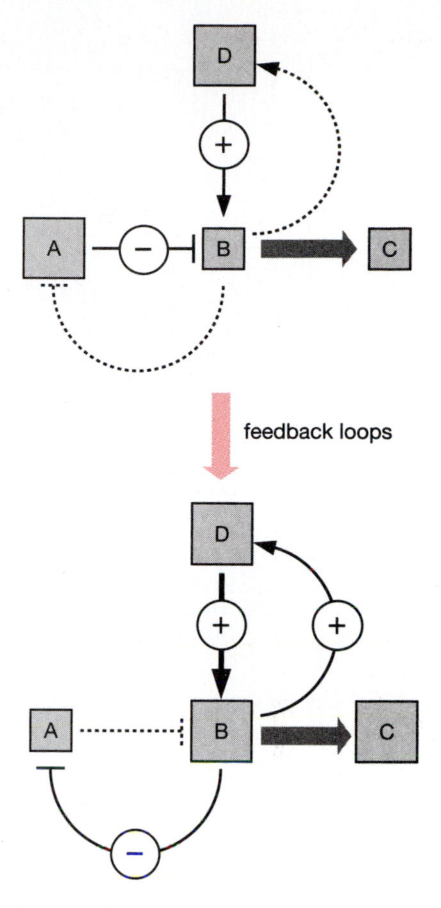

Figure 12-6. Positive- and negative-feedback loops. Protein A blocks either the production or activity of B which itself is an inhibitor of A. Protein D stimulates either the production or activity of B which itself is a stimulator of D. Thus, when D appears on the scene, the level of B and its downstream products or targets (represented as C) increases dramatically at the expense of A.

Driving The $G_2 \rightarrow M$ Transition

The $G_2 \rightarrow M$ transition requires two protein kinase activities, Cdk1 complexed with cyclin B (Cdk1•CcnB1) and with cyclin A2 (Cdk1•CcnA2). CcnB1 begins to accumulate during late S phase, forms a complex with Cdk1 (Cdk1•CcnB1), and this complex is localized to the nucleus during the $G_2 \rightarrow M$ transition. However, the ability of Cdk1 to phosphorylate proteins is itself stringently regulated (Figure 12-7).

Cyclins induce specific conformational changes in CDKs that expose their catalytic cleft and a conserved threonine between amino acids 160 and 170 to phosphorylation by the CDK-activating kinase (CAK). CAKs are expressed constitutively throughout the cell cycle, thereby maintaining all CDK•cyclins in an active form. CAKs are monomeric proteins in yeast, but heterotrimeric complexes of Cdk7, cyclin H, and Mat1 in metazoa. The Cdk1•CcnB1 complex, however, is maintained in an inactive state through Wee1 phosphorylation of Tyr-15, and Myt1 phosphorylation of both Tyr-15 and Thr-14 adjacent to the CDK catalytic site.

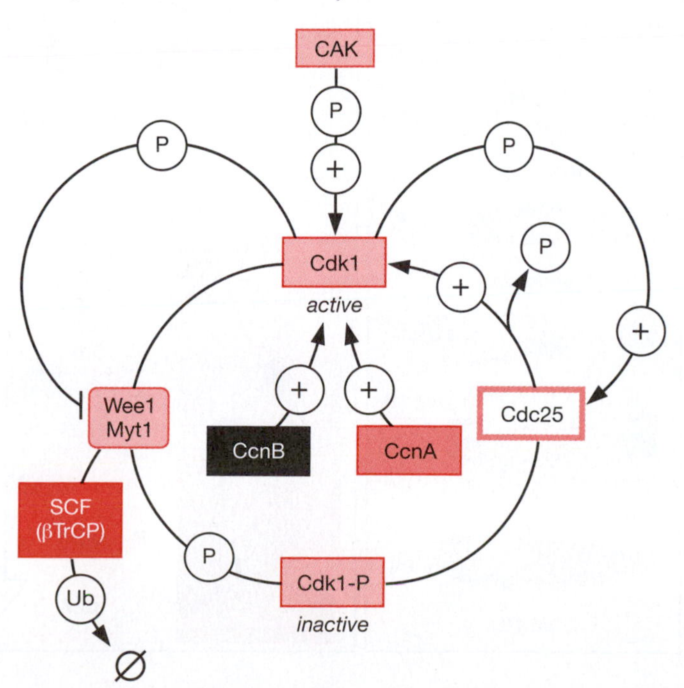

Figure 12-7. Activation of Cdk1, the engine that drives the $G_2 \rightarrow M$ transition. Arrows indicate protein phosphorylation (P), ubiquitination (Ub), or protein–protein associations that either activate (+→) or inhibit (⊥) the target. Polyubiquitinated proteins become substrates for degradation by the 26S proteasome (∅). In mammals, the protein phosphatase Cdc25 is primarily responsible for converting the inactive Cdk1-P enzyme into its active Cdk1 form. Colors are as in Figure 12-5. Species-specific analogs are provided in Table 12-3.

Prophase begins when the stockpile of Thr-14/Tyr-15-phosphorylated Cdk1 is abruptly activated by Cdc25-mediated dephosphorylation of Cdk1-P. As Cdk1•CcnB1 activity appears, it catalyzes its own activation by two feedback loops. Wee1 is phosphorylated by CDKs, facilitating its degradation by the ubiquitin ligases SCF(βTrCP) and SCF(Tome1). Conversely, Cdc25 is activated by multiple CDK phosphorylations. This allows Cdk1–CcnB1 to convert graded inputs into switch-like, irreversible responses once a critical portion of the enzyme becomes active. CDKs phosphorylate serine (S) and threonine (T) residues in proteins at [S/T]PX[K/R] where S/T is either serine or threonine, P is proline, X is any amino acid, and K/R is either lysine (K) or arginine (R). Although Cdc25C is frequently associated with the conversion of inactive Cdk1 to active Cdk1, Cdc25A appears to be the predominant isoform used throughout the cell cycle. Cdc25A functions at multiple cell-cycle transitions, and its protein level oscillates throughout the cell cycle. Moreover, Cdc25A is essential for embryonic development, whereas Cdc25B and Cdc25C are not.

Regulating preRC Assembly

Given the cell-cycle-dependent changes in CDK activities (Figure 12-5B), there is a single window of opportunity for assembly of preRCs during cell proliferation that occurs from anaphase to late G_1 (Table 12-4; Figure 12-8). After this window, Cdk2 is active from S through G_2, first in association with CcnE and then with CcnA. Prior to this window, Cdk1 is active from prophase through metaphase in conjunction with CcnB and CcnA. At metaphase, the mitotic cyclins A and B are targeted for ubiquitin-dependent degradation by APC(Cdc20), and Cdk1 is inactivated by Wee1- and Myt1-dependent phosphorylation. In addition, all eukaryotic cells produce at least one CKI

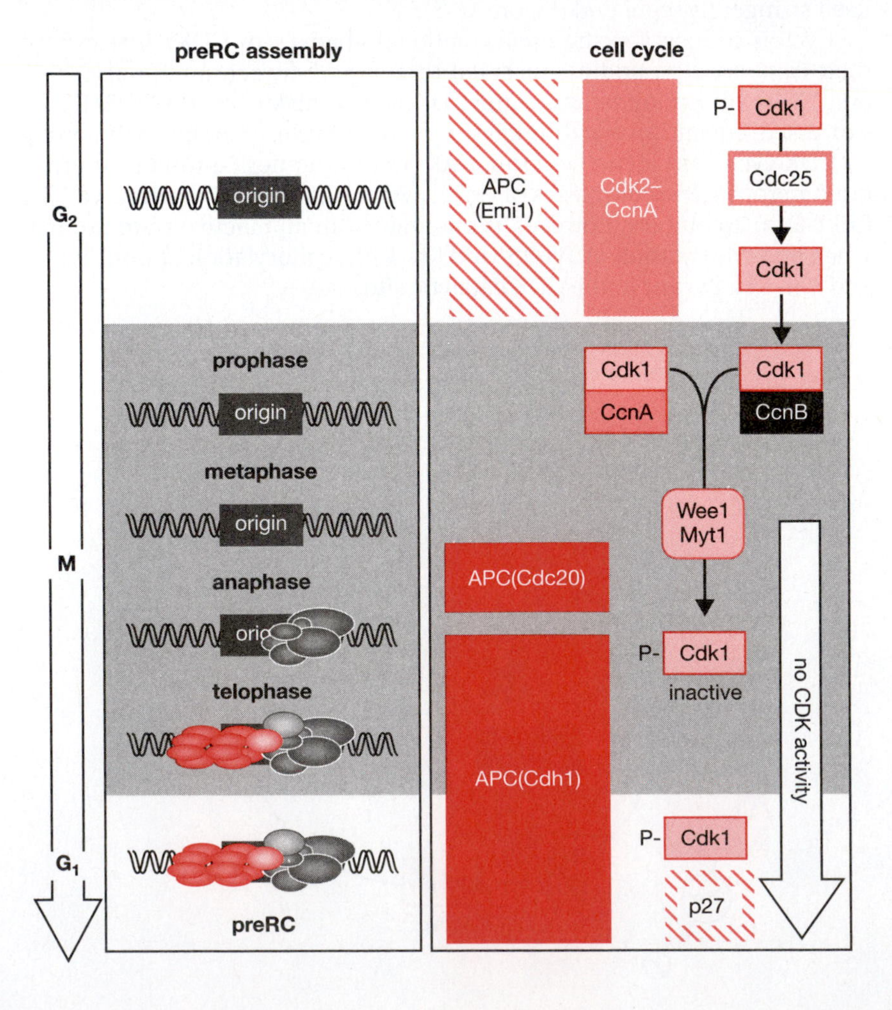

Figure 12-8. Regulation of preRC assembly in mammalian cells. Cdk1, which is maintained in an inactive phosphorylated form during most of the cell cycle, is converted into an active form by the protein phosphatase Cdc25 during G_2 phase. This allows formation of active Cdk1•mitotic cyclin complexes that drive cells into mitosis (M). The mitotic cyclins are degraded following APC-dependent ubiquitination, and Cdk1 is inactivated by Wee1 and Myt1 during the metaphase→anaphase transition. Suppression of CDK activity is enforced by the CKI p27 (red cross-hatch). The APC is present throughout cell division, but Emi1 (red cross-hatch) inhibits its activity during interphase. Emi1 is degraded as cells enter mitosis, and APC(Cdc20) assembly is activated by Cdk1•CcnA, CcnB. However, APC(Cdc20) activity is suppressed by the spindle-assembly checkpoint (Figure 13-19) until metaphase is complete. Upon inactivation of Cdk1, the APC changes its targeting subunit from Cdc20 to Cdh1 so that it can continue to function in the absence of CDK activity. Species-specific analogs are given in Table 12-3. Colors are as in Figure 12-5. preRC consists of ORC (dark gray), Cdc6 (light gray), Cdt1 (light red), and two MCM helicases (dark red) proteins.

during the M→G_1 transition (e.g. p27/Kip1 in mammals; Table 12-3). CKIs bind both subunits of the cyclin–Cdk complex distorting the active site and blocking the ATP-binding site, thus completely disrupting the catalytic function of the enzyme. The net result is a complete shutdown of CDK activity during the metaphase to G_1-phase transition.

preRC assembly during mitotic cell cycles begins with assembly of an ORC•chromatin complex (Figure 11-11) at or near the origin of bidirectional DNA replication (Figure 9-1). In single-cell eukarya such as yeast, ORC(1-6) binds to new replication origins as soon as they have been duplicated during S phase, and ORC•chromatin complexes that were assembled during the preceding G_1 phase remain intact. This means that, in yeast, replication origins are associated with ORC throughout cell division, including mitosis. This is not the case with metazoa.

In flies, frogs, and mammals, one or more of the ORC subunits is subject to cell-cycle-dependent modifications that can result either in the selective degradation of Orc1 (flies and mammals), or in the displacement of ORC from chromatin (frogs). Since Orc1 is essential for ORC assembly on chromatin, inactivation of Orc1 can account for the absence of ORC from metazoan chromosomes during early mitosis (prophase→metaphase). The status of ORC during G_2 is not clear. ORC cannot re-associate with chromosomes until the mitotic cyclins are degraded by the APC during the metaphase→anaphase transition, and CDK activity is suppressed. This permits the dephosphorylation of ORC subunits, and the assembly of ORC•chromatin complexes as early as anaphase. However, the rate at which ORC associates with chromatin during the M→G_1 transition, and therefore the rate of preRC assembly at replication origins, is determined by the cellular levels of ORC, and in particular by the level of Orc1. High levels of Orc1, such as those in cancer cells, drives association of ORC with chromatin. This means that the rate of preRC assembly can vary among different cell types as well as among different species of eukarya. Once ORC is bound to chromatin, preRC assembly can be completed, but the rate of preRC assembly will depend on the availability of the remaining preRC components. Chromatin-bound Mcm proteins appear as early as telophase in both fission yeast and mammals, but the amount of chromatin-bound Mcm protein increases steadily until late G_1. In mammals, it appears that formation of a nuclear membrane is a prerequisite to the appearance of functional preRCs.

Driving The M→G_1 Transition

The Cdk1 feedback loops (Figure 12-7) ensure that Cdk1 is inactive during interphase as a result of site-specific phosphorylation by the Wee1 and Myt1 protein kinases, an event that occurs during the M→G_1 transition (Figure 12-8). When the cell has completed S phase (i.e. entered G_2 phase), Cdc25 protein phosphatase and cyclin B translocate from the cytoplasm to the nucleus. Cdc25 removes the inhibitory phosphorylations from Cdk1. Then, the active form of Cdk1 partners with mitotic cyclins A and B to produce an active enzyme that then drives the cell into mitosis. Both Cdk1•CcnA and Cdk1•CcnB activities are required for the G_2→prophase→prometaphase→metaphase transitions.

Regulating APC activity

APC activity is present only during the anaphase→late G_1 period (Figures 12-5A and 12-8). Although the APC is present physically throughout S and G_2 phase, its activity is suppressed by association with the APC-specific inhibitor Emi1 (Figures 12-5C and 12-9) and by Cdk2-dependent phosphorylation of Cdh1 (Figures 12-5B and 12-10). Cdh1 is one of the two APC substrate-targeting proteins. Cdh1 is expressed throughout the mitotic cell cycle, but formation of APC(Cdh1) is prevented from S through

G_2 phase by Cdk2-dependent phosphorylation of Cdh1 (Figure 12-10). This targets Cdh1 for ubiquitination by SCF(Skp2) and then degradation by the 26S proteasome.

During the $G_2 \rightarrow M$ transition, Cdk1•CcnB activates the APC by phosphorylating several of its subunits, including the substrate-targeting protein Cdc20. At the same time, Emi1 is rapidly phosphorylated by the polo-like protein kinase (Plk1) whose activity appears during early mitosis. The phosphorylated form of Emi1 is then ubiquitinated by SCF(βTrCP) and degraded by the 26S proteasome. Destruction of Emi1 occurs during prometaphase shortly before CcnA degradation is initiated by APC(Cdc20), implying that Emi1 destruction is required for APC(Cdc20) activation. These two events, inactivation of Emi1 and CDK-dependent phosphorylation of APC(Cdc20), result in a fully active APC(Cdc20) that is prevented from acting by the spindle-assembly checkpoint until metaphase is completed (Chapter 13). APC(Cdc20) then targets cyclins A and B for ubiquitin-dependent proteolysis, thereby suppressing Cdk1 activity and triggering the metaphase→anaphase transition.

Suppressing CDK activity

To ensure that preRC assembly occurs after mitosis is finished, three separate pathways converge to suppress CDK activities from anaphase until late G_1. First, Wee1 and Myt1 phosphorylate Cdk1, converting it into a form that cannot be activated by cyclins (Figures 12-7 and 12-8). This prevents a return to the mitotic state, because Cdc25 activity will not reappear until another round of DNA replication has occurred. In addition, a CKI appears and inhibits any CDK•cyclin complexes it encounters (Figures 12-8 and 12-9).

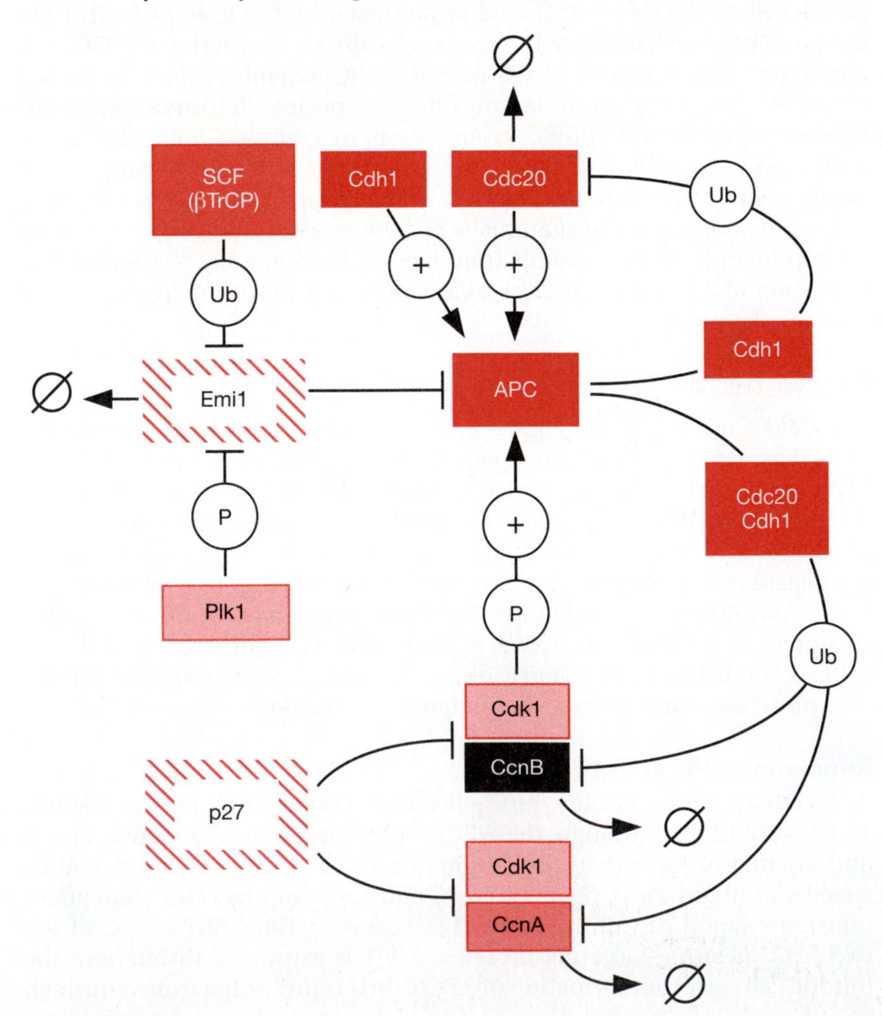

Figure 12-9. Regulation of Cdk1 activity in mammals by the APC and p27/Kip1. Cdk1•CcnB activates the APC by phosphorylating it. The APC inactivates Cdk1 by targeting mitotic cyclins for ubiquitin-dependent degradation. p27 inhibits CDK activities by binding to both the cyclin and Cdk1 components. Species-specific analogs are given in Table 12-3. Colors are as in Figure 12-5. Symbols are as in Figure 12-7.

Finally, with the disappearance of Cdk1 activity, APC(Cdc20) undergoes dephosphorylation with the consequence that Cdc20 (whose association with the APC is dependent on phosphorylation by a CDK activity) is replaced by Cdh1 (Figure 12-8). APC(Cdh1) then ubiquitinates Cdc20, resulting in its degradation by the 26S proteasome (Figure 12-9). Cdc20 is not expressed again until S phase. This completes a feedback loop that converts APC(Cdc20), an enzyme that functions only when CDK activity is high, into APC(Cdh1), an enzyme that functions only in the absence of CDK activity. Both forms of the APC target cyclins A and B for degradation. APC(Cdc20) triggers the metaphase→G_1 transition. APC(Cdh1) continues this transition and maintains the cell in G_1 by preventing expression of Skp2 and Dbf4, two proteins that are required to drive cells into S phase (Figure 12-10).

Driving The G_1→S Transition

The transition from G_1 to S phase is driven by the concerted action of two protein kinases, Cdc7•Dbf4 (often referred to as a Dbf4-dependent protein kinase, DDK) and Cdk2•CcnE. Both are required for preIC assembly (Figure 11-15), and both are regulated by similar, albeit independent, mechanisms. Both enzymes consist of one catalytic subunit and one regulatory subunit. The catalytic subunits are present at constant levels throughout the cell cycle, but the regulatory subunits are regulated by ubiquitin-dependent degradation (Figure 12-10).

Dbf4 regulates Cdc7•Dbf4 activity. During G_1, Dbf4 is absent, because it is targeted for degradation by the APC(Cdh1). As cells enter S phase, APC is inactivated by Cdk2•CcnE-dependent phosphorylation of Cdh1 and by the appearance of the APC-specific inhibitor Emi1. This stabilizes Dbf4 (which remains until metaphase) with concomitant assembly of preICs.

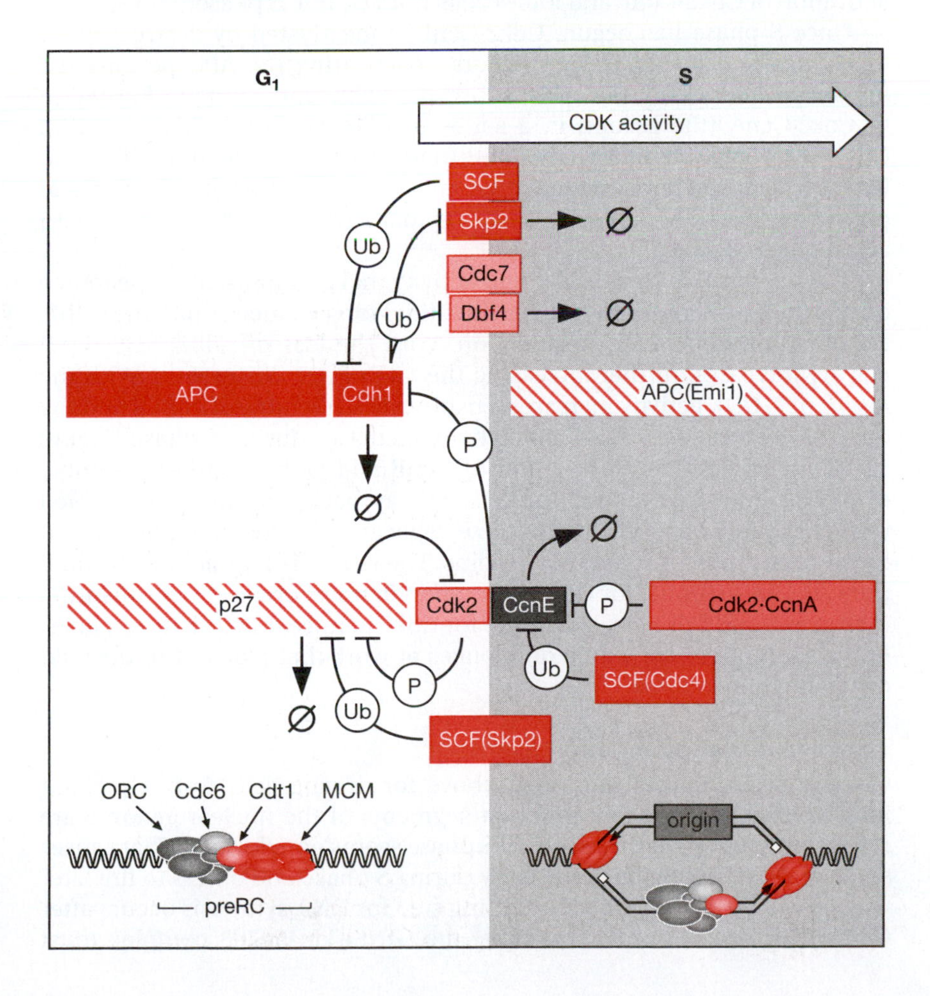

Figure 12-10. Regulation of the G_1→S transition in mammalian cells depends on feedback loops involving Cdk2 and Cdc7, the two engines that drive the G_1→S transition. preRC assembly is first detected as cells progress from anaphase to G_1 phase, the time when APC(Cdc20) and then APC(Cdh1) target cyclins A and B for ubiquitin-dependent degradation, and p27 ensures that Cdk1 and Cdk2 activities remain suppressed. Both APC(Cdh1) and APC(Cdc20) target Skp2 for degradation, thereby restricting SCF(Skp2) activity to S phase through prometaphase. During S phase, SCF(Skp2) targets a phosphorylated form of Cdh1 for destruction, thereby delaying APC(Cdh1) activity until Cdh1 appears again during late mitosis. Symbols are as described in Figure 12-7.

With two small variations, Cdk2 is converted into an active protein kinase by the same pathways that activate Cdk1 (Figure 12-7). The inactive form of Cdk2-P is activated by Cdc25A instead of Cdc25C, and the active form of Cdk2 utilizes cyclins E and A instead of A and B. CcnE is not a substrate for the APC, and therefore it can be expressed late during G_1 phase without being degraded (Figure 12-10). Cdk2•CcnE then phosphorylates the CKI p27, thereby making p27 a target for ubiquitination by SCF(Skp2) and subsequent degradation by the 26S proteasome. Loss of p27 results in increased levels of Cdk2•CcnE activity with concomitant activation of preIC assembly.

No way back

Cdk2•CcnE suppresses p27 activity by targeting p27 for ubiquitination by SCF(Skp2). However, during G_1 phase, APC(Cdh1) suppresses SCF(Skp2) activity by targeting Skp2 for ubiquitin-dependent degradation. Therefore, for SCF(Skp2) to carry out its attack on p27, APC(Cdh1) must be silenced. This occurs in two ways. First, Cdk2•CcnE phosphorylates Cdh1. Cdh1-P does not bind to the APC, and Cdh1-P is targeted by SCF(Skp2) and hence degraded. Second, the APC-specific inhibitor Emi1 appears with the onset of S phase (Figures 12-5C and 12-10) and shuts down the APC. These events increase the level of SCF(Skp2) activity so that it can eliminate p27 from the cell. This, in turn, increases the level of Cdk2•CcnE activity and completes the activation of preICs. It is the activation of preICs that generates DNA replication forks at replication origins, and it is the DNA synthesis at these replication forks that defines S phase. Thus, we have a feedback loop: Cdk2•CcnE facilitates inhibition of APC(Cdh1), which facilitates activation of SCF(Skp2), which facilitates inactivation of p27, which facilitates activation of Cdk2•CcnE and forces cells from G_1 into S phase!

Once S phase has begun, Cdk2•CcnE is inactivated by destruction of CcnE. Again, a feedback loop occurs. Inactivating the APC permits the appearance of CcnA (an APC substrate) with concomitant formation of Cdk2•CcnA (the second in a series of CDK activities; Figure 12-5B). Cdk2•CcnA now phosphorylates CcnE, thereby targeting it for ubiquitination by SCF(Cdc4) and hence degradation. In addition, Cdk2•CcnA continues to phosphorylate Cdh1 and p27, ensuring that neither the APC nor the CKI awaken prematurely.

The net result is that Cdk2•CcnE activity makes a transient appearance during the $G_1 \rightarrow S$ transition where it identifies targets, such as p27 and Cdh1, for ubiquitin-dependent degradation. With the loss of Cdh1, SCF(Skp2) activity increases, further increasing the rate of p27 destruction. With the loss of p27, CDK activity increases, ensuring the assembly of preICs and the onset of S phase. As Cdk2•CcnA activity increases during S phase (Figure 12-5B), it triggers the destruction of CcnE and with it further assembly of preICs. With one caveat, there is no way back; the newly assembled replication forks are left to complete replication of the genome without further activation of replication origins. The caveat is that neither cyclin E nor Cdk2 are essential for cell proliferation in mice; cyclin A can substitute for cyclin E and Cdk1 can substitute for Cdk2 in mice that lack these genes! Presumably, there are subtle differences at work that play out in the wild over many generations.

Temporal order of initiation events

Given the mechanism described above for driving cells from G_1 phase into S phase, the fact that different segments of the nuclear genome are replicated at different times during S phase seems paradoxical. What causes some replication origins to fire early during S phase and others to fire late? One possibility is that the rate-limiting step for DNA synthesis occurs after Cdk2•CcnE activation. For example, the GINS•Cdc45•Sld3 complex must bind to the activated preIC (Figure 11-15) before DNA unwinding by the GINS•Cdc45•MCM complex can occur (Figure 5-4). Alternatively, some replication origins may be activated by a different protein kinase. For example, Cdk1/Cdc28•Clb5 can activate both early- and late-firing origins in budding yeast, whereas Cdk1/Cdc28•Clb6 suffices for initiation at early origins but not at late origins. Finally, some origins may simply be more accessible to DNA replication proteins than others (e.g. euchromatin versus heterochromatin regions of the genome).

PREVENTING DNA RE-REPLICATION

DNA re-replication occurs when one or more of the normal controls that prevent preRC and preIC assembly during S phase are circumvented. New replication forks are assembled in regions that have already replicated before the first set of replication forks has completed duplication of the chromosome. This results in replication bubbles within replication bubbles (Figure 12-11). For example, DNA re-replication can be induced in some metazoan cells either by overexpression of Cdt1, or by suppression of the Cdt1-specific inhibitor geminin. Both changes promote loading of the MCM helicase at replication origins. DNA re-replication results in stalled replication forks and DNA damage, thereby triggering a DNA damage response that arrests cell proliferation (Figure 13-13). However, if the situation is not corrected, cells soon adapt to the arrest signal and trigger apoptosis. To prevent this catastrophe, eukarya contain several mechanisms that inactivate existing preRCs and prevent new preRC assembly until mitosis is complete and a nuclear membrane is present.

When eukaryotic cells enter S phase, several independent pathways converge to prevent DNA re-replication. These pathways are based on phosphorylation, ubiquitination, and protein–protein associations that inhibit either the assembly or function of preRCs (Figure 12-12). However, not all of these pathways are active in all eukarya, and different organisms may target some preRC proteins more stringently than others. Comparison of data from yeast, flies, frogs, and mammals reveals the presence of five different mechanisms that prevent DNA re-replication (Table 12-5). The

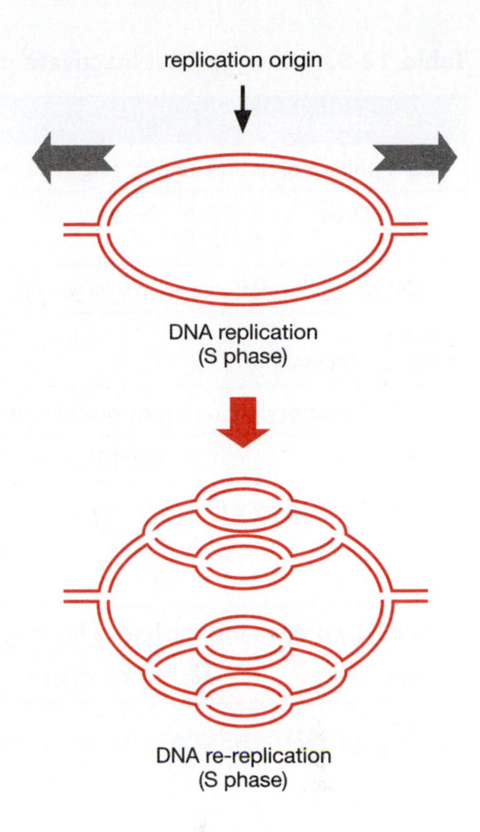

Figure 12-11. DNA re-replication is an aberrant event that occurs during S phase when replication origins are activated more than once.

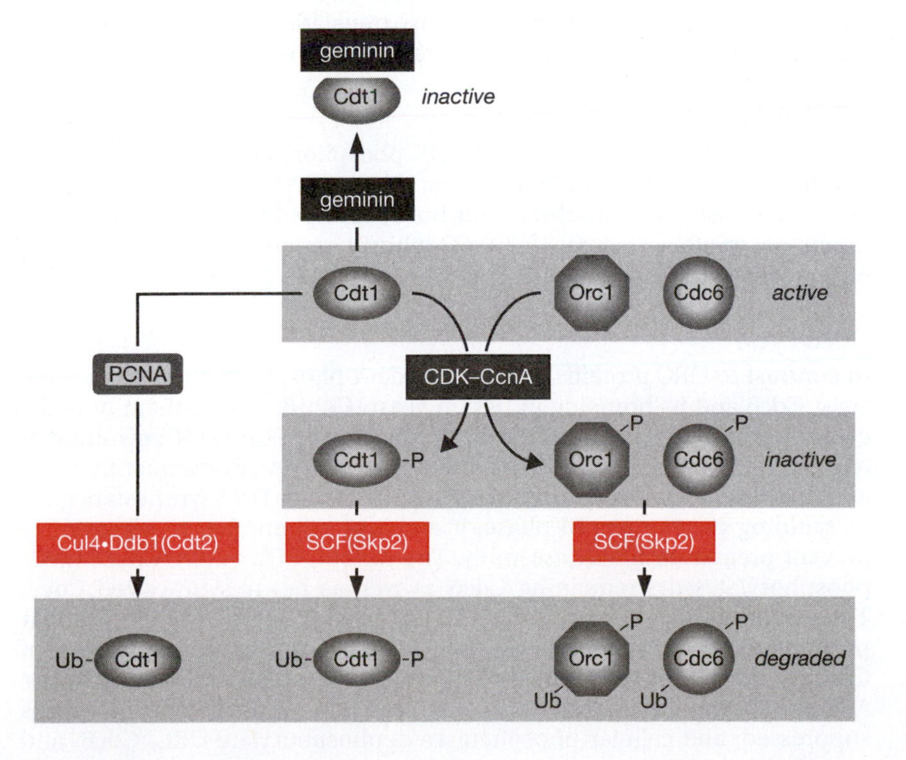

Figure 12-12. Pathways that prevent DNA re-replication in eukarya. In metazoa, geminin binds and inhibits Cdt1. In both single-cell and multicellular eukarya, CDK•CcnA analogs phosphorylate Cdt1, Orc1, and Cdc6, reducing their activities to varying extents. SCF can ubiquitinate the phosphorylated forms of Cdt1, Orc1, and Cdc6. Cul4 can ubiquitinate the unphosphorylated form of Cdt1 when Cdt1 is associated with the sliding clamp protein PCNA. Polyubiquitinated forms of Cdt1, Orc1, and Cdc6 are subject to degradation by the 26S proteasome.

Table 12-5. Pathways That Inactivate preRC Proteins

Budding yeast	Fission yeast	Flies	Frogs	Mammals
CDK phosphorylation inhibits target protein's function				
Orc2, Orc6	Orc2		Orc1, Orc2	Orc1 Mcm2, Mcm4
CDK phosphorylation exports target protein from the nucleus				
Cdc6 Mcm3, Mcm4			Cdc6	Cdc6
CDK phosphorylation targets protein for SCF ubiquitination				
Cdc6	Cdc18/Cdc6		Cdt1	Orc1, Cdt1
Ubiquitination by a non-CDK-dependent ubiquitin ligase				
		Orc1 by APC(Fzr)	Cdt1 by Cul4•Ddb1(Cdt2)	Cdt1 by Cul4•Ddb1(Cdt2)
Binding a protein that inhibits its function				
Cdc6⊢Clb2	ORC⊢Cdc13	Cdt1⊢geminin	Cdt1⊢geminin	Cdt1⊢geminin

An empty box simply indicates the absence of evidence for that mechanism in that organism.

goal, however, is the same: restrict genome duplication to once and only once each time a cell divides.

Single-Cell Eukarya

In single-cell eukarya such as budding yeast and fission yeast all six ORC subunits bind to new replication origins as soon as they are produced during S phase, and ORC•chromatin complexes assembled during the preceding G_1 phase remain intact. Thus, ORC remains associated with yeast replication origins throughout their cell cycle (including mitosis). How then does the cell prevent DNA re-replication? The solution lies in four convergent pathways that regulate post-translational modifications and subcellular localization of individual preRC proteins (Table 12-5).

ORC cycle

As cells enter S phase, the S phase CDK phosphorylates Orc2 and Orc6 in budding yeast and Orc2 in fission yeast. Phosphorylation of ORC subunits does not release ORC from chromatin, but does impede its ability to assemble functional preRCs (Figure 12-13). Dephosphorylation of ORC subunits occurs during the metaphase to anaphase transition (Figure 12-15).

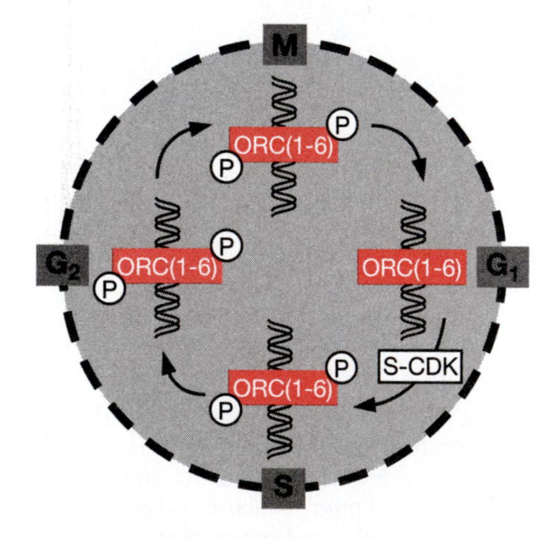

Figure 12-13. The ORC cycle in budding yeast. ORC function is suppressed when budding yeast enter S phase by Cdk1/Cdc28•Clb5,6 (S-CDK) phosphorylation of the Orc2 and Orc6 subunits of the ORC•chromatin complex. A similar pathway occurs in fission yeast. Shaded area represents the nucleus. Broken circle represents the nuclear membrane. Yeast nuclei do not break down during mitosis.

Cdc6 cycle

In contrast to ORC proteins, CDK-dependent phosphorylation of budding yeast Cdc6 and its homolog in fission yeast, Cdc18, targets them both for export to the cytoplasm, and for ubiquitination by SCF(Cdc4), resulting in its degradation by the 26S proteasome (Figure 12-14A). Consequently, Cdc6 and Cdc18 are remarkably unstable ($t_{1/2} \leq 5$ min) once DNA synthesis begins. As budding yeast enters M phase, it employs still another mechanism to prevent premature preRC assembly. The mitotic CDK (Cdk1/Cdc28•Clb2) phosphorylates the remaining Cdc6 as well as newly synthesized Cdc6. The resulting Cdc6-P forms a tight complex with mitotic cyclin Clb2 that prevents Cdc6 from assembling preRCs, and prevents ubiquitination of Cdc6 by the APC as cells pass through mitosis. As Clb2 is degraded by the APC during the metaphase→anaphase transition, Cdk1/Cdc28 activity is suppressed, and cellular phosphatases dephosphorylate ORC, Cdc6, and

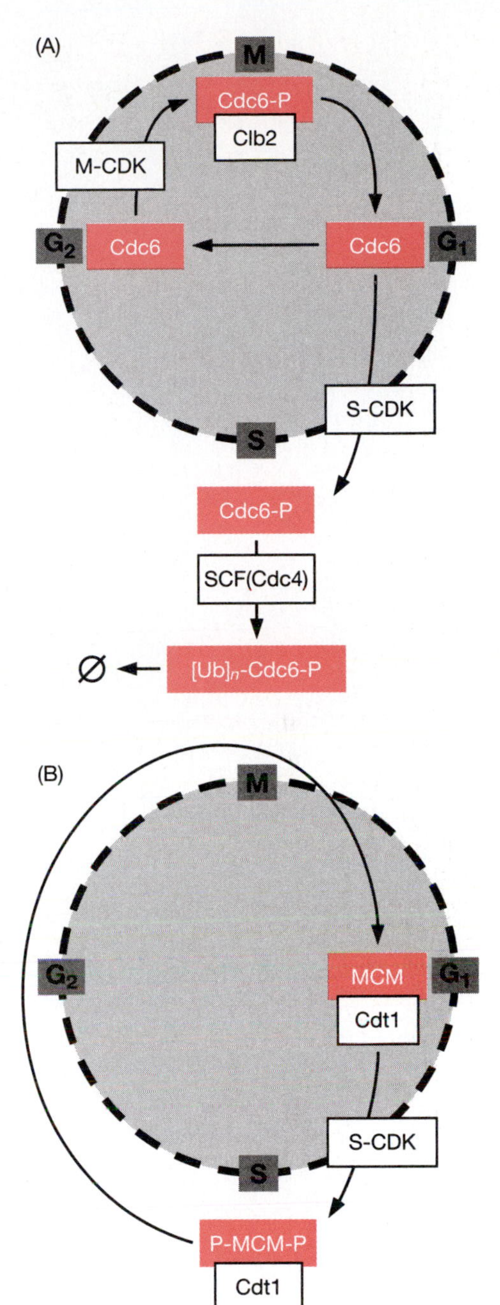

Figure 12-14. The Cdc6 and MCM cycles in budding yeast. (A) Cdc6 is phosphorylated by Cdk1/Cdc28•Clb5,6 (S-CDK) as cells enter S phase and by Cdk1/Cdc28•Clb2 (M-CDK) as cells enter mitosis. Inactivation of Cdc6-P results from three subsequent events: export of Cdc6-P from the nucleus (shaded area), ubiquitination-dependent degradation of Cdc6-P by the 26S proteasome (Ø), and formation of a Clb2•Cdc6-P complex. (B) MCM subunits also are phosphorylated as cells enter S phase. The phosphorylated MCM helicase localizes to the cytoplasm as soon as it is no longer bound to chromatin. Cdt1 is bound to MCM and thereby exported along with it. Dephosphorylation of both Cdc6 and MCM occurs during the M→G_1 transition with concomitant nuclear localization of MCM and Cdt1. Shaded area represents the nucleus. Broken circle represents the nuclear membrane.

MCM proteins. In this way, preRC assembly is prevented during cell division until the genome has been duplicated and mitosis is complete, and at least some preRCs can be assembled as soon as cells exit mitosis (Figure 12-15).

MCM cycle

In contrast to Cdc6, Cdt1 protein levels remain essentially the same throughout the yeast cell cycle. However, the fates of the helicase loader Cdt1 and its cargo, the MCM helicase, are closely linked by virtue of the fact that free MCM is bound to Cdt1. In budding yeast, the Cdt1•MCM helicase complex enters the nucleus at the end of mitosis, remains nuclear during G_1 phase, and is then transported to the cytoplasm as MCM helicase is released from replication forks (Figure 12-14B). The accumulation of MCM in the nucleus during G_1 does not require Cdc6, and therefore does not require loading the helicase onto chromatin. Neither Cdt1 nor MCM can localize to the nucleus alone. Export of Cdt1•MCM requires CDK-dependent phosphorylation of at least two subunits, Mcm3 and Mcm4.

Multicellular Eukarya

Although the same sequence of events ensures that genome duplication in both yeast and mammals always precedes cell division, these organisms emphasize different pathways to restrict genome duplication to once per cell division. Whereas yeast rely primarily on regulating Cdc6 activity to prevent premature preRC assembly, metazoa rely primarily on regulating the activities of ORC and Cdt1. Whereas ORC remains intact and tightly bound to chromatin throughout the yeast cell cycle, metazoa release ORC from chromatin during the G_2→M transition. Whereas yeast do not modify Cdt1 during cell division, metazoa phosphorylate it, ubiquitinate it, and bind it to a specific inhibitor termed geminin. Metazoa employ

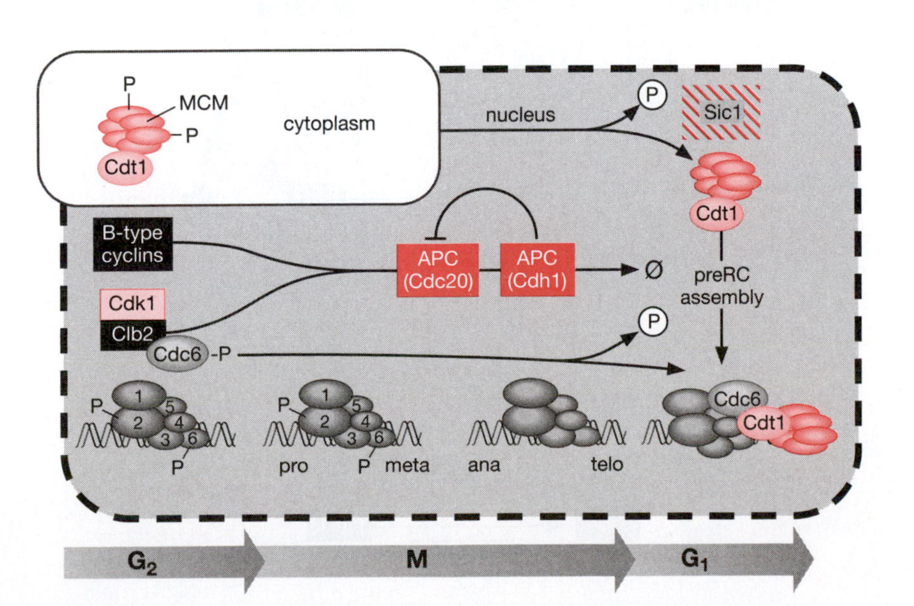

Figure 12-15. preRC assembly in budding yeast is coordinated with completion of mitosis. Dephosphorylation of preRC proteins is a consequence of the suppression of Cdk1/Cdc28 activity by APC-mediated ubiquitination of cyclins during the metaphase to anaphase transition, and the concomitant appearance of the budding yeast CKI, Sic1. ORC subunits are indicated by numbered gray ovals. Only Orc2 and Orc6 are phosphorylated during S phase in *S. cerevisiae*.

five convergent pathways (Figure 12-12; Table 12-5) to ensure that DNA re-replication does not occur.

ORC cycle

As mammalian cells enter S phase, the association between Orc1 and the remaining ORC•chromatin complex appears to change by virtue of the fact that Orc1 can be eluted from cells under conditions where Orc2 and other ORC(2-5) core proteins cannot (Figure 12-16). Moreover, Orc2 can be cross-linked to DNA in S-phase cells, whereas Orc1 cannot. This change may be induced by DNA unwinding, and it may account for the fact that Orc1 can now be ubiquitinated selectively by SCF(Skp2), the same ubiquitin ligase that targets p27 for destruction. Monoubiquitinated Orc1 has been detected in some mammalian cells, and this modification can cause Orc1 to be localized to the cytoplasm, which would prevent it from initiating preRC assembly. In human cancer cells, Orc1 is targeted for ubiquitin-dependent degradation when cells enter S phase, a reaction that depends on site-specific binding of cyclin A to Orc1. Orc1 is subject to phosphorylation by Cdk2•CcnA and targeted for ubiquitination by SCF(Skp2). Orc1 reappears during the $G_2 \rightarrow M$ transition where it is associated with Cdk1•CcnA, the enzyme responsible for hyperphosphorylation of Orc1 in mitotic cells. At this point, none of the ORC subunits in mammalian cells are associated with chromatin. However, they rapidly re-associate with chromatin during anaphase as Cdk1 activity is suppressed by the mechanisms described above (Figures 12-7, 12-8, and 12-9). Mutating serine and threonine residues in Orc1 to aspartic acid in order to mimic protein phosphorylation prevents Orc1 from binding to DNA *in vitro* and to chromatin *in vivo*.

Embryos from the fruit fly *Drosophila* also selectively target Orc1 for ubiquitin-dependent proteolysis, except that it is carried out by APC(Fzr/Cdh1) as cells exit mitosis instead of SCF(Skp2) as cells enter S phase. Orc1 reappears during G_1 phase, and is then hyperphosphorylated during the $G_2 \rightarrow M$ transition, thereby releasing ORC from chromatin. As with mammalian cells, ORC re-associates with *Drosophila* chromatin during

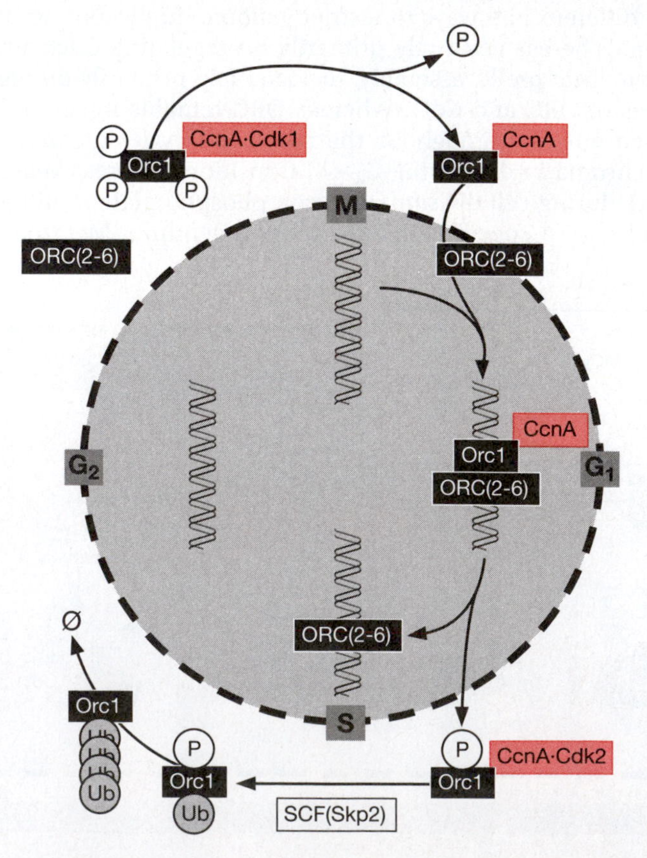

Figure 12-16. The mammalian ORC cycle. In contrast to yeast, mammalian ORC is systematically displaced from chromatin as cells enter S phase and transit into mitosis. This process begins with as yet undefined changes in the ORC•chromatin complex that allows targeting of Orc1 for ubiquitin-dependent degradation. This requires site-specific binding of cyclin A to Orc1, and apparently CDK phosphorylation of Orc1 converts it to a substrate for the SCF ubiquitiin ligase. During the $G_2 \rightarrow M$ transition, Orc1 is hyperphosphorylated and the remaining ORC subunits are displaced from chromatin. As with yeast, the ability of ORC to bind to chromatin is restored during the anaphase→G_1 transition.

anaphase. Cleavage-stage embryos from the frog *Xenopus* do not appear to ubiquitinate ORC, but they do hyperphosphorylate Orc1 and Orc2 as cells enter mitosis, thereby releasing ORC from chromatin. As cells transit from metaphase to anaphase, cell-cycle-dependent inactivation of Cdk1 allows dephosphorylation of ORC and rebinding to DNA. Taken together, it is clear that ORC activity is regulated in all eukarya, although the precise mechanism can vary from one species to another.

The fact that ORC is released from chromatin during the G_2→metaphase transition, and then rebinds to chromatin during the anaphase→G_1 transition means that cells could change both the number and locations of replication origins each time they divide (Figure 12-18). This would allow them to change the pattern of replication origins to accommodate changes in the pattern of gene expression during cell differentiation and animal development. In fact, the cellular level of Orc1 determines the rate at which ORC binds to chromatin when human cells exit mitosis, and the amount of DNA replication that occurs in cells during *Drosophila* development.

Regulation of Cdc6 activity

Cdc6 is not a primary target for regulation in metazoa. Both nematode and human Cdc6, like yeast Cdc6, are subject to CDK-dependent phosphorylation (in this case, by Cdk2•CcnA), but in contrast to yeast Cdc6, neither nemotode nor human Cdc6-P are degraded during S phase, although it can be exported to the cytoplasm (Figure 12-17). However, the amount exported appears to depend on the cellular concentration of Cdc6, suggesting that nuclear export is used primarily to prevent an excess of Cdc6-P from accumulating in the nucleus.

Human Cdc6 is targeted by APC(Cdh1) for ubiquitin-mediated proteolysis during the M→G_1 transition. Thus, while the half-life of human Cdc6 protein is 2-3 h in asynchronously proliferating cells, its half-life in G_1 cells is 15–30 min. Intriguingly, Cdc6-P is not an APC substrate (Figure 12-18). Therefore, CDK phosphorylation of human Cdc6 actually protects it from degradation rather than targeting it for destruction. As soon as Cdc6-P is dephosphorylated during the anaphase→G_1 transition, Cdc6 is either incorporated into preRCs, or it is destroyed by APC(Cdh1). This limits the number of preRCs that can be assembled as cells exit mitosis.

Regulation of Cdt1 activity

Cdt1 is not an APC substrate. Therefore, Cdt1 levels during G_1 are unlikely to be rate limiting in preRC assembly. However, once DNA synthesis begins,

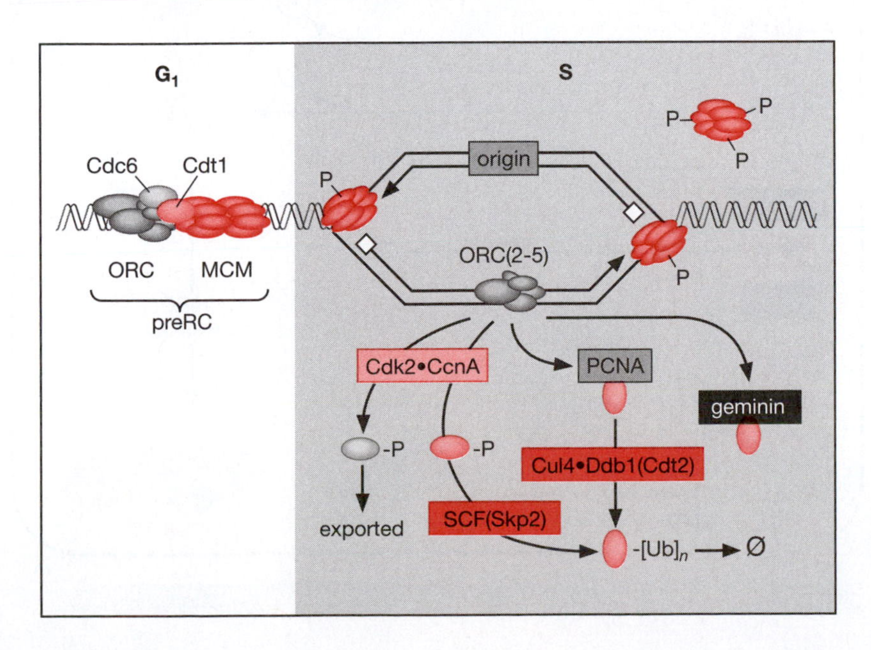

Figure 12-17. Regulation of MCM helicase loading in metazoa. In addition to the ORC cycle (Figure 12-16), mammals and other metazoa inactivate Cdt1 activity in their nuclei by three independent mechanisms. Cdt1 is degraded by a CDK-dependent, SCF-dependent pathway, as well as by a CDK-independent, Cul4•Ddb1-dependent pathway, and geminin inhibits Cdt1 binding to the MCM helicase. In addition, the same enzyme that phosphorylates Cdt1 (Cdk2•CcnA) also phosphorylates Cdc6 and Orc1, thereby modifying three proteins essential to loading the MCM helicase onto chromatin. Mcm proteins also are phosphorylated during S phase.

Cdt1 is rapidly inactivated by three independent pathways (Figure 12-17), marking it as a primary target for preventing DNA re-replication during S phase. First, it is phosphorylated by Cdk2•CcnA which makes it a substrate for ubiquitination by SCF(Skp2) and degradation by the 26S proteasome. Thus, the same ubiquitin ligase that targets p27 and Orc1 also targets Cdt1. In addition, unphosphorylated Cdt1 is targeted by a different ubiquitin ligase, Cul4•Ddb1(Cdt2), that is involved in DNA repair as well as cell proliferation. In this case, Cdt2 is the substrate receptor (analogous to an F-box protein) that targets Cul4•Ddb1 to Cdt1. Cdt2 is recruited to replication forks via Cdt1 and chromatin-bound PCNA. Mutant Cdt1 proteins that cannot be phosphorylated by CDK and therefore cannot interact with Skp2 are still degraded normally at the onset of S phase by the Cul4•Ddb1(Cdt2) pathway. Therefore, Cdt1 is degraded during cell proliferation by both CDK-dependent and CDK-independent pathways. In the third pathway, Cdt1 is both inactivated and stabilized by binding to geminin, a 33-kDa protein that binds specifically to Cdt1. In doing so, geminin interferes with Cdt1 binding to the MCM helicase without preventing Cdt1 binding to DNA. In addition, geminin protects Cdt1 from ubiquitination and therefore degradation. Since geminin is selectively targeted by the APC for ubiquitination and subsequent degradation, that fraction of Cdt1 bound to geminin during S phase will become available for preRC assembly upon completion of mitosis (Figure 12-18).

Which of these three pathways constitutes the principal regulatory mechanism for Cdt1 activity can be both species-specific and cell-type specific. In *C. elegans*, Cul4•Ddb1(Cdt2) negatively regulates Cdt1 to prevent DNA re-replication during S phase. In human cancer cells, the ratio of geminin to Cdt1 is the principal mechanism that prevents re-replication; but, in normal mammalian cells both geminin and cyclin A must be suppressed in order to induce DNA re-replication. Therefore, normal mammalian cells suppress DNA re-replication by at least one cyclin A-dependent pathway as well as by geminin.

MCM helicase modifications

MCM proteins are released from chromatin during S phase, presumably a consequence of replication-fork termination. However, in contrast to budding yeast, Mcm(2-7) in both fission yeast and metazoa are not

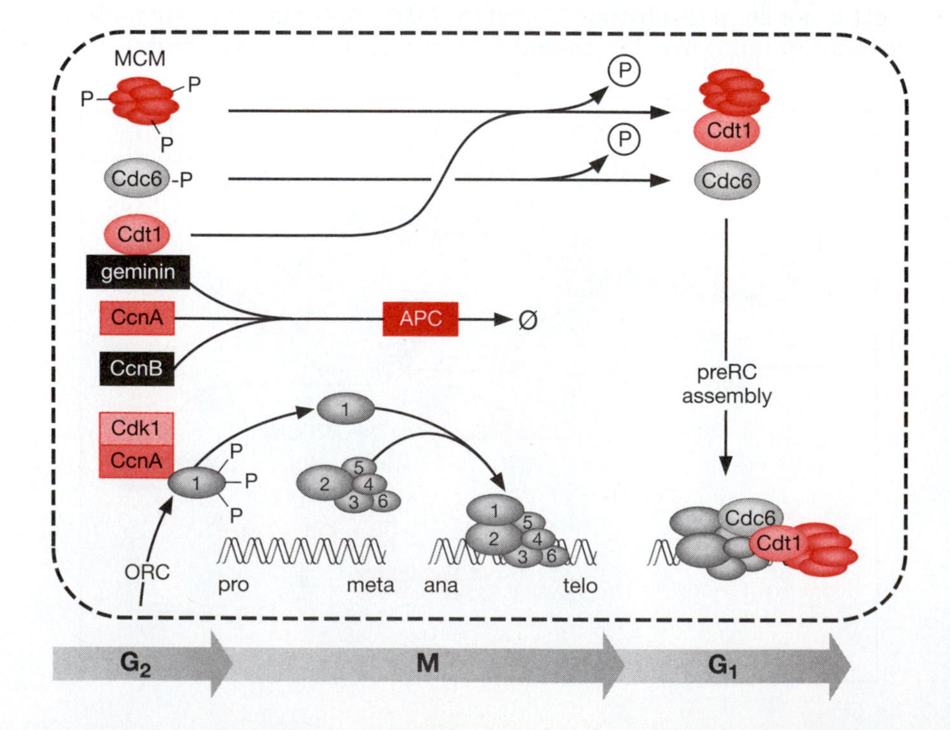

Figure 12-18. Coordination of preRC assembly with completion of mitosis in mammals. As in yeast, dephosphorylation of preRC proteins in mammals is mediated by the APC. Therefore, preRC assembly is delayed until the anaphase→G_1 transition. Ubiquitin-dependent degradation by the 26S proteasome is indicated as →Ø. Phosphorylation at one or more sites is indicated by -P. Mcm4 is hyperphosphorylated in metaphase-arrested cells and may require dephosphorylation at some sites and phosphorylation at others before preRC assembly.

exported from the nucleus into the cytoplasm. Instead, they remain in the nucleus throughout interphase. Although both Mcm2 and Mcm4 in the chromatin-unbound fraction become phosphorylated during S phase, and Mcm4 is hyperphosphorylated in metaphase-arrested cells, the evidence that these events help prevent DNA re-replication is circumstantial. Cdk2•CcnA phosphorylation of Mcm4 *in vitro* inhibits Mcm(4,6,7) helicase activity, and premature expression of Cdk2•CcnA in G_1-phase cells inhibits preRC assembly. Cdk1•CcnB appears to be responsible for hyperphosphorylation of Mcm4 in metaphase-arrested cells, where MCM is not associated with chromatin (Figure 12-18). While such results suggest that CDK phosphorylation of MCM contributes to preventing premature preRC assembly, the primary reason that MCM helicase does not rebind to chromatin during S phase can be attributed to the modifications of Orc1, Cdc6, and Cdt1 described above. Since the intracellular levels of Mcm proteins are generally at least 10-fold greater than other preRC proteins, the levels of Orc1, Cdc6, or Cdt1 must limit the rate at which the MCM helicase can be loaded onto chromatin. Overexpression either of stable, biologically active, Cdc6 mutants, or of wild-type Cdc6 does not induce DNA re-replication, but overexpression of Cdt1 alone does. Therefore, at least in some cell types, Cdt1 is the rate-limiting component for preRC assembly. These include human cells derived from malignant cancers, *Drosophila* embryos, *Xenopus* eggs, and some plant cells.

PARALLEL PATHWAYS

All of the pathways summarized in Figure 12-12 contribute to preventing DNA re-replication in eukarya, although the contribution of any one mechanism clearly varies among species and even among different cell types. For example, inhibition of CDK activity in fission yeast inhibits mitosis and induces a second round of DNA replication, but the comparable condition in budding yeast inhibits mitosis without inducing DNA replication. All three components – Cdc6, MCM, and ORC – must be deregulated in budding yeast in order to induce substantial re-replication. This was demonstrated by providing the cells with a Cdc6 protein that cannot be degraded, along with an Mcm protein to which a nuclear localization signal has been added, and mutant forms of Orc2 and Orc6 that lack CDK phosphorylation sites. Only when regulation of two of the three preRC components had been suppressed did DNA re-replication occur in budding yeast. A similar story exists in mammalian cells where suppression of geminin expression can induce DNA re-replication in human cancer cells but not in normal human cells. However, if expression of both geminin and cyclin A are suppressed, then normal as well as cancer cells can be forced to re-replicate their DNA. Similarly, over-expression of Cdc6 and Cdt1 can force DNA re-replication in some metazoan cells, but not in others. These examples reveal that multiple regulatory pathways prevent DNA re-replication.

Sometimes different species appear to regulate cell division by different mechanisms, when in fact, the same basic mechanism is simply applied to different DNA replication proteins, thereby creating distinct but parallel regulatory pathways. For example, Cdc6 and Cdt1 are both essential for preRC assembly in eukarya, but in budding yeast, only Cdc6 is targeted for CDK-dependent phosphorylation and ubiquitin-dependent proteolysis, whereas in mammals, the same two regulatory pathways target Cdt1 instead of Cdc6. Moreover, in budding yeast Cdc6 is both inhibited and protected from proteolysis by Clb, a B-type cyclin, whereas in mammals Cdt1 is both inhibited and protected from proteolysis by geminin, a protein unique to metazoa. As yeast and mammalian cells exit mitosis, both Clb and geminin are targeted for degradation by APC-dependent ubiquitination, thereby allowing preRC assembly under conditions of low CDK activity.

Another example is the regulation of Cdc6/Cdc18 in yeast, and the parallel regulation of Orc1 in mammalian cells. Both proteins are essential to preRC assembly. Both proteins are targets of CDK-dependent phosphorylation and ubiquitin-dependent proteolysis. Both proteins are released from chromatin after completing their role in preRC assembly, and both are responsible for initiating assembly of preRCs during the M→G_1 transition. Different targets in different species are regulated by essentially the same sequence of events.

FUNCTIONAL REDUNDANCY

Functional redundancy simply means that the function of one gene can be carried out by another gene. For example, the DNA Pol-ε catalytic activity can be replaced by that of DNA Pol-δ, although the physical presence of Pol-ε is still required for assembly of a preIC (Figure 11-15). Such gene swaps reveal the presence of functionally redundant proteins; proteins that normally serve a specific function in wild-type organisms, but that can substitute for another protein of similar function, if necessary. Functional redundancy appears to have exploded with the evolution of metazoa. For example, of the three Cdc25 genes in multicellular eukarya, only one is essential. Any two of the three can be deleted in mice and embryos will still produce a viable, fertile adult animal. Of the four mammalian CDKs that regulate cell division, only Cdk1 is required throughout animal development; none of the other CDKs can substitute for Cdk1. Thus, Cdk1 can substitute for Cdk2 in cells lacking a functional Cdk2 gene, but Cdk2 cannot substitute for Cdk1. This eliminates the danger that cells expressing high levels of Cdk2•cyclin activities might accidentally enter mitosis before completing genome duplication. Thus, mouse embryos lacking Cdk2 can develop from a fertilized egg into a viable adult, and cells derived from these animals continue to proliferate *in vitro*.

Inactivation of either cyclin E1 or cyclin E2 does not prevent embryos from developing into viable, fertile adults, but inactivation of both genes does prevent embryonic development. CcnE is required only during the G_0→S transition and during endocycles, because CcnA can substitute for CcnE during cell proliferation. Again, this is analogous to budding yeast where cyclins Clb5 and Clb6 are required for timely DNA replication. Cells lacking both of these cyclins enter the cell cycle normally except that the initiation of DNA synthesis is delayed and now depends on the 'mitotic cyclins' Clb1 to Clb4 (Table 12-3). However, functional redundancy among the cyclins is not universal. Mice lacking CcnB2 develop normally and are fertile whereas mouse embryos lacking CcnB1 die *in utero*. Inactivation of the CcnA2 gene results in embryonic lethality shortly after implantation. Although loss of CcnA1, whose expression is normally restricted to **germ cells**, results in some tissue- and gender-specific defects, CcnA1 cannot substitute for the loss of CcnA2. Thus, CcnA2 appears to play a unique role in the G_1→S and G_2→M transitions as well as during DNA replication.

Functional redundancy allowed the evolution of multicellular organisms. As cells differentiated in order to perform a myriad of specialized tasks, the functions of particular genes were adapted to these tasks through gene duplication with mutations arising in the duplicated copy that could be selected as required to carry out new tasks. An additional advantage to functional redundancy is that complex eukaryotes contain backup mechanisms for essential functions, such as cell proliferation, in the event that something goes wrong with the primary pathway.

DEVELOPMENTALLY PROGRAMMED POLYPLOIDY

All cell cycles are not the same. Most cell proliferation in eukaryotic organisms employs the mitotic cell cycle described above, but early phases of embryonic development often require more rapid cell divisions and employ the cleavage cell cycle where Gap phases are too brief to measure (Table

12-1). However, these too use the same regulatory pathways employed by mitotic cell cycles and produce the same diploid cells containing a single nucleus with two complete copies of the genome (2N). Other cell cycles, however, have been modified in order to produce polyploid cells.

Cells in G_2 or M phase are tetraploid, because they contain 4N DNA. Cells with greater than 4N DNA are termed polyploid. Polyploidy results either from DNA re-replication (Figure 12-11), an aberration in cell division, or as a normal part of animal and plant development. DNA re-replication can be distinguished from developmentally regulated polyploidy in two ways. First, DNA re-replication produces cells whose DNA content varies continuously from 2N to much greater values, whereas polyploidy produces cells in which the DNA content is an integral multiple of the normal diploid DNA content (e.g. 4N, 8N, 16N, 32N, etc.). Second, DNA re-replication triggers the DNA damage response (Figure 13-13) followed by apoptosis (programmed cell death). In contrast, terminally differentiated polyploid cells remain viable for long periods of time, because they disconnect the link between DNA damage sensors and apoptosis.

Natural polyploidy is especially common among ferns and flowering plants. It also occurs in arthropods (e.g. flies) and vertebrates (e.g. fish, and salamanders), but it is rare among mammals. In contrast to DNA re-replication, developmentally programmed polyploidy is the result of multiple S phases in the absence of cytokinesis under conditions that prevent apoptosis. Such cells are terminally differentiated; they grow in size, but they no longer proliferate. These cells arise by one of four mechanisms (Table 12-6).

Acytokinetic mitosis is a mitotic cell cycle without cytokinesis. A well-characterized example is the syncytial blastoderm produced at the beginning of invertebrate development where *Drosophila*, for example, undergoes 10 rapid cleavage cell cycles in the absence of cytokinesis to produce a single cell (syncytium) filled with 1024 diploid nuclei. A more modest example occurs during postnatal liver development with the gradual appearance of tetraploid and octoploid hepatocytes containing one or two nuclei. These cells undergo mitosis (separate their homologous chromosomes into two groups) but fail to form a cleavage furrow in the center of the cell and undergo cytokinesis. The block to cytokinesis is unknown.

Cell fusion occurs during myogenesis. Mononucleated myoblasts withdraw from the cell cycle when they express the CDK-specific inhibitors p21 and p57 during G_1 phase. This event is linked directly to expression of MyoD, one of the earliest myogenic regulatory proteins associated with muscle differentiation. MyoD induces expression of p21 and p57 which then prevent CDK-dependent phosphorylation of MyoD, thereby preventing its ubiquitin-dependent degradation, as well as preventing the myoblasts from entering S phase. This induces muscle-specific gene expression followed by cell fusion to form multinucleated myofibers.

Endoreduplication occurs when a cell undergoes multiple S phases without an intervening mitosis and without undergoing cytokinesis. The result is a giant cell with a single giant nucleus. Examples of endoreduplication are found among protozoa, arthropods, mollusks, and plants. The clearest example of endoreduplication in mammals occurs during differentiation of trophoblast stem cells into the trophoblast giant cells that are required for implantation of blastocysts into the uterine endothelium and placental development. The DNA content of these giant cells generally ranges from 8N to 64N, although levels as high as 1000N have been reported.

Endomitosis is similar to endoreduplication, except that endoreduplication results from arresting cells in G_2 before they enter mitosis, whereas endomitosis results from arresting cells after they begin mitosis, but before they complete mitosis. The clearest example of endomitosis occurs in the bone marrow when megakaryoblasts differentiate into megakaryocytes, the cells that give rise to blood thrombocytes (platelets).

Table 12-6. Eukaryotic Cell-Cycle Variations And Their Effect On Ploidy

Cell cycle	Examples	Sequence of events	Cell anatomy
Mitotic cell cycle	Most somatic cells	$G_1 \rightarrow S \rightarrow G_2 \rightarrow M \rightarrow G_1$	2N 2N
Cleavage cell cycle	Early embryos of amphibians, fish, and echinoderms	$S \rightarrow M \rightarrow S \rightarrow M \rightarrow S$	2N 2N 2N 2N
Acytokinetic mitosis	Fly blastoderm, slime mold plasmodium	$S \rightarrow M(-C) \rightarrow S \rightarrow M(-C) \rightarrow S$	2N 2N
	Liver hepatocyte development	$G_1 \rightarrow S \rightarrow G_2 \rightarrow M(-C) \rightarrow G_1$	
Cell fusion	Skeletal muscle Myoblast→myotube Syncytiotrophoblasts	$G_1 \rightarrow S \rightarrow G_2 \rightarrow M \rightarrow G_0 \rightarrow$ fuse	2N 2N
Endoreduplication	Placenta Trophoblast stem cell→giant cell	$G_1 \rightarrow S \rightarrow G_2 \rightarrow S \rightarrow G$	4N
Endomitosis	Bone marrow Megakaryoblast→megakaryocyte	$G_1 \rightarrow S \rightarrow G_2 \rightarrow M^* \rightarrow S \rightarrow G \rightarrow M^*$	2N 2N

M(–C) indicates mitosis in the absence of cytokinesis. M* indicates anaphase arrest.

The result is a single giant cell containing a single multilobulated nucleus with a genome content of up to 64N. Each lobe presumably contains a diploid genome. With time, individual lobes may separate from one another to produce a multinucleated cell.

Initiating Endocycles

Endocycles are multiple rounds of nuclear DNA replication without an intervening mitosis. Endocycles begin in most, perhaps all, cells whenever they are arrested in G_2 phase under conditions that permit assembly and subsequent activation of preRCs. However, most eukaryotic cells eventually induce apoptosis under these conditions. What is unique about cells that are developmentally programmed to endoreduplicate their genome is that they have disconnected the normal linkage between the DNA damage-response pathways and apoptosis.

In the fission yeast *S. pombe*, for example, either overexpression of the CKI Rum1 or deletion of the mitotic cyclin Cdc13 induces repeated rounds of genome duplication to produce enormous polyploid cells. Both conditions inhibit mitotic CDK (Cdk1/Cdc2•Cdc13) activity. Analogous results have also been demonstrated in the budding yeast *S. cerevisiae*, but only after it was realized that *S. cerevisiae* S-phase cyclins (Clb5, Clb6) and M-phase cyclins (Clb1 to Clb4) are functionally redundant. When Clb5 and Clb6 are absent, then Clb1, 2, 3, and 4 can function in their place to initiate DNA replication, albeit less efficiently. So, in single-cell eukarya, preventing the onset of mitotic CDK activity prevents cells from entering mitosis and induces cells to assemble preRCs.

Endoreduplication in yeast, however, is an anomaly; it is not part of the organism's normal life cycle. Some single-cell eukaryotes, such as the ciliated protozoan *Tetrahymena*, are developmentally programmed to endoreduplicate their genomes. *Tetrahymena thermophila* contains two types of nuclei, a polyploid somatic macronucleus and a diploid germ-line micronucleus. After mating, the old macronucleus is destroyed and a new polyploid macronucleus develops from the diploid micronucleus by repeated endocycles, resulting in a final DNA content of about 45N. During macronuclear development, a 10 kb DNA minichromosome containing the rDNA locus is produced (Figure 10-22) and then amplified to 5000 copies during a period of time that follows the endocycles. These cells then proliferate by mitotic cell cycles, during which the copy number of the rDNA minichromosome is maintained by initiating replication once and only once during each subsequent S phase. Unfortunately, the mechanism that switches from mitotic to endocycles and back again has not yet been elucidated.

The best-characterized mechanisms by which cells switch from mitotic cell cycles to endocycles are found in flies and in mammals where one additional hurdle – geminin – must be overcome. Geminin is present throughout S, G_2, and early M phases of mitotic cell cycles where it prevents Cdt1 from loading the MCM helicase onto replication origins. Therefore, not only must CDK activity be inhibited following S phase, geminin must be inactivated to allow preRC assembly without passing through mitosis.

In *Drosophila* (and presumably other arthropods), entry into developmentally programmed endocycles is triggered by a decrease in Cdk1 activity that is accompanied by an increase in APC(Fzr/Cdh1) activity (Figure 12-19A). In the follicle cells of the *Drosophila* ovary, these events are under control of the Notch signaling pathway. During development, this pathway upregulates Fzr gene expression and downregulates the expression of genes that activate mitosis. Transcriptional repression of the String/Cdc25 phosphatase prevents conversion of the inactive form of Cdk1 into its active form, while transcriptional repression of the genes that encode mitotic cyclins A and B prevents any active Cdk1 from phosphorylating its substrates. The suppression of cyclin-driven CDK activity is reinforced by increasing transcription of the Fzr gene to promote assembly of APC(Fzr/Cdh1), the ubiquitin ligase that operates only in the absence of CDK activity and targets both mitotic cyclins and geminin for polyubiquitination and subsequent degradation by the 26S proteasome. The net result is that mitosis is blocked and cells enter a G_1-like state where preRC assembly can occur.

Mammals have a specialized mechanism for triggering endoreduplication. Trophoblast stem (TS) cells differentiate into trophoblast giant cells in response to the absence of fibroblast growth factor 4 (FGF4) and factors provided by fibroblast-conditioned medium (Figure 12-19B). FGF4 (produced by the inner cell mass) binds to the Fgfr2 receptor protein in the membrane of adjacent trophoblast cells. This event results in repression of two CKIs, p21 and p57, both of which are unique to mammals.

When TS cells are deprived of FGF4, they rapidly express p21/Cip1, a protein linked to the disappearance of Chk1 protein kinase, a key intermediate in the DNA damage response checkpoint (Chapter 13). The mechanism by which p21 accomplishes this is unknown, but it likely involves inhibition of a CDK activity that phosphorylates Chk1. The level of Chk1 kinase activity is determined by the extent to which it is phosphorylated, and the abundance of Chk1 protein. The level of Chk1 protein is regulated by the SCF(Fbx6) and Cul4A•Ddb1 ubiquitin ligases.

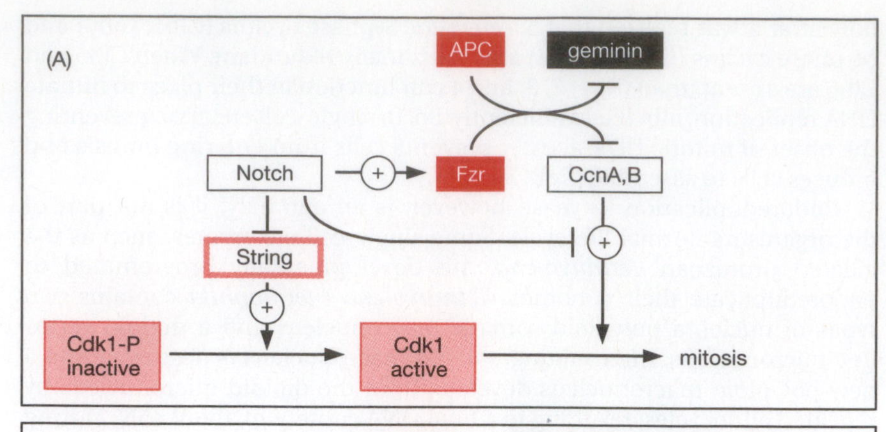

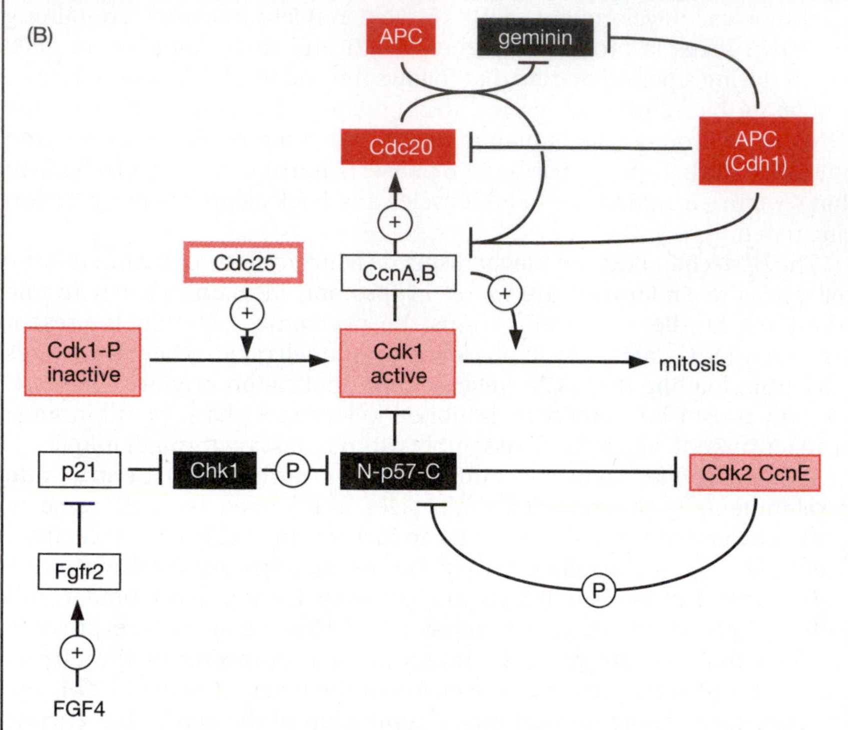

Figure 12-19. The switch from mitotic cell cycles to endocycles in *Drosophila* follicle cells (A) and in mouse trophoblast stem cells (B) requires preventing Cdk1•mitotic cyclin complexes from driving cells into mitosis and replacing APC(Cdc20/Fzy) with APC(Cdh1/Fzr). The Notch signaling pathway in flies and the FGF signaling pathway in mammals triggers endoreduplication in cells that are developmentally programmed for this event. Flies rely on transcriptional downregulation (⊥) of genes that trigger mitosis and upregulation (+→) of the APC substrate-targeting protein Fzr. APC(Cdh1/Fzr) activity is essential for endoreduplication, because it targets both mitotic cyclins and geminin for proteolysis. Mammals induce a unique CDK-specific inhibitor (p57/Kip2) to trigger a sequence of events that blocks mitosis under conditions that induce preRC assembly. Note the feedback loops between p57 and Cdk2•CcnE, and between Cdk1•CcnA,B and the APCs.

Chk1 kinase activity in TS cells prevents p57 expression by phosphorylating a site in the N-terminal region of p57 that targets p57 for ubiquitin-dependent degradation by the 26S proteasome. p57 also contains a CDK phosphorylation site near its C-terminus. Phosphorylation by Cdk2•CcnE converts p57 into a substrate for SCF(Skp2) ubiquitin ligase. Thus, the appearance of p57 protein in TS cells is suppressed by Chk1 during the metaphase to G_1 phase of cell division when CDK activity is low, and by Cdk2 and Cdk1 during the S phase to prometaphase period when CDK activity is high. This allows p57 to be expressed under conditions of low CDK activity that are essential for preRC assembly, and then degraded during genome duplication and spindle assembly when high CDK activity is required. Periodic cycling of p57 is essential to permit multiple rounds of DNA replication without an intervening mitosis.

Since Cdk1•CcnA,B activity is required to activate APC(Cdc20), inhibition of Cdk1 by p57 not only prevents cells from entering mitosis, but it also prevents assembly of APC(Cdc20), while allowing premature assembly of APC(Cdh1). APC(Cdh1) reinforces this situation by targeting both Cdc20 and the mitotic cyclins A and B for ubiquitin-dependent proteolysis. Since APC(Cdh1) also targets geminin for destruction, preRC assembly can now begin even though cells have not passed through mitosis. Mouse embryos lacking Cdh1 die shortly after implantation due to defects in endoreduplication of trophoblast cells and placental malfunction.

Sustaining Endocycles

Endocycles are successive waves of G and S phases that appear remarkably similar in both flies and mammals (Figure 12-20). Cdk2•CcnE and APC(Fzr/Cdh1) activities oscillate inversely during endocycles and both cyclin E (CcnE) and APC(Fzr/Cdh1) are essential to sustaining endocycles. Of critical importance is the fact that Cdk2•CcnE inactivates Fzr/Cdh1, suggesting that APC(Fzr/Cdh1) oscillations are driven by periodic inhibition of Fzr/Cdh1 by Cdk2•CcnE. Low Cdk2•CcnE activity allows high APC(Fzr/Cdh1) activity. High APC(Fzr/Cdh1) activity degrades mitotic cyclins and geminin, thereby allowing assembly of preRCs. High Cdk2•CcnE activity inactivates APC(Fzr/Cdh1), thereby allowing the onset of S phase in the presence of geminin. Cdk2•CcnE also phosphorylates Cdc6, thereby preventing its degradation by the APC and allowing Cdc6 to participate in preRC assembly. Thus, premature expression of APC(Fzr/Cdh1) activity, a consequence of inhibition of Cdk1 activity, triggers endoreduplication by preventing mitosis under conditions that allow preRC assembly.

The cell now responds as if it were in G_1 phase by expressing p27/Dacapo, the same CKI that suppresses CDK activity during G_1 of mitotic cell cycles. In trophoblast giant cells, p57 levels as well as p27 levels oscillate during endocycles so that preRC assembly can occur in the absence of CDK activity during G phases, and then DNA replication can occur in the presence of CDK activity during S phases (Figure 12-20B). Both p27 and p57 are targeted for ubiquitin-dependent degradation by the SCF ubiquitin ligase during the G→S transition in endocycles, just as p27 is targeted by the same enzyme during the G→S transition in mitotic cell cycles. Endocycles,

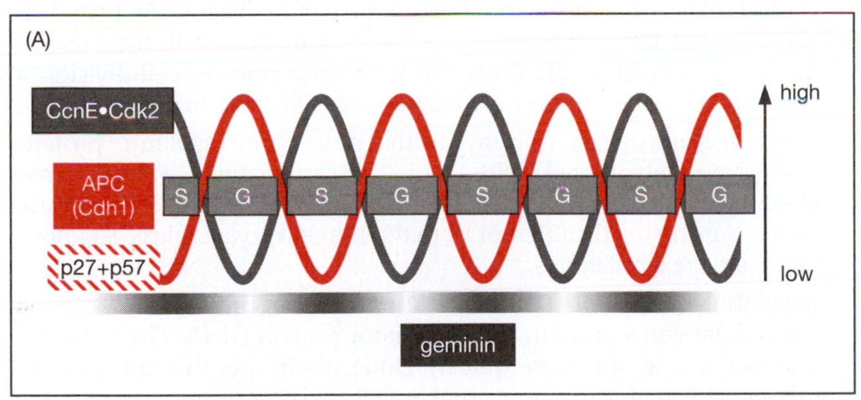

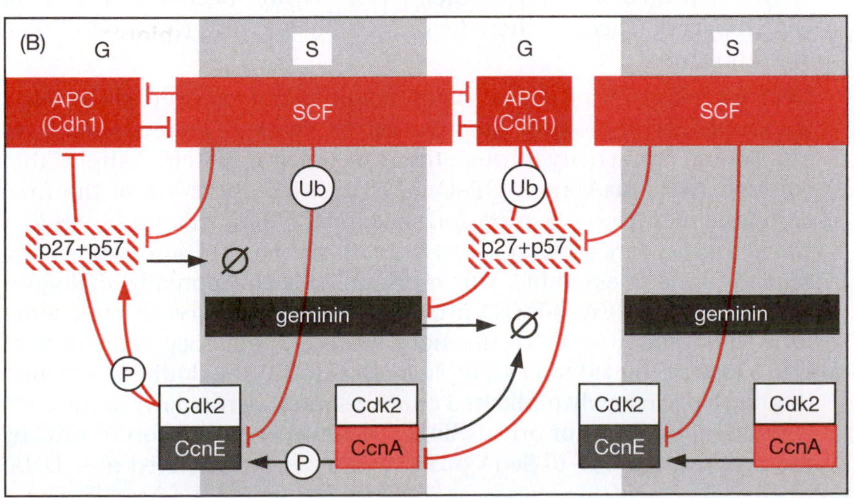

Figure 12-20. Endocycles in flies and mammals consist of repeated Gap (G) phases and S phases in the absence of mitosis. (A) Once initiated, endocycles appear similar in flies and mammals with one exception. p57 is unique to mammals and is required to initiate endoreduplication in trophoblast stem cells. The levels of CKIs p27/Dacapo and p57, as well as APC(Cdh1/Fzr) activity, are high during G phase and low during S phase. Conversely, the levels of geminin and Cdk2•CcnE are high during S phase and low during G phase. (B) The levels of individual proteins oscillate during endocycles in response to the same pathways that drive the G_1→S and G_2→G_1 transitions in mitotic cell cycles. The role of Cdk2•CcnA is based on mammalian cells; it has not been demonstrated in flies.

like mitotic cell cycles, use Cdk2 to initiate and regulate S phases. Cdk1 can substitute for Cdk2 in this function, although Cdk2 cannot replace Cdk1's role in mitosis. This difference allows cells to suppress M phase without also suppressing S phase. Oscillation of Cdk2•CcnE activity likely drives the progression of endocycles in much the same way that it drives progression of mitotic cell cycles.

Two APCs, Two Cell Cycles

Eukaryotic cells contain two forms of the APC, one that functions only when CDK activity is high [APC(Cdc20)] and one that functions only when CDK activity is low [APC(Cdh1)]. It is the existence of these two forms of the APC that allows both mitotic cell cycles and endocycles to occur within different cells of the same organism. APC(Cdc20) is required for mitotic cell cycles in both plants and animals, but not for endocycles, whereas APC(Cdh1/Fzr) is required to initiate and sustain endocycles in both plants and animals, but it is dispensable for mitotic cell cycles. Switching from mitotic cell cycles to endocycles involves two events: (1) inhibition of Cdk1 activity which prevents mitosis, and (2) activation of APC(Cdh1) in place of APC(Cdc20) which degrades mitotic cyclins and geminin in the absence of CDK activity thereby allowing preRC assembly. Cdk1 activity in flies is inhibited by downregulating the mitotic activators, whereas Cdk1 activity in mammals is inhibited by upregulating the CKI p57. These events, in turn, prevent assembly of APC(Cdc20), and in the absence of APC-specific inhibitors such as Emi1, allow assembly of APC(Cdh1).

SUMMARY

- Cell-division cycles are a series of pathways that regulate genome duplication (discussed here) and cell division (discussed in Morgan 2007, Lutkenhaus 2007, and Graumann 2007). There are just three simple rules: DNA replication must always precede cell division, and the genome must be duplicated at least once prior to cell division in bacteria, but once and only once prior to cell division in eukarya.
- Genomic sequence analysis reveals that none of the regulatory proteins that drive eukaryotic cell cycles (Table 12-3) are encoded by the genomes of either bacteria or archaea. Therefore, cell cycles in bacteria and archaea are distinctly different from those in eukarya. Little is yet known about archaeal cell cycles.
- Regulation of bacterial DNA replication is linked directly to changes in the cellular concentration of the initiator protein DnaA. The richer the nutrient source, the more quickly DnaA levels rise, the more quickly oriC is activated, the more quickly cell division occurs. Initiation of DNA replication occurs when the ratio of DnaA to oriC reaches a critical 'initiation mass'.
- DNA re-replication in bacteria is suppressed by three concerted pathways (Figure 12-3). Regulatory inactivation of DnaA is linked directly to replication-fork activity through the Hda•β-clamp protein complex that converts ATP–DnaA into ADP–DnaA. Thus, incorporation of the final component of the replication-fork machinery, the β-clamp, serves as a signal that the fork apparatus has been assembled and acts to prevent a second round of assembly. This mechanism is conceptually analogous to the Cul4•Ddb1(Cdt2)•PCNA protein complex at eukaryotic replication forks ubiquitinating Cdt1 during S phase. Regulatory reduction of DnaA levels is linked to DNA replication through the doubling of DnaA binding sites in newly replicated chromosomes, particularly at the *datA* locus. Sequestration of oriC is linked directly to replication of oriC by the rapid association of SeqA protein with hemimethylated oriC DNA.

The SeqA•oriC complex is refractory to initiation by DnaA. The *seqA* and *dam* methyltransferase genes are not commonly found among bacteria, although bacteria may utilize analogous DNA methylation mechanisms to regulate origin utilization.

- Regulation of eukaryal nuclear DNA replication (mtDNA replication occurs throughout the cell cycle) is linked directly to changes in CDK activity. preRC assembly occurs only when CDK activity is low, but preIC assembly occurs only when CDK activity is high. The appearance of cyclin E following cell division is the critical event that triggers the $G_1 \rightarrow S$ transition. In metazoa, preRC assembly is further restricted by the presence of geminin, a specific inhibitor of Cdt1-dependent loading of the MCM helicase. Given these restrictions, preRC assembly occurs only during the period from anaphase through early G_1 phase, and preIC assembly occurs only during late G_1 phase. Thus, eukarya employ a two-step strategy to regulate initiation of DNA replication. First, they regulate the ability of initiator [ORC+Cdc6] and Cdt1 to load the MCM helicase. Then, they regulate the ability of DDK and CDK to activate the MCM helicase and initiate the process of replication-fork assembly.
- The eukaryal cell cycle is driven by a series of feedback loops that regulate CDK activities through changes in the levels of cyclins and CKIs. These changes are regulated by a series of feedback loops that regulate ubiquitin ligases SCF ($S \rightarrow G_2$) and APC ($M \rightarrow G_1$). SCF activities are regulated by changes in F-box proteins and CDK activities, and APC activities are regulated by changes in substrate-targeting proteins Cdc20 and Cdh1, and by the APC-specific inhibitor Emi1.
- Two feedback loops dominate the $G_1 \rightarrow S$ transition: (1) APC(Cdh1) inhibits SCF(Skp2) which inhibits Cdk2•CcnE which inhibits APC(Cdh1), and (2) Cdk2•CcnE inhibits APC(Cdh1) which increases SCF activity and allows CcnA to accumulate. Cdk2•CcnE phosphorylates p27 which targets it for ubiquitination by SCF and subsequent degradation by the 26S proteasome. This allow CDK activities to increase further. Cdk2•CcnA phosphorylates CcnE, thereby converting it into an SCF substrate and targeting it for proteolysis. The cell has now moved from low CDK and high APC activities to high CDK, low APC but high SCF activities. Low APC activity during S phase is reinforced by the synthesis of Emi1.
- DNA re-replication from S through early M phase in eukarya is suppressed by five concerted pathways that inhibit either the assembly or function of preRCs through phosphorylation, ubiquitination, and protein–protein associations (Table 12-5). Not all of these pathways are used in each organism, and different organisms may target some preRC proteins more stringently than others. The final result is that genome duplication is restricted to once and only once each time a cell divides. (1) CDK phosphorylates selected ORC subunits thereby inhibiting the ability of ORC to initiate preRC assembly. (2) CDK phosphorylates Cdc6 and selected MCM subunits resulting in their export from the nucleus. (3) CDK phosphorylation of Cdc6 and Cdt1 targets them for ubiquitination by SCF and subsequent degradation by the 26S proteasome. (4) Non-CDK-dependent ubiquitination of Orc1 and Cdt1 targets them for degradation. (5) Binding of cyclins to selected ORC subunits and to Cdc6 in yeast, and binding of geminin to Cdt1 in metazoa, inhibits their function.
- Eukarya, and particularly the metazoa, have evolved families of proteins with closely related functions. Consequently, the proteins within a single family tend to be functionally redundant. When the gene for one protein is inactivated, another member of the same family can carry out its task. Cdc25, CDK, cyclin, and CKI are examples of functionally redundant gene families.

- Most cell divisions occur by the mitotic cell cycle, but organisms that require rapid early development to avoid predators have evolved rapid cycles composed only of S and M phases. Elimination of G phases is possible because large stores of the necessary replication proteins are already present. Nevertheless, DNA re-replication is still suppressed during these rapid cell cycles by the same mechanisms employed by mitotic cell cycles.
- Developmentally programmed polyploidy occurs in metazoan organisms by four different mechanisms (Table 12-6), all of which include a block to cytokinesis and a block to apoptosis. In addition, some cells (e.g. myoblasts) are developmentally programmed to arrest proliferation in G_1 phase by expressing the CKIs p21 and p57, whereas others (e.g. trophoblast stem cells) are programmed to arrest proliferation in G_2 phase by expression of the same CKIs. Still others (e.g. megakaryoblasts) are programmed to arrest proliferation in anaphase by a mechanism as yet unknown. Endoreduplication and endomitosis (Table 12-6) are possible because APC(Cdh1) can degrade geminin after CDK activity has terminated.

REFERENCES

Additional Reading

Bernander, R. (2007) The cell cycle of *Sulfolobus*. *Mol. Microbiol.* 66: 557–562.

Blow, J.J. and Gillespie, P.J. (2008). Replication licensing and cancer–a fatal entanglement? *Nat. Rev. Cancer* 8: 799–806.

DePamphilis, M.L., Blow, J.J., Ghosh, S., Saha, T., Noguchi, K. and Vassilev, A. (2006) Regulating the licensing of DNA replication origins in metazoa. *Curr. Opin. Cell Biol.* 18: 231–239.

Diffley, J.F. (2004) Regulation of early events in chromosome replication. *Curr. Biol.* 14: R778–786.

Graumann, P.L. (2007) Cytoskeletal elements in bacteria. *Annu. Rev. Microbiol.* 61: 589–618.

Kaguni, J.M. (2006) DnaA: controlling the initiation of bacterial DNA replication and more. *Annu. Rev. Microbiol.* 60: 351–375.

Lutkenhaus, J. (2007) Assembly dynamics of the bacterial MinCDE system and spatial regulation of the Z ring. *Annu. Rev. Biochem.* 76: 539–562.

Morgan, D.O. (2007) *The Cell Cycle, Principles of Control*. OUP/New Science Press, Oxford.

Murray, A. and Hunt, T. (1993) *The Cell Cycle, An Introduction*. Oxford University Press, Oxford.

Pesin, J.A. and Orr-Weaver, T.L. (2008) Regulation of APC/C activators in mitosis and meiosis. *Annu. Rev. Cell Dev. Biol.* 24: 475–499.

Sclafani, R.A. and Holzen, T.M. (2007) Cell cycle regulation of DNA replication. *Annu. Rev. Genet.* 41: 237–280.

Tada, S. (2007) Cdt1 and geminin: role during cell cycle progression and DNA damage in higher eukaryotes. *Front. Biosci.* 12: 1629–1641.

Thanbichler, M. and Shapiro, L. (2008) Getting organized--how bacterial cells move proteins and DNA. *Nat. Rev. Microbiol.* 6: 28–40.

Ullah, Z., Lee, C.Y. and DePamphilis, M.L. (2009) Cip/Kip cyclin-dependent protein kinase inhibitors and the road to polyploidy. *Cell Div.* 4: 10.

Waldminghaus, T. and Skarstad, K. (2009) The *Escherichia coli* SeqA protein. *Plasmid* 61: 141–150.

Chapter 13
CHECKPOINTS

In the previous chapter, we learned that cell-division cycles are controlled by a series of feedback loops that ensure a linear progression of events from cell growth to genome duplication to cell division. In addition, eukaryotic cells employ at least four mechanisms (Figure 12-12) to prevent DNA from undergoing more than one round of replication in a single cell cycle, and at least four mechanisms (Table 12-6) that permit some eukaryotic cells to bypass these controls and duplicate their nuclear genomes more than once without undergoing cytokinesis (termed developmentally regulated polyploidy). Here we describe the cell's police force, mechanisms that check the cell's progress at various points along the road to cell division. These **cell-cycle checkpoints**, as Leland Hartwell called them in 1989, are surveillance mechanisms that arrest the cell cycle at specific stages in response to problems either in chromosome replication or in cell division. They ensure that cells do not begin genome duplication before they have produced the components required for its completion, that DNA damage will not derail the replication process, that cells will complete genome duplication before attempting to divide, and that each cell receives a complete set of chromosomes during cytokinesis. Checkpoints presumably exist in all living organisms, although only those in bacteria and eukarya (primarily yeast and mammals) have been characterized extensively.

Cell-cycle checkpoints sense an imbalance in the normal flow of events and then respond by blocking the transition to the next event (Figure 13-1). Based on the event they sense and on the transition they block, cell-cycle checkpoints fall into three groups: those that restrict cell growth, those that sense changes in DNA structure, and those that sense changes in mitotic structures. Although checkpoints are not essential for cell division, they are essential for the existence of living organisms. When checkpoints are eliminated by mutation or bypassed by other means, the results are infidelity of chromosome transmission, increased susceptibility to agents that damage DNA, and cell death. Inactivation of checkpoint pathways leads directly to genome instability, and genome instability leads to cancer. Given the differences in their genetic complexity (Table 1-4), it comes as no surprise that the complexity of checkpoints increases from bacteria to yeast to simple metazoa to complex metazoa such as humans. Nevertheless, common principles are evident. Both bacteria and eukarya have a restriction checkpoint and at least one DNA damage checkpoint (Figure 13-2). Moreover, these two checkpoints sense the same defects and respond in analogous ways, even though the pathways and proteins involved differ markedly (Table 13-1). Among the eukarya, single-cell and multicellular organisms employ the same checkpoints using a similar (but not homologous) cast of players.

The **restriction checkpoint** prevents conversion of pre-replication complexes (preRCs) into pre-initiation complexes (preICs; Figure 11-15) until conditions are sufficient to support cell division. Otherwise, cells

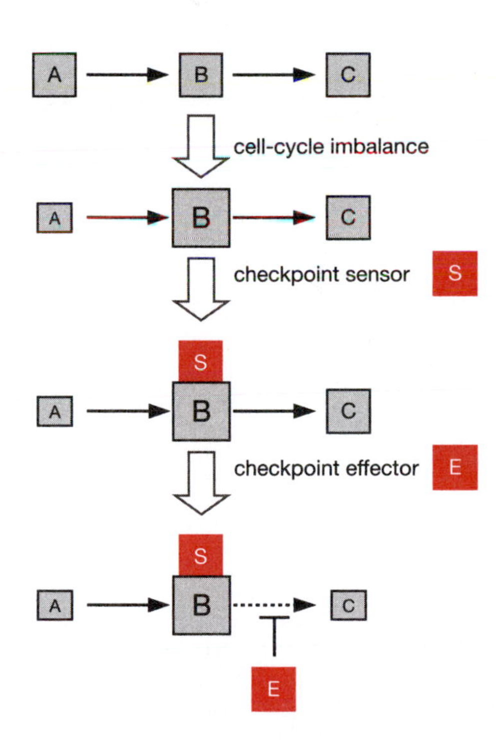

Figure 13-1. Checkpoints sense an imbalance in the normal sequence of events in a cell cycle, as represented by A→B→C. The checkpoint sensor (S) responds to changes either in the concentration or in the activity of a specific target molecule (B). This event activates the checkpoint effector (E), which then modifies its target molecule to prevent the cell from progressing to the next cell-cycle event.

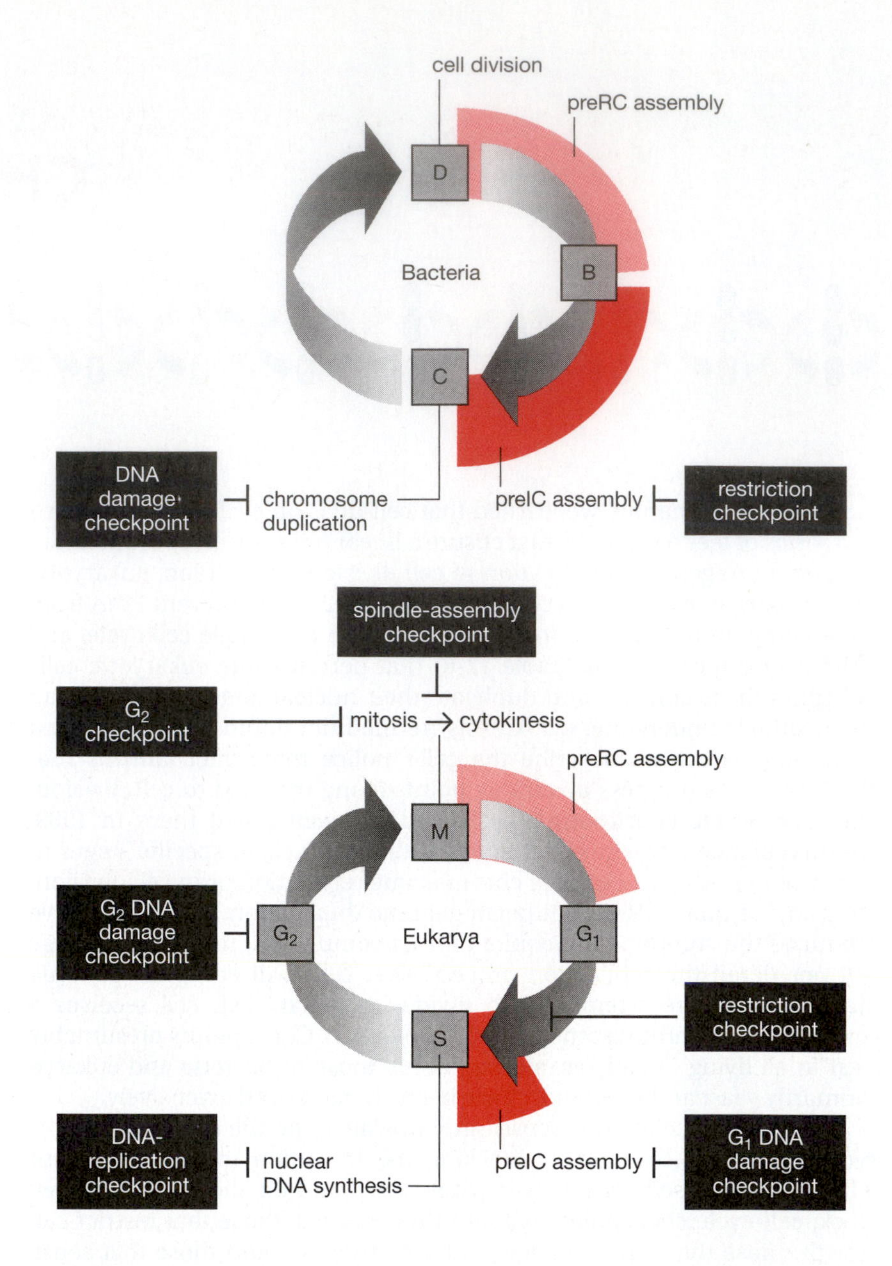

Figure 13-2. Cell-cycle checkpoints in bacteria and eukarya block (⊥) specific events at specific points during genome duplication and cell division. Light orange and dark orange regions indicate preRC and preIC assembly, respectively.

that lack sufficient nutrients to increase their mass and to support genome duplication would die in a futile attempt to divide. Arthur Pardee introduced the term **restriction point** or simply R-point in 1974 to describe the arrest of mammalian cells in G_1 phase when deprived of required nutrients or mitogens. Cells that lack the nutritional environment to synthesize adequate proteins and nucleotide precursors soon slide into a quiescent state (G_0) that supports life without proliferation, or in the case of certain bacteria or fungi, a dormant state referred to as sporulation. Consequently, the restriction checkpoint blocks the B→C transition in bacteria by delaying oriC-dependent DNA unwinding, and the G_1→S transition in eukarya by inhibiting transcription of genes whose products are required for genome duplication (Figure 13-2; Table 13-1).

The DNA damage checkpoint in bacteria is active only during DNA replication (C period). It senses DNA damage and then activates transcription of genes whose products are required for DNA repair and recombination. In addition, one of those gene products inhibits expression of a protein (FtsZ) essential for cell division. In contrast to bacteria, the DNA damage response in eukarya exerts its effects throughout interphase by preventing progression through the cell cycle if the cell's genome contains

Table 13-1. Cell-Cycle Checkpoints

Checkpoint	Senses	Sensor	Effector	Target	Inhibits	Cell-cycle block
BACTERIA						
Restriction	DnaA/oriC ratio	DnaA	DnaA	oriC	DNA unwinding	B→C
	Unloaded tRNA	RelA	ppGpp			
DNA damage	SSB–ssDNA	RecA	RecA•ssDNA	SOS genes	FtsZ	C→D
EUKARYA						
Restriction	Nutrient, mitogen deficiency	CcnD gene	CcnD•Cdk4,6	Rb	S-phase gene transcription	G_1→S
G_1 DNA damage	RPA•ssDNA	Atm•MRN	Chk2	Tp53	CcnE•Cdk2	G_1→S
				Cdc25A	Cdk2	
DNA replication		Atr•Atrip	Chk1	Cdc25A	Cdk2	S phase
G_2 DNA damage		Atr•Atrip	Chk1	Cdc25A, C	Cdk1	G_2→M
G_2	Replication forks	Atr•Atrip	Chk1	Cdc25A, C	Cdk1	G_2→M
	Cdc6-P	Chk1	Chk1			
Spindle assembly	Improper chromosome-spindle attachment	CPC	MPC	Cdc20	APC	Metaphase→ anaphase
	Meiotic maturation	?	Emi2	Cdc20	APC	Metaphase II→ G_1 phase

Bacterial genes are those for *E. coli.* Eukaryal genes are those for humans. The distinction between pathways driven by Atm and Atr sensor kinases and between Chk2 and Chk1 effector kinases is preferential, not absolute. The targets listed are ones that clearly affect cell division. A large-scale proteomic analysis for human proteins that are phosphorylated on consensus sites recognized by Atm and Atr in response to DNA damage identified more than 700 proteins (Matsuoka *et al.*, 2007). Thus, the pathways described here represent the beginning of the story, not the end.

either damaged DNA or stalled replication forks. Consequently, there are three manifestations of the eukaryotic DNA damage response. The G_1 DNA damage checkpoint, like the restriction checkpoint, blocks the G_1→S transition by inhibiting Cdk2 activity, which is required for preIC assembly. Once cells have entered S phase, the same pathways that detect DNA damage also detect stalled replication forks. Therefore, the S-phase DNA damage response is frequently termed the DNA-replication checkpoint, because it inhibits replication-fork progress and suppresses further activation of preICs by inhibiting Cdk2 activity and presumably other, as yet unidentified, targets. Once DNA replication is completed, the G_2 DNA damage checkpoint prevents entrance into mitosis by inhibiting Cdk1 activity in response to DNA damage. Keep in mind that these boundaries are artificial; the response to various types of DNA damage untilizes the same proteins regardless of when the damage appears during cell division. The DNA replication and G_2 DNA damage checkpoints senses primarily the presence of ssDNA gaps and stalled replication forks, whereas the G_1 DNA damage checkpoint senses primarily dsDNA breaks. Nevertheless, when dsDNA breaks occur during S or G_2 phase, the G_1 DNA damage checkpoint responds, as evidenced by the appearance of its signature proteins p53 and p21.

Even in the absence of DNA damage, the G_2 checkpoint prevents eukaryotic cells from beginning mitosis until their nuclear DNA has completed replication. This checkpoint is essentially an extension of the DNA damage-response pathway operating at normal replication forks in the absence of experimentally induced DNA damage. It also appears to sense the level of Cdc6-P, a byproduct of origin activation during S phase (Figure 12-17).

The **spindle-assembly checkpoint** (**SAC**) operates during both mitosis and meiosis. The mitotic SAC prevents proliferating cells from advancing beyond metaphase until all of the chromosomes are attached properly on the metaphase plate. The meiotic form of this checkpoint prevents female gametes from advancing beyond meiosis II. Both forms act by inhibiting the **anaphase-promoting complex** (**APC**), but they do so by different mechanisms.

RESTRICTION CHECKPOINT

The **restriction checkpoint** (R-point) is a transition in the cell cycle after which cells can proliferate independently of mitogenic stimulation. Prior to the R-point, cells are sensitive to the presence or absence of mitogenic stimuli and nutrients. After the R-point, simply reducing the level of a required nutrient cannot arrest cell division. Thus, the transition from growth factor dependence to growth factor independence represents a commitment to a new cell-division cycle. Once cells pass through the R-point, there is no turning back; failure to complete cell division results in cell death.

Bacteria

The restriction checkpoint in bacteria is commonly referred to as the **stringent response**, and it accounts for a phenomenon known as the **initiation mass**.

Initiation mass

As early as 1961, Ole Maaløe and Phil Hanawalt found that when bacterial cells are deprived of an essential amino acid, they complete replication of chromosomes already in the process of replicating, but they do not initiate new rounds of DNA replication. Thus, if bacterial cells do not reach a critical mass during the B period of their cell cycle (Figures 11-1 and 13-2), they will not initiate DNA replication. This critical mass later became known as the 'initiation mass' and understood as the concentration of DnaA protein required to initiate DNA replication at oriC (Figure 12-2).

Stringent response

The stringent response in bacteria is triggered by nutritional deprivation, such as amino acid starvation, and is accompanied by a switch in gene expression from cell proliferation to biosynthesis. Nutritional deprivation results in the accumulation of uncharged tRNA molecules, which then cause ribosomes to stall as they attempt to translate mRNAs and encounter a codon for which the cognate charged aminoacyl-tRNA is missing from the ribosome tRNA acceptor site (Figure 13-3). Stalled ribosomes activate a ppGpp synthetase encoded by the RelA gene that is essential to the stringent response. RelA produces the effector nucleotides guanosine tetraphosphate (5′-ppGpp-3′) and pentaphosphate (5′-pppGpp-3′) by ATP-dependent phosphorylation of GTP or GDP. The predominant and more stable of the two effector molecules is ppGpp. Nutritional stress stimulates production of ppGpp, and ppGpp induces changes in gene expression in most bacteria and plants. The cellular level of ppGpp is determined both by RelA and by SpoT, the sole hydrolase for ppGpp degradation. Hydrolysis of ppGpp regenerates GTP or GDP and pyrophosphate. SpoT functions both to degrade and to synthesize ppGpp. Another event that induces ppGpp synthesis is fatty acid starvation. Uncharged Acyl carrier protein (Acp) accumulates that then binds to SpoT (Figure 13-3). This event changes SpoT from a hydrolase into a synthetase, thereby increasing the cellular level of ppGpp. Other stress conditions provoke a similar shift by unknown mechanisms.

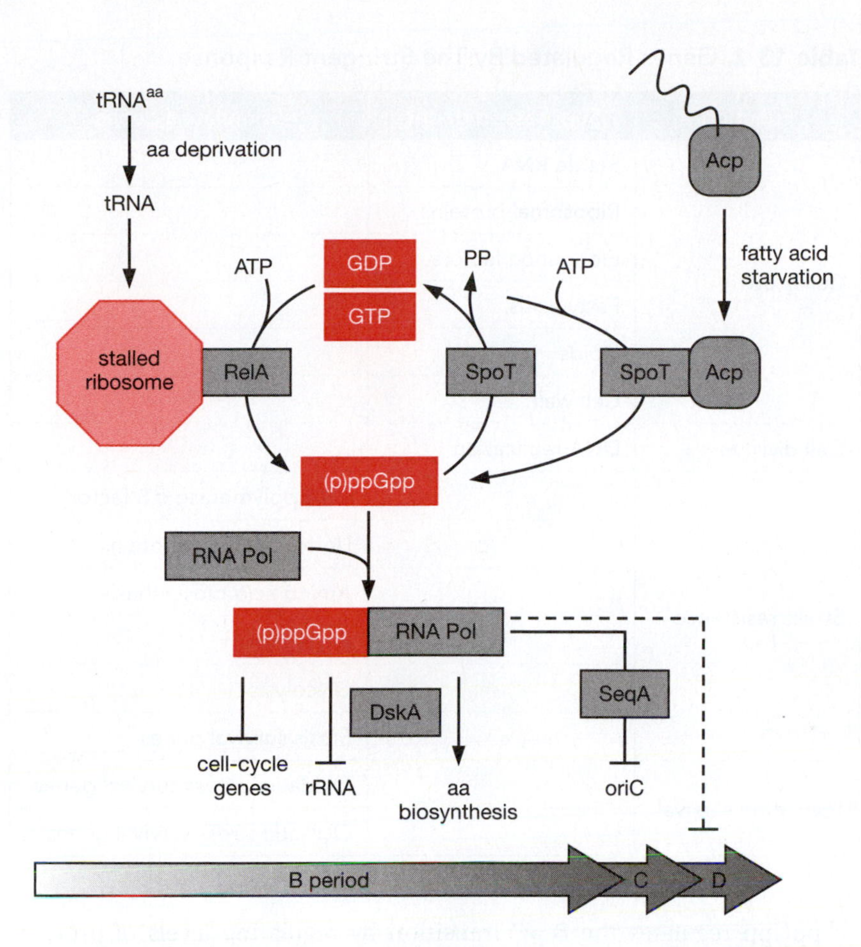

Figure 13-3. Stringent response in *E. coli*. → indicates conversion to product; ⊥ indicates inhibition of target. aa, amino acid. (Data from Ferullo, D.J. and Lovett, S.T. (2008) *PLoS Genet.* 4: e1000300, and Potrykus, K. and Cashel, M. (2008) *Annu. Rev. Microbiol.* 62: 35–51.)

The stringent response ultimately involves changing the expression pattern of almost 500 of the cell's approximately 4700 genes. Genes that are suppressed include those encoding rRNA, flagella, chemotaxis, metabolic functions, and transporters (Table 13-2). Conversely, a significantly larger number of genes are upregulated. These include amino acid biosynthetic genes, as well as regulators and effectors of bacterial stress responses. However, one-half of the upregulated genes have unknown functions. ppGpp regulates gene transcription by binding to RNA polymerase and, with the help of transcription factor DskA, suppresses expression of genes involved in regulating cell proliferation and ribosome synthesis, while activating genes involved in amino acid synthesis (Figure 13-3). The net effect on cell proliferation is to prevent initiation events at oriC (a process requiring the SeqA protein; Figure 12-3) while allowing replicating chromosomes to complete replication. Most cells are arrested in the B period, although some are arrested in the D period after completing DNA replication.

The stringent response arrests *E. coli* cell proliferation after the C period is completed, and the cells contain an integral number of chromosomes with a ratio of oriC/Ter = 1 (Figure 12-1). The growth rate prior to starvation determines the number of chromosomes upon arrest (Figure 12-2). Nucleoids of these cells are decondensed. Chromosomal regions containing oriC segregate, but the termini remain colocalized. The majority of cells arrest in the B period, prior to initiation, and contain an integer number of chromosomes appropriate for the growth medium prior to arrest. A minority of cells arrest proliferation in the D period after replication is completed but prior to chromosome segregation. Thus, bacterial cells are arrested after they have completed DNA replication already in progress, but before they initiate new rounds of DNA replication. Release from starvation causes rapid nucleoid reorganization, chromosome segregation, and resumption of replication.

Table 13-2. Genes Regulated By The Stringent Response

Function	Suppressed genes	Stimulated genes
Cell growth	Stable RNA	
	Ribosomal proteins	
	Elongation factors	
	Fatty acids	
	Lipids	
	Cell wall	
Cell division	DNA replication	
Stress resistance		RNA polymerase σ S factor
		Universal stress proteins
		Amino acid biosynthesis
		Proteolysis
		Glycolysis
		Stasis survival genes
Starvation survival		Oxidative stress survival genes
		Osmotic stress survival genes

ppGpp regulates the B→C transition by regulating levels of proteins required for oriC activation. Expression of DnaA, IHF, and Fis is repressed by ppGpp, as is transcription from the mioC and gidA genes that flank oriC and facilitate oriC activity. Moreover, an *E. coli* mutant devoid of ppGpp has a reduced initiation mass in rich culture medium. Hence, high levels of ppGpp inhibit oriC activity in *E. coli*. The effect of ppGpp on the B→C transition in *E. coli* is relieved by mutations that inactivate SeqA and deoxyadenosine methyltransferase (Dam), both of which are involved in regulation of oriC activity (Figure 12-3). Since neither SeqA nor the methyltransferase bind guanine nucleotides, this effect is likely regulated through ppGpp-sensitive transcription either of these genes, or genes that regulate SeqA or Dam activity. An additional ppGpp-regulated checkpoint just beyond oriC has been detected in *B. subtilis* where increased levels of ppGpp arrest replication forks at specific sites located 130 kb from one side of oriC and 230 kb from the other side. Release of the stringent response allows these forks to complete chromosome replication.

Archaea

So far, there have been no reports that archaea contain a restriction checkpoint, but it would be surprising if they didn't. One likely pathway would involve a small molecule such as (p)ppGpp. Although there is no indication from genomic sequences that archaea possess the ability to synthesize (p)ppGpp, some thermophilic archaea contain enzymes that can degrade (p)ppGpp.

Eukarya

Stringent response genes in plants

Animal cells do not encode (p)ppGpp synthetases, but they have been recovered in genetic screens for stress resistance in yeast, and they have been identified in plant cells. The genes are located in plant cell nuclei, and the (p)ppGpp synthetases appear in chloroplasts. This presumably reflects the evolutionary origins of the chloroplast, an organelle generally believed to be the result of an endosymbiotic event involving a cyanobacterium-like organism. Remarkably, the plant genes mimic the bacterial genes by responding to stress when transferred to bacteria. Moreover, the (p)ppGpp signal is triggered in plants in response to pathogens, wounding, ultraviolet light, heat shock, salt, and drought. Taken together, these observations suggest that the function of (p)ppGpp synthetases in plants parallels their function in bacteria; to produce (p)ppGpp in response to stress.

Parallel pathways for restriction checkpoints in eukarya

A restriction checkpoint must exist in multicellular animals, because at some point, cell proliferation must cease in order to form specific tissues and organs. In fact, most of the cells in our body are in a quiescent G_0 state in which cells shut down expression of genes required for genome duplication and cell division, but retain the ability to proliferate again in response to external mitogens or internal differentiation signals. Cells in a G_0 state should not be confused with terminally differentiated cells that can never again proliferate. Cells can remain in G_0 for years, and still respond to mitogenic signals that drive them back into a mitotic cell cycle. Liver cells, for instance, divide only once or twice a year. In mammals, progression from M phase to the R-point (Figure 11-1) requires a continuous mitogenic signal and a high rate of protein synthesis. Interrupting either of these requirements rapidly results in a transition from G_1 to the G_0 state. This usually occurs approximately 3 h after cells have exited mitosis (about 50% of their normal G_1 phase) and after preRCs are assembled at replication origins.

The proteins that coordinate cell growth with cell division in mammals have analogs in both budding yeast and fission yeast (Table 13-3), and these analogs follow parallel pathways in preventing expression of genes required for S phase until the proper nutrients and mitogens are present. In the budding yeast *S. cerevisiae*, where the restriction checkpoint is termed start, gene expression at the $G_1 \rightarrow S$ transition is attributed to SBF and

Table 13-3. Restriction Checkpoint Proteins In Eukarya

Yeast		Mammals	Function
S. cerevisiae	*S. pombe*	*H. sapiens*	
SBF/Swi6•Swi4		E2f1, E2f2, E2f3a	Transcription activation
MBF/Swi6•Mbp1	MBF/Res1•Res2•Cdc10	E2f3b, E2f4, E2f5	Transcription repression
Swi4, Mbp1	Res1, Res2	E2f1, E2f2, E2f3a, E2f3b, E2f4, E2f5 (E2f6, 7, and 8 do not bind Rb)	Sequence-specific DNA binding
Swi6	Cdc10	Dp1, Dp2, Dp3	Transcription activation
Whi5, Nrm1	Nrm1, Yox1	Rb1, Rbl1/p107, Rbl2/p130	Transcription repression, MBF corepressor
Cdk1/Cdc28	Cdk1/Cdc2	Cdk4, Cdk6	Inactivation of repressor proteins
Cln3	Puc1	Cyclin D (CcnD)	
		p16/INK4a, p15/INK4b, p18/INK4c, p19/INK4d/Arf	Inhibit Cdk4, Cdk6
	Cds1/Chk2		Inhibit SpNrm1

/ indicates different names for the same protein or protein complex. SBF, Swi4,6 cell-cycle box (SCB)-binding factor; MBF, Mul1 cell-cycle box (MCB)-binding factor.

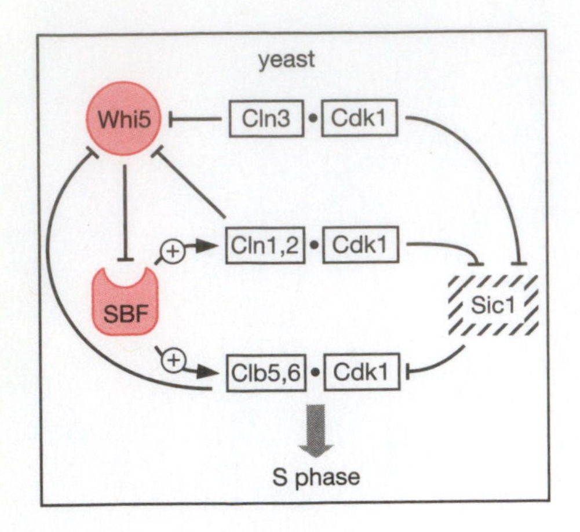

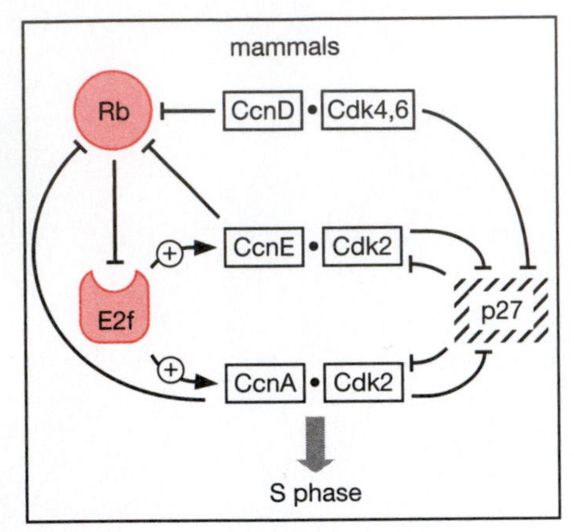

Figure 13-4. Restriction checkpoints in budding yeast ('start') and mammals ('restriction point') use different proteins to achieve the same ends. This is an example of the evolution of parallel pathways in the regulation of genome duplication. +→ indicates activation of target; ⊥ indicates inhibition of target. (Data from Sherr, C.J. and Roberts, J.M. (2004) *Genes Dev.* 18: 2699–2711.)

MBF, two transcription factors comprised of the Swi6 protein and one of the two sequence-specific proteins that bind DNA, called Swi4 and Mbp1. SBF functions primarily as a transcriptional activator during G_1 phase, whereas MBF functions primarily as a transcriptional repressor outside of G_1 phase. Genes involved in the timing or efficiency of cell-cycle events are regulated mainly by SBF. Genes involved in the DNA damage response, as well as those involved in the initiation of DNA replication and in nucleotide synthesis, are controlled largely by MBF. Two corepressor proteins, Whi5 and Nrm1, regulate SBF and MBF activities. Whi5 represses SBF-dependent transcription in early G_1 phase prior to commitment of the cell to enter S phase. The Nrm1 protein acts together with MBF to confine transcription of its target genes to the G_1 phase of the cell cycle by repressing expression of target genes outside of G_1. In the fission yeast *S. pombe*, only the MBF transcription complex (Res1•Res2•Cdc10) is found, and its activity is regulated by two proteins, Nrm1 and Yox1.

Although the mammalian E2f family of transcription factors and their repressors, the retinoblastoma (Rb) family of proteins, has no recognizable sequence homology with their yeast counterparts, they play analogous roles in their respective restriction checkpoints. The E2f family of transcription factors can be subdivided into transcriptional activators (E2f1 through E2f3a) comparable to SBF in budding yeast, and transcriptional repressors (E2f3b, E2f4 through E2f8) similar to MBF in budding and fission yeast. Moreover, just as Whi5 and Nrm1 repress SBF and MBF in yeast, the Rb proteins repress most (but not all) of the E2f transcription factors in mammals. Many of the genes regulated by E2f during the $G_1 \rightarrow S$ transition in human cells are homologs of genes regulated by MBF in yeast. In summary, Rb and E2f in mammals play the roles of Whi5 and SBF in yeast; Whi5 and Rb repress SBF- and E2f-dependent transcription, respectively. Conversely, Whi5 and Rb activity is suppressed by various CDK•cyclin complexes. The net result is a series of feedback loops that drive cells from G_1 into S phase (Figure 13-4).

G_1 cyclins together with their CDK partners promote cell-cycle progression by inactivating proteins that inhibit S-phase CDK activity. In budding yeast, the CKI Sic1 is targeted by Cdk1/Cdc28•cyclins. In mammals, the CKI p27 is targeted by Cdk2, Cdk4, and Cdk6 and their cyclin activators. In addition, the same CDK•cyclin complexes phosphorylate and thereby inactivate proteins that inhibit transcription of genes required for S phase (Whi5 in budding yeast, and Rb proteins in mammals). Inactivation of both classes of inhibitory proteins is initiated by cyclins that are expressed at the beginning of G_1 phase (Cln3 in yeast; cyclin Ds in mammals) and then maintained by cyclins that are expressed later during G_1 phase (Clns 1 and 2 in yeast; cyclin Es in mammals). This transition from early G_1 cyclins to late G_1 cyclins switches the cell from dependence on nutrients and mitogens to independence. These pathways converge on one goal: expression of the S-phase cyclins (Clbs 5 and 6 in yeast; CcnE and CcnA2 in mammals) that trigger the $G_1 \rightarrow S$ transition.

The E2f pathway in mammals

Given the striking similarities between restriction checkpoint pathways in single-cell and multicellular eukarya, the well-characterized E2f pathway in mammals serves as a paradigm for the R-point in all eukarya (Figure 13-5). The initial response of G_0 cells to mitogenic stimuli is induction of a set of genes that encode transcription factors such as Fos, Jun, Ets, and Myc. This induces expression of cyclin D (CcnD), the protein that specifically regulates Cdk4 and Cdk6 protein kinase activities. Thus, as CcnD levels rise, CcnD•Cdk4 and CcnD•Cdk6 activities increase, reaching a maximum in mid-G_1, and phosphorylate Rb. The activities of CcnD•Cdk4,6 complexes are kept in check by two CDK-specific protein inhibitors from the INK4 family (Tables 12-3 and 13-3), commonly referred to as p15

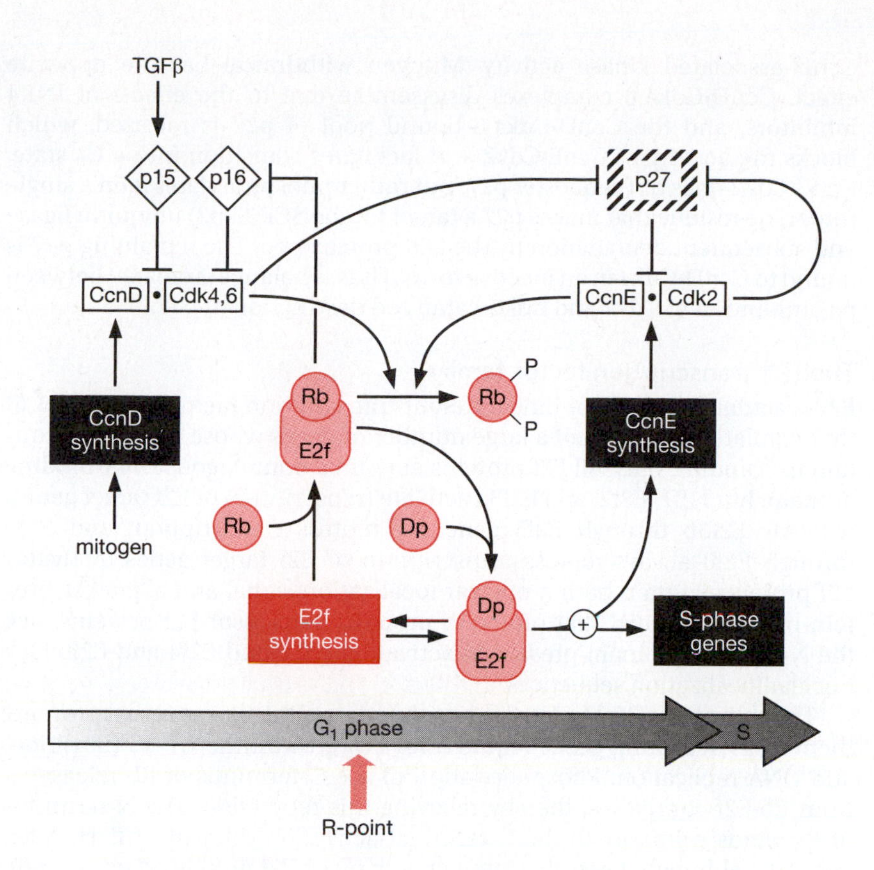

Figure 13-5. The eukaryotic restriction checkpoint (R-point) marks the transition from an inactive form of transcription factor E2f (the E2f•Rb complex) to an active form (E2f). This transition occurs when Rb is phosphorylated by cyclin D (CcnD) complexed with either Cdk4 or Cdk6. E2f then activates transcription of genes such as cyclin E (CcnE) to trigger S phase. Both CcnD•Cdk4,6 and CcnE•Cdk2 inactivate the CKI p27, but by totally different mechanisms.

and p16. Transforming growth factor β (TGFβ), a secreted protein that regulates cell proliferation and differentiation, stimulates production of p15. INK4 proteins block the $G_0 \rightarrow G_1$-phase transition by binding specifically to either Cdk4 or Cdk6, thereby preventing their activation by CcnD. Eventually, the level of CcnD is sufficient to overcome the inhibitory effects of INK4 inhibitors, and allow CcnD•Cdk4,6 complexes to inactivate Rb by phosphorylating it. This transition is effected by a feedback loop in which E2f•Rb represses expression of p16, thereby increasing CcnD•Cdc4,6 activity with concomitant phosphorylation of Rb and release of E2f. The E2f1 promoter itself creates a second feedback loop. This promoter has several E2F binding sites, thereby allowing E2f1 protein to activate its own expression in a cell-cycle-dependent manner. This increases the amount of E2f protein during late G_1 phase and activates transcription of E2f target genes. One of those genes is cyclin E (CcnE).

S phase is triggered by increased levels of CcnE and Dbf4, resulting in the appearance of CcnE•Cdk2 and Cdc7•Dbf4 protein kinase activities that are required for preIC assembly (Figure 11-15). Hence, CcnE levels peak during the $G_1 \rightarrow S$ transition as a result of E2f activation of *CcnE* gene transcription. CcnE•Cdk2 then forms another feedback loop by maintaining Rb in a hyperphosphorylated state during late G_1 and S phases when CcnD levels fall. In this way, CcnD and CcnE operate sequentially to drive the $G_1 \rightarrow S$ transition forward. Ectopic expression of CcnD causes phosphorylation of Rb, whereas ectopic expression of CcnE does not. Ectopic expression of both cyclins in the same cell shortens G_1 phase. Thus, they control different events that together determine the time required for entry into S phase. Once DNA replication begins, the activities of CcnD-dependent kinases are no longer required until cells complete M phase and re-enter G_1 phase.

CcnD•Cdk4,6 complexes also play a noncatalytic role in regulating the $G_1 \rightarrow S$ transition by sequestering p27/Kip1 and p21/Cip1, two potent inhibitors of Cdk2. Binding of Cip/Kip proteins to CcnD•Cdk4 stabilizes the complex and facilitates its nuclear import, without inhibiting

CcnD-associated kinase activity. Mitogen withdrawal has the opposite effect. CcnD•Cdk4,6 complexes disassemble due to the effects of INK4 inhibitors, and the CcnD•Cdk4,6-bound pool of p27 is released, which blocks the activity of CcnE•Cdk2 and facilitates transition into a G_0 state. CcnE•Cdk2 does not sequester p27, but rather phosphorylates it on a single threonine residue that makes p27 a target for the SCF(Skp2) ubiquitin ligase and subsequent degradation by the 26S proteasome. The remaining p27 is bound to CcnD•Cdk4 in an inactive form. Thus, a balance is created between p27 inhibition of Cdk2 and Cdk2-catalyzed destruction of p27.

The E2f transcription factor family

E2f is actually a family of nine different transcription factors (Figure 13-6) that regulate expression of a large number of genes whose promoters contain E2f binding sites. All E2f proteins contain a homologous DNA-binding domain, but E2f1, E2f2, and E2f3a activate transcription of E2f target genes, whereas E2f3b through E2f5 generally repress transcription, and E2f6 through E2f8 always repress transcription of E2f target genes. Activator E2f proteins contain both a nuclear localization signal and a 'pocket-protein-binding domain' that binds Rb proteins. Repressor E2f proteins lack the N-terminal domain present in activating E2fs, and E2f4 and E2f5 lack nuclear localization sequences.

The activating E2fs bind hypophosphorylated Rb1, an event that prevents them from activating genes required for cell growth, the G1→S transition, and DNA replication. Phosphorylation of the C-terminus of Rb releases it from Rb•E2f complexes, thereby relieving this repression. The N-terminus of Rb binds primarily to the C-terminal activation domain of E2f1, E2f2, and E2f3, although it can also interact with other E2f proteins (Figure 13-6). Rb-like protein p130 binds to either E2f4 or E2f5, and p107 binds only E2f4. E2f6, 7, and 8 lack sequences required for transactivation and Rb binding. This allows them to repress transcription without interference from Rb.

E2f1 through E2f6 contain conserved dimerization domains that allow them to form DNA-binding heterodimers with a family of three co-activator proteins termed differentiation-regulated transcription factor polypeptide (Dp1, 2, and 3). This interaction is required for sequence-specific DNA binding by E2f proteins and for activation of E2f target genes. E2f7 and E2f8 lack a Dp-binding domain, but they contain the tandem repeats of an E2f DNA-binding domain that apparently increase the specificity of DNA binding.

Functionally redundant proteins

Multicellular eukarya often contain families of genes that encode proteins with common homologies and similar biochemical properties. Each member

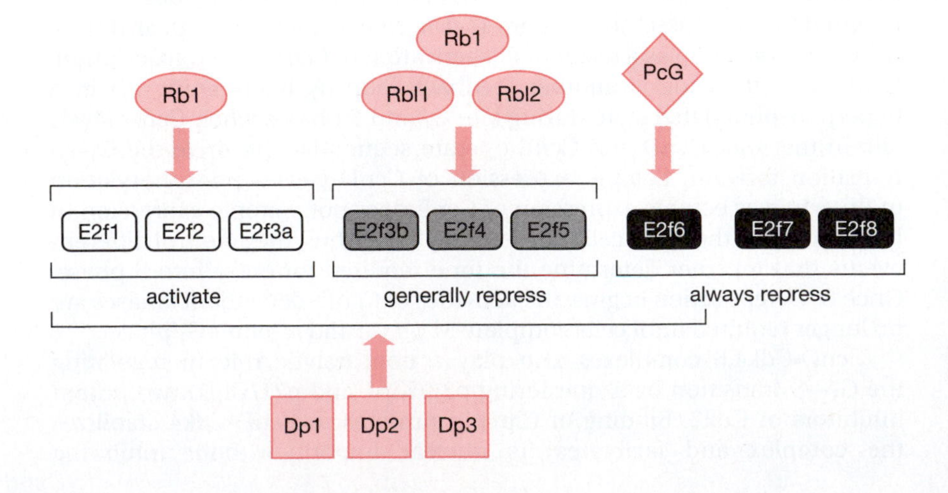

Figure 13-6. Mammals encode eight E2f transcription factors, as well as three Rb proteins that repress E2f activities, and three Dp proteins that serve as co-activators of E2f. E2f3a and E2f3b are different products of the E2f3 locus. E2f6 interacts with polycomb-group (PcG) proteins characteristically associated with heterochromatin. (Data from van den Heuvel, S. and Dyson, N.J. (2008) *Nat. Rev. Mol. Cell Biol.* 9: 713–724.)

of a **gene family** presumably has specific tasks to which it is best suited over many generations, but in the short run, one member can often substitute for another. The R-point is an example of this **functional redundancy**. Both Cdk4 and Cdk6 require CcnD to regulate the R-point. Inactivation of any one of the three CcnD genes does not prevent mouse embryos from developing into viable, fertile adults, but inactivation of any two of them blocks embryonic development. Conversely, embryos lacking either Cdk4 or Cdk6 can develop from a fertilized egg into a viable adult, and cells derived from these animals continue to proliferate *in vitro,* whereas embryos lacking both Cdk4 and Cdk6 die during post-implantation development. Prior to that stage, either Cdk1 or Cdk2 can substitute for Cdk4 and Cdk6.

A similar relationship exists between E2f, Rb, and Dp proteins. Either all three activating E2f proteins or all three Rb proteins must be knocked out in mice in order to inactivate the R-point. Inactivation of any one of the three activator E2f genes has little effect on proliferation of cultured cells, but inactivation of all three genes severely reduces DNA replication and cell proliferation. Similarly, Dp1 is largely dispensable for embryonic development, although it is essential for development of extra-embryonic tissues, suggesting that the other two Dp proteins can substitute for Dp1 during embryonic development. The Rb relationships are a bit more complex. Mouse embryos that lack Rb1 die *in utero* between days 13 and 15 of gestation with defects in erythropoiesis and placental development. However, the fact that development can extend this far reveals functional redundancy within the Rb family. Quiescent $Rb1^{-/-}$ cells have elevated levels of Rbl1/p107, and embryos lacking both Rb1 and Rbl1/p107 die 2 days earlier than their $Rb1^{-/-}$ counterparts, and they die with the same phenotype. When all three Rb family members are inactivated, the resulting cells are completely refractory to the induction of quiescence by any means; they are unable to produce normal mouse embryos. Thus, all three Rb proteins collaborate to regulate the $G_0 \rightarrow S$ transition.

E2f target genes

E2f transcription factors regulate genes that affect a wide range of biological processes, some of which are required directly for genome duplication (Table 13-4). Quiescent cells do not express genes encoding preRC components Orc1, Cdc6, Cdt1, and Mcm2 through Mcm7, as well as Cdk1, cyclins A and B, and Rbl1/p107. As cells transit from $G_0 \rightarrow G_1$, Orc1, Cdc6, Cdt1, and Mcm2 through Mcm7 are upregulated, and their expression is dependent on E2f. Since expression of ORC subunits 2 through 6 is not E2f-dependent, Orc1 appears to regulate the first step in preRC assembly, the formation of ORC•chromatin complexes. Genes required for preventing DNA re-replication during S phase appear to be downregulated during this period, and expression of other E2f-dependent genes required for DNA replication appear to be upregulated later during the $G_1 \rightarrow S$ transition.

E2f repressor complexes suppress the transcription of their target genes in G_0 cells, in differentiated cells, and during the early G_1 phase of the cell cycle. In proliferating cells, the repressor E2f proteins are either disrupted or replaced by activator E2f proteins that promote gene expression. Rb proteins inhibit E2f-mediated activation and augment E2f-mediated repression. The ability of Rb proteins to repress E2f-dependent transcription is regulated by periodic CDK activity, thereby converting temporal changes in CDK activity into temporal patterns of gene expression.

DNA DAMAGE RESPONSE

Cell division in all cellular organisms depends both on completion of DNA replication and on the presence of undamaged DNA. The problem is that all cellular organisms are exposed to a multitude of agents that damage DNA. **Chemical lesions** occur through oxidation, alkylation, and hydrolysis

Table 13-4. Genes Regulated By E2f That Are Required For Genome Duplication

Function	Gene	Function
preRC assembly	Orc1	Initiator protein
	Cdc6	Initiator protein
	Cdt1	Helicase loader
	Geminin	Initiation inhibitor
	Mcm2, Mcm3, Mcm4, Mcm5, Mcm6, Mcm7	MCM helicase
preIC assembly	Cdc45	MCM activity, Pol-α loading
dNTP synthesis	Dihydrofolate reductase	Dihydrofolate→Tetrahydrofolate
	Thymidylate synthase	dUMP→dTMP
	Thymidine kinase	dT→dTMP
	Ribonucleotide reductase	NDPs→dNDPs
DNA replication	DNA Pol-α catalytic subunit	RNA-primed DNA synthesis
	DNA Pol-δ catalytic subunit	Lagging-strand DNA synthesis
	PCNA	Sliding clamp
Cell cycle	Cdk1	Regulate G_2→M transition
	Cdkn1B/p27/Kip1	Regulate G_1→S transition
	Cdkn2D/p19/INK4d	Regulate G_0→S transition
	Cdkn2C/p18/INK4c	Regulate G_0→S transition
	Cyclin A	Regulate Cdk2 and Cdk1 activities, maintain S phase, regulate G_2→M transition
	Cyclin E	Regulate Cdk2 activity, regulate G_1→S transition
	Dbf4/Ask	Regulate Cdc7 kinase activity, regulate G_1→S transition

At least 130 genes have been identified as E2f-activated. For literature references, see Bracken *et al.* (2004) and Ozono *et al.* (2009).

of bases to produce a variety of modified bases such as 7-methylguanine, 1-methyladenine, and O^6-methylguanine, as well as deamination of cytosine, adenine, and guanine. In addition, chemical lesions can result in both depurination and depyrimidation of DNA. Elevated temperatures increase the rate of depurination, and increase the frequency of phosphodiester bond interruptions (single-strand breaks) that result from depurination as well as generation of reactive oxygen species. A huge diversity of **DNA adducts** result from exposure to smoke, soot, and tar. Ultraviolet light (sunlight) causes cross-linking between adjacent cytosine and thymine bases to create pyrimidine dimers, as well as formation of free radicals that can damage other molecules. Ionizing radiation (radioactive decay or cosmic rays) causes breaks in dsDNA. Add to this list misincorporation of the wrong base during DNA replication, and it is not surprising that living organisms had to evolve DNA damage response systems that sense the presence of DNA damage, arrest cell proliferation, and activate genes that encode DNA-repair proteins.

The same mechanisms that respond to the presence of chemical lesions also respond to stalled DNA replication forks and to dsDNA breaks. **Stalled replication forks** occur when DNA synthesis is arrested under conditions that allow DNA unwinding to continue. This can occur when a polymerase encounters a DNA lesion in the template, or a noncanonical DNA structure (Chapter 1), or a replication-fork barrier (Chapter 6), or inadequate nucleotide pools, or the absence of a required protein. For example, drugs that reduce nucleotide pools or inhibit replicative DNA polymerases (Table 12-2) cause replication forks to stall and trigger a DNA damage response. This can be a serious problem in synchronizing cells for experimental studies. Collectively, these problems are referred to as replicative stress.

In eukarya, most dsDNA breaks occur in genetically encoded replication-fork **slow zones** (yeast) or **fragile sites** (mammals) where replication forks are more likely to stall. Stalled forks become fragile sites where cleavage of a single phosphodiester bond in one of the exposed templates creates a dsDNA break (Figure 14-7). Therefore, the importance of mechanisms for repairing dsDNA breaks cannot be overstated; they are the most lethal form of DNA damage, and they generate gross chromosomal rearrangements. In bacteria, the mechanism for dsDNA break repair plays a critical role in restarting stalled replication forks. Eukarya, on the other hand, place more emphasis on stabilizing stalled forks so that they do not collapse and are therefore easier to restart. **Collapsed replication forks** occur when the replisome dissociates from the fork, generating extensive ssDNA regions due to the nucleolytic processing of nascent chains, formation of branched DNA structures resembling recombination intermediates, and DNA breaks. Stalled forks collapse when DNA-replication checkpoints are defective.

DNA damage-response pathways in bacteria and eukarya all sense the same signal: extensive regions of ssDNA complexed with a unique protein that binds single-stranded DNA (SSB in bacteria, RPA in eukarya; Table 13-1). This simple fact implies that active replication forks contain little ssDNA, otherwise they could not be distinguished from stalled replication forks. In fact, analysis of replicating chromosomes by electron microscopy, two-dimensional gel electrophoresis, and nuclease sensitivity confirms that significant ssDNA at one or both arms of a replication fork is a rare event.

The SOS Response In Bacteria

The prototype of a DNA damage response was discovered in *E. coli* by Miroslav Radman in 1975 and named the SOS response after the international distress call 'save our souls'. The SOS response arrests cell division in response to excess ssDNA, and sets into motion the machinery required either to repair the DNA damage, or to restart the replication fork. At least 10 gene products are directly involved in the SOS response in bacteria (Table 13-5) but the principal player is the RecA protein. It senses DNA damage, and it is required when a DNA segment is transferred from one DNA molecule to another DNA molecule by virtue of their homologous DNA sequences, an event termed **homologous recombination**. Several different pathways for homologous recombination exist in *E. coli*, and all of them require RecA, because RecA is the only protein that promotes base-pairing between an ssDNA molecule and a homologous sequence in another DNA molecule. The importance of RecA is emphasized by the fact that while archaea and eukarya often contain proteins with the same function as ones in bacteria, only two bacterial DNA damage-response proteins, RecA and RecQ, have homologs in eukarya (Table 13-5).

Response to DNA damage

Damaged DNA in bacteria is quickly converted into regions of ssDNA that are rapidly complexed with SSB (Figure 13-7). These SSB•ssDNA regions are 'sensed' by RecA protein, which displaces the SSB to form a RecA•ssDNA filament. The relationship between the SOS response, DNA recombination, and DNA repair is best illustrated by the mechanism for repairing breaks in dsDNA. When a dsDNA break occurs at a stalled replication fork, repair is initiated by resection of the ends of the DNA by RecBCD (Figure 13-7). RecBCD consists of three different proteins, RecB (3′→5′ helicase), RecC [recognizes short sequences called **chi(χ)-sites** (5′-GCTGGTGG-3′)], and

Table 13-5. Bacterial DNA Damage-Response Proteins And Their Analogs In Other Organisms

Bacteria	Archaea	Yeast		Metazoa	Function
E. coli		*S. cerevisiae*	*S. pombe*	*H. sapiens*	
RecBCD complex RecB RecC RecD	Mre11 Rad50	MRX complex Mre11 Rad50 Xrs2 Sae2	MRN complex Rad32 Rad50 Nbs1 Ctp1	MRN complex Mre11A Rad50 Nbn/Nbs1 Bcl11a/Ctip1	Resects 5′-end of dsDNA breaks to create 3′-ssDNA termini suitable for DNA recombination
SSB•ssDNA		RPA•ssDNA	RPA•ssDNA	RPA•ssDNA	ssDNA-binding protein complexed with ssDNA is the DNA damage signal
LexA					Represses DNA-repair gene expression
RecA	RadA	Rad51	Rad51	Rad51	Binds SSB/RPA•ssDNA, initiates homologous recombination, activates DNA-repair gene expression
RecFOR RecF RecO RecR		Rad52	Rad52	Rad52	Facilitates SSB/RPA displacement from ssDNA by RecA/Rad51
RecX					Blocks RecA filament extension
DinI					Stabilizes RecA filaments
RecQ UvrD		Sgs1 Srs2	Rqh1	Blm, Wrn, RecQL, RecQL4, RecQL5 Rad54L	DNA helicases, displace RecA/Rad51 from 5′-end of resected DNA, facilitate D-loop formation during homologous recombination
RuvC	Hjc Hje	Yen1	Yen1	Gen1	Resolves Holliday junctions

/ separates two names for the same gene. The first name is the one proposed by a genome nomenclature organization. By convention, abbreviated names of human proteins are in upper case, but for simplicity the names of all proteins are in title case; upper case is used for common multiprotein complexes. Although Mre11 and Rad50 are conserved in all three domains of life, they are not homologs of RecBCD. Their homologs in bacteria are SbcC and SbcD, two proteins involved in recombination at palindrome sequences. Blm (Bloom's), Wrn (Warner's), and RecQL4 are homologs of *E. coli* RecQ. *E. coli* UvrD and RecQ are from different helicase families. Only budding yeast has a homolog (Srs2) of UvrD.

RecD (5′→3′ helicase). RecBCD efficiently degrades the ends of linear dsDNA, cleaving off both strands in the process. When RecBCD encounters a χ-site, the RecD subunit is altered such that the dsDNA exonuclease is inactivated, and the RecBC complex becomes a helicase with 5′→3′ ssDNA nuclease activity. RecBC now unwinds the DNA with concomitant degradation of the 5′-terminated strand. The resulting 3′-ssDNA overhangs are converted into RecA•ssDNA filaments.

To avoid excess DNA recombination, interaction between RecA and SSB•ssDNA is stringently regulated by a group of **mediator proteins** (Figure 13-8). RecA does not bind well to ssDNA, because long regions of ssDNA contain secondary structures (hairpins). Association of ssDNA with SSB not only eliminates these structures, but also prevents RecA filament nucleation, thereby avoiding unwanted recombination events. The RecO, RecR, and RecF proteins (termed RecFOR) load RecA protein onto SSB-coated ssDNA by exchanging SSB for RecA. Assembly of RecA filaments proceeds in the 5′→3′ direction with one RecA molecule bound to every three DNA bases.

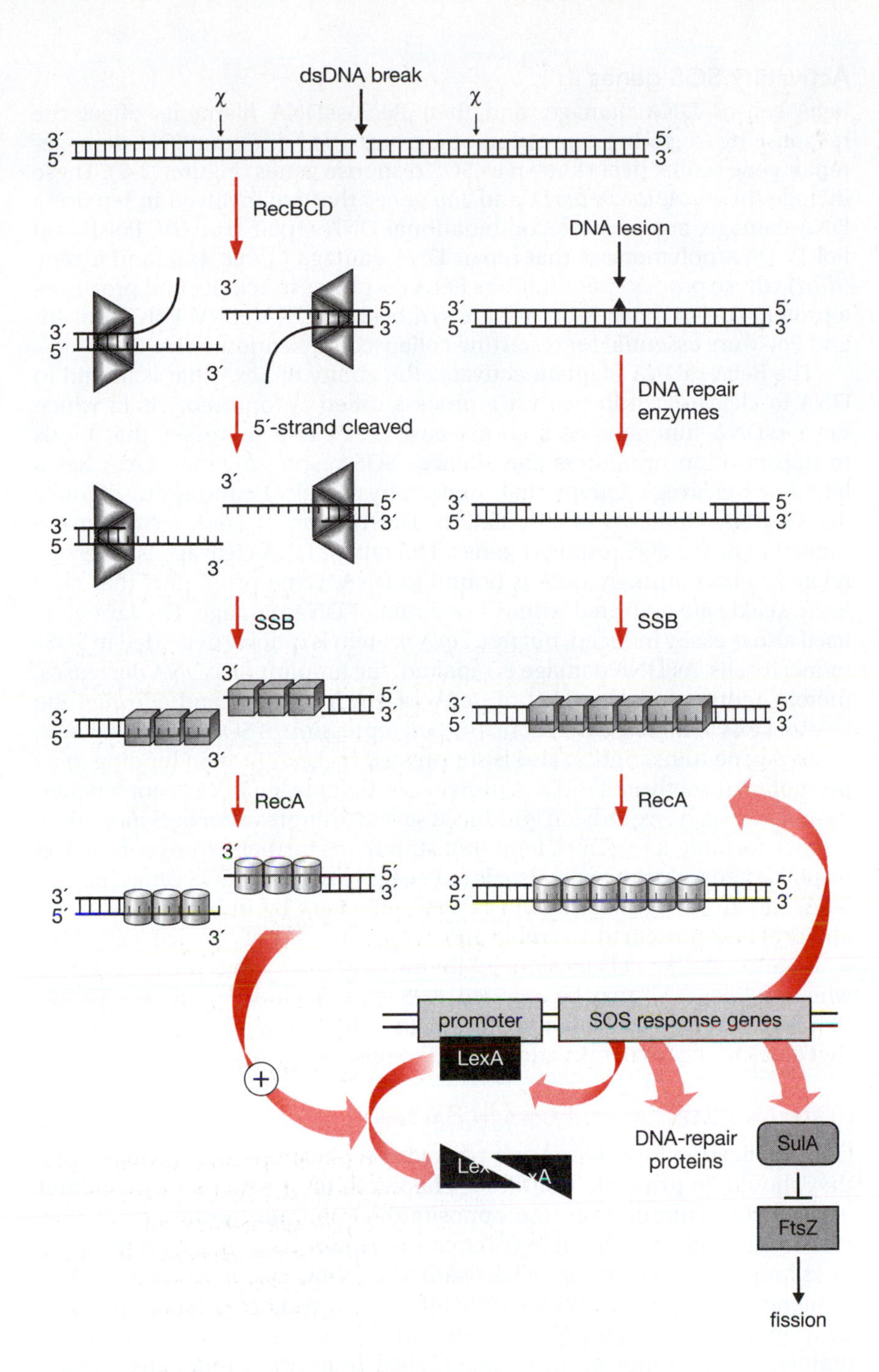

Figure 13-7. The SOS response in *E. coli*. DNA damage in the form of dsDNA breaks or DNA lesions is detected by recruitment of RecA protein to SSB•ssDNA complexes that result from the process of DNA repair. Repair of dsDNA breaks is initiated by RecBCD to provide 3′-ssDNA overhangs that begin at a χ-site. Repair of DNA lesions involves excision of the modified nucleotide to produce a ssDNA gap within the duplex DNA. RecA•ssDNA filaments trigger proteolysis of the transcriptional repressor LexA to allow expression of SOS response genes.

Assembly requires binding of ATP to RecA, and disassembly requires ATP hydrolysis. The DinI protein that is induced very early in the SOS response stabilizes RecA filaments. RecX protein is also produced in order to limit the extension of RecA filaments by binding to the 3′-terminal RecA monomer. Thus, DinI and RecX are antagonists; DinI stabilizes the RecA filament, whereas RecX destabilizes it. The SOS response is eventually terminated by the appearance of UvrD, a DNA helicase induced as part of the SOS response. UvrD dismantles RecA filaments. Together, these proteins determine where and when RecA induces DNA recombination.

dsDNA break repair begins with resection of the break by RecBCD (Figure 13-7) to provide 3′-ssDNA overhangs that are converted into RecA•ssDNA filaments (Figure 13-8). Not only do these filaments activate expression of the SOS response genes (Figure 13-7), but also they initiate 'strand invasion' to allow homologous recombination repair of the dsDNA break (Figure 13-9).

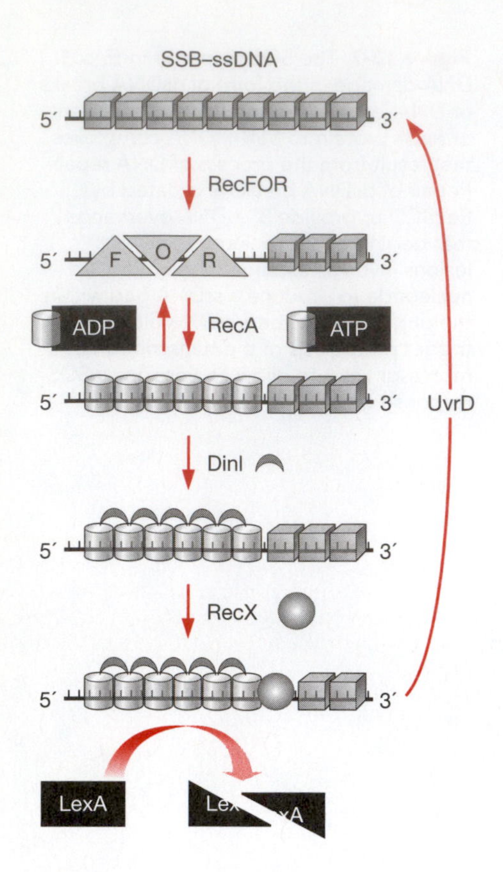

Figure 13-8. Assembly of RecA filaments in *E. coli*. (Data from Cox, M.M. (2007) *Crit. Rev. Biochem. Mol. Biol.* 42: 41–63.)

Activating SOS genes

RecA senses DNA damage, and then RecA•ssDNA filaments effect the response by triggering expression of a group of 43 different DNA damage-repair genes collectively known as SOS response genes (Figure 13-7). These include the *uvrA*, *uvrB*, *uvrD*, and *cho* genes that are involved in repairing DNA damage, a gene for recombinational DNA repair (*ruvAB*), Pol-II and Pol-IV DNA polymerases that repair DNA damage (Table 4-3), and a gene (*dinI*) whose product both inhibits RecA co-protease activity and processes a protein (UmuD) necessary for the synthesis of DNA Pol-V. Pol-II, Pol-III, and Pol-V are essential for restarting collapsed replication forks.

The RecA•ssDNA filament activates the ability of LexA that is bound to DNA to cleave itself in two via a process called autoproteolysis in which RecA–ssDNA functions as a co-protease. LexA is a repressor that binds to transcription promoters and silences SOS response genes. LexA has a latent self-cleavage activity that, under physiological conditions, requires RecA•ssDNA filaments to stimulate it. Destruction of LexA results in the induction of the SOS response genes. The rate of LexA cleavage is inversely related to how strongly LexA is bound to DNA. Gene promoters that bind LexA weakly are activated within 1 or 2 min of DNA damage. The *lexA* gene itself also is easily induced, but free LexA protein is quickly degraded in SOS-induced cells. As DNA damage is repaired, the amount of ssDNA decreases, thereby reducing the amount of RecA•ssDNA in the cell and allowing the level of LexA to increase with concomitant repression of SOS response genes.

recA gene transcription also is suppressed by LexA protein binding to its promoter and activated by LexA proteolysis. If damaged DNA is not repaired rapidly, *recA* gene expression is induced several minutes after SOS induction, thereby forming a feedback loop that stimulates further expression of SOS response genes due to higher levels of RecA (Figure 13-7). SOS genes include SulA, which prevents formation of FtsZ polymers by interfering with the ability of FtsZ protein to assemble into a ring at the site of cell cleavage. Since FtsZ is essential for cell division, inhibition of FtsZ prolongs the time during which cellular DNA may be repaired. SOS genes include UmuD and UmuC that function in **translesion synthesis**, the process by which DNA lesions that block replication forks are repaired so that forks can continue.

Restarting Replication Forks In Bacteria

Once replication forks have been created at replication origins (Chapter 9), they should, in principle, continue (Chapter 3) until either they encounter another fork coming from the opposite direction and terminate, or they run off the end of a telomere (Chapter 6). However, in reality, replication forks frequently stall during cell division when they encounter atypical DNA sequences (Chapter 1) that are difficult to copy, modified bases produced by agents that damage DNA that are difficult to bypass, or breaks in the template that cannot be traversed. Stalled replication forks are a major source of genomic instability in all cellular organisms. To counter this threat, bacteria evolved with mechanisms that recognize the structure of stalled replication forks, and then set about restoring the replisome (Figure 4-21) in order to restart replication where it had left off.

Replication-restart proteins

Replication-restart proteins were discovered through biochemical reconstitution of bacteriophage ϕX174 DNA replication *in vitro*. The PriA, PriB, PriC, and DnaT proteins, along with DnaB, DnaC, and DnaG, are required for the initial steps in the conversion of the circular ϕX174 ssDNA genome into its dsDNA replicative form (Figure 8-8). Although this mechanism was initially thought to mirror the workings of Okazaki-fragment synthesis at chromosomal replication forks, subsequent discoveries revealed that only DnaB, DnaC, and DnaG are required to assemble a replisome at oriC

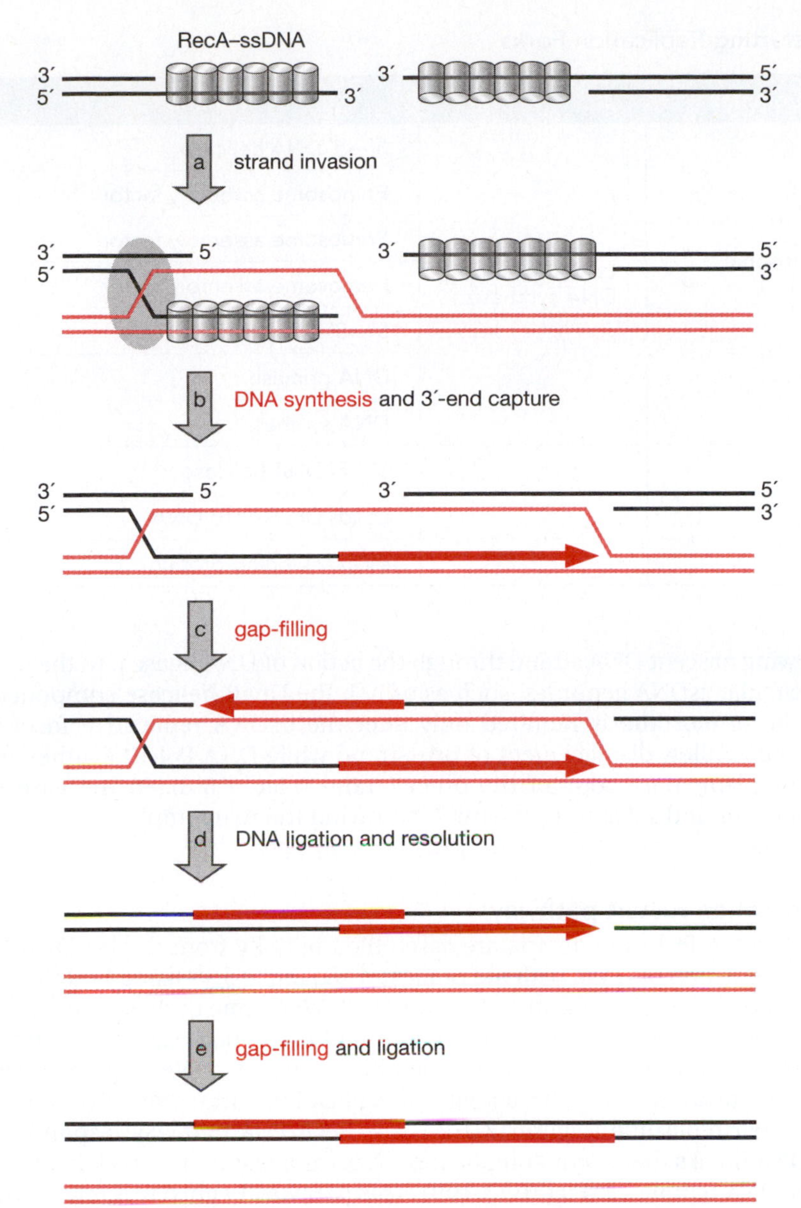

Figure 13-9. dsDNA break repair in bacteria. (a) RecA•SSB filaments catalyze strand invasion into a homologous sequence where base-pairing occurs between complementary strands. A transient Holliday junction (named for its originator, Robin Holliday) forms between four strands of DNA (shaded oval). This junction can migrate up or down the DNA molecule. (b) The 3′-OH end of the invading strand acts as a primer for DNA polymerase to extend the invading strand as well as the displaced loop of DNA. Nascent DNA is indicated by red arrows. This event allows 'capture' of the other 3′-ssDNA•RecA filament. Again, RecA facilitates base pairing between complementary sequences. (c) DNA polymerase fills in ssDNA gap. (d) DNA ligase seals parental DNA strands. Holliday junctions are then resolved by specific endonucleases that recognize the unique structure of this DNA junction. (e) Remaining ssDNA gap is filled in and sealed by DNA ligase.

(Figure 11-6). The role of PriA, PriB, PriC, and DnaT proteins – termed replication-restart proteins – is to assemble a primosome at replication forks that have lost their replisome.

The **primosome** is a complex of six proteins that synthesizes RNA primers on a ssDNA template (Table 13-6). Initially, the complex formed by PriA, PriB, and PriC binds to a SSB•ssDNA complex. Then the DnaB helicase is escorted to this site by DnaT to form a pre-primosome. Finally, DnaG, the bacterial DNA primase, binds to the pre-primosome to form the primosome. At this point, DNA synthesis is essentially the same as previously described for Okazaki fragments on the lagging-strand template of replication forks (Table 3-2). DnaG synthesizes oligoribonucleotides of approximately 10 residues on the ssDNA template to create a hybrid RNA:DNA duplex.

This complex of RNA-primer–DNA template associated with a primosome recruits the DNA Pol-III holoenzyme that then carries out DNA synthesis on both arms of the fork. RNA primers are ultimately replaced with DNA through the actions of RNase H, the 5′→3′ exonuclease of DNA Pol-I, and either the Pol-I or Pol-III DNA polymerase. The 3′-end of the short nascent DNA chain is then covalently joined to the 5′-end of the long

Table 13-6. Bacterial Proteins Involved In Restarting Replication Forks

<table>
<tr><th>Protein</th><th colspan="3">Complex</th><th>Function</th></tr>
<tr><td>PriA</td><td rowspan="5">Pre-primosome</td><td rowspan="6">Primosome</td><td rowspan="7">Leading-strand DNA synthesis</td><td>3′→5′ DNA helicase</td></tr>
<tr><td>PriB</td><td>Primosome assembly factor</td></tr>
<tr><td>PriC</td><td>Primosome assembly factor</td></tr>
<tr><td>DnaT</td><td>Primosome assembly factor</td></tr>
<tr><td>DnaB</td><td>5′→3′ DNA helicase</td></tr>
<tr><td>DnaG</td><td></td><td>DNA primase</td></tr>
<tr><td>Pol-III holoenzyme</td><td></td><td></td><td>DNA synthesis</td></tr>
<tr><td>Rep</td><td></td><td></td><td></td><td>3′→5′ DNA helicase</td></tr>
<tr><td>DnaC</td><td></td><td></td><td></td><td>Loads DnaB onto DNA</td></tr>
<tr><td>SSB</td><td></td><td></td><td></td><td>ssDNA-binding protein</td></tr>
</table>

growing nascent DNA strand through the action of DNA ligase I. In the case of circular ssDNA genomes, such as ϕX174, the DnaB helicase component of the primosome is required only after the dsDNA replicative form is nicked to allow displacement of one strand while DNA Pol-III synthesizes a complementary copy of the other strand (Figure 8-8). In the case of replication forks, DnaB is essential to unwind the two template strands at the fork.

Replication-restart pathways

Replication forks in bacteria are assembled initially from duplex DNA by the interaction of DnaA with the replicator sequence oriC using 11 different proteins (Table 11-4, Figures 11-5 and 11-6). When one of these forks stalls or collapses, bacteria utilize one of four additional pathways to assemble a replication fork that does not depend on a specific DNA sequence. The PriA pathway restores the activity of replication forks that have simply lost their replisome (Figure 13-10A). The RecA-PriA pathway responds to dsDNA breaks that occur at replication forks as a result of a break in one of its ssDNA templates (Figure 13-10B). This pathway begins by applying the same SOS-induced dsDNA break repair-mechanism involving homologous recombination (Figure 13-7). Here the RecA-PriA pathway converts the D-loop that forms as an intermediate structure in dsDNA break repair (Figure 13-9) into a replication fork. The PriC pathway restores replication when leading-strand DNA synthesis is arrested but DNA unwinding continues, thereby producing a large ssDNA gap on the leading-strand template (Figure 13-10C). When replication forks encounter a DNA lesion, an enzyme complex called RecFOR loads RecA onto the damaged DNA. This event triggers either a DNA repair or a DNA damage-bypass pathway, and then the fork can be restarted by the RecA-PriC pathway (Figure 13-10D). Cells with mutations in either *priA* or *priC* rapidly accumulate DNA damage, revealing that bacterial replication forks stall frequently and therefore require replication-restart mechanisms to forestall DNA damage. Mutations in both *priA* and *priC* are lethal, consistent with two independent restart mechanisms.

PriA pathway

The PriA pathway requires PriA, PriB, DnaT, and DnaB (Table 13-6). PriA is a 3′→5′ DNA helicase that not only interacts specifically with SSB, but also has a high affinity for branched DNA junctions, including D-loop and

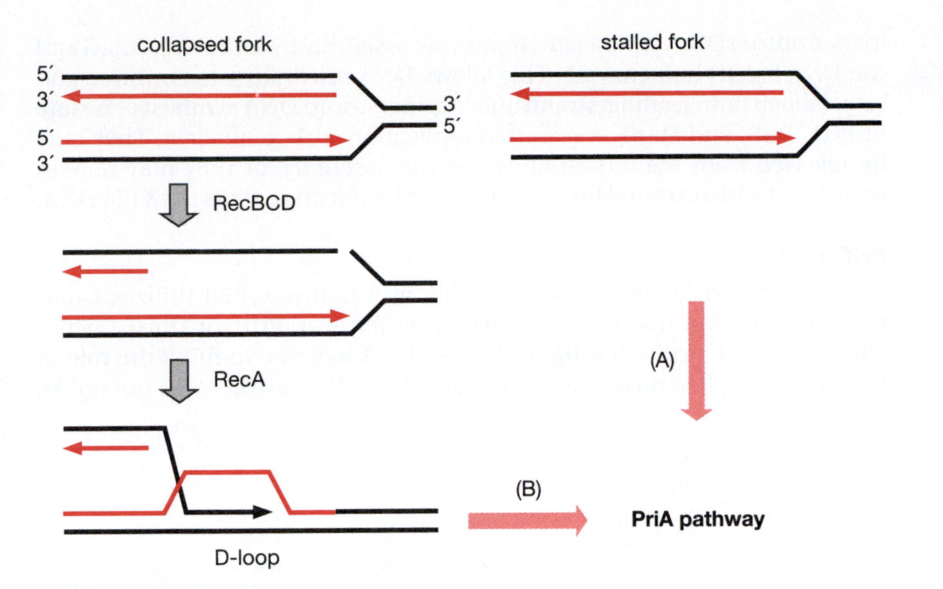

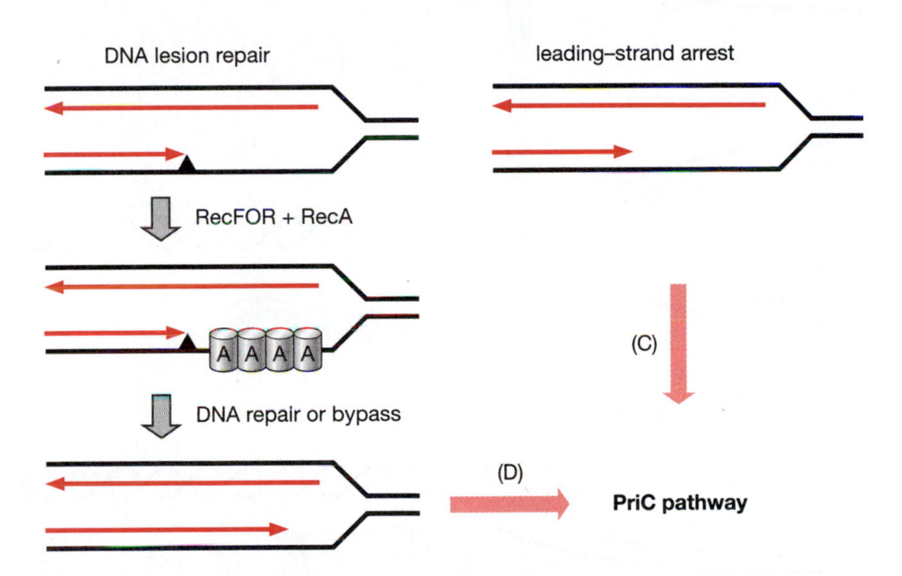

Figure 13-10. Four pathways for restarting replication forks in bacteria. The PriA pathway is used when a fork either stalls because of replisome malfunction (A), or collapses from a break in one of its ssDNA template strands (B). The PriC pathway is used either when leading-strand synthesis is arrested, but DNA unwinding continues (C), or when leading-strand DNA synthesis encounters a DNA lesion (D).

stalled fork structures with a nascent 3′-OH terminus near the fork junction (Figure 13-11). As the distance increases, the affinity of PriA decreases, as does its ability to nucleate replisome assembly. Small gaps of 3 to 5 nucleotides are tolerated, but larger gaps severely reduce PriA-dependent replisome assembly. Thus, PriA-dependent replisome assembly can restart forks following homologous recombination of the type driven by RecA protein, but larger ssDNA gaps require the PriC-dependent pathway.

Binding of PriA to DNA rapidly induces binding of PriB, which stabilizes the PriA•DNA complex (Figure 13-11). PriB has two oligonucleotide-binding (OB) folds, suggesting that PriB facilitates replisome assembly through its ability to form a complex with ssDNA. This complex facilitates association with DnaT, which brings the DnaB helicase with it to the replication fork. Thus, the role of PriA is analogous to that of DnaA, except that where DnaA recognizes a specific DNA sequence (oriC), PriA recognizes a specific DNA structure (branched DNA). Moreover, the role of PriB is analogous to that of SSB, and the role of DnaT is analogous to that of DnaC. The PriA•PriB•DnaT•DNA complex functions as a 'helicase loader' for DnaB at branched DNA structures in much the same way that DnaA and DnaC load DnaB at oriC (compare Figure 13-11 with Figure 11-6). Once DnaB has been loaded onto ssDNA, it nucleates replisome assembly by recruiting DnaG and the DNA Pol-III holoenzyme. This allows DNA unwinding to continue and re-establish both leading-strand and lagging-strand DNA synthesis. The fate of PriA, PriB, and DnaT at restarted replication forks is unclear. They may be released from the fork upon replisome assembly, or they may remain associated with restarted forks, as they do at replication forks in ϕX174 DNA.

PriC pathway

PriC is required for replisome assembly in a pathway that utilizes DnaC to load DnaB, but that does not involve either PriA, PriB, or DnaT (Figure 13-12). Instead, PriC relies upon the Rep DNA helicase to fulfill the role of PriA. Like PriA, Rep possesses an intrinsic 3′→5′ helicase activity, but unlike

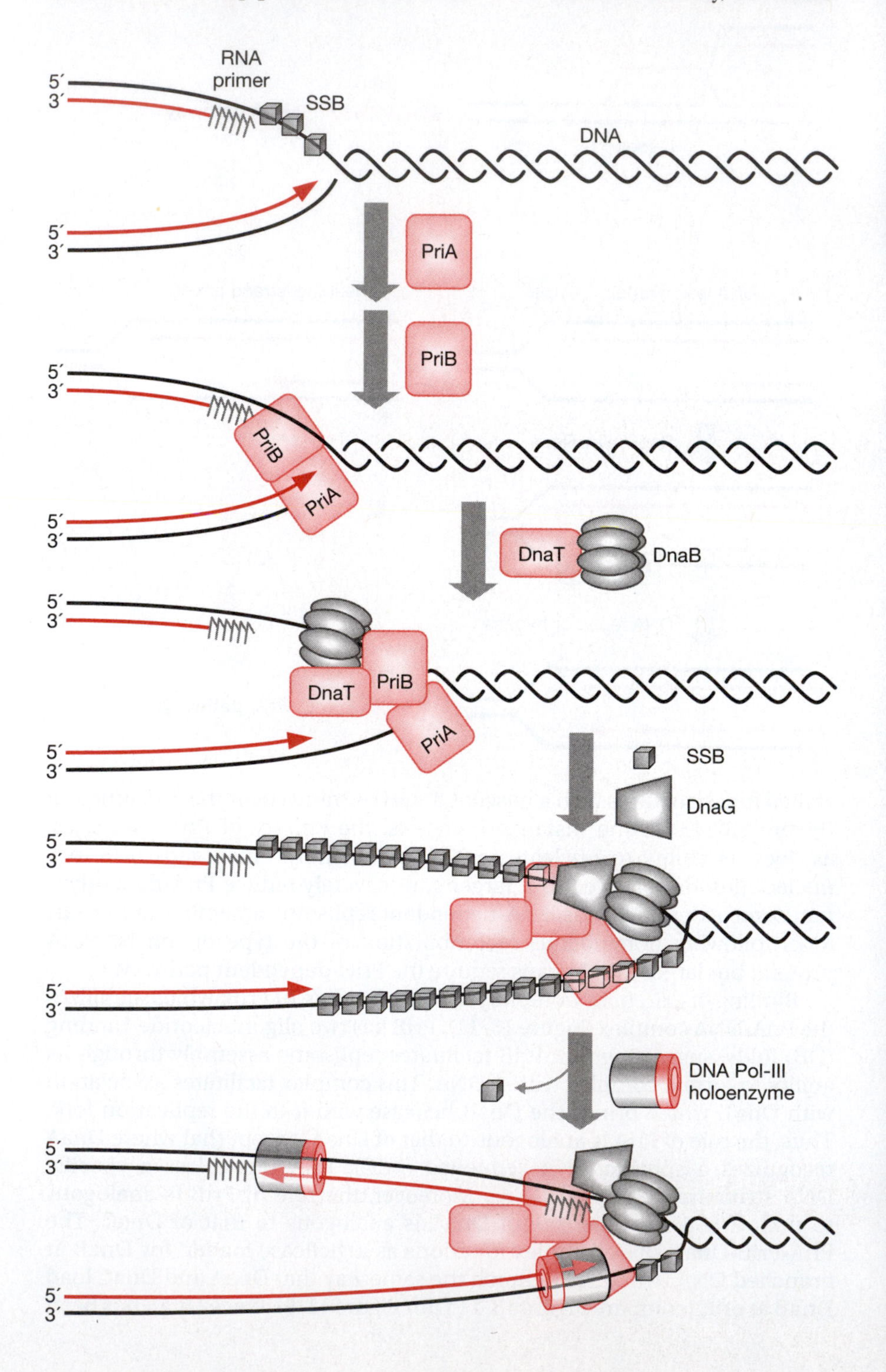

Figure 13-11. PriA-dependent replisome assembly at a stalled replication fork. Black is template DNA strand; red is nascent strand; gray bars are RNA primer.

PriA its helicase activity is stimulated by PriC. Otherwise, it appears that PriC•Rep functions much the way that PriB•PriA does, except that the two protein complexes target different DNA structures.

Whereas optimum PriA-dependent replisome assembly requires a gap size of less than six or seven nucleotides between the 3′-terminus of the nascent leading strand and the fork junction, PriC-dependent replisome assembly requires a gap size of seven nucleotides or greater. Therefore, PriC targets forks in which leading-strand DNA synthesis stopped, but DNA unwinding continued. *In vitro*, purified PriA protein in the company of PriB, DnaT, and DnaC effectively recruits DnaB only when a 3′-DNA end is nearby. In contrast, PriC in the company of DnaC recruits DnaB most effectively when there is a ssDNA gap at the fork. Rep appears to facilitate the PriC pathway by unwinding sufficient lagging-strand DNA template to allow loading of DnaB during replisome assembly.

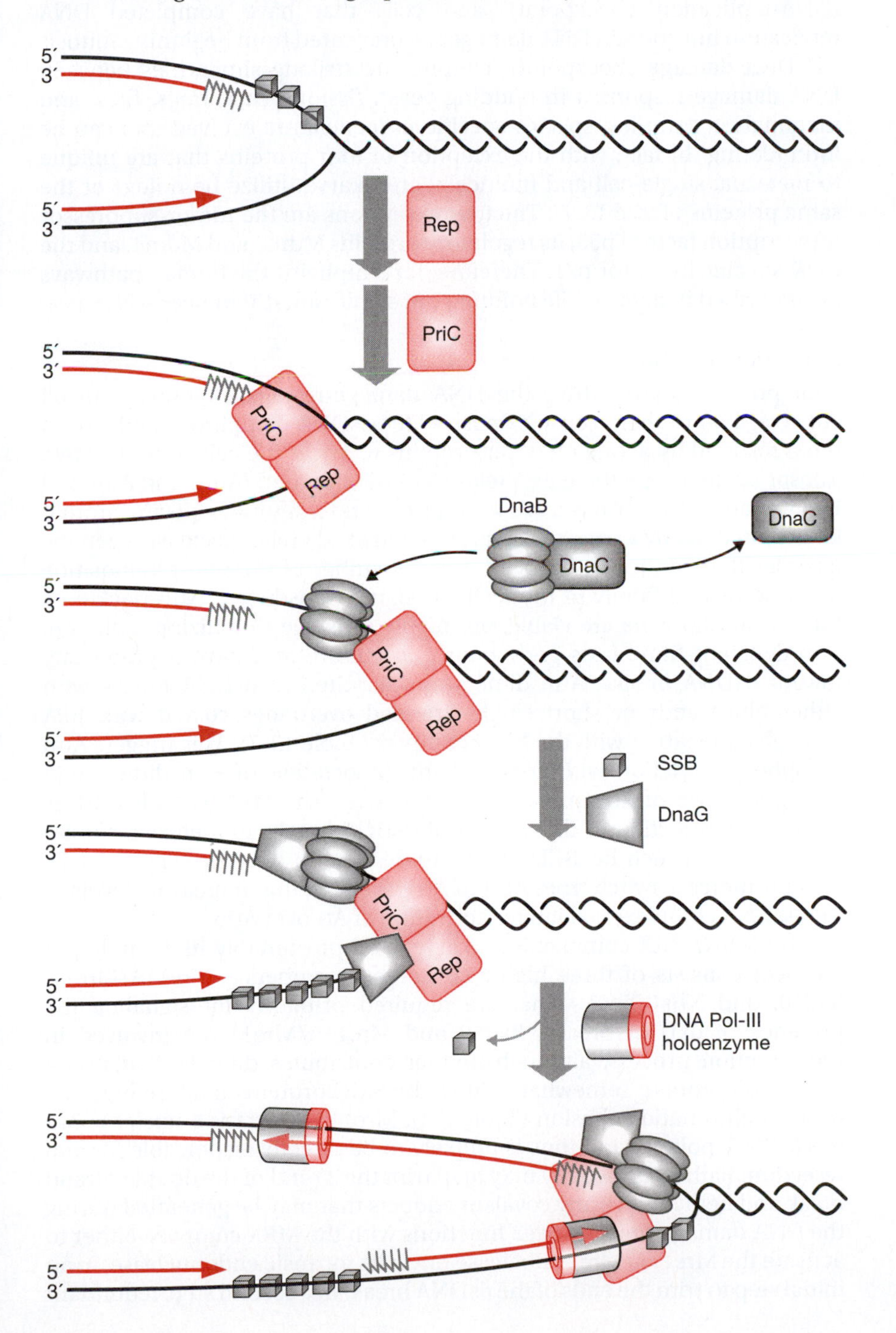

Figure 13-12. PriC-dependent replisome assembly at a stalled replication fork. DnaC is released from the fork when DnaB is loaded onto the Rep•PriC-SSDNA complex. The events in recruiting SSB, DnaG, and DNA Pol-III holoenzyme are the same as in the PriA pathway (Figure 13-11).

Archaea

Like bacteria and eukarya, archaea must have mechanisms that recognize damaged DNA and stalled replication forks. Archaea do encode a Rad51/RecA ortholog (called RadA) and orthologs of the Mre11 and Rad50 subunits of the eukaryotic MRN complex (Table 13-5), but as yet little is known about the DNA damage response in these unusual forms of life.

Eukarya

In contrast to bacteria, the eukaryotic DNA damage response operates throughout interphase, not just during chromosome duplication. As a result, cells with damaged DNA are prevented from entering S phase (G_1 DNA damage checkpoint), S-phase cells with damaged DNA or stalled replication forks are prevented from continuing the replication process (DNA-replication checkpoint), and cells that have completed DNA replication but contain DNA damage are prevented from beginning mitosis (G_2 DNA damage checkpoint). Despite the striking similarities between DNA damage responses in budding yeast, fission yeast, frogs, flies, and mammals, a complex, species-specific nomenclature evolved that can be intimidating. In fact, with the exception of four proteins that are unique to metazoa, single-cell and multicellular eukarya utilize homologs of the same proteins (Table 13-7). The four exceptions are the tumor-suppressor transcription factor Tp53, its regulatory proteins Mdm2 and Mdm4, and the CDK-specific inhibitor p21. Therefore, for simplicity, the human pathways are described in detail while pointing out significant differences with yeast.

Sensors and effectors

Four protein kinases drive the DNA damage-response pathways in all eukarya. Two of them sense extensive RPA•ssDNA complexes, and two of them respond by acting on other proteins to arrest the cell cycle. The two 'sensor proteins' are the **a**taxia telangiectasia mutated (Atm) and Atm and Rad3-related (Atr). Atm is a 351 kDa serine- and threonine-specific protein kinase encoded by a gene that is mutated in ataxia telangiectasia, a genetic disorder that can lead to cancer. It is a member of the phosphoinositide 3-kinase-related family of intracellular signal transducers. Mice lacking a functional *Atm* gene are viable, but highly sensitive to ionizing radiation. Atm-deficient fibroblasts grow poorly in culture and have a propensity toward dsDNA breaks. Atm dimers are recruited to dsDNA breaks with either blunt ends or short single-stranded overhangs coated with RPA through association with the MRN complex (Table 13-7). This triggers Atm autophosphorylation with concomitant dissociation of Atm dimers and the appearance of Atm protein kinase activity. The MRN complex drives progressive resection of the 5′-ends of dsDNA breaks in metazoa, similar to the way in which RecBCD resects these ends in bacteria (Figure 13-7). This promotes a switch from Atm to Atr, revealing that increasing levels of RPA•ssDNA complexes facilitates activation of Atr over Atm.

The MRN/MRX complex in eukarya (and presumably its homolog in archaea) consists of three highly conserved members – Mre11A/Mre11, Rad50, and Nbs1/Xrs2 – that are required primarily for signaling the presence of dsDNA breaks. Rad50 and Mre11A/Mre11 are involved in the resection process, although neither contributes directly. Rad50 is a coiled-coil protein, somewhat akin to the SMC proteins that are involved in sister chromatid cohesion (Table 7-4). Mre11A/Mre11 is a nuclease but has a 3′→5′ polarity and thus is unlikely to be directly responsible for end resection. Rather the activity may help trim the 3′-end of the double-strand break and help remove any covalent adducts that may be generated during the DNA damage process. Sae2 functions with the MRX complex, either to activate the Mre11A/Mre11 nuclease or via its intrinsic endonuclease, in an initial step to trim the ends of the dsDNA break. In a second step, redundant

Table 13-7. DNA Damage-Response Proteins In Eukarya

Yeast		Metazoa	Function
S. cerevisiae	*S. pombe*	*H. sapiens*	
Mec1	Rad3	Atr	Protein kinase, senses RPA•ssDNA
Ddc2	Rad26	Atrip	Atr interactive protein
Tel1	Tel1	Atm	Protein kinase, senses RPA•ssDNA
Tel2 *MRX complex* Mre11 Rad50 Xrs2 Sae2	Tel2	*MRN complex* Mre11A Rad50 Nbn/Nbs1 Bcl11A/Ctip1	Atm interactive proteins; Tel2 is a homolog of Ddc2/Atrip
Rad24•RFC(2-5)	Rad17•RFC(2-5)	Rad17•RFC(2-5)	Checkpoint sliding clamp loader
Ddc1 Rad17 Mec3	*9-1-1 complex* Rad9 Rad1 Hus1	*9-1-1 complex* Rad9 Rad1 Hus1	Checkpoint sliding clamp
Dpb11	Rad4	Topbp1	Activates Atr
Pol-ε	Pol-ε	Pol-ε	Leading-strand DNA synthesis
Tof1	Swi1	Tim/Timeless	Fork stability
Csm3	Swi3	Tipin	Fork stability
Mrc1	Mrc1	Clspn/Claspin	Fork stability and mediator
Ino80		Ino80	Chromatin remodeling
Rad9	Rhp9	Brca1	Mediator
Rad53	Cds1	Chek2/Chk2	Effector protein kinase
Chk1	Chk1	Chek1/Chk1	Effector protein kinase
		Tp53/p53	Tumor-suppressor transcription factor
		Mdm2	Ubiquitin ligase, inhibits Tp53
		Mdm4/MdmX	Mdm2-associated protein
		Cdkn1A/p21/Cip1	CDK-specific inhibitor

Multiple names for the same gene are separated by / with the first name recommended by the HUGO Gene Nomenclature Committee. The second and third names are ones used more commonly. Abbreviations for human proteins are generally written in capitals, but for simplicity all protein names are written in title case.

systems remove long tracts of DNA to reveal extensive 3′ single-stranded tails. One system is dependent on the helicase Sgs1 and the nuclease DNA2, and the other on the 5′→3′ exonuclease Exo1.

Atr is a 301 kDa protein kinase that is homologous to the Atm gene product in human cells and the Rad3/Mec1 proteins in yeast. In contrast to Atm, Atr is essential for cell viability in mammals. Mouse embryos lacking a functional *Atr* gene accumulate chromosome breaks and die prior to embryonic day 7.5. Inactivating the *Atr* gene in mouse embryonic fibroblasts leads to accumulation of dsDNA breaks during S phase, even in the absence of induced DNA damage, suggesting that replication forks in eukarya frequently stall, as they do in bacteria. The Atr kinase is activated by a broader spectrum of genomic insults than the Atm kinase, especially DNA damage that interferes with DNA replication.

The two 'effectors' in eukaryotic DNA damage-response pathways are the 54 kDa checkpoint kinase 1 (Chek1/Chk1) and the 61 kDa checkpoint kinase 2 (Chek2/Chk2). Both are nuclear proteins that phosphorylate specific serine or threonine residues in other proteins. Although both Chk1 and Chk2 kinase activities are activated by phosphorylation by either Atm or Atr, they are structurally unrelated proteins. Chk1 contains a protein kinase domain in its N-terminal half and a regulatory domain in its C-terminal half. These are reversed in Chk2; the regulatory domain is in the N-terminal portion and the protein kinase domain is in the C-terminal portion. *In vivo*, Atm activates primarily Chk2, and Atr activates primarily Chk1, but the distinction is not absolute. The substrate specificity for all four of these protein kinases may rely on as yet unidentified co-activator proteins whose expression may be dependent on the cell cycle and that may respond to different genotoxic insults.

The primary target of Chk1 and Chk2 is the Cdc25 phosphatase, of which there are three isoforms in mammals, but only one in yeast and insects (Table 12-3). These enzymes dephosphorylate two residues within the catalytic site of CDKs, converting them from an inactive form into an active form. Cdc25A activates both Cdk1 and Cdk2, thereby regulating the $G_1 \rightarrow S$ transition, S phase, and mitosis. The primary, if not exclusive, role for Cdc25C *in vivo* is activation of Cdk1. Thus, Cdc25C regulates the $G_2 \rightarrow M$ transition. Cdc25B appears to activate CDK1 specifically at the centrosome. Either Chk1 or Chk2 can phosphorylate all three isoforms of Cdc25. In each case, phosphorylation of Cdc25 inhibits CDK activity and arrests cell division.

Required versus essential genes

Some genes are essential to the viability of an organism, whereas other genes contribute to a particular pathway, but their role can be carried out by another gene with similar activity. For example, mouse cells lacking both Cdc25B and Cdc25C display minimum defects in cell-cycle progression, because Cdc25A is sufficient. Cdc25A, however, is essential for mouse development. Therefore, Cdc25A is required for entrance into mitosis as well as into S phase. Atr, Atm, Chk1, and Chk2 elicit specific DNA damage responses in various eukaryotic organisms (Table 13-8). The Atm/Tel1–Chk2/Rad53 pathway detects DNA damage in G_1-phase cells, but the Atr/Mec1–Chk1 pathway detects DNA damage in G_2-phase cells. Chk2 is the preferred effector kinase in budding yeast, whereas Chk1 is the preferred effector kinase in mammals. Depending upon the species in question, their function is not always essential for the viability of the organism (Table 13-9). Viability refers to an organism's ability to live and develop. In the case of mammals, this means outside the uterus. For example, *Chk1* is an essential gene in flies and mammals, but not in yeast. In mammals, *Chk1*$^{-/-}$ embryos die *in utero*. On the other hand, *Chk2* is an essential gene in budding yeast, but not in fission yeast, flies, or mammals. These differences reflect the evolutionary divergence among organisms as they develop functionally redundant pathways where one gene can substitute for another, and multiple concerted pathways where different mechanisms can achieve the same result. Such strategies make complex life forms possible.

Table 13-8. Species Specificity For DNA Damage-Response Genes

Cell-cycle block	Budding yeast (*S. cerevisiae*)		Mammals (*H. sapiens*)	
	Sensor	Effector	Sensor	Effector
$G_1 \rightarrow S$	Tel1•MXR	Rad53	Atm•MRN	Chk2
S phase	Mec1•Ddc2	Rad53	Atr•Atrip	Chk1
$G_2 \rightarrow M$	Mec1•Ddc2	Chk1	Atr•Atrip	Chk1

Table 13-9. DNA Damage-Response Genes And Their Effect On Viability

Gene	Mammals (mice)	Budding yeast (*S. cerevisiae*)	Fission yeast (*S. pombe*)	*Insects* (*Drosophila*)
Atr/Mec1/Rad3 /Mei41	Essential	Essential	None	Essential
Atm/Tel1/Tel1/Tefu	None	None	None	Essential
Chk1/ Chk1/Chk1/Grp	Essential	None	None	Essential
Chk2/Rad53/Cds1 /Loki	None	Essential	None	None

Gene names are for mice/*S. cerevisiae*/*S. pombe*/*D. melanogaster*

G_1 DNA Damage Checkpoint

Exposing mammalian cells either to ionizing radiation or to radiomimetic agents causes dsDNA breaks. When this happens prior to S phase, the G_1 DNA damage checkpoint is induced. dsDNA breaks are initially sensed by Atm, but Atr may amplify the signal as the MRN complex resects the ends of the DNA break. Embryonic fibroblasts derived from *Chk2*$^{-/-}$ mice are deficient in this checkpoint, but not in the DNA replication or G_2 DNA damage checkpoints, suggesting that Chk2 is the primary effector kinase for dsDNA break repair. However, two independent pathways are involved in preventing the onset of S phase, one that requires a Chk kinase and one that does not (Figure 13-13).

Atm→Chk2⊣Cdc25A→Cdk2•CcnE

One pathway targets Cdc25A phosphatase, the enzyme that converts the inactive Thr-14 and Tyr-15-phosphorylated form of Cdk2 into an active form. In the company of CcnE, Cdk2 drives the G_1→S transition by triggering the conversion of preRCs into preICs (Figure 11-15). Either Chk2 or Chk1 phosphorylation of Cdc25A inactivates this phosphatase and converts it into a substrate for the SCF(βTrCP) ubiquitin ligase and subsequent proteolysis by the 26S proteasome. This action prevents the appearance of an active form of Cdk2 and inhibits the G_1→S-phase transition (Figure 13-13).

Atm or Chk2→Tp53→p21⊣Cdk2•CcnE

A second pathway is unique to metazoa, because only metazoa contain the Tp53 protein. Tp53 is a transcription factor that regulates the cell cycle and functions as a tumor suppressor by preventing genomic instability. The Atm-activated Chk2 kinase phosphorylates the C-terminal end of Tp53, thereby activating Tp53-dependent transcription (Figure 13-13). One of the genes whose transcription is Tp53-dependent is the CDK-specific inhibitor p21. p21 then inhibits Cdk2•CcnE, thereby preventing it from triggering the G_1→S-phase transition. The Atm kinase also stabilizes Tp53 by preventing its proteolysis and export into the cytoplasm. In unstressed cells, Tp53 protein is restrained by the ubiquitin ligase Mdm2. Mdm2 binds to the N-terminal transactivation domain of Tp53 and prevents Tp53-dependent transcription. In addition, Mdm2 transports Tp53 out of the nucleus by the virtue of its nuclear export signal and tags it for ubiquitin-mediated proteolysis. In normal cells, Tp53 increases transcription of Mdm2 over basal levels, thereby insuring that Tp53 activity is held in check.

In stressed conditions, the function of the Mdm2 is blocked by phosphorylation, protein-binding events, and enhanced degradation. When Atm is activated, it stimulates the cellular level and transcriptional activity of Tp53 by phosphorylating the N-terminal region of Tp53 as well as the C-terminal region of Mdm2, thereby preventing Mdm2 from interacting with Tp53. In addition, Atm activates Chk2, which then phosphorylates several sites within the N-terminal transactivation domain of Tp53, further

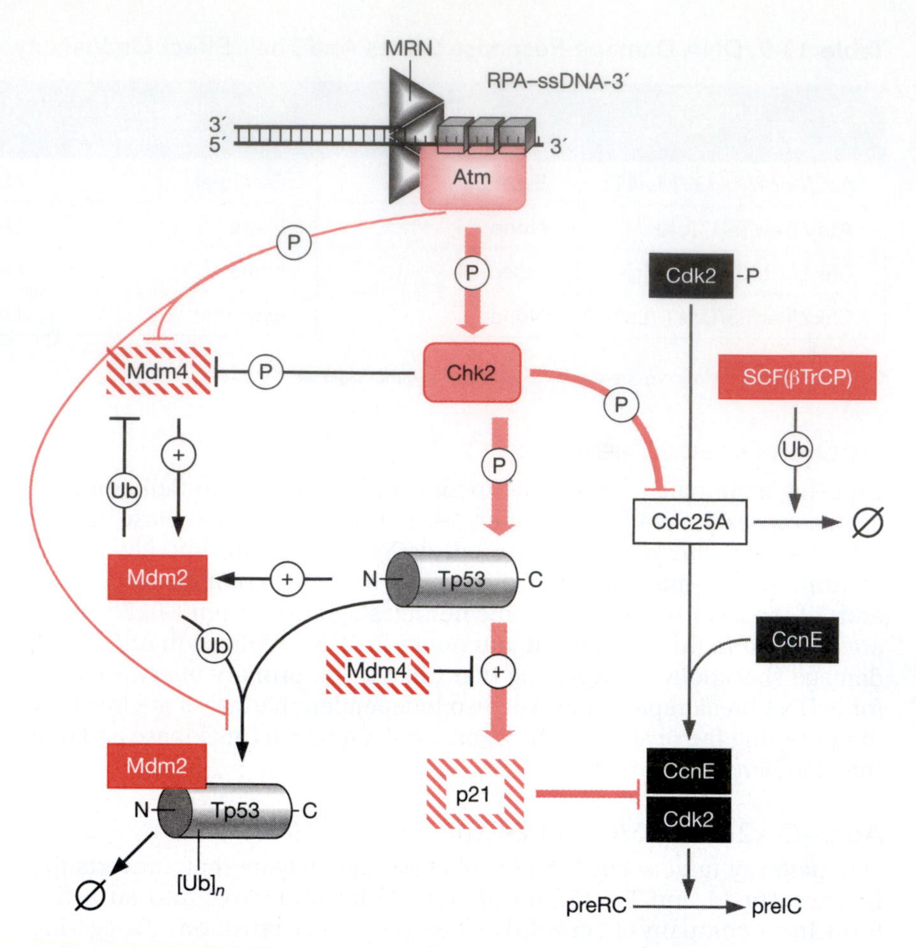

Figure 13-13. The G_1 DNA damage checkpoint responds to breaks in dsDNA. The MRN complex binds to the dsDNA break and promotes resection of the 5′-end. The resulting 3′-ssDNA tail is coated with RPA, which activates the Atm 'sensor' kinase. Atm kinase then activates the Chk2 'effector' kinase. Thus begins a cascade of phosphorylation events that culminate in the inhibition of the CDK activity required to trigger preIC assembly and the onset of S phase. +, activation; P, phosphorylation; Ub, ubiquitination; $[Ub]_n$, polyubiquitination; Ø, proteolysis; ⊥, inhibition; →, conversion of substrate into product. Heavy orange-colored arrows and inhibition bars indicate primary DNA damage-response pathway in G_1-phase cells.

inhibiting the interaction between Tp53 and Mdm2. These events stabilize Tp53, which in turn, increases its cellular concentration. The fact that *Chk2*$^{-/-}$ cells can still give a robust Tp53 response following exposure to ionizing radiation attests to the existence of multiple regulatory pathways for regulating Tp53 activity *in vivo*.

Mdm4, an Mdm2-related protein, binds to Tp53 and inhibits its transcriptional activity without targeting Tp53 for ubiquitination. In the absence of stress, Mdm4 binds and stabilizes Mdm2 by preventing Mdm2 from auto-ubiquitination and subsequent degradation. When Atm is activated, Mdm4 is phosphorylated both by Atm and Chk2. These phosphorylation events enhance the ability of Mdm2 to target Mdm4 for ubiquitin-dependent degradation. Removal of Mdm4 in turn destabilizes Mdm2 and promotes the accumulation and activation of Tp53.

DNA-Replication Checkpoint

Detecting stalled replication forks and DNA damage

When RPA•ssDNA complexes appear during S phase, they can recruit Atr via its cofactor, the Atr-interacting protein (Atrip) (Figure 13-14). However, since RPA bound to ssDNA is also a normal transient intermediate at replication forks, the DNA damage signal must in some way be distinguished from the signal to initiate Okazaki-fragment synthesis on the lagging-strand template and possibly the signal to continue DNA synthesis on the leading-strand template. One way this could be accomplished is if RPA•ssDNA at active replication forks is always in contact with the replisome. At bacterial replication forks, for example, the χ-subunit of the DNA Pol-III holoenzyme links it to SSB•ssDNA (Figure 4-21), and the bacterial DNA helicase and DNA primase form a tight complex (Figure 5-3) that appears to be linked dynamically to the rest of the replisome via the clamp loader complex (Figure 5-16). Similar interactions between components of the replisome at a eukaryotic replication fork may prevent RPA•ssDNA from recruiting Atr•Atrip complexes. Alternatively, the DNA damage response may be related to the relative amounts of RPA•ssDNA present at any one time or at any one location. Such complexes will accumulate as a result of stalled replication forks and DNA damage of the type illustrated in Figure 13-10. Note that inhibition of DNA polymerase uncouples DNA unwinding from DNA synthesis, thereby increasing the amount of RPA•ssDNA at replication foci, which triggers the DNA-replication checkpoint.

RPA-coated ssDNA alone is not sufficient for Atr to activate Chk1; it must also contain a 3′-OH polynucleotide primer, and in the case of ssDNA produced by nucleolytic processing of dsDNA breaks, a free 5′-terminus (Figure 13-14A). In addition, the **checkpoint sliding clamp**, a heterotrimer consisting of Rad9, Rad1, and Hus1 (termed the 9-1-1 complex), must be present at the site of action. The 9-1-1 complex is structurally similar to PCNA, and it interacts with proteins that repair DNA much the way PCNA interacts with and organizes proteins at DNA replication forks. The 9-1-1

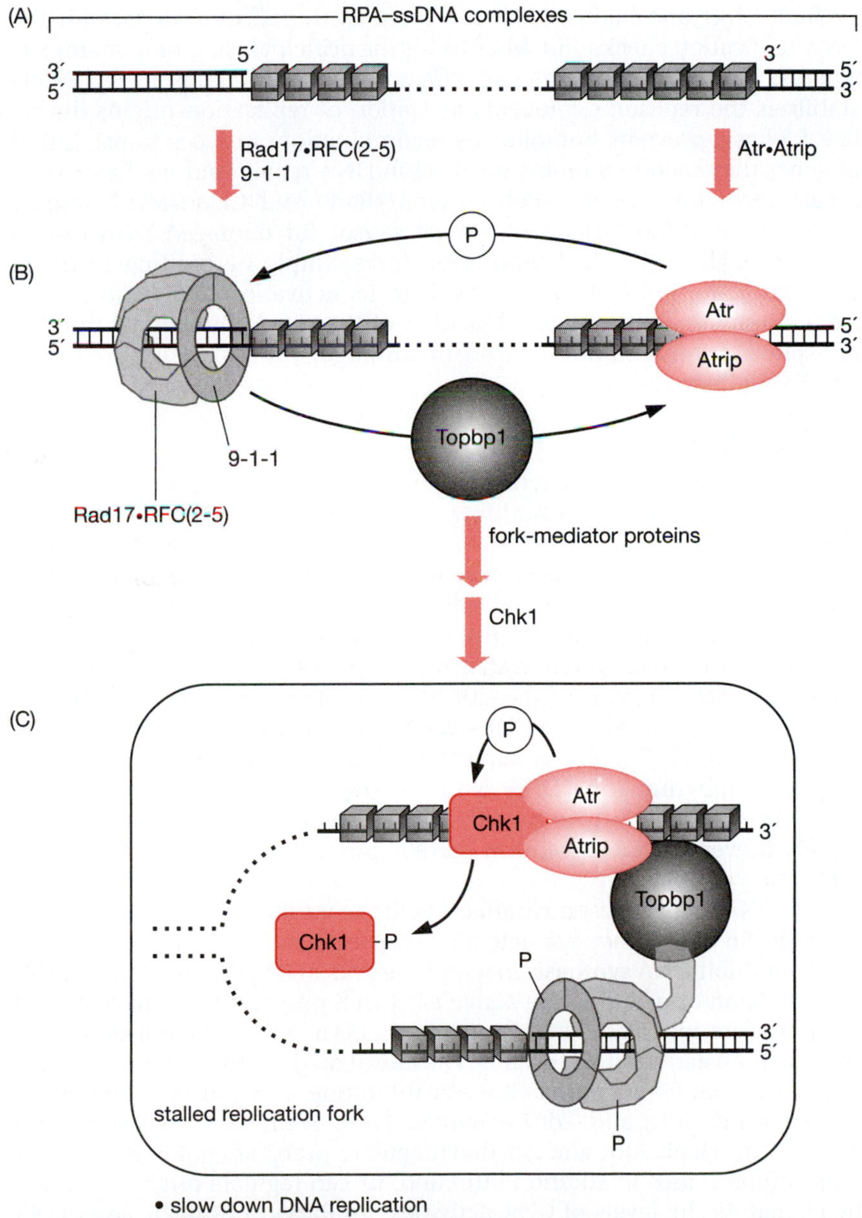

Figure 13-14. The DNA-replication checkpoint recognizes RPA•ssDNA complexes (A) produced by stalled replication forks, dsDNA breaks, and DNA lesion repair. Rad17•RFC(2-5) loads the 9-1-1 complex onto 5′-ends of duplex DNA with a 3′-ssDNA extension (B). This recruits Topbp1 which in turn recruits the sensor Atr•Atrip to the 3′-ends of duplex DNA with a 5′-ssDNA extension. Atr then phosphorylates Rad1 and Hus1, two subunits of the 9-1-1 complex, an event that is required for Dpb11 function (Figure 11-15) in the DNA damage response. (C) Atr•Atrip•ssDNA complexes recruit the 'effector' Chk1 kinase and activate it by phosphorylation, an event that releases Chk1 into the cell to amplify the signal by phosphorylating its target proteins.

complex is loaded onto primer:template junctions adjacent to RPA•ssDNA complexes by the **checkpoint sliding clamp loader** Rad17•RFC(2-5). This is the same protein complex used to load PCNA onto primed DNA templates (Figure 5-8), but with the Rfc1 subunit replaced by Rad17. Thus, the 9-1-1 and Atr•Atrip complexes are recruited independently to RPA•ssDNA sites, and together they constitute a DNA damage signal (Figure 13-14B).

The preference of the 9-1-1 complex for 5′-termini and the Atr•Atrip complex for 3′-termini place them at opposite ends of a ssDNA gap in duplex DNA. For the same reasons, the 9-1-1 complex would reside on the lagging-strand template arm of a stalled replication fork, whereas the Atr–Atrip would reside on the leading-strand template arm (Figure 13-14C). The 9-1-1 complex then recruits topoisomerase-binding protein-1 (Topbp1) that binds Atrip and contributes to activation of the Atr kinase activity. Since colocalization of the 9-1-1 and Atr–Atrip complexes appears sufficient to activate the DNA damage response, even in the absence of induced DNA damage, the extent to which this checkpoint arrests cell proliferation depends on the amount of Chk1 kinase activity produced.

Both Chk1 and Rad53/Chk2 can serve as the effector kinases for the DNA-replication checkpoint, but Chk1 is the principal effector in mammals whereas Rad53 is the principal effector in yeast (Table 13-8). Rad53 stabilizes the replisome, prevents activation of replication origins during late S phase, prevents homologous recombination, induces transcription of genes that encode proteins involved in DNA repair, and mediates DNA repair. Mediator proteins such as Brca1/Rad9 and Clspn/Mrc1 amplify the signal, and Mec1•Ddc2 (the yeast sensor for damaged DNA during S phase) stabilizes stalled replication forks simply by binding to them. Phosphorylation of Chk1 by Atr leads to its activation and rapid release from chromatin, allowing amplification of the checkpoint signal through phosphorylation of multiple downstream targets, two of which are Cdc25 and Tp53 (Table 13-1).

Arresting S phase

The principal targets for inactivation by the Chk1 kinase in mammals and the Chk2/Rad53 kinase in budding yeast appear to be Cdc25, Dbf4 and Sld3. Phosphorylation of Cdc25 isoforms at specific sites inactivates them, thereby arresting any process that requires either Cdk2 or Cdk1 kinase activity (Figure 13-15). Cdk2 and Cdk1 are the only CDKs that can regulate the activity of replication origins in metazoa. Normally, Cdk2 handles this job, but if Cdk2 is inactivated then Cdk1 can take its place. Dbf4 is an essential component of the Cdc7•Dbf4 protein kinase (also known as DDK) that, together with the CDK•cyclin kinase, is required for assembly of preICs (Figure 11-15). Sld3 is required for loading DNA Pol-α•primase into preICs, thereby initiating DNA synthesis (Figure 11-16). Thus, activation of the Atr-Chk1 pathway in humans, or the Mec1•Rad53 pathway in yeast, rapidly prevents preIC assembly, thereby preventing further activation of replication origins.

The DNA damage response affects both the number of active origins and the order in which they are activated. For example, when hydroxyurea is used to inhibit DNA synthesis in wild-type yeast, DNA Pol-α•primase binds to replication origins that are active early in S phase, but not to ones that are active late, whereas in the absence of Rad53 activity, DNA Pol-α•primase binds to both early and late origins. The rate of origin activation in a *Xenopus* egg extract can be increased either by inhibiting Atm and Atr activities, or by increasing Cdk2 and Cdc7 activities. These results are consistent with a model in which Atm and Atr downregulate preIC assembly at selective replication origins. In addition, Atm and Atr can regulate origin activation by regulating the levels of CDK activities. Inhibition of Cdc25 slows DNA replication by reducing conversion of the inactive forms of Cdk1 and Cdk2 into their active forms, thereby reducing origin activation. Conversely, inactivating Chk1 increases Cdc25 activities, which in turn, increases

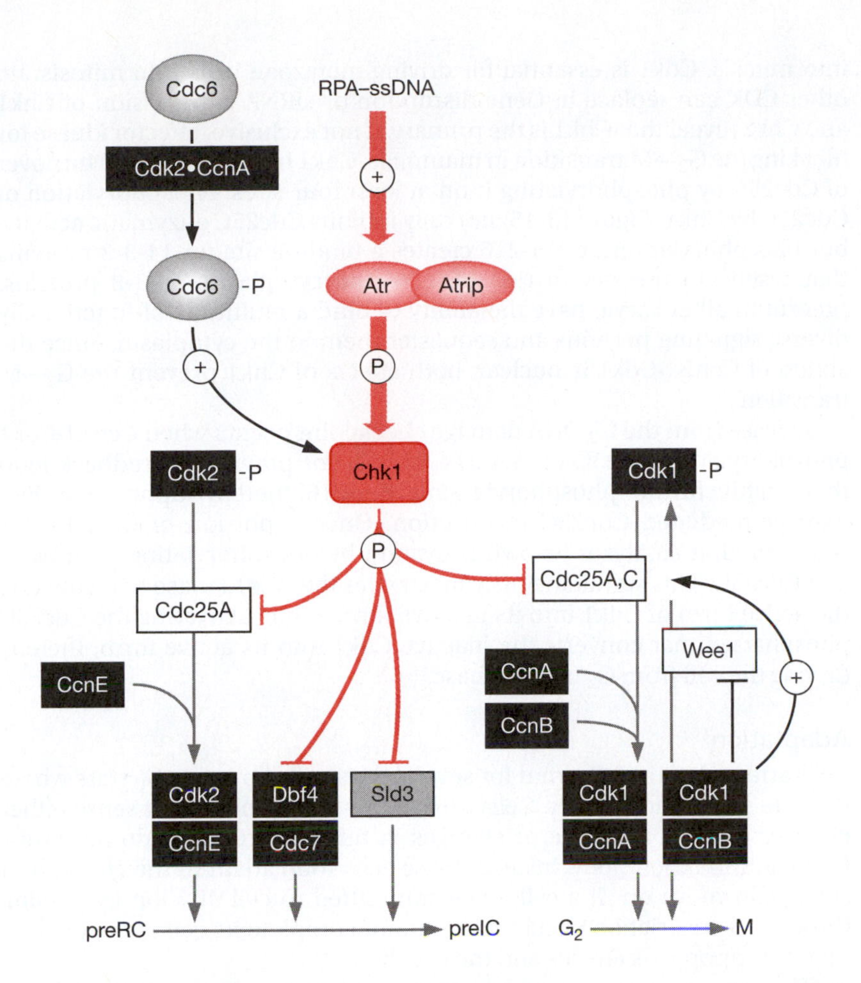

Figure 13-15. The DNA-replication and G_2 DNA damage checkpoints are two sides of the same coin. Stalled replication forks and damaged DNA are both detected in mammals by Atr•Atrip (Figure 13-14), and then a signal is transmitted to the Chk1 effector through phosphorylation. High levels of Cdc6-P (a preRC protein that is phosphorylated during S phase by Cdk2•CcnA) facilitate this reaction, thereby preventing premature entrance into mitosis. Chk1 kinase phosphorylates both Cdc25 isoforms and Dbf4, the activator for Cdc7 protein kinase. These events inhibit the protein kinases required for preIC assembly and mitosis. As the G_2 DNA damage checkpoint fades, two feedback loops (phosphorylation of both Cdc25A, C and Wee1) promote the transition to mitosis.

CDK activities. Increased S-phase CDK activity in yeast stimulates DNA replication as a consequence of premature replication of chromosomal regions normally replicated late during S phase.

In mammalian cells, inactivating Chk1 reduces the global rate of replication-fork progression by about 50% while increasing the density of active origins by two- to threefold. Thus, Chk1 both contributes to the efficiency of replication forks and links fork velocity to the activation of nearby **dormant origins**. Two mechanisms appear to be at work here. First, preIC assembly is affected directly by DNA damage-response pathways, and second, Chk1 and Chk2/Rad53 facilitate replication-fork progress, and in doing so prevent less efficient replication origins (i.e. late origins, dormant origins) from being activated, because replication forks from efficient origins disrupt replication complexes they may encounter as they duplicate the genome.

Other mechanisms for slowing down DNA replication in response to DNA damage or stalled replication forks also exist, but specific pathways have not yet been established. For example, both Atm and Atr can directly phosphorylate the MCM helicase, although the physiological significance of this event is not evident. Atm and Atr prevent dsDNA breaks from accumulating in *Xenopus* egg extracts by promoting restart of collapsed replication forks that arise during DNA replication. Collapsed forks appear to retain the MCM helicase, but lose DNA Pol-ε. Reloading Pol-ε requires Atm and Atr. Cdt1, an essential component of preRC assembly, also is degraded in response to agents that damage DNA, and in human cells this degradation is sensitive to inhibitors of Atr kinase activity.

G_2 DNA Damage Checkpoint

In the presence of damaged DNA or stalled replication forks, maintaining Cdk1 in its inactive Thr-14, Tyr-15-phosphorylated state blocks entrance into mitosis. Cdk1 is essential for driving metazoan cells into mitosis; no other CDK can replace it. Gene disruption or siRNA suppression of Chk1 and Chk2 reveal that Chk1 is the primary, if not exclusive, effector kinase for blocking the $G_2 \rightarrow M$ transition in mammals. Chk1 triggers the rapid turnover of Cdc25A by phosphorylating it on at least four sites. Phosphorylation of Cdc25C by Chk1 (Figure 13-15) not only inhibits Cdc25C enzymatic activity, but phosphorylation on Ser-216 creates a binding site for 14-3-3 proteins that results in transfer of Cdc25C-P to the cytoplasm. 14-3-3 proteins, present in all eukarya, have the ability to bind a multitude of functionally diverse signaling proteins and sequester them in the cytoplasm. Since the action of CcnB1•Cdk1 is nuclear, both effects of Chk1 prevent the $G_2 \rightarrow M$ transition.

Release from the G_2 DNA damage checkpoint occurs when CcnB1•Cdk1 phosphorylates Cdc25C on Ser-214. This event provides a feedback loop that inhibits further phosphorylation at Ser-216, thereby suppressing DNA damage-mediated Cdc25C inactivation. Once a portion of CcnB1•Cdk1 is activated, it catalyzes its own activation by phosphorylating both Wee1 and Cdc25C. This simultaneously inactivates the Wee1 kinase that converts the active form of Cdk1 into its inactive form, while activating the Cdc25C phosphatase that converts the inactive Cdk1 into its active form, thereby driving the cell from G_2 into M phase.

Adaptation

Cells arrested at their R-point for several days will slip into a G_0 state where they are stable indefinitely. Cells arrested by checkpoints that sense either changes in DNA structure or changes in mitotic structures do not arrest proliferation indefinitely. Instead, these cells soon adapt to the checkpoint signal and move on. If a cell, once committed to cell division by passing through the restriction checkpoint, cannot complete its cell cycle, then, in metazoa, apoptosis ensues and the cell dies.

The mechanism by which cells adapt to checkpoints (termed '**adaptation**') remains pretty much of a mystery. Cdk1 phosphorylation of Wee1 and Cdc25C provides one example, although what exactly tips the balance between inactivation and activation of Cdc25C is not clear. Another example is the role of the polo-like kinase-1 (**Plk1**), an activity required for progression through mitosis (Figure 12-9). Plk1 protein itself is expressed throughout the cell cycle. When replication forks stall in *Xenopus* egg extracts, Atr-dependent phosphorylation of Mcm2 induces binding of Plk1 to chromatin. This event leads to activation of replication origins. In addition, Plk1 can inactivate Claspin. Since Claspin mediates the interaction of Atr with RPA•ssDNA, this event could release the cell from its DNA damage checkpoint. Since depletion of Plk1 leads to accumulation of chromosomal breakage, Plk1 may promote genome stability by regulating DNA replication under stressful conditions.

THE EUKARYOTIC REPLISOME REVISITED

The MCM helicase does not travel alone at replication forks, but in the company of other proteins. The Cdc45•MCM•GINS complex is required for efficient DNA unwinding at replication forks (Figure 5-4), but purification of GINS from extracts of budding yeast cells suggests the presence of a larger assembly of proteins at replication forks that have been termed the **replisome progression complex** (Figure 13-16). In metazoa, this complex contains Cdc45, MCM, and GINS as well as Clspn (Claspin), Tim (Timeless), and Tipin (Tim-interacting protein). The latter three proteins increase fork stability, permitting forks to pause at replication-fork barriers (Figure 6-8) without collapsing. Deficiency in Tim and Tipin results in replication-fork instability, increased sister chromatid exchange, and chromatid breaks during otherwise unperturbed DNA replication. The **checkpoint mediator** Clspn links DNA Pol-ε to the MCM helicase.

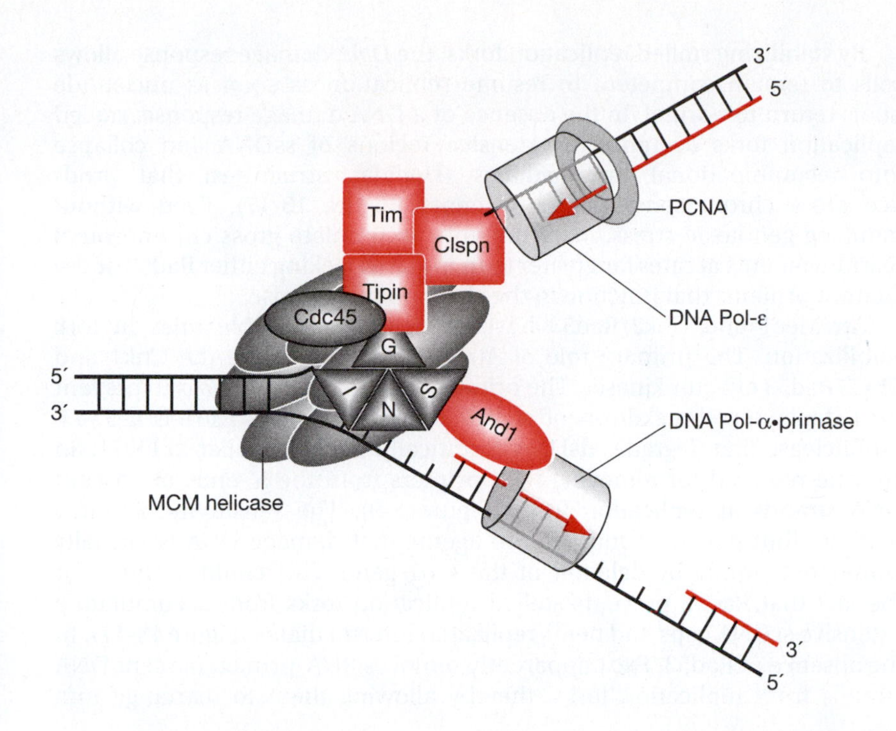

Figure 13-16. The replisome progression complex in eukarya is analogous to the bacterial replisome (Figure 4-21) with the GINS complex possibly playing a role analogous to that of the central organizing tau factor in *E. coli*. Human proteins indicated in orange are required for replication-fork stability and for the DNA damage response. Their yeast equivalents are in Table 13-7. (Data from Gambus, A., van Deursen, F., Polychronopoulos, D. et *al.* (2009) *EMBO J.* 28: 2992–3004.)

The replisome progression complex also contains the histone chaperone complex FACT that facilitates nucleosome disassembly at forks (Figure 7-16), the type-I topoisomerase Topo I that relaxes supercoiled DNA generated by DNA replication (Table 4-8), the preIC assembly factor Mcm10 (Figure 11-15), and a protein called And1. Both Mcm10 and And1 bind DNA Pol-α. And1 stabilizes the replication fork by linking DNA Pol-α to the MCM helicase via the GINS complex. Not shown in Figure 13-16 for the sake of clarity are Mcm10, Topo I, and FACT, all of which can bind DNA Pol-α.

The replisome progression complex does not appear in cells until the MCM helicase is loaded onto chromatin at replication origins during G_1 phase, and then it persists in cells throughout S phase, even when DNA synthesis is inhibited by reduction of nucleotide pools. Thus, the replisome progression complex appears to represent the physical state of the MCM helicase *in vivo*.

Stabilization Of Replication Forks

When replication forks stall, they are in danger of collapsing. Stabilizing of replication forks requires that the replicative DNA polymerases and the replicative MCM helicase remain associated with the replication fork, but inactive, and that homologous recombination is inhibited. Analysis of mutations in budding yeast reveals that Mec1 (a homolog of Atr) has two functions at stalled replication forks. One is to recruit Chk1 to RPA•ssDNA complexes as described above, and the other is to stabilize the replisome by keeping the replicative DNA polymerases engaged. For example, treating wild-type cells with hydroxyurea reduces dNTP pools in budding yeast, which causes replication forks to stall. Mec1 accumulates at replication forks, and DNA Pol-α and Pol-ε remain with replication forks for about one hour [50% of a yeast cell cycle (Table 12-1)]. In the absence of either Mec1 or Rad53 (a homolog of Chk2), Pol-α and Pol-ε rapidly disengage when nucleotide pools are depleted and the cell dies. Thus, Mec1 appears to stabilize the replisome through phosphorylation of the MCM helicase, RFC clamp loader, and DNA polymerases. The importance of fork stabilization in the face of DNA damage is demonstrated by the discovery that conditional mutations in Mcm proteins that inactivate MCM helicase activity results in DNA damage and genome instability similar to that seen with mutations in checkpoint proteins.

By stabilizing stalled replication forks, the DNA damage response allows cells to remain competent to resume replication as soon as nucleotide pools return to normal. In the absence of a DNA damage response, stalled replication forks accumulate extensive regions of ssDNA and collapse into recombinational intermediates (Holliday structures) that produce gross chromosomal rearrangements (Figure 13-17). Even without inducing genotoxic stress, *Mec1* mutants accumulate gross chromosomal rearrangements at rates far greater than mutants lacking either Rad53 or the adaptor proteins that function in the checkpoint response.

Atr/Mec1 and Chk2/Rad53 have genetically separable roles in fork stabilization. The primary role of Atr/Mec1 is to activate the Chk1 and Chk2/Rad53 effector kinases. The primary role of Chk2/Rad53 is to prevent Exo1-dependent breakdown of stalled replication forks. Exo1 is a 5′→3′ exonuclease that degrades dsDNA specifically. Exo1 is similar to FEN1, an enzyme required for removing RNA primers from the 5′-ends of nascent DNA strands at replication forks (Figure 5-9). The sensitivity of *rad53* mutants (but not *mec1* mutants) to agents that damage DNA is virtually eliminated simply by deletion of the *exo1* gene. This could account for the fact that Rad53 prevents stalled replication forks from accumulating extensive ssDNA gaps and hemi-replicated intermediates (Figure 13-17). In the absence of Rad53, Exo1 apparently removes RNA-primed nascent DNA strands from replication forks, thereby allowing them to rearrange into

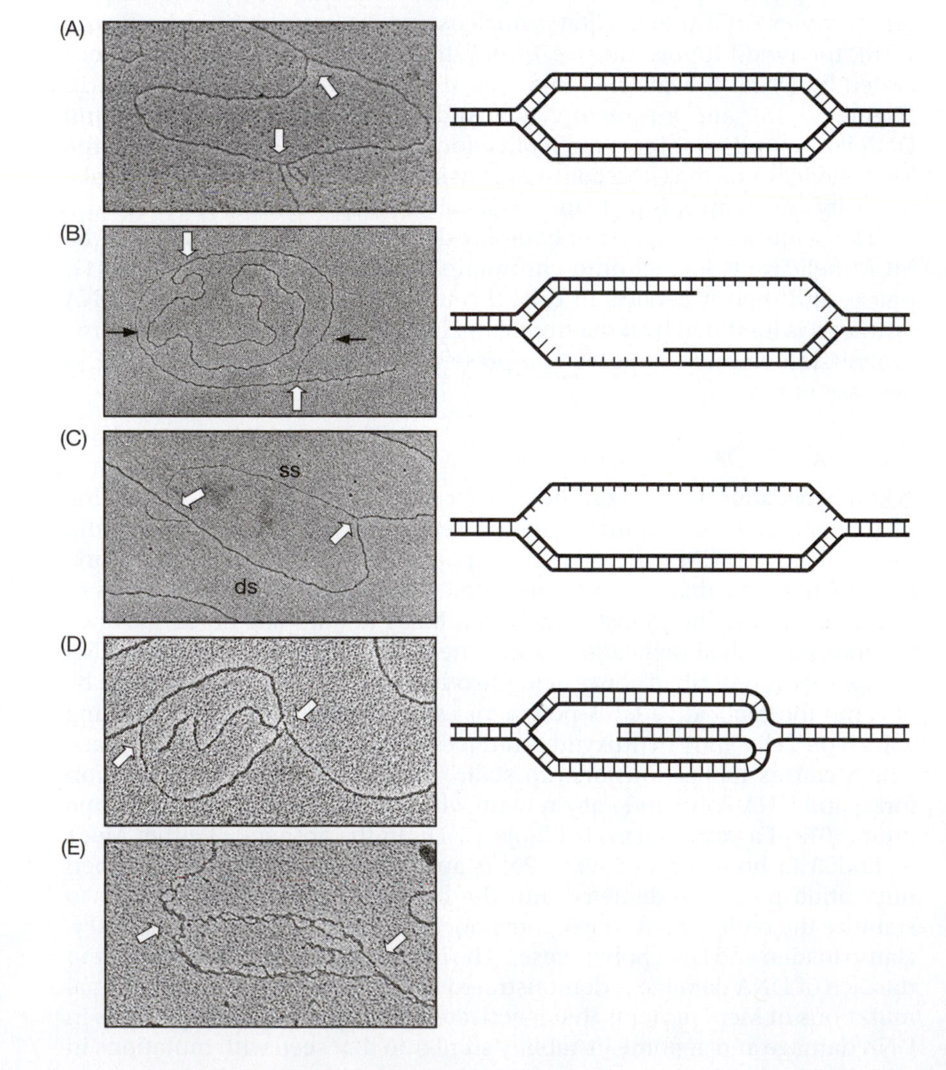

Figure 13-17. Stalled replication forks in *S. cerevisiae* mutants lacking Rad53. DNA replication forks in wild-type cells (A) and in Rad53 mutant cells (B–E) cultured briefly in the presence of hydroxyurea to arrest DNA synthesis. Replicating DNA intermediates were isolated from *S. cerevisiae* after treating the cells with psoralen to cross-link the two complementary DNA strands in duplex DNA. The DNA was then viewed by electron microscopy either under nondenaturing (A–D) or denaturing conditions (E). Denaturing conditions reveal the presence of nucleosomes *in vivo* as rows of ssDNA bubbles of approximately 150 nucleotides connected by short dsDNA segments. The dsDNA corresponds to intranucleosomal linker regions that were accessible to psoralen *in vivo*. Open arrows identify the replication forks. Arrowheads in B identify the transition from ssDNA to dsDNA. ssDNA (ss) and dsDNA (ds) strands are indicated in C. In D and E, one fork is rearranged into a Holliday junction. (From Sogo, J.M., Lopes, M. and Foiani, M. (2002) *Science* 297: 599–602. With permission from the American Association for the Advancement of Science.)

Holliday junctions. To counter this problem in wild-type cells, helicases of the RecQ family (Table 13-5) are targeted for phosphorylation by Atr/Mec1, an event required for efficient replication-fork restart. The role of RecQ helicases is to unravel abnormal DNA structures that can occur at replication forks and to prevent improper recombination events. Chk2/Rad53 also stabilizes the replisome itself, because even though replication forks can be stabilized by deletion of Exo1, their ability to restart is impaired in the absence of Chk2/Rad53.

Mediators

Mediators are proteins that facilitate a reaction, but that are not required for the reaction to take place. For example, in the absence of the Tim•Tipin complex, ssDNA accumulates at replication forks and dramatically increases their dependence on Atr to maintain genome stability and permit the continuation of DNA synthesis. When the sensor kinase Atr•Atrip is activated, it phosphorylates various downstream targets, including Brca1 and Clspn, that contribute to the activation of the downstream effector kinases Chk1 and Chk2. Brca1 and Clspn (Claspin) mediate the DNA damage response. Assembly of Brca1 oligomers occurs at sites of DNA damage as a consequence of its phosphorylation by Atr, and this oligomerization is required to sustain checkpoint signaling. Claspin stabilizes replication forks even in an unperturbed S phase, and it is a key phosphorylation target of Atr during the DNA damage response. In both normal and stalled replication forks, the N-terminus of Claspin interacts with the C-terminus of the catalytic subunit of the leading-strand DNA polymerase, Pol-ε (Figure 13-18). Claspin appears to stabilize the link between Pol-ε and the MCM helicase, thereby helping to maintain coordination between DNA unwinding and DNA synthesis. In the absence of Claspin, ssDNA accumulates. Phosphorylation of Claspin by Atr•Atrip disrupts the interaction between Claspin and Pol-ε, an event that could account for the observed reduction in replication-fork velocity in response to DNA damage. Moreover, the link mediated by Claspin between MCM and Pol-ε helps to prevent release of Pol-ε when replication forks stall.

Replication-Fork Restart In Eukarya

Replication-fork restart in eukarya is complicated by the fact that multiple concerted pathways exist to inactivate preRCs assembled during G_1 phase and to prevent assembly of new preRCs during S, G_2, and M phases (Figure 12-12). Therefore, *de novo* assembly of replisomes during S phase would be a challenging task. This could account for the absence of specific replication-fork-restart pathways in eukarya analogous to those in bacteria (Figure 13-10). In fact, Srs2 is the only eukaryotic homolog of a bacterial replication-restart protein found so far (Table 13-5). Srs2 promotes strand annealing that is dependent on DNA synthesis by unwinding the elongating invading strand from the donor strand. Instead, eukarya focus on stabilizing replication forks in order to prevent their collapse. This requires Atr and Chk1, because loss of these proteins results in replication-fork collapse and gross chromosomal rearrangements. The advantage is that once the barrier is either removed by DNA repair, or bypassed by allowing time for the replicative polymerases to progress through a difficult sequence, genome duplication can resume without the need to reassemble a replisome. If the fork remains blocked, replication may be completed by a fork that originated from an adjacent origin or by homologous recombination. Rad51, the eukaryotic homolog of *E. coli* RecA, is essential for proliferation of vertebrate cells. Rad51 and RecA are essential for homologous DNA recombination. They are both required to repair spontaneously occurring chromosome breaks in proliferating cells that presumably occur as a consequence of stalled replication forks.

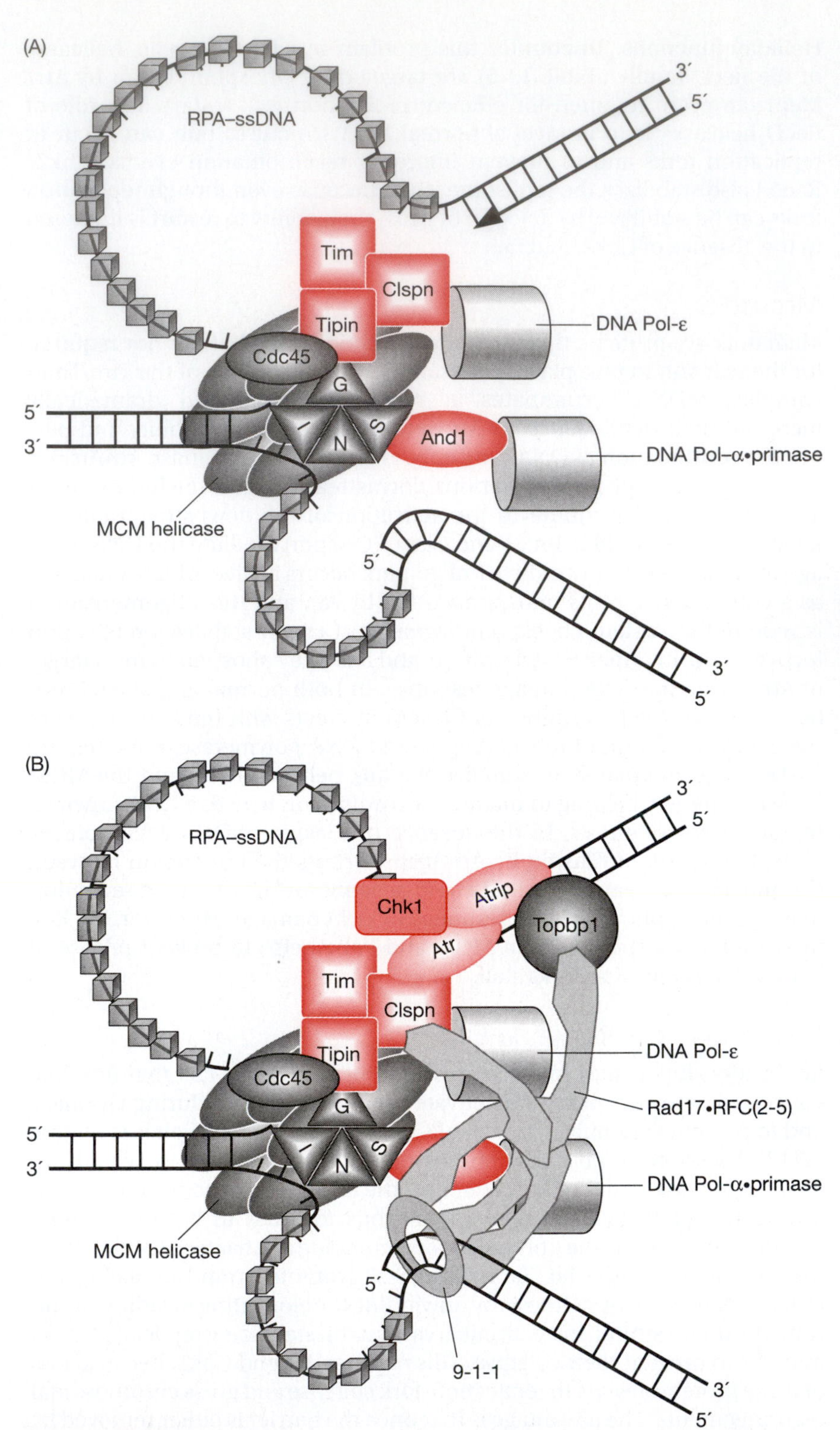

Figure 13-18. Stalled replication forks in a human cell before (A) and after (B) the DNA damage response. GINS (a four-subunit complex) and Cdc45 optimize MCM helicase (six-subunit complex) stability and activity. Tim, Tipin, and Claspin (Clspn) anchor the leading-strand DNA Pol-ε to the MCM helicase complex, while And1 fulfills the same function for the lagging-strand DNA Pol-α•primase. These mediator proteins stabilize the replisome until help arrives. Atr, Atrip, and Chk1 are recruited to the 3′-end of the leading strand, and the sliding clamp 9-1-1 complex and checkpoint sliding clamp loader [Rad17•RFC(2-5)] at the 5′-end of the lagging strand (Figure 13-14). Topbp1 and the sliding clamp loader link the two arms of the fork together. Yeast homologs (Table 13-7) carry out the same functions in yeast cells.

G_2 CHECKPOINT

Even in the absence of 'replication stress', all three domains of life must prevent cell division or mitosis from occurring before genome duplication has been completed. Otherwise, sister chromatids could not separate, because they would be linked together by one or more unreplicated regions. The G_2 checkpoint prevents premature entry of bacteria into cell division and eukarya into mitosis. This problem may not be as serious in single-cell organisms as it is in multicellular organisms, because multicellular organisms depend on a developmental program to produce the specialized cells, tissues, and organs required for viability and fertility. Whether or not chromosomes must be completely devoid of replication bubbles before beginning cell division depends on whether or not additional checkpoints exist downstream to allow completion of DNA replication. Nevertheless, one would expect evolution to select for cells that delay the onset of cell division until genome duplication has finished. In fact, this job appears to be carried out by the DNA-replication checkpoint during normal (unstressed) cell cycles.

Bacteria

As early as 1968, Charles Helmstetter demonstrated that when DNA synthesis is inhibited by ultraviolet irradiation, *E. coli* cells that have not completed DNA replication cannot divide, whereas cells that have completed DNA replication (i.e. contain two complete genomes) can divide. Therefore, DNA replication must be completed before cell division can proceed. Since this is analogous to the G_2 checkpoint in eukaryotic cells where the failure to complete DNA replication prevents mitosis, one can infer that a C/D-checkpoint exists in bacteria that prevents binary fission until DNA replication has been completed.

Cells past the midpoint of their cell cycle contain two separate nucleoids, because they have completed one round of DNA replication. Cells past the midpoint at the time of ultraviolet irradiation divide, whereas cells prior to the midpoint do not divide. Since ultraviolet irradiation presumably damages DNA in all cells, the SOS DNA damage response appears to arrest cell division only in those cells that have not yet completed DNA replication. In *E. coli*, at least nine gene products are required for proper septation, all of which localize to the division plane (the Z-ring). One of these, FtsZ, appears to act at the earliest step and is required throughout cytokinesis. If the SOS response is active at some basal level in normal unstressed cells, then the SOS response gene SulA may prevent FtsZ from acting until replication forks have completed genome duplication (Figure 13-7). Rather than FtsZ levels, it is the actual ability of FtsZ to form a ring at midcell that is critical in regulating the timing of Z-ring formation in both *E. coli* and *B. subtilis.*

Eukarya

The G_2 checkpoint as a manifestation of the DNA-replication checkpoint

What is commonly referred to as the G_2 checkpoint in eukarya is most likely the DNA-replication checkpoint recognizing the presence of replication-fork structures in the genome. This could occur in two ways. Either an Atr/Mec1 protein is a normal component of the replisome, or more likely, the frequency of replication-fork stalling under normal growth conditions is sufficient to prevent premature mitosis. In either case, S-phase cells would emit a constant signal to Chk1 to inhibit Cdc25A, C and prevent entrance into mitosis (Figure 13-15). The strength of this signal would reflect the number of replication forks and the amount of Atr/Mec1 per fork.

In fact, both the Atr and Chk1 genes are essential for normal mouse development, and Chk1 is required to maintain normal rates of replication-fork progression in both human and avian cells, even in the absence of DNA damage. When Chk1 is suppressed or its gene deleted in cultured cells by conditional knockout techniques, then replication forks stall and dsDNA breaks appear. These observations imply that the Atr-Chk1 pathway is required for normal cell proliferation, even in the absence of agents that induce DNA damage or stalled replication forks.

The same conclusion applies to yeast. From 10% to 20% of *S. cerevisiae* cells spontaneously accumulate Rad52 foci during normal cell cycles. Rad52 is an essential protein in dsDNA break repair and homologous recombination (Table 13-5). The fact that these spontaneous Rad52 foci are most frequently observed in S-phase cells and only rarely in G_1 cells implies a coupling between recombinational repair and DNA replication. Therefore, these spontaneous Rad52 foci likely represent replication forks that have stalled during the normal course of genome duplication. This seems a common occurrence in eukaryotic cells. *S. cerevisiae* mutants in Rad51, an essential gene in dsDNA break repair (Table 13-5), are viable, but extremely sensitive to gamma-rays due to defective repair of dsDNA breaks. In contrast, multicellular eukarya are far less tolerant of stalled forks than single-cell eukarya. Rad51 is essential for early mouse embryonic development. Loss of Rad51 in vertebrates causes cells to arrest in G_2 with numerous dsDNA breaks and then succumb to apoptosis.

The G_2 checkpoint as a manifestation of DNA re-replication control

An alternative mechanism for preventing premature mitosis through activation of Chk1 involves the accumulation of phosphorylated Cdc6 protein, one of the pathways that prevents DNA re-replication during S phase (Figures 12-14 and 12-15). In both *Xenopus* egg extracts and mammalian cells, the presence of Cdc6 during S phase is required to activate Chk1 in response to inhibition of DNA replication. The role of Cdc6 in activating Chk1 is independent of its role in preRC assembly, and phosphorylated Cdc6 can inhibit the onset of mitosis by inducing phosphorylation of Chk1 by a mechanism as yet unknown (Figure 13-15). Reduction in Cdc6-P levels would then release the G_2 checkpoint. In both single-cell and multicellular eukarya, a portion of the Cdc6 from S phase survives until mitosis and then it is degraded by the APC. In yeast, where Cdc6 is targeted for degradation, a portion is stabilized during the G_2 and M phase as a complex with cyclin Clb2. In mammalian cells, only excess Cdc6 is targeted for degradation, the rest survives until metaphase when it is ubiquitinated by the APC as the cell advances to anaphase. Therefore, Cdc6-P could play a role in preventing premature mitosis by activating the Chk1 kinase.

SPINDLE-ASSEMBLY CHECKPOINT

When a cell divides, its chromosomes must be delivered flawlessly to the two daughter cells, because missing or extra chromosomes can result in developmental catastrophe, birth defects, and cancer. Therefore, both single-cell and multicellular eukaryotic cells employ the **spindle-assembly checkpoint** (**SAC**) to prevent premature separation of sister chromatids during mitosis and meiosis. Although the existence of SAC was anticipated from the sensitivity of cell division to inhibitors of microtubule assembly, it remained hypothetical until 1991 when Andrew Hoyt and Andrew Murray independently discovered genes that defined it.

Mitotic Cells

SAC monitors the attachment of chromosomes to the mitotic spindle, and it blocks the transition from metaphase to anaphase by inhibiting the APC until the kinetochore on each chromosome is properly attached to the mitotic spindle (Figure 13-19). In that way, each daughter cell is assured of receiving one and only one copy of each chromosome. Unattached kinetochores trigger SAC. In yeast, a lack of tension is sensed, and then converted into a lack of attachment that is sensed by the chromosomal passenger complex (CPC) that consists of aurora kinase B and three cofactors (Table 13-10). Exactly how these events activate aurora kinase B is not yet clear, but the primary sequence of events downstream of the CPC 'sensor' are well established.

SAC prevents Cdc20 (one of the two APC subunits that target substrate) from activating the APC-mediated polyubiquitylation of cyclin B and securin, thereby preventing their destruction by the 26S proteasome (Figure 13-19). Cyclin B (CcnB) is required for Cdk1 to drive the cell from G_2 into metaphase; degradation of CcnB inactivates CDK activity and allows the

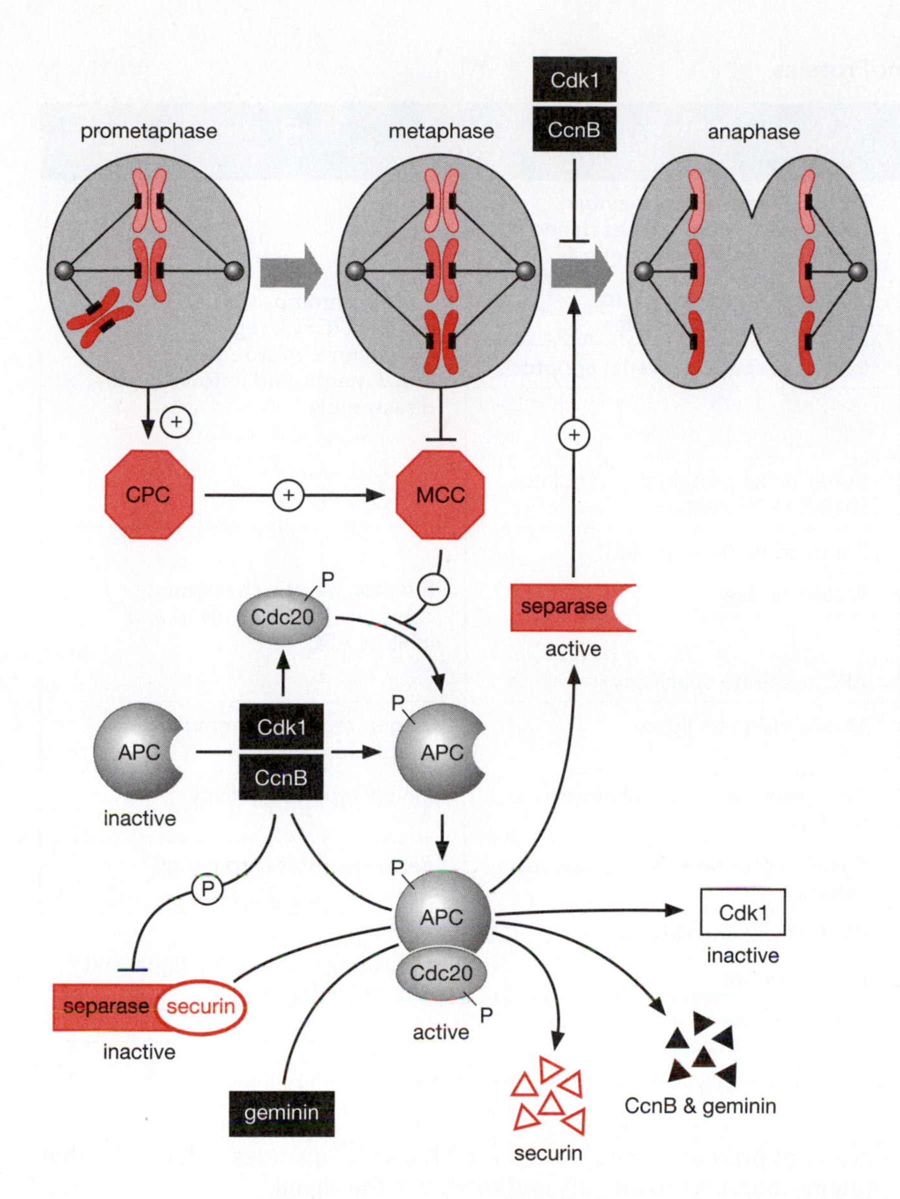

Figure 13-19. In mitotic cells, the spindle-assembly checkpoint (SAC) operates during prometaphase to prevent the transition to anaphase before all of the chromosomes are attached correctly to the spindle fibers. Sister chromatids are indicated in shades of orange, kinetochores as a black bar on each chromatid, and spindle fibers as lines with one end attached to the kinetochore and the other end attached to the centrosome (gray sphere) at the spindle poles. Figure 12-4 contains images of the human mitotic cell cycle. MCC, mitotic checkpoint complex.

metaphase to anaphase transition (Figure 12-8). Securin is a protein that binds to and inhibits a protease called separase. Separase is required to cleave the cohesin complex that binds sister chromatids together (Figure 7-17). Separase activity also is required for the metaphase to anaphase transition. Therefore, SAC prevents this chain of events by keeping Cdc20 in check. SAC prolongs prometaphase until all chromosomes have become bi-orientated between separated spindle poles on the metaphase plate. For proper chromosome segregation, sister kinetochores must attach to microtubules extending from opposite spindle poles prior to the onset of anaphase (termed chromosome bi-orientation). This is the condition that finally extinguishes SAC, relieving the mitotic arrest and allowing anaphase to proceed. Note the feedback loop in which Cdk1•CcnB is required to activate the APC that then targets CcnB for ubiquitin-dependent proteolysis.

During prometaphase, Cdc20 as well as all of the SAC proteins are concentrated at **kinetochores**, a complex of proteins on each mitotic chromosome at the site of the spindle fiber attachment. This attachment is essential in order to pull the sister chromatids apart during anaphase. Kinetochores are thought to catalyze assembly of the SAC 'effector', the **mitotic checkpoint complex** (**MCC**). The MCC in yeast consists of Mad2, Mad3, Bub3, and Cdc20 (Table 13-10). It binds the APC and prevents it from ubquitinating CcnB and securin. Both Mad2 and Mad3 bind Cdc20 directly, although exactly how the MCC is assembled and how it inhibits the APC are

Table 13-10. Spindle-Assembly Checkpoint Proteins

Mammals (human)	Budding yeast (*S. cerevisiae*)	Activity	Function
Aurkb/aurora kinase B	Ipl1	Protein kinase, helps maintain condensed chromosomes during anaphase and early telophase	Sensor, chromosomal passenger complex (CPC), regulates kinetochore-microtubule attachments, and mitotic spindle disassembly
CdcA8/borealin		Targets aurora kinase B to centromeres and spindles	
Birc5/survivin	Bir1	Inhibits caspase; prevents apoptosis	
Incenp/inner centromere protein	Sli15		
	Nbl1	Borealin-like protein that mediates Sli15•Bir1 interaction	
Mad2L1	Mad2	Forms complex with Mad1	Effector, mitotic checkpoint complex (MCC), binds to and inhibits APC
Bub1b/Bubr1	Mad3	Protein kinase	
Bub3	Bub3		
Cdc20	Cdc20	APC substrate specificity	
APC	APC	M→G_1 ubiquitin ligase	Target, regulates metaphase to anaphase transition
Pttg1/securin	Pds1	Binds separase and transports it to nucleus	Inhibit separase activity
Espl1/separase	Esp1	Cysteine protease that hydrolyzes cohesin	Separate sister chromatids
Mad1L1	Mad1	Binds Mad2 and Hdac1	Mediators that amplify signal and rate of MCC assembly
Bub1	Bub1	Protein kinase	
Mps1	Mps1	Protein kinase	

With names separated by /, the first name is the official one, and the second is the common one. Hdac1, histone deacetylase-1.

unclear at present. In addition to the MCC, SAC includes at least 17 other proteins that serve to amplify and modulate the signal.

SAC monitors the attachment of kinetochores to microtubule plus-ends during an unperturbed mitosis, and SAC remains active at sister kinetochores until bi-orientation is achieved, the only condition that ensures accurate segregation at anaphase. Any interference with microtubule assembly or dynamics immediately activates SAC. CcnB and securin start to be degraded after the last chromosome has aligned, and they are essentially absent just prior to anaphase. Mad1 and Mad2 accumulate at unattached kinetochores, and they are removed upon their attachment to spindle fibers. APC-dependent ubiquitination of Mps1 irreversibly inactivates SAC at anaphase.

Meiotic Cells

Females are born with a large store of **oocytes** (the female gametes) arrested in prophase of the first stage in meiosis (meiosis I). Each oocyte contains high concentrations of mRNAs and proteins that are consumed during the first rounds of cell cleavage following fertilization. Oocytes in *Xenopus*, *Drosophila*, and mice contain all the proteins required for cell proliferation, but one; they are missing Cdc6, an essential component of preRC assembly (Table 11-3). In this way, oocytes can remain dormant from months to years, because the capacity to initiate DNA replication is bestowed upon them by Cdc6 only during ovulation when the cells undergo **meiotic maturation**. They complete meiosis I and then arrest again in metaphase II to await fertilization. At this point, they are called eggs. DNA licensing, however, still cannot begin, because eggs are arrested in metaphase and therefore contain large amounts of active CcnB2•Cdk1 that prevents preRC assembly (Table 12-4). Once again, SAC prevents premature transition from metaphase to anaphase, although the mechanism in meiotic cells differs somewhat from the mechanism in mitotic cells.

Emi2, an oocyte-specific APC inhibitor, prevents activation of the APC in eggs and thus prevents passage beyond metaphase II (Figure 13-20). The APC is required to degrade the mitotic cyclins in order to inactivate Cdk1 and allow cells to exit metaphase, and in multicellular eukarya, to degrade geminin in order to activate Cdt1 (Figure 12-18). Hence, preRC assembly is prevented as long as APC activity is inhibited. Fertilization activates preRC assembly (DNA licensing) by inducing a rapid influx of Ca^{2+} that activates a calmodulin-dependent protein kinase II (CamK2). CamK2 functions as a 'priming kinase' by phosphorylating Emi2 at a specific motif that results in a strong interaction with Plk1. This allows Plk1 to phosphorylate a different site within Emi2, one that converts Emi2 into a substrate for the ubiquitin ligase SCF(βTrCP). The net result is ubiquitin-dependent proteolysis of Emi2 with concomitant activation of APC(Cdc20) and transition from metaphase II in meiosis to G_1 phase of the cell-division cycle.

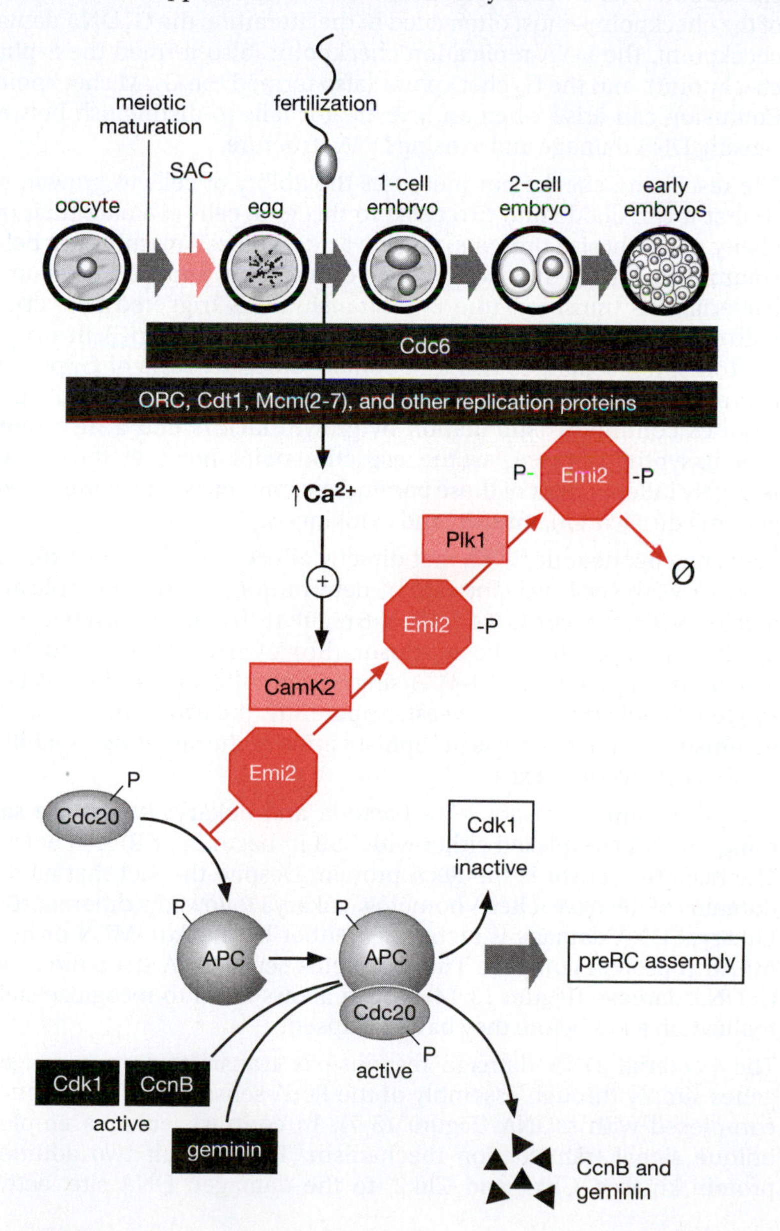

Figure 13-20. The spindle-assembly checkpoint operates during oocyte maturation in *Xenopus*, *Drosophila*, and mice to prevent initiation of genome duplication before fertilization.

SUMMARY

- Cell cycles are driven ever forward by the feedback loops involving CDKs, Cdc7, and ubiquitin ligases described in Chapter 12. Cell-cycle checkpoints are not required to drive the cell cycle, but to ensure that preceding events are completed successfully before moving on to the next event. Thus, checkpoints are not essential for cell division, because cell division can occur without checkpoints. However, checkpoints are essential for the organism's viability, because without them mistakes will occur. Even if the mistakes are not lethal, the progeny of these cells will be defective. If the mistakes are serious, the progeny will no longer be able to divide. Eventually, the organism will become extinct. It should come as no surprise in Chapter 14 to learn that mutations in checkpoint genes are frequently associated with cancer .
- Checkpoints are always active, not just when cells are under stress.
- Checkpoints are satisfied in one of two ways: either the cell corrects the condition, or it adapts to the condition and moves on.
- Checkpoints fall into four groups: the restriction checkpoint, the DNA damage response, the G_2-phase checkpoint, and the spindle-assembly checkpoint. The DNA damage response is responsible for at least three of the checkpoints most often cited in the literature: the G_1 DNA damage checkpoint, the DNA-replication checkpoint (also termed the S-phase checkpoint), and the G_2 checkpoint (also termed the G_2/M checkpoint). Confusion can arise when an investigator fails to distinguish between sensing DNA damage and sensing DNA structure.
- The restriction checkpoint measures the ability of cells to grow in size so that it can divide into two cells. To this end, cells evaluate their own ability to synthesize proteins, nucleic acids, and cell membranes before committing themselves to genome duplication and cell division. In bacteria, this transition (the B→C transition) is triggered by a critical ratio of initiator protein (DnaA) to replicator sequence (oriC). In eukarya, the transition separates two functionally different parts of G_1 phase in continuously cycling cells. Progression through the restriction point requires continuous stimulation by growth factors and a high rate of protein synthesis. Once past the restriction point, however, the cell cycle is largely independent of these parameters; the cell is then committed to genome duplication, mitosis, and cytokinesis.
- Metazoa encode four CDKs that directly affect the cell cycle. Cdk1 and Cdk2 drive the cell cycle forward by determining when DNA replication and mitosis can occur. Cdk4 and Cdk6 regulate the restriction checkpoint by determining when the repressor (Rb) of E2f transcription factor activity is active (Figure 13-4). A single CDK fulfills all of these roles in single-cell eukarya such as yeast. Apparently, the evolution of complex organisms required increased sophistication in the targeting capabilities of CDK•cyclin complexes.
- The DNA damage response in bacteria and eukarya *senses* the same thing: ssDNA complexed either with SSB in bacteria or RPA in eukarya. The bacterial sensor is the RecA protein. Despite the fact that all three domains of life have a RecA homolog, eukarya followed a different route. Eukaryal DNA damage is recognized either by the Atm•MRN or by the Atr•Atrip protein complex. These proteins sense DNA structure as well as DNA damage (Figure 13-14), which allows them to recognize stalled replication forks before they have collapsed.
- The bacterial DNA damage response is transmitted to a range of genes simply through assembly of the RecA sensor into long filaments complexed with ssDNA (Figure 13-7). In contrast, eukarya employ a unique signal transduction mechanism. They recruit two additional protein kinases, Chk1 and Chk2, to the damaged DNA site, activate

them, and then release them into the nucleoplasm to activate other cellular proteins. Such a mechanism may have evolved in response to increases in genome size.

- A significant difference between the bacterial and eukaryotic DNA damage response is that the bacterial response operates only during DNA replication (C period), whereas the DNA damage response in eukarya operates throughout interphase. This gives rise to the impression that there are three different eukaryotic DNA damage checkpoints, one for G_1, one for S, and one for G_2. In fact, there is only one, and it inhibits either Cdk1 or Cdk2, depending on the phase of the cell cycle in which the damage is detected.
- A single dsDNA break, if not repaired, is lethal. Therefore, the primary role of the G_1 DNA damage response is to prevent cells with dsDNA breaks from initiating DNA replication. To this end, the sensor (Atm•MRN) and the effector (Chk2) conspire to inhibit Cdk2 activity by two independent means (Figure 13-13). Cdc25A (the phosphatase that activates Cdk2) is inhibited, and p21 (a CDK-specific inhibitor) is expressed. The goal of the G_1 DNA damage checkpoint is to prevent Cdk2•CcnE from triggering preIC assembly and DNA replication. Intriguingly, as the 5′-end of a dsDNA break is resected further, the longer 3′-ssDNA tails recruit Atr rather than Atm. This would allow dsDNA breaks that occur during S phase to signal the same types of response elicited by the DNA-replication checkpoint.
- Replication forks in bacteria and eukarya stall frequently, even under optimal growth conditions. The greatest danger from stalled replication forks is their vulnerability to cleavage at their ssDNA sites. This would produce a potentially lethal dsDNA break. In response to this threat, bacteria evolved specialized pathways for restarting forks, even if they have collapsed, or suffered a DNA lesion or a runaway helicase (Figure 13-10). These mechanisms are critical to bacterial viability. Eukarya, however, strive to prevent stalled forks from collapsing. To this end, additional proteins are included with the replisome to facilitate its ability to unwind and replicate chromosomal DNA without dissociating from the DNA template (Figures 13-16 and 13-18).
- Once a cell enters S phase, it is the job of the DNA-replication checkpoint to prevent cells with either damaged DNA or stalled replication forks from continuing genome duplication until the problem is corrected (Figure 13-15). This checkpoint is essentially the same as the G_1 DNA damage checkpoint, except that a different sensor (Atr•Atrip) conspires with a different effector (Chk1) to prevent Cdc25 isoforms from activating either Cdk2 or Cdk1, as well as to prevent Dbf4 from activating Cdc7 (another protein kinase required for preIC assembly). Such actions not only prevent activation of late-firing replication origins and reduce fork velocity, but also stabilize the replisome at stalled forks, prevent forks from undergoing homologous DNA recombination, and induce expression of genes encoding proteins that repair DNA. Should a dsDNA break occur, the G_1 DNA damage checkpoint can deal with it.
- For mitosis to succeed, sister chromatids cannot be linked by regions of unreplicated DNA. It is the job of the G_2 checkpoint to make sure that DNA replication is complete before mitosis begins. For this purpose, the DNA-replication checkpoint is ideal (Figure 13-15). What is often termed the G_2 checkpoint is likely the DNA-replication checkpoint acting to prevent cells from entering mitosis with replication forks present. An independent G_2 checkpoint may also exist as a pathway that detects post-translationally modified preRC proteins. For example, Cdc6-P appears to arrest the cell cycle by activating Chk1 kinase, the same effector used in the DNA-replication checkpoint.

- The spindle-assembly checkpoint prevents transit through either mitosis or meiosis until all sister chromatids are correctly attached to the metaphase plate before anaphase begins (Figures 13-19 and 13-20). Although the proteins involved differ, both the mitotic and the meiotic checkpoints inhibit APC activity by preventing its association with Cdc20, one of two APC substrate-targeting subunits. When active, the APC ubiquitin ligase targets mitotic cyclins and geminin for ubiquitin-dependent protein degradation, thereby allowing the cell to once again assemble preRCs.

REFERENCES

Additional Reading

Brown, C.J., Lain, S., Verma, C.S., Fersht, A.R. and Lane, D.P. (2009) Awakening guardian angels: drugging the p53 pathway. *Nat. Rev. Cancer* 9: 862–873.

Chen, Y. and Poon, R.Y.C. (2008) The multiple checkpoint functions of CHK1 and CHK2 in maintenance of genome stability. *Front. Biosci.* 13: 5016–5029.

Cooper, S. (2006) Checkpoints and restriction points in bacteria and eukaryotic cells. *BioEssays* 28: 1035–1039.

de Bruin, R.A. and Wittenberg, C. (2009) All eukaryotes: before turning off G_1-S transcription, please check your DNA. *Cell Cycle* 8: 214–217.

Friedberg, E.C., Walker, G.C., Siede, W., Wood, R.D., Schultz, R.A. and Ellenberger, T. (2005) *DNA Repair and Mutagenesis*, 2nd edn. American Society for Microbiology, Washington DC.

Friedel, A.M., Pike, B.L. and Gasser, S.M. (2009) Atr/Mec1: coordinating fork stability and repair. *Curr. Opin. Cell Biol.* 21: 237–244.

Heller, R.C. and Marians, K.J. (2006) Replisome assembly and the direct restart of stalled replication forks. *Nat. Rev. Mol. Cell Biol.* 7: 932–943.

Janion, C. (2008) Inducible SOS response system of DNA repair and mutagenesis in Escherichia coli. *Int. J. Biol. Sci.* 4: 338–344.

Maya-Mendoza, A., Tang, C.W., Pombo, A. and Jackson, D.A. (2009) Mechanisms regulating S phase progression in mammalian cells. *Front. Biosci.* 14: 4199–4213.

Musacchio, A. and Salmon, E.D. (2007) The spindle-assembly checkpoint in space and time. *Nat. Rev. Mol. Cell Biol.* 8: 379–393.

Paulsen, R.D. and Cimprich, K.A. (2007) The Atr pathway: fine-tuning the fork. *DNA Repair* 6: 953–966.

Potrykus, K. and Cashel, M. (2008) (p)ppGpp: still magical? *Annu. Rev. Microbiol.* 62: 35–51.

Zetterberg, A., Larsson, O. and Wiman, K.G. (1995) What is the restriction point? *Curr. Opin. Cell Biol.* 7: 835–842.

Literature Cited

Bracken, A.P., Ciro, M., Cocito, A. and Helin, K. (2004) E2F target genes: unraveling the biology. *Trends Biochem. Sci.* 29: 409–417.

Matsuoka, S., Ballif, B.A., Smogorzewska, A. *et al.* (2007) ATM and ATR substrate analysis reveals extensive protein networks responsive to DNA damage. *Science* 316: 1160–1166.

Ozono, E., Komori, H., Iwanaga, R., Ikeda, M.A., Iseki, S. and Ohtani, K. (2009) E2F-like elements in p27(Kip1) promoter specifically sense deregulated E2F activity. *Genes Cells* 14: 89–99.

Chapter 14
HUMAN DISEASE

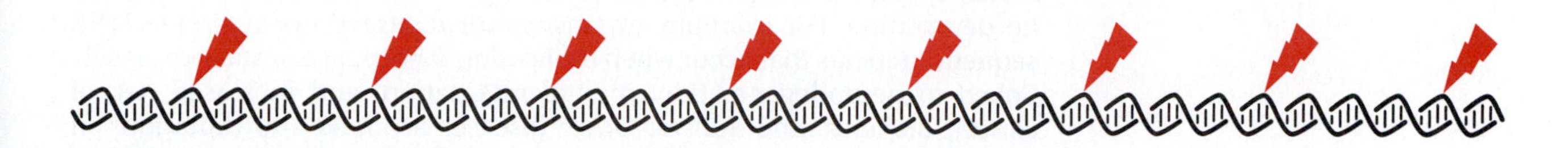

All human disease stems from four root causes: infection by viruses, microorganisms, prions, and genetic mutations. Remarkably, with the exception of diseases caused by prions (Creutzfeldt–Jakob disease) and RNA viruses (e.g. influenza, measles, and polio), all of the remaining infectious diseases are caused either by microorganisms, DNA viruses, or retroviruses. Retroviruses are RNA viruses that encode a **reverse transcriptase**, a DNA polymerase that transcribes ssRNA into a complementary copy of ssDNA and then into a dsDNA molecule. The dsDNA is used to express viral genes and to integrate the viral genome into the host's genome. Thus, all of these infectious agents must replicate either their DNA genome or a DNA copy of their genome in order to infect their host and produce a pathological state. This means that all of these contagions are susceptible to drugs that inhibit a critical enzyme required to replicate its DNA or a copy of its DNA. In fact, drugs targeted against viral DNA replication are among the most potent weapons against human immunodeficiency virus (HIV), the causative agent for acquired immune deficiency syndrome (AIDS), and hepatitis B virus (HBV), the causative agent for the liver disease called hepatitis B. About a third of the world's population (~2 billion people) are infected with HBV, although only a fraction of them exhibit the disease. In contrast, HIV infects about 0.6% of the world's population, but AIDS has already killed more than 25 million people.

Infections are not the only human diseases susceptible to drugs targeting DNA replication. About 600 000 people in the USA alone died from cancer in 2008, making cancer the second leading cause of death in the USA. Cancer is a disease in which cells continue to proliferate under conditions where they normally would not. In fact, malignant tumors are so rich in proliferating cells that proteins specific to DNA replication can be used as biomarkers for early detection of cancer. Given these facts, it is not surprising that cancer chemotherapy commonly targets DNA replication in an attempt to kill the cancer without killing normal tissues.

The causes of cancer are many and varied, but one of them is a viral infection. Of the seven viruses known to cause cancer in humans, five are DNA viruses and one is a retrovirus. Inhibiting a protein essential to duplication of the viral genome but not to the host genome would prevent these viruses from proliferating without harming the host cell. This could be a tremendous help in reducing the number of people infected by the virus. Unfortunately, it would not kill the cancer cell produced by the virus, because these cancer cells contain only a replication-defective copy of the viral genome that is carried along as part of the host cell's genome. Although the defective viral genome cannot replicate independently of its host, it does express one or more viral genes that transforms the host cell into one that ignores signals that arrest cell proliferation. Since most of the cells in our body are not actively proliferating, drugs that preferentially interfere with genome duplication provide effective cancer chemotherapy.

Noninfectious diseases such as cancers and hereditary diseases such as Huntington's are the result of genetic mutations. These genetic mutations result from errors that occur during DNA replication, from DNA damage induced by chemicals and radiation in the environment, and from integration of foreign DNA, particularly viral DNA. Fortunately for the development of complex life forms, most genetic alterations are silent; they produce no obvious phenotype. However, those that do can be devastating. For example, expansions or contractions of simple DNA sequence repeats that occur when replication forks pause or stall are closely linked to neurodegeneration, mental retardation, and increased risk of cancer, diabetes, and autism. Other diseases are linked to mutations in genes required for either DNA replication or repair. Still others are linked to mutations in mitochondrial DNA. Our genomes are more likely to be sensitive to agents that damage DNA when chromatin is disassembled and ssDNA is generated at replication forks. The mitochondria, in particular, lack DNA repair and proofreading mechanisms for mtDNA replication. Therefore, mtDNA accumulates mutations over time. Even the ultimate human disease, aging, can be viewed as the consequence of genetic and epigenetic changes that accumulate during the trillions of cell-proliferation events required to produce and then maintain an adult animal. These errors occur from DNA replication, from failure to repair DNA damage, and from changes in DNA methylation patterns that regulate gene expression.

Given the simple fact that nothing is more fundamental to living things than the ability to reproduce, it is not hyperbole to state that understanding how genomes are duplicated, how the process of genome duplication is regulated, and how genome duplication differs among the various forms of life is critical to the understanding and treatment of human disease. To this end, one must identify and target features that distinguish viruses and bacteria from mammals, cancer cells from normal cells, and heritable gene disorders from normal gene functions.

INFECTIOUS DISEASES

Infectious human diseases are caused either by viruses, bacteria, protozoa, or fungi. All three microorganisms contain a DNA genome, and therefore can, in principle, be targeted for destruction by drugs directed against their DNA replication proteins. The efficacy of this approach depends on how well the drug distinguishes the replication machinery in the microbial pathogen from that of its human counterpart.

Viruses

In addition to two retroviruses, seven families of DNA viruses are known to cause a multitude of human diseases. In each case, the viral genome encodes at least one protein essential for duplication of its own genome, and the host cell provides the remaining proteins. Therefore, diseases caused by these viruses can, in principle, be treated by selectively inhibiting the action of a unique viral protein. In fact, most antiviral drugs are targeted specifically at enzymes essential to viral genome duplication.

Small nuclear DNA viruses

There are four families of small DNA viruses that replicate in the nucleus of their host cell, some of which are responsible for one or more human diseases (Table 14-1). Since the viral-encoded protein(s) must interact functionally as well as physically with the proteins provided by its host, the fewer proteins provided by the virus (i.e. the smaller the viral genome), the more restricted is the virus' host range. Not only are they specific for a particular species, but often for particular cell types within that species. With the exception of hepatitis B virus (HBV), antiviral drugs have not targeted DNA replication proteins encoded by small nuclear DNA viruses.

Table 14-1. Small Nuclear DNA Viruses That Cause Human Disease

Virus	Strain	Disease	Replication proteins
Hepadnavirus	Hepatitis B virus (HBV)	Acute and chronic hepatitis, fulminant hepatitis, liver cirrhosis, primary hepatocellular carcinoma	Reverse transcriptase
Parvovirus	B19	Fifth disease, non-immune hydrops fetalis, pure red cell aplasia, congenital anemia, acute or chronic arthropathy, aplastic crisis, persistent anemia, chronic arthritis	NS1
Polyomavirus	JC	Progressive multifocal leukoencephalopathy	T-ag
	BK	Nephropathy	
	Merkel cell	Merkel skin cancer	
Human Papillomavirus (HPV)	6, 11	Genital warts, respiratory papillomatosis→pulmonary cancer	E1, E2
	16, 18, 31, 33, 35, 39, 45, 51, 52, 56, 58, 59, 68, 69	Cervical dysplasias, cancer (cervical, penile, anal, pharyngeal, tonsillar)	
	1, 2, 4, 5, 8	Warts (plantar, palmar, cutaneous, genital), epidermaldysplasia verruciformis, ultraviolet-induced skin cancers	

HBV is a member of the *Hepadnaviridae* family. They replicate primarily in hepatocytes, the major parenchymal cell of the liver, although each of the five hepadnaviruses infects a different mammalian host. Infection and virus replication do not lead to hepatocyte death; that results from the host's immune response. Hepadnaviruses circulate in the bloodstream and are poorly transmittable except by contact with blood or products contaminated with blood from infected individuals. Nevertheless, the human HBV infects about a third of the world's population, 350 million of whom are chronic carriers of the virus.

Hepadnaviruses are the smallest DNA viruses that infect mammals. They replicate to high levels, and maintain a chronic, productive infection, often in the face of a vigorous immune response. They contain an unusual genome consisting of a 3 kbp relaxed, circular, partially duplex DNA molecule whose replication requires reverse transcription of a viral RNA intermediate termed the pregenome. Accordingly, these viruses encode their own reverse transcriptase essential for viral genome duplication. In virtually all other details, however, the pathway for HBV genome duplication differs from the reverse transcription pathway evolved by retroviruses. In fact, the HBV provirus is a covalently closed, circular dsDNA episome, rather than an integrated copy of viral DNA like that employed by retroviruses. Drugs that target reverse transcriptase are effective against HBV.

The *Parvoviridae* contain a linear ssDNA genome of 4–6 kb with short imperfect terminal palindromes that fold back on themselves to form duplex hairpin telomeres (Table 8-1). It duplicates its genome by a gap-filling single-strand displacement mechanism (Figure 8-9) that requires the parvovirus NS1 protein, a site-specific ssDNA nuclease with DNA helicase activity. Parvovirus B19 is the only known human pathogenic parvovirus. Approximately 50% of all children age 15 are seropositive for this virus, a prevalence that rises to more than 90% in the elderly. In healthy individuals, the major consequence of B19 infection is erythema infectiosum (fifth disease), a rash on the cheeks, arms, and legs. In patients with underlying hemolytic disorders, infection is the primary cause of temporary suppression of erythropoiesis. In immunocompromised individuals, persistent infection may result in chronic anemia, and *in utero* infection may result in an accumulation of fluid (hydrops fetalis) or congenital anemia.

The *Polyomaviridae* contain a circular dsDNA genome of approximately 5 kb in length that replicates bidirectionally from a single replicator (Figure 8-1) using the replication-fork mechanism (Figure 3-8). These viruses encode only one protein, large T-antigen (T-ag; Figures 10-5 and 11-3), essential for viral genome duplication. Children and young adults are commonly infected with polyomaviruses, but most of these infections produce few if any symptoms. For example, approximately 80% of the adult population in the USA is seropositive for polyomaviruses BK and JC, but serious diseases are common only among those who become immunosuppressed either by AIDS, by old age, or following organ transplantation. Only three polyomaviruses are associated with human disease, all of which are associated with respiratory infections, suggesting that polyomavirus infection occurs primarily by inhaling the virus. JC virus can infect the brain where it can cause demyelination and death. BK virus promotes kidney malfunction. The Merkel cell polyomavirus causes a rare and highly aggressive cancer where malignant cells develop on or just beneath the skin and in hair follicles.

The *Papillomaviridae* are similar to the Polyomaviridae. In fact, they were once considered members of a single family called the Papovaviridae. Human papillomaviruses (HPVs) have a circular dsDNA genome of approximately 8 kb that replicates bidirectionally from a single replicator (Figure 8-1) using the replication-fork mechanism. Papillomavirus DNA replication requires only two HPV-encoded proteins, the replicative DNA helicase (E1) and the viral origin-binding protein (E2) (Figure 10-5). HPVs infect the squamous epithelium. Not only can they cause warts in virtually any epithelial tissue, but HPVs are the primary cause of cervical cancer. The mucosotropic groups are the most thoroughly investigated because they are prevalent, they can be sexually transmitted, and they cause significant morbidity and mortality worldwide. HPV infections can be asymptomatic or they may go into a productive phase. Productive infections often regress into subclinical latency after a year or so of activity, but overt symptoms may reappear years later during periods of immune suppression.

Large nuclear DNA viruses

The large nuclear DNA viruses are comprised of the *Adenoviridae* and the *Herpesviridae*, most of which cause diseases in humans and other mammals (Table 14-2). These viruses encode three or more proteins essential for duplication of the viral genome. Consequently, their host range is more extensive than the host range for smaller viruses. Fortunately, a number of inhibitors targeted against the DNA polymerases encoded by these viruses are effective antiviral agents.

The *Adenoviridae* (AdV) are nonenveloped icosahedral viruses that replicate in the nucleus by a single-strand displacement mechanism (Figures 8-7 and 10-2). Human AdV are grouped as six species (formerly subgroups) termed AdV-A through F, in the genus *Mastadenovirus* (one of four *Adenoviridae* genera). Human AdV contain a 36 kb linear, dsDNA genome with inverted terminal repeats of approximately 100 bp. Both 5′-termini are covalently attached to a 55 kDa terminal protein. AdV encodes only three proteins essential for duplication of the AdV genome, the AdV initiator protein 'pre-terminal protein', AdV DNA polymerase, and AdV DNA-binding protein.

AdV are implicated in a wide variety of human ailments such as nasal congestion, coughs, pneumonia, respiratory disease, conjunctivitis, epidemic kerato-conjunctivitis, hemorrhagic cystitis, and diarrhea (Table 14-2). The gastrointestinal disease caused by AdV-F is considered a major contributing factor in childhood diarrhea in underdeveloped and

Table 14-2. Large Nuclear Pathogenic DNA Viruses

Virus	Strain	Disease	Replication proteins
Adenovirus (Ad)	Many serotypes (HAdV-A to E)	Respiratory illness	DNA polymerase, pre-terminal protein, DNA-binding protein
	Serotypes 3, 7, and 14 (HAdV-B)	Conjunctivitis	
	Serotypes 8, 19, and 37 (HAdV-D)	Epidemic keratoconjunctivitis	
	Serotypes 3, 7, 16, and 21 (HAdV-B1), and 4 (HAdV-E)	Acute respiratory syndrome	
	Serotypes 40 and 41 (HAdV-F)	Gastroenteritis	
	Serotypes 9, 17, 20, 22, 23, 26, 27, 42–49, and 51 (HAdV-D)	AIDS-associated adenoviruses	
Herpes simplex virus (HSV)	Type 1	Oral herpes, keratitis, encephalitis, mucocutaneous infections	DNA polymerase, helicase•primase, SSB, ori-binding protein
	Type 2	Genital herpes	
Varicella-zoster virus		Chickenpox (children), shingles (adults)	
Cytomegalovirus (CMV)		Infectious mononucleosis, congenital deafness, mental retardation	
Herpesvirus 6, 7		Roseola infantum	
Herpesvirus 8 (KSV)		Kaposi's sarcoma, pleural effusion lymphoma	
Epstein–Barr virus (EBV)		Infectious mononucleosis, post-transplant lymphoproliferative disease, Hodgkin disease, Burkitt's lymphoma, nasopharyngeal carcinoma, gastric carcinoma	EBNA1, BZLF1, DNA polymerase, helicase•primase, SSB

KSV, Kaposi sarcoma virus, is a common name for herpesvirus 8. SSB is ssDNA-binding protein.

overpopulated regions. In addition, AdV is the most common infectious cause of ocular disease worldwide. AdV is isolated frequently from patients with a compromised immune system, and it is a frequent cause of morbidity and mortality after allogeneic stem cell transplantation in children. The incidence of AdV infection is now about 23%. Another group at risk is HIV-infected individuals. In fact, many patients with AIDS shed AdV serotypes rarely found in immunocompetent individuals.

At least eight of the viruses that comprise the *Herpesviridae* family are known to infect humans (Table 14-2), all of which fall under two paradigms, the herpes simplex virus (HSV) paradigm and the Epstein–Barr virus (EBV) paradigm. HSV contains a 152 kbp dsDNA molecule that consists of two unique regions (UL and US) flanked by inverted repeat sequences. It encodes seven proteins essential for viral DNA replication as well as several nonessential replication proteins. Six of the essential proteins have homologs in other herpesviruses: an ssDNA-binding protein, a two-subunit DNA polymerase, and a three-subunit helicase•primase complex. HSV also encodes a protein that binds the HSV origin of replication. The overall strategy of lytic viral DNA replication appears to be conserved in all herpesviruses; however, the mechanism and regulation of initiation of viral DNA synthesis differs sharply between the HSV and EBV paradigms.

EBV virions contain a linear, dsDNA molecule of about 175 kbp that is characterized by a number of different sequence repetitions. The termini consist of tandem repeats of about 540 bp that allow the EBV genome to circularize when released into the nucleus of its host cell. EBV persists in its host through its ability to establish a latent state from which it can periodically explode into a lytic state. In the latent state, a small number of EBV genomes replicate under control of the cellular replication machinery. The only viral protein required for EBV DNA replication during the latent period is EBNA1, the initiator protein that recruits cellular ORC to bind to oriP and thereby establish EBV as a cellular episome (Figures 10-18 and 10-19). In the lytic state, multiple copies of the genome are produced under control of the virus's replication machinery. EBV enters a lytic state when it produces BZLF1, the EBV oriLyt initiator protein, as well as the EBV two-subunit DNA polymerase, three-subunit DNA helicase•primase complex, and ssDNA-binding protein.

Herpesviruses are second only to influenza and cold viruses as a leading cause of virus-induced human disease. They can either cause disease or remain silent for many years only to be reactivated (e.g. shingles). In underdeveloped countries, for example, antibodies against HSV are found in more than 90% of children. In the USA, more than 90% of the population is seropositive for varicella-zoster. From 90% to 95% of the world's population is infected with EBV and these people, although usually asymptomatic, shed the virus from time to time throughout life. The virus is spread by close contact (infectious mononucleosis is known as the kissing disease). Up to 80% of students entering college in the USA are seropositive for the virus and many of those that are negative will become positive while at college. Cytomegalovirus (CMV) can infect virtually any organ and cause disease in immunocompromised people. CMV is transmitted through saliva, urine, vaginal secretions, and semen and therefore their frequency within the population increases with age. About 15% of the college-age population in the USA is infected, and this rises to about half by 35 years of age. In developing countries with more crowded conditions these viruses are found in a much higher proportion of the population than in Western countries. Herpesvirus 6 is found worldwide in the saliva of most adults (>90%).

The hallmark of herpes infection is the ability to infect epithelial mucosal cells or lymphocytes. The virus then travels up peripheral nerves to a nucleated neuron where it may remain dormant for years before it is reactivated. HSV enters humans either through wounds or by infecting the mucoepithelia. HSV then frequently sets up latent infections in neuronal cells, macrophages, and lymphocytes. Once epithelial cells are infected, there is replication of the virus around the lesion and entry into the innervating neuron. The virus travels along the neuron to the ganglion. It also can travel in the opposite direction to arrive at the mucosa that was initially infected. Vesicles containing infectious virus are formed on the mucosa and the virus spreads.

Cytoplasmic DNA viruses

Poxviruses are large, enveloped, DNA viruses that replicate entirely in the cytoplasm. Their genomes are linear, unmethylated dsDNA that varies in length from 130 to 230 kbp. Notably, the two DNA strands are connected at their termini to form a continuous polynucleotide chain with no interruptions. These hairpin termini exist in inverted and complementary forms that are incompletely base-paired and AT-rich. Although poxviruses encode 11 proteins that are involved in duplicating the viral genome, only six are essential (DNA polymerase, NTPase, uracil DNA glycosylase, processivity factor, protein kinase, and ssDNA-binding protein). Host-cell proteins can substitute for the remaining five viral proteins (thymidine kinase, thymidylate kinase, ribonucleotide diphosphate reductase, dUTPase, and DNA ligase).

The variola and molluscum contagiosum poxviruses are specific for humans. The former causes smallpox, a severe disease with high mortality that was eradicated more than two decades ago. The latter is distributed worldwide and produces discrete benign skin lesions in infants but more extensive disease in immunocompromised individuals. In addition, other poxviruses that normally infect nonhuman species can be transmitted to humans (Table 14-3). Some of the inhibitors targeted against the DNA polymerases encoded by large nuclear DNA virus also are effective against poxviruses.

Retroviridae

Retroviruses are RNA viruses that encode a reverse transcriptase to produce a dsDNA copy of the virus' RNA genome, and an **integrase** that incorporates the dsDNA copy into the host's genome. The resulting 'provirus' DNA is then replicated as part of the host cell's genome. Two retroviruses cause human disease. Human T-cell leukemia virus 1 (HTLV-1) is responsible for adult T-cell leukemia and T-cell lymphoma. For every 25 people infected with HTLV-1, one will develop cancer. HIV is responsible for AIDS. In 2008, an estimated 33.4 million people worldwide carried HIV; more than 90% of them either have or will develop AIDS. Drugs that inhibit reverse transcriptase are effective against HIV proliferation.

Bacteria

Standard medical textbooks list at least 82 different human diseases caused by bacteria. Some of the more infamous ones and their causative agents include Lyme disease (*Borrelia burgdorferi*), syphilis (*Treponema pallidum*), plague (*Yersinia pestis*), leprosy (*Mycobacterium leprae*), pneumonia (*Streptococcus pneumoniae*), cholera (*Vibrio cholerae*), tuberculosis (Mycobacteria), and common food poisoning (*Salmonella* and other Enterobacteria). Given the fact that genome duplication is highly conserved among the three domains of life, one might assume that pharmacological agents that inhibit a protein essential for bacterial DNA replication would also inhibit its human counterpart. However, while the DNA replication proteins in bacteria are analogous to those in humans, they evolved independently, and therefore exhibit significant differences. In fact, several inhibitors targeted against the replicative DNA polymerase of Gram-positive bacteria, and topoisomerase II of both Gram-positive and Gram-negative bacteria, show promise in the treatment of bacterial infections.

Archaea

Remarkably, no pathogenic archaea have been reported. Whether archaea are innocent of attacks on humans, or simply as yet unrecognized agents of human disease, remains to be determined. Certainly, human

Table 14-3. Cytoplasmic Pathogenic DNA Viruses

Virus	Strain	Disease	Replication proteins
Orthopoxvirus	Variola Monkeypox Cowpox Vaccinia	Smallpox Human monkeypox (zoonosis) Human cowpox (zoonosis) Vaccinia (vaccine-derived)	DNA polymerase, NTPase, uracil DNA glycosylase, processivity factor, protein kinase, SSB
Parapoxvirus	Orf virus Pseudocowpox	Human orf (zoonosis) Milker's nodule (zoonosis)	
Molluscipoxvirus	Molluscum contagiosum	Molluscum contagiosum	
Yatapoxvirus	Tanapoxvirus	Tanapox (zoonosis)	

A zoonosis is a disease that can be transmitted from other vertebrates to humans.

physiology does not exclude archaea, because at least 50% of the human population has methanogenic archaea (*Methanobrevibacter smithii* and *Methanosphaera stadtmanae*) as a component of their gut microflora. In addition, there has been a positive correlation observed between the presence of *Methanobrevibacter oralis* in subgingival sites and the presence of periodontal disease. However a direct causal relationship has not been established. Thus, it is possible that archaea do indeed play as yet undiscovered roles in disease; alternatively, the archaea may harbor important clues to living in harmonious coexistence with their metazoan hosts.

NONINFECTIOUS DISEASES

You can catch a cold from your neighbor, but you can't catch a cancer. That's the difference between infectious and noninfectious diseases. Agents that cause infectious diseases such as viruses or bacteria can pass from one person to another. Cancer is not infectious, because the agents that cause it (radiation, chemical carcinogens, and oncogenic viruses) are inactivated in the process.

The Road To Cancer

The cancer phenotype is invariably associated with a loss of growth control leading to inappropriate or excessive proliferation of cells and eventually malignancy. Simply stated, cancer cells are not defective in genome duplication; they are defective in *regulating* genome duplication. Cancer cells continue to proliferate under conditions where normal cells do not. To accomplish this feat, cancer cells ignore the chemical signals emitted by cells (cytokines) or present on cell surfaces (cell surface markers) that either stimulate or arrest the proliferation of neighboring cells. Such signals are essential to normal tissue formation and morphology. Ignoring these signals allows cancer to spread throughout the body, a process termed **metastasis**. Finally, cancer cells often must be able to form into a mass of cells (tumor) that continues to take up nutrients and remain metabolically active. The host organism eventually dies when the tumor directly or indirectly disrupts the function of a normal tissue, or when the cancer destroys the ability of its host to protect itself from infection.

To accomplish these feats – that is, to circumvent the controls that regulate cell proliferation – cancer cells suppress the checkpoint pathways described in Chapter 13. They do this through the accumulation of genetic mutations, thereby steadily progressing from a normal healthy cell to a malignant cell. For this, cancer cells need to be genetically unstable compared with normal cells. Hence, mutations in genes that directly affect the replication or repair of DNA are fundamental to driving cells down the road to cancer. Thus, mutations in some genes do not 'cause' cancer; they simply increase the 'risk' of cancer by predisposing the cell toward developing into a cancer cell. Any genetic change that increases genomic instability increases the likelihood of cancer. That is why the incidence of cancer increases sharply with age; we accumulate mutations over time.

Mutations in restriction checkpoint genes

Cancer cells frequently harbor mutations in pathways that govern the $G_1 \rightarrow S$ transition in the cell-division cycle: the restriction checkpoint that senses nutrient and mitogen deficiencies (Figure 13-5), and the G_1 DNA damage checkpoint that senses the accumulation of RPA•ssDNA complexes (Figure 13-13). During early stages of many types of cancer, quiescent cells inappropriately re-enter the cell cycle, a phenomenon that reflects the high frequency with which upstream elements of pathways that regulate origin licensing or function are mutated in cancer cells (Table 14-4). Mutations in the *Ras* and *Myc* genes, for example, constitutively activate mitogenic

Table 14-4. Restriction Checkpoint Genes Associated With Cancer And Other Disorders

Gene	Function	Disease
Rb1	Inhibits certain E2f transcription factors, thereby suppressing E2f-dependent gene expression	Retinoblastoma, osteosarcoma, bladder cancer, pinealoma
Cdkn2A/p16/Ink4a	Inhibits Cdk4 and Cdk6 kinase activity	Melanoma, Li–Fraumeni syndrome, neural system tumor syndrome, pancreatic cancer, orolaryngeal cancer
CcnD1	Activates Cdk4 and Cdk6 kinase activity	Colorectal cancer, von Hippel–Lindau disease
Cdk4	Inhibits Rb and p27	Melanoma
Cdkn1B/p27/Kip1	Inhibits Cdk1 and Cdk2	Multiple endocrine neoplasia
Cdkn1C/p57/Kip2	Inhibits Cdk1 and Cdk2	Beckwith–Wiedemann syndrome

signaling pathways leading to the inappropriate activation of CcnD•Cdk4, CcnD•Cdk6, and CcnE•Cdk2 (Figure 13-4). The *Myc* gene, which was originally identified in the avian myelocytomatosis virus, is disrupted in numerous hematopoietic neoplasias. The *Ras* gene was first identified as the rat sarcoma virus gene responsible for causing cancers in animals infected with this virus. *Ras* is one of the most frequently disrupted genes in colorectal carcinomas.

Constitutive activation of mitogenic signaling also occurs in association with inactivation, or amplification, or rearrangements of genes that regulate E2f (Table 14-4), the transcription factor that is required to transcribe genes needed to support S phase. Loss of Rb contributes to a wide range of cancers. After Rb, the most frequent mutations in cancers that disrupt the restriction checkpoint are those that inactivate p16. This increases CcnD•Cdk4 activity, which leads to Rb phosphorylation and the release of E2f transcription factor. Similarly, deregulation of CcnD or Cdk4 stimulates E2f activity. Inactivation of the CDK inhibitors p27 and p57 likewise stimulates the $G_1 \rightarrow S$ transition. Notable by their absence are mutations that deregulate expression of E2f genes. Such mutations may be too pleiotropic to allow cell survival.

Mutations in DNA damage-response genes

Once cells enter S phase, the susceptibility to DNA damage of unwound DNA at replication forks can lead to genome instability, a hallmark of cancer cells. Genome instability arises from several sources, all of which occur specifically during genome duplication. They include decreased fidelity in DNA synthesis (Table 4-3), improperly repaired DNA damage (see Table 14-7, below), bypassing DNA lesions instead of repairing them (Figure 14-6), defects in DNA damage checkpoints (Table 13-1), inactivation of proteins that stabilize replication forks (Figure 13-18), premature entrance into S phase that reduces the number of functional origins of replication (Table 9-2), and failing to restrict the activation of replication origins to once per origin per cell cycle (Figure 12-12).

Defects in proteins that regulate checkpoints and/or replication-fork stability are commonly found in cancers (Tables 14-4 and 14-5). The best example is the tumor suppressor Tp53, a major component of the G_1-phase DNA damage response (Figure 13-13). Tp53 is mutated in at least 50% of all human malignancies. In addition, two negative regulators of Tp53, Mdm2 and Mdm4/Mdmx, are overexpressed in many human tumors, making Tp53 the most frequently suppressed activity in the majority of cancers. The importance of Tp53 regulation is demonstrated by the fact that overexpression of Mdm2 suppresses Tp53 and promotes tumor formation.

Tp53 activity also is modulated by the DNA damage sensor Atm•MRN complex (Figure 13-13), and mutations in the *Atm* gene are responsible

Table 14-5. DNA Damage-Response Genes Associated With Cancer And Other Disorders

Gene	Function	Disease
Tp53	Regulate expression of p21	Li–Fraumeni syndrome, colorectal cancer, hepatocellular carcinoma, osteosarcoma, choroid plexus papilloma, nasopharyngeal carcinoma, pancreatic cancer, adrenal cortical carcinoma, breast cancer
Mdm2	Regulate Tp53 activity	
Mdm4	Regulate Mdm2 activity	
Atm	Detect DNA damage and stalled forks	Ataxia telangiectasia, B-cell non-Hodgkin lymphoma, mantle cell lymphoma, T-cell prolymphocytic leukemia, breast cancer
Mre11A	Meiotic recombination, rescue stalled forks, repair DNA damage	Ataxia telangiectasia-like disorder
Nbn/Nbs1	Rescue stalled forks, repair DNA damage	Nijmegen breakage syndrome
Atr	Detect DNA damage and stalled forks, other checkpoint/replication genes may also be involved	Seckel syndrome
Chek2/Chk2	Respond to DNA damage and stalled forks	Li–Fraumeni-like syndrome, osteosarcoma, prostate cancer, increased risk of breast and colorectal cancer
Brca1	Amplify the DNA damage-response signal and stabilize stalled replication forks	Breast cancer, ovarian cancer
Brca2	Regulate function of Rad51 in DNA repair by homologous recombination	
Mcph1	Required for expression of Brca1 and Chk1, and for phosphorylation of Nbn	Microcephaly, mental retardation

for the predisposition to cancer observed in ataxia telangiectasia. People homozygous for these mutations frequently develop leukemias and lymphomas, and their heterozygous relatives have a greater risk than normal individuals of acquiring the disease. Mutations in the MRS genes (*Mre11•Rad50•Nbs1*) exhibit similar phenotypes. *Nbs1* gene mutations cause the cancer-predisposing Nijmegen breakage syndrome, and *Mre11* mutations lead to an ataxia telangiectasia-like disorder. Brca1 protein also mediates the DNA damage response, and mutations in *Brca1* and *Brca2* genes predispose individuals to early onset of breast and ovarian cancer, and to a lesser extent, ovary, prostate, and pancreatic cancer. Mutations in Atr, the other DNA damage sensor protein in mammals, are associated with a birth defect called 'Seckel syndrome' that is characterized by severe short stature, low birth weight, and a very small head.

Humans have two DNA damage effector proteins, Chk1 and Chk2, that operate downstream of Atr and Atm (Table 13-1). Both Atm and Chk2 modulate Tp53 function in response to dsDNA breaks (Figure 13-13). Chk1 mutations are rarely found in tumor cells (presumably because Chk1 is essential for cell proliferation), but Chk2 mutations appear in a subpopulation of families with the cancer-predisposing Li–Fraumeni syndrome. This syndrome is often associated with germ-line mutations in Tp53. Mutations in Chk2 also are associated with an increased risk for breast and colon cancer, and mutations in Chk2 appear in a variety of cell lines derived from different types of tumors.

The spindle-assembly checkpoint prevents the transition from metaphase to anaphase until all of the chromosomes are attached correctly to the spindle fibers (Figure 13-19). Therefore, it is not surprising that mutations in spindle-assembly checkpoint genes appear to create a predisposition to cancer. For example, the *Bub1* gene (Table 13-10) is mutated or expressed at low levels in colorectal cancers with chromosomal instability. Moreover, when expression of Bub1 is gradually reduced in mice by genetic manipulations, the mice develop spontaneous tumors with reduced latency and increased incidence in a dose-dependent fashion. Mutations in the *Mad1L1* gene (Table 13-10) have been found in lymphomas and prostate cancer. Such mutations could account for the aneuploidy that is characteristic of the majority of human cancers.

Mutations in DNA helicases

The human genome encodes at least 25 different DNA helicases, several of which are essential in the repair of dsDNA breaks, ssDNA gaps, and stalled replication forks, as well as in homologous DNA recombination. Mutations in these enzymes give rise to various syndromes, some of which increase a cell's propensity toward cancer.

Xpb and Xpd are two DNA helicases associated with xeroderma pigmentosum (XP), a disease characterized by ultraviolet sensitivity, skin cancers, and neurological abnormalities (Table 14-6). They are also associated with XP-Cockayne syndrome, which includes dwarfism, mental retardation, and retinal and skeletal abnormalities, and with trichothiodystrophy, which is characterized by brittle hair, mental retardation, reduced stature, ichthyotic skin, and other unusual facies. Many mutations in the *Xpd* gene have been reported, but no identical substitutions are associated with these three hereditary disorders. Thus, it appears that the site of the mutation determines the clinical outcome.

In normal human cells, Xpb and Xpd play a critical role in the **nucleotide excision repair** pathway by removing the vast majority of ultraviolet-induced DNA damage in the form of pyrimidine dimers (Figure 14-1). Failure to eliminate these lesions can lead to oncogenesis (development of tumors), developmental abnormalities, and accelerated aging. Hence, the ability of patients to tolerate chemotherapy relies on the capacity of their normal cells to eliminate drug-induced DNA lesions.

Xpb and Xpd are part of the ten-subunit transcription factor IIH (TFIIH) complex that functions in both transcription and in nucleotide excision repair. The role of Xpb and Xpd is to stabilize binding of TFIIH to damaged DNA and to open the DNA to allow excision of the lesion and repair of the resulting ssDNA gap. Xpb is indispensable for normal TFIIH-dependent

Table 14-6. DNA Helicases Associated With Cancer And Other Disorders

Gene	Function	Disease
Xpb; Xpd	Nucleotide excision repair	Xeroderma pigmentosum, XP-Cockayne syndrome, trichothiodystrophy
Rad54L	dsDNA-break repair. Dissociates Rad51 from nucleoprotein filaments formed on dsDNA. Could be involved in the turnover of Rad51 protein-dsDNA filaments. Component of Swi2/Snf2 chromatin-remodeling machine	Non-Hodgkin lymphoma, breast cancer, colon adenocarcinoma
RecQ2/Blm	dsDNA-break repair, homologous recombination, and recovery of stalled replication forks. RecQ dissociates Rad51 from nucleoprotein filaments formed on dsDNA	Bloom syndrome
RecQ3/Wrn		Werner syndrome
RecQ4/Rts		Rothmund–Thomson syndrome, Baller–Gerold syndrome
RecQ4		Rapadilino syndrome
Peo1/ Twinkle	mtDNA helicase required for mtDNA replication	Progressive external ophthalmoplegia, mtDNA depletion syndrome

transcription, and therefore mutations in the *Xpb* gene are generally lethal. Xpd, however, maintains the stability of the TFIIH complex, so it is required only for optimal transcription.

The Rad54-like and RecQ-like DNA helicases (Table 13-5) are involved in dsDNA break repair, homologous recombination, and recovery of stalled replication forks associated with the expansion and contraction of repeated DNA sequences (discussed below). Mutations in these DNA helicases are associated with a predisposition to cancer and a variety of other physical abnormalities (Table 14-6). Bloom syndrome causes a predisposition to many cancers including non-Hodgkin lymphoma and leukemia. This syndrome is characterized by dwarfism, sun-induced erythema, type II diabetes, and infertility. Werner syndrome causes a predisposition to mesenchymal tumors such as sarcomas, melanoma, and thyroid cancer. Werner syndrome is characterized by premature aging features such as cataracts, type II diabetes, osteoporosis, atherosclerosis, and graying and loss of hair. Rothmund–Thomson syndrome also creates a cancer predisposition, but chiefly to osteosarcoma. This syndrome is characterized by juvenile cataracts, skin atrophy, and pigmentation changes and skeletal abnormalities. Only Rapadilino syndrome, characterized by developmental and skeletal abnormalities, is not associated with increased cancer risk.

Human cancer viruses

Viruses cause approximately 20% of cancers worldwide. These include cervical cancer, the second most frequent cancer in women, and about 80% of liver cancers. Some cancers are associated with viruses, although the virus has not yet been proven to be the causative agent. Of the seven viruses that do cause cancer in humans, five are DNA viruses (HPV, EBV, human herpesvirus type 8, Merkel cell polyomavirus, and HBV), one is a retrovirus (HTLV-1), and one is an RNA virus (hepatitis C virus). Viruses that

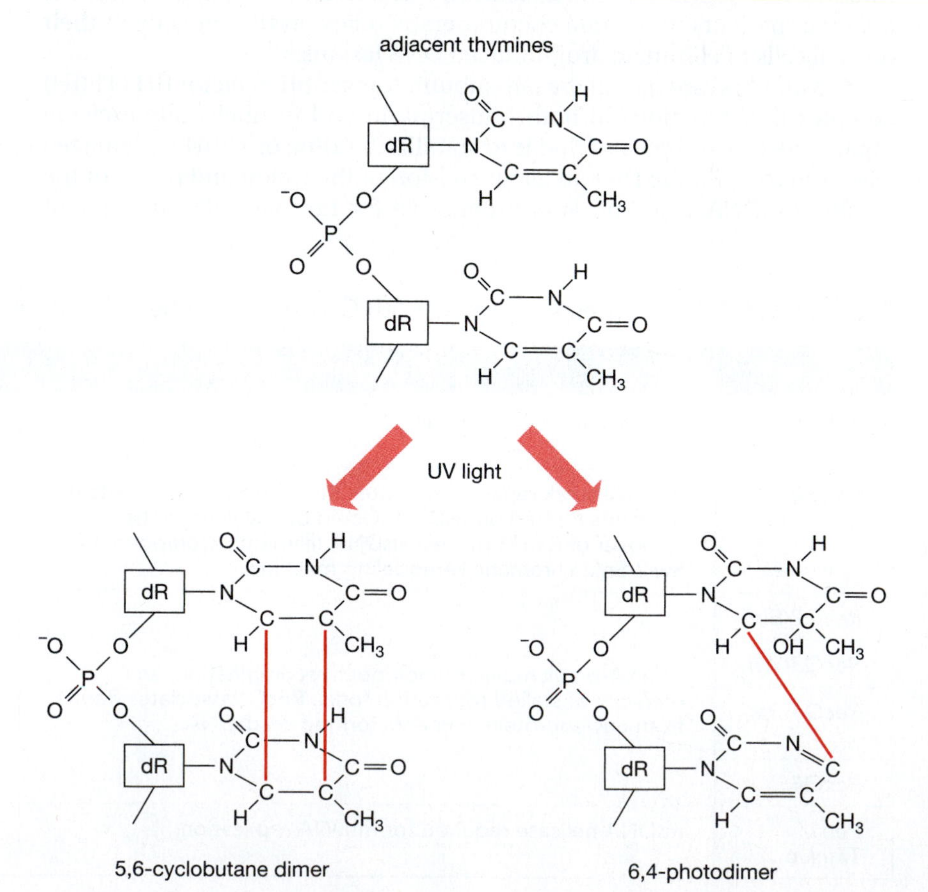

Figure 14-1. Ultraviolet (UV) light induces two types of cross-linking between adjacent pyrimidine bases in the same strand of DNA. Shown are T-T dimers (numbering system as in Figure 1-1). T-C dimers also form. Deoxyribose is indicated as dR.

cause cancer are termed oncogenic viruses; they each encode one or more **oncogenes** that drive their host cell into S phase. This ensures that the virus has all of the nucleotides and proteins needed to replicate its own genome, thereby allowing the virus to propagate efficiently even though its genome is but a tiny fraction the size of its host's genome. With the exception of hepatitis C virus, proliferation of the remaining six human oncogenic viruses requires replication of the viral DNA genome or a DNA copy of the viral RNA genome.

How does an infectious agent cause a noninfectious disease? Cancer viruses have both a lytic cycle, in which more viruses are produced, and a lysogenic cycle, in which the viral genome does not replicate but does express one or more viral genes. The lytic cycle allows the virus to spread to other cells and to other organisms; the lysogenic cycle allows the virus to transform its host into a cancer cell. When a DNA virus enters a cell, its genome is uncoated and the first viral genes expressed are transcribed by the host cell's RNA polymerases. When an RNA virus that encodes a reverse transcriptase enters a cell, some of that enzyme is carried with it inside their virions. This is how Howard Temin and Satoshi Mizutani discovered the first RNA-dependent DNA polymerase in 1970; they released it from the virions of the avian Rous sarcoma virus, an experiment for which Dr. Temin shared the 1975 Nobel Prize in Physiology or Medicine. The reverse transcriptase rapidly converts the ssRNA viral genome into a dsDNA molecule that can be transcribed by the cell and that can be integrated into the host's chromosomal DNA. It is only after the viral DNA becomes part of the cell's genome that oncogenesis can take place.

If the integrated viral genome continues to express its oncogene(s), the cell is transformed into one that can proliferate under conditions in which normal cells cannot. For example, transformed cells can proliferate when nutrient conditions are poor or growth stimulatory factors are absent. Transformed cells are no longer restricted to forming a monolayer in culture, but can pile up upon one another to form dense populations of cells. They also can form colonies in soft agar (termed anchorage-independence) whereas normal cells cannot. Moreover, these virally-transformed cells can produce malignant tumors in animals, the consequence of the viral oncogene's ability to alter the expression of multiple cellular genes, and at least in some cases to induce genomic instability. It is important to note that cell transformation by oncogenic viruses is a very-low frequency event.

The HPV paradigm

The mechanism by which viruses induce cancer in humans is understood best for the small nuclear DNA viruses represented by HPV. In 2008, Harald zur Hausen received the Nobel Prize in Physiology or Medicine for his discovery that cervical cancer is caused by HPV. This human cancer virus expresses two oncogenes: E6 and E7. In cervical cancers and cell lines derived from cervical cancers, HPV DNA is integrated into the cell's DNA in more than 99.7% of the samples. In these cells, HPV genes E6 and E7 are expressed, but the ability of the HPV DNA to replicate on its own has been lost. The ability of integrated viral genomes to initiate replication at their own replicator using their own initiator protein is selected against during cell transformation by the simple fact that unregulated origin activation events would result in DNA re-replication (Figure 12-11) and the creation of fragile sites (described below) where chromosomes are easily broken and replication forks stall. Therefore, in the cancer cell that eventually emerges, either the HPV replication origin or one or both of its initiator proteins are lost.

Oncogenesis occurs by virtue of the E6 protein binding to and inactivating its host cell's Tp53 protein, thereby preventing Tp53 from participating in the DNA damage-response pathway that arrests cells in G_1 phase (Figure 13-13). Moreover, Tp53-dependent gene expression is suppressed, with the consequence that Tp53-dependent growth-arrest and apoptosis-induction functions are inactive in the transformed cells. Concurrently, the oncogenic E7 protein binds to its host cell's Rb protein, thereby preventing it from inhibiting E2f-dependent transcription that is necessary for moving cells from G_0 into S phase (Figure 13-4). By this means, E6 and E7 operate in concert to inactivate at least two mechanisms that govern the $G_0 \rightarrow G_1 \rightarrow S$ transition in normal cell cycles. Both of these viral genes are necessary to produce epithelial neoplasia characteristic of those produced by HPV-16 and HPV-18, the two viral strains that account for more than 99% of cervical carcinomas (Table 14-1). The transformed cell is now at liberty to mutate more easily due to the suppression of a critical DNA damage checkpoint. Additional mutations increase the probability that this cell will eventually become a cancer cell.

The HPV paradigm extends easily to the polyomaviruses, each of which encode two or three tumor antigens, as a result of alternate mRNA splicing sites. Large tumor antigen (T-ag) serves as a viral-specific initiator protein and DNA helicase during a lytic virus infection. In addition, T-ag stimulates the host cell to enter S phase by associating both with the DNA-binding domain of Tp53, and with Rb proteins, preventing these proteins from carrying out their normal regulatory functions. However, the mechanisms by which other human cancer viruses operate are less clear.

The EBV paradigm

EBV is an enormously successful human parasite, having infected more than 6.5 billion people worldwide, and its success is intimately linked to its ability to establish itself as an extrachromosomal element by recruiting its host cell's DNA replication machinery to initiate and carry out EBV DNA replication (Figures 10-18 and 10-19), a function that requires the EBV EBNA1 protein. However, only three EBV genes (EBNA-2, EBNA-3C, and LMP-1) are essential for cell transformation, and the prime suspects for EBV oncogenes are two small, highly abundant, noncoding nuclear RNAs (EBER1 and EBER2) expressed consistently during EBV latency. Their role, however, remains speculative.

The HBV paradigm

Chronic infection by HBV is one of the major causes of hepatocellular carcinoma in the world. The viral genome (3–3.3 kb) encodes four proteins: a viron coat protein, a DNA polymerase, a cell-surface antigen, and the X protein. HBV X protein is a multifunctional regulator that modulates transcription, signal transduction, cell-cycle progress, protein-degradation pathways, apoptosis, and genetic stability by directly or indirectly interacting with host factors, making it the likely HBV oncogene.

The HTLV paradigm

HTLV-1 is a retrovirus that causes T-cell leukemia. HTLV encodes five regulatory proteins, all of which are important for infecting cells and stimulating DNA replication. The HTLV-1 Tax protein appears to be an oncogene. Tax expression in cultured cells leads to leukogenesis and immortalization of T-lymphocytes through the stimulation of interleukins 2 and 15, which, in turn, promote cell transformation.

DNA Replication Proteins As Cancer Biomarkers

Most cells in our body exist either in a quiescent (G_0), reversibly arrested state (e.g. hepatocytes), or in a terminally differentiated state that does not undergo cell division (e.g. neurons and muscle cells). Cancers, on the other hand, are characterized by uncontrolled cell growth and therefore contain a high proportion of rapidly proliferating cells. Thus, genes involved specifically in DNA replication and cell proliferation can provide biomarkers for a variety of cancers. The better the biomarker, the earlier a cancer can be detected, the earlier it is detected, the more likely it can be treated successfully. The usefulness of a particular replication protein as a cancer biomarker depends largely on the sensitivity of the assay in a clinical setting, on the stability of the protein or mRNA in tissue biopsies, and on the specificity of the assay for proliferating cells as compared to cells suffering from DNA damage.

One popular cancer biomarker has been proliferating cell nuclear antigen (PCNA), the sliding clamp for replicative DNA polymerases δ and ε (Figure 4-23). PCNA is easy to detect, but it has two drawbacks. Its expression is restricted largely to S phase, and it is used in DNA repair as well as DNA replication. In addition, PCNA exhibits striking fluctuations in abundance among cells, making interpretation of stained samples difficult. Nevertheless, two isoforms of PCNA have been detected in breast cancer cells, one of which is associated specifically with breast cancer. The cancer-specific PCNA isoform functions in cancer cell DNA replication and interacts with DNA Pol-δ. Hence, it may prove useful as a breast cancer biomarker.

In some cases, oncogenic viruses produce their own biomarkers, in addition to activating cellular biomarkers. For example, persistent infection with oncogenic strains of HPV is the major cause of cervical cancer and its pre-invasive lesions. Viral DNA replication and integration in the cervix result in overexpression of viral-specific proteins. These proteins serve as biomarkers for HPV-induced cancer. In fact, HPV DNA, HPV E6, and E7 mRNA, as well as human PCNA and Mcm5 proteins, are currently the most useful biomarkers for early diagnosis of cervical carcinoma.

Two of the most promising cancer biomarkers are the replicative MCM DNA helicase and the preRC assembly inhibitor geminin. Both are unique to the process of genome duplication. Of these, the Mcm(2-7) proteins are the most abundantly expressed, because they are present throughout cell division, whereas geminin expression is restricted to S, G_2, and early M phases (see Figure 14-2 in color plate section). Thus, an Mcm protein reveals the fraction of proliferating cells (non-G_0 cells), whereas geminin reveals the fraction of cells committed to cell division. Together, they provide a more accurate assessment of the status of a neoplasia. For example, the level of geminin protein in normal breast tissue is low or absent, but in invasive breast carcinomas, geminin is present in a higher proportion of epithelial cells (see Figure 14-3 in color plate section). High levels of cells expressing geminin correlate with both poor overall patient survival and the development of distant metastases. In other words, as the fraction of geminin-expressing cells increases, the prognosis for successful treatment of invasive breast cancer decreases. The *geminin* gene shows no evidence of amplification, deletion, or mutation in breast cancer, and there is no deregulation of its cell-cycle-dependent expression. Thus, the increased levels of geminin result from failure to arrest cells in G_1 or G_0 phase, rather than changes in the *geminin* gene itself. A study of melanomas reached the same conclusion; as the fraction of cells expressing geminin and Mcm proteins increased, patient survival decreased.

Geminin, Mcm7, and the Ki-67 antigen are generally coexpressed in colorectal cancer, suggesting that a combination of markers will be the most reliable method of diagnosis. Ki-67 is a commonly used cancer biomarker. It is a nonhistone nuclear protein of unknown function that is expressed throughout the cell cycle in proliferating cells but not in quiescent cells. Direct comparisons of these three biomarkers reveal that the Mcm and geminin labeling indices (fraction of positive cells) are superior to the Ki-67 labeling index and to the widely used clinicopathological parameters of tumor grade, stage, and size. Ki-67 is absent from a portion of cells in each phase of the cell cycle, including S phase, and it is expressed in fewer malignant and premalignant cells than Mcm proteins (see Figures 14-3 and 14-4 in color plate section).

Oral squamous-cell carcinoma represents the majority of malignant lesions of the oral cavity. A combination of Mcm, geminin, and Ki-67 cytology can detect the pre-cancerous state called oral epithelial dysplasia as well as the cancerous state called oral squamous-cell carcinoma. Labeling indices increase progressively from normal oral mucosa, to oral epithelial dysplasia (OED) to oral squamous-cell carcinoma. In all the OED cases, the levels of Mcm2, geminin, and Ki-67 were elevated indicating a constant cell-cycle re-entry. When the OED groups were compared, there was a significant increase in the ratios of Mcm2 to Ki-67 and geminin to Ki-67. Thus, Mcm2 and geminin proteins are novel biomarkers of cell proliferation that may be useful prognostic tools for oral epithelial dysplasia.

Mcm proteins also allow normal and cancer tissues to be distinguished clinically, because Mcm proteins in normal cells are broken down before cells are shed from tissues into the body fluids, whereas malignant and premalignant (dysplastic) cells lost from the tumor surface retain their Mcm proteins. This critical observation allows early detection of colorectal cancer, lung cancer, oral cancer, and anal cancer (see Figure 14-5 in color plate section).

HERITABLE DISEASES

Heritable diseases are noninfectious diseases that result from genetic disorders, and genetic disorders arise from mutations in the DNA sequence. Carcinogenic chemicals and radiation mutate DNA by altering nucleotide bases so that they are either mistaken for other nucleotide bases, or they actually change into another base (e.g. deamination converts cytosine into uracil; Figure 1-1). Frightening is the realization that the number of mutations that occur during a single cell division is amplified by the number of cell divisions (10–100 trillion) required to produce an adult human from a fertilized egg. Fortunately, the human genome contains, on average, only 7 genes per million base pairs (Table 1-4), which means that most mutations are unlikely to affect either development or viability. Moreover, only those mutations that appear in our germ lines (sperm and eggs) can be passed on from one generation to the next. With this in mind, consider the dangers inherent in genome duplication.

The 'Seven Deadly Sins' Of DNA Replication

The very act of DNA replication exposes the 3.08 billion base pairs that constitute the nuclear portion of the human genome to 'seven deadly sins' that could lead to cell death (Table 14-7). This places tremendous stress on the mechanisms that prevent or repair the damage. First, the very act of synthesizing a complementary DNA strand results in a demonstrable frequency of copy errors. Base substitution error rates for replicative DNA polymerases with an intrinsic 3′→5′ exonuclease proofreading activity are typically $\geq 10^{-6}$ (greater than or equal to one in 10^6 bases). Since their exonuclease-deficient derivatives are considerably less accurate, proofreading improves replication fidelity, on average, by 10 to 100-fold (Table 4-3). In addition to proofreading of mismatched base pairs during replication, most organisms also have DNA mismatch repair systems so that the spontaneous mutation rate in human cells is 10^{-9} to 10^{-11} bp. This allows the human species to continue to evolve through natural selection while remaining fertile and viable.

The second and third sins are that DNA replication invariably produces ssDNA during DNA unwinding and ssDNA interruptions during discontinuous DNA synthesis on the lagging-strand template (Figure 3-8). If this ssDNA is not immediately complexed with protein (a problem that arises when forks stall) it can form complex secondary structures through internal base pairing. These structures can lead to either addition or deletion of DNA sequences (described below). Even more dangerous, however, is the possibility that one or both of the ssDNA sites in replication forks will break, either from physical stress or from enzymatic activity. dsDNA breaks not only arrest cell division (Figure 13-13), but they induce homologous DNA recombination in eukarya as well as bacteria (Figure 13-9). The presence of a ssDNA interruption between the 3′-end of an Okazaki fragment and the 5′-end of the downstream nascent DNA strand can induce DNA damage-response pathways, particularly when replication forks stall.

Finally, there is the simple fact that genome duplication can disrupt gene expression patterns in four ways (Table 14-7). DNA replication disrupts the DNA methylation pattern (Figure 7-24), a mechanism frequently used in mammals to maintain genes in a silent state. Failure to remethylate these sites during genome duplication can result in inappropriate gene expression (a hallmark of cancer cells). Similarly, DNA replication disrupts chromatin structure (Figures 7-15 and 7-16). Nucleosomes not only compact and protect DNA, they also regulate gene expression through their ability to restrict access of replication and transcription proteins to DNA. Thus, disruption of chromatin structure can affect subsequent patterns of DNA

Table 14-7. The 'Seven Deadly Sins' Of DNA Replication

'Sin'	Potential consequence	Replication fork
Mismatch base pairing	Generate genetic mutations	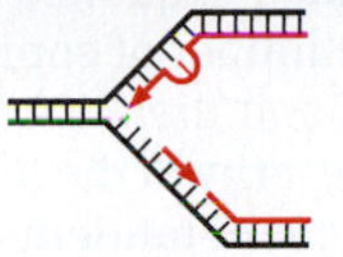
Produce ssDNA template	Generate secondary structures	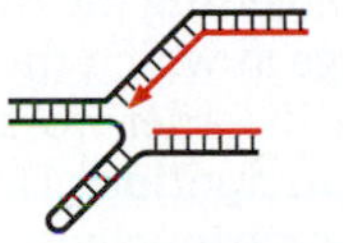
	Promote dsDNA breaks Promote DNA recombination	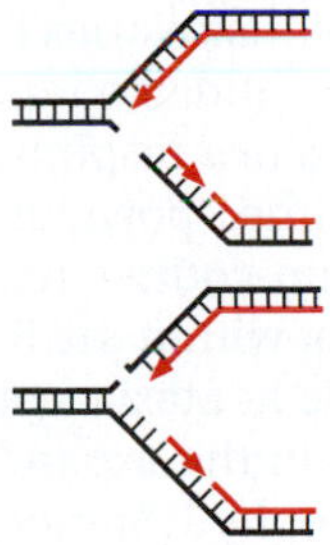
Produce ssDNA interruptions	Activate DNA damage response	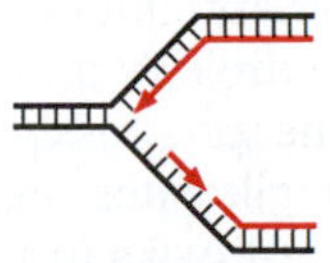
Disrupt DNA methylation pattern	Alter gene expression	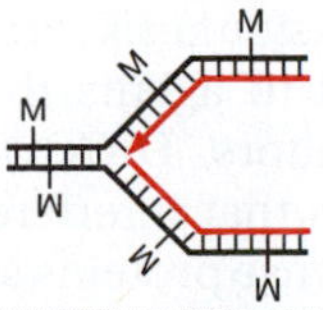 M M M M M
Disrupt • chromatin structure • nucleosome phasing • histone-modification		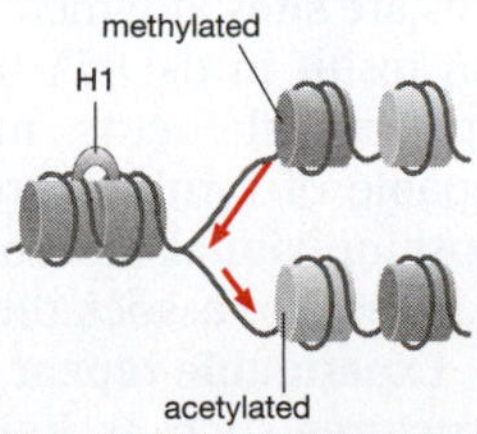 methylated H1 acetylated

replication and gene expression. Finally, newly assembled nucleosomes are hyperacetylated, and newly synthesized histones lack post-translational modifications to their N-terminal and C-terminal tails (Figures 7-21 and 7-22). Post-translational modifications of chromosomal proteins can strongly affect gene expression. All of these disruptions to the cell's gene expression pattern have the potential to alter the gene expression pattern. They may trigger developmentally regulated cell differentiation, or they may open the door to deregulation of cell proliferation.

The Road To Nucleotide-Repeat Disorders

Huntington's disease is a neurodegenerative genetic disorder that affects muscle coordination and some cognitive functions, typically becoming noticeable during middle age. Spinocerebellar ataxia is a group of 10 distinct genetic disorders characterized by slowly progressive lack of coordination in motor skills often associated with gait, hands, speech, and eye movements. Fragile X syndrome, the most common form of inherited mental retardation in males, is a genetic disorder that results in characteristic physical, intellectual, emotional, and behavioral features whose manifestations range from severe to mild. These are just 12 of the 37 known nucleotide-repeat disorders that result either in a heritable disease or an increased risk of disease (Table 14-8). These repeats occur either within a particular gene itself, or within the noncoding sequences of a particular gene. Each repeated sequence usually, but not always, consists of three nucleotides. The number of copies of the repeat (**'repeat length'**) is frequently polymorphic, and there is often a direct correlation between the repeat length and the severity of the disease. Since the instability of the repeat increases as the number of tandem copies of the repeat increases (**'repeat expansion'**), the risk of getting the disorder is greater in successive generations. In addition, the age at which the symptoms appear is likely to be lower and the severity of the disorder worse with increasing repeat length, a phenomenon known as '**anticipation**'. Healthy individuals harbor about 5 to 50 tandem repeats of these trinucleotides, whereas the number of tandem repeats in diseased individuals varies from 2 to 1000-times greater.

Expandable repeats are found in various regions of their resident gene. The genes responsible for numerous hereditary diseases contain codons for either polyglutamine, polyalanine, or polyaspartate runs within the protein-coding region. However, some diseases contain the expanded repeat within the 5′-untranslated region of the gene (those associated with fragile X, ataxia syndrome, and spinocerebellar ataxia-12), whereas others contain the expanded repeat within the 3′-untranslated region (myotonic dystrophy 1, spinocerebellar ataxia-8, and Huntington's disease-like 2), and still others contain the expanded repeat within introns (myotonic dystrophy 2, Friedreich's ataxia, and spinocerebellar ataxia-10) or within promoter regions (progressive myoclonus epilepsy 1).

Fragile sites are specific chromosomal loci that preferentially exhibit gaps or breaks in metaphase chromosomes following partial inhibition of DNA synthesis. Fragile sites can be induced in cultured cells by exposing them to agents that perturb DNA replication such as DNA polymerase inhibitors, DNA chain elongation terminators, nucleotide analogs, and drugs that interfere with either nucleotide biosynthesis or DNA damage-response proteins such as Atm and Atr (see Figure 14-7, below). Thus, fragile sites are sites at which replication forks are prone to stalling, an event that can result in dsDNA breaks. Fragile sites are enriched with interspersed repetitive elements, nuclear matrix attachment regions, and sequences capable of forming unusual DNA structures. Fragile sites are present in most individuals, but of more than 100 reported, only a handful have been mapped and associated with disease.

Expandable repeat sequences all have the ability to form unusual DNA structures such as imperfect **hairpins** (cruciforms) that can arise when

Table 14-8. Human Diseases Associated With Expanded Repeats

Repeat	Affected Gene	Disease
Single-amino acid polypeptide tracts		
$(CAG)_n$/polyGln	*Htt*	Huntington's disease
$(CAG)_n$/polyGln	*Atn1*	Dentatorubral-pallidoluysian atrophy
$(CAG)_n$/polyGln	*Atxn1, 2, 3, 7 Cacna1a, Tbp*	Spinocerebellar ataxias (six variants)
$(CAG)_n$/polyGln	*Ar*	Spinal and bulbar muscular atrophy
$(GCG)_n$/polyAla	*Pabpn1*	Oculopharyngeal muscular dystrophy
$(GCX)_n$/polyAla	*Hoxd13*	Synpolydactyly
$(GCX)_n$/polyAla	*Runx2*	Cleidocranial dysplasia
$(GCX)_n$/polyAla	*Zic2*	Holoprosencephaly
$(GCX)_n$/polyAla	*Arx*	Infantile spasm syndrome
$(GCX)_n$/polyAla	*Hoxa13*	Hand-foot-genital syndrome
$(GCX)_n$/polyAla	*Foxl2*	Blepharophimosis syndrome
$(GAC)_n$/polyAsp	*Comp*	Pseudoachondroplasia (PSACH) and multiple epiphyseal dysplasia (MED)
$(CAX)_n$/polyGln	*Ncoa3*	Increased risk of prostate cancer
Polypeptide repeat		
24 bp repeat octapeptide	*Prnp*	Creutzfeldt–Jakob disease, prion protein
3.3 kb D4Z4 repeat	*Fshmd1a*	Facioscapulohumeral muscular dystrophy 1A
Noncoding repeats		
$(CAG)_n$	*Jph3*	Huntington's disease-like 2
$(CTG)_n$	*Atxn8os*	Spinocerebellar ataxia-8
$(ATTCT)_n$	*Atxn10*	Spinocerebellar ataxia-10
$(CAG)_n$	*Ppp2r2b*	Spinocerebellar ataxia-12
$(TGGAA)_n$	*Tk2/Bean*	Spinocerebellar ataxia-31
$(CCGCC)_n$	*Smyd3*	Increased risk of cancer
14 bp-VNTR	*Ins*	Increased risk of diabetes
28 bp-VNTR	*Hras*	Increased risk of ovarian cancer
$(CGG)_n$	*Reln*	Increased risk of autism
$(C_4GC_4GCG)_n$	*Cstb*	Progressive myoclonus epilepsy (PME)
$(CTG)_n$	*Dmpk*	Myotonic dystrophy type 1
$(CCTG)_n$	*Znf9*	Myotonic dystrophy type 2
$(GAA)_n$	*Fxn*	Friedreich's ataxia
Rare fragile site-associated repeats		
$(CGG)_n$, ^{me}CpG	*Fraxa/Fmr1*	Fragile X mental retardation syndrome (FRAXA)
$(CGG)_n$	*Fraxa/Fmr1*	Fragile X-associated tremor/ataxia syndrome (FXTAS)
$(CGG)_n$, ^{me}CpG	*Jbs*	Jacobsen syndrome
$(GGC)_n$	*Fmr2*	Fragile X-associated mental retardation (FRAXE)

Adapted from Castel, L. et al (2010) *Nat. Rev. Mol. Cell Biol.*. 11: 165–170. VNTR is variable number of tandem repeats.

one strand of an inverted repeat folds back upon itself (Figure 1-10). Under physiological conditions, the stability of such imperfect hairpins decreases according to the sequence of the triplet, in the order CGG>CTG>CAG = CCG. In addition to hairpins, single-stranded $(CGG)_n$, $(CCG)_n$, and $(CGCG_4CG_4)_n$ repeats can fold into G-quadruplexes (Figure 1-13). Denaturation and renaturation of dsDNA fragments containing expandable repeats also can promote formation of **slipped-strand DNA** in which an out-of-register realignment of the complementary repetitive strands gives rise to 'slipouts' that are folded into hairpin-like structures (Figure 1-15). $(GAA)_n \cdot (TTC)_n$ (homopurine–homopyrimidine) repeats can convert into intramolecular **triplex DNA** structures (Figures 1-11 and 1-12). Intriguingly, an (A+T)-rich repeat that is responsible for spinocerebellar ataxia-10 [$(ATTCT)_n \cdot (AGAAT)_n$] has the thermodynamic properties of an easily unwound DNA sequence of the type associated with DNA replication origins (DNA-unwinding element; Figure 9-1).

Expansion and contraction of nucleotide repeats

Triplet repeats are normal components of the human genome. The problem arises when they expand in such a way as to alter either the expression or the function of a particular gene. Invariably, these expansions are limited to just one repeat at a unique locus for each disease. Therefore, they are not likely the result of mutations in proteins involved in DNA replication, repair, or recombination, because such mutations would affect numerous repeats at multiple genetic loci. Instead, they appear to be caused by aberrant events at replication forks that are enhanced by specific sequence contexts.

The stability of expandable repeats is strongly affected by their orientation relative to replication origins. Furthermore, contraction of these repeats (i.e. their instability) is most striking when the same strand of a repeated sequence that most readily folds into secondary structures also is the lagging-strand template of the replication fork passing through this region. For example, when repeats of 5′-CGG-3′ base-pair with 3′-GCC-5′, one strand will be G-rich and one strand C-rich. Since CGG repeats form stable DNA secondary structures, such as hairpins, more readily than CCG repeats, G-rich secondary structures are more likely to form in the lagging-strand template than C-rich structures. Since the location of the nearest replication origin determines which DNA strand will be the lagging-strand template, the location of a replication origin relative to the repeated DNA sequences will influence the stability of those repeats and therefore the probability that changes in those repeats will affect expression of a gene that may reside within the same locus. One example is fragile X syndrome which arises when the normally stable 5 to 50 CGG repeats in the 5′-untranslated region of the fragile X mental retardation protein 1 (Fmr1) gene expand to over 200 copies. This triplet repeat expansion leads to DNA methylation and silencing of the *Fmr1* promoter. Analysis of a 35 kb region centered at the *Fmr1* promoter revealed a single replication origin located adjacent to the *Fmr1* promoter and CGG repeats. The position of the *Fmr1* origin relative to the CGG repeats is consistent with a role in repeat maintenance; the CGG repeats reside in the lagging-strand template. The *Fmr1* origin is active in fibroblasts both from healthy individuals and from patients with fragile X syndrome, as well as in fetal cells as early as 8 weeks old.

The hypothesis that aberrant DNA replication-fork structures lead to either expansion or contraction of triplet repeats is supported by mutations in yeast replication proteins. Inactivation of Srs2, a helicase involved in restarting stalled replication forks, results in expansion of CTG, CAG, and CGG repeats by a mechanism that does not require DNA-repair proteins. Srs2 functions to block expansions by working in concert with DNA Pol-δ, the enzyme that carries out the bulk of lagging-strand DNA synthesis. Mutations in DNA ligase I, the enzyme that joins Okazaki fragments to the 5′-ends of downstream nascent DNA strands, elevate the frequency of expansions, particularly when either CAG or CTG serves as the lagging-strand template. In wild-type cells, repeats tend to contract in this orientation. Similarly, inactivation of FEN1, the nuclease that excises RNA primers from Okazaki fragments, also causes expansion of DNA sequence repeats. Thus, actions that retard the joining of Okazaki fragments promote triplet repeat expansion. The actions of Pol-δ, lig I, and FEN1 are coordinated by PCNA (Figure 5-12), the sliding clamp that endows DNA Pol-δ and DNA Pol-ε with processivity. Orthologous mutations in either lig I, FEN1, or PCNA that interrupt the association between the two enzymes and PCNA reveal that the interactions among PCNA, lig I, and FEN1 are critical in maintaining triplet repeat stability and in preventing a series of triplet repeats from expanding.

Mechanism for expanding and contracting nucleotide repeats

Expansion or contraction of most triplet repeats appears to result primarily, if not exclusively, from aberrant events at DNA replication forks. Based on the ability of ssDNA to form hairpins, a simple model can account for either the contraction or expansion of repeated DNA sequences at replication forks during genome duplication. Contractions occur when a replication fork unwinds through an inverted repeat and the ssDNA lagging-strand template folds into a cruciform (hairpin) structure (Figure 1-10) before the ssDNA-binding protein RPA (Figure 4-11) can prevent it (Figure 14-6A). As the fork advances, Okazaki-fragment synthesis on the lagging-strand template may simply bypass the hairpin, thereby creating a nascent DNA strand in which the sequence of the bypassed repeat is absent. When the two sister chromatids are segregated into daughter cells, and the daughter cells undergo DNA replication, one of them will produce a normal genome and one of them will produce a genome in which one or more copies of the repeated sequence are missing. This process is referred to as a contraction or instability of the repeated sequence. The threshold length for repeat instability in either yeast or humans is approximately 150 bp for repeats of 3 to 12 nucleotides. Since this is the median length of an Okazaki fragment initiation zone (Figure 7-13), it suggests that formation of secondary structures in the lagging-strand template prior to Okazaki-fragment synthesis gives rise to repeat instability.

Expansions occur if a replication fork stalls when it encounters a repeated DNA sequence, and then reverses itself by collapsing into a Holliday structure (Figure 14-6B). The structure associated with fork reversal has been observed by electron microscopy when yeast chromosomes replicate in the absence of Rad53/Chk2 (Figure 13-17D). Under these conditions, the nascent DNA strand on the leading-strand template can form a hairpin from an inverted repeat. When the fork is restarted, then the nascent DNA strand on the leading-strand template will contain one or more additional copies of the repeated sequence. Thus, when the sister chromatids are replicated again during the next cell cycle, one of them will produce a fully normal genome, and one will produce a genome with additional copies of the repeat.

In this model, the question of contraction or expansion of DNA sequences is determined by the response of the replication machine to the secondary structure possibilities inherent in the DNA templates. *In vitro*, purified DNA polymerases pause when they encounter a hairpin structure in the ssDNA template, a behavior that appears to be recapitulated *in vivo* when replication forks encounter unusual DNA sequences or replication-fork barriers (Figure 6-8). In addition, the propensity to pause increases markedly as two forks approach one another to terminate replication (Figures 6-1 and 6-2). Therefore, whether or not DNA Pol-δ successfully bypasses a hairpin in the lagging-strand template depends on its association with PCNA and the ability of FEN1 and lig I to quickly link the new Okazaki fragment to the longer nascent DNA strand, thereby stabilizing the bypassed conformation. In addition, hairpins can be eliminated by the action of a maintenance helicase such as Srs2 in budding yeast or one of the RecQ-like helicases in

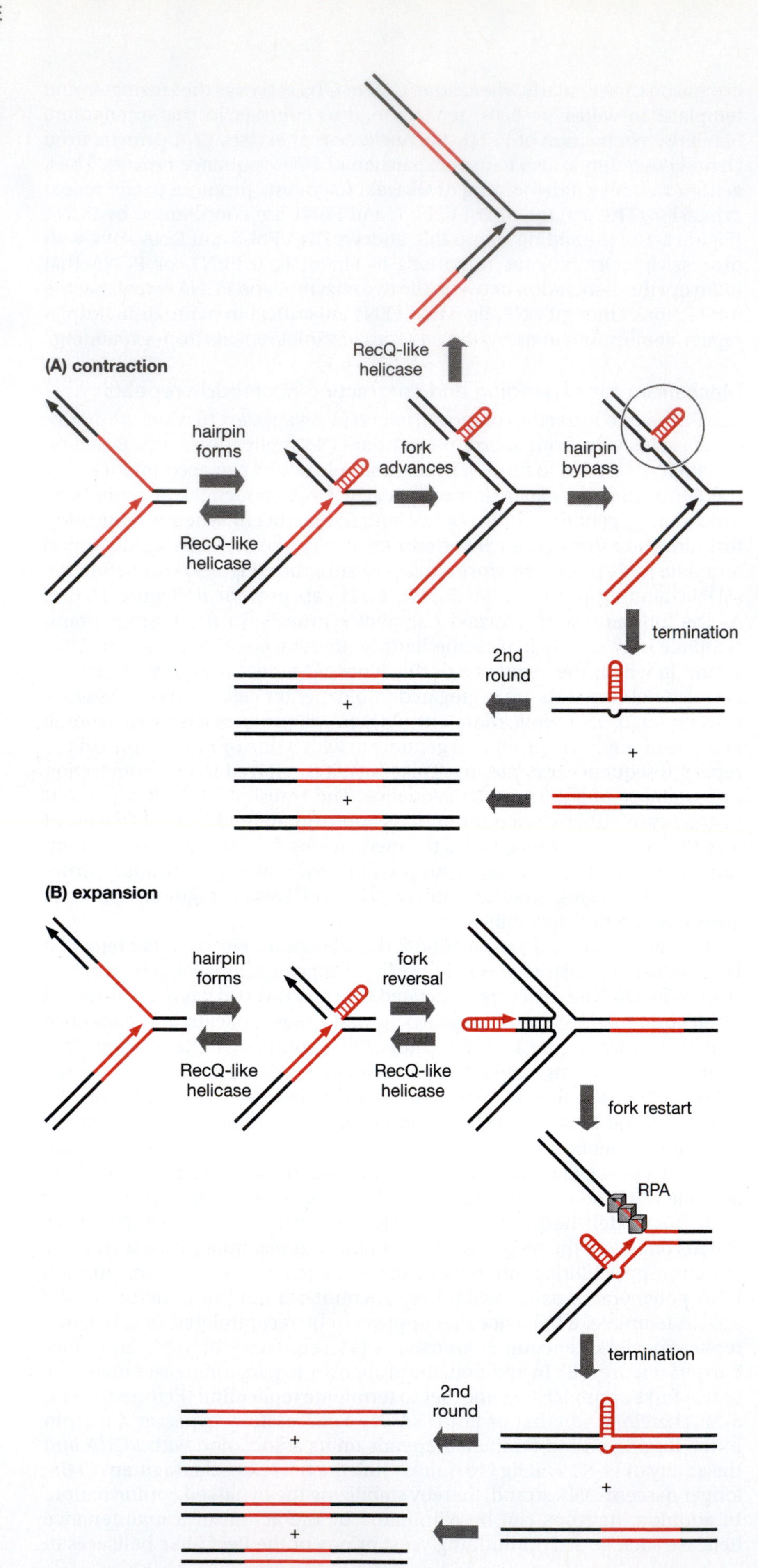

Figure 14-6. Model for contracting (A) and expanding (B) inverted repeats (red lines). (A) Contraction (instability) of an inverted repeat results from formation of hairpins in the ssDNA lagging-strand template, followed by a second round of DNA replication in which one of the template strands is now missing one or more copies of the repeated sequence. During normal replication, either hairpin structures are prevented from forming by association of ssDNA with the ssDNA-binding protein RPA, or they are removed by the action of RecQ-like maintenance DNA helicases. (B) Expansion of an inverted repeat results when the fork does not continue to unwind through the repeat, but reverses itself by collapsing into a Holliday structure. Under these conditions, the nascent strand on the leading-strand template can form a hairpin. When the fork is restarted, then the nascent DNA strand will contain one or more additional copies of the repeat. When the sister chromatids are replicated again during the next cell cycle, one of them will produce a fully normal genome and the other a genome with one additional copy of the repeat. (Data from Cleary, J.D., Pearson, C.E. and La Spada, A.R. (2006) DNA replication, repeat instability, and human disease. In *DNA Replication and Human Disease* (DePamphilis, M.L., ed.), pp. 461–480. Cold Spring Harbor Laboratory Press, and Mirkin, S.M. (2007) *Nature* 447: 932–940.)

human cells (Table 13-5), thereby allowing DNA Pol-δ to continue copying the template without interruption. Similarly, whether or not a restarted replication fork contains a hairpin within the nascent DNA strand on the leading-strand template depends on whether or not a maintenance helicase eliminates the hairpin in the 'reversed-fork' conformation.

Since the response may be different each time the genome is duplicated, sequences are likely to experience many expansions and contractions during the trillions of cell divisions that occur as a fertilized egg develops into a viable adult. Thus, what amounts to a low probability event in any single round of genome duplication can be amplified greatly during human development. Errors introduced into the germ line of a species are then carried over to the next generation where the problem can be amplified further. 'Anticipation' likely results from progressively increased instability due to consecutive stalling and restarting of replication forks within longer and longer tracts of DNA repeat.

Mutations In DNA Replication Genes

At least half of the human diseases known to result from genetic mutations contain mutations in genes encoding proteins involved either in regulation of genome duplication or in DNA repair. Some of them have been described above (Tables 14-4 to 14-6). Others include genes involved directly in DNA replication (Table 14-9), DNA repair (Table 14-10), chromatin modification

Table 14-9. Human Diseases Associated With Mutations In DNA Replication Proteins

Gene	Function	Disease
PolG	DNA Pol-γ replicates mtDNA	Progressive external ophthalmoplegia, Alpers syndrome, spinocerebellar ataxia with epilepsy, sensory ataxic neuropathy, dysarthria, and ophthalmoparesis, MNGIE without leukoencephalopathy
PolG2	DNA Pol-γ accessory subunit	Progressive external ophthalmoplegia
RNaseH2A, B, or C	Ribonuclease H2	Aicardi–Goutieres syndrome
Lig1	DNA ligase, joins Okazaki fragments to long nascent strand; DNA repair	DNA ligase I deficiency
Contiguous haploinsufficiency (~20 genes) on 17p13.3, including *Rpa1*	Required for both DNA replication and repair	Miller–Dieker lissencephaly syndrome[a]
Contiguous haploinsufficiency (~30 genes) on 22q11, including *Cdc45*	Required for loading DNA Pol-α at replication origins and for MCM helicase activity at replication forks	DiGeorge syndrome[a]
Contiguous haploinsufficiency (~24 genes) on 7q11.23, including *Rfc2*	Required for loading PCNA	Williams–Beuren syndrome[a]
PPP2R2B	Brain-specific regulatory subunit of protein phosphatase 2A (PP2A). PP2A regulates binding of Cdc45 to the pre-replication complex	Spinocerebellar ataxia type 12[a]
Tinf2	A component of the shelterin telomere-protection complex	Dyskeratosis congenita
Tert	Telomerase maintains telomere length	Decreased telomerase activity correlates with increased cell senescence

[a]Syndromes that may be related to DNA replication or repair proteins. MNGIE, mitochondrial neurogastrointestinal encephalopathy syndrome.

(Table 14-11), or nucleotide metabolism (Table 14-12). Diseases resulting from defective DNA replication genes are rare, because failure to complete nuclear DNA replication leads to apoptosis and embryonic lethality. In contrast, proteins that repair damaged DNA and proteins that modify chromatin structure often only affect expression of specific genes, and therefore are more likely to result in a disease state rather than cell death.

Table 14-10. Human Diseases Associated With Mutations In DNA-Repair Genes

Gene	Function	Disease
Msh2, Msh3, Msh6, Mlh1	'Mismatch repair genes' repair base-pairing mistakes and simple repeat errors made during DNA replication	Hereditary non-polyposis colorectal cancer (Lynch syndrome)
Brca1, Brca2	Repair radiation-induced breaks in dsDNA	Familial breast cancer, ovarian cancer, prostate cancer, pancreatic cancer
Eleven *FA* genes, *D1* is *Brca2*		Fanconi anemia
Xpa, Xpc, Xpf, Xpg	Repair nucleotide excisions	Xeroderma pigmentosum
PolH	DNA Pol-η, carries out trans-lesion DNA synthesis	Xeroderma pigmentosum variant
Csa, Csb	Repair DNA damage	Cockayne syndrome
Lig4	DNA ligase IV is required for non-homologous DNA end-joining	LIG4 syndrome
Tdp1	Tyrosyl-DNA phosphodiesterase, repairs Topo I isomerase DNA adducts and 3′-phosphoglycolate ends	Spinocerebellar ataxia with axonal neuropathy

Table 14-11. Human Diseases Associated With Defects In Chromatin Modification

Gene	Function	Disease
Dnmt3b	DNA (cytosine-5-)-methyltransferase 3β is a *de novo* CpG DNA methyltransferase that regulates epigenetic phenomena at specific sites	ICF syndrome (immunodeficiency, centromeric instability, and facial anomalies)
Mecp2	^{me}CpG-binding protein that is involved in assembly of specialized chromatin structure	Rett syndrome; mental retardation, X-linked; encephalopathy, neonatal severe; increased risk of autism; Angelman syndrome
Ehmt1	Euchromatic histone-lysine N-methyltransferase 1	Chromosome 9q subtelomeric deletion syndrome

Table 14-12. Human Diseases Associated With Reduction In Nucleotide Pools

Gene	Function	Disease
Tk2	Mitochondrial thymidine kinase	mtDNA depletion syndromes
Dguok	Deoxyguanosine kinase	
Slc25a4	Mitochondrial adenine nucleotide translocator	Autosomal dominant progressive external ophthalmoplegia
Tymp	Thymidine phosphorylase	Mitochondrial neurogastrointestinal encephalopathy
Slc25a19	Mitochondrial thiamine pyrophosphate carrier	Amish microcephaly

The frequency of hereditary diseases in response to genetic mutations is limited by two related facts: many genes are functionally redundant, and some genes are required but not essential. Mutations in Cdk1, for example, are lethal in mice, because none of the other CDKs in mammals can substitute for Cdk1 in the $G_2 \rightarrow M$ transition, whereas loss of Cdk2 activity is not lethal, because Cdk1 can replace Cdk2 in the $G_1 \rightarrow S$ transition. Similarly, Chk1 is essential to cell proliferation in mammals, whereas Chk2 is not. Therefore, mutations in Chk2 can result in a hereditary disease, whereas mutations in Chk1 are likely to be lethal.

Mitochondria are a story in themselves. At least 25 human diseases result from defects in mitochondrial proteins encoded either by nuclear DNA or by mtDNA. Mutations in mtDNA accumulate more easily than those in nuclear DNA, because mitochondria lack an efficient DNA damage-repair system. In addition, mutations that affect mtDNA replication are more likely to be pathological rather than lethal, because the number of mitochondria per cell can be reduced without causing the death of the organism. A similar argument applies to genes that affect nucleotide pool levels. Changes in the concentrations of NTPs and dNTPs can adversely affect the ability of cells to replicate and repair DNA efficiently without killing the cell.

PHARMACOLOGICAL AGENTS THAT TARGET DNA REPLICATION

Given the fact that most infectious diseases are caused by microorganisms and viruses with DNA genomes, considerable effort has gone into identifying pharmacological agents that target proteins essential to DNA replication and even agents that induce DNA damage. To be effective, such agents must be administered either orally or by injection into blood or muscle, and they must inactivate their target with little or no harm to the patient. Many are prodrugs, small molecules that are administered in an inactive form that is subsequently metabolized into an active form. The metabolic transition occurs in specific tissues such as the liver or in the target cell itself. With infectious diseases, the most successful drugs take advantage of viral or bacterial proteins with features that distinguish them from their human counterpart. In the case of noninfectious diseases such as cancer, success depends largely on the simple fact that cancer cells are proliferating whereas most of the cells in our bodies are not.

Drugs that target DNA replication fall into eight groups (Figure 14-7). One group inhibits DNA synthesis by reducing nucleotide pools; another terminates DNA chain elongation. A third group blocks DNA replication forks either by damaging the DNA template itself or by inhibiting DNA damage-repair systems. Still other drugs inhibit the activities of DNA polymerases, DNA topoisomerases, or cyclin-dependent protein kinases (CDKs). More recent targets include DNA damage-response proteins such as Atr, and suppressors of DNA re-replication such as geminin. Inhibitors of DNA replication are among the most effective antiviral agents, particularly against herpesviruses, HIV, and HBV. Some antibiotics also are targeted against DNA replication, but the greatest impact of anti-DNA replication drugs has been in the field of cancer chemotherapy.

Antiviral Pharmaceuticals

Drugs that target herpesviruses

Most of the antiviral agents developed to combat herpesvirus infections target the viral DNA polymerase. The most successful of these, acycloguanosine, is one of several drugs designed by Gertrude Elion and George Hitchings, for which they received the 1988 Nobel Prize in Physiology or Medicine. Commonly termed aciclovir or acyclovir, this nucleoside analog is rapidly converted into a DNA replication inhibitor only in cells infected with HSV

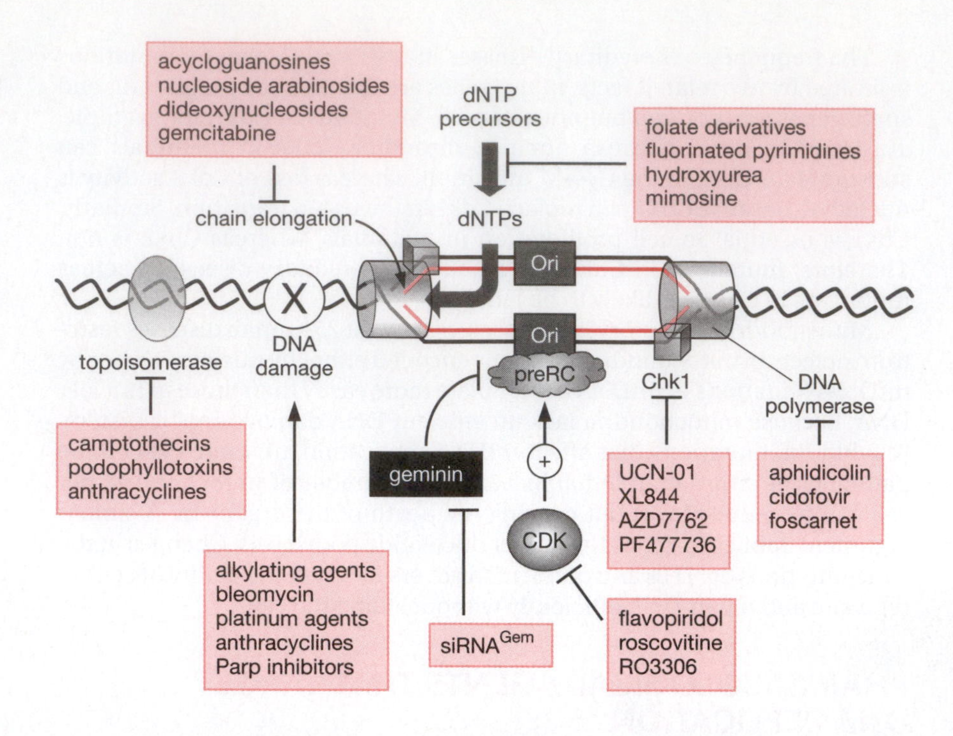

Figure 14-7. Pharmacological agents that target DNA replication. Most agents inhibit (⊥) their target. Other agents either cause DNA damage or prevent DNA damage from being repaired (Parp inhibitors). (Data from Pommier, Y. and Diasio, R.B. (2006) Pharmacological agents that target DNA replication. In *DNA Replication and Human Disease* (DePamphilis, M.L. ed.), pp. 519–546. Cold Spring Harbor Laboratory Press.)

or herpes varicella-zoster virus. Aciclovir is a derivative of guanosine and deoxyguanosine in which the sugar ring is replaced by an open-chain structure (Figure 14-8). It is selectively converted into acycloguanosine monophosphate (acyclo-GMP) by the HSV thymidine kinase, which is about 3000 times more effective in phosphorylating thymidine than the cellular thymidine kinase (Figure 14-9). Thymidine kinase catalyzes the reaction: Thd+ATP→TMP+ADP. Acyclo-GMP is further phosphorylated by cellular kinases into its active triphosphate form, acyclo-GTP, which

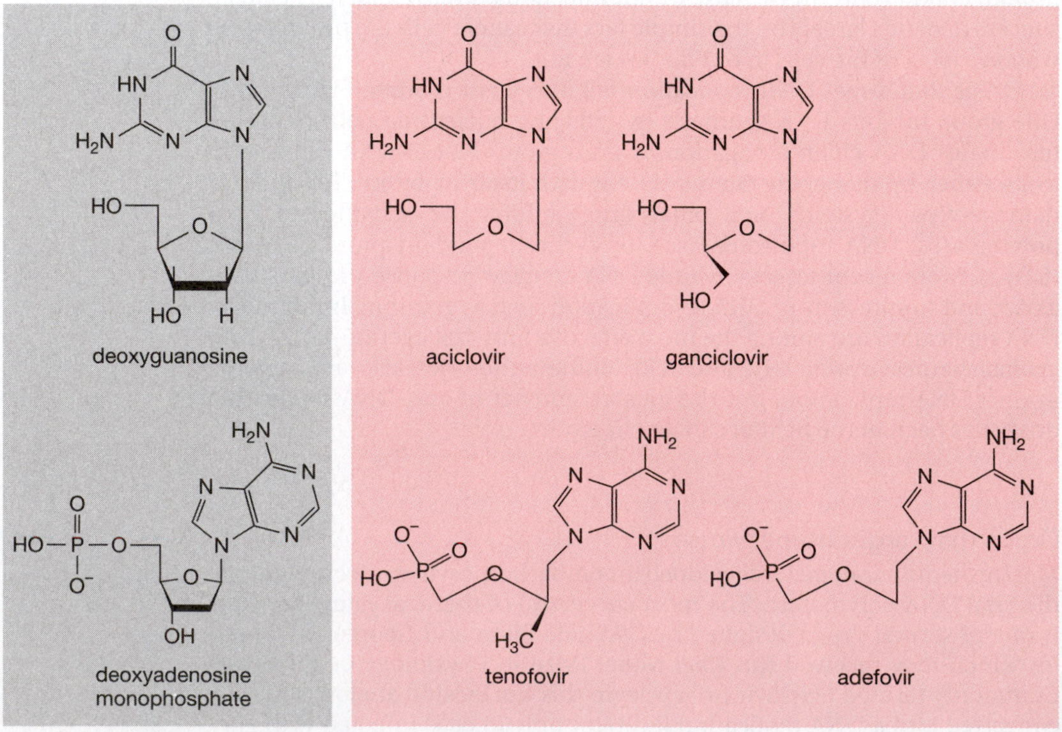

deoxyguanosine

aciclovir

ganciclovir

deoxyadenosine monophosphate

tenofovir

adefovir

Figure 14-8. Drugs that inhibit DNA polymerases by competing with dGTP or dATP (red box). Normal substrates and products are shown in the gray box.

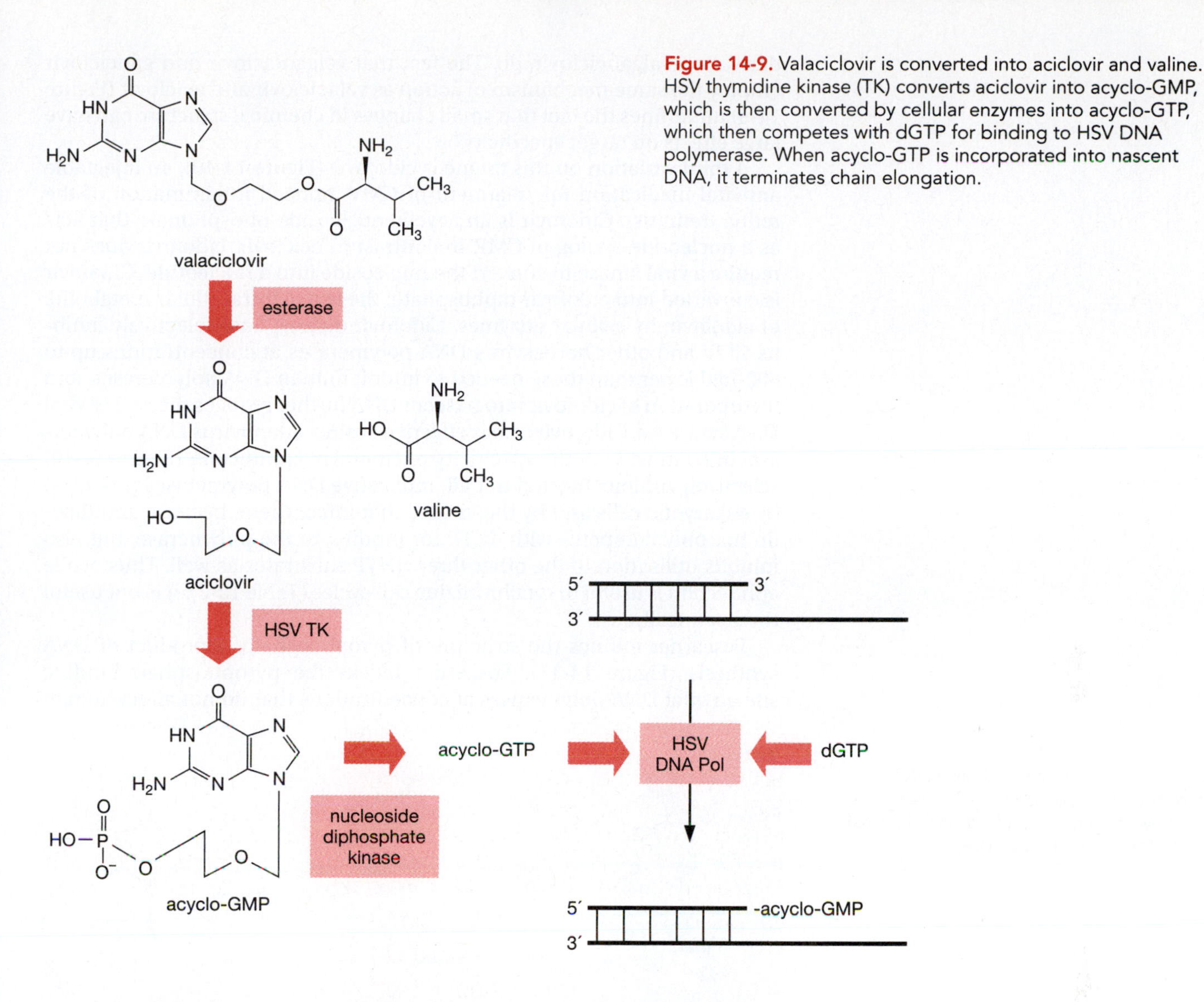

Figure 14-9. Valaciclovir is converted into aciclovir and valine. HSV thymidine kinase (TK) converts aciclovir into acyclo-GMP, which is then converted by cellular enzymes into acyclo-GTP, which then competes with dGTP for binding to HSV DNA polymerase. When acyclo-GTP is incorporated into nascent DNA, it terminates chain elongation.

then competes with dGTP for binding to DNA polymerases. When acyclo-GTP is incorporated into DNA, it terminates elongation of nascent DNA chains, because it lacks a ribose 3′-OH group. Since viral enzymes cannot remove acyclo-GTP from nascent DNA, HSV DNA replication is effectively terminated. Acyclo-GTP effectively inhibits HSV DNA replication under conditions where it has no significant effect on cellular DNA replication. Not only is unincorporated acyclo-GTP rapidly metabolized by cellular phosphatases, the affinity of acyclo-GTP for HSV DNA polymerase is about 100 times greater than for human replicative DNA polymerases.

Valaciclovir, an esterified version of aciclovir (Figure 14-9), is more potent than aciclovir. Esterases in the intestine and liver convert valaciclovir into aciclovir plus the innocuous amino acid valine. The advantage of valaciclovir is that approximately 55% of an orally administered dose of this prodrug enters the bloodstream compared to approximately 10% of aciclovir. The problem with valaciclovir and aciclovir is that they are not very effective against other types of virus infections. Aciclovir inhibits EBV DNA synthesis during EBV lytic infections, but not in latent infections. Aciclovir given to patients with infectious mononucleosis causes a reduction in the level of oropharyngeal viral replication, but after the treatment is stopped, EBV DNA replication returns to the pre-aciclovir treatment levels. Aciclovir does not inhibit human CMV replication, but other acycloguanosines, such as ganciclovir (Figure 14-8) and its esterified derivative, valganciclovir, do. The fact that valganciclovir and ganciclovir employ the same mechanism of action as valaciclovir and aciclovir (Figure 14-9) underlines the fact that small changes in chemical structure can have large effects on target specificity.

A third variation on this theme is cidofovir (Figure 14-10), an injectable antiviral medication for treatment of CMV-induced inflammation of the retina (retinitis). Cidofovir is an acyclic nucleoside phosphonate that acts as a nucleotide analog of CMP. In contrast to aciclovir, cidofovir does not require a viral kinase to convert the nucleoside into a nucleotide. Cidofovir is converted into cidofovir diphosphate, the active intracellular metabolite of cidofovir, by cellular enzymes. Cidofovir diphosphate selectively inhibits CMV and other herpesvirus DNA polymerases at concentrations up to 600-fold lower than those needed to inhibit human DNA polymerases, and incorporation of cidofovir into nascent DNA further reduces the rate of viral DNA synthesis. Cidofovir also is effective against adenovirus DNA polymerase. In contrast with the specificity of cidofovir, aphidicolin (Figure 14-10) selectively inhibits most, if not all, replicative DNA polymerases produced by eukaryotic cells and by the viruses that infect them, because aphidicolin not only competes with dCTP for binding to the polymerase, but also inhibits utilization of the other three dNTP substrates as well. Thus, while aphidicolin is useful in synchronizing cell cycles (Table 12-2), it is not useful therapeutically.

Foscarnet mimics the structure of pyrophosphate, a product of DNA synthesis (Figure 14-11). Foscarnet blocks the pyrophosphate-binding site on viral DNA polymerases at concentrations that do not affect human

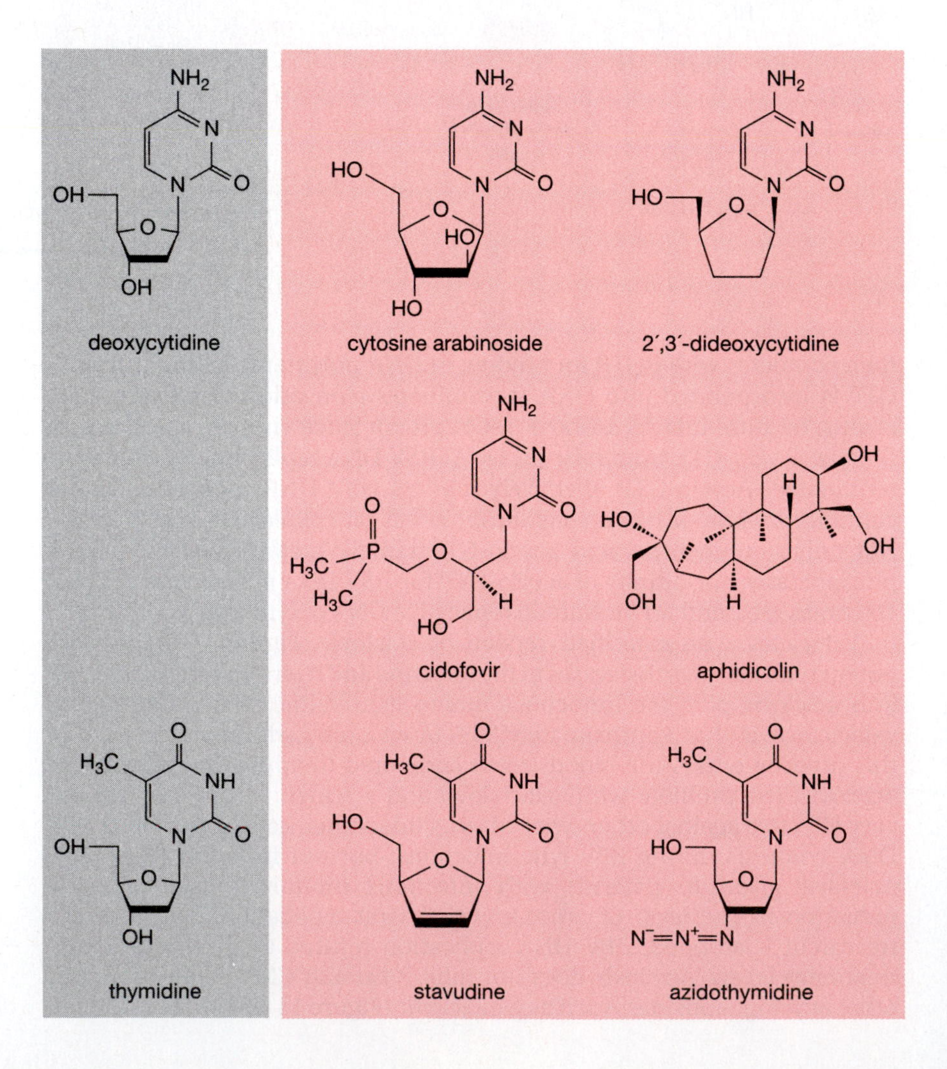

Figure 14-10. Drugs that inhibit DNA polymerases by competing with their deoxypyrimidine analog (red box). Normal substrates are in the gray box.

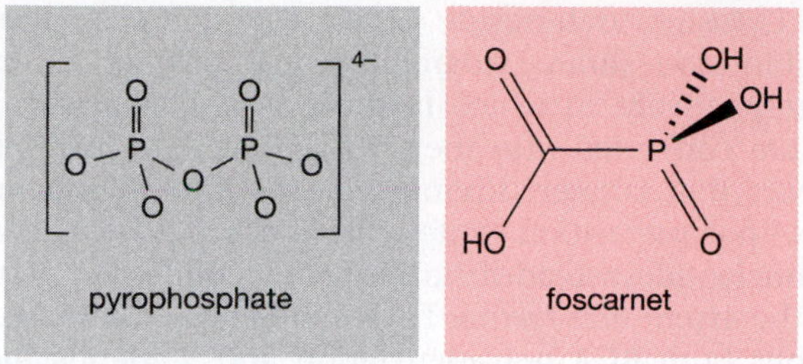

pyrophosphate

foscarnet

Figure 14-11. Foscarnet is a mimetic of pyrophosphate.

DNA polymerases, thereby preventing cleavage of pyrophosphate from nucleoside triphosphates and blocking DNA synthesis. Its antiviral activity is equivalent to that of ganciclovir, making it useful for HSV infections that are resistant to the aciclovir family of drugs. Foscarnet is not metabolized, and therefore is excreted unchanged.

Another class of non-nucleoside compounds that inhibit herpesvirus DNA polymerase is 4-oxo-dihydroquinolines. Studies of resistant mutants taken together with the crystal structure of HSV-1 DNA polymerase suggest that these compounds act by binding near the dNTP-binding site. Other replication proteins have also been targeted. In particular, three related classes of compounds (2-amino-thiazole, thiazolyl phenyl, and thiazolylsulfonamide derivatives) inhibit helicase•primase activities by binding to the interface shared by these two proteins. These agents exhibit potent *in vitro* anti-herpes activity, and show excellent efficacy against HSV in animal models.

Drugs that target HIV and HBV

A number of nucleoside and nucleotide analogs, beginning with azidothymidine (Figure 14-10), are designed to selectively inhibit the reverse transcriptase enzymes encoded by HIV and HBV. In human cells, only telomerase (the enzyme required for telomere maintenance) synthesizes DNA from an RNA template, although DNA Pol-γ (the mitochondrial replicative DNA polymerase) exhibits reverse transcriptase activity *in vitro*. Thus, both telomere maintenance and mitochondrial replication are possible secondary targets of drugs targeting viral reverse transcriptase.

Reverse transcriptase inhibitors are of two types. Some are nucleotides, such as tenofovir and adefovir (Figure 14-8), that compete with the normal substrate (in this example, dATP) for binding to the enzyme's active site. They are quite similar structurally, but they have different side effects and different preferences for HIV and HBV. Both drugs in combination bypass cases where the viral enzyme is resistant to one drug or the other. Non-nucleoside reverse transcriptase inhibitors, such as tetrazole, triazole thioacetanilides, and tetrahydroquinoline derivatives, bind to a different site on the protein that indirectly (allosterically) regulates reverse transcriptase activity. Combinations of these inhibitors are very effective in the treatment of HIV.

Other reverse transcriptase inhibitors are nucleosides that are converted into nucleotide competitive substrate inhibitors as well as DNA chain terminators. Examples include 2′,3′-dideoxycytidine, 2′,3′-dideoxyinosine, and stavudine, an analog of thymidine (Figure 14-10). These nucleosides are phosphorylated by cellular kinases into nucleoside triphosphates that not only compete with their dNTP analog, but that are incorporated onto the 3′-ends of nascent DNA chains where they prevent further nucleotide addition due to the absence of a 3′-OH group. Dideoxynucleotides are so effective at terminating DNA synthesis, that they are widely used to sequence DNA, a method developed by Fred Sanger for which he received his second Nobel Prize in Chemistry in 1980.

DNA chain termination also can be achieved by incorporation of arabinoside analogs of the normal ribonucleoside substrates for RNA polymerase. Cytosine arabinoside (araC, cytarabine; Figure 14-10) and adenine arabinoside (araA, vidarabine) are analogs of cytosine and adenosine, respectively, in which D-ribose is replaced with D-arabinose. Both drugs are converted into their 5′-triphosphate forms by the same cellular kinases that convert ribonucleosides into ribonucleotides. Both drugs are DNA polymerase substrates, but as nascent DNA accumulates the arabinoside nucleotide instead of the ribose nucleotide, its structure becomes distorted to the extent that further DNA synthesis is inhibited. In addition, araATP inhibits both RNA polyadenylation and *S*-adenosylhomocysteine hydrolase, the enzyme that catalyzes hydrolysis of *S*-adenosylhomocysteine to adenosine and homocysteine. The latter effect causes an accumulation of adenosylhomocysteine, which inhibits methylation reactions that use adenosylmethionine as methyl donor. Methylation of viral mRNA (5′-capping reaction) is essential for the maturation and translation of viral mRNA. Both araADP and araCDP inhibit ribonucleotide diphosphate reductase, thereby reducing dNTP pools, and both araATP and araCTP inhibit RNA polymerases. These multiple effects are felt by both viral and host replication machines, but since most cells in an adult are not replicating DNA, these drugs can inhibit virus production dramatically with minimum side effects on the host.

Despite the remarkable similarity of these two drugs, their therapeutic applications differ significantly. AraC is used mainly in the treatment of hematological malignancies such as acute myeloid leukemia and non-Hodgkin lymphoma, whereas araA is used primarily as an antiviral agent. It is generally effective against herpesviruses, poxviruses, rhabdoviruses, and hepadnaviruses, although many of its previous applications have been superseded by aciclovir. Nevertheless, araA still finds application in the treatment of acute keratoconjunctivitis and recurrent superficial keratitis caused by HSV-1 and HSV-2, and in the treatment of varicella-zoster infections (shingles) suffered by AIDS patients.

Antibiotic Pharmaceuticals

Several bacterial DNA replication proteins have proven reliable targets for antibiotics. Trimethoprim (Figure 14-12) is a synthetic analog of methotrexate (an anti-cancer drug discussed below) that inhibits the bacterial dihydrofolate reductase (DHFR) more effectively than the corresponding human enzyme. Trimethoprim is effective against a wide variety of bacteria. It is used mainly in the treatment of urinary tract infections.

Quinolones such as the fluoroquinolone ciprofloxacin (Figure 14-12) are drugs that inhibit bacterial DNA gyrase and topoisomerase IV activities (Table 4-9) and block septation in addition to DNA replication. Since these enzymes are unique to bacteria, ciprofloxacin arrests bacterial cell proliferation without disturbing human cells. Quinolones provide a broad spectrum of antibiotics that are effective against many different types of Gram-negative and Gram-positive bacteria.

6-Anilinouracils (Figure 14-12) appear to inhibit specifically DNA polymerase IIIC (Pol-IIIC), the replicative DNA polymerase in Gram-positive bacteria that are commonly associated with infections acquired during treatment in a hospital. The uracil component is thought to mimic the guanine moiety of dGTP, while the anilino moiety (aryl domain) presumably targets a hydrophobic pocket present only in the Gram-positive DNA Pol-IIIC, resulting in a nonproductive, reversible protein•DNA complex leading to growth inhibition and cell death. Only the aryl ring of the aniline moiety and the 3-position of the uracil can be modified without reducing the potency of these compounds. Some 6-anilinouracils have an IC_{50} of 5 nM against DNA Pol-IIIC. This is the concentration required for 50% inhibition *in vitro*. DNA Pol-IIIC shares little homology either with mammalian DNA Pol-α, or with the Gram-negative DNA Pol-III. Consequently, DNA Pol-IIIC inhibitors should selectively kill Gram-positive bacteria without either

trimethoprim

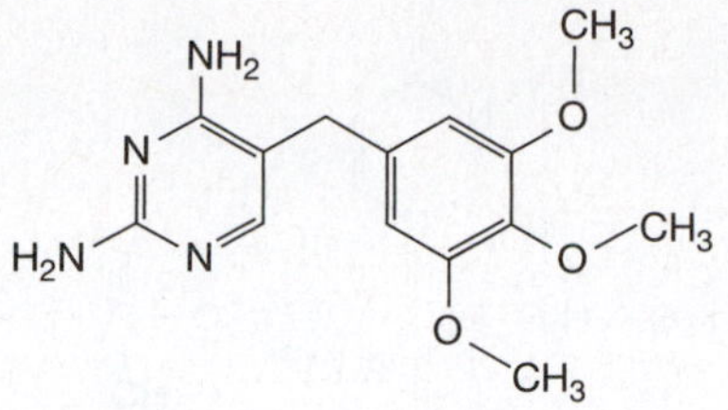

ciprofloxacin

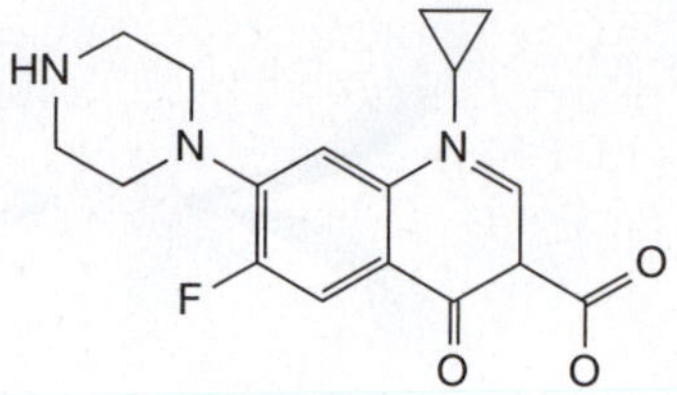

anilinouracils

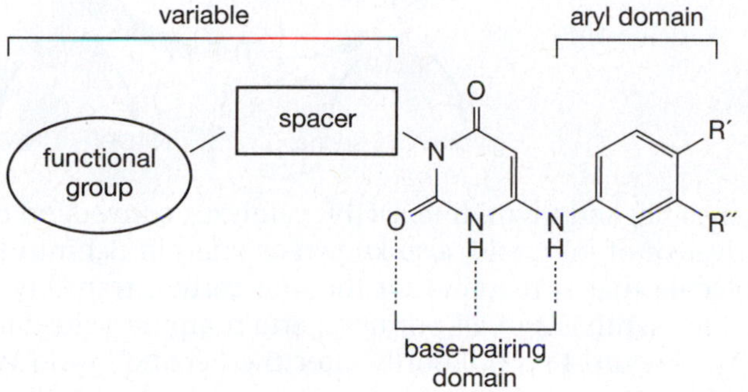

novobiocin

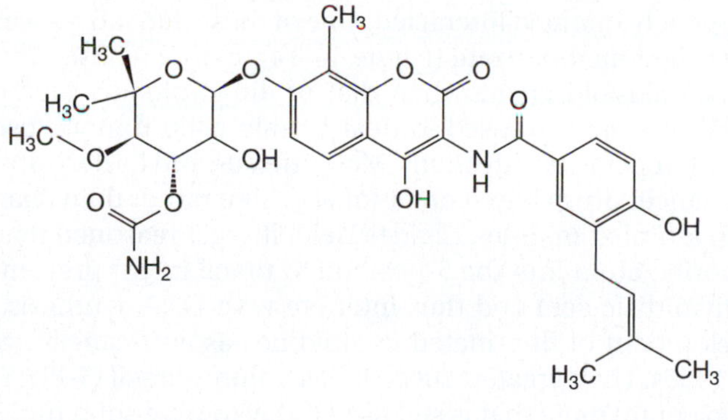

Figure 14-12. Antibiotics targeted against bacterial DNA replication proteins.

arresting mammalian cell proliferation, or harming the intestinal bacterial flora. In an attempt to target simultaneously both Pol-IIIC and DNA gyrase, the 3-position of selected anilinouracils has been substituted with various fluoroquinolones. These anilinouracil-fluoroquinolone hybrids are 15 times more potent at inhibiting bacterial DNA synthesis than anilinouracils alone.

Aminocoumarin antibiotics such as novobiocin (Figure 14-12), clorobiocin, and coumermycin A1 are potent inhibitors of bacterial DNA gyrase. They exert their action by competing with ATP for binding to the B subunit of this enzyme, thereby inhibiting ATP-dependent DNA supercoiling. Their affinity for gyrase is considerably higher than that of fluoroquinolones, which target the A subunit.

Anti-Cancer Pharmaceuticals

Nucleotide synthesis inhibitors

Replication-fork velocity is sensitive to nucleotide pool levels (Table 3-5), marking the enzymes in nucleotide metabolism as primary targets for chemotherapy. For example, DHFR is critical to the synthesis of adenosine,

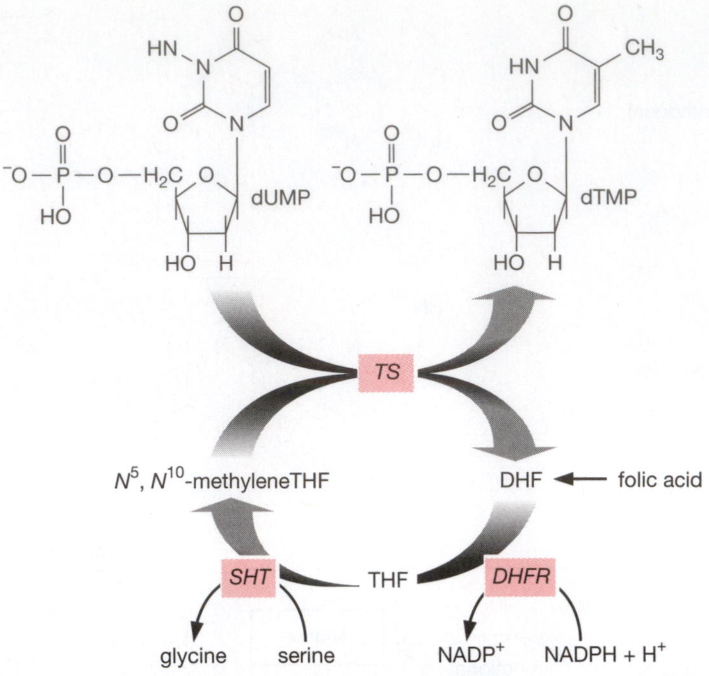

Figure 14-13. Thymidylate synthase (TS), dihydrofolate reductase (DHFR), and serine hydroxymethyltransferase (SHT) are required to convert deoxyuridine monophosphate (dUMP) into deoxythymidine monophosphate (dTMP) (Table 1-1). Dihydrofolate (DHF) and tetrahydrofolate (THF) are derived from folic acid.

guanosine, and thymidine. DHFR catalyzes conversion of dihydrofolate (a derivative of folic acid, also known as vitamin B_9) into tetrahydrofolate, a molecule that is required for the one-carbon transfers that occur during *de novo* synthesis of all purines, certain amino acids, and the pyrimidine dTMP (Figure 14-13). Shortly after the Second World War, Sidney Farber and his colleagues realized that cell proliferation required folic acid, and therefore designed derivatives of folic acid that might inhibit the protein(s) with which folic acid interacted. One of these derivatives was amethopterin, now called methotrexate (Figure 14-14), a drug whose affinity for DHFR is about 1000-fold greater than that of dihydrofolate. Methotrexate and its derivatives are now used to treat certain solid tumors, psoriasis, malaria, and other parasitic infections. Methotrexate and its derivatives were the first anti-cancer drugs based on rational design rather than chance discovery.

In a similar manner, Charles Heidelberger reasoned that substitution of a fluorine atom into the 5-position of uracil might prevent its metabolism to thymidylic acid and thus interfere with DNA synthesis. This led to the development of fluorinated pyrimidines as anti-cancer agents during the late 1950s. Their greatest success was 5-fluorouracil (5-FU; Figure 14-15), an analog of thymine that is still used today to treat solid tumors that occur in the colon, stomach, and breast tissues, as well as to treat fungal infections.

5-FU has multiple effects on cell metabolism, but its primary toxicity results from inhibiting thymidylate synthase, the enzyme that uses the N^5,N^{10}-methylenetetrahydrofolate produced by DHFR and serine hydroxymethyltransferase to convert dUMP into dTMP (Figure 14-13). This reaction is essential to providing the dTTP necessary for DNA synthesis. 5-FU is converted by the cell into the active metabolite fluorodeoxyuridine monophosphate (FdUMP). FdUMP rapidly inactivates thymidylate synthase by binding to its nucleotide-binding site and forming the stable ternary complex FdUMP•TS•N^5,N^{10}-methylenetetrahydrofolate. This complex prevents access of dUMP to the nucleotide-binding site, thereby inhibiting dTMP synthesis.

Inhibition of thymidylate synthase not only reduces the cellular concentration of dTTP, but simultaneously increases the concentration of dUMP which is rapidly converted into dUTP. Since most DNA polymerases do not discriminate well between dTTP and dUTP, this results in significant incorporation of uracil into DNA (Figure 14-15). FdUMP also is converted

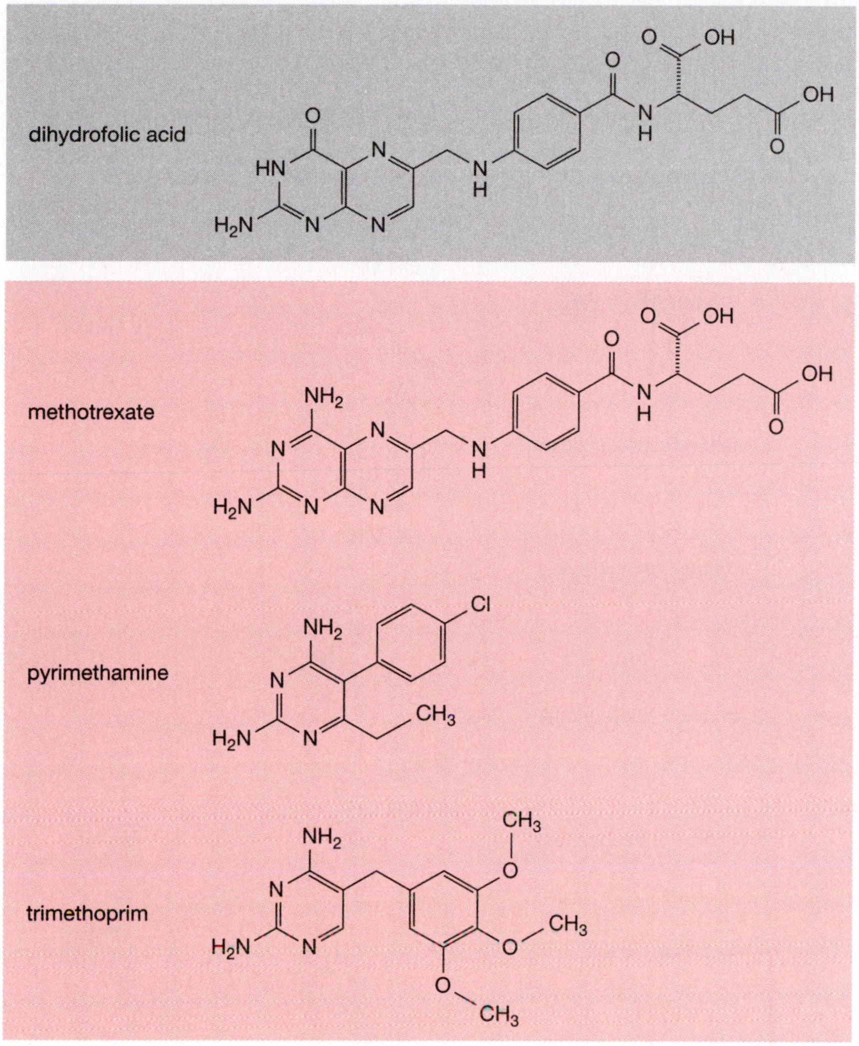

Figure 14-14. Inhibitors of DHFR, an enzyme essential to maintenance of dNTP pools (Figure 14-13). Methotrexate, pyrimethamine, and trimethoprim are inhibitors of DHFR (red box). Dihydrofolate (gray box), is the natural substrate for DHFR. They are all analogs of folic acid, an essential vitamin.

into FdUTP which is then incorporated into DNA where it produces genetic mutations. To deal with these problems, the cell excises these misincorporated uracil bases from DNA with uracil glycosylase and the DNA backbone in this strand is broken by abasic endonuclease. The 5′-deoxyribose phosphate is then removed by a flap endonuclease, and the resulting gap is filled in through the action of a repair DNA polymerase followed by DNA ligation. Although uracil excision is not essential for the cytotoxic activity of 5-FU, incomplete repair can result in the accumulation of toxic repair intermediates including abasic sites and DNA strand breaks that induce apoptosis. In fact, repair of DNA containing uracil and 5-FU becomes futile in the presence of high (F)dUTP/dTTP ratios and only results in multiple cycles of misincorporation, excision, and repair that eventually lead to DNA strand breaks and cell death. 5-FU also is converted into FUTP, a substrate for RNA synthesis. This leads to protein miscoding, inhibition of pre-rRNA processing, inhibition of post-transcriptional modification of tRNAs, as well as inhibition of polyadenylation and splicing of mRNA. Although these effects do not kill cells immediately, they are disadvantageous for cell proliferation.

The anti-cancer activity of 5-FU has since been improved over the years by modulating the drug's metabolism. For example, up to 80% of administered 5-FU is broken down by dihydropyrimidine dehydrogenase in the liver, but administration of uracil inhibits this pathway. Tetrahydrofolate is required for purine and TMP biosynthesis. Blocking purine biosynthesis by

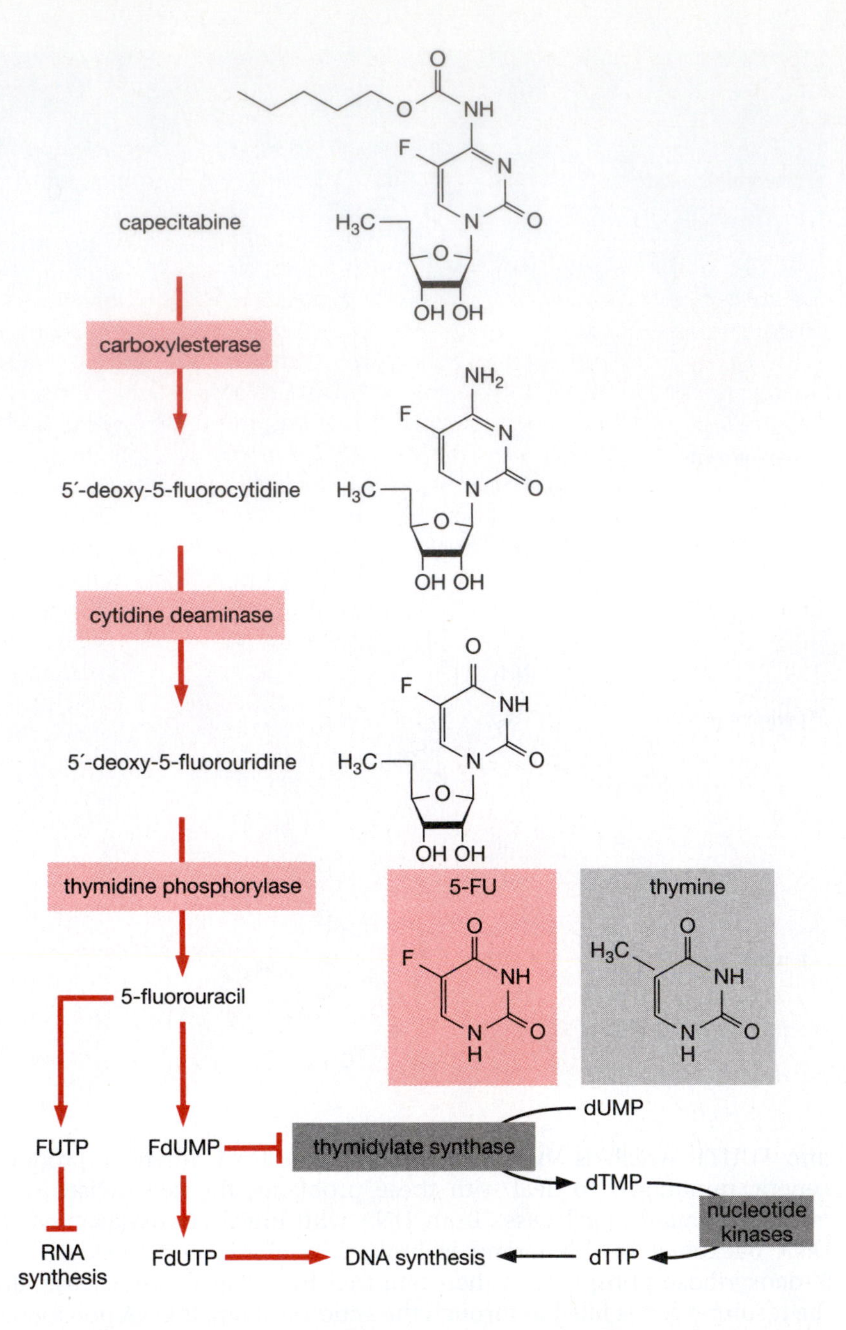

Figure 14-15. 5-Fluorouracil (5-FU) targets thymidylate synthase. Capecitabine is a 5-FU prodrug. The effects of 5-FU are mediated through 5-fluoro-2′-deoxyuridine-5′-monophosphate (FdUMP), FdUTP, and FUTP. 5-FU is converted into FdUMP and FUMP by the normal pyrimidine synthesis pathway by first attaching the pyrimidine base to phosphoribosyl pyrophosphate, a reaction catalyzed by orotate phosphoribosyl transferase. The resulting NDP is then converted into dNDP by ribonucleotide diphosphate reductase. Thymine (shaded gray) is included for comparison.

inhibiting DHFR with methotrexate increases the levels of phosphoribosyl pyrophosphate, the cofactor required for the conversion of 5-FU to FUMP. In this manner, pretreatment of patients with methotrexate increases conversion of 5-FU into FdUTP. Capecitabine is a 5-FU prodrug that is converted into 5′-deoxy-5-fluorouridine in the liver by the sequential action of carboxylesterase and cytidine deaminase, and 5′-deoxy-5-fluorouridine is converted to 5-FU in all cells by thymidine phosphorylase. Thus, low doses of capecitabine deliver higher concentrations of 5-FU to the target tissues than would be obtained by administering comparable doses of 5-FU directly.

Another established target for anti-cancer therapy is **ribonucleotide diphosphate reductase**, an enzyme essential to the conversion of ribonucleotides into deoxyribonucleotides (Figure 3-12). Hydroxyurea (HU) and mimosine (Figure 14-16) inhibit this enzyme by scavenging a tyrosine radical at the enzyme's active site and by chelating iron, thereby reducing the cell's dNTP pools and inhibiting DNA synthesis. HU is currently used to treat hematologic malignancies and sickle cell anemia. Mimosine is effective against ovarian cancer cells *in vitro*. Both drugs are commonly used to synchronize cells at their G_1/S boundary (Table 12-2). HU is used with yeast, and mimosine is used with cultured mammalian cells.

hydroxyurea

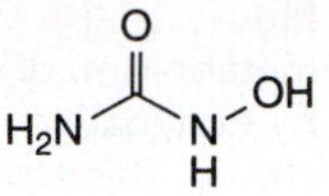

mimosine

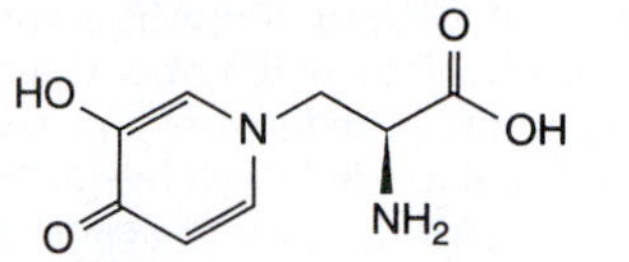

Figure 14-16. Inhibitors of ribonucleotide diphosphate reductase.

Fluorinated pyrimidines are proving effective anti-cancer agents. Gemcitabine (Figure 14-17) is used to treat carcinomas such as lung cancer, pancreatic cancer, bladder cancer, and breast cancer. It is less debilitating than some other forms of chemotherapy, and it inhibits DNA polymerase activity as well as dNTP synthesis. Gemcitabine (dFdC) is phosphorylated first to dFdCMP by deoxycytidine kinase and then to gemcitabine diphosphate (dFdCDP) and gemcitabine triphosphate (dFdCTP) by nucleoside monophosphate (UMP/CMP) and diphosphate kinases (Figure 14-18). dFdCTP is incorporated at the end of nascent DNA chains where it allows addition of only one more nucleotide, before terminating chain elongation. Proofreading enzymes are unable to remove gemcitabine from this position. In addition, the gemcitabine metabolites inhibit ribonucleoside diphosphate reductase (thereby reducing dNTP pools) as well as deoxycytidine monophosphate deaminase, an enzyme that converts dFdCMP into dFdUMP. These two actions promote a high dFdCTP/dCTP ratio, thereby enhancing the drug's overall ability to inhibit cell proliferation. On the other hand, more than 90% of administered gemcitabine is deaminated (and thus inactivated) by cytidine deaminase.

Clofarabine (Figure 14-17) is a fluorinated analog of adenosine that is used for treating refractory acute lymphoblastic leukemia. Its mechanism of

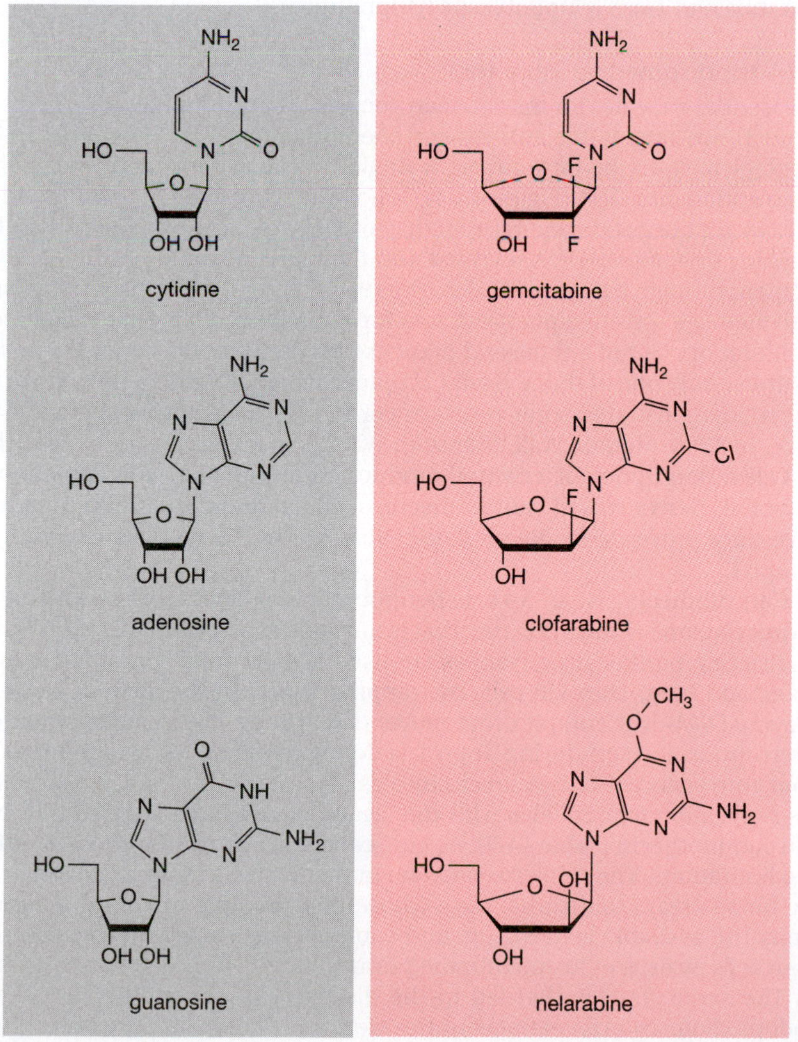

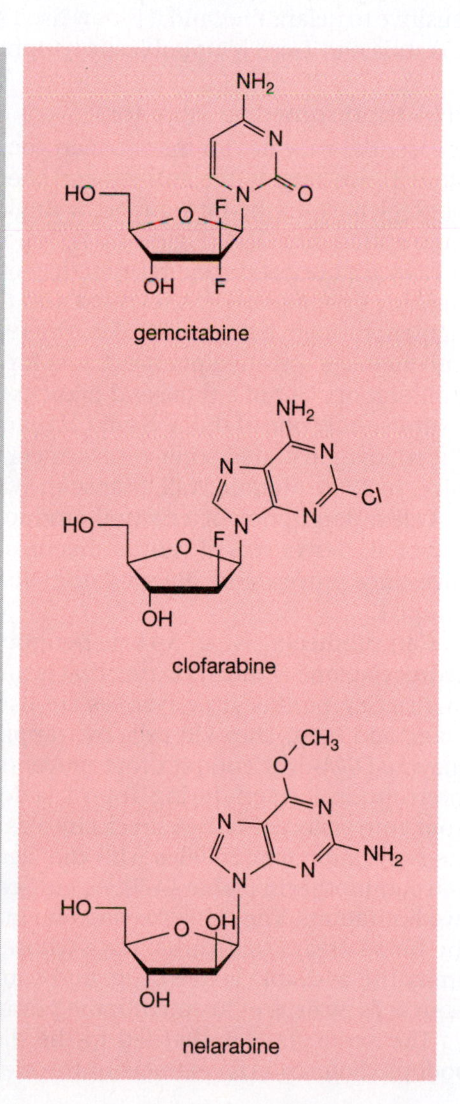

Figure 14-17. Ribonucleoside analogs that inhibit DNA polymerase activity or dNTP synthesis.

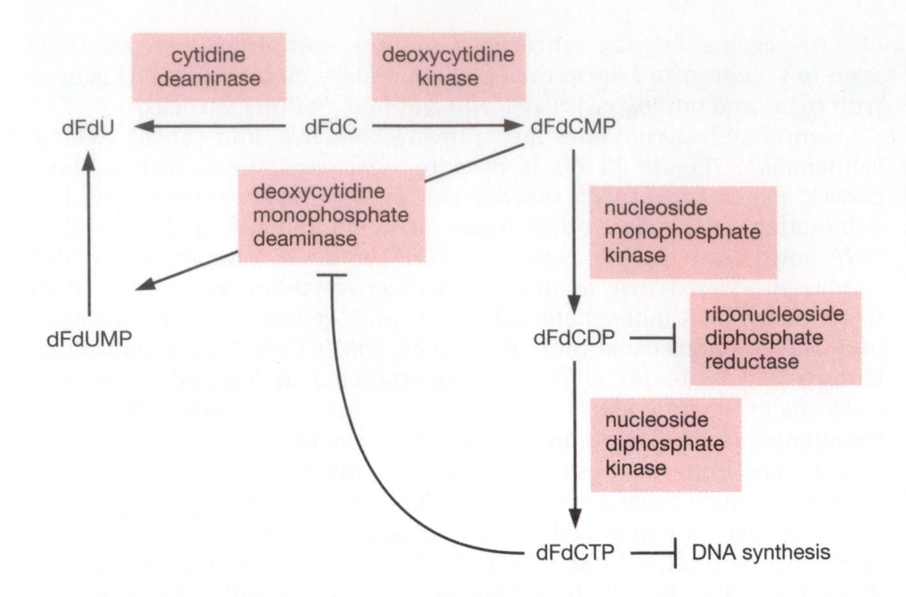

Figure 14-18. Gemcitabine (dFdC) metabolism. dFdCTP inhibits DNA synthesis by terminating elongation of nascent DNA chains. This action is enhanced by the ability of dFdCTP to inhibit conversion of dFdCMP into dFdUMP, and the ability of dFdCDP to inhibit ribonucleotide diphosphate reductase, an enzyme essential for dNTP synthesis. dFdC is inactivated by cytidine deaminase to form 2′-deoxy-2′,2′-difluorouridine (dFdU) (see Figure 14-15).

action is probably similar to that of gemcitabine. **Nelarabine** (Figure 14-17) is a prodrug of arabinosylguanosine that is converted into araGTP. As with araCTP and araATP (Figure 14-11), araGTP incorporation into nascent DNA strands retards and eventually halts DNA synthesis. T-cells are particularly sensitive to nelarabine, and it is now used to treat T-cell acute lymphoblastic leukemia and T-cell lymphoblastic lymphoma.

DNA topoisomerase inhibitors

As essential enzymes for mammalian genome duplication, topoisomerases I and II are targets for anti-cancer chemotherapy. Topo I relieves DNA topological stress by introducing a single transient phosphodiester bond interruption into DNA (Figure 4-24). Topo II removes catenated intertwines in DNA that occur during termination of DNA replication (Figure 6-2) by breaking one duplex DNA segment and then passing another duplex DNA segment through it (Figure 4-25). In the 1960s, teams of biologists, botanists, and chemists screened plant extracts for natural products that inhibit cell proliferation. Out of a thousand plant extracts sent to the National Cancer Institute, only one of them, from *Camptotheca acuminata* (a tree native to China), demonstrated potent anti-cancer activity against tumors planted in mice. In 1966, Monroe Wall, Mansukh Wani, and colleagues published the first clear description of a natural product, camptothecin, with anti-cancer potential. Derivatives of camptothecin, such as irinotecan and topotecan, have since proven to be potent anti-cancer agents. Their only known target is Topo I.

Camptothecin is a 5-ring heterocyclic alkaloid that contains a hydroxylactone within its E-ring that is unstable at physiological pH (Figure 14-19). Camptothecin derivatives are used against solid tumors in colon, ovary, and lung. They kill cells by trapping Topo I in the form of a stable enzyme•DNA•drug complex that prevents religation of the single-stranded interruption formed during Topo I action. Camptothecin toxicity results from conversion of ssDNA breaks into dsDNA breaks during S phase when the replication fork collides with the cleavage complexes formed by DNA and camptothecin (Table 14-7). As anti-cancer drugs, camptothecins suffer two limitations. The α-hydroxylactone in the E-ring is rapidly converted in the bloodstream to a carboxylate whose tight binding to serum albumin limits the available active drug, and camptothecins are actively exported from cells by drug efflux membrane pumps.

The same strategy that led to the discovery of camptothecin yielded podophyllotoxin from extracts of the stem of *Podophyllum peltatum* (the

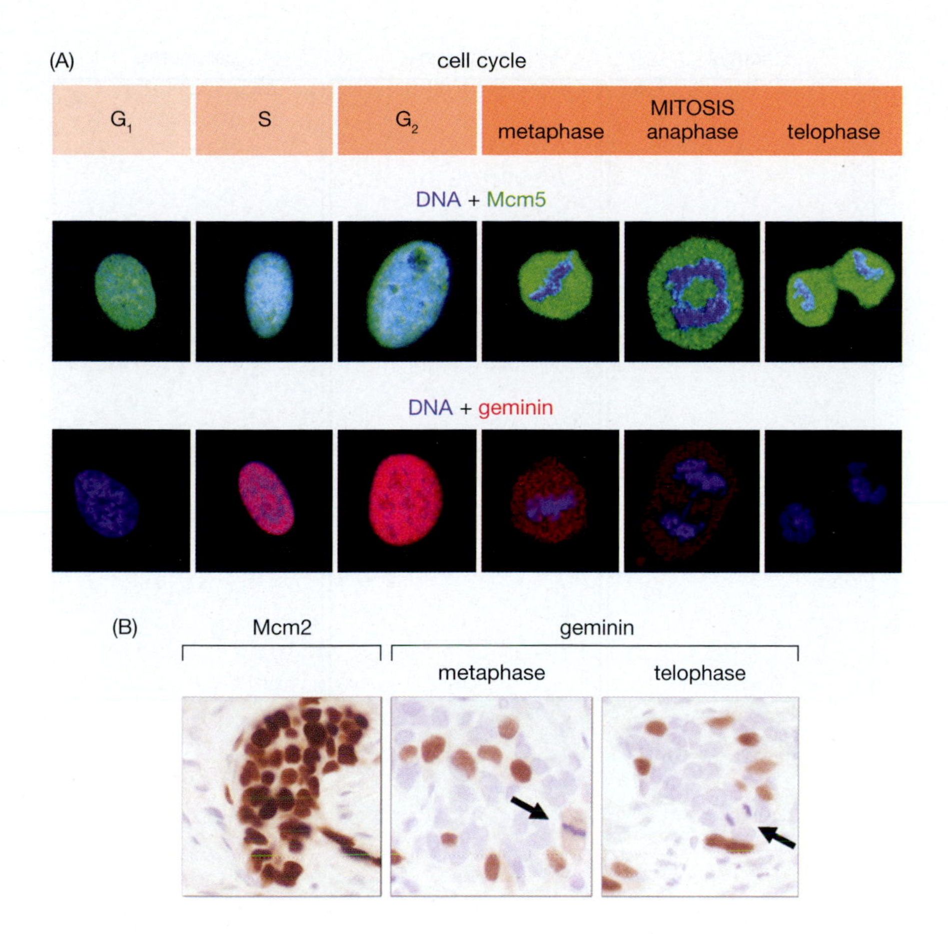

Figure 14-2. Expression of Mcm and geminin proteins at each stage of cell division in human cells. (A) Human U2OS cells cultured *in vitro* were stained with DAPI to visualize nuclear DNA (blue) and with specific antibodies to visualize Mcm5 (green) and geminin (red). Where DNA (blue) and Mcm5 (green) are coincident, the color is turquoise. Where DNA and geminin (red) are coincident, the color is pink. (B) Mcm2 immunohistochemical labeling identifies all cycling cells in an invasive breast carcinoma (formalin-fixed, paraffin-embedded tissue). In contrast, geminin labeling detects only cells in S, G_2, and early mitotic phases. Arrows indicate a geminin-positive cell in metaphase and a geminin-negative cell in telophase. (From Gonzalez, M.A., Tachibana, K.E., Laskey, R.A. and Coleman, N. (2005) *Nat. Rev. Cancer* 5: 135–141. With permission from Macmillan Publishers Ltd.)

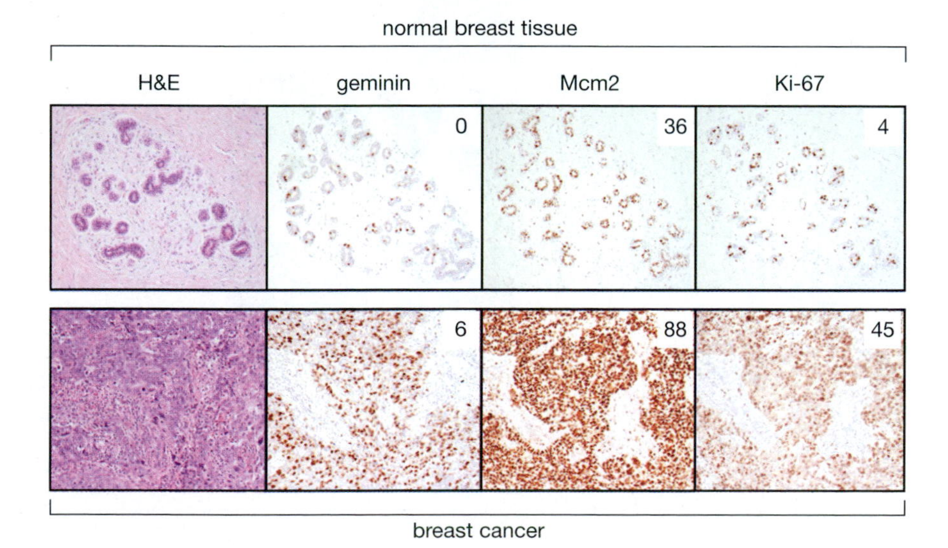

Figure 14-3. Comparison of geminin, Mcm2, and Ki-67 biomarkers in normal human breast tissue (top panel) and a breast cancer (bottom panel). For comparison, a control tissue slice was stained with hematoxylin and eosin (H&E). Hematoxylin stains nucleic acid structures, such as nuclei and ribosomes, blue-purple. Eosin Y stains protein-rich areas, such as cytoplasm, pink. The percentage of epithelial nuclei that stained positively for each protein (the labeling index) is indicated. (From Gonzalez, M.A., Tachibana, K.E., Chin, S.F. *et al.* (2004) *J. Pathol.* 204: 121–130. With permission from John Wiley and Sons, Inc.)

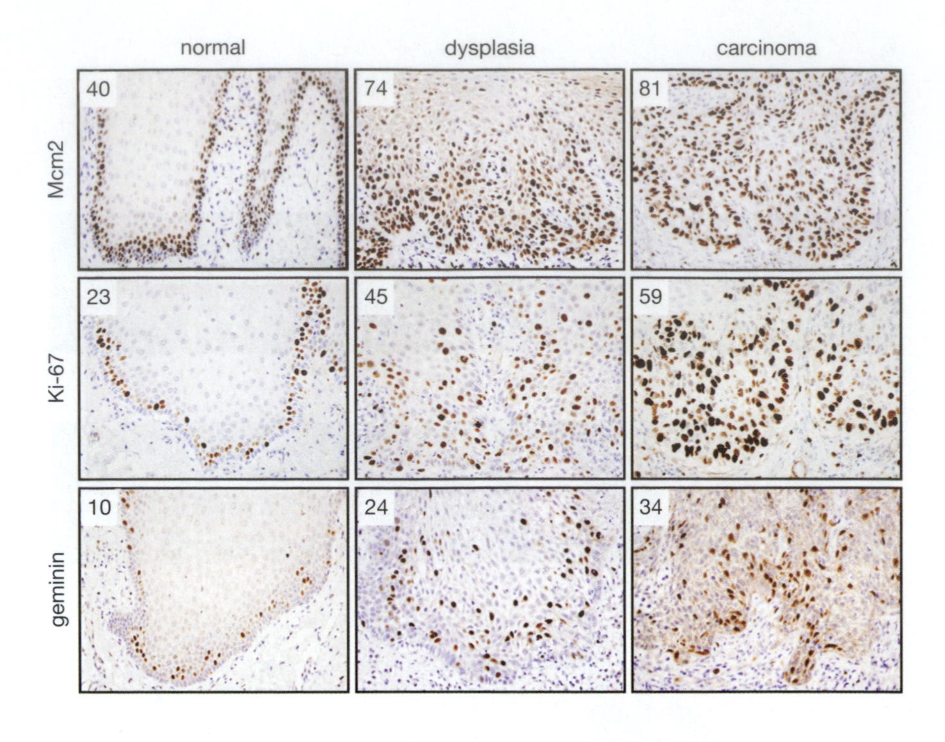

Figure 14-4. Comparison of Mcm2, Ki-67, and geminin expression in normal oral mucosa (normal), oral epithelial dysplasia that progressed to oral squamous-cell carcinoma (dysplasia), and oral squamous-cell carcinoma (carcinoma). Numbers are the mean labeling indices. (From Torres-Rendon, A., Roy, S., Craig, G.T. and Speight, P.M. (2009) *Br. J. Cancer* 100: 1128–1134. With permission from Macmillan Publishers Ltd.)

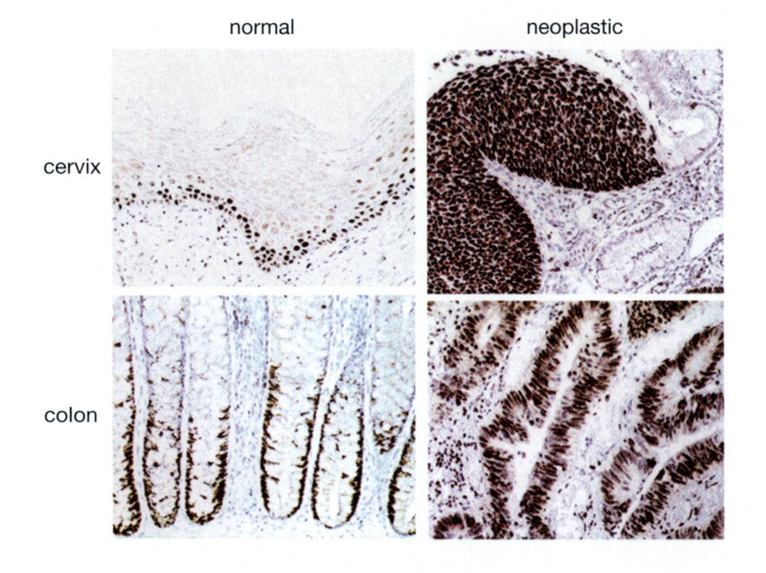

Figure 14-5. Expression of Mcm5 in normal and neoplastic tissue from cervix and colon. The protein is restricted to basal proliferative compartments in normal cervical squamous epithelium and normal large bowel crypts (brown immunoperoxidase staining) and is lost as cells differentiate as they move toward the epithelial surface. In contrast, neoplastic epithelia at the same sites (both malignant and premalignant) show full-thickness expression of Mcm5. Numerous immunopositive cells are present at the epithelial surface, from which they can be sampled either actively or passively. Expression of Mcm2 through Mcm7 is essentially the same in all tissues examined. (From Freeman, A., Morris, L.S., Mills, A.D. et al. (1999) *Clin. Cancer Res.* 5: 2121–2132. With permission from the American Association for Cancer Research.)

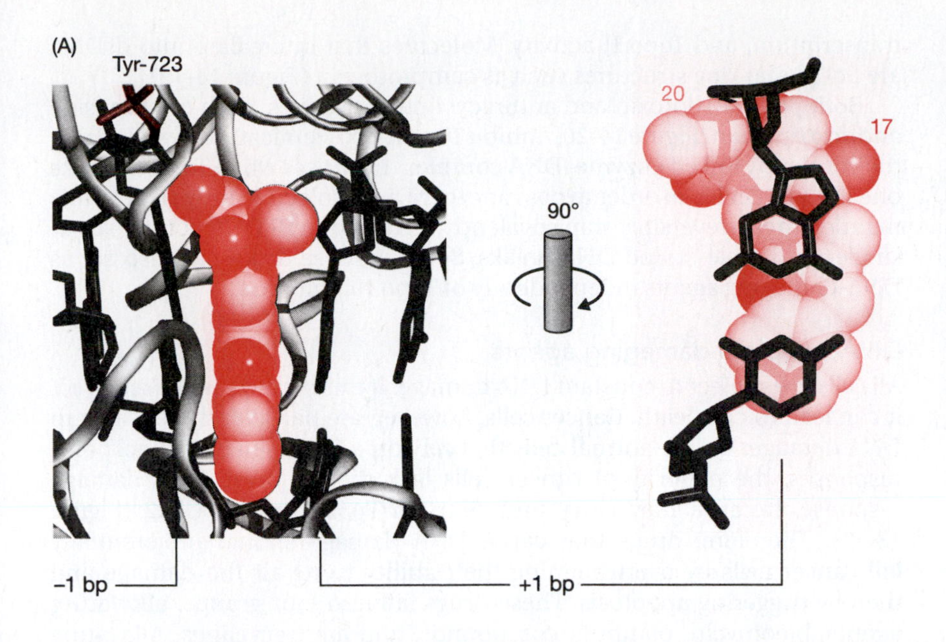

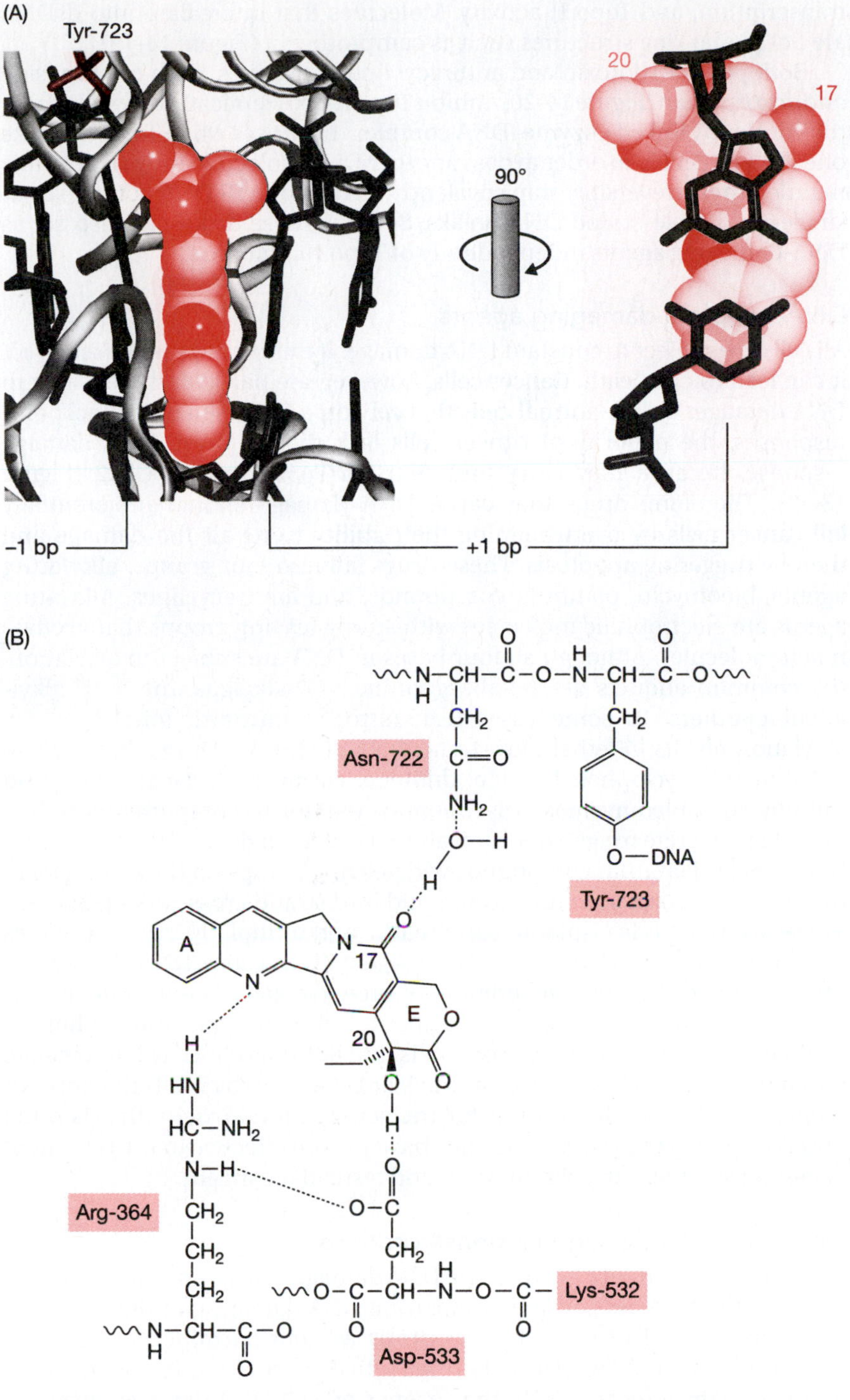

(B)

Asn-722

Tyr-723

Arg-364

Lys-532

Asp-533

Figure 14-19. The topoisomerase I-DNA–camptothecin complex. (A) The camptothecin molecule intercalates between the two DNA base pairs (–1 and +1) flanking the topoisomerase I cleavage complex. Space-filling spheres are camptothecin. Black sticks are the DNA structure; gray ribbon is Topo I backbone. The left panel is oriented with the DNA viewed from the minor groove and the right panel is rotated 90° and shows the +1 nucleotide covering the camptothecin molecule. The catalytic Topo I Tyr-723 is shown. Colored numbers correspond to the camptothecin atoms in (B). (B) The natural camptothecin forms three hydrogen bonds with topoisomerase I covalently linked to DNA at Tyr-723. The B ring is bonded to arginine, the D ring to Asn-722, and the E ring to Asp-533. (Adapted from Marchand, C., Antony, S., Kohn, K.W. *et al.* (2006) *Mol. Cancer Therapeut.* 5: 287–295. With permission from the American Association for Cancer Research.)

American mayapple). Its anti-proliferative properties have proven useful in the treatment of genital warts, but it is too toxic to be used as an anti-cancer drug. Nevertheless, podophyllotoxin is the precursor of the highly effective anti-cancer drugs, etoposide and teniposide (Figure 14-20A).

Anthracyclines are another class of anti-cancer drugs isolated from soil bacteria of the genus *Streptomyces*, a major source of antibiotics. The anthracyclines include daunorubicin (daunomycin), doxorubicin (adriamycin), epirubicin, and idarubicin (Figure 14-20B). They are used to treat a wide range of cancers, including leukemias and lymphomas, as well as cancers in breast, uterus, ovary, and lungs. Anthracyclines exert their effects on cancer cells by intercalating between the base pairs in DNA through base stacking interactions. This results in the inhibition of DNA synthesis, DNA

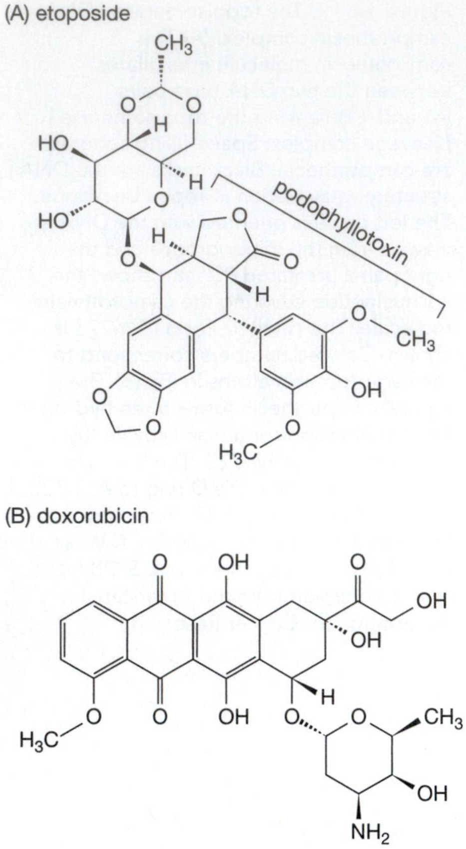

(A) etoposide

(B) doxorubicin

Figure 14-20. Topoisomerase II inhibitors and DNA-damaging agents. Etoposide and doxorubicin are both potent anti-cancer drugs.

transcription, and Topo II activity. Molecules that intercalate into dsDNA are flat planar ring structures such as camptothecin (Figure 14-19).

Both podophyllotoxin and anthracycline derivatives, such as etoposide and doxorubican (Figure 14-20), inhibit Topo II DNA replication by stabilizing the covalently linked enzyme–DNA complex that arises when Topo II breaks one DNA molecule in order to pass another DNA molecule through it. In this manner, they irreversibly trap covalently linked Topo II–DNA complexes as single- or double-strand DNA breaks. Some of these drugs can also act as DNA-damaging agents independently of Topo II inhibition.

DNA-template-damaging agents

All cells are subject to constant DNA damage. If this damage is not repaired, it can lead to cell death. Cancer cells, however, are particularly sensitive to DNA damage. Unlike normal cells that rely on a full arsenal of checkpoint responses, the majority of cancer cells lack the G_1-phase DNA damage response, because they carry mutations in Tp53, Atm, or Chk2 (Figure 13-13). Therefore, drugs that cause DNA damage should preferentially kill cancer cells by overwhelming their ability to repair the damage and thereby triggering apoptosis. These drugs fall into four groups: **alkylating agents,** bleomycin, platinum compounds, and anthracyclines. Alkylating agents are electrophilic molecules with strong leaving groups that alkylate macromolecules. Although all four bases in DNA are subject to alkylation, the common adducts are N^7-alkylguanine>>O^6-alkylguanine = N^3-alkyladenine>others. In some cases (e.g. nitrogen mustard, mitomycin C), alkylation results in lethal interstrand DNA crosslinks. Drugs that alkylate DNA include cyclophosphamide, thiotepa, carmustine, dacarbazine, and mitomycin C. Bleomycin is a glycopeptide that forms complexes with Cu^{2+} or Fe^{2+} to generate oxygen free radicals that cause single- and double-strand DNA breaks. Platinum compounds with leaving groups on the same side of the molecule cause both intrastrand and interstrand cross-linking of DNA. These drugs include cisplatin, carboplatin, and oxaliplatin. Anthracyclines (Figure 14-20B) are DNA intercalating agents that cause DNA damage by the NADPH-dependent generation of free radicals. Daunorubicin and doxorubicin are used in cancer treatment, but their application is limited by their clinical toxicity. One problem is that anthracyclines will intercalate into mitochondrial DNA as well as nuclear DNA, thereby inhibiting mtDNA replication. This could account for the severe cardiotoxicity that is often observed. Many of these compounds have proven effective in the treatment of certain solid tumors, lymphomas, and testicular cancer.

Targeting DNA damage-response proteins

Given the sensitivity of cancer cells to DNA damage, inactivating proteins that are essential to either sensing or repairing DNA damage can preferentially induce cancer cells to undergo apoptosis without harming normal cells. Such drugs should be particularly effective in conjunction with DNA-template-damaging agents. In the absence of a G_1 DNA damage response, cancer cells must rely on the S and G_2 DNA damage response (Figure 13-15). The DNA damage response relies on two checkpoint effector proteins called checkpoint kinase 1 (Chk1) and checkpoint kinase 2 (Chk2). Depletion of Chk1 by RNA interference provokes premature entry into mitosis, whereas depletion of Chk2 does not. Thus, Chk1 is the more promising target for therapeutic intervention.

The antibiotic staurosporine was originally isolated in 1977 from the bacterium *Streptomyces staurosporeus.* Later studies revealed that 7-hydroxystaurosporine (UCN-01) inhibits serine and threonine kinases, in particular Chk1, by preventing ATP from binding to the protein. Chk1 is essential to mammalian cell proliferation where it acts as the effector kinase in both the DNA replication and G_2 DNA damage checkpoints (Figure 13-15).

UCN-01 arrests cell proliferation and induces apoptosis. Accordingly, it is proving useful in chemotherapy against a variety of cancers, particularly in conjunction with other anti-cancer drugs. Newer drugs such as XL844, AZD7762, and PF477736 are at least 100-fold more specific than UCN-01 for Chk1, and these drugs are currently under evaluation as chemotherapeutic agents. Drugs that block ATP binding specifically to Atm also have been discovered, but none that are specific for Atr.

Another promising target is poly(ADP-ribose) polymerase (Parp). Of the 17 Parp enzymes in humans, Parp-1 and Parp-2 are the only ones involved in the repair of ssDNA breaks. Parp cleaves NAD^+ to generate ADP and ADP-ribose. The ADP-ribose is then covalently linked to lysines on acceptor proteins to produce proteins linked to polymers of ADP-ribose in much the way ubiquitin ligases produce proteins linked to polymers of ubiquitin. Inhibition of Parp-1-dependent DNA repair results in the accumulation of stalled replication forks, single-strand breaks, and the conversion of single-strand breaks into potentially lethal dsDNA breaks.

Parp inhibitors block the catalytic domain of Parp by competing with NAD^+ for binding to the enzyme. These inhibitors shows promise as anti-cancer agents either by promoting selective killing of tumors with pre-existing DNA repair defects, or in conjunction with DNA-damaging agents, including radiation therapy. Proof of principle is revealed by the observation that both Brca1- and Brca2-deficient cells are exquisitely sensitive to Parp inhibition in comparison with heterozygous mutant or wild-type cells. The Brca1 tumor-suppressor gene is involved in DNA damage signaling, cell-cycle checkpoint regulation, and recruitment of repair enzymes, while Brca2 binds and translocates Rad51 (Table 13-5) to the site of DNA damage to initiate repair. Since breast cancers frequently are defective in either Brca1 or Brca2, Parp-specific inhibitors now under development may prove effective as anti-cancer agents.

Targeting cell-cycle-regulatory proteins

Protein kinases that regulate the cell-division cycle are obvious targets for anti-cancer drugs, and 11 of them show promise as therapeutic targets. Cdk4 and Cdk6 drive the cell out of a G_0 state and beyond the restriction point, resulting in an irreversible commitment to genome duplication (Figure 13-4). Cdk1, Cdk2, Plk1, and aurora kinase A are involved in regulating centrosome assembly. Either Cdk2 or Cdk1 are essential for entrance into S phase (Figure 12-10). Cdk1 is essential for entrance into mitosis (Figure 12-7). Plk1 activity is required for progression through mitosis (Figure 12-9). Aurora kinase B, Bub1, Bub1b, Bub3, and Mps1 are components of the spindle-assembly checkpoint (Table 13-10). All of the inhibitors for these protein kinases compete with ATP for binding to the protein kinase. This results in a lack of target specificity, and therefore increased toxicity due to inhibition of more than one protein kinase.

Flavopiridol is a broad-spectrum CDK inhibitor inhibiting the CDKs controlling both the cell cycle and transcription. Roscovitine has a slightly narrower CDK inhibitory profile with selectivity for Cdk1, Cdk2, Cdk7, and Cdk9. At low concentrations, RO3306 selectively inhibits Cdk1. However, if UCN-01 is an example, a broad target range can sometimes give surprisingly good results against particular cancers.

Targeting proteins that prevent DNA re-replication

Pre-replication complexes (preRCs) are assembled during the $M \rightarrow G_1$ transition, and then converted into pre-initiation complexes (preICs) during the $G_1 \rightarrow S$ transition. Assembly and activation of preRCs is suppressed during the S, G_2, and early M phases in order to prevent DNA re-replication (Figure 12-11), an event that results in DNA damage and eventually apoptosis. Of the 14 proteins involved in preRC assembly in eukarya, only Cdt1, Orc1, and Cdc6 are targeted for cell-cycle-dependent regulation of preRC activity (Figure 12-12). In higher eukarya Cdt1 is inhibited specifically by the protein geminin. In addition, Cdt1, Orc1, and Cdc6 are targeted for phosphorylation by Cdk2•CcnA, an event that not only inhibits their activity, but converts them into substrates for the SCF(Skp2) ubiquitin ligase and eventual degradation. Therefore, if one or more of the proteins that prevents DNA re-replication could be inactivated, proliferating cells could be destroyed.

Geminin is normally expressed only during S, G_2, and early mitosis (Figure 14-2) where it inhibits preRC assembly. Therefore, it is not surprising that overexpression in cells of a non-degradable form of geminin inhibits loading of the MCM helicase onto chromatin, and prevents cells from entering S phase, thereby arresting cell proliferation. This anti-proliferative effect of stable geminin occurs *in vivo* as well. Nude mice bearing xenografts of cells expressing stable geminin exhibit a dramatic reduction in tumor induction and growth compared to cells expressing wild-type geminin. However, overexpression of stable geminin arrests proliferation of both normal cells and cancer cells, although cells lacking Tp53 and Rb (typical of cancer cells) are more likely to trigger apoptosis than are normal cells, suggesting that overexpression of stable forms of geminin might be a strategy to selectively kill cancer cells *in vivo*. Remarkably, the opposite strategy also looks promising as an anti-cancer therapy. Depletion of geminin by siRNA induces DNA re-replication in cells derived from human cancers, but not in cells derived from normal human tissues, or even when normal cells have been immortalized with tumor virus genes. Moreover, induction of DNA re-replication rapidly induces DNA damage, and the DNA damage triggers apoptosis. In contrast, induction of DNA re-replication in normal cells requires suppression of both geminin and cyclin A expression, revealing that normal cells employ multiple pathways to prevent DNA re-replication (Figure 12-12), whereas some, perhaps most, cancer cells rely solely on geminin. Therefore, drugs that inhibit geminin action may kill cancer cells *in vivo* without harming normal tissues.

SUMMARY

- Replication and repair of DNA lie at the core of human disease, both in understanding it and in treating it.
- Most infectious human diseases are caused by pathogens that must either duplicate their DNA genome in order to proliferate (DNA viruses, bacteria, protozoa, and fungi), or produce a DNA copy of their RNA genome (retroviruses). In principle, drugs can be designed to inhibit one or more proteins essential for replicating a pathogen's DNA, but with little, if any, effect on DNA replication of its human host.
- The best examples are the antiviral agents that either directly or indirectly inhibit the activity of viral DNA polymerases. Some of these agents take advantage of the differences between viral nucleoside kinases and the corresponding cellular enzyme, as well as differences between the viral DNA polymerase and that of its host.
- Although most antibiotics are directed at cell-wall formation and protein biosynthesis, those that are targeted at bacterial DNA replication proteins appear quite effective. The best defense against drug resistance in bacteria is targeting drugs to completely different pathways.
- Noninfectious diseases such as cancer (of which there are numerous manifestations) and heritable diseases such as Huntington's, spinocerebellar ataxia, fragile X syndrome, and many others are the result of genetic mutations. These include base-pair changes, deletions, insertions, and rearrangements of DNA sequences, such as the expansion and contraction of nucleotide repeats. Most genetic mutations occur during DNA replication as a consequence of proofreading errors, failure

to respond to DNA damage checkpoints, and stalled replication forks. Mutations from radiation, free radicals, and chemical carcinogens are passed on during cell proliferation only if they are not repaired. Thus, the effectiveness by which DNA is replicated and repaired determines the rate at which mutations accumulate, and this, in turn, determines the frequency of noninfectious and heritable diseases.

- Cancer cells continue to proliferate under conditions where normal cells do not. Therefore, proteins unique to DNA replication are excellent biomarkers for the early detection of cancer.
- Cancer arises either from the accumulation of genetic mutations, or from transformation by an oncogenic virus.
- Cancer is not caused by mutations in genes required for nuclear DNA replication, but rather by mutations in genes required for cell-cycle checkpoints and for repair of DNA damage. Mutations in genes required for nuclear DNA replication are likely to be lethal, whereas mutations in genes that regulate nuclear DNA replication can produce a disease state. Since the number of mitochondria per cell is less critical than the number of nuclei, interference with mtDNA replication can result in a genetic disease.
- Genetic instability is a hallmark of cancer cells. Much of this instability results from suppression of their DNA damage response. Errors that go undetected go uncorrected. This results in more mutations from which those cells that proliferate the fastest, ignore stop signals, and metastasize efficiently flourish at the expense of normal cells. Notably, the Rb–E2f and Mdm2–Tp53 pathways are both defective in most, if not all, human tumors. This underscores the crucial role of the restriction and G_1 DNA damage checkpoints in regulating cell-cycle progression and viability.
- Oncogenic viruses in cancer cells exist either as an episome, such as EBV, or more commonly as a replication-defective viral DNA integrated into its host's chromosomes (e.g. HPV). Viral genes expressed in these cells allow the cell to bypass one or more checkpoints. Remarkably, papillomaviruses and polyomaviruses produce proteins that inactivate both Rb and Tp53 proteins, again pointing to the importance of the restriction and G_1 DNA damage checkpoints in regulating cell proliferation.
- Most chemotherapy is based on preferentially killing proliferating cells, regardless of whether or not they are cancer cells. This strategy works, because most cells in an adult human are either quiescent or terminally differentiated.
- The drugs commonly used as antiviral or anti-cancer agents are those that target DNA replication and mitosis. The DNA replication drugs fall into six categories: inhibitors of dNTP synthesis, DNA synthesis, topoisomerases, the DNA damage response, or cell-cycle regulation, as well as agents that promote DNA damage. Drugs that arrest cells during mitosis do so by inhibiting either assembly (vinblastine) or dissociation (paclitaxel) of microtubules.
- Despite similar mechanisms of action, various nucleoside and nucleotide analogs frequently exhibit different efficacies with different pathogens or with different cancers. Thus, subtle differences in protein configuration apparently can result in significant differences in the ability of a drug to interact with its target.
- While drugs that target DNA replication can be extremely effective in blocking the lytic cycle of an oncogenic virus, they do not selectively prevent viral replication in transformed cells.
- An ounce of prevention is worth a pound of cure. Smoking is linked directly to cancer of the lung, cervix, larynx, mouth, esophagus, bladder, pancreas, and kidney. If we simply stopped inhaling tobacco smoke, the incidence of lung cancer would decrease by approximately 90%!

REFERENCES

Additional Reading

Abeloff, M.D., Armitage, J.O., Niederhuber, J.E., Kastan, M.B. and McKenna, W.G. (eds) (2008) *Clinical Oncology*, 4th edn. Elsevier, Amsterdam.

Brunton, L.L., Lazo, J.S. and Parker K.L. (2005) *Goodman and Gilman's The Pharmacological Basis of Therapeutics*, 11th edn. Elsevier, Amsterdam.

Chabner, B.A. and Longo, D.L. (eds) (2006) *Cancer Chemotherapy and Biotherapy, Principles and Practice*, 4th edn. Lippincott Williams and Wilkins, Philadelphia.

DePamphilis, M.L. (ed.) (2006) *DNA Replication and Human Disease*. Cold Spring Harbor Laboratory Press, Cold Spring Harbor, NY.

Genes and Disease. http://www.ncbi.nlm.nih.gov/books/bv.fcgi?rid=gnd.

Golan, D.E., Tashjian, Jr, A.H., Armstrong, E.J. and Armstrong, A,W. (eds) (2008) *Principles of Pharmacology: The Pathophysiologic Basis of Drug Therapy*, 2nd edn. Lippincott Williams and Wilkins, Philadelphia.

Lapenna, S. and Giordano, A. (2009) Cell cycle kinases as therapeutic targets for cancer. *Nat. Rev. Drug Discov.* 8: 547–566.

Petropoulou, C., Kotantaki, P., Karamitros, D. and Taraviras, S. (2008) Cdt1 and Geminin in cancer: markers or triggers of malignant transformation? *Front. Biosci.* 13: 4485–4494.

Weinberg, R.A. (2007) *The Biology of Cancer*. Garland Science, New York.

zur Hausen, H. (2006) *Infections Causing Human Cancer*. John Wiley and Sons, New York.

Literature Cited

Castel, L.A., Cleary, J.D., and Pearson, C.E. (2010) Repeat instability as the basis for human diseases and as a potential target for threapy. *Nat. Rev. Mol. Cell Biol.* 11: 165–170.

Chapter 15
EVOLUTION OF CELLULAR REPLICATION MACHINES

The ability to replicate DNA is central to the propagation of all life on the planet. However, from the preceding chapters it will be apparent that while common features exist, a great many mechanisms and a bewildering array of proteins can be employed to duplicate the double helix. At the gross organizational level, certain common features can be gleaned for the replication of cellular DNA. For example, replication starts at defined sites and the genomes of eukaryotic nuclei, bacteria, and archaea are replicated by the replication-fork mechanism. Somehow, all of these cellular DNA replication machines evolved from a common ancestor. The question is how?

REPLICATION MACHINES IN BACTERIA, ARCHAEA, AND EUKARYA

Examination of the sequences of the proteins that carry out replication-fork-directed replication reveals a striking split in evolution. More specifically, while a clear relationship is observed between archaeal and eukaryotic replication factors, the machinery of bacteria is clearly non-orthologous to that of archaea and eukarya (Figure 15-1).

The reader will recall that **orthologs** are proteins of related sequence that have the same function in disparate organisms. For example, archaeal Orc1/Cdc6 proteins are related to eukaryal Orc1 and Cdc6, and like Orc1 they bind to replication origins. Thus, these proteins are orthologs. **Paralogs** are proteins of related sequence but that have distinct functions. Mcms are a good example of paralogs. Most archaea have a single Mcm gene, the protein product of which forms a homohexamer. However, in eukarya there are multiple related Mcm genes. All eukarya possess Mcm(2–7) and most also have Mcm8 and Mcm9. Mcm(2–7) form a heterohexamer, and Mcm8 and Mcm9 are believed to form homohexamers. These distinct Mcm complexes appear to play distinct roles in replication and are therefore paralogous. **Homolog** is a catch-all term for proteins of related sequence that embraces both orthologs and paralogs. Finally, **analogs** have unrelated sequences but carry out the same function. DnaA, the bacterial replication initiator, like archaeal Orc1/Cdc6 and eukaryal Orc1 [in the company of ORC(2-6) and Cdc6] binds to replication origins, but DnaA is essentially unrelated either to Orc1/Cdc6 or to Orc1 at the level of its amino acid sequence. Thus, DnaA is an analog of its archaeal and eukaryal counterparts.

Virtually all of the archaeal proteins are orthologs of eukaryotic proteins. The only exception is the D-family DNA polymerase found in some archaea (Table 4-4); none are present in eukarya. However, eukarya contain at least 10 replication proteins not found in archaea (Orc2, Orc3, Orc4, Orc5, Orc6, Mcm10, Cdc45, Sld2, Sld3, and Dbp11). Moreover, although the DNA polymerases in eukarya, like those in some archaea, are B-family enzymes, DNA polymerases α, δ, and ε are unique to the eukarya. These uniquely eukaryotic proteins represent an expansion of the stage in fork assembly

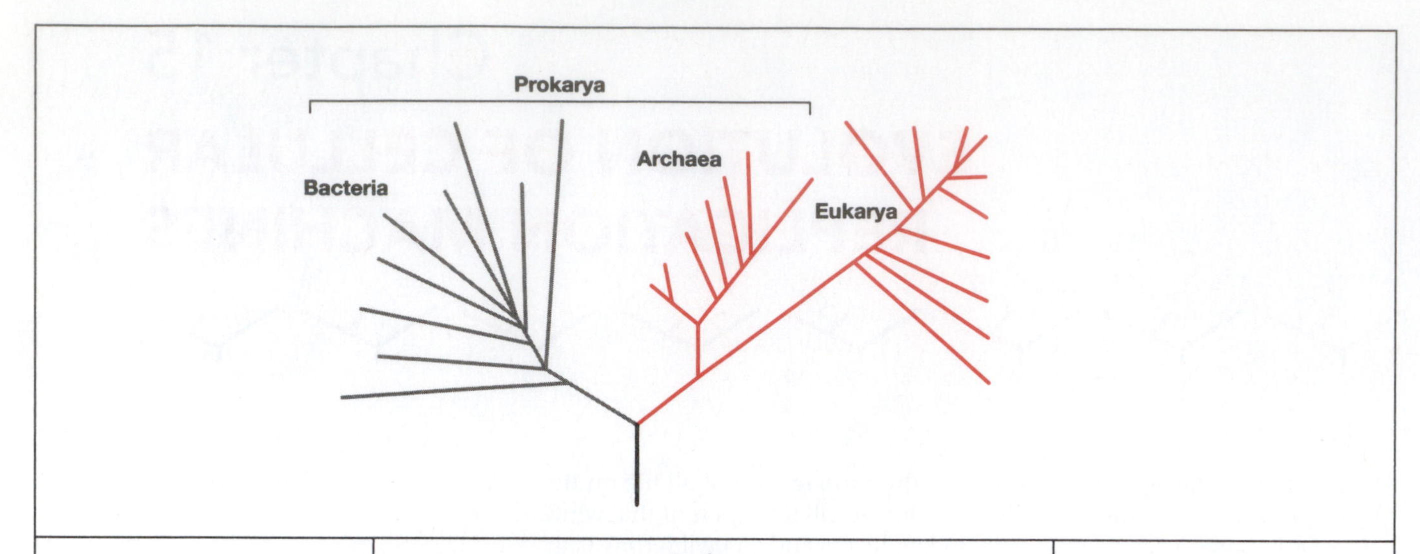

Function	Prokarya		Eukarya
	Bacteria	Archaea	
Origin recognition	DnaA	Orc1/Cdc6 WhIP	ORC (Orc1, 2, 3, 4, 5, 6)
Helicase loading (Pre-replication complex)	DnaC DnaB	Orc1/Cdc6 WhiP Mcm hexamer	Cdc6 Cdt1 Mcm(2-7)
Pre-initiation complex		GINS (Gins23 + Gins15)	GINS (Psf1, 2, 3 and Sld5) Mcm10 Cdc45 Sld2 Sld3 Dbp11
DNA priming	DnaG	Primase (PriS + PriL)	DNA Pol-α•primase (Pol-α B-subunit, PriS + PriL)
DNA synthesis	DNA Pol-III core τ-subunit γ-complex (clamp loader) β-clamp (sliding clamp β	B-family DNA Pol D-family DNA Pol RFC (clamp loader) PCNA (sliding clamp)	B-family DNA Pol (Pol-δ, Pol-ε) RFC PCNA
Primer excision	DNA Pol-I RNase H DNA ligase (+ NAD^+)	Fen1 [Dna2] RNase HII (?) DNA ligase I (+ ATP)	Fen1 Dna2 RNase HII (?) DNA ligase I (+ ATP)

Figure 15-1. The evolutionary relationship between DNA replication machines in the three domains of life. Both bacteria and archaea lack nuclei and are, therefore, prokarya. The core replication proteins in bacteria, archaea, and eukarya are grouped according to their function. The function of bacterial replication components (shaded in gray) is analogous to those listed in parallel for archaea and eukarya, but the bacterial proteins are not orthologous to the archaeal and eukaryal proteins. Archaeal and eukaryotic replication proteins that are orthologous to one another are indicated in black. Replication proteins that are unique to either archaea or eukarya are indicated in red. The Cdt1 homolog, WhiP, is restricted to the crenarchaeal kingdom of archaea; its role in origin binding has been established but it is not yet known if it interacts with MCM. Dna2 in archaea is phylogenetically restricted to a few species where its role is uncertain; therefore, it is indicated in brackets. RNase HII can facilitate primer excision in both archaea and eukarya, but it is not essential. DNA ligase in bacteria uses NAD^+ as its cofactor, whereas DNA ligases in archaea and eukarya use ATP. Data and illustrations are adapted from Woese, C.R., Kandler, O. and Wheelis, M.L. (1990) *Proc. Natl. Acad. Sci. USA* 87: 4576–4579 (with permission from the National Academy of Sciences), and from McGeoch, A.T. and Bell, S.D. (2008) *Nat. Rev. Mol. Cell Biol* 9: 569–574 (with permission from Macmillan Publishers Ltd.).

termed the pre-initiation complex (Table 11-4) in which the DNA helicase is activated, the replication origin is thought to be melted, and the initiator DNA polymerase-α•primase is loaded onto the forks (Figure 11-15). This is the stage in eukarya that requires CDK activity, one of the cell-cycle regulators that is absent from archaea. Thus, a number of the more significant differences between the archaeal and eukaryotic machines lie not in the core replication engines, but rather in the apparatus that transduces signals from the cell cycle control circuits to the core apparatus via a range of eukaryal-specific adaptors, such as Cdc45, Sld2, Sld3, and Dbp11. In addition, three different DNA polymerases are used to replicate DNA in eukarya whereas a single DNA polymerase is sufficient in both archaea and bacteria (only the DNA Pol-I 5′→3′ exonuclease is required; it's a component of RNA primer excision). Clearly, the DNA replication machines in archaea evolved earlier than those in eukarya, but independently from those in bacteria.

The fact that the bacterial replication machinery is distinct from that in both archaea and eukarya is truly remarkable. Moreover, it has significant implications for the evolution of life on Earth. The obvious conclusion is that cellular DNA replication machines evolved twice, once in the lineage that led to bacteria and once in the lineage that gave rise to modern-day archaea and eukarya. However, it is generally accepted that all extant organisms share a common ancestor, the **L**ast **U**niversal **C**ommon **A**ncestor (LUCA). So what sort of genome did LUCA have? As discussed in Chapter 2, it is generally believed that RNA was the first genetic material. Perhaps LUCA had an RNA genome and the switch to the more stable DNA genome occurred after the emergence of the two lineages leading to bacteria on the one hand, and to archaea and eukarya on the other. In principle, this proposition could account for the two separate DNA replication machines by positing that they evolved independently in the two post-LUCA branches, concomitant with the switches to DNA-based genomes (Figure 15-2).

However, whereas the DNA replication machinery of bacteria is clearly non-orthologous to that of archaea and eukarya, the same is not the case for certain key components of the DNA repair machinery. More specifically, the central players in homologous recombination, proteins such as the strand exchange proteins, bacterial RecA, archaeal RadA, and eukaryotic Rad51, and the bacterial SbcCD complex and its counterpart in archaea and eukarya, the Rad50•Mre11 complex, show strong sequence and structural conservation in all three domains of life (Table 13-5). Homologous recombination is thus a conserved pathway that mediates essentially error-free DNA repair. That this apparatus is conserved in all three domains of life indicates that LUCA possessed a DNA repair apparatus that is based on homologous recombination. It therefore follows that if LUCA had a homologous recombination DNA repair apparatus then it must have been performing DNA repair. This conservation of DNA repair pathways therefore strongly supports the idea that LUCA had a DNA-based genome.

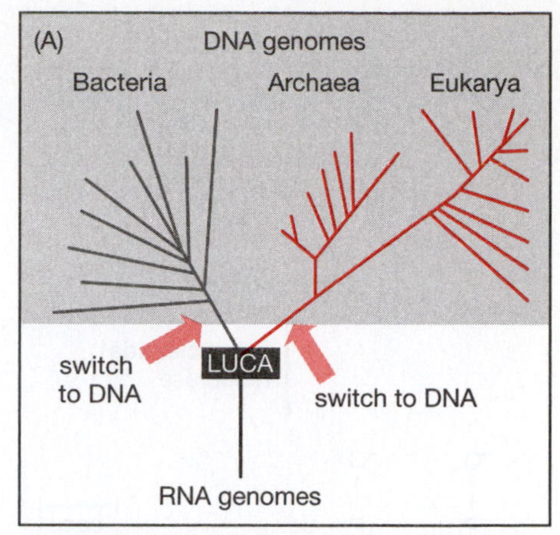

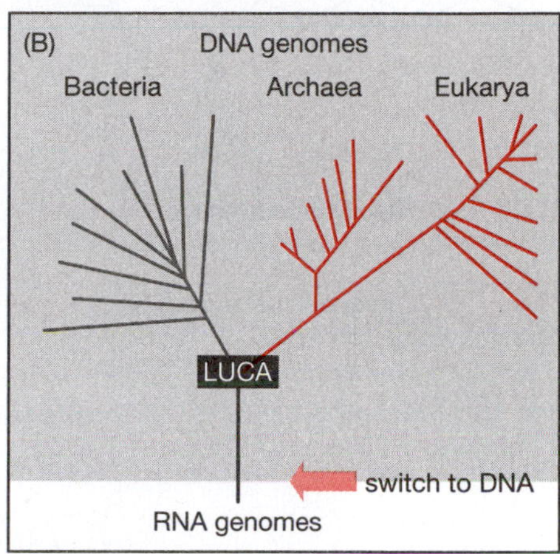

Figure 15-2. Assuming that the earliest forms of life had RNA genomes, at some point in evolutionary history a transition took place where DNA took over as the genetic material. (A) If that transition took place after the time of LUCA, once in the lineage leading to bacteria (gray) and once in the lineage eventually leading to archaea and eukarya (red), then two distinct replication machines could have evolved in these lineages. In panel (B), a scenario is depicted where the switch from RNA to DNA genomes took place prior to the time of LUCA. In the example depicted the ancestral DNA replication machinery was related to that of present-day archaea and eukarya (indicated by the red line). At or just following the time of LUCA an event took place in which a novel DNA replication machinery supplanted the ancestral form in the lineage that gave rise to present-day bacteria.

MAKING THE JUMP FROM RNA TO DNA GENOMES

The current way in which cells make DNA precursors hints at the ancestral transitions that occurred during the switch from RNA to DNA genomes. In fact, all dNTPs are synthesized in a pathway which starts with the reduction of a NMP or NDP precursor to dNMP or dNDP by a ribonucleotide reductase (RNR; see Figure 15-3). However, DNA contains dTMP, not dUMP, and so a second enzyme, thymidylate synthase, converts the uracil base in dUMP into the thymine base in dTMP (Figure 15-4).

Assuming that these two steps occurred at distinct times, one could invoke a period in evolutionary history after the acquisition of RNR where DNA contained uracil in place of thymine. The subsequent appearance of thymidylate synthase could account for the switch to DNA that contains

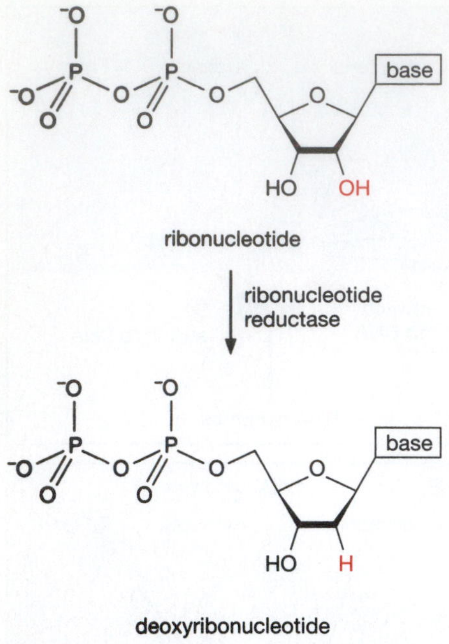

ribonucleotide

ribonucleotide reductase

deoxyribonucleotide

Figure 15-3. Reaction catalyzed by ribonucleotide reductase.

thymidine instead of uracil. In principle, one might expect that, by tracing the phylogenetic relationships between either RNR or thymidylate synthase enzymes, one could elucidate whether these enzymes predate or postdate LUCA. Such analyses might provide support for the hypothesis that LUCA had a DNA-based genome. Remarkably, however, there are three distinct classes of RNR (although these do share a common core fold) as well as two entirely non-orthologous thymidylate synthase enzymes. Furthermore, these enzymes show a patchy distribution across the three Domains of Life. This distribution prevents the drawing of conclusions about the ancestral state of these enzymes. What might be causing this scrambling of the phylogenetic signature of these genes? Interestingly, many viruses encode their own RNRs or thymidylate synthases, presumably as a way of fine-tuning the metabolism of infected cells for optimal virus production. It is possible therefore that some of the scrambling of the distribution of these genes may be accounted for by replacement of endogenous cellular genes with ones donated by viruses.

A ROLE OF VIRUSES IN THE EVOLUTION OF DNA GENOMES?

Viruses are engaged in an arms race with their cellular hosts and are thus under huge selection pressure to evolve strategies to evade host defense systems. This battle is likely almost as old as life itself. Recent studies have revealed common features in the organization of the viral capsid proteins of diverse viruses that infect organisms from all three Domains of Life. More specifically, the bacteriophage PRD1, the archaeal virus STIV, and the eukaryotic virus adenovirus share a common capsid structure in the form of the 'double jelly-roll' fold. The extreme conservation of this fold indicates that it is a descendant of an ancestral fold that was present in a virus that existed at, or even preceded, the time of LUCA. Thus, the evolution of all cellular life on this planet has taken place in the context of constant exposure to viral infection. It is inconceivable therefore, that viruses have not impacted on the evolution of their hosts and vice versa. Indeed, as discussed below, it is likely that viruses have played pivotal roles in some of the major transitions in evolution.

DID DNA EVOLVE IN VIRUSES AS A WAY TO EVADE HOST DEFENSES?

One of the principal ways that host cells can eliminate the potential of viral infection is by recognizing the viral genome as foreign and targeting it for elimination. This strategy is employed by plants and invertebrates in the

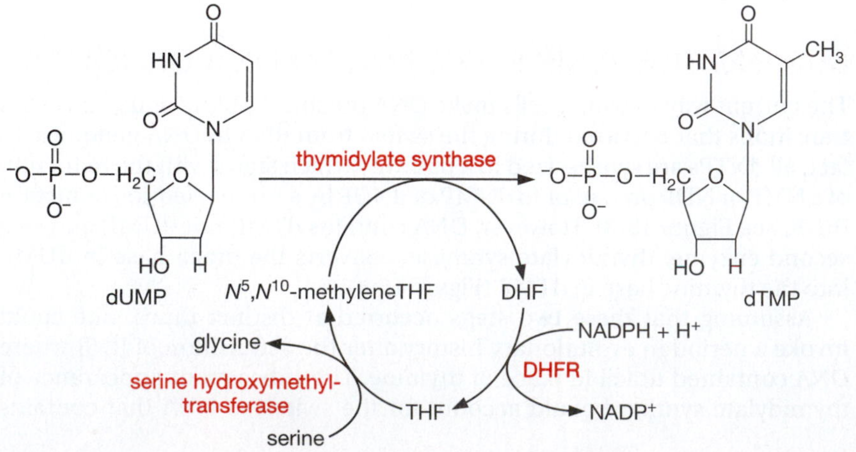

thymidylate synthase
dUMP
dTMP
N^5,N^{10}-methyleneTHF
DHF
NADPH + H^+
DHFR
$NADP^+$
THF
glycine
serine hydroxymethyl-transferase
serine

Figure 15-4. The reaction catalyzed by thymidylate synthase. DHF is dihydrofolate, THF is tetrahydrofolate, and DHFR is dihydrofolate reductase.

form of the RNA interference (RNAi) machinery. RNAi acts as a defense against RNA-based virus infection. In this pathway, the double-stranded RNA intermediate that is generated during viral replication is recognized by the RNAi machinery, resulting in the cleavage of the dsRNA into shorter fragments. These are then used to generate additional copies of short interfering RNA (siRNA) that can in turn interfere with virus production. The key feature of this process is that the host recognizes the extensive dsRNA form of the viral genome as a foreign molecule.

The restriction and modification systems of bacteria and archaea typify another mechanism by which foreign nucleic acid is detected and eliminated. In this case, the host cell encodes a bipartite system composed of a DNA methyltransferase and a restriction endonuclease. The methyltransferase specifically methylates bases in a short sequence; the restriction endonuclease recognizes this same sequence but cannot cleave it if the DNA is methylated. Thus, thanks to the presence of the methyltransferase in the host cell, its own genome is protected. In contrast, bacteriophage DNA, upon injection into the host, will not be methylated and so will be attacked by the restriction enzyme and thus degraded. However, some bacteriophages have come up with a way of circumventing host restriction systems. For example, bacteriophage T4 encodes the enzyme dCMP-hydroxymethylase that converts cytosine into 5-hydroxymethylcytosine, resulting in T4 DNA that contains hydroxymethylated cytosine (HMC; Figure 15-5).

HMC-containing DNA can be modified further by addition of glucose residues. These modifications of T4 DNA serve two purposes. First, they protect the DNA from the host restriction enzyme systems. Second, they allow the phage to turn the tables on the host by targeting normal cytosine-containing DNA for degradation by viral enzymes, thereby increasing the pool of DNA precursors available for making new viruses.

Patrick Forterre has proposed that a conceptually analogous situation could account for the appearance of DNA during biological evolution. It is conceivable that early forms of life with RNA genomes could have been preyed upon by viruses and so may have evolved antiviral systems that would degrade RNA-based viruses. Therefore, a virus that had a modified genome containing deoxyribonucleotides instead of ribonucleotides

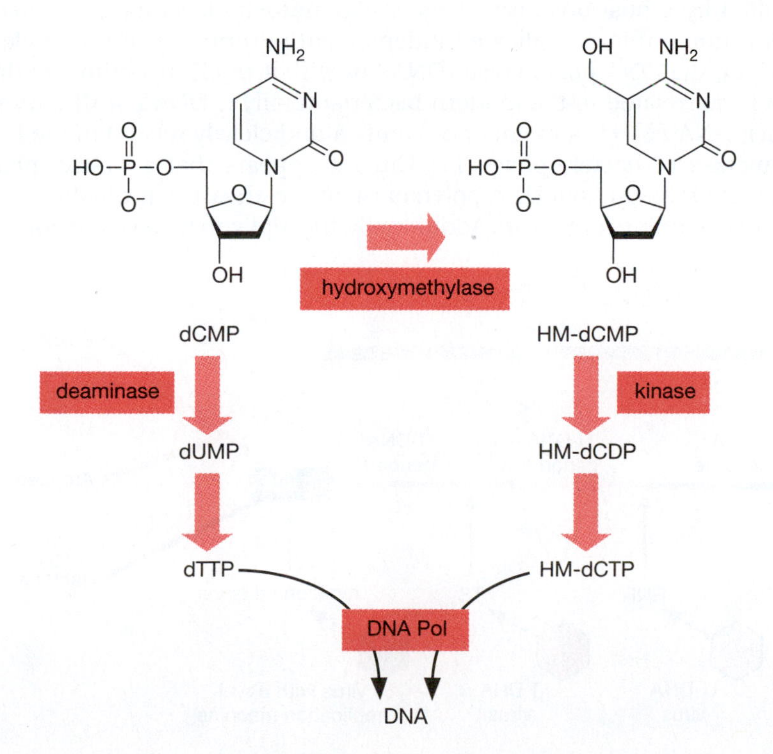

Figure 15-5. T4 as well as other T-even bacteriophage encode a hydroxymethylase that converts dCMP into 5-hydroxymethyl-dCMP (HM-dCMP). Another bacteriophage enzyme, nucleoside monophosphate kinase, converts HM-dCMP into HM-dCDP, and then host enzymes convert it into HM-dCTP, a substrate for T4 DNA polymerase. At the same time, another bacteriophage enzyme deaminates dCMP into dUMP, thereby ensuring that phage DNA will contain primarily, if not exclusively, HM-dCMP. Bacteriophage T4 enzymes are indicated in orange.

in its chromosome (Figure 15-3) would evade the host defense strategy. Subsequent capture of a gene for the RNR-like enzyme by a host genome could have led to the generation of a DNA-based cellular genome. A similar argument could be applied to the switch from U-containing DNA to T-containing DNA: T-DNA may have initially arisen as an evasion strategy (Figure 15-6). Modern cells contain surveillance mechanisms that ensure that U-containing DNA is eliminated by excision of U bases (Figure 3-12). Perhaps this is reflective of an ancient viral strategy akin to T4 phage's degradation of host DNA.

REPLICON TAKEOVER

Another potential role for viruses in driving evolutionary transitions has been proposed by Forterre in the context of the replicon takeover hypothesis. This proposal provides a simple and elegant mechanism to account for the dichotomous distribution of non-orthologous replication proteins between bacteria and archaea and eukarya. In its simplest form we assume that LUCA had a DNA-based genome with a replication machinery that was related either to the current archaeal system or to the bacterial system. If a virus infected a host cell, and in doing so integrated into the chromosome such that it inactivated an essential host-replication function, then the resultant hybrid viral/cellular chromosome would be dependent entirely on the viral replication machinery for its propagation. Suppose that an archaeal-like cell was infected with a virus with a bacterial-like replication machinery, or conversely, a bacterial-like host infected with a virus replicated by an archaeal-like system. The result would be conversion of replication machines and establishment of that machinery in all of the descendants of that organism. Over subsequent generations the original and now redundant cellular machinery either would be lost gradually or else subverted to ancillary functions such as DNA repair.

While such a scenario may seem unlikely, there is in fact precedent for precisely such events taking place. Remarkably, that evidence is found in our own cells: in the genomes of mitochondria. Mitochondria are derived from an endosymbiosis event early in the evolution of the eukaryotic lineage wherein a bacterium, related to modern proteobacteria, was engulfed by a host organism. Present-day mitochondria have retained a DNA genome that is replicated independently from that of the nucleus. However, the DNA polymerase (DNA Pol-γ) that replicates mitochondrial DNA is not related to the modern bacterial family C DNA Pol-III enzyme. Rather, DNA Pol-γ is a member of family A and closely related to the DNA polymerase of bacteriophage T7. Thus, it appears that a T7-like phage supplanted the original DNA polymerase of the ancestral proteobacterial-like endosymbiont organism. Additionally, the replicon takeover hypothesis

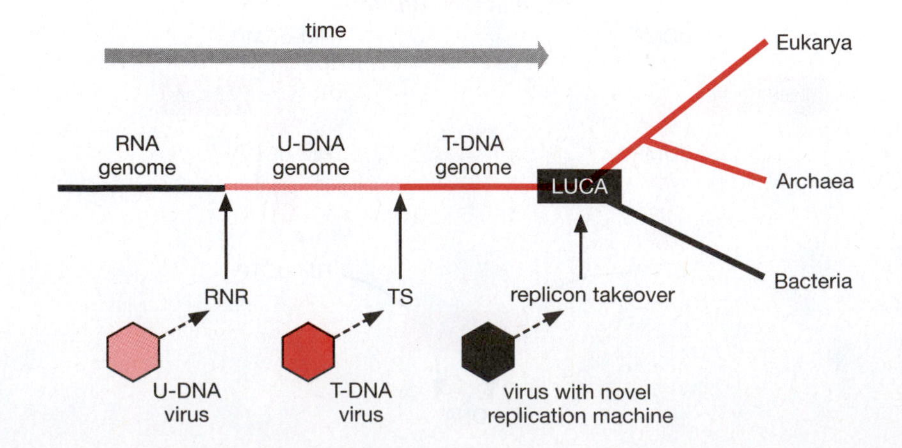

Figure 15-6. The transitions that have occurred during the evolution of DNA genomes indicate the potential role for viruses in mediating these changes. Assuming that the earliest organisms had RNA genomes, with the acquisition of ribonucleotide reductase (RNR), these RNA genomes became DNA genomes composed of uracil-containing DNA (U-DNA) instead of thymine-containing DNA (T-DNA). Here it is envisaged that RNR was captured by a host cell from an invading U-DNA virus (pink hexagon) that had discovered RNR as a defense against cellular mechanisms directed against RNA viruses. The next transition was from U-DNA to T-DNA, a transition dependent upon the enzyme thymidylate synthase (TS). The source of TS is envisaged as a T-DNA virus (red hexagon) that had discovered TS as a defense against cellular mechanisms directed against U-DNA viruses. For simplicity these events are diagrammed as taking place before the time of LUCA. Finally, at the time of LUCA a replicon takeover event took place in which a DNA virus (black hexagon) supplanted the host DNA replication machinery in the lineage that led to bacteria.

is compatible with observations in *E. coli* revealing that, albeit under artificially selected conditions, it is possible to drive host chromosome replication in the absence of a functional oriC by integration of a plasmid into the chromosome.

A prerequisite for the replicon takeover hypothesis is the existence either of a bacteriophage with an archaeal-like replication machinery, or of an archaeal virus with a bacterial-like apparatus. Is there any evidence within extant viruses and phage for such interdomain relationships? Examination of the genome of *Bacillus cereus* reveals an integrated phage genome. Within that phage genome there exists a gene that encodes a functional primase–helicase fusion protein. The primase domain is in the N-terminal half, and the C-terminal half of the protein contains the helicase domain. Remarkably, these primase and helicase domains are clearly related to the archaeal and eukaryal primase and MCM helicase, respectively, rather than their bacterial analogs. While it is highly unlikely that this particular phage is a molecular fossil of the replicon takeover event that occurred at the time of LUCA, its existence nevertheless demonstrates that phage can deliver components of an archaeal/eukaryotic-like DNA replication machinery into bacterial chromosomes.

THE ACQUISITION OF MULTIPLE REPLICATION ORIGINS

One of the fundamental organizational differences between bacterial and eukaryal genomes lies in the number of replication origins. All bacterial chromosomes have a single replication origin, oriC, whereas all eukaryotic organisms have many origins per chromosome. Intriguingly, archaea represent a half-way house. Some archaea, such as *Pyrococcus*, have a single oriC; others have more than one origin, for example *Sulfolobus* species have three origins per chromosome. Furthermore, all three origins in every cell appear to fire once during each cell cycle.

The presence of multiple origins can confer advantages: clearly the more active origins that are present, the smaller the inter-origin distance will be and therefore the more quickly a given chromosome can be replicated. Indeed, it seems likely that the evolution of chromosomes with multiple origins was an important step in permitting large genomes to evolve. However, with increasing origin numbers comes increasing organizational complexities, most obviously in coordinating the firing of multiple origins, although issues with termination events also may arise.

What is the nature of the mechanism that leads to amplification of origin numbers? For example, are the three origins of *Sulfolobus* derived from duplication of a single ancestral origin, giving rise to three related origins? Inspection of the sequence environment and architecture of the three *Sulfolobus* origins suggests that this is not the case. Rather, it points to yet another role that viruses or extrachromosomal elements may have played in shaping the cellular DNA replication process. The three *Sulfolobus* origins show only very modest relationships to each other, and they bind distinct subsets of the Orc1/Cdc6 initiator proteins, thereby forming distinct nucleoprotein architectures at the three origin loci. Examination of the genes flanking these origins indicates the presence of a range of genes normally associated with plasmids or viruses, including genes that control the number of genome copies and integrases, enzymes that catalyze integration of viruses into chromosomes. Furthermore, examination of the genome of a distantly related archaeon, *Aeropyrum pernix*, reveals two origins of replication. One of these is flanked by genes which in *Sulfolobus* have been split apart into two clusters, one that is associated with oriC2 and the other with oriC3. This suggests that an integration event has occurred in the lineage that gave rise to *Sulfolobus* and which split these gene clusters apart while concomitantly adding an extra origin of replication.

The combination of evidence for integration and the presence of viral or plasmid-derived genes in the immediate vicinity of the replication origins leads to the proposal that these origins have been acquired by integration of plasmids or viruses into the host chromosome and adoption of these formerly extrachromosomal replication origins for replication of the host chromosome. Thus, integration of extrachromosomal elements into host genomes may have played roles, not just in changing the information content of the genome, but also in shaping the very manner in which that host molecule is replicated. This model is of course contingent on the ability of the 'new' replication origins to be appropriately controlled in the context of the host's cell cycle. Clearly, a situation such as this is found in present-day eukarya, where the cell-cycle machinery post-translationally regulates the progression of preRCs to preICs to full elongation complexes at replication forks and clearly is capable of coordinately regulating many replication origins. Whether or not a simple analog of the eukaryotic cell-cycle regulatory machinery exists in organisms like *Sulfolobus* remains unknown.

VIRUSES AS A FOSSIL RECORD

One of the key questions about LUCA, assuming that it had a DNA-based genome, is whether its replication machinery was archaeal-like or bacterial-like. Eugene Koonin has described an interesting way of addressing this issue, and once again viruses are involved. What Koonin proposed was that the relative distribution of DNA polymerase families among DNA-based viruses would be reflective of their temporal history of appearance during evolution. Thus, the most widely distributed polymerase would likely be the most ancient, and similarly more recent polymerases would be less widely distributed in the 'virosphere'. The analysis is of course potentially limited by the range of viral sequences available in databases and could also be influenced by the relative efficiencies of given families of DNA polymerases for replicating viral genome-sized molecules. However, despite these caveats, the findings by Koonin are striking. By far the most widely distributed polymerases are members of family B, the family responsible for replicating present-day archaeal and eukaryotic genomes. Examples of family B polymerases are found in viruses and phage that infect all three domains of life. In contrast, family A and family C DNA polymerases show much narrower distributions. These data indicate that LUCA possessed a DNA replication system akin to that of present-day archaea.

SUMMARY

- Archaea and eukarya share a common core replication machinery.
- The bacterial machinery is distinct from that in the other two domains of life.
- The last universal common ancestor probably had a DNA genome.
- Viruses may have made significant contributions to the replication machines of their cellular hosts.

REFERENCES

Additional Reading

Edgell, D.R. and Doolittle, W.F. (1997) Archaea and the origin(s) of DNA replication proteins. *Cell* 89: 995–998.

Forterre, P. (2002) The origin of DNA genomes and DNA replication proteins. *Curr. Opin. Microbiol.* 5: 525–532.

Forterre, P. (2006) Three RNA cells for ribosomal lineages and three DNA viruses to replicate their genomes: a hypothesis for the origin of cellular domain. *Proc. Natl. Acad. Sci. USA* 103: 3669–3674.

Koonin, E.V. (2006) Temporal order of evolution of DNA replication systems inferred by comparison of cellular and viral DNA polymerases. *Biol. Direct* 1: 39.

McGeoch, A.T. and Bell, S.D. (2008) Extra-chromosomal elements and the evolution of cellular DNA replication machineries. *Nat. Rev. Mol. Cell Biol.* 9: 569–574.

Woese, C.R., Kandler, O., and Wheelis, M.L. (1990) Towards a natural system of organisms: proposal for the domains Archaea, Bacteria, and Eucarya. *Proc. Natl. Acad. Sci. USA* 87: 4576–4579.

GLOSSARY

AAA+ domain A 200–250 amino acid ATP-binding domain (AAA domain) containing Walker A and Walker B motifs, as well as other motifs, that distinguish it from classic P-loop nucleoside triphosphate hydrolases.

active replicator Specific *cis*-acting DNA sequences that are required both for the binding and function of its cognate initiator protein.

acytokinetic mitosis Mitosis in the absence of cytokinesis.

adaptation DNA damage persists but all or part of the DNA damage-response pathway becomes inactivated.

A-domain Essential element in autonomously replicating sequences that binds initiator protein.

A-form DNA A right-handed double helix similar to B-form DNA, but with a shorter, more compact helical structure.

alkylating agents Transfer an alkyl group from one molecule to another.

allele Any one of a number of copies of a unique genetic locus on a chromosome.

amplification origins Replication origins that initiate DNA replication multiple times in the absence of cell division.

analogs Proteins that have a similar function but unrelated sequences.

anaphase That stage of mitosis in which sister chromatids separate from one another.

anaphase-promoting complex (APC) An E3 ubiquitin ligase that targets proteins for degradation by the 26S proteasome.

angstrom (Å) 0.1 nanometers or 10^{-10} meters.

anticipation The tendency in trinucleotide-repeat disorders for progressively earlier or more severe expression of a disease to appear in successive generations.

antiparallel Complementary strands in DNA, DNA:RNA hybrids, or duplex regions of DNA and RNA molecules run in opposite directions (5′→3′:3′→5′).

apoptosis Programmed cell death.

archaea A major division of prokaryotes that differ from bacteria in the composition of their cell membrane, cell wall, and in the genes required for DNA replication, DNA transcription, and RNA translation.

ARS elements Autonomously replicating sequence found in plasmids or episomes.

A:T-rich element DNA sequence in which adenines are preferentially, if not exclusively, in one strand and thymines in the complementary strand.

AT-rich element DNA sequence composed preferentially, if not exclusively, of adenines and thymines.

autonomously replicating sequence (ARS) A *cis*-acting sequence that confers on DNA the ability to replicate autonomously when placed in a responsive environment. For example, *S. cerevisiae* ARS elements allow plasmids to replicate in an *S. cerevisiae* cell.

auxiliary components DNA sequences at replication origins that are not essential for initiation of DNA replication, but whose presence facilitates origin activity.

bacteria Single-cell prokaryotes that are genetically and evolutionarily distinct from archaea.

bacteriophage Viruses that replicate in bacterial cells.

base pairing Complementary bases form hydrogen bonds [A:T(U); G:C].

base stacking Interaction between nucleotide bases within the same DNA or RNA strand.

B-domain DNA sequence in ARS elements that facilitates initiation of DNA replication.

B-form DNA A right-handed helix with 10.5 base pairs per turn is the form of DNA that occurs at the hydration levels present in living cells.

bidirectional replication The consequence of two replication forks that originate at a common site but travel in opposite directions.

binary fission The process by which a prokaryotic cell divides into two cells, each with the potential to grow to the size of the original cell.

catenated DNA Circular DNA molecules linked together in a chain.

cell-cycle Series of events leading to cell division.

cell cycle checkpoint Mechanism that ensures the fidelity of cell division by verifying that events in one phase of the cell cycle have been completed correctly before progression onto the next phase.

cell fusion Merging together of two or more cells to form a single cell with two or more nuclei.

centrifugal elutriation Method for separating cells into G_1, S, and G_2/M-phase cells based on their size and density.

centriole Barrel-shaped structure in animal cells generally composed of nine triplets of microtubules.

centrosome Two orthogonally arranged centrioles surrounded by an amorphous mass of protein that serves as the microtubule-organizing center of eukaryotic cells as well as a regulator of cell-cycle progression.

C-form DNA A right-handed double helix similar to B-form DNA, but with a less compact helical structure.

checkpoint mediator Facilitates recruitment of DNA-repair proteins to damaged sites in response to a DNA damage checkpoint.

checkpoint sliding clamp Unique heterotrimeric protein ring structure produced in response to DNA damage (e.g. Rad9•Hus1•Rad1 in vertebrates) that functions as a sliding clamp for DNA polymerases in DNA-repair reactions.

checkpoint sliding clamp loader Unique heteropentameric protein complex produced in response to DNA damage [e.g. Rad17•RFC(2-5) in vertebrates] that loads the checkpoint sliding clamp onto damaged DNA.

chemical lesion Chemically modified base in DNA.

chi(χ)-site 5′-GCTGGTGG-3′ DNA sequence found at recombination hot spots in some bacterial genomes.

chromatid One of the two newly replicated homologous chromosomes.

chromatin DNA and its associated proteins.

chromatin assembly Organization of DNA into a DNA–protein complex in which DNA is associated with basic proteins such as histones in a periodic manner.

chromatosome A nucleosome with one bound linker histone such as H1.

chromosome Single linear or circular DNA molecule and its associated proteins.

***cis*-acting elements** DNA sequences such as replicators, promoters, and enhancers whose action is experienced within the DNA molecule in which they are located.

cleavage cell cycles Multiple cell divisions in the absence of cell growth typified by the absence of gap phases separating the S and M phases.

closed mitosis Mitosis in the absence of nuclear membrane breakdown.

cohesin Protein complex that maintains cohesion between sister chromatids.

collapsed replication fork Stalled replication fork in which the replisome has dissociated from the fork.

concatemer Multiple copies of a DNA sequence arranged end to end in tandem.

conditional replicator Passive replicator whose activity depends on epigenetic factors, such as chromatin structure, as well as on the sequence or composition of the DNA.

conservative replication Duplication of the parental DNA molecule without changing its structure or composition.

conserved sequence block (CSB) DNA sequence that is found at the same genetic or functional location in related organisms (e.g. CSBs at mitochondrial replication origins).

core histones H2A, H2B, H3, and H4.

covalent bond Covalent bond between two atoms formed by the sharing of one, two, or three pairs of electrons to produce a single, double, or triple bond.

CpG island Region of at least 200 bp with a GC content that is greater than 50% and with an observed/expected CpG ratio that is greater than 60%.

cytokinesis The process by which a cell divides into two approximately equal portions.

denaturation Process by which proteins or nucleic acids lose their tertiary and secondary structure as a result of some external stress such as heat, acid, base, or organic solvent.

diploid Cell containing two copies (homologs) of each chromosome, usually one from the mother and one from the father.

direct repeat Two copies of a nucleotide sequence with the same strand polarity (5′→→3′).

dispersive replication Newly replicated sections of DNA are dispersed among old unreplicated DNA such that the newly replicated portions are covalently linked to the unreplicated portions.

distributive Describes a nucleic acid polymerase or nuclease that carries out only one (or a few) catalytic events before dissociating and then rebinding randomly to the same or another nucleic acid molecule.

D-loop (displacement loop) A DNA structure where the two strands of a double-stranded DNA molecule are separated for a stretch and held apart by a third strand of DNA.

DNA adduct A piece of DNA covalently bonded to a chemical.

DNA-binding domain Specific region of a protein that binds to DNA.

DNA-binding protein (DBP) A protein that binds to dsDNA or ssDNA.

DNA damage response checkpoint Cell's response to chromosomal insults by arresting progress through the cell cycle, activating DNA-repair networks, and eventually inducing apoptosis.

DNA gap Short region of single-strand DNA within an otherwise duplex DNA molecule.

DNA gyrase A type II topoisomerase that introduces negative supercoils into DNA by looping the template so as to form a crossing, then cutting one of the dsDNA molecules and passing the other through it before releasing the break, thereby changing the linking number by two in each enzymatic step.

DNA helicase Enzyme that unwinds DNA by deriving the energy required from hydrolysis of ATP or other nucleoside triphosphate.

DNA ligase Enzyme that forms a covalent phosphodiester bond between the 3′-OH of one nucleotide and the 5′-P of an adjacent nucleotide in DNA by utilizing the energy from hydrolyzing either ATP (archaea, eukarya) or NAD^+ (bacteria).

DNA methylation Addition of methyl groups to DNA, specifically to the number 5 carbon in cytosine, the number 6 nitrogen in adenine, and less commonly to the number 4 nitrogen in cytosine.

DNA methyltransferase Enzyme that catalyzes the transfer of a methyl group from *S*-adenosylmethionine to DNA.

DNA polymerase Enzyme that catalyzes the polymerization of deoxyribonucleotides into a DNA strand.

DNA-replication checkpoint/S phase DNA damage checkpoint Detects either DNA damage or stalled replication fork during S phase and then inhibits replication-fork progress and suppresses further activation of preICs by inhibiting Cdk2 activity and presumably other, as yet unidentified, targets.

DNA unwinding Separating the two complementary strands in a duplex DNA molecule.

DNA-unwinding element (DUE) A *cis*-acting DNA sequence whose intrinsic helical instability facilitates the activity of replication origins by allowing the DNA to be easily unwound.

dNMP, dNDP, dNTP Deoxyribonucleotide containing one, two, or three phosphates, respectively.

dormant origin Replication origin that is silent under normal S-phase conditions, but active under conditions that inhibit replication-fork progress.

double-strand DNA break repair Re-establishing a single DNA molecule with the correct nucleotide sequence from the two ends of a broken duplex DNA molecule.

duplex DNA Alternative name for double-stranded DNA.

dyad symmetry element Two regions of a single DNA strand whose nucleotide sequences are inverted repeats of each other.

early mitosis inhibitor (Emi) Protein that binds to the anaphase-promoting complex and inhibits its activity, thereby preventing premature entrance into mitosis.

ectopic site activity DNA sequence that functions as a DNA replication origin when moved to a new chromosomal locus.

endomitosis Two or more rounds of genome duplication without completing mitosis or undergoing cytokinesis. Both endoreduplication and endomitosis result in cells with multiple copies of their genome contained within a single nucleus.

endoreduplication Two or more rounds of genome duplication without an intervening mitosis or cytokinesis.

endosymbiont Any organism that lives within the body or cells of another organism.

endosymbiont hypothesis Proposal that chloroplasts and mitochondria evolved from bacteria that were engulfed by a larger, presumably eukaryotic cell, through endophagocytosis.

epigenetic Reversible heritable changes in genomic functions (e.g. gene expression and replication) that occur without a change in the sequence of DNA. One example of an epigenetic change is DNA methylation.

episome A DNA molecule that can replicate autonomously outside of the cell's genome.

essential origin component DNA sequence, such as the origin core element, that is required for origin function.

eukaryote Organism with one or more cells that have visible membrane-bound nuclei and other cellular organelles.

expandable repeat sequences Repeated DNA sequences (usually trinucleotide repeats) that can expand or contract during cell proliferation, and that are often associated with specific inherited human disorders.

family of repeats ~20 imperfect copies of a 30 bp repeat to which the Epstein–Barr virus EBNA1 protein binds specifically and with high affinity.

feedback loops When a biochemical reaction generates a product that either enhances or represses the reaction's activity.

fidelity The degree of exactness with which DNA is copied or reproduced.

fragile sites Chromosomal loci that are especially sensitive to forming gaps or breaks when DNA replication is perturbed.

functional redundancy Two different genes that are capable of fulfilling the same function.

G_1 DNA damage checkpoint Blocks the $G_1 \rightarrow S$ transition by inhibiting Cdk2 activity, which is required for preIC assembly.

G_1 phase The gap in time between mitosis and DNA replication.

G_2 checkpoint Prevents entrance into mitosis in response to the presence of replication forks (may be same mechanism as G_2 DNA damage checkpoint).

G_2 DNA damage checkpoint Prevents entrance into mitosis by inhibiting Cdk1 activity in response to DNA damage.

G_2 phase The gap in time between DNA replication and mitosis.

gamete A mature sexual reproductive haploid cell.

gene Sequence of DNA or RNA that encodes the information to build either a protein or an RNA molecule.

gene amplification DNA replication events that are restricted to a specific chromosomal locus.

gene family Set of genes that share a known homology, usually in their protein products.

genetic Inheritable changes in the genome.

genome The total genetic information required to produce an organism, a virus or, an episome.

germ cell The haploid reproductive cells produced by metazoa termed eggs in the female and sperm in the male.

Gram-negative bacteria Bacteria that do not stain with crystal violet dye, a chemical that cannot penetrate the cell wall of Gram-negative bacteria.

Gram-positive bacteria Defined by the fact that they are stained dark blue or violet by Gram staining due to the high peptidoglycan content of their cell wall. These bacteria typically lack the outer membrane found in Gram-negative bacteria.

hairpins A structure generated by a region of single-stranded DNA or RNA that folds back and base-pairs with itself resulting in a double-stranded stem adjacent to a single-stranded loop.

hairpin structure Intramolecular base pairing in single-stranded DNA or RNA occurs when two complementary regions of the same molecule form a double helix that ends in an unpaired loop.

haploid Single set of unpaired chromosomes.

heavy (H)-strand synthesis Leading-strand DNA synthesis in mtDNA replication.

histone chaperones Protein that assists the assembly of histone octamers and nucleosome assembly.

histone modification Chemical modification of a histone such as acetylation, methylation, phosphorylation, or ubiquitination.

histone octamer A stable complex of core histones consisting of an H3•H4 tetramer sandwiched between two H2A•H2B dimers.

histones Highly basic proteins in the nuclei of eukaryotic cells that package DNA into nucleosomes.

holoenzyme Active, complex enzyme consisting of an apoenzyme and one or more coenzymes.

homologous chromosomes Two different copies of the same chromosome that diploid organisms inherit, one from each parent. Homologous chromosomes contain the same genes but two different copies of each allele, which might or might not be identical.

homologous recombination Exchange of nucleotide sequences between two similar or identical strands of DNA.

homologs Genes related by descent from a common ancestral DNA sequence, regardless of whether they have retained the same function (orthologs) or evolved a different function (paralogs).

Hoogsteen base pairs Hoogsteen base pair uses the N7 position of the purine base as a hydrogen-bond acceptor and C6 amino group as a donor to bind the Watson–Crick N3–N4 face of the pyrimidine base.

hydrogen bond Weak, primarily electrostatic, bond between a hydrogen atom bound to a highly electronegative element in one molecule and a second highly electronegative atom somewhere else.

immature chromatin Chromatin containing newly synthesized DNA that is hypersensitive to nonspecific endonucleases.

initiation mass Cell mass per replication origin at the time when DNA replication begins.

initiation region Region within a chromosome wherein initiation events have been detected by nascent strand analysis techniques.

initiation zone Region within a chromosome wherein initiation events have been detected by two-dimensional gel electrophoresis techniques.

initiator *trans*-acting protein or protein complex that binds to the replicator, thereby triggering DNA replication.

initiator-specific motif The α-helical helix-loop-helix motif found at the DNA interface within the conserved nucleotide-binding and catalytic site termed the AAA+ domain present in all initiator proteins produced by bacteria, archaea, and eukarya.

integrase Enzyme produced by a virus that enables its genetic material to be integrated into the DNA of the infected cell.

interphase Period of time during cell division that is not occupied by mitosis.

inverted repeat Two copies of the same nucleotide sequence but with opposite strand polarities so that they are read by RNA polymerase as a palindrome (5′→3′:3′← 5′).

iterons Directly repeated sequences that regulate plasmid copy number in bacteria by binding the initiator protein that activates the plasmid's replicator.

jumping-back mechanism The newly synthesized pre-terminal protein-trinucleotide intermediate in the replication of linear dsDNA genomes with terminal replicators jumps backward and pairs with the first of two identical triplets within ori-core.

kinetochore Protein structure on chromosomes where the spindle fibers attach during mitosis to pull the two chromatids apart.

lagging-strand synthesis Nascent DNA synthesized in the direction opposite to the direction of DNA unwinding.

lagging-strand template DNA template whose 5′→3′ polarity demands that the direction of nascent DNA synthesis on this template is opposite to the direction of DNA unwinding.

leading-strand synthesis Nascent DNA synthesized in the same direction as DNA unwinding.

leading-strand template DNA template whose 5′→3′ polarity demands that the direction of nascent DNA synthesis on this template is the same as the direction of DNA unwinding.

light (L)-strand DNA synthesis Lagging-strand DNA synthesis in mtDNA replication.

linking number Number of times that one strand winds around the other, including both helical turns and superhelical turns, when a DNA molecule is forced to lie in a flat plane.

macronuclear rDNA The larger of the two nuclei in ciliated protozoans that contains only small linear DNA molecules encoding the rRNA genes.

matrix-attachment regions (MARs) DNA sequences in eukaryotic chromosomes that attach to nuclear matrix proteins, thereby anchoring the DNA and organizing chromatin into structural domains.

mature chromatin Chromatin whose histone composition, modification, and sensitivity to nonspecific endonucleases are typical of bulk chromatin.

mediator proteins Proteins that detect changes in proteins that sense a signal and then act on proteins that execute an action in response to that signal (sensor→mediator→effector).

meiotic maturation The transition between diakinesis and metaphase of meiosis I that is characterized by nuclear envelope breakdown, rearrangement of the cortical cytoskeleton, and meiotic spindle assembly.

metaphase That stage during mitosis in which condensed, highly coiled chromosomes align in the middle of the cell before being separated into each of the two daughter cells.

metastasis Spread of a disease from one site within an organism to another non-adjacent site.

metazoan Multicellular animal containing different types of cells.

methyl-CpG-binding proteins Proteins that bind specifically to methylated CpG sites in DNA.

micronucleus The smaller of two nuclei in ciliate protozoans that contains the organism's complete chromosomes and that functions in reproduction.

mirror repeats The same sequence is repeated with mirror-image polarity (5′→←3′).

mismatch repair A system for strand-specific recognition and repair of erroneous insertion, deletion, and incorporation of bases that arise during DNA replication and recombination.

mitochondrial DNA (mtDNA) The DNA located in mitochondria.

mitosis The process by which a eukaryotic cell separates the chromosomes in its nucleus into two identical sets of chromosomes, each sequestered in its own nucleus.

mitotic cell cycle Cell-division cycles that include four sequential stages termed G_1→S→G_2→M.

mitotic checkpoint complex (MCC) A complex of four proteins that binds to and inhibits the anaphase-promoting complex, thereby preventing the metaphase to anaphase transition.

monoubiquitination Protein with no more than one ubiquitin molecule attached to a lysine residue.

negative superhelical turns Duplex DNA wound about itself in a right-handed direction.

nelarabine A purine nucleoside analog that is used in chemotherapeutic treatment of T-cell acute lymphoblastic leukemia.

N-glycosidic bond A bond between the hemiacetal group of a saccharide (sugar) molecule and a nitrogen in another molecule such as a purine or pyrimidine.

nick Single phosphodiester bond interruption in one of the two strands of duplex DNA.

nucleoid The bacterial equivalent of a eukaryotic nucleus, but lacking a nuclear membrane.

nucleosome Histone octamer consisting of two $[H3 \cdot H4]_2$ and two $[H2A \cdot H2B]_2$ dimers around which is coiled 146 base pairs of DNA.

nucleosome core particle (NCP) Two copies each of histone proteins H2A, H2B, H3, and H4 in complex with 146 base pairs of DNA.

nucleotide excision repair A mechanism for repairing damage to bases in DNA that results from a variety of sources including chemicals and ultraviolet (UV) light.

oncogene A gene that, when mutated or overexpressed, promotes the transition from a normal cell into a cancer cell.

oocyte The female germ cell that matures into an egg (meiosis II) which, in vertebrates, can be fertilized to begin animal development.

open mitosis The nuclear envelope breaks down and then re-forms around the two sets of separated chromosomes.

origin density Number of replication origins per unit of DNA.

origin of bidirectional replication (OBR) A chromosomal locus where bidirectional DNA replication begins.

origin-recognition box (ORB) The term used for 'origin recognition element' in archaeal replication origins.

origin-recognition complex (ORC) Heterohexameric protein complex that binds to DNA at or near the site where bidirectional DNA replication begins and initiates formation of a pre-replication complex in eukaryotes.

origin-recognition element (ORE) Replicator DNA sequence that determines where the initiator protein binds.

origin-core component DNA sequence essential to replicator function.

orthologs Genes that evolved from a common ancestral gene and retain the same function in different species.

paradox Statement that is seemingly contradictory or opposed to common sense and yet appears to be true.

paralogs Genes related by duplication within a genome, but that have evolved a new function.

passive replicator DNA binding site for initiator proteins that is not required for initiator protein function.

peptide bond Amide linkage joining amino acids to form peptides consists of a covalent bond between the carboxyl group of one amino acid and the amino group of another.

phosphoanhydride bond Covalent linkage between two phosphoric acid molecules in which an OH group from each molecule reacts to form an O linkage between the two molecules.

phosphodiester bond Covalent linkage between a phosphate group and two five-carbon ring carbohydrates (pentoses) over two ester bonds.

plasmid Autonomously replicating, extrachromosomal circular DNA molecules, distinct from the normal bacterial genome and nonessential for cell survival under nonselective conditions.

ploidy The number of sets of chromosomes within a cell.

poly-cistronic mRNA Single mRNA molecule that encodes several different genes that are translated into several different proteins.

polyploid Containing more than two homologous sets of chromosomes.

polyubiquitination Adding multiple ubiquitin molecules onto the primary amine in a single lysine residue within a target protein.

positive superhelical turns Duplex DNA wound about itself in a left-handed direction.

prenucleosomal DNA Region of DNA at replication forks that has not yet been assembled into nucleosomes.

pre-replication complex (preRC) Protein complex in eukaryotes consisting of the origin-recognition complex (ORC), Cdc6, Cdt1, and the MCM DNA helicase [Mcm(2-7)] that determines where DNA replication will begin.

pre-terminal protein (pTP) The initiator protein for linear dsDNA genomes with terminal replicators.

primer excision Removal of the RNA primer and the RNA-p-DNA phosphodiesterase linkage from nascent DNA chains.

primosome Protein complex capable of initiating DNA synthesis on a ssDNA template in the absence of a preformed RNA or DNA primer.

processivity The extent to which a polymerase or nuclease continues to synthesize or degrade a nucleic acid molecule before dissociating from it and then rebinding to the same or another molecule.

prokaryote Any cell that lacks a nucleus.

prometaphase The phase of mitosis following prophase and preceding metaphase during which the nuclear envelope breaks into fragments and disappears.

proofreading The 3′→5′ exonuclease activity associated with most replicative DNA polymerases excises an incorrect base-pair insertion, thereby allowing the polymerase to re-insert the correct base and continue DNA synthesis.

prophase Stage of mitosis in which the chromatin condenses into chromosomes visible by light microscopy.

proteasome Macromolecular protein complexes in eukarya and archaea that degrade unneeded or damaged proteins by proteolysis, a chemical reaction that breaks peptide bonds.

purine Heterocyclic aromatic organic compound, consisting of a pyrimidine ring fused to an imidazole ring.

pyrimidine Heterocyclic aromatic organic compound similar to benzene and pyridine, containing two nitrogen atoms at positions 1 and 3 of the six-member ring.

quadruplex DNA Square planar arrangement of four guanine bases (termed a G-tetrad) that can form remarkably stable stacks of G-tetrads, referred to as G-quadruplexes.

renaturation Process by which proteins or complementary strands of nucleic acids re-establish their native conformations.

repeat expansion Increased number of tandemly arranged copies of a particular repeated sequence (usually three nucleotides) at a particular genomic locus.

repeat length Number of copies of a particular tandemly repeated sequence.

replication-fork barrier (RFB) DNA site that causes replication forks to pause or arrest.

replication origin Genomic site where DNA replication begins.

replicative DNA helicase The DNA helicase responsible for unwinding DNA at replication forks [DnaB in bacteria, Mcm in archaea, Mcm(2-7) in eukarya].

replicative DNA polymerase DNA polymerase that is required for replicating a DNA genome.

replicator *cis*-acting sequence required for initiation of DNA replication.

replicon A unit of DNA replication. It contains a replicator where replication begins and a termination region where replication ends.

replicon clusters A subset of origins within a single chromosomal locus within a single cell that are activated both reliably and efficiently in each successive cell cycle (different cells may activate different replicon clusters at different times during S phase).

replisome A protein complex that replicates DNA, commonly consisting of helicase, primase, polymerase, sliding clamp, sliding clamp loader, and single-strand-specific DNA-binding protein (activities required for joining Okazaki fragments onto longer nascent DNA strands may or may not be physically part of the replisome).

replisome progression complex A protein complex required for maintaining association of the replicative MCM helicase with other components of the replisome.

restriction checkpoint/restriction point/R-point/start point The G_1-phase checkpoint in eukaryotic cell cycles that prevents S phase in the absence of adequate cell growth, commonly termed the 'restriction point' in animal cells and the 'start point' in yeast.

restriction endonucleases Enzymes that recognize specific short DNA sequences and cleave the sugar-phosphate backbone.

restriction enzyme An enzyme that cuts dsDNA or ssDNA at specific nucleotide sequences known as restriction sites, apparently evolved in bacteria and archaea as a defense against viruses.

restriction-modification system Host DNA is methylated by a modification enzyme (usually a DNA methyltransferase) to protect it from restriction endonuclease activity.

reverse gyrase A type I topoisomerase fused to a helicase that introduces positive superhelical turns into duplex DNA instead of negative superhelical turns.

reverse transcriptase A DNA polymerase that transcribes single-stranded RNA into double-stranded DNA.

ribonucleotide reductase/ribonucleoside diphosphate reductase Enzyme that converts ribonucleoside diphosphates into deoxyribonucleoside diphosphates.

ribosomal RNA (rRNA) RNA with the structural role of organizing a specific group of proteins into a ribosome.

ribosome Macromolecular complex of RNA and proteins that translates mRNA into protein.

ribozyme RNA molecule possessing a well-defined tertiary structure that enables it to catalyze a chemical reaction.

RNA polymerase Enzyme that synthesizes RNA with a sequence complementary to a DNA template.

RNA primer RNA annealed to a DNA template that allows DNA polymerase to initiate DNA synthesis by addition of dNMPs to the 3′-OH end of the RNA chain.

rolling-circle mechanism Unidirectional replication of a circular ssDNA or ssRNA molecule, resulting in long DNA or RNA products that are then processed into genome-length molecules that can be packaged into bacteriophage or viruses.

scaffold-attachment regions (SARs) DNA sequences in eukaryotic chromosomes that attach to topoisomerase II and at least one 'structural maintenance of chromosomes' (Smc2) protein, thereby anchoring the DNA and organizing chromatin into structural domains.

semi-conservative replication The two strands of the double helix are unwound, and a complementary copy of each strand is synthesized.

sequence context The DNA and protein environment in which a specific nucleotide sequence resides.

single-strand-specific DNA-binding protein Protein that binds specifically to single-stranded DNA.

sister chromatid cohesion Force holding together the two newly replicated sister chromatids.

sister chromatids Identical copies of a chromosome containing the same genes and alleles of each gene.

sliding clamp Homodimer (β-clamp in bacteria) or homo- or heterotrimer (PCNA in archaea and eukarya) that serves as a processivity promoting factor in DNA replication by preventing release of a DNA polymerase from its primer: template.

sliding clamp loader Proteins that load 'sliding clamps' onto DNA template strands and that disassemble the clamps after replication is completed.

slipped-strand DNA Regions of repetitive DNA (often containing trinucleotide repeats) where the complementary strands have aligned out of register, creating single-stranded bulges in the DNA.

slow zone A genetically encoded region that causes slower fork progression and tends to accumulate convergent replication forks.

somatic cell Any cell of the body with the exception of germ cells.

S phase That period of time during cell division when the nuclear genome is duplicated.

spindle-assembly checkpoint (SAC) The onset of anaphase is delayed when some chromosomes are not properly attached to the mitotic spindle apparatus during metaphase.

ssDNA-binding protein (SSB) Protein that binds specifically to single-stranded DNA.

stalled replication fork Replication fork that stops functioning before it terminates by colliding with another active replication fork during genome duplication.

stringent response Bacterial checkpoint in response to poor growth conditions, equivalent to restriction checkpoint in eukarya.

structural maintenance of chromosomes (SMC) proteins Six highly conserved eukaryotic genes encoding proteins that are required for chromosome segregation, and therefore, in most cases, essential for cell viability.

supercoiling A duplex DNA molecule relieving helical stress by twisting around itself.

syncytial cell cycle Nuclear genome duplication and mitosis in the absence of cytokinesis and therefore occurring within a common cytoplasm.

telomerase Enzyme that extends the 3′-end of telomeres.

telomere A series of tandemly repeated direct repeats at the ends of eukaryotic chromosomes.

telomere-repeat binding factors (TRFs) Double-stranded telomere-repeat binding proteins TRF1 and TRF2.

telophase Following anaphase, nuclear envelopes form around each set of sister chromatids from the fragments of the nuclear envelope of the parent cell, usually with concurrent cytokinesis (a separate process).

terminal deoxyribonucleotidyl transferase A DNA polymerase that can extend the 3′-OH end of a DNA chain in the absence of a DNA template.

termination When two oncoming replication forks collide, thereby resolving into two separate DNA molecules and terminating DNA synthesis.

termination site Defined chromosomal locus at which two oncoming replication forks pause and then terminate replication simply by completing DNA unwinding and then DNA synthesis.

terminator sequence See 'replication-fork barrier.'

tetrasome A stable complex of two H3•H4 dimers $[(H3\bullet H4)_2]$.

topoisomerase Enzyme that can either increase or decrease the superhelical density of DNA, or that can either catenate or decatenate two different DNA molecules.

toroidal coil A DNA molecule that wraps about a center axis as though it were a coil of wire which, if it were a circular DNA molecule, would resemble a doughnut.

***trans*-acting factor** Diffusible substance whose action is experienced anywhere within its boundaries, regardless of whether or not its target is located in the same DNA molecule.

translesion synthesis A mechanism that replication forks to bypass DNA lesions such as thymine dimers or apurinic sites by ubiquitinating PCNA, thereby releasing Pol-δ and replacing it with Pol-ζ or Pol-η, two DNA polymerases that are extremely efficient at inserting correct bases opposite specific types of damage.

triplex DNA Triple-strand DNA structure that forms when either a polypyrimidine or a polypurine sequence occupies the major groove of dsDNA by forming Hoogsteen base pairs with purines in the normal dsDNA.

ubiquitin Small (76 amino acid) highly conserved polypeptide ubiquitously expressed in eukarya where it regulates the stability, function, and intracellular localization of a wide variety of proteins.

ubiquitin ligase Protein that, in combination with an E2 ubiquitin-conjugating enzyme, links the C-terminal residue in ubiquitin to the primary amine on a lysine in the target protein via an isopeptide bond.

ubiquitination The process by which one or more copies of the 76 amino acid protein ubiquitin are covalently attached to a target protein. The process requires three enzyme complexes, E1, E2, and E3, that sequentially activate the ubiquitin and target it to a substrate protein.

ubiquitylation Post-translational modification of a protein by the covalent attachment of one or more ubiquitin molecules via an isopeptide bond.

virion A complete virus particle with its DNA or RNA core and protein coat as it exists outside the cell.

virus A protein-encapsulated nucleic acid molecule that can reproduce itself but only in living cells.

yeast artificial chromosome (YAC) A chromosome containing the telomeric, centromeric, and replication origin sequences required for replication in yeast cells and used in cloning DNA fragments >100 kb ≤3000 kb.

INDEX

Notes: All genes mentioned in the text are listed in the index under 'Genes (list)'. Page numbers followed by 'f', 't' and 'b' refer to figures, tables and boxes respectively. Page numbers followed by 'ff" (or 'bb' or 'tt') refer to figures (or boxes or tables) which span consecutive pages. Plate refers to color plates which can be found between pages 82 and 83, 114 and 115, 210 and 211, 274 and 275, 402 and 403.

B

D

E

I

N

S

T

U

V

W

X

Y

Z